Lecture Notes in Computer Scie

Commenced Publication in 1973
Founding and Former Series Editors:
Gerhard Goos, Juris Hartmanis, and Jan van Leeuwen

T0238601

Erzsébet Csuhaj-Varjú
Martin Dietzfelbinger   Zoltán Ésik (Eds.)

Mathematical Foundations of Computer Science 2014

39th International Symposium, MFCS 2014
Budapest, Hungary, August 25-29, 2014
Proceedings, Part II

 Springer

Volume Editors

Erzsébet Csuhaj-Varjú
Eötvös Loránd University
Faculty of Informatics
Budapest, Hungary
E-mail: csuhaj@inf.elte.hu

Martin Dietzfelbinger
Technische Universität Ilmenau
Fakultät für Informatik und Automatisierung
Ilmenau, Germany
E-mail: martin.dietzfelbinger@tu-ilmenau.de

Zoltán Ésik
Szeged University
Institute of Informatics
Szeged, Hungary
E-mail: ze@inf.u-szeged.hu

ISSN 0302-9743   e-ISSN 1611-3349
ISBN 978-3-662-44464-1   e-ISBN 978-3-662-44465-8
DOI 10.1007/978-3-662-44465-8
Springer Heidelberg New York Dordrecht London

Library of Congress Control Number: 2014945809

LNCS Sublibrary: SL 1 – Theoretical Computer Science and General Issues

Typesetting: Camera-ready by author, data conversion by Scientific Publishing Services, Chennai, India

Printed on acid-free paper

Springer is part of Springer Science+Business Media (www.springer.com)

Preface

The series of MFCS symposia has a long and well-established tradition. The MFCS conferences encourage high-quality research into all branches of theoretical computer science. Their broad scope provides an opportunity to bring together researchers who do not usually meet at specialized conferences. The first symposium was held in 1972. Until 2012 MFCS symposia were organized on a rotating basis in Poland, the Czech Republic, and Slovakia. The 2013 edition took place in Austria, and in 2014 Hungary joined the organizing countries. The 39th International Symposium on Mathematical Foundations of Computer Science (MFCS 2014) was held in Budapest during August 25–29, 2014.

Due to the large number of accepted papers, the proceedings of the conference were divided into two volumes on a thematical basis: Logic, Semantics, Automata and Theory of Programming (Vol. I) and Algorithms, Complexity and Games (Vol. II). The 95 contributed papers were selected by the Program Committee (PC) out of a total of 270 submissions. All submitted papers were peer reviewed and evaluated on the basis of originality, quality, significance, and presentation by at least three PC members with the help of external experts. The PC decided to give the Best Paper Award, sponsored by the European Association of Theoretical Computer Science (EATCS), to the paper "Zero Knowledge and Circuit Minimization" written by Eric Allender and Bireswar Das. In addition, the paper entitled "The Dynamic Descriptive Complexity of k-Clique" by Thomas Zeume earned the Best Student Paper Award.

The scientific program of the symposium included seven invited talks by:

- Krishnendu Chatterjee (IST Austria, Klosterneuburg, Austria)
- Achim Jung (University of Birmingham, UK)
- Dániel Marx (MTA SZTAKI, Hungary)
- Peter Bro Miltersen (Aarhus University, Denmark)
- Cyril Nicaud (Université Paris-Est Marne-la-Vallé, France)
- Alexander Sherstov (University of California, Los Angeles, USA)
- Christian Sohler (Technische Universität Dortmund, Germany)

We are grateful to all invited speakers for accepting our invitation and for their excellent presentations at the symposium. We thank all authors who submitted their work for consideration to MFCS 2014. We deeply appreciate the competent and timely handling of the submissions of all PC members and external reviewers.

The members of the Organizing Committee were Erzsébet Csuhaj-Varjú (chair, Budapest), Zsolt Gazdag (Budapest), Katalin Anna Lázár (Budapest) and Krisztián Tichler (Budapest).

The website design and maintenance were carried out by Zoltán L. Németh (University of Szeged). The publicity chair was Szabolcs Iván (University of Szeged).

The editors express their gratitude to Zsolt Gazdag, Katalin Anna Lázár, and Krisztián Tichler for their valuable contribution to the technical edition of the two volumes of proceedings.

We thank Andrej Voronkov for his EasyChair system, which facilitated the work of the PC and the editors considerably.

June 2014

Erzsébet Csuhaj-Varjú
Martin Dietzfelbinger
Zoltán Ésik

Conference Organization

The organization of the scientific part of the conference was supported by the Faculty of Informatics, Eötvös Loránd University, Budapest, and the Institute of Computer Science, Faculty of Science and Informatics, University of Szeged.

Some parts of the local arrangements were taken care of by the ELTE-Soft Non-Profit Organization and Pannonia Tourist Service.

Program Committee Chairs

Zoltán Ésik, chair	University of Szeged, Hungary
Erzsébet Csuhaj-Varjú, co-chair	Eötvös Loránd University, Hungary
Martin Dietzfelbinger, co-chair	Technische Universität Ilmenau, Germany

Program Committee

Albert Atserias	Technical University of Catalonia, Spain
Giorgio Ausiello	Sapienza University of Rome, Italy
Jos Baeten	CWI, The Netherlands
Therese Biedl	University of Waterloo, Canada
Mikołaj Bojańczyk	University of Warsaw, Poland
Gerth Stølting Brodal	Aarhus University, Denmark
Christian Choffrut	Paris Diderot University, France
Rocco De Nicola	IMT Institute for Advanced Studies Lucca, Italy
Manfred Droste	Leipzig University, Germany
Robert Elsässer	University of Salzburg, Austria
Uli Fahrenberg	IRISA/Inria Rennes, France
Fedor V. Fomin	University of Bergen, Norway
Fabio Gadducci	University of Pisa, Italy
Anna Gál	The University of Texas at Austin, USA
Dora Giammarresi	University of Tor Vergata, Rome, Italy

Roberto Grossi	University of Pisa, Italy
Anupam Gupta	Carnegie Mellon University, USA
Michel Habib	Paris Diderot University, France
Kristoffer Arnsfelt Hansen	Aarhus University, Denmark
Edith Hemaspaandra	Rochester Institute of Technology, USA
Kazuo Iwama	Kyoto University, Japan
Yoshihiko Kakutani	The University of Tokyo, Japan
Juhani Karhumäki	University of Turku, Finland
Bakhadyr Khoussainov	University of Auckland, New Zealand
Elias Koutsoupias	University of Oxford, UK
Jan Kratochvíl	Charles University, Prague, Czech Republic
Stefan Kratsch	Technical University of Berlin, Germany
Rastislav Královič	Comenius University in Bratislava, Slovakia
Amit Kumar	Indian Institute of Technology Delhi, India
Kim G. Larsen	Aalborg University, Denmark
Frédéric Magniez	Paris Diderot University, France
Ralph Matthes	IRIT (CNRS and University of Toulouse), France
Madhavan Mukund	Chennai Mathematical Institute, India
Jean-Éric Pin	LIAFA CNRS and Paris Diderot University, France
Alexander Rabinovich	Tel Aviv University, Israel
Peter Rossmanith	RWTH Aachen University, Germany
Jan Rutten	CWI and Radboud University Nijmegen, The Netherlands
Wojciech Rytter	Warsaw University and Copernicus University in Torun, Poland
Luigi Santocanale	Aix-Marseille University, France
Christian Scheideler	University of Paderborn, Germany
Thomas Schwentick	TU Dortmund University, Germany
Alex Simpson	University of Edinburgh, UK
Mohit Singh	Microsoft Research Redmond, USA
Klaus Sutner	Carnegie Mellon University, USA
Gábor Tardos	Rényi Institute, Hungary
György Turán	University of Illinois at Chicago, USA
Peter Widmayer	ETH Zürich, Switzerland
Philipp Woelfel	University of Calgary, Canada

Steering Committee

Juraj Hromkovič	ETH Zürich, Switzerland
Antonín Kučera (chair)	Masaryk University, Czech Republic
Jerzy Marcinkowski	University of Wrocław, Poland
Damian Niwinski	University of Warsaw, Poland
Branislav Rovan	Comenius University in Bratislava, Slovakia
Jiří Sgall	Charles University, Prague, Czech Republic

Additional Reviewers

Abel, Andreas
Aceto, Luca
Acher, Mathieu
Aghazadeh, Zahra
Ahrens, Benedikt
Allender, Eric
Ambos-Spies, Klaus
Andrews, Matthew
Anselmo, Marcella
Bacci, Giorgio
Bacci, Giovanni
Baertschi, Andreas
Balbiani, Philippe
Barceló, Pablo
Bartha, Miklós
Barto, Libor
Basavaraju, Manu
Basold, Henning
Baumeister, Dorothea
Béal, Marie-Pierre
Ben-Amram, Amir
Berenbrink, Petra
Berthé, Valérie
Bilò, Davide
Bilò, Vittorio
Blanchet-Sadri, Francine
Bliznets, Ivan
Boasson, Luc
Bodlaender, Hans L.
Böhmova, Katerina
Bollig, Benedikt
Bonchi, Filippo
Bonelli, Eduardo
Bonfante, Guillaume
Bonifaci, Vincenzo
Bonsma, Paul
Boudjadar, A. Jalil
Bozzelli, Laura
Breveglieri, Luca
Brinkmann, André
Broadbent, Christopher
Brotherston, James
Brunet, Paul

Buss, Sam
Calì, Andrea
Carayol, Arnaud
Carpi, Arturo
Caucal, Didier
Cechlárová, Katarína
Chang, Richard
Christophe, Reutenauer
Chroboczek, Juliusz
Colcombet, Thomas
Colin de Verdière, Éric
Cook, Stephen
Cosme Llópez, Enric
Cranen, Sjoerd
Crescenzi, Pierluigi
Crespi Reghizzi, Stefano
Cyriac, Aiswarya
Czeizler, Elena
D'Alessandro, Flavor
D'Souza, Deepak
Dalmau, Victor
Damaschke, Peter
Datta, Samir
de Boysson, Marianne
de Mesmay, Arnaud
de Rougemont, Michel
de Wolf, Ronald
Dell, Holger
Demetrescu, Camil
Diekert, Volker
Dinneen, Michael
Drange, Pål Grønås
Dube, Simant
Dück, Stefan
Dürr, Christoph
Duparc, Jacques
Durand, Arnaud
Eikel, Martina
Elahi, Maryam
Elbassioni, Khaled
Elberfeld, Michael
Epifanio, Chiara
Espírito Santo, José

Facchini, Alessandro
Faliszewski, Piotr
Feldmann, Andreas Emil
Fernau, Henning
Ferreira, Francisco
Fertin, Guillaume
Fici, Gabriele
Finocchi, Irene
Firmani, Donatella
Fitzsimmons, Zack
Forisek, Michal
Fortier, Jérôme
Forys, Wit
Fournier, Hervé
François, Nathanaël
Fratani, Séverine
Fredriksson, Kimmo
Freivalds, Rusins
Frid, Anna
Frigioni, Daniele
Frougny, Christiane
Fusy, Éric
Gairing, Martin
Gasieniec, Leszek
Gaspers, Serge
Gastin, Paul
Gavalda, Ricard
Gawrychowski, Pawel
Gharibian, Sevag
Giustolisi, Rosario
Gliozzi, Valentina
Gmyr, Robert
Goldwurm, Massimiliano
Golovach, Petr
Grigorieff, Serge
Gualà, Luciano
Guillon, Bruno
Gupta, Sushmita
Habermehl, Peter
Harju, Tero
Hatami, Pooya
Hellerstein, Lisa
Hemaspaandra, Lane A.

Henglein, Fritz
Hertrampf, Ulrich
Hill, Cameron
Hirvensalo, Mika
Holzer, Stephan
Honda, Kentaro
Horn, Florian
Hovland, Dag
Huang, Chien-Chung
Huang, Sangxia
Huang, Zhiyi
Huber, Stefan
Hung, Ling-Ju
Hyvernat, Pierre
Imreh, Csanád
Inaba, Kazuhiro
Iván, Szabolcs
Ivanyos, Gábor
Jancar, Petr
Jansen, Bart M.P.
Jansen, Klaus
Jeřábek, Emil
Kaaser, Dominik
Kabanets, Valentine
Kaminski, Michael
Kamiński, Marcin
Kanté, Mamadou
  Moustapha
Kara, Ahmet
Kari, Jarkko
Kawamoto, Yusuke
Kerenidis, Iordanis
Kieronski, Emanuel
Kimura, Daisuke
Klasing, Ralf
Klein, Philip
Klimann, Ines
Klin, Bartek
Kling, Peter
Kniesburges, Sebastian
Kociumaka, Tomasz
Kollias, Konstantinos
Komm, Dennis
Konrad, Christian
Kontchakov, Roman

Kosowski, Adrian
Koutris, Paraschos
Koutsopoulos, Andreas
Královič, Richard
Kratsch, Dieter
Krebs, Andreas
Krokhin, Andrei
Křetinský, Jan
Kučera, Petr
Kufleitner, Manfred
Kuhnert, Sebastian
Kuperberg, Denis
Kutrib, Martin
Kuznets, Roman
La Torre, Salvatore
Labarre, Anthony
Labbé, Sébastien
Labella, Anna
Ladra, Susana
Laura, Luigi
Le Gall, François
Lecroq, Thierry
Leroy, Julien
Liaghat, Vahid
Limouzy, Vincent
Liu, Jiamou
Lluch Lafuente, Alberto
Lodaya, Kamal
Lohrey, Markus
Lokshtanov, Daniel
Lombardy, Sylvain
Lomonaco, Sam
Lonati, Violetta
Loreti, Michele
Lubiw, Anna
Luttik, Bas
Madelaine, Florent
Madonia, Maria
Malcher, Andreas
Maletti, Andreas
Mamageishvili, Akaki
Mamcarz, Antoine
Manlove, David
Markovski, Jasen
Martens, Wim

Martin, Barnaby
Marx, Dániel
Mathieson, Luke
Matoušek, Jiří
Maudet, Nicolas
Mayr, Richard
Mendler, Michael
Mihal'ák, Matúš
Mikulski, Lukasz
Mio, Matteo
Miyazaki, Shuichi
Momigliano, Alberto
Monmege, Benjamin
Montanari, Sandro
Morton, Jason
Mouawad, Amer
Mühlenthaler, Moritz
Müller, Moritz
Nagy-György, Judit
Narayanaswamy, N.S.
Nederlof, Jesper
Negri, Sara
Neven, Frank
Nichterlein, André
Niewerth, Matthias
Nishimura, Naomi
Nisse, Nicolas
Niwinski, Damian
Nonner, Tim
Nyman, Ulrik
Ochremiak, Joanna
Okhotin, Alexander
Olesen, Mads Chr.
Ouaknine, Joël
Palfrader, Peter
Pandey, Omkant
Panella, Federica
Paperman, Charles
Parberry, Ian
Pardubska, Dana
Parekh, Ojas
Paulusma, Daniel
Pavan, Aduri
Perevoshchikov, Vitaly
Perrin, Dominique

Petersen, Holger
Piątkowski, Marcin
Pignolet, Yvonne-Anne
Pinsker, Michael
Piperno, Adolfo
Plandowski, Wojciech
Poulsen, Danny Bøgsted
Pozzato, Gian Luca
Praveen, M.
Pröger, Tobias
Proietti, Guido
Quaas, Karin
Quyen, Vuong Anh
Radoszewski, Jakub
Raffinot, Mathieu
Reidl, Felix
Rettinger, Robert
Reutenauer, Christophe
Ribichini, Andrea
Ricciotti, Wilmer
Riscos-Núñez, Agustín
Riveros, Cristian
Ronchi Della Rocca,
    Simona
Rossman, Benjamin
Rubin, Sasha
Saarela, Aleksi
Sabharwal, Yogish
Sadakane, Kunihiko
Salo, Ville
Salomaa, Kai
Salvati, Sylvain
Santini, Francesco
Sau, Ignasi
Sauerwald, Thomas
Saurabh, Saket
Savicky, Petr
Schabanel, Nicolas
Schaefer, Marcus
Schmidt, Jens M.
Schmidt, Johannes

Schwartz, Roy
Schweitzer, Pascal
Segoufin, Luc
Selečéniová, Ivana
Serre, Olivier
Seto, Kazuhisa
Setzer, Alexander
Sgall, Jirí
Shallit, Jeffrey
Shen, Alexander
Siebertz, Sebastian
Silva, Alexandra
Skrzypczak, Michał
Spoerhase, Joachim
Srivathsan, B.
Stacho, Juraj
Stephan, Frank
Stiebitz, Michael
Straubing, Howard
Strozecki, Yann
Studer, Thomas
Sun, He
Suresh, S.P.
Svensson, Ola
Szörényi, Balázs
Tadaki, Kohtaro
Talebanfard, Navid
Tamaki, Suguru
Tan, Li-Yang
Tanabe, Yoshinori
Tantau, Till
Telikepalli, Kavitha
ten Cate, Balder
Thierry, Éric
Thomas, Rick
Tillich, Jean-Pierre
Törmä, Ilkka
Trinker, Horst
Tschager, Thomas
Tuosto, Emilio
Tzameret, Iddo

Umboh, Seeun
van 't Hof, Pim
van Breugel, Franck
van Hulst, Allan
van Leeuwen, Erik Jan
van Raamsdonk, Femke
Vanier, Pascal
Variyam, Vinodchandran
Vatan, Farrokh
Velner, Yaron
Vigo, Roberto
Villaret, Mateu
Volkov, Mikhail
Vrt'o, Imrich
Wagner, Uli
Waldmann, Johannes
Walen, Tomasz
Walter, Tobias
Wanka, Rolf
Warnke, Lutz
Weidner, Thomas
Weiner, Mihály
Wieder, Udi
Willemse, Tim
Williams, Ryan
Winter, Joost
Witkowski, Piotr
Wu, Zhilin
Xiao, David
Xue, Bingtian
Ye, Deshi
Zantema, Hans
Zdanowski, Konrad
Zemek, Petr
Zeume, Thomas
Zhang, Liyu
Ziegler, Martin
Zielonka, Wieslaw
Živný, Stanislav

Table of Contents – Part II

Algorithms, Complexity and Games

Table of Contents – Part I

On r-Simple k-Path

Hasan Abasi, Nader H. Bshouty, Ariel Gabizon*, and Elad Haramaty**

Department of Computer Science Technion, Haifa, Israel

Abstract. An r-simple k-path is a path in the graph of length k that passes through each vertex at most r times. The r-SIMPLE k-PATH problem, given a graph G as input, asks whether there exists an r-simple k-path in G. We first show that this problem is NP-Complete. We then show that there is a graph G that contains an r-simple k-path and no simple path of length greater than $4 \log k / \log r$. So this, in a sense, motivates this problem especially when one's goal is to find a short path that visits many vertices in the graph while bounding the number of visits at each vertex.

We then give a randomized algorithm that runs in time

$$\text{poly}(n) \cdot 2^{O(k \cdot \log r / r)}$$

that solves the r-SIMPLE k-PATH on a graph with n vertices with one-sided error. We also show that a randomized algorithm with running time $\text{poly}(n) \cdot 2^{(c/2)k/r}$ with $c < 1$ gives a randomized algorithm with running time $\text{poly}(n) \cdot 2^{cn}$ for the Hamiltonian path problem in a directed graph - an outstanding open problem. So in a sense our algorithm is optimal up to an $O(\log r)$ factor in the exponent.

The crux of our method is to use *low degree testing* to efficiently test whether a polynomial contains a monomial where all individual degrees are bounded by a given r.

1 Introduction

Let G be a directed graph on n vertices. A path ρ is called *simple* if all the vertices in the path are distinct. The SIMPLE k-PATH problem, given G as input, asks whether there exists a simple path in G of length k. This is a generalization of the well known HAMILTONIAN-PATH problem that asks whether there is a simple path passing through *all* vertices, i.e., a simple path of length n in G. As HAMILTONIAN-PATH is NP-complete, we do not expect to find polynomial time algorithms for SIMPLE k-PATH for general k. Moreover, we do not even expect to find good approximation algorithms for the corresponding

---

* The research leading to these results has received funding from the European Community's Seventh Framework Programme (FP7/2007-2013) under grant agreement number 257575.
** This research was partially supported by the Israel Science Foundation (grant number 339/10).

E. Csuhaj-Varjú et al. (Eds.): MFCS 2014, Part II, LNCS 8635, pp. 1–12, 2014.
© Springer-Verlag Berlin Heidelberg 2014

optimization problem: the *longest path problem*, where we ask what is the length of the longest simple path in G. This is because Björklund et al. [6] showed that the longest path problem cannot be approximated in polynomial time to within a multiplicative factor of $n^{1-\epsilon}$, for any constant $\epsilon > 0$, unless P=NP. This (in addition to being a natural problem in parameterized complexity) motivates finding algorithms for SIMPLE k-PATH with running time whose dependence on k is as small as possible. The first result in this venue by Monien [17] achieved a running time of $k! \cdot \text{poly}(n)$. Since then, there has been extensive research on constructing algorithms for SIMPLE k-PATH running in time $f(k) \cdot \text{poly}(n)$, for a function $f(k)$ as small as possible [5,2,16,10,13]. The current state of the art is $2^k \cdot poly(n)$ by Williams [19] for directed graphs and $O(1.657^k) \cdot \text{poly}(n)$ by Björklund [7] for undirected graphs.

1.1   Our Results

In this paper we look at a further generalization of SIMPLE k-PATH which we call r-SIMPLE k-PATH. In this problem instead of insisting on ρ being a simple path, we allow ρ to visit any vertex a fixed number of times. We now formally define the problem r-SIMPLE k-PATH.

Definition 1. *Fix integers $r \leq k$. Let G be a directed graph.*

- *We say a path ρ in G is r-simple, if each vertex of G appears in ρ at most r times. Obviously, ρ is a simple path if and only if it is a 1-simple path.*
- *The r-SIMPLE k-PATH problem, given G as input, asks whether there exists an r-simple path in G of length k.*

At first, one may wonder whether for some fixed $r > 1$, r-SIMPLE k-PATH always has a polynomial time algorithm. We show this is unlikely by showing that for any r, for some k r-SIMPLE k-PATH is NP-complete. See Theorem 3 in Section 3 for a formal statement and proof of this. Thus, as in the case of SIMPLE k-PATH, one may ask what is the best dependency of the running time on r and k that can be obtained in an algorithm for r-SIMPLE k-PATH.

Our main result is

Theorem 2. *Fix any integers r, k with $2 \leq r \leq k$. There is a randomized algorithm running in time*

$$\text{poly}(n) \cdot O\left(r^{\frac{2k}{r}+O(1)}\right) = \text{poly}(n) \cdot 2^{O(k \cdot \log r / r)}$$

solving r-SIMPLE k-PATH on a graph with n vertices with one-sided error.

One may ask how far from optimal is the dependency on k and r in Theorem 2. Theorem 3 implies that a running time of $\text{poly}(n) \cdot 2^{o(k/r)}$ would give an algorithm with running time $2^{o(n)}$ for HAMILTONIAN-PATH Moreover, even a running time of $\text{poly}(n) \cdot 2^{c \cdot k/r}$, for a small enough constant $c < 1/2$, would imply a better algorithm for HAMILTONIAN-PATH than those of [19,7] which are the best currently known. So, in a sense our algorithm is optimal up to an

$O(\log r)$ factor. We find closing this $O(\log r)$ gap, e.g. by a better reduction to HAMILTONIAN-PATH or a better algorithm for r-SIMPLE k-PATH, to be an interesting open problem.

1.2 r-Monomial Detection

As will be explained later on, the main step of our algorithm is, given a circuit computing a multivariate polynomial f, to efficeintly check whether f contains a monomial where all individual degrees are at most r. Let us refer to this problem as r-*monomial detection*. As the problem of 1-*monomial* detection, or multilinear monomial detection, has proven to be central to parameterized complexity (cf. Koutis and Williams [14]), we believe the r-monomial detection problem will prove useful.

1.3   Comparision to the Method of Koutis and Williams

Koutis [13] uses group algebras over $\mathbb{F}_2$ in a clever way to solve multilinear monomial detection. An extension of his method by Williams [19] to group algebras over extension fields of the form $\mathbb{F}_{2^t}$ gives the current best randomized algorithm for SIMPLE k-PATH on directed graphs. It is possible that working with the group algebra $\mathbb{F}_{r^t}[\mathbb{Z}_r^{O(k/r)}]$ in a similar way could give an algorithm for r-monomial detection with running times comparable to the algorithm of this paper. However, this seems difficult to analyze: [13] and [19] only need to distiniguish between products in the algebra corresponding to independent and dependent vectors; whereas here the situation will be more complex. Moreover, this method has the potential of working only for prime r, whereas our algorithm works for all r (albeit giving the best running time when $r + 1$ is prime).

2   Overview of the Proof of Theorem 2

We give an informal sketch of Theorem 2. We are given a directed graph G on n vertices, and integers $r \leq k$. We wish to decide if G contains an r-simple path of length k. There are two main stages in our algorithm. The first is to reduce the task to another one concerning multivariate polynomials. This part, described below, is very similar to [1].

Reduction to a question about polynomials. We want to associate our graph G with a certain multivariate polynomial p_G.

We associate with the i'th vertex a variable x_i. The monomials of the polynomial will correspond to the paths of length k in G. So we have

$$p_G(\mathbf{x}) = \sum_{i_1 \to i_2 \to \cdots \to i_k \in G} x_{i_1} \cdots x_{i_k},$$

where $i_1 \to i_2 \to \cdots \to i_k \in G$ means that $i_1, i_2, \cdots, i_k$ is a directed path in G. An important issue is *over what field $\mathbb{F}$ is p_G defined?* A central part of the algorithm is indeed choosing the appropriate field to work over. Another issue is how efficiently p_G can be evaluated? (Note that it potentially contains n^k different monomials.) Williams shows in [19] that using the adjacency matrix of G it can be computed in poly(n)-time. See Section 5. For now, think of p_G as defined over $\mathbb{Q}$, i.e., having integer coefficients. It is easy to see that G contains an r-simple path of length k if and only if p_G contains a monomial such that the individual degrees of all variables are at most r. Let us call such a monomial an r-*monomial*. Thus our task is reduced to checking whether a homogenous polynomial of degree k contains an r-monomial.

Checking whether p_G contains an r-monomial. Let us assume in this overview for simplicity that $p = r + 1$ is prime. Let us view p_G as a polynomial over $\mathbb{F}_p$. One problem with doing this is that if we have p directed paths of length k passing through the same vertices in different order, this translates in p_G to p copies of the same monomial summing up to 0. To avoid this we need to look at a variant of p_G that contains auxiliary variables that prevent this cancelation. For details on this issue see [1] and Section 5. For this overview let us assume this does not happen. Recall that we have the equality $a^p = a$ for any $a \in \mathbb{F}_p$. Let us look at a monomial that *is not* an r-monomial, say $x_1^{r+1} \cdot x_2 = x_1^p \cdot x_2$. The equality mentioned implies this monomial is equivalent as a function from $\mathbb{F}_p^n$ to $\mathbb{F}_p$ to the monomial $x_1 \cdot x_2$. By the same argument, any monomial that is *not* an r-monomial will be 'equivalent' to one of smaller degree. More generally, p_G that is homogenous of degree k over $\mathbb{Q}$ will be equivalent to a polynomial of degree smaller than k as a function from $\mathbb{F}_p^n$ to $\mathbb{F}_p$ if and only if it *does not* contain an r-monomial. Thus, we have reduced our task to the problem of *low-degree testing*. In this context, this problem is as follows: Given black-box access to a function $f : \mathbb{F}_p^n \rightarrow \mathbb{F}_p$ of degree at most k, determine whether it has degree exactly k or less than k, using few queries to the function. Here, for a function $f : \mathbb{F}_p^n \rightarrow \mathbb{F}_p$, by its degree we mean the total degree of the lowest-degree polynomial $p \in \mathbb{F}_p[x_1, \ldots, x_n]$ representing it. Haramati, Shpilka and Sudan [12] gave an optimal solution (in terms of the number of queries) to this problem for prime fields. An important observation is that for the right choice of parameters, the test of [12] can be performed in *linear time* in the number of queries. See Section 6 for details. For details on dealing with the case that $r + 1$ is not prime, see Section 7.

3   Definitions and Preliminary Results

In this section we give some definitions and preliminary results that will be used throughout this paper.

Let $G(V, E)$ be a directed graph where V is the set of vertices and $E \subseteq V \times V$ the set of edges. We denote by $n = |V|$ the number of vertices in the graph and by $m = |E|$ the number of edges in the graph. A k-*path* or a *path of length k* is a sequence $\rho = v_1, \ldots, v_k$ such that (v_i, v_{i+1}) is an edge in G for all $i = 1, \ldots, k-1$. A *path* is a k-path for some integer $k > 0$. A path ρ is called *simple* if all the vertices in the path are distinct. We say that a path ρ in G is r-*simple*, if each vertex of G appears in ρ at most r times. Obviously, a simple path is a 1-simple path.

Given as input a directed graph G on n vertices, the r-SIMPLE k-PATH problem asks for a given G whether it contains an r-simple path of length k. When $r = 1$ then the problem is called SIMPLE k-PATH. The r-SIMPLE PATH problem asks for a given G and integer k whether G contains an r-simple k-path of length k.

We show that r-SIMPLE PATH is NP-complete.

Theorem 3. *For any r the decision problem r-SIMPLE PATH is NP-complete.*

The proof of the theorem is deferred to the full version. The idea is to create a copy v' of each vertex v, accessible only from v. A Hamiltonian path in the original graph translates into a long r-simple path by repeatedly going from v to v' back and forth after reaching v in the Hamiltonian path. The above result implies

Corollary 1. *If r-SIMPLE k-PATH can be solved in $T(r, k, n, m)$ time then HAMILTONIAN-PATH can be solved in $T(r, 2rn - n + 2, 2n, m + 2n)$.*

In particular, if there is an algorithm for r-SIMPLE k-PATH that runs in time $poly(n) \cdot 2^{(c/2)(k/r)}$ then there is an algorithm for HAMILTONIAN-PATH that runs in time $poly(n) \cdot 2^{cn}$.

4 Gap

In this section we show that the gap between the longest simple path and the longest r-simple path can be exponentially large even for $r = 2$.

We first give the following lower bound for the gap

Theorem 4. *If G contains an r-simple path of length k then G contains a simple path of length $\lceil \frac{\log k}{\log r} \rceil$.*

Proof: Let $t = \lfloor \frac{\log k - 1}{\log r} \rfloor$. Let ρ be an r-simple path whose first vertex is v_0. We will use ρ to construct a simple path $\bar{\rho}$ of length $\lfloor \frac{\log k}{\log r} \rfloor$. We denote $\rho_0 = \rho$. As v_0 appears at most r times in ρ_0, there must be a subpath ρ_1 of ρ_0 of length at least $(k - r)/r$ where v_0 does not appear. Let v_1 be the first vertex of ρ_1. Similarly, for $1 < i \leq t$, we define the subpath ρ_i of ρ_{i-1} to be a subpath of length at least

$$(k - r - \ldots - r^i)/r^i \geq (k - r^{i+1})/r^i,$$

where $v_1, \ldots, v_{i-1}$ do not appear, and define v_i to be the first vertex of ρ_i. Note that we can always assume there is an edge from v_{i-1} to v_i as we can start ρ_i just after an appearance of v_{i-1} in ρ_{i-1}. Note that for $1 \leq i \leq t$, such a v_i as defined indeed exists as $(k - r^{i+1})/r^i \geq 1$ when

$$k \geq 2 \cdot r^{i+1} \leftrightarrow i + 1 \leq (\log k - 1)/\log r$$

Thus, $v_0 \cdots v_{t-1}$ is a simple path of the desired length. $\square$

Before we give the upper bound we give the following definition. A *full r-tree* is a tree where each vertex has r children and all the leaves of the tree are in the same level. The root is on level 1.

Theorem 5. *There is a graph G that contains an r-simple path of length k and no simple path of length greater than $4 \log k / \log r$.*

Proof: We first give the proof for $r \geq 3$. Consider a full $(r-1)$-tree of depth $\lceil \log n / \log(r-1) \rceil$. Remove vertices from the lowest level (leaves) so the number of vertices in the graph is n. Obviously there is an r-simple path of length $k \geq n$. Any simple tour in this tree must change level at each step and if it changes from level ℓ to level $\ell + 1$ it cannot go back in the following step to level ℓ. So the longest possible simple path is $2\lceil \log n / \log(r-1) \rceil - 2 \leq 3.17(\log k / \log r)$.

For $r = 2$ we take a full binary tree (2-tree) and add an edge between every two children of the same vertex. The 2-simple path starts from the root v, recursively makes a tour in the left tree of v then moves to the root of the right tree of v (via the edge that we added) then recursively makes a tour in the right tree of v and then visit v again. Obviously this is a 2-simple path of length $k > n$. A simple tour in this graph can stay in the same level only twice, can move to a higher level or can move to a lower level. Again here if it moves from level ℓ to $\ell + 1$ it cannot go back in the following step to level ℓ. Therefore the longest simple path is of length at most $4 \log n \leq 4 \log k$.   □

5   From *r*-Simple *k*-Path to Multivariate Polynomial

The purpose of this section is to reduce the question of whether a graph G contains an r-simple k-path, to that of whether a certain multivariate polynomial *contains an r-monomial*, as defined below.

Definition 6 (r-monomial). *Fix a field $\mathbb{F}$. Fix a monomial $M = M(z) = z_1^{i_1} \cdots z_t^{i_t}$.*

- *We say M is an r-monomial if no variable appears with degree larger than r in M. That is, for all $1 \leq j \leq t$, $i_j \leq r$.*
- *Let $f(z)$ be a multivariate polynomial over $\mathbb{F}$. We say f contains an r-monomial, if there is an r-monomial $M(z)$ appearing with a nonzero coefficient $c \in \mathbb{F}$ in f.*

We now describe this reduction.

Let $G(V, E)$ be a directed graph where $V = \{1, 2, \ldots, n\}$. Let A be the adjacency matrix and B be the $n \times n$ matrix such that $B_{i,j} = x_i \cdot A_{i,j}$ where x_i, $i = 1, \ldots, n$ are indeterminates. Let $\mathbf{1}$ be the row n-vector of 1s and $\mathbf{x} = (x_1, \ldots, x_n)^T$. Consider the polynomial $p_G(\mathbf{x}) = \mathbf{1} \cdot B^{k-1} \cdot \mathbf{x}$. It is easy to see

$$p_G(\mathbf{x}) = \sum_{i_1 \to i_2 \to \cdots \to i_k \in G} x_{i_1} \cdots x_{i_k}$$

where $i_1 \to i_2 \to \cdots \to i_k \in G$ means that $i_1, i_2, \cdots, i_k$ is a directed path in G.

Obviously, for field of characteristic zero there is an r-simple k-path if and only if $p_G(\boldsymbol{x})$ contains an r-monomial. For other fields the later statement is not true. For example, in undirected graph, $k = 2$, and $r = 1$ if $(1, 2) \in E$ and the field is of characteristic 2 then the monomial $x_1 x_2$ occurs twice and will vanish in $p_G(\boldsymbol{x})$. We solve the problem as follows.

Let $B^{(m)}$ be an $n \times n$ matrices, $m = 2, \ldots, k$, such that $B_{i,j}^{(m)} = x_i \cdot y_{m,i} \cdot A_{i,j}$ where x_i and $y_{m,i}$ are indeterminates. Let, $\boldsymbol{y} = (\boldsymbol{y}_1, \ldots, \boldsymbol{y}_k)$ and $\boldsymbol{y}_m = (y_{m,1}, \ldots, y_{m,n})$. Let $\boldsymbol{x} \boldsymbol{.} \boldsymbol{y} = (x_1 y_{1,1}, \ldots, x_n y_{1,n})$. Consider the polynomial $P_G(\boldsymbol{x}, \boldsymbol{y}) = 1 B^{(k)} B^{(k-1)} \cdots B^{(2)}(\boldsymbol{x} \boldsymbol{.} \boldsymbol{y})$. It is easy to see that

$$P_G(\boldsymbol{x}, \boldsymbol{y}) = \sum_{i_1 \to i_2 \to \cdots \to i_k \in G} x_{i_1} \cdots x_{i_k} y_{1,i_1} \cdots y_{k,i_k}$$

Obviously, no two paths have the same monomial in P_G. Note that as P_G contains only $\{0, 1\}$ coefficients, we can define it over any field $\mathbb{F}$. It will actually be convenient to view it as a polynomial $P_G(\boldsymbol{x})$ whose coefficients are in the field of rational functions $\mathbb{F}(\boldsymbol{y})$. Therefore, for any field, there is an r-simple k-path if and only if $P_G(\boldsymbol{x}, \boldsymbol{y})$ contains an r-monomial in $\boldsymbol{x}$. We record this fact in the lemma below.

Lemma 1. *Fix any field* $\mathbb{F}$. *The graph* G *contains an* r-simple k-path *if and only if the polynomial* P_G, *defined over* $\mathbb{F}(\boldsymbol{y})$, *contains an* r-monomial $M(\boldsymbol{x})$.

6   Low Degree Tester

In this section we present a tester that determines whether a function $f : \mathbb{F}_p^n \to \mathbb{F}_p$ of degree *at most* d has, in fact, degree *less than* d. The important point is that the tester will be able to do this using few black-box queries to f. The results of this section essentially follow from the work of Haramaty, Sudan and Shpilka [12]. A crucial observation is that for a certain choice of parameters the low-degree test of [12] can be performed in linear time in the number of queries.

First, let us say precisely what we mean by the *degree* of a function $f : \mathbb{F}_p^n \to \mathbb{F}_p$. We define this to be the degree of the lowest degree polynomial $f' \in \mathbb{F}_p[\mathbf{x}]$ that agrees with f as a function from $\mathbb{F}_p^n$ to $\mathbb{F}_p$. It is known from the theory of finite fields that there is a unique such f', and that the individual degrees of all variables in f' are smaller than p. Moreover, given any polynomial $g \in \mathbb{F}_p[\boldsymbol{x}]$ agreeing with f as a function from $\mathbb{F}_p^n$ to $\mathbb{F}_p$, f' can be derived from g by replacing, for any $1 \leq i \leq n$, occurrences of x_i^t with $x_i^t \mod x_i^p - x_i$ (i.e., $x_i^{((t-1) \mod (p-1))+1}$ when $t \neq 0$). We do not prove these basic facts formally here. They essentially follow from the fact that $a^p = a$ for $a \in \mathbb{F}_p$.

This motivates the following definition.

Definition 7. *Fix positive integers* n, d *and a prime* p. *Let* $f \in \mathbb{F}_p[\boldsymbol{x}] = \mathbb{F}_p[x_1, \ldots, x_n]$. *We define* $\deg_p(f)$ *to be the degree of the polynomial* f *when replacing, for* $1 \leq i \leq n$, x_i^t *by* $(x_i^t \mod x_i^p - x_i)$. *More formally,* $\deg_p(f) \triangleq \deg(f')$ *where*

$$f'(x_1, \ldots, x_n) \triangleq f(x_1, \ldots, x_n) \bmod x_1^p - x_1, \ldots, \bmod x_n^p - x_n.$$

Moreover, for a function $g : V \to \mathbb{F}_p$ where $V \subseteq \mathbb{F}_p^n$ is a subspace of dimension k, we define $\deg_p(g) = \min_f \deg_p(f)$ where $f \in \mathbb{F}_p[x_1, \ldots, x_n]$ and $f|_V = g$. Here g can be regarded as a function in $\mathbb{F}_p[x_1, \ldots, x_k]$.

We note that this notion of degree is affine invariant, i.e doesn't change after affine transformations. In addition it has the property that for any affine subspace V, $\deg_p(f|_V) \leq \deg_p(f)$.

We now present the main result of this section.

Lemma 2. *There is a randomized algorithm A running in time $\mathrm{poly}(n) \cdot p^{\lceil \frac{d}{p-1} \rceil + 1}$ that determines with constant one-sided error whether a function f of degree at most d has degree less than d. More precisely, given black-box access to a function $f : \mathbb{F}_p^n \to \mathbb{F}_p$ with $\deg_p(f) \leq d$,*

- *If $\deg_p(f) = d$, A accepts with probability at least $99/100$.*
- *If $\deg_p(f) < d$, A rejects with probability one.*

The proof of Lemma 2 is can be found in the full version.

7   Testing if P_G Contains an r-Monomial

In this section we present a method for testing whether the polynomial P_G, described in Section 5, contains an r-monomial. This is done using the low-degree tester from the previous section.

As stated in Lemma 1, this is precisely equivalent to whether G contains an r-simple k-path. Recall we viewed P_G as a polynomial over a field of rational functions $\mathbb{F}_p(\mathbf{y})$. To obtain efficient algorithms, we first reduce the question to checking whether a different polynomial defined over $\mathbb{F}_p$ rather than $\mathbb{F}_p(\mathbf{y})$ contains an r-monomial. It is important in the next Lemma that we are able to do this reduction for *any* p, in particular a 'small' one.

Lemma 3. *Fix any integers r, k, with $r \leq k$. Let p be any prime and $t = \lceil \log_p 10k \rceil$. Let G be a directed graph on n vertices. Given an adjacency matrix A_G for G, we can return in $\mathrm{poly}(n)$-time $\mathrm{poly}(n)$-size circuits computing polynomials $f_G^1, \ldots, f_G^t : \mathbb{F}_p^n \to \mathbb{F}_p$ on inputs in $\mathbb{F}_p^n$ such that*

- *For $1 \leq i \leq t$, f_G^i is (either the zero polynomial or) homogenous of degree k.*
- *If G contains an r-simple k-path then with probability at least $9/10$, for some $1 \leq i \leq t$, f_G^i contains an r-monomial.*
- *If G does not contain an r-simple k-path, for all $1 \leq i \leq t$, f_G^i does not contain an r-monomial.*

Proof: Note that the discussion in Section 5 implies we can compute P_G in $\mathrm{poly}(n)$-time over inputs in $\mathbb{F}_p^{2n}$. We choose random $\mathbf{b} \in \mathbb{F}_{p^t}^n$ and let

$$f_G(\mathbf{x}) \triangleq P_G(\mathbf{x}, \mathbf{b}).$$

Suppose P_G, as a polynomial over $\mathbb{F}(\mathbf{y})$, contains an r-monomial $M'(\mathbf{x})$. The coefficient $c_{M'}(\mathbf{y})$ of M' in P_G is a nonzero polynomial of degree k. So, by the Schwartz-Zippel Lemma, $c_{M'}(\mathbf{b}) = 0$ with probability at most $k/p^t \le 1/10$. In the event that $c_{M'}(\mathbf{b}) \ne 0$, $f_G(\mathbf{x})$ is a homogenous polynomial of degree k in $\mathbb{F}_{p^t}[\mathbf{x}]$ containing an r-monomial. Let us assume from now on, we chose a $\mathbf{b}$ such that indeed $a_M \triangleq c_{M'}(\mathbf{b}) \ne 0$. We now discuss how to end up with polynomials having coefficients in $\mathbb{F}_p$ rather than $\mathbb{F}_{p^t}$.

Let $T_1, \ldots, T_t : \mathbb{F}_{p^t} \to \mathbb{F}_p$ be independent $\mathbb{F}_p$-linear maps. Suppose $f_G = \sum_M a_M \cdot M(\mathbf{x})$. For $1 \le i \le t$, define a polynomial $f_G^i \in \mathbb{F}_p[\mathbf{x}]$ by

$$f_G^i(\boldsymbol{x}) \triangleq \sum_M T_i(a_M) \cdot M(\boldsymbol{x}).$$

Note that for all $1 \le i \le t$, f_G^i is the zero polynomial or homogenous of degree k. As $a_{M'} \ne 0$, for some i, $T_i(a_{M'}) \ne 0$. For this i, f_G^i is homogenous of degree k and contains an r-monomial, specifically the r-monomial $a_{M'} \cdot M'(\mathbf{x})$. We claim that for all $1 \le i \le t$, f_G^i can be computed by a $poly(n)$-size circuit on inputs $\mathbf{a} \in \mathbb{F}_p^n$. This is because f_G and T_i are efficiently computable, and because for $\mathbf{a} \in \mathbb{F}_p^n$,

$$T_i(f_G(\mathbf{a})) = T_i\left(\sum_M a_M \cdot M(\mathbf{a})\right) = \sum_M T_i(a_M) \cdot M(\mathbf{a}) = f_G^i(\mathbf{a}),$$

where the second equality is due to the $\mathbb{F}_p$-linearity of T_i. $\qquad\square$

The above lemma implies

Corollary 2. *Fix any prime p. Suppose that given black-box access to a polynomial $g \in \mathbb{F}_p[\mathbf{x}]$ that is homogenous of degree k, we can determine in time $poly(n) \cdot S$ if it contains an r-monomial. Then we can also determine in time $poly(n) \cdot S$ whether P_G as a polynomial over $\mathbb{F}_p(\mathbf{y})$ contains an r monomial.*

Our reduction to low-degree testing is based on the following simple observation that, for the right p and for homogenous polynomials, containing an r-monomial is equivalent to having a certain $\deg_p$-degree.

Lemma 4. *Suppose $g \in \mathbb{F}_p[\mathbf{x}]$ is a homogenous polynomial of degree k. Suppose $r = p - 1$. Then $\deg_p(g) = k$ if and only if g contains an r-monomial.*

Proof: If g contains an r-monomial M then, as $r < p$, $\deg_p(M) = k$, which implies that $\deg_p(g) = k$. If g does not contain an r-monomial, then for every monomial M in g there is an $i \in [n]$ such that the degree of x_i in M is at least $r + 1 = p$. So replacing x_i^p by x_i will reduce the degree of M and therefore $\deg_p(M) < k$. Since this happens for all monomials of g, $\deg_p(g) < k$. $\qquad\square$

We introduce another element on notation that will be convenient in the rest of this section.

Definition 8. *Fix integers n, d and prime p. Let $f \in \mathbb{F}_p[\mathbf{x}]$ be an n-variate polynomial of degree at most d. We define $LDT(f, n, d, p)$ to be 1 if $\deg_p(f) = d$, and 0 otherwise.*

Before proceeding, we note that the results of Section 6 imply that given n, d, p and black-box access to f, $LDT(f, n, d, p)$ can be computed in time $\mathrm{poly}(n) \cdot O(p^{\lceil d/(p-1) \rceil + 1})$. In particular, if given $\mathbf{a} \in \mathbb{F}_p^n$, we can compute $f(\mathbf{a})$ in $\mathrm{poly}(n)$-time, then we can compute $LDT(f, n, d, p)$ in time $\mathrm{poly}(n) \cdot O(p^{\lceil d/(p-1) \rceil + 1})$. The following lemma is an easy corollary of Lemma 4.

Lemma 5. *Fix integers r, k with $r \leq k$. Suppose $p = r + 1$ is prime. Let $g \in \mathbb{F}_p[\mathbf{x}]$ be homogenous of degree k and computable in $\mathrm{poly}(n)$-time. There is a randomized algorithm running in time*

$$\mathrm{poly}(n) \cdot O((r+1)^{\lceil \frac{k}{r} \rceil + 1})$$

determining whether g contains an r-monomial.

Proof: The algorithm simply returns $LDT(g, n, d = k, p = r + 1)$. The running time follows from the discussion above. The correctness follows from Lemma 4. $\square$

We wish to have a similar result when $r + 1$ is not a prime.

Lemma 6. *Fix integers r, k with $r \leq k$. Let p be the smallest prime such that $\frac{p-1}{r} \in \mathbb{Z}$. Let $g \in \mathbb{F}_p[\mathbf{x}]$ be homogenous of degree k and computable by a $\mathrm{poly}(n)$-size circuit. There is a randomized algorithm running in time $\mathrm{poly}(n) \cdot O(p^{\lceil \frac{k}{r} \rceil + 1})$ determining whether g contains an r-monomial.*

Proof: Denote $l \triangleq \frac{p-1}{r}$ and define

$$h(x_1, x_2, \ldots, x_n) := g(x_1^l, x_2^l, \ldots, x_n^l).$$

The algorithm returns $LDT(h, n, d = k \cdot l, p)$.

Note that h is homogenous of degree $k \cdot l$. Note also that h contains an $r \cdot l$-monomial if and only if g contains an r-monomial. As $r \cdot l + 1 = p$ correctness now follows from Lemma 4.

$\square$

The best known bound for the smallest prime number p that satisfies $r | p - 1$ is $r^{5.5}$ due to Heath-Brown [18]. This gives a randomized algorithm running in time

$$\mathrm{poly}(n) \cdot O(r^{\frac{5.5k}{r} + O(1)}).$$

Schinzel, Sierpinski, and Kanold have conjectured the value to be 2 [18]. In the following Theorem we give a better bound. We first give the following

Lemma 7. *Fix integers r, k with $r \leq k$. Let p be the smallest prime such that there is an $l \in \mathbb{Z}$ for which $r \cdot l \leq p - 1$ and $(r + 1) \cdot l > p - 1$. Let $g \in \mathbb{F}_p[\mathbf{x}]$ be homogenous of degree k and computable by a $\mathrm{poly}(n)$-size circuit. There is a randomized algorithm running in time*

$$\mathrm{poly}(n) \cdot O\left(p^{\lceil \frac{l \cdot k}{p-1} \rceil + 1}\right)$$

determining whether g contains an r-monomial.

Proof: As in the proof of Lemma 6, we define $h(x_1, x_2, ..., x_n) \triangleq g(x_1^l, x_2^l, ..., x_n^l)$. The algorithm returns $LDT(h, n, d = k \cdot l, p)$. As in the proof of Lemma 6, h is homogenous of degree $k \cdot l$ and contains an $(r \cdot l)$-monomial if and only if g contains an r-monomial. Furthermore, as $r \cdot l \le p-1$ and $(r+1) \cdot l \ge p$, h contains a $(p-1)$-monomial if and only if g contains an r-monomial. Correctness now follows from Lemma 4.

$\square$

The main result of this section contains two results. The first is unconditional. The second is true if Cramer's conjecture is true. Cramer's conjecture states that the gap between two consecutive primes $p_{n+1} - p_n = O(\log^2 p_n)$, [9].

Theorem 9. *(Unconditional Result) Fix any integers r, k with $2 \le r \le k$. Let $g \in \mathbb{F}_p[\mathbf{x}]$ be homogenous of degree k and computable by a poly(n)-size circuit. There is a randomized algorithm running in time*

$$\text{poly}(n) \cdot O\left(r^{\frac{2k}{r} + O(1)}\right)$$

determining whether g contains an r-monomial.

(Conditional Result) If Cramer's Conjecture is true then the time complexity of the algorithm is

$$\text{poly}(n) \cdot O\left(r^{\frac{k}{r} + o\left(\frac{k}{r}\right)}\right).$$

Proof: We will find p and l as required in Lemma 7. Fix a prime p such that $r^2 + r + 1 < p < 2r^2 + 2r \le 3r^2$. (This can be done as for any positive integer $t > 3$, there is always a prime between t and $2t$.)

Define $l \triangleq \lfloor \frac{p-1}{r} \rfloor$. We have

$$r \cdot l = r \cdot \lfloor \frac{p-1}{r} \rfloor \le p-1$$

$$(r+1) \cdot l \ge (r+1) \cdot (\frac{p-1}{r} - 1) = (p-1) + \frac{p-1}{r} - r - 1 > (p-1)$$

The first claim now follows from Lemma 7 and Corollary 2.

If Cramer's conjecture is true then there is a constant c such that for every integer x there is a prime number in $[x, x + c \log^2(x)]$. Then there is a prime number p in the interval $[2cr \log^2 r, 2cr \log^2 r + c \log^2(2cr \log^2 r)]$ and we can choose $l = 2c \log^2 r$. Then the time complexity will be

$$\text{poly}(n) \cdot O\left(r^{\frac{k}{r} + o\left(\frac{k}{r}\right)}\right).$$

$\square$

In the full version we summarize the running times attained for $r \le 11$.

References

1. Abasi, H., Bshouty, N.H.: A simple algorithm for undirected hamiltonicity. Electronic Colloquium on Computational Complexity (ECCC) 20, 12 (2013)
2. Alon, N., Yuster, R., Zwick, U.: Color-Coding. J. ACM 42(4), 844–856 (1995)
3. Bellman, R.: Dynamic programming treatment of the travelling salesman problem. J. Assoc. Comput. Mach. 9, 61–63 (1962)
4. Bellman, R.: Combinatorial processes and dynamic programming, Combinatorial Analysis. In: Bellman, R., Hall, M. (eds.) Proceedings of Symposia in Applied Mathematics 10, American Mathematical Society, pp. 217–249 (1960)
5. Bodlaender, H.L.: On linear time minor tests with depth-first search. J. Algorithm. 14(1), 1–23 (1993)
6. Björklund, A., Husfeldt, T., Khanna, S.: Approximating Longest Directed Paths and Cycles. In: Díaz, J., Karhumäki, J., Lepistö, A., Sannella, D. (eds.) ICALP 2004. LNCS, vol. 3142, pp. 222–233. Springer, Heidelberg (2004)
7. Björklund, A., Husfeldt, T., Kaski, P., Koivisto, M.: Narrow sieves for parameterized paths and packings. CoRR abs/1007.1161 (2010)
8. Baker, R.C., Harman, G., Pintz, J.: The Difference between Consecutive Primes. II. Proc. London Math. Soc. 83(3), 532–562 (2001)
9. Cramer, H.: On the order of magnitude of the difference between consecutive prime numbers. Acta Arithmetica 2, 23–46 (1936)
10. Chen, J., Lu, S., Sze, S.-H., Zhang, F.: Improved algorithms for path, matching, and packing problems. In: Proc. 18th Annual ACM SIAM Symposium on Discrete Algorithms, SODA 2007, Philadelphia, PA, USA, pp. 298–307 (2007)
11. Gabow, H.N., Nie, S.: Finding Long Paths, Cycles and Circuits. In: Hong, S.-H., Nagamochi, H., Fukunaga, T. (eds.) ISAAC 2008. LNCS, vol. 5369, pp. 752–763. Springer, Heidelberg (2008)
12. Haramaty, E., Shpilka, A., Sudan, M.: Optimal Testing of Multivariate Polynomials over Small Prime Fields. SIAM J. Comput. 42(2), 536–562 (2013)
13. Koutis, I.: Faster algebraic algorithms for path and packing problems. In: Aceto, L., Damgård, I., Goldberg, L.A., Halldórsson, M.M., Ingólfsdóttir, A., Walukiewicz, I. (eds.) ICALP 2008, Part I. LNCS, vol. 5125, pp. 575–586. Springer, Heidelberg (2008)
14. Koutis, I., Williams, R.: Limits and applications of group algebras for parameterized problems. In: Albers, S., Marchetti-Spaccamela, A., Matias, Y., Nikoletseas, S., Thomas, W. (eds.) ICALP 2009, Part I. LNCS, vol. 5555, pp. 653–664. Springer, Heidelberg (2009)
15. Karger, D.R., Motwani, R., Ramkumar, G.D.S.: On Approximating the Longest Path in a Graph. Algorithmica 18(1), 82–98 (1997)
16. Kneis, J., Mölle, D., Richter, S., Rossmanith, P.: Divide-and-Color. In: Fomin, F.V. (ed.) WG 2006. LNCS, vol. 4271, pp. 58–67. Springer, Heidelberg (2006)
17. Monien, B.: How to find long paths efficiently. Annals of Discrete Mathematics 25, 239–254 (1985)
18. Ribenboim, P.: The New Book of Prime Number Records. Springer, New York (1996)
19. Williams, R.: Finding paths of length k in $O^*(2^k)$. Inform. Process Lett. 109(6), 301–338 (2009)

Low-Depth Uniform Threshold Circuits and the Bit-Complexity of Straight Line Programs

Eric Allender[1], Nikhil Balaji[2], and Samir Datta[2]

[1] Department of Computer Science, Rutgers University, USA
allender@cs.rutgers.edu
[2] Chennai Mathematical Institute, India
{nikhil,sdatta}@cmi.ac.in

Abstract. We present improved uniform TC^0 circuits for division, matrix powering, and related problems, where the improvement is in terms of "majority depth" (as studied by Maciel and Thérien). As a corollary, we obtain improved bounds on the complexity of certain problems involving arithmetic circuits, which are known to lie in the counting hierarchy.

1 Introduction

How hard is it to compute the 10^{100}-th bit of the binary expansion of $\sqrt{2}$? Datta and Pratap [6], and Jeřábek [15] considered the question of computing the m-th bit of an algebraic number. Jeřábek [15] showed that this problem has uniform TC^0 circuits[1] of size polynomial in m (which is not so useful when $m = 10^{100}$). Earlier, Datta and Pratap showed a related result: when m is expressed in *binary*, this problem lies in the counting hierarchy. More precisely, Datta and Pratap showed that this problem is reducible to the problem of computing certain bits of the quotient of two numbers represented by arithmetic circuits of polynomial size.[2] Thus, we are led to the problem of evaluating arithmetic circuits. **In this paper, we focus on arithmetic circuits *without input variables*.** Thus an arithmetic circuit is a (possibly very compact) representation of a number.

Arithmetic circuits of polynomial size can produce numbers that require exponentially-many bits to represent in binary. The problem[3] known as BitSLP ($= \{(C, i, b) :$ the i-th bit of the number represented by arithmetic circuit C is $b\}$) is known to be hard for #P [3]. It was known that BitSLP lies in the counting hierarchy [3], but the best previously-known bound for this problem is the bound mentioned in [3] and credited there to [2]: $\mathsf{PH}^{\mathsf{PP}^{\mathsf{PP}^{\mathsf{PP}^{\mathsf{PP}}}}}$. That bound follows via a straightforward translation of a uniform TC^0 algorithm presented in [12].

---

[1] For somewhat-related TC^0 algorithms on sums of radicals, see [14].

[2] It is mistakenly claimed in [6] that this problem lies in $\mathsf{PH}^{\mathsf{PP}^{\mathsf{PP}}}$. In this paper, we prove the weaker bound that it lies in $\mathsf{PH}^{\mathsf{PP}^{\mathsf{PP}^{\mathsf{PP}}}}$.

[3] "SLP" stands for "straight-line program": a model equivalent to arithmetic circuits. Throughout the rest of the paper, we will stick with the arithmetic circuit formalism.

E. Csuhaj-Varjú et al. (Eds.): MFCS 2014, Part II, LNCS 8635, pp. 13–24, 2014.
© Springer-Verlag Berlin Heidelberg 2014

In this paper, we improve this bound on the complexity of BitSLP to $\mathsf{PH}^{\mathsf{PP}^{\mathsf{PP}^{\mathsf{PP}^{\mathsf{PP}}}}}$. In order to do this, we present improved uniform TC^0 algorithms for a number of problems that were already known to reside in uniform TC^0. The improvements that we provide are related to the *depth* of the TC^0 circuits. There are several possible variants of "depth" that one could choose to study. For instance, several papers have studied circuits consisting only of majority gates, and tight bounds are known for the depth required for several problems, in that model. (See, for instance [10,24,27,23] and other work referenced there.) Since our motivation comes largely from the desire to understand the complexity of problems in the counting hierarchy, it turns out that it is much more relevant to consider the notion of *majority depth* that was considered by Maciel and Thérien [20]. In this model, circuits have unbounded-fan-in AND, OR, and MAJORITY gates (as well as NOT gates). The class $\widehat{\mathsf{TC}}^0_d$ consists of functions computable by families of threshold circuits of polynomial size and constant depth such that no path from an input to an output gate encounters more than d MAJORITY gates. Thus the class of functions with majority depth zero, $\widehat{\mathsf{TC}}^0_0$, is precisely AC^0. In order to explain the connection between $\widehat{\mathsf{TC}}^0_d$ and the counting hierarchy, it is necessary to define the levels of the counting hierarchy.

Define $\mathsf{CH}_1 = \mathsf{PP}$, and $\mathsf{CH}_{k+1} = \mathsf{PP}^{\mathsf{CH}_k}$.

Proposition 1. *(Implicit in [3, Theorem 4.1].) Let A be a set such that for some k, some poly-time function f and some dlogtime-uniform $\widehat{\mathsf{TC}}^0_d$ circuit family C_n:*
$$x \in A \text{ iff } C_{|x|+2^{|x|^k}}(x, f(x,1)f(x,2)\dots f(x, 2^{|x|^k})) \text{ accepts. Then } A \in \mathsf{PH}^{\mathsf{CH}_d}.$$

(One important part of the proof of Proposition 1 is the fact that, by Toda's theorem [25], for every oracle A, $\mathsf{PP}^{\mathsf{PH}^A} \subseteq \mathsf{P}^{\mathsf{PP}^A}$. Thus all of the AC^0 circuitry inside the $\widehat{\mathsf{TC}}^0_d$ circuit can be swallowed up by the PH part of the simulation.)

Note that the dlogtime-uniformity condition is crucial for Proposition 1. Thus, for the remainder of this paper, all references to $\widehat{\mathsf{TC}}^0_d$ will refer to the dlogtime-uniform version of this class, unless we specifically refer to nonuniform circuits. Table 1 compares the complexity bounds that Maciel and Thérien obtained in the *nonuniform* setting with the bounds that we are able to obtain in the uniform setting. (Maciel and Thérien also considered several problems for which they gave uniform circuit bounds; the problems listed in Table 1 were not known to lie in dlogtime-uniform TC^0 until the subsequent work of [12].) All previously-known dlogtime-uniform TC^0 algorithms for these problems rely on the CRR-to-binary algorithm of [12], and thus have at *least* majority-depth 4 (as analyzed by [2]); no other depth analysis beyond $O(1)$ was attempted.

In all of the cases where our uniform majority-depth bounds are worse than the nonuniform bounds given by [20], our algorithms also give rise to nonuniform algorithms that match the bounds of [20] (by hardwiring in some information that depends only on the length), although in all cases the algorithms differ in several respects from those of [20].

Problem	Nonuniform Majority-Depth [20]	Uniform Majority-Depth [This Paper]
Iterated multiplication	3	3
Division	2	3
Powering	2	3
CRR-to-binary	1	3
Matrix powering	$O(1)$ [21,12]	3

All of the TC^0 algorithms that are known for the problems considered in this paper rely on partial evaluations or approximations. The technical innovations in our improved algorithms rely on introducing yet another approximation, as discussed in Lemmas 2 and 3.

Table 1 also lists one problem that was not considered by Maciel and Thérien: given as input 1^m and a $k \times k$ matrix A, produce A^m. For any fixed k, this problem was put in nonuniform TC^0 [21]; and by [12] it is also in dlogtime-uniform TC^0. The corresponding problem of computing *large* powers of a $k \times k$ matrix (i.e., when m is given in *binary*) has been discussed recently; see the final section of [22]. We show that this version of matrix powering is in $\mathsf{PH}^{\mathsf{PP}^{\mathsf{PP}^{\mathsf{PP}}}}$.

In addition to BitSLP, there has also been interest in the related problem PosSLP ($= \{C :$ the number represented by arithmetic circuit C is positive$\}$) [9,18,17,19]. PosSLP $\in \mathsf{PH}^{\mathsf{PP}^{\mathsf{PP}}}$, and is not known to be in PH [3], but in contrast to BitSLP, it is not known (or believed [9]) to be NP-hard. Our theorems do not imply any new bounds on the complexity of PosSLP, but we do conjecture that BitSLP and PosSLP both lie in $\mathsf{PH}^{\mathsf{PP}}$. This conjecture is based mainly on the heuristic that says that, for problems of interest, if a nonuniform circuit is known, then corresponding dlogtime-uniform circuits usually also exist. Converting from CRR to binary can be done nonuniformly in majority-depth one, and there is no reason to believe that this is not possible uniformly – although it seems clear that a different approach will be needed, to reach this goal.

The well-studied Sum-of-Square-Roots problem reduces to PosSLP [3], which in turn reduces to BitSLP. But the relationship between PosSLP and the matrix powering problem (given a matrix A and n-bit integer j, output the jth bit of a given entry of A^j) is unclear, since matrix powering corresponds to evaluating *very restricted* arithmetic circuits. Note that some types of arithmetic involving large numbers *can* be done in P; see [13]. Might matrix powering also lie in PH?

We provide a very weak "hardness" result for the problem of computing the bits of large powers of 2-by-2 matrices, to shed some dim light on this question. We show that the Sum-of-Square-Roots problem reduces to this problem via $\mathsf{PH}^{\mathsf{PP}}$-Turing reductions. Due to lack of space, we defer the proof to the full version of this paper.

2   Preliminaries

Given a list of primes $\Pi = (p_1, \ldots, p_m)$ and a number X, the CRR_Π representation of X is the list $(X \bmod p_1, \ldots, X \bmod p_m)$. We omit the subscript Π if context makes clear. For more on complexity classes such as $\mathsf{AC}^0, \mathsf{TC}^0, \mathsf{NC}^1$, as well as a discussion of dlogtime uniformity, see [26].

3   Uniform Circuits for Division

Theorem 1. *The function taking as input $X \in [0, 2^n), Y \in [1, 2^n)$, and 0^m and producing as output the binary expansion of X/Y correct to m places is in $\widehat{\mathsf{TC}}_3^0$.*

Proof. This task is trivial if $Y = 1$; thus assume that $Y \geq 2$. Computing the binary expansion of Z/Y correct to m places is equivalent to computing $\lfloor 2^m Z/Y \rfloor$. Thus we will focus on the task of computing $\lfloor X/Y \rfloor$, given integers X and Y.

Our approach will be to compute $\widetilde{V}(X, Y)$, a strict underestimate of X/Y, such that $X/Y - \widetilde{V}(X, Y) < 1/Y$. Since $Y > 1$, we have that $\lfloor X/Y \rfloor \neq \lfloor (X + 1)/Y \rfloor$ if and only if $(X+1)/Y = \lfloor X/Y \rfloor + 1$. It follows that in *all* cases $\lfloor X/Y \rfloor = \lfloor \widetilde{V}(X + 1, Y) \rfloor$, since

$$\left\lfloor \frac{X}{Y} \right\rfloor \leq \frac{X}{Y} = \frac{X+1}{Y} - \frac{1}{Y} < \widetilde{V}(X + 1, Y) < \frac{X+1}{Y}.$$

Note: in order to compute $\lfloor \frac{X}{Y} \rfloor$, we compute an approximation to $(X + 1)/Y$.

The approximation $\widetilde{V}(X, Y)$ is actually defined in terms of another rational approximation $W(X, Y)$, which will have the property that $\widetilde{V}(X, Y) \leq W(X, Y) < X/Y$. We postpone the definition of $\widetilde{V}(X, Y)$, and focus for now on $W(X, Y)$, an under approximation of $\frac{X}{Y}$ with error at most $2^{-(n+1)}$.

Using AC^0 circuitry, we can compute a value t such that $2^{t-1} \leq Y < 2^t$.

Let $u = 1 - 2^{-t}Y$. Then $u \in (0, \frac{1}{2}]$. Thus, $Y^{-1} = 2^{-t}(1 - u)^{-1} = 2^{-t}(1 + u + u^2 + \ldots)$. Set $Y' = 2^{-t}(1 + u + u^2 + \ldots + u^{2n+1})$, then

$$0 < Y^{-1} - Y' \leq 2^{-t} \sum_{j>2n+1} 2^{-j} < 2^{-(2n+1)}$$

Define $W(X, Y)$ to be XY'. Hence, $0 < \frac{X}{Y} - W(X, Y) < 2^{-(n+1)}$.

We find it useful to use this equivalent expression for $W(X, Y)$:

$$W(X, Y) = \frac{X}{2^t} \sum_{j=0}^{2n+1} \left(1 - \frac{Y}{2^t}\right)^j = \frac{1}{2^{2(n+1)t}} \sum_{j=0}^{2n+1} X(2^t - Y)^j 2^{(2n+1-j)t};$$

$W(X, Y) = \frac{1}{2^{2(n+1)t}} \sum_{j=0}^{2n+1} W_j(X, Y)$, where $W_j(X, Y) = X(2^t - Y)^j (2^{(2n+1-j)t})$.

Lemma 1. *(Adapted from [6]) Let Π be any set of primes such that the product M of these primes lies in $(2^{n^c}, 2^{n^d})$ for some $d > c \geq 3$. Then, given X, Y, Π we can compute the CRR_Π representations of the $2(n + 1)$ numbers $W_j(X, Y)$ (for $j \in \{0, \ldots, 2n + 1\}$) in $\widehat{\mathsf{TC}}_1^0$.*

Proof. Using AC^0 circuitry, we can compute $2^t - Y$, $2^j \bmod p$ for each prime $p \in \Pi$ and various powers j, as well as finding generators mod p. In $\widehat{\mathsf{TC}}_1^0$ we can compute $X \bmod p$ and $(2^t - Y) \bmod p$ (each of which has $O(\log n)$ bits). Using those results, with AC^0 circuitry we can compute the powers $(2^t - Y)^j \bmod p$ and then do additional arithmetic on numbers of $O(\log n)$ bits to obtain the product $X(2^t - Y)^j(2^{(2n+1-j)t}) \bmod p$ for each $p \in \Pi$. (The condition that $c \geq 3$ ensures that the numbers that we are representing are all less than M.) □

Having the CRR_Π representation of the number $W_j(X, Y)$, our goal might be to convert the $W_j(X, Y)$ to binary, and take their sum. For efficiency, instead we compute an approximation (in binary) to $W(X, Y)/M$ where $M = \prod_{p \in \Pi} p$. In Lemma 3 we build on this to compute our approximation $\widetilde{V}(X, Y)$ to $W(X, Y)$.

Recall that $W(X, Y) = \frac{1}{2^{2(n+1)t}} \sum_{j=0}^{2n+1} W_j(X, Y)$. Thus $2^{2(n+1)t}W(X, Y)$ is an integer with the same significant bits as $W(X, Y)$.

Lemma 2. *Let Π be any set of primes such that the product M of these primes lies in $(2^{n^c}, 2^{n^d})$ for a fixed constant $d > c \geq 3$, and let b be any natural number. Then, given X, Y, Π we can compute the binary representation of a good approximation to $\frac{2^{2(n+1)t}W(X,Y)}{M}$ in $\widehat{\mathsf{TC}}_2^0$ (where by good we mean that it under-estimates the correct value by at most an additive term of $1/2^{n^b}$).*

Proof. Let $h_p^\Pi = (M/p)^{-1} \bmod p$ for each prime $p \in \Pi$.

If we were to compute a good approximation $\tilde{A}_\Pi$ to the fractional part of:

$$A_\Pi = \sum_{p \in \Pi} \frac{(2^{2(n+1)t}W(X, Y) \bmod p)h_p^\Pi}{p}$$

i.e. if $\tilde{A}_\Pi$ were a good approximation to $A_\Pi - \lfloor A_\Pi \rfloor$, then $\tilde{A}_\Pi M$ would be a good approximation to $2^{2(n+1)t}W(X, Y)$. This follows from observing that the fractional part of A_Π is exactly $\frac{2^{2(n+1)t}W(X,Y)}{M}$ (as in [12,3]).

Instead, we will compute a good approximation $\widetilde{A'}_\Pi$ to the fractional part of

$$A'_\Pi = \sum_{p \in \Pi} \sum_{j=0}^{2n+1} \frac{(W_j(X, Y) \bmod p)h_p^\Pi}{p}.$$

Note: the two quantities A_Π, A'_Π are not equal but their *fractional parts* are. Since we are adding $2(n + 1)|\Pi|$ approximate quantities it suffices to compute each of them to $b_m = 2n^b + 2(n + 1)|\Pi|$ bits of accuracy to ensure:

$$0 \leq \frac{W(X, Y)}{M} - \widetilde{A'}_\Pi < \frac{1}{2^{n^b}}.$$

Now we analyze the complexity. By Lemma 1, we obtain in $\widehat{\mathsf{TC}}_1^0$ the CRR_Π representation of $W_j(X, Y) \in [0, 2^n)$ for $j \in \{0, \ldots, O(n)\}$. Each h_p^Π is computable in $\widehat{\mathsf{TC}}_1^0$, and poly-many bits of the binary expansion of $1/p$ can be obtained in AC^0. Using AC^0 circuitry we multiply together the $O(\log n)$-bit numbers

h_p^{Π} and $W_j(X,Y) \bmod p$, to obtain the binary expansion of $((W_j(X,Y) \bmod p)h_p^{\Pi}) \cdot (1/p)$ (since multiplying an n-bit number by a $\log n$ bit number can be done in AC^0). Thus, with one more layer of majority gates, we compute

$$A_{\Pi}' = \sum_{p \in \Pi} \sum_{j=0}^{2n+1} \frac{(W_j(X,Y) \bmod p)h_p^{\Pi}}{p}$$

and strip off the integer part, to obtain the desired approximation.   □

Corollary 1. *Let Π be any set of primes such that the product M of these primes lies in $(2^{n^c}, 2^{n^d})$ for a fixed constant $d > c \geq 3$. Then, given Z in CRR_{Π} representation and the numbers h_p^{Π} for each $p \in \Pi$, we can compute the binary representation of a good approximation to $\frac{Z}{M}$ in $\widehat{\mathsf{TC}}_1^0$*

Now, finally, we present our desired approximation. $\tilde{V}(X,Y)$ is $2^{n^c} \cdot V'(X,Y)$, where $V'(X,Y)$ is an approximation (within $1/2^{n^{2c}}$) of

$$V(X,Y) = \frac{W(X,Y)\prod_{i=1}^{n^c}(M_i-1)/2}{\prod_{i=1}^{n^c}M_i}.$$

$$W(X,Y) - 2^{n^c}V(X,Y) = W(X,Y) - 2^{n^c}\frac{W(X,Y)\prod_{i=1}^{n^c}(M_i-1)/2}{\prod_{i=1}^{n^c}M_i}$$

$$= W(X,Y) - \frac{W(X,Y)\prod_{i=1}^{n^c}(M_i-1)}{\prod_{i=1}^{n^c}M_i}$$

$$< W(X,Y)\frac{n^c}{2^{n^c}} < \frac{2^{2n}n^c}{2^{n^c}}$$

and

$$2^{n^c}V(X,Y) - \tilde{V}(X,Y) = 2^{n^c}V(X,Y) - 2^{n^c}V'(X,Y)$$

$$= 2^{n^c}(V(X,Y) - V'(X,Y))$$

$$\leq 2^{n^c}(\frac{1}{2^{n^{2c}}}) = \frac{2^{n^c}}{2^{n^{2c}}}.$$

Thus $X/Y - \tilde{V}(X,Y) = (X/Y - W(X,Y)) + (W(X,Y) - 2^{n^c}V(X,Y)) + (2^{n^c}V(X,Y) - \tilde{V}(X,Y)) < 2^{-(n+1)} + n^c2^{2n}/2^{n^c} + 2^{n^c}/2^{n^{2c}} < 1/Y$.

Lemma 3. *Let Π_i for $i \in \{1, \ldots, n^c\}$ be n^c pairwise disjoint sets of primes such that $M_i = \prod_{p \in \Pi_i} p \in (2^{n^c}, 2^{n^d})$ (for some constants $c, d : 3 \leq c < d$). Let $\Pi = \cup_{i=1}^{n^c}\Pi_i$. Then, given X,Y and the Π_i, we can compute $\tilde{V}(X,Y)$ in $\widehat{\mathsf{TC}}_3^0$.*

Proof. In $\widehat{\mathsf{TC}}_1^0$ we compute the CRR_{Π} representation of each M_i, and the numbers $W_j \bmod p$ (using Lemma 1). Also, as in Lemma 2, we get the values h_p^{Π}.

Then, with one more layer of majority gates we can compute the CRR representation of $\prod_i (M_i - 1)/2$ and of $2^{2(n+1)t}W(X,Y) = \sum_{j=0}^{2n+1} W_j(X,Y)$. The CRR representation of the product $2^{2(n+1)t}W(X,Y) \cdot \prod_i (M_i - 1)/2$ can then be computed with AC^0 circuitry to obtain the CRR representation of the numerator of the expression for $V(X,Y)$. (It is important to note that $2^{2(n+1)t}W(X,Y) \cdot \prod_i (M_i - 1)/2 < \prod_i M_i$, so that it is appropriate to talk about this CRR representation. Indeed, that is the reason why we divide each factor $M_i - 1$ by two.)

This value can then be converted to binary with one additional layer of majority gates, via Corollary 1, to obtain $\widetilde{V}(X,Y)$. $\square$

This completes the proof of Theorem 1. $\square$

Corollary 2. *Let Π be any set of primes such that the product M of these primes lies in $(2^{n^c}, 2^{n^d})$ for a fixed constant $d > c \geq 3$. Then, given Z in CRR_Π representation, the binary representation of Z can be computed in $\widehat{\mathsf{TC}}_3^0$*

Proof. Recall from the proof of Theorem 1 that, in order to compute the bits of $Z/2$, our circuit actually computes an approximation to $(Z+1)/2$. Although, of course, it is trivial to compute $Z/2$ if Z is given to us in binary, let us consider how to modify the circuit described in the proof of Lemma 3, if we were computing $\widetilde{V}(Z+1, 2)$, where we are given Z in CRR representation.

With one layer of majority gates, we can compute the CRR_Π representation of each M_i and the values h_p^Π for each prime p. (We will not need the numbers $W_j \bmod p$.)

Then, with one more layer of majority gates we can compute the CRR representation of $\prod_i (M_i - 1)/2$. In place of the gates that store the value of the CRR representation of $2^{2(n+1)t}W(X,Y)$, we insert the CRR representation of Z (which is given to us as input) and using AC^0 circuitry store the value of $Z+1$. The CRR representation of the product $Z + 1 \cdot \prod_i (M_i - 1)/2$ can then be computed with AC^0 circuitry to obtain the CRR representation of the numerator of the expression for $V(Z+1, 2)$.

Then this value can be converted to binary with one additional layer of majority gates, from which the bits of Z can be read off. $\square$

It is rather frustrating to observe that the input values Z are not used until quite late in the $\widehat{\mathsf{TC}}_3^0$ computation (when just one layer of majority gates remains). However, we see no simpler uniform algorithm to convert CRR to binary.

For our application regarding problems in the counting hierarchy, it is useful to consider the analog to Theorem 1 where the values X and Y are presented in CRR notation.

Theorem 2. *The function taking as input $X \in [0, 2^n), Y \in [1, 2^n)$ (in CRR) as well as 0^m, and producing as output the binary expansion of X/Y correct to m places is in $\widehat{\mathsf{TC}}_3^0$.*

Proof. We assume that the CRR basis consists of pairwise disjoint sets of primes M_i, as in Lemma 3.

The algorithm is much the same as in Theorem 1, but there are some important differences that require comment. The first step is to determine if $Y = 1$, which can be done using AC^0 circuitry (since the CRR of 1 is easy to recognize). The next step is to determine a value t such that $2^{t-1} \leq Y < 2^t$. Although this is trivial when the input is presented in binary, when the input is given in CRR it requires the following lemma:

Lemma 4. *(Adapted from [1,7,3]) Let X be an integer from $(-2^n, 2^n)$ specified by its residues modulo each $p \in \Pi_n$. Then, the predicate $X > 0$ is in $\widehat{\mathsf{TC}}^0_2$*

Since we are able to determine inequalities in majority-depth two, we will carry out the initial part of the algorithm from Theorem 1 using *all* possible values of t, and then select the correct value between the second and third levels of MAJORITY gates.

Thus, for each t, and for each j, we compute the values $W_{j,t}(X + 1, Y) = (X + 1)(2^t - Y)^j (2^{(2n+1-j)t})$ in CRR, along with the desired number of bits of accuracy of $1/p$ for each p in our CRR basis.

With this information available, as in Lemma 3, in majority-depth one we can compute h_p^{Π}, as well as the CRR representation of each M_i, and thus with AC^0 circuitry we obtain $(W_{j,t}(X + 1, Y)$ and the CRR for each $(M_i - 1)/2$.

Next, with our second layer of majority gates we sum the values $W_{j,t}(X + 1, Y)$ (over all j), and at this point we also will have been able to determine which is the correct value of t, so that we can take the correct sum, to obtain $2^{2(n+1)t}W(X, Y)$.

Thus, after majority-depth two, we have obtained the same partial results as in the proof of Lemma 3, and the rest of the algorithm is thus identical. □

Proposition 2. *Iterated product is in uniform $\widehat{\mathsf{TC}}^0_3$.*

Proof. The overall algorithm is identical to the algorithm outlined in [20], although the implementation of the basic building blocks is different. In majority-depth one, we convert the input from binary to CRR. With one more level of majority gates, we compute the CRR of the product.

Simultaneously, in majority-depth two we compute the bottom two levels of our circuit that computes from CRR to binary, as in Corollary 2.

Thus, with one final level of majority gates, we convert the answer from CRR to binary. □

3.1   Consequences for the Counting Hierarchy

Corollary 3. $\mathsf{BitSLP} \in \mathsf{PH}^{\mathsf{PP}^{\mathsf{PP}^{\mathsf{PP}^{\mathsf{PP}}}}}$.

Proof. This is immediate from Proposition 1 and Corollary 2.

Let f be the function that takes as input a tuple $(C, (p, j))$ and if p is a prime, evaluates the arithmetic circuit C mod p and outputs the j-th bit of the result. This function f, taken together with the $\widehat{\mathsf{TC}}_3^0$ circuit family promised by Corollary 2, satisfies the hypothesis of Proposition 1. (There is a minor subtlety, regarding how to partition the set of primes into the groupings M_i, but this is easily handled by merely using all of the primes of a given length, at most polynomially-larger than $|C|$.)   □

Via essentially identical methods, using Theorem 2, we obtain:

Corollary 4. $\{(C_X, C_Y, i)$: the i-th bit of the quotient X/Y, where X and Y are represented by arithmetic circuits C_X and C_Y, respectively, is in $\mathsf{PH}^{\mathsf{P}^{\mathsf{P}^{\mathsf{P}^{\mathsf{P}^{\mathsf{P}^{\mathsf{P}}}}}}}$.

4   Integer Matrix Powering

Theorem 3. The function $MPOW(A, m, p, q, i)$ taking as input a $(d \times d)$ integer matrix $A \in \{0,1\}^{d^2 n}$, $p, q, 1^i$, where $p, q \in [d]$, $i \in [O(n)]$ and producing as output the i-th bit of the (p,q)-th entry of A^m is in $\widehat{\mathsf{TC}}_3^0$.

For a $(d \times d)$ matrix, the characteristic polynomial $\chi_A(x) : \mathbb{Z} \to \mathbb{Z}$ is a univariate polynomial of degree at most d. Let $q, r : \mathbb{Z} \to \mathbb{Z}$ be univariate polynomials of degree at most $(m - d)$ and $(d - 1)$ such that $x^m = q(x)\chi_A(x) + r(x)$. By the Cayley-Hamilton theorem, we have that $\chi_A(A) = 0$. So, in order to compute A^m, we just have to compute $r(A)$.

Lemma 5. Given a $(d \times d)$ matrix A of n-bit integers, the coefficients of the characteristic polynomial of A in CRR can be computed in $\widehat{\mathsf{TC}}_1^0$.

Proof. We convert the entries of A to CRR and compute the determinant of $(xI - A)$. This involves an iterated sum of $O(2^d d!)$ integers each of which is an iterated product of d n-bit integers. The conversion to CRR is in $\widehat{\mathsf{TC}}_1^0$. Since addition, multiplication, and powering of $O(1)$ numbers of $O(\log n)$ bits is computable in AC^0, it follows that the coefficients of the characteristic polynomial can be computed in $\widehat{\mathsf{TC}}_1^0$.

Lemma 6. Given the coefficients of the polynomial r, in CRR, and given A in CRR, we can compute A^m in CRR using AC^0 circuitry.

Proof. Recall that $A^m = r(A)$. Let $r(x) = r_0 + r_1 x + \ldots + r_{d-1} x^{d-1}$. Computing any entry of $r(A)$ in CRR involves an iterated sum of $O(1)$ many numbers which are themselves an iterated product of $O(1)$ many $O(\log n)$-bit integers.   □

Lemma 7. (Adapted from [11]) Let p be a prime of magnitude $\mathsf{poly}(m)$. Let $g(x)$ of degree m and $f(x)$ of degree d be monic univariate polynomials over GF_p, such that $g(x) = q(x)f(x) + r(x)$ for some polynomials $q(x)$ of degree $(m - d)$ and $r(x)$ of degree $(d - 1)$. Then, given the coefficients of g and f, the coefficients of r can be computed in $\widehat{\mathsf{TC}}_1^0$.

Proof. Following [11], let $f(x) = \sum_{i=0}^{d} a_i x^i$, $g(x) = \sum_{i=0}^{m} b_i x^i$, $r(x) = \sum_{i=0}^{d-1} r_i x^i$ and $q(x) = \sum_{i=0}^{m-d} q_i x^i$. Since f, g are monic, we have $a_d = b_m = 1$. Denote by $f_R(x), g_R(x), r_R(x)$ and $q_R(x)$ respectively the polynomial with the i-th coefficient $a_{d-i}, b_{m-i}, r_{d-i-1}$ and q_{m-d-i} respectively. Then note that $x^d f(1/x) = f_R(x)$, $x^m g(1/x) = g_R(x)$, $x^{m-d} q(1/x) = q_R(x)$ and $x^{d-1} r(1/x) = r_R(x)$.

We use the Kung-Sieveking algorithm (as implemented in [11]). The algorithm is as follows:

1. Compute $\tilde{f}_R(x) = \sum_{i=0}^{m-d}(1 - f_R(x))^i$ via interpolation modulo p.
2. Compute $h(x) = \tilde{f}_R(x) g_R(x) = c_0 + c_1 x + \ldots + c_{d(m-d)+m} x^{d(m-d)+m}$. from which the coefficients of $q(x)$ can be obtained as $q_i = c_{d(m-d)+m-i}$.
3. Compute $r(x) = g(x) - q(x) f(x)$.

To prove the correctness of our algorithm, note that we have $g(1/x) = q(1/x) f(1/x) + r(1/x)$. Scaling the whole equation by x^m, we get $g_R(x) = q_R(x) f_R(x) + x^{m-d+1} r_R(x)$. Hence when we compute $h(x) = \tilde{f}_R(x) g_R(x)$ in step 2 of our algorithm, we get

$$h(x) = \tilde{f}_R(x) g_R(x) = \tilde{f}_R(x) q_R(x) f_R(x) + x^{m-d+1} \tilde{f}_R(x) r_R(x).$$

Note that $\tilde{f}_R(x) f_R(x) = \tilde{f}_R(x)(1 - (1 - f_R(x))) = \sum_{i=0}^{m-d}(1 - f_R(x))^i - \sum_{i=0}^{m-d}(1 - f_R(x))^{i+1} = 1 - (1 - f_R(x))^{m-d+1}$ (a telescoping sum). Since f is monic, f_R has a constant term which is 1 and hence $(1 - f_R(x))^{m-d+1}$ does not contain a monomial of degree less than $(m - d + 1)$. This is also the case with $x^{m-d+1} \tilde{f}_R(x) r_R(x)$, and hence all the monomials of degree less than $(m-d+1)$ belong to $q_R(x)$.

To see that this can be done in $\widehat{\mathsf{TC}}_1^0$, first note that given $f(x)$ and $g(x)$, the coefficients of $f_R(x)$ and $g_R(x)$ can be computed in NC^0. To compute the coefficients of $\tilde{f}_R(x)$, we use interpolation via the discrete Fourier transform (DFT) using arithmetic modulo p. Find a generator w of the multiplicative group modulo p and substitute $x = \{w^1, w^2, \ldots, w^{p-1}\}$ to obtain a system of linear equations in the coefficients F of $\tilde{f}_R(x) : V \cdot F = Y$, where Y is the vector consisting of $\tilde{f}_R(w^i)$ evaluated at the various powers of w. Since the underlying linear transformation $V(w)$ is a DFT, it is invertible; the inverse DFT $V^{-1}(w)$ is equal to $V(w^{-1}) \cdot (p-1)^{-1}$, which is equivalent to $-V(w^{-1}) \bmod p$. We can find each coefficient of $\tilde{f}_R(x)$ evaluating $V^{-1} Y$, i.e., by an inner product of a row of the inverse DFT-matrix with the vector formed by evaluating $\sum_{i=1}^{(m-d+1)}(1 - f_R(x))^{i-1}$ at various powers of w and dividing by $p - 1$. The terms in this sum can be computed in AC^0, and then the sum can be computed in majority-depth one, to obtain the coefficients of $\tilde{f}_R(x)$. The coefficients of $h(x)$ in step 2 could be obtained by iterated addition of the product of certain coefficients of $\tilde{f}_R$ and g_R, but since the coefficients of $\tilde{f}_R$ are themselves obtained by iterated addition of certain terms t, we roll steps 1 and 2 together by multiplying these terms t by the appropriate coefficients of g_R. Thus steps 1 and 2 can be accomplished in majority-depth 1. Then step 3 can be computed using AC^0 circuitry.   □

Proof. (of Theorem 3)

Our $\widehat{\mathsf{TC}}_3^0$ circuit C that implements the ideas above is the following:

0. At the input, we have the d^2 entries A_{ij}, $i,j \in [d]$ of A, and a set Π of short primes (such that Π can be partitioned in to n^c sets Π_i that are pairwise disjoint, i.e., $\Pi = \cup_{i=1}^{n^c} \Pi_i$).
1. In majority-depth one, we obtain (1) A_{ij} mod p for each prime p in our basis, and (2) $M_i = \prod_{p \in \Pi_i} p$ for all the n^c sets that constitute Π, and (3) the CRR of the characteristic polynomial of A (via appeal to Lemma 5).
2. In the next layer of threshold gates, we compute (1) $\prod_i^{n^c} (M_i - 1)/2$ in CRR, and (2) the coefficients of the polynomial r in CRR, by appeal to Lemma 7.
3. At this point, by Lemma 6, AC^0 circuitry can obtain $r(A) = A^m$ in CRR, and with one more layer of MAJORITY gates we can convert to binary, by appeal to Corollary 2.

$\square$

5 Open Questions and Discussion

Is conversion from CRR to binary in dlogtime-uniform $\widehat{\mathsf{TC}}_1^0$? This problem has been known to be in P-uniform $\widehat{\mathsf{TC}}_1^0$ starting with the seminal work of Beame, Cook, and Hoover [4], but the subsequent improvements on the uniformity condition [5,12] introduced additional complexity that translated into increased depth. We have been able to reduce the majority-depth by rearranging the algorithmic components introduced in this line of research, but it appears to us that a fresh approach will be needed, in order to decrease the depth further.

Is BitSLP *in* PH^PP? An affirmative answer to the first question implies an affirmative answer to the second, and this would pin down the complexity of BitSLP between $\mathsf{P}^{\#\mathsf{P}}$ and $\mathsf{PH}^{\mathsf{PP}}$.

Is PosSLP *in* PH? Some interesting observations related to this problem were announced recently [8,16].

Is it easy to compute bits of large powers of small matrices? Recall that some surprising things about large powers of *integers* can be computed [13].

Acknowledgments. The first author acknowledges the support of NSF grants CCF-0832787 and CCF-1064785.

References

1. Agrawal, M., Allender, E., Datta, S.: On TC⁰, AC⁰, and Arithmetic circuits. Journal of Computer and System Sciences 60(2), 395–421 (2000)
2. Allender, E., Schnorr, H.: The complexity of the BitSLP problem (2005) (unpublished manuscript)
3. Allender, E., Bürgisser, P., Kjeldgaard-Pedersen, J., Miltersen, P.B.: On the complexity of numerical analysis. SIAM J. Comput. 38(5), 1987–2006 (2009)

4. Beame, P.W., Cook, S.A., Hoover, H.J.: Log depth circuits for division and related problems. SIAM Journal on Computing 15, 994–1003 (1986)
5. Chiu, A., Davida, G.I., Litow, B.: Division in logspace-uniform NC^1. Informatique Théorique et Applications 35(3), 259–275 (2001)
6. Datta, S., Pratap, R.: Computing bits of algebraic numbers. In: Agrawal, M., Cooper, S.B., Li, A. (eds.) TAMC 2012. LNCS, vol. 7287, pp. 189–201. Springer, Heidelberg (2012)
7. Dietz, P., Macarie, I., Seiferas, J.: Bits and relative order from residues, space efficiently. Information Processing Letters 50(3), 123–127 (1994)
8. Etessami, K.: Probability, recursion, games, and fixed points. talk presented at Horizons in TCS: A Celebration of Mihalis Yannakakis' 60th Birthday (2013)
9. Etessami, K., Yannakakis, M.: On the complexity of Nash equilibria and other fixed points. SIAM J. Comput. 39(6), 2531–2597 (2010)
10. Goldmann, M., Karpinski, M.: Simulating threshold circuits by majority circuits. SIAM J. Comput. 27(1), 230–246 (1998)
11. Healy, A., Viola, E.: Constant-depth circuits for arithmetic in finite fields of characteristic two. In: Durand, B., Thomas, W. (eds.) STACS 2006. LNCS, vol. 3884, pp. 672–683. Springer, Heidelberg (2006)
12. Hesse, W., Allender, E., Barrington, D.: Uniform constant-depth threshold circuits for division and iterated multiplication. Journal of Computer and System Sciences 65, 695–716 (2002)
13. Hirvensalo, M., Karhumäki, J., Rabinovich, A.: Computing partial information out of intractable: Powers of algebraic numbers as an example. Journal of Number Theory 130, 232–253 (2010)
14. Hunter, P., Bouyer, P., Markey, N., Ouaknine, J., Worrell, J.: Computing rational radical sums in uniform TC^0. In: FSTTCS, pp. 308–316 (2010)
15. Jeřábek, E.: Root finding with threshold circuits. Theoretical Computer Science 462, 59–69 (2012)
16. Jindal, G., Saranurak, T.: Subtraction makes computing integers faster. CoRR abs/1212.2549 (2012)
17. Kayal, N., Saha, C.: On the sum of square roots of polynomials and related problems. TOCT 4(4), 9 (2012)
18. Koiran, P., Perifel, S.: The complexity of two problems on arithmetic circuits. Theor. Comput. Sci. 389(1-2), 172–181 (2007)
19. Koiran, P., Perifel, S.: Interpolation in Valiant's theory. Computational Complexity 20(1), 1–20 (2011)
20. Maciel, A., Thérien, D.: Threshold circuits of small majority-depth. Inf. Comput. 146(1), 55–83 (1998)
21. Mereghetti, C., Palano, B.: Threshold circuits for iterated matrix product and powering. ITA 34(1), 39–46 (2000)
22. Ouaknine, J., Worrell, J.: Positivity problems for low-order linear recurrence sequences. In: SODA, pp. 366–379 (2014)
23. Sherstov, A.A.: Powering requires threshold depth 3. Inf. Process. Lett. 102(2-3), 104–107 (2007)
24. Siu, K.Y., Roychowdhury, V.P.: On optimal depth threshold circuits for multiplication and related problems. SIAM J. Discrete Math. 7(2), 284–292 (1994)
25. Toda, S.: PP is as hard as the polynomial time hierarchy. SIAM J. Comput. 20, 865–877 (1991)
26. Vollmer, H.: Introduction to Circuit Complexity. Springer (1999)
27. Wegener, I.: Optimal lower bounds on the depth of polynomial-size threshold circuits for some arithmetic functions. Inf. Process. Lett. 46(2), 85–87 (1993)

Zero Knowledge and Circuit Minimization

Eric Allender[1] and Bireswar Das[2]

[1] Department of Computer Science, Rutgers University, USA
allender@cs.rutgers.edu
[2] IIT Gandhinagar, India
bireswar@iitgn.ac.in

Abstract. We show that every problem in the complexity class SZK (Statistical Zero Knowledge) is efficiently reducible to the Minimum Circuit Size Problem (MCSP). In particular Graph Isomorphism lies in $\mathsf{RP}^{\mathsf{MCSP}}$.

This is the first theorem relating the computational power of Graph Isomorphism and MCSP, despite the long history these problems share, as candidate NP-intermediate problems.

1 Introduction

For as long as there has been a theory of NP-completeness, there have been attempts to understand the computational complexity of the following two problems:

- Graph Isomorphism (GI): Given two graphs G and H, determine if there is permutation τ of the vertices of G such that $\tau(G) = H$.
- The Minimum Circuit Size Problem (MCSP): Given a Boolean function f on n variables, represented by its truth table of size 2^n, and a number i, determine if f has a circuit of size i. (There are different versions of this problem depending on precisely what measure of "size" one uses (such as counting the number of gates or the number of wires) and on the types of gates that are allowed, etc. For the purposes of this paper, any reasonable choice can be used.)

Cook [Coo71] explicitly considered the graph isomorphism problem and mentioned that he "had not been able" to show that GI is NP-complete. Similarly, it has been reported that Levin's original motivation in defining and studying NP-completeness [Lev73] was in order to understand the complexity of GI [PS03], and that Levin delayed publishing his work because he had hoped to be able to say something about the complexity of MCSP [Lev03]. (Trakhtenbrot has written an informative account, explaining some of the reasons why MCSP held special interest for the mathematical community in Moscow in the 1970s [Tra84].)

For the succeeding four decades, GI and MCSP have been prominent candidates for so-called "NP-Intermediate" status: neither in P nor NP-complete. No connection between the relative complexity of these two problems has been established. Until now.

E. Csuhaj-Varjú et al. (Eds.): MFCS 2014, Part II, LNCS 8635, pp. 25–32, 2014.

It is considered highly unlikely that GI is NP-complete. For instance, if the polynomial hierarchy is infinite, then GI is not NP-complete [BHZ87]. Many would conjecture that GI ∈ P; Cook mentions this conjecture already in [Coo71]. However this is still very much an open question, and the complexity of GI has been the subject of a great deal of research. We refer the reader to [KST93, AT05] for more details.

In contrast, comparatively little was written about MCSP, until Kabanets and Cai revived interest in the problem [KC00], by highlighting its connection to the so-called Natural Proofs barrier to circuit lower bounds [RR97]. Kabanets and Cai provided evidence that MCSP is not in P (or even in P/poly); it is known that BPP^{MCSP} contains several problems that cryptographers frequently assume are intractable, including the discrete logarithm, and several lattice-based problems [KC00, ABK+06]. The integer factorization problem even lies in ZPP^{MCSP} [ABK+06].

Is MCSP complete for NP? Krajíček discusses this possibility [Kra11], although no evidence is presented to suggest that this is a likely hypothesis. Instead, evidence has been presented to suggest that it will be difficult to reduce SAT to MCSP. Kabanets and Cai define a class of "natural" many-one reductions; after observing that most NP-completeness proofs are "natural" in this sense, they show that any "natural" reduction from SAT to MCSP yields a proof that EXP ⊄ P/poly. Interestingly, Vinodchandran studies a problem called SNCMP, which is similar to MCSP, but defined in terms of strong nondeterministic circuits, instead of deterministic circuits [Var05]. (SNCMP stands for Strong Non-deterministic Circuit Minimization Problem.) Vinodchandran shows that any "natural" reduction from graph isomorphism to SNCMP yields a nondeterministic algorithm for the complement of GI that runs in subexponential time for infinitely many lengths n.

We show that GI ∈ RP^{MCSP}; our proof also shows that GI ∈ RP^{SNCMP}. Thus, although it would be a significant breakthrough to give a "natural" reduction from GI to SNCMP, no such obstacle prevents us from establishing an RP-Turing reduction.

One of the more important results about GI is that GI lies in SZK: the class of problems with statistical zero-knowledge interactive proofs [GMW91]. After giving a direct proof of the inclusion GI ∈ RP^{MCSP} in Section 3, we give a proof of the inclusion SZK ⊆ BPP^{MCSP} in Section 4. We conclude with a discussion of additional directions for research and open questions.

But first, we present the basic connection between MCSP and resource-bounded Kolmogorov complexity, which allows us to invert polynomial-time computable functions (on average), using MCSP.

2   Preliminaries and Technical Lemmas

A small circuit for a Boolean function f on n variables constitutes one form of a short description for the bit string of length 2^n that describes the truth table of f. In fact, as discussed in [ABK+06, Theorem 11], there is a version of time-bounded Kolmogorov complexity (denoted KT) that is roughly equivalent to circuit size. That is, if x is a string of length m representing the truth table of a function f with minimum circuit size s, it holds that

$$\left(\frac{s}{\log m}\right)^{1/4} \leq \mathrm{KT}(x) \leq O(s^2(\log s + \log \log m)).$$

The connection with Kolmogorov complexity is relevant, because of this simple observation: The output of a pseudorandom generator consists of strings with small time-bounded Kolmogorov complexity. Thus, with an oracle for MCSP, one can take as input a string x and accept iff x has no circuits of size, say, $\sqrt{|x|}$, and thereby ensure that one is accepting a very large fraction of all of the strings of length n (since most x encode functions that require large circuits), and yet accept no strings x such that $\mathrm{KT}(x) \leq n^\epsilon$. Such a set is an excellent test to distinguish the uniform distribution from the distribution generated by a pseudorandom generator. Using the tight connection between one-way functions and pseudorandom generators [HILL99], one obtains the following result:

Theorem 1. *[ABK$^+$06, Theorem 45] Let L be a language of polynomial density such that, for some $\epsilon > 0$, for every $x \in L$, $\mathrm{KT}(x) \geq |x|^\epsilon$. Let $f(y,x)$ be computable uniformly in time polynomial in $|x|$. There exists a polynomial-time probabilistic oracle Turing machine N and polynomial q such that for any n and y*

$$\Pr_{|x|=n,s}[f(y, N^L(y, f(y,x), s)) = f(y,x)] \geq 1/q(n),$$

where x is chosen uniformly at random and s denotes the internal coin flips of N.

Here, "polynomial density" means merely that L contains at least $2^n/n^k$ strings of each length n, for some k. That is, let f_y be a collection of functions indexed by a parameter y, where $f_y(x)$ denotes $f(y,x)$. Then, if one has access to an an oracle L that contains many strings but no strings of small KT-complexity, one can use the probabilistic algorithm N to take as input $f_y(x)$ for a randomly-chosen x, and with non-negligible probability find a $z \in f_y^{-1}(f_y(x))$, that is, a string z such that $f_y(z) = f_y(x)$.

Note that such a set L can be recognized in deterministic polynomial time with an oracle for MCSP, as well as with an oracle for SNCMP. One could also use an oracle for R_{KT}, the KT-random strings: $R_{\mathrm{KT}} = \{x : \mathrm{KT}(x) \geq |x|\}$.

3 Graph Isomorphism and Circuit Size

Theorem 2. $\mathrm{GI} \in \mathrm{RP}^{\mathrm{MCSP}}$.

Proof. We are given as input two graphs G and H, and we wish to determine whether there is an isomorphism from G to H.

Consider the polynomial-time computable function $f(G, \tau)$ that takes as input a graph G on n vertices and a permutation $\tau \in S_n$ and outputs $\tau(G)$. We will use the notation $f_G(\tau)$ to denote $f(G, \tau)$. That is, f_G takes a permutation τ as input, and produces as output the adjacency matrix of the graph obtained by permuting G according to τ. Observe that f_G is uniformly computable in time polynomial in the length of τ.

Thus, by Theorem 1, there is a polynomial-time probabilistic oracle Turing machine N and polynomial q such that for any n and G

$$\Pr_{\tau \in S_{n,s}}[f_G(N^{\mathsf{MCSP}}(G, f_G(\tau), s)) = f_G(\tau)] \geq 1/q(n),$$

where τ is chosen uniformly at random and s denotes the internal coin flips of N.

Now, given input (G, H) to GI, our $\mathsf{RP}^{\mathsf{MCSP}}$ algorithm does the following for $100q(n)$ independent trials:

1. Pick τ and probabilistic sequence s uniformly at random.
2. Compute $\tau(G)$.
3. Run $N^{\mathsf{MCSP}}(H, \tau(G), s)$ and obtain output π.
4. Report "success" if $\pi(H) = \tau(G)$.

The $\mathsf{RP}^{\mathsf{MCSP}}$ algorithm will accept if at least one of the $100q(n)$ independent trials are successful.

Note that if H and G are not isomorphic, then there is no possibility that the algorithm will succeed.

On the other hand, if H and G are isomorphic, then $\tau(G)$ does appear in the image of f_H. In fact, the distributions $\tau(G)$ and $\tau(H)$ are identical over τ picked uniformly at random. Thus, with probability at least $1/q(n)$ (taken over the choices of τ and s), the algorithm will succeed in any given trial. Thus the expected number of trials that will succeed is at least 100, and hence, by the Chernoff bounds, the probability of having at least one success is well over $1/2$. □

Since truth-tables that require large strong nondeterministic circuits also require large deterministic circuits, it is immediate that this reduction can be carried out also with SNCMP.

Corollary 1. $\mathsf{GI} \in \mathsf{RP}^{\mathsf{SNCMP}} \cap \mathsf{RP}^{R_{\mathrm{KT}}}$.

4   Zero Knowledge

In this section, we show $\mathsf{SZK} \subseteq \mathsf{BPP}^{\mathsf{MCSP}}$. Note that SZK is best defined not as a class of languages but as a class of "promise problems". A promise problem consists of a pair of disjoint languages (Y, N) where Y consists of "yes-instances" and N consists of "no-instances". Thus the inclusion $\mathsf{SZK} \subseteq \mathsf{BPP}^{\mathsf{MCSP}}$ is perhaps more properly stated in terms of "promise" $\mathsf{BPP}^{\mathsf{MCSP}}$. That is, we will show that, for every $(Y, N) \in \mathsf{SZK}$ there is a probabilistic polynomial time oracle Turing machine M with the property that $x \in Y$ implies $M(x)$ accepts with probability at least $2/3$ when given oracle MCSP, and $x \in N$ implies $M(x)$ accepts with probability at most $1/3$ when given oracle MCSP. M may exhibit any behavior on inputs outside of $N \cup Y$.

It was shown by Chailloux et al. [CCKV08] that SZK is equal to a class that Ben-Or and Gutfreund [BOG03] defined and called NISZK$|_h$. Importantly for us, Ben-Or and Gutfreund showed that a promise problem they called IID (Image Intersection Density) is complete for NISZK$|_h$ (and thus, by [CCKV08], IID is also complete for SZK). The yes-instances of IID consist of pairs of circuits (C_0, C_1), each of size n, taking m-bit inputs, such that the distributions $C_0(x)$ and $C_1(x)$ (where x is chosen uniformly at random) have statistical distance at most $1/n^2$. The no-instances of IID consist of pairs of circuits (C_0, C_1) with the property that $\Pr_{|x|=m}[\exists y \ C_1(y) = C_0(x)] < 1/n^2$.

We will not work directly with IID, but rather with a related problem that is shown to be complete for NISZK$|_h$ in [BOG03, Lemma 20], which is just like IID but with different parameters. Let us call this problem PIID for "polarized IID". The yes-instances of PIID consist of triples (n, D_0, D_1), where each D_i is an m-input circuit of size at most n^k (for some fixed k), such that the distributions $D_0(x)$ and $D_1(x)$ (where x is chosen uniformly at random) have statistical distance at most $1/2^n$. The no-instances of PIID consist of triples (n, D_0, D_1) with the property that $\Pr_{|x|=m}[\exists y \ D_1(y) = D_0(x)] < 1/2^n$.

Furthermore, we need to make use of the fact that we can assume that the length m of the inputs to the circuits D_0 and D_1 may be assumed without loss of generality to be at least n^δ for some fixed $\delta > 0$. This can be accomplished by simply adding dummy input variables. It is easy to check that adding dummy variables to both circuits does not change the statistical difference. Similarly, this does not alter the probability that the output produced by a random input to the first circuit is in the support of the second circuit.

Theorem 3. SZK $\in$ BPP^{MCSP}

Proof. It will suffice to show that PIID $\in$ BPP^{MCSP}.

Consider the polynomial-time computable function $F(C, x)$ that takes as input a Boolean circuit C (on m-bit inputs), and a string x of length m, and outputs $C(x)$. We will use the notation $F_C(x)$ to denote $F(C, x)$. Since the length of x is polynomially-related to the size of C in the instances of PIID that we consider, it follows that F_C is uniformly computable in time polynomial in the length of x.

Thus, by Theorem 1, there is a polynomial-time probabilistic oracle Turing machine N and polynomial q such that for any m and C

$$\Pr_{|x|=m,s} [F_C(N^{\mathsf{MCSP}}(C, F_C(x), s)) = F_C(x)] \geq 1/q(m),$$

where x is chosen uniformly at random and s denotes the internal coin flips of N.

Now, given input (n, D_0, D_1) to PIID, our BPP^{MCSP} algorithm does the following for n^ℓ independent trials (for an ℓ to be determined later):

1. Pick m-bit input x and probabilistic sequence s uniformly at random.
2. Compute $z = D_0(x)$.
3. Run $N^{\mathsf{MCSP}}(D_1, z, s)$ and obtain output y.
4. Report "success" if $D_1(y) = z$.

The $\mathsf{BPP}^{\mathsf{MCSP}}$ algorithm will accept if at least $\log n$ of the n^ℓ independent trials are successful.

If (n, D_0, D_1) is a no-instance of PIID, then the probability that any given trial succeeds is at most $1/2^n$. Thus, for all large n the expected number of the n^ℓ trials that will succeed is at most $n^\ell/2^n < 1$. By the Chernoff bounds, the probability that $\log n$ trials will succeed is less than $1/3$.

If (n, D_0, D_1) is a yes-instance of PIID, then $D_0(x)$ and $D_1(x)$ have statistical distance at most $1/2^n$.

Note that

$$\Pr_{|x|=m,s}[F_{D_1}(N^{\mathsf{MCSP}}(D_1, F_{D_0}(x), s)) = F_{D_0}(x)]$$

$$= \sum_z \Pr_{|x|=m,s}[F_{D_1}(N^{\mathsf{MCSP}}(D_1, z, s)) = z | z = F_{D_0}(x)] \Pr[z = F_{D_0}(x)]$$

$$= \sum_z \Pr_{|x|=m,s}[F_{D_1}(N^{\mathsf{MCSP}}(D_1, z, s)) = z | z = F_{D_1}(x)] \Pr[z = F_{D_0}(x)]$$

Also,

$$\Pr_{|x|=m,s}[F_{D_1}(N^{\mathsf{MCSP}}(D_1, F_{D_1}(x), s)) = F_{D_1}(x)]$$

$$= \sum_z \Pr_{|x|=m,s}[F_{D_1}(N^{\mathsf{MCSP}}(D_1, z, s)) = z | z = F_{D_1}(x)] \Pr[z = F_{D_1}(x)]$$

Thus the difference of these two probabilities is

$$\sum_z \Pr_{|x|=m,s}[F_{D_1}(N^{\mathsf{MCSP}}(D_1, z, s)) = z | z = F_{D_1}(x)] \times$$

$$(\Pr[z = F_{D_0}(x)] - \Pr[z = F_{D_1}(x)])$$

$$\leq \sum_z 1 \cdot (\Pr[z = F_{D_0}(x)] - \Pr[z = F_{D_1}(x)])$$

$$\leq 1/2^n$$

Since $\Pr_{|x|=m,s}[F_{D_1}(N^{\mathsf{MCSP}}(D_1, F_{D_1}(x), s)) = F_{D_1}(x)] > 1/q(m) > 1/q(n^k)$, it follows that each trial has probability at least $1/q(n^k) - 1/2^n$ of success. Thus, the expected number of the n^ℓ trials that will succeed is at least $n^\ell(1/q(n^k) - 1/2^n)$. Picking ℓ so that n^ℓ is enough greater than $q(n^k)$ guarantees that this expected value is at least n. Thus, by the Chernoff bounds the probability that at least $\log n$ trials succeed is greater than $2/3$.   □

In the above proof, notice that we obtain one-sided error on those instances (n, D_0, D_1) of PIID where $\Pr_{|x|=m}[\exists y \; D_1(y) = D_0(x)] = 0$, instead of merely being bounded by $1/2^n$. In particular, the promise problem known as $\overline{\mathsf{SD}^{1,0}}$ (consisting of pairs of circuits (D_0, D_1) where, for the yes-instances, D_0 and D_1 represent identical distributions, and the no-instances have disjoint images) is in $\mathsf{RP}^{\mathsf{MCSP}}$. It was shown in [KMV07] that this problem is complete for the class of problems that have "V-bit" perfect zero knowledge protocols; this class contains most of the problems that are known to have perfect zero-knowledge protocols, including the problems studied in [AD08].

5   Conclusions and Open Problems

We are the first to admit that there appears to be no reason why these results could not have been proved earlier. The techniques involved have been available to researchers for years, and the proofs have much the same flavor as the reductions of factoring, discrete logarithm, and other cryptographic problems to MCSP that were presented in [ABK+06]. Perhaps the only missing ingredient is that the earlier work involved using MCSP (or, equivalently, R_{KT}) to break pseudorandom generators that were constructed from one-way functions that people actually believed *were* cryptographically secure. In contrast, the functions f_G considered here have never seemed like promising candidates to use, in constructing pseudorandom generators.

It is natural to wonder if better reductions are also possible. Is $\mathsf{GI} \in \mathsf{P}^{\mathsf{MCSP}}$? Or in $\mathsf{ZPP}^{\mathsf{MCSP}}$?

Equally temptingly, is it possible to build on these ideas to reduce larger classes to MCSP? The Wikipedia article on "NP-Intermediate Problems" (as of April 10, 2014) says "...MCSP is believed to be NP-complete" [Wik14]. We are unaware of much evidence for this "belief" being very widespread in the complexity theory community, but it is certainly an intriguing possibility.

Alternatively, is it possible to tie MCSP more closely to SZK? For instance, what is the complexity of the promise problem whose yes-instances consist of strings with KT-complexity at most $\sqrt{n}$, and whose no-instances consist of strings with KT-complexity $> n/2$?

Acknowledgments. We acknowledge helpful comments from Lance Fortnow, Valentine Kabanets, Rahul Santhanam, Bruce Kapron, and Salil Vadhan. The first author acknowledges the support of NSF grants CCF-0832787 and CCF-1064785. This work was performed while the second author was a DIMACS postdoctoral fellow at Rutgers University, under support provided by the Indo-US Science and Technology Forum.

References

[ABK+06]   Allender, E., Buhrman, H., Koucký, M., van Melkebeek, D., Ronneburger, D.: Power from random strings. SIAM Journal on Computing 35, 1467–1493 (2006)

[AD08]   Arvind, V., Das, B.: SZK proofs for black-box group problems. Theory Comput. Syst. 43(2), 100–117 (2008)

[AT05]   Arvind, V., Torán, J.: Isomorphism testing: Perspective and open problems. Bulletin of the EATCS 86 (2005)

[BHZ87]   Boppana, R.B., Håstad, J., Zachos, S.: Does co-NP have short interactive proofs? Information Processing Letters 25(2), 127–132 (1987)

[BOG03]   Ben-Or, M., Gutfreund, D.: Trading help for interaction in statistical zero-knowledge proofs. J. Cryptology 16(2), 95–116 (2003)

[CCKV08] Chailloux, A., Ciocan, D.F., Kerenidis, I., Vadhan, S.P.: Interactive and noninteractive zero knowledge are equivalent in the help model. In: Canetti, R. (ed.) TCC 2008. LNCS, vol. 4948, pp. 501–534. Springer, Heidelberg (2008)

[Coo71]   Cook, S.A.: The complexity of theorem-proving procedures. In: ACM Symposium on Theory of Computing (STOC), pp. 151–158 (1971)

[GMW91]   Goldreich, O., Micali, S., Wigderson, A.: Proofs that yield nothing but their validity for all languages in NP have zero-knowledge proof systems. Journal of the ACM 38(3), 691–729 (1991)

[HILL99] Håstad, J., Impagliazzo, R., Levin, L., Luby, M.: A pseudorandom generator from any one-way function. SIAM Journal on Computing 28, 1364–1396 (1999)

[KC00]   Kabanets, V., Cai, J.-Y.: Circuit minimization problem. In: ACM Symposium on Theory of Computing (STOC), pp. 73–79 (2000)

[KMV07]   Kapron, B., Malka, L., Srinivasan, V.: A characterization of non-interactive instance-dependent commitment-schemes (NIC). In: Arge, L., Cachin, C., Jurdziński, T., Tarlecki, A. (eds.) ICALP 2007. LNCS, vol. 4596, pp. 328–339. Springer, Heidelberg (2007)

[Kra11]   Krajíček, J.: Forcing with Random Variables and Proof Complexity. Cambridge University Press (2011)

[KST93]   Köbler, J., Schöning, U., Torán, J.: The Graph Isomorphism Problem: Its Structural Complexity. Birkhauser Verlag, Basel (1993)

[Lev73]   Levin, L.A.: Universal sequential search problems. Problems of Information Transmission 9, 265–266 (1973)

[Lev03]   Levin, L.: Personal communication (2003)

[PS03]   Pemmaraju, S., Skiena, S.: Computational Discrete Mathematics: Combinatorics and Graph Theory with Mathematica. Cambridge University Press, New York (2003)

[RR97]   Razborov, A., Rudich, S.: Natural proofs. Journal of Computer and System Sciences 55, 24–35 (1997)

[Tra84]   Trakhtenbrot, B.A.: A survey of Russian approaches to perebor (brute-force searches) algorithms. IEEE Annals of the History of Computing 6(4), 384–400 (1984)

[Var05]   Variyam, V.N.: Nondeterministic circuit minimization problem and derandomizing Arthur-Merlin games. Int. J. Found. Comput. Sci. 16(6), 1297–1308 (2005)

[Wik14]   Wikipedia (2014), http://en.wikipedia.org/wiki/NP-intermediate

A Tight Lower Bound on Certificate Complexity in Terms of Block Sensitivity and Sensitivity⋆

Andris Ambainis and Krišjānis Prūsis

Faculty of Computing, University of Latvia, Raina bulv. 19, Rīga, LV-1586, Latvia

Abstract. Sensitivity, certificate complexity and block sensitivity are widely used Boolean function complexity measures. A longstanding open problem, proposed by Nisan and Szegedy [7], is whether sensitivity and block sensitivity are polynomially related. Motivated by the constructions of functions which achieve the largest known separations, we study the relation between 1-certificate complexity and 0-sensitivity and 0-block sensitivity.

Previously the best known lower bound was $C_1(f) \geq \frac{bs_0(f)}{2s_0(f)}$, achieved by Kenyon and Kutin [6]. We improve this to $C_1(f) \geq \frac{3bs_0(f)}{2s_0(f)}$. While this improvement is only by a constant factor, this is quite important, as it precludes achieving a superquadratic separation between $bs(f)$ and $s(f)$ by iterating functions which reach this bound. In addition, this bound is tight, as it matches the construction of Ambainis and Sun [3] up to an additive constant.

1 Introduction

Determining the biggest possible gap between the sensitivity $s(f)$ and block sensitivity $bs(f)$ of a Boolean function is a well-known open problem in the complexity of Boolean functions. Even though this question has been known for over 20 years, there has been quite little progress on it.

The biggest known gap is $bs(f) = \Omega(s^2(f))$. This was first discovered by Rubinstein [8], who constructed a function f with $bs(f) = \frac{s^2(f)}{2}$, and then improved by Virza [9] and Ambainis and Sun [3]. Currently, the best result is a function f with $bs(f) = \frac{2}{3}s^2(f) - \frac{1}{3}s(f)$ [3]. The best known upper bound is exponential: $bs(f) \leq s(f)2^{s(f)-1}$ [2] which improves over an earlier exponential upper bound by Kenyon and Kutin [6].

In this paper, we study a question motivated by the constructions of functions that achieve a separation between $s(f)$ and $bs(f)$. The question is as follows:

⋆ This research has received funding from the EU Seventh Framework Programme (FP7/2007-2013) under projects QALGO (No. 600700) and RAQUEL (No. 323970) and ERC Advanced Grant MQC. Part of this work was done while Andris Ambainis was visiting Institute for Advanced Study, Princeton, supported by National Science Foundation under agreement No. DMS-1128155. Any opinions, findings and conclusions or recommendations expressed in this material are those of the author(s) and do not necessarily reflect the views of the National Science Foundation.

E. Csuhaj-Varjú et al. (Eds.): MFCS 2014, Part II, LNCS 8635, pp. 33–44, 2014.

Let $s_z(f)$, $bs_z(f)$ and $C_z(f)$ be the maximum sensitivity, block sensitivity and certificate complexity achieved by f on inputs x: $f(x) = z$. What is the best lower bound of $C_1(f)$ in terms of $s_0(f)$ and $bs_0(f)$?

The motivation for this question is as follows. Assume that we fix $s_0(f)$ to a relatively small value m and fix $bs_0(f)$ to a substantially larger value k. We then minimize $C_1(f)$. We know that $s_1(f) \leq C_1(f)$ (because every sensitive bit has to be contained in a certificate). We have now constructed an example where both $s_0(f)$ and $s_1(f)$ are relatively small and $bs_0(f)$ large. This may already achieve a separation between $bs_0(f)$ and $s(f) = \max(s_0(f), s_1(f))$ and, if $s_1(f) > s_0(f)$, we can improve this separation by composing the function with OR (as in [3]).

While this is just one way of achieving a gap between $s(f)$ and $bs(f)$, all the best separations between these two quantities can be cast into this framework. Therefore, we think that it is interesting to explore the limits of this approach.

The previous results are as follows:

1. Rubinstein's construction [8] can be viewed as taking a function f with $s_0(f) = 1$, $bs_0(f) = k$ and $C_1(f) = 2k$. A composition with OR yields [3] $bs(f) = \frac{1}{2}s^2(f)$;
2. Later work by Virza [9] and Ambainis and Sun [3] improves this construction by constructing f with $s_0(f) = 1$, $bs_0(f) = k$ and $C_1(f) = \lfloor \frac{3k}{2} \rfloor + 1$. A composition with OR yields $bs(f) = \frac{2}{3}s^2(f) - \frac{1}{3}s(f)$;
3. Ambainis and Sun [3] also show that, given $s_0(f) = 1$ and $bs_0(f) = k$, the certificate complexity $C_1(f) = \lfloor \frac{3k}{2} \rfloor + 1$ is the smallest that can be achieved. This means that a better bound must either start with f with $s_0(f) > 1$ or use some other approach;
4. For $s_0(f) = m$ and $bs_0(f) = k$, it is easy to modify the construction of Ambainis and Sun [3] to obtain $C_1(f) = \lfloor \frac{3\lceil k/m \rceil}{2} \rfloor + 1$ but this does not result in a better separation between $bs(f)$ and $s(f)$;
5. Kenyon and Kutin [6] have shown a lower bound of $C_1(f) \geq \frac{k}{2m}$. If this was achievable, this could result in a separation of $bs(f) = 2s^2(f)$.

The gap between the construction $C_1(f) = \frac{3k}{2m} + O(1)$ and the lower bound of $C_1(f) \geq \frac{k}{2m}$ is only a constant factor but the constant here is quite important. This gap corresponds to a difference between $bs(f) = (\frac{2}{3}+o(1))s^2(f)$ and $bs(f) = 2s^2(f)$, and, if we achieved $bs(f) > s^2(f)$, iterating the function f would yield an infinite sequence of functions with a superquadratic separation $bs(f) = s(f)^c$, where $c > 2$.

In this paper, we show that for any f

$$C_1(f) \geq \frac{3}{2}\frac{bs_0(f)}{s_0(f)} - \frac{1}{2}.$$

This matches the best construction up to an additive constant and shows that no further improvement can be achieved along the lines of [8,9,3]. Our bound is shown by an intricate analysis of possible certificate structures for f.

Since we now know that $bs_0(f) \leq (\frac{2}{3} + o(1))C_1(f)s_0(f)$, it is tempting to conjecture that $bs_0(f) \leq (\frac{2}{3} + o(1))s_1(f)s_0(f)$. If this was true, the existing separation between $bs(f)$ and $s(f)$ would be tight.

2   Preliminaries

Let $f : \{0,1\}^n \to \{0,1\}$ be a Boolean function on n variables. The i-th variable of input x is denoted by x_i. For an index set $S \subseteq [n]$, let x^S be the input obtained from an input x by flipping every bit x_i, $i \in S$. Let a z-input be an input on which the function takes the value z, where $z \in \{0,1\}$.

We briefly define the notions of sensitivity, block sensitivity and certificate complexity. For more information on them and their relations to other complexity measures (such as deterministic, probabilistic and quantum decision tree complexities), we refer the reader to the surveys by Buhrman and de Wolf [4] and Hatami et al. [5].

Definition 1. *The* sensitivity complexity $s(f, x)$ *of* f *on an input* x *is defined as* $|\{i \mid f(x) \neq f(x^{\{i\}})\}|$. *The* z-sensitivity $s_z(f)$ *of* f, *where* $z \in \{0,1\}$, *is defined as* $\max\{s(f, x) \mid x \in \{0,1\}^n, f(x) = z\}$. *The* sensitivity $s(f)$ *of* f *is defined as* $\max\{s_0(f), s_1(f)\}$.

Definition 2. *The* block sensitivity $bs(f, x)$ *of* f *on input* x *is defined as the maximum number* b *such that there are* b *pairwise disjoint subsets* $B_1, \ldots, B_b$ *of* $[n]$ *for which* $f(x) \neq f(x^{B_i})$. *We call each* B_i *a* block. *The* z-block sensitivity $bs_z(f)$ *of* f, *where* $z \in \{0,1\}$, *is defined as* $\max\{bs(f, x) \mid x \in \{0,1\}^n, f(x) = z\}$. *The* block sensitivity $bs(f)$ *of* f *is defined as* $\max\{bs_0(f), bs_1(f)\}$.

Definition 3. *A* certificate c *of* f *on input* x *is defined as a partial assignment* $c : S \to \{0,1\}, S \subseteq [n]$ *of* x *such that* f *is constant on this restriction. If* f *is always 0 on this restriction, the certificate is a* 0-certificate. *If* f *is always 1, the certificate is a* 1-certificate.

We denote specific certificates as words with $*$ in the positions that the certificate does not assign. For example, $01****$ denotes a certificate that assigns 0 to the first variable and 1 to the second variable.

We say that an input x *satisfies* a certificate c if it matches the certificate in every assigned bit.

The number of *contradictions* between an input and a certificate or between two certificates is the number of positions where one of them assigns 1 and the other assigns 0. For example, there are two contradictions between $0010**$ and $100***$ (in the 1st position and the 3rd position).

The number of *overlaps* between two certificates is the number of positions where both have assigned the same values. For example, there is one overlap between $001***$ and $*0000$ (in the second position). We say that two certificates *overlap* if there is at least one overlap between them.

We say that a certificate remains *valid* after fixing some input bits if none of the fixed bits contradicts the certificate's assignments.

Definition 4. *The* certificate complexity $C(f, x)$ *of* f *on input* x *is defined as the minimum length of a certificate that* x *satisfies. The* z-certificate complexity $C_z(f)$ *of* f, *where* $z \in \{0,1\}$, *is defined as* $\max\{C(f, x) \mid x \in \{0,1\}^n, f(x) = z\}$. *The* certificate complexity $C(f)$ *of* f *is defined as* $\max\{C_0(f), C_1(f)\}$.

3 Background

We study the following question:

Question: Assume that $s_0(g) = m$ and $bs_0(g) = k$. How small can we make $C_1(g)$?

Example 1. Ambainis and Sun [3] consider the following construction.

They define $g_0(x_1, \ldots, x_{2k}) = 1$ if and only if $(x_1, \ldots, x_{2k})$ satisfies one of k certificates $c_0, \ldots, c_{k-1}$ with c_i ($i \in \{0, 1, \ldots, k-1\}$) requiring that

(a) $x_{2i+1} = x_{2i+2} = 1$;
(b) $x_{2j+1} = 0$ for $j \in \{0, \ldots, k-1\}$, $j \neq i$;
(c) $x_{2j+2} = 0$ for $j \in \{i+1, \ldots, i+\lfloor k/2 \rfloor\}$ (with $i+1, \ldots, i+\lfloor k/2 \rfloor$ taken $\mathrm{mod}\, k$).

Then, we have:

- $s_0(g_0) = 1$ (it can be shown that, for every 0-input of g_0, there is at most one c_i in which only one variable does not have the right value);
- $s_1(g_0) = C_1(g_0) = \lfloor 3k/2 \rfloor + 1$ (a 1-input that satisfies a certificate c_i is sensitive to changing any of the variables in c_i and c_i contains $\lfloor 3k/2 \rfloor + 1$ variables);
- $bs_0(g_0) = k$ (the 0-input $x_1 = \cdots = x_{2k} = 0$ is sensitive to changing any of the pairs (x_{2i+1}, x_{2i+2}) from $(0,0)$ to $(1,1)$).

This function can be composed with the OR-function to obtain the best known separation between $s(f)$ and $bs(f)$: $bs(f) = \frac{2}{3}s^2(f) - \frac{1}{3}s(f)$ [3]. As long as $s_0(g) = 1$, the construction is essentially optimal: any g with $bs_0(g) = k$ must satisfy $C_1(g) \geq s_1(g) \geq \frac{3k}{2} - O(1)$.

In this paper, we explore the case when $s_0(g) > 1$. An easy modification of the construction from [3] gives

Theorem 1. *There exists a function g for which $s_0(g) = m$, $bs_0(g) = k$ and* $C_1(f) = \left\lfloor \frac{3\lceil k/m \rceil}{2} \right\rfloor + 1$.

Proof. To simplify the notation, we assume that k is divisible by m. Let $r = k/m$.

We consider a function $g(x_{m1}, \ldots, x_{m,2r})$ with variables $x_{i,j}$ ($i \in \{1, \ldots, m\}$ and $j \in \{1, \ldots, 2r\}$) defined by

$$g(x_{11}, \ldots, x_{m,2r}) = \vee_{i=1}^m g_0(x_{i,1}, \ldots, x_{i,2r}). \tag{1}$$

Equivalently, $g(x_{11}, \ldots, x_{m,2r}) = 1$ if and only if at least one of the blocks $(x_{i,1}, \ldots, x_{i,2r})$ satisfies one of the certificates $c_{i,0}, \ldots, c_{i,r-1}$ that are defined similarly to $c_0, \ldots, c_{k-1}$ in the definition of g_0.

It is easy to see [3] that composing a function g_0 with OR gives $s_0(g) = m\,s_0(g_0)$, $bs_0(g) = m\,bs_0(g_0)$ and $C_1(g) = C_1(g_0)$, implying the theorem. ∎

While this function does not give a better separation between $s(f)$ and $bs(f)$, any improvement to Theorem 1 could give a better separation between $s(f)$ and $bs(f)$ by using the same composition with OR as in [3].

On the other hand, Kenyon and Kutin [6] have shown that

Theorem 2. *For any f with $s_0(g) = m$ and $bs_0(g) = k$, we have $C_1(f) \geq \frac{k}{2m}$.*

4   Separation between $C_1(f)$ and $bs_0(f)$

In this paper, we show that the example of Theorem 1 is optimal.

Theorem 3. *For any Boolean function f the following inequality holds:*

$$C_1(f) \geq \frac{3}{2} \frac{bs_0(f)}{s_0(f)} - \frac{1}{2}. \tag{2}$$

Proof. Without loss of generality, we can assume that the maximum bs_0 is achieved on the all-0 input denoted by 0. Let $B_1, ..., B_k$ be the sensitive blocks, where $k = bs_0(f)$. Also, we can w.l.o.g. assume that these blocks are minimal and that every bit belongs to a block. (Otherwise, we can fix the remaining bits to 0. This can only decrease s_0 and C_1, strengthening the result.)

Each block B_i has a corresponding minimal 1-certificate c_i such that the word $(\{0\}^n)^{B_i}$ satisfies this certificate. Each of these certificates has a 1 in every position of the corresponding block (otherwise the block would not be minimal) and any number of 0's in other blocks.

We construct a complete weighted graph G whose vertices correspond to certificates $c_1, ..., c_k$. Each edge has a weight that is equal to the number of contradictions between the two certificates the edge connects. *The weight of a graph is just the sum of the weights of its edges.* We will prove

Lemma 1. *Let w be the weight of an induced subgraph of G of order m. Then*

$$w \geq \frac{3}{2} \frac{m^2}{s_0(f)} - \frac{3}{2}m. \tag{3}$$

Proof. The proof is by induction. As a basis we take induced subgraphs of order $m \leq s_0(f)$. In this case,

$$\frac{3}{2} \frac{m^2}{s_0(f)} - \frac{3}{2}m \leq 0 \tag{4}$$

and $w \geq 0$ is always true, as the number of contradictions between two certificates cannot be negative.

Let $m > s_0(f)$. We assume that the relation holds for every induced subgraph of order $< m$. Let G' be an induced subgraph of order m. Let $H \subset G'$ be its induced subgraph of order $s_0(f)$ with the smallest total weight.

Lemma 2. *For any certificate $c_i \in G' \setminus H$ in G' not belonging to this subgraph H the weight of the edges connecting c_i to H is ≥ 3.*

Proof. Let t be the total weight of the edges in H. Let us assume that there exists a certificate $c_j \notin H$ such that the weight of the edges connecting c_j to H

is ≤ 2. Let H' be the induced subgraph $H \cup \{c_j\}$. Then the weight of H' must be $\leq t + 2$.

We define the weight of a certificate $c_i \in H'$ as the sum of the weights of all edges of H' that involve vertex c_i. If there exists a certificate $c_i \in H'$ such that its weight in H' is ≥ 3, then the weight of $H' \setminus \{c_i\}$ would be $< t$, which is a contradiction, as H was taken to be the induced subgraph of order $s_0(f)$ with the smallest weight. Therefore the weight of every certificate in H' is at most 2.

In the next section, we show

Lemma 3. *Let f be a Boolean function for which the following properties hold: $f(\{0\}^n) = 0$ and f has such k minimal 1-certificates that each has at most 2 contradictions with all the others together. Furthermore, for each input position, exactly one of these certificates assigns the value 1. Then, $s_0(f) \geq k$.*

This lemma implies that $s_0(f) \geq |H'|$ which is in contradiction with $|H'| = s_0(f) + 1$. Therefore no such c_j exists. ∎

We now examine the graph $G' \setminus H$. It consists of $m - s_0(f)$ certificates and by the inductive assumption has a weight of at least

$$\frac{3}{2} \frac{(m - s_0(f))^2}{s_0(f)} - \frac{3}{2}(m - s_0(f)). \tag{5}$$

But there are at least $3(m - s_0(f))$ contradictions between H and $G' \setminus H$, thus the total weight of G' is at least

$$\frac{3}{2} \frac{(m - s_0(f))^2}{s_0(f)} - \frac{3}{2}(m - s_0(f)) + 3(m - s_0(f)) \tag{6}$$

$$= \frac{3}{2} \frac{m^2 - 2m s_0(f) + s_0(f)^2}{s_0(f)} + \frac{3}{2}m - \frac{3}{2}s_0(f) \tag{7}$$

$$= \frac{3}{2} \frac{m^2}{s_0(f)} - 3m + \frac{3}{2}s_0(f) + \frac{3}{2}m - \frac{3}{2}s_0(f) \tag{8}$$

$$= \frac{3}{2} \frac{m^2}{s_0(f)} - \frac{3}{2}m. \tag{9}$$

This completes the induction step. ∎

By taking the whole of G as G', we find a lower bound on the total number of contradictions in the graph:

$$\frac{3}{2} \frac{k^2}{s_0(f)} - \frac{3}{2}k. \tag{10}$$

Each contradiction requires one 0 in one of the certificates and each 0 contributes to exactly one contradiction (since for each position exactly one of c_i assigns a 1). Therefore, by the pigeonhole principle, there exists a certificate with at least

$$\frac{3}{2} \frac{k}{s_0(f)} - \frac{3}{2} \tag{11}$$

zeroes. As each certificate contains at least one 1, we get a lower bound on the size of one of these certificates and $C_1(f)$:

$$C_1(f) \geq \frac{3}{2}\frac{bs_0(f)}{s_0(f)} - \frac{1}{2}. \tag{12}$$

∎

5   Functions with $s_0(f)$ Equal to Number of 1-certificates

In this section we prove Lemma 3.

5.1   General Case: Functions with Overlaps

Let $c_1, \ldots, c_k$ be the k certificates. We start by reducing the general case of Lemma 3 to the case when there are no overlaps between any of $c_1, \ldots, c_k$.

Note that certificate overlaps can only occur when two certificates assign 0 to the same position. Then a third certificate assigns 1 to that position. This produces 2 contradictions for the third certificate, therefore it has no further overlaps or contradictions. For example, here we have this situation in the 3rd position (with the first three certificates) and in the 6th position (with the last three certificates):

$$\begin{pmatrix} 1 & 1 & 0 & * & * & * & * & * & * & * \\ * & * & 1 & * & * & * & * & * & * & * \\ * & * & 0 & 1 & 1 & 0 & * & * & * & * \\ * & * & * & * & 1 & 1 & 1 & * & * \\ 0 & * & * & * & * & 0 & * & * & 1 & 1 \end{pmatrix}. \tag{13}$$

Let t be the total number of such overlaps. Let D be the set of certificates assigning 1 to positions with overlaps, $|D| = t$. We fix the position of every overlap to 0. Since the remaining function contains the word $\{0\}^n$, it is not identically 1. Every certificate not in D is still a valid 1-certificate, as they assigned either nothing or 0 to the fixed positions. If they are no longer minimal, we can minimize them, which cannot produce any new overlaps or contradictions.

The certificates in D are, however, no longer valid. Let us examine one such certificate $c \in D$. We denote the set of positions assigned to by c by S. Let i be the position in S that is now fixed to 0. We claim that certificate c assigns value 1 to all $|S|$ positions in c. (If it assigned 0 to some position, there would be at least 3 contradictions between c and other certificates: two in position i and one in position where c assigns 0.)

If $|S| = 1$, then the remaining function is always sensitive to i on 0-inputs, as flipping x_i results in an input satisfying c.

If $|S| > 1$, we examine the $2^{|S|-1}$ subfunctions obtainable by fixing the remaining positions of S. We fix these positions to the subfunction that is not identically 1 with the highest number of bits fixed to 1, we will call this the *largest non-constant subfunction*. If it fixes 1 in every position, it is sensitive to

i on 0-inputs, as flipping it produces a word which satisfies c. Otherwise it is sensitive on 0-inputs to every other bit fixed to 0 in S besides i, as flipping them would produce a word from a subfunction with a higher amount of bits fixed to 1. But that subfunction is identically 1 or we would have fixed it instead.

In either case we obtain at least one sensitive bit in S on 0-inputs in the remaining function. Furthermore, every certificate not in D is still valid, if not minimal. But we can safely minimize them again.

We can repeat this procedure for every certificate in D. The resulting function is not always 1 and, on every 0-input, it has at least t sensitive bits among the bits that we fixed. Furthermore, we still have $k-t$ non-overlapping valid minimal 1-certificates with no more than 2 contradictions each. In the next section, we show that this implies that it has 0-sensitivity of at least $k - t$ (Lemma 4). Therefore, the original function has a 0-sensitivity of at least $k - t + t = k$.

5.2   Functions with No Overlaps

Lemma 4. *Let f be a Boolean function, such that f is not always 1 and f has such k non-overlapping minimal 1-certificates that each has at most 2 contradictions with all the others together. Then, $s_0(f) \geq k$.*

Proof. To prove this lemma, we consider the weighted graph G on these k certificates where the weight of an edge in this graph is the number of contradictions between the two certificates the edge connects.

We examine the connected components in this graph, not counting edges with weight 0. There can be only 4 kinds of components – individual certificates, two certificates with 2 contradictions between them, paths of 2 or more certificates with 1 contradiction between every two subsequent certificates in the path and cycles of 3 or more certificates with 1 contradiction between every two subsequent certificates in the cycle. As there are no overlaps between the certificates, each position is assigned to by certificates from at most one component.

We will now prove by induction on k that we can obtain a 0-input with as many sensitive bits in each component as there are certificates in it.

As a basis we take $k = 0$. Since f is not always 1, $s_0(f)$ is defined, but obviously $s_0(f) \geq 0$.

Then we look at each graph component type separately.

Individual Certificates. We first examine individual certificates. Let us denote the examined certificate by c and the set of positions it assigns by S. We fix all bits of S except for one according to c and we fix the remaining bit of S opposite to c. The remaining function cannot be always 1, as otherwise the last bit in S would not be necessary in c, but c is minimal. Therefore on 0-inputs the remaining function is also sensitive to this last bit, as flipping it produces a word which satisfies c.

Afterwards the remaining certificates might no longer be minimal. In this case we can minimize them. This cannot produce any more contradictions and no certificate can disappear, as the function is not always 1. Therefore the remaining function still satisfies the conditions of this lemma and has $k - 1$ minimal 1-certificates, with each certificate having at most 2 contradictions with the others.

Then by induction the remaining function has a 0-sensitivity of $k-1$. Together with the sensitive bit among the fixed ones, we obtain $s_0(f) \geq k$.

Certificate Paths. We can similarly reduce certificate paths. A certificate path is a structure where each certificate has 1 contradiction with the next one and there are no other contradictions. For example, here is an example of a path of length 3:

$$\begin{pmatrix} i \\ 1\,1\,0\,*\,*\,*\,* \\ *\,*\,1\,1\,0\,*\,* \\ *\,*\,*\,*\,1\,1\,1 \end{pmatrix}. \tag{14}$$

We note that every certificate in a path assigns at least 2 positions, otherwise its neighbours would not be minimal.

We then take a certificate c at the start of a path, which is next to a certificate d. Let S be the set of positions c assigns. Let i be the position where c and d contradict each other.

We then fix every bit in S but i according to c, and we fix i according to d. The remaining function cannot be always 1, as otherwise i would not be necessary in c, but c is minimal. But on 0-inputs the remaining function is also sensitive to i because flipping it produces a word which satisfies c.

We note that in the remaining function the rest of d (not all of d was fixed because d assigns at least 2 positions) is still a valid certificate, since it only assigns one of the fixed bits and it was fixed according to d. Similarly to the first case we can minimize the remaining certificates and obtain a function with $k - 1$ certificates satisfying the lemma conditions.

Then by induction the remaining function has a 0-sensitivity of $k-1$. Together with the sensitive bit i, we obtain $s_0(f) \geq k$.

Two Certificates with Two Contradictions. Let us denote these 2 certificates as c and d and the two positions where they contradict as i and j. For example, we can have 2 certificates like this:

$$\begin{pmatrix} i\,j \\ 1\,1\,1\,0\,* \\ *\,*\,0\,1\,1 \end{pmatrix}. \tag{15}$$

Let S be the set of positions c assigns and T be the set of positions d assigns. We then fix every bit in S except j according to c but we fix j according to d. The remaining function cannot be always 1 because, otherwise, j would not be necessary in c, but c is minimal. But on 0-inputs the remaining function is also sensitive to j, as flipping it produces a word which satisfies c.

If $|T| = 2$, then on 0-inputs the remaining function is also sensitive to i because flipping the ith variable produces a word which satisfies d.

If $|T| > 2$, we examine the $2^{|T|-2}$ subfunctions obtainable by fixing the remaining positions of T. We can w.l.o.g. assume that d assigns the value 1 to each of these. Similarly to section 5.1, we find the largest non-constant subfunction among these – the subfunction that is not identically 1 with the highest number of bits fixed to 1. Then on 0-inputs we obtain a sensitive bit either at i if this subfunction fixes all these positions to 1 or at a fixed 0 otherwise.

Therefore we can always find at least one additional sensitive bit among T.

Again we can minimize the remaining certificates and obtain a function with $k - 2$ certificates satisfying the conditions of the lemma.

Then by induction the remaining function has a 0-sensitivity of $k-2$. Together with the two additional sensitive bits found, we obtain $s_0(f) \geq k$.

Certificate Cycles. A certificate cycle is a sequence of at least 3 certificates where each certificate has 1 contradiction with the next one and the last one has 1 contradiction with the first one. For example, here is a cycle of length 5:

$$\begin{pmatrix} j_{5,1} & j_{1,2} & j_{2,3} & & j_{3,4} & j_{4,5} & \\ 1 & 1 & 0 & * & * & * & * & * \\ * & * & 1 & 0 & * & * & * & * \\ * & * & * & 1 & 1 & 1 & * & * \\ * & * & * & * & * & 0 & 0 & * \\ 0 & * & * & * & * & * & 1 & 1 \end{pmatrix}. \tag{16}$$

Every certificate in a cycle assigns at least 2 positions, otherwise its neighbours in the cycle would overlap. We denote the length of the cycle by m. Let $c_1, \ldots, c_m$ be the certificates in this cycle, let $S_1, \ldots, S_m$ be the positions assigned by them, and let $j_{1,2}, \ldots, j_{m,1}$ be the positions where the certificates contradict.

We assign values to variables in $c_2, \ldots, c_m$ in the following way. We first assign values to variables in S_2 so that the variable $j_{2,3}$ contradicts c_2 and is assigned according to c_3, but all other variables are assigned according to c_2.

We have the following properties. First, the remaining function cannot be always 1, as otherwise $j_{2,3}$ would not be necessary in c_2, but c_2 is minimal. Second, any 0-input that is consistent with the assignment that we made is sensitive to $j_{2,3}$ because flipping this position produces a word which satisfies c_2. Third, in the remaining function $c_3, \ldots, c_m$ are still valid 1-certificates because we have not made any assignments that contradict them. Some of these certificates c_i may no longer be minimal. In this case, we can minimize them by removing unnecessary variables from c_i and S_i.

We then perform a similar procedure for $c_i \in \{3, \ldots, m\}$. We assume that the variables in S_2, ..., S_{i-1} have been assigned values. We then assign values to variables in S_i. If c_i and c_{i+1} contradict in the variable $j_{i,i+1}$, we assign it according to c_{i+1}. (If $i = m$, we define $i + 1 = 1$.) If c_i and c_{i+1} no longer contradict (this can happen if $j_{i,i+1}$ was removed from one of them), we choose a variable in S_i arbitrarily and assign it opposite to c_i. All other variables in S_i are assigned according to c_i.

We now have similar properties as before. The remaining function cannot be always 1 and any 0-input that is consistent with our assignment is sensitive to changing a variable in S_i. Moreover, $c_{i+1}, \ldots, c_m$ are still valid 1-certificates and, if they are not minimal, they can be made minimal by removing variables.

At the end of this process, we have obtained $m-1$ sensitive bits on 0-inputs: for each of $c_2, \ldots, c_m$, there is a bit, changing which results in an input satisfying c_i. We now argue that there should be one more sensitive bit. To find it, we consider the certificate c_1.

During the process described above, the position $j_{1,2}$ where c_1 and c_2 contradict was fixed opposite to the value assigned by c_1. The position $j_{m,1}$ where c_1 and c_m contradict is either unfixed or fixed according to c_1. All other positions of c_1 are unfixed.

If there are no unfixed positions of c_1, then changing the position $j_{1,2}$ in a 0-input (that satisfies the partial assignment that we made) leads to a 1-input that satisfies c_1. Hence, we have m sensitive bits.

Otherwise, let $T \subset S_1$ be the set of positions in c_1 that have not been assigned and let $p = |T|$. W.l.o.g, we assume that c_1 assigns the value 1 to each of those positions. We examine the 2^p subfunctions obtainable by fixing the positions of T in some way. Again we find the largest non-constant subfunction among these – the subfunction that is not identically 1 with the highest number of bits fixed to 1. Then on 0-inputs we obtain a sensitive bit either at $j_{1,2}$ if this subfunction fixes all these positions to 1 or at a fixed 0 otherwise.

Similarly to the first three cases, we can minimize the remaining certificates and obtain a function with $k - m$ certificates satisfying the conditions of the lemma. By induction, the remaining function has a 0-sensitivity of $k - m$. Together with the m additional sensitive bits we found, we obtain $s_0(f) \geq k$.   ∎

6   Conclusions

In this paper, we have shown a lower bound on 1-certificate complexity in relation to the ratio of 0-block sensitivity and 0-sensitivity:

$$C_1(f) \geq \frac{3}{2} \frac{bs_0(f)}{s_0(f)} - \frac{1}{2}. \tag{17}$$

This bound is tight, as the function constructed in Theorem 1 achieves the following equality:

$$C_1(f) = \frac{3}{2} \frac{bs_0(f)}{s_0(f)} + \frac{1}{2}. \tag{18}$$

The difference of 1 appears as the proof of Theorem 3 requires only a single 1 in each certificate but the construction of Theorem 1 has two.

Thus, we have completely solved the problem of finding the optimal relationship between $s_0(f)$, $bs_0(f)$ and $C_1(f)$. For functions with $s_1(f) = C_1(f)$, such as those constructed in [3,8,9], this means that

$$bs_0(f) \leq \left(\frac{2}{3} + o(1) \right) s_0(f) s_1(f). \tag{19}$$

That is, if we use such functions, there is no better separation between $s(f)$ and $bs(f)$ than the currently known one.

For the general case, it is important to understand how big the gap between $s_1(f)$ and $C_1(f)$ can be. Currently, we only know that

$$s_1(f) \leq C_1(f) \leq 2^{s_0(f)-1} s_1(f), \tag{20}$$

with the upper bound shown in [2]. In the general case (17) together with this bound implies only

$$bs_0(f) \leq \left(\frac{2}{3} + o(1) \right) 2^{s_0(f)-1} s_0(f) s_1(f). \tag{21}$$

However, there is no known f that comes even close to saturating the upper bound of (20) and we suspect that this bound can be significantly improved.

There are some examples of f with gaps between $C_1(f)$ and $s_1(f)$, though. For example, the 4-bit non-equality function of [1] has $s_0(NE) = s_1(NE) = 2$ and $C_1(NE) = 3$ and it is easy to use it to produce an example $s_0(NE) = 2$, $s_1(NE) = 2k$ and $C_1(NE) = 3k$. Unfortunately, we have not been able to combine this function with the function that achieves (18) to obtain a bigger gap between $bs(f)$ and $s(f)$.

Because of that, we conjecture that (19) might actually be optimal. Proving or disproving this conjecture is a very challenging problem.

References

1. Ambainis, A.: Polynomial degree vs. quantum query complexity. J. Comput. Syst. Sci. 72(2), 220–238 (2006)
2. Ambainis, A., Bavarian, M., Gao, Y., Mao, J., Sun, X., Zuo, S.: Tighter relations between sensitivity and other complexity measures. In: Esparza, J., Fraigniaud, P., Husfeldt, T., Koutsoupias, E. (eds.) ICALP 2014. LNCS, vol. 8572, pp. 101–113. Springer, Heidelberg (2014)
3. Ambainis, A., Sun, X.: New separation between $s(f)$ and $bs(f)$. CoRR, abs/1108.3494 (2011)
4. Buhrman, H., de Wolf, R.: Complexity measures and decision tree complexity: a survey. Theor. Comput. Sci. 288(1), 21–43 (2002)
5. Hatami, P., Kulkarni, R., Pankratov, D.: Variations on the Sensitivity Conjecture. Graduate Surveys, vol. 4. Theory of Computing Library (2011)
6. Kenyon, C., Kutin, S.: Sensitivity, block sensitivity, and l-block sensitivity of Boolean functions. Inf. Comput. 189(1), 43–53 (2004)
7. Nisan, N., Szegedy, M.: On the degree of Boolean functions as real polynomials. Computational Complexity 4, 301–313 (1994)
8. Rubinstein, D.: Sensitivity vs. block sensitivity of Boolean functions. Combinatorica 15(2), 297–299 (1995)
9. Virza, M.: Sensitivity versus block sensitivity of Boolean functions. Inf. Process. Lett. 111(9), 433–435 (2011)

$\tilde{O}(\sqrt{n})$-Space and Polynomial-Time Algorithm for Planar Directed Graph Reachability

Tetsuo Asano[1], David Kirkpatrick[2], Kotaro Nakagawa[3], and Osamu Watanabe[3]

[1] Japan Advanced Institute of Technology, Japan
[2] Department of Computer Science, University of British Columbia, Canada
[3] Department of Math. and Comput. Sci., Tokyo Institute of Technology, Japan

Abstract. The directed graph reachability problem takes as input an n-vertex directed graph $G = (V, E)$, and two distinguished vertices v_0, and vertex v_*. The problem is to determine whether there exists a path from v_0 to v_* in G. The main result of this paper is to show that the directed graph reachability problem restricted to planar graphs can be solved in polynomial time using only $\tilde{O}(\sqrt{n})$ space[1].

1 Introduction and Motivation

For a directed graph $G = (V, E)$, its *underlying graph* is the undirected graph $`G = (V, `E)$, where the vertex pair $\{u, v\}$ belongs to $`E$ if and only if at least one of (u, v) or (v, u) belongs to E. The *planar directed graph reachability problem* is a special case of the directed graph reachability problem where we restrict attention to input graphs whose underlying graph is *planar*.

The general directed graph reachability problem is a core problem in computational complexity theory. It is a canonical complete problem for nondeterministic log-space, NL, and the famous open question L = NL is essentially asking whether the problem is solvable deterministically in log-space. The standard breadth first search algorithm and Savitch's algorithm are two of the most fundamental algorithms known for solving the directed graph reachability problem. The former uses space and time linear in the number of edges of the input graph and the latter uses only $O((\log n)^2)$-space but requires $\Theta(n^{\log n})$ time. Hence a natural and significant question is whether we can design an algorithm for directed graph reachability that is efficient in both space and time. In particular, can we design a polynomial-time algorithm for the directed graph reachability problem that uses only $O(n^\epsilon)$-space for some small constant $\epsilon < 1$? This question was asked by Wigderson in his excellent survey paper [13], and it remains unsettled. The best known result in this direction is the two decades old bound due to Barns, Buss, Ruzzo and Schieber [4], who showed a polynomial-time algorithm for the problem that uses $O(n/2^{\sqrt{\log n}})$ space. Note that this space bound is only

---

[1] In this paper "$\tilde{O}(s(n))$-space" means $O(s(n))$-words intuitively and precisely $O(s(n) \log n)$-space.

E. Csuhaj-Varjú et al. (Eds.): MFCS 2014, Part II, LNCS 8635, pp. 45–56, 2014.
© Springer-Verlag Berlin Heidelberg 2014

slightly sublinear, and improving this bound remains a significant open question. In fact, there are indications that it may be difficult to improve this bound because there are matching *lower bounds* known for solving the directed graph reachability problem on a certain model of computation known as NNJAG; see, e.g., [5]. Though NNJAG is a restrictive model, all the known algorithms for the directed reachability can be implemented in NNJAG without significant blow up in time and space.

Some important progress has been made for restricted graph classes. The most remarkable one is the log-space algorithm of Reingold (which we will refer as UReach) for the undirected graph reachability [12]. Recently, Asano and Doerr [2] gave a $\widetilde{O}(n^{1/2+\varepsilon})$-space and $n^{O(1/\epsilon)}$-time algorithm for the reachability problem restricted to directed grid graphs. Inspired by this result Imai et al. proposed [8] an $\widetilde{O}(n^{1/2+\varepsilon})$-space and $n^{O(1/\epsilon)}$-time algorithm for the planar graph reachability problem. It has been left open to design an $\widetilde{O}(n^{1/2})$-space and yet polynomial-time algorithm. More recently, Asano and Kirkpatrick [3] introduced a more efficient way to control the recursion, thereby succeeding to obtain an $\widetilde{O}(\sqrt{n})$-space and polynomial time algorithm for the reachability problem restricted to directed grid graphs. The main result of this paper is to show that this technique can be adapted to design an $\widetilde{O}(\sqrt{n})$-space and polynomial-time algorithm for the planar graph reachability problem.

2   Background

The Planar Separator Theorem, shown first by Lipton and Tarjan [10], asserts that, for any n-vertex undirected planar graph G, there is a polynomial-time algorithm for computing an $O(\sqrt{n})$-size "separator" for G, i.e. a set of $O(\sqrt{n})$ vertices whose removal separates the graph into two subgraphs of similar size. The key idea of the $\widetilde{O}(n^{1/2+\varepsilon})$-space algorithm of Imai et al. is to use an space-constrained algorithmic version of the Planar Separator Theorem. For a given input instance (G, v_0, v_*), they first compute a separator S of the underlying planar graph of G that separates G into two smaller subgraphs G^0 and G^1. They then consider a new directed graph H on $S \cup \{v_0, v_*\}$; it has a directed edge (a, b) if and only if there is a path from a to b in either G^0 or G^1. Clearly, reachability (of v_* from v_0) in H is equivalent to the original reachability. On the other hand, since S has $O(\sqrt{n})$ vertices, the standard linear-space and polynomial-time algorithm can be used to solve the reachability problem in H by applying the algorithm recursively to G^0 and G^1 whenever it is necessary to know if an edge (a, b) exists in H. It should be mentioned that the idea of using separators to improve algorithms for the reachability and related problems is natural, and in fact it has been proposed by several researchers; see, e.g., [7]. The main contribution of [8] is to show how to implement this idea by giving a space efficient separator algorithm based on the parallel separator algorithms of Miller [9] and Gazit and Miller [6].

The algorithm of Asano and Kirkpatrick uses a recursive separation of a grid graph; at each level a separator is formed by the set of vertices on one of the grid center lines. In order to get a polynomial time bound, Asano and Kirkpatrick introduce a kind of budgeted recursion, controlled by a "universal sequence", that restricts the time complexity of each recursive execution on smaller grids. Here we use the same idea for the general planar graph reachability. To mimic the grid-based algorithm of Asano and Kirkpatrick we need a simple separator that allows us to express/identify a hierarchy of subgraphs succinctly/simply. The main technical contribution of this paper is to describe a space efficient way to construct such a simple separator together with a succinct way to express the separated subgraphs. (We note that it is not immediately clear whether the sublinear-space algorithm of [8] always yields such a simple separator. Here instead of analyzing the separator algorithm of [8], we show how to modify a given separator to obtain a suitable cycle-separator.)

3   Planar Graph Reachability Algorithm

3.1   Preliminaries

Let $G = (V, E)$ denote an arbitrary directed graph. For any subset U of V we use $G[U]$ to denote the subgraph of G induced by U. For two graphs G_1 and G_2, by $G_1 \cup G_2$ we mean the graph $(V(G_1) \cup V(G_2), E(G_1) \cup E(G_2))$. For any graph G, consider any subgraph H of G and any vertex v of G that is not in H; then by $H \sqcup_G v$ we denote an induced subgraph of G obtained from H by adding vertex v; that is, $H \sqcup_G v = G[V(H) \cup \{v\}]$. Similarly, for an arbitrary set $A \subset V$, we use $H \sqcup_G A$ to denote $G[V(H) \cup A]$.

Recall that a graph is *planar* if it can be drawn on a plane so that the edges intersect only at end vertices. Such a drawing is called a *planar embedding*. Here we use the standard way to specify a *planar embedding* of G; that is, a sequence of vertices adjacent to v in a clockwise order around v under the planar embedding, for all $v \in V$. We use $N(v)$ to denote this sequence for v, which is often regarded as a set. For a planar graph and its planar embedding, its *triangulation* (w.r.t. this planar embedding) means to add edges to the planar graph until all its faces under the planar embedding (including the outer one) are bounded by three edges. We note that, in assuming that the input of our algorithm is a planar graph, we assume only that a planar embedding exists, not that it is given as part of the input.

3.2   The Algorithm

We now describe our algorithm for planar graph reachability. To illustrate the idea of the algorithm we consider first the case in which the input graph G is a subgraph of a bi-directed grid graph like Figure 1(a). Here we assume that the original bi-directed grid graph is a $(2^h - 1) \times (2^h - 1)$ square grid. Let 'G denote the undirected version of this original grid. Note that both G and 'G have $n = (2^h - 1) \times (2^h - 1)$ vertices.

Although reachability is determined by the edges of G, the computation is designed based on the underlying graph 'G. Consider a set S of vertices that are

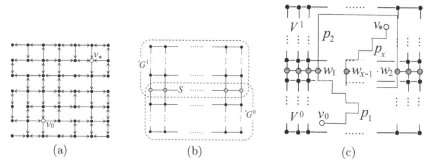

Fig. 1. Example grid graph G and a path from v_0 to v_*

on the horizontal center line. We call S a *grid-separator* because 'G is separated to two disconnected subgraphs by removing S. Our strategy is to determine, for every vertex $v \in G$, the reachability from v_0 in each subgraph independently, thereby saving the space for the computation. More specifically, we consider a subgraph 'G^0 (resp., 'G^1) of 'G consisting of vertices below (resp., above) S including S (*cf.* Figure 1(b)), and compute the reachability from v_0 on G in the area defined by each subgraph. A crucial point is that we need to keep the reachability information only for vertices in S in order to pass the reachability information to the next computation on the opposite subarea. Also for showing the $\widetilde{O}(\sqrt{n})$-space bound, it is important that S consists of $2^h - 1 = \sqrt{n}$ vertices, and that both subgraphs 'G^0 and 'G^1 are almost half of 'G.

Suppose that v_* is reachable from v_0 in G. We explain concretely our strategy to confirm this by identifying some directed path that witnesses the reachability from v_0 to v_*. Notice here that such a path p is divided into some x subpaths $p_1, \ldots, p_x$ such that the following holds for each $j \in [x-1]$ (*cf.* Figure 1(c)): (i) the end vertex w_j of p_j (that is the start vertex of p_{j+1}) is on S, and (ii) all inner vertices of p_j are in the same V^b for some $b \in \{0, 1\}$, where V^0 and V^1 are respectively the set of vertices below and above the separator S. Then it is easy to see that we can find that w_1 is reachable from v_0 by searching vertices in S that are reachable from v_0 in $G[S \cup V^0]$. Next we can find that w_2 is reachable (from v_0) by searching vertices in S that are reachable in $G[S \cup V^1]$ from some vertex in S for which we know already its reachability; in fact, by the reachability from w_1 we can confirm that w_2 is reachable from v_0. Similarly, the reachability of $w_3, \ldots, w_{x-1}$ is confirmed, and then by considering the subgraph $G[S \cup V^1]$ we confirm that v_* is reachable from v_0 because it is reachable from w_{x-1}. This is our basic strategy. Note that the reachability in each subgraph can be checked recursively.

Since a path can cross the separator $\Theta(\sqrt{n})$ times, we cannot avoid making as many recursive calls at each level of recursion as there are vertices in the separator at that level. Consequently, without some modification the algorithm as described cannot hope to terminate in time bounded by some polynomial in n. In order to achieve polynomial-time computability, we introduce the idea of "budgeted recursion": the computation time allocated to individual recursive calls for checking the reachability in each subgraph is restricted in accordance with a predetermined sequence. Since the time required to trace a connecting path within an individual subgrid is not known in advance, we rely on a universality property of the sequence: eventually every subproblem will be allocated a budget sufficiently large to complete the required computation within that subproblem.

Asano and Kirkpatrick [3] describe the construction of a "universal sequence" suitable for this purpose. A similar universal sequence has been used in the context of oblivious algorithms; see [11], for example. Here we consider the following version.

For any $s \geq 0$, the *universal sequence* σ_s *order* s is defined inductively by

$$\sigma_s = \begin{cases} \langle 1 \rangle & \text{if } s = 0, \text{ and} \\ \sigma_{i-1} \diamond \langle 2^i \rangle \diamond \sigma_{i-1} & \text{otherwise,} \end{cases}$$

where $\diamond$ signifies concatenation of sequences. By the definition, each element of the sequence is a power of 2. The length of the sequence σ_s is $2^{s+1} - 1$. For example, $\sigma_2 = \langle 1, 2, 1, 4, 1, 2, 1 \rangle$. We will use the following properties of universal sequences in the design and analysis of our algorithm. Their proof is straightforward, and is omitted here due to space constraints.

Lemma 1. *(a) The sequence $\sigma_s = \langle c_1, \ldots, c_{2^{s+1}-1} \rangle$ is 2^s-universal in the sense that for any positive integer sequence $\langle d_1, \ldots, d_x \rangle$ such that $\sum_{i \in [x]} d_i \leq 2^s$, there exists a subsequence $\langle c_{i_1}, \ldots, c_{i_x} \rangle$ of σ_s such that $d_j \leq c_{i_j}$ holds for all $j \in [x]$; (b) the sequence σ_s contains exactly 2^{s-i} appearances of the integer 2^i, for all $i \in [s]$, and nothing else; and (c) the sequence σ_s is computable in $O(2^s)$-time and $\tilde{O}(1)$-space.*

Now we define our reachability algorithm following the strategy explained above. The technical key point here is to define a sequence of separators dividing subgraphs into two parts in a way that we can specify a current target subgraph succinctly. For this we introduce the notion of "cycle-separator." Intuitively, a cycle-separator S of a graph G is a set of cycles $S = \{C_1, \ldots, C_h\}$ that separates G into two subgraphs, those consisting of vertices located left (resp., right) of the cycles (including cycle vertices). In section 4 we outline how such a simple cycle-separator can be efficiently computed from any given separator. Based on this we have the following lemma.

Lemma 2. *There exists an $\tilde{O}(\sqrt{n})$-space and polynomial-time algorithm (which we refer as* CycleSep*) that computes a cycle-separator S for a given undirected graph $G = (V, E)$ with its triangulated planar embedding. The size of the separator is at most $c_{\text{sep}}\sqrt{n}$. Furthermore, there is a way to define subsets V^0 and V^1 of V with the following properties: (a) $V^0 \cup V^1 = V$, $V^0 \cap V^1 = S$, and (b) $|V^b| \leq (2/3)|V| + c_{\text{sep}}\sqrt{n}$ for each $b \in \{0, 1\}$.*

Intuitively we can use cycle-separators like grid-separators to define a sequence of progressively smaller subgraphs of a given planar directed graph G. (Note that

its underlying graph 'G is used for defining the subgraphs.) From technical reason[2], however, we need to add some edges to the outer faces of '$G[V^b]$ to get it triangulated under the current embedding. We can show that the number of added
edges is bounded by $O(|S|)$ and that there is an algorithm AddTri that computes
these edges and their planar embedding in $\widetilde{O}(|S|)$-space and polynomial-time.
We refer to this information as an *additional triangulation edge list* T and consider it with a cycle-separator S and a Boolean label b. By $['G]^b_{S,T}$ we mean both
a graph obtained from '$G[V^b]$ by adding those triangulation edges specified by
T and its planar embedding obtained by modifying the original planar embed-
ding by T. In general, for any sequence $\mathbf{S} = \langle (b_1, S_1, T_1), \ldots, (b_t, S_t, T_t) \rangle$ of such
triples of a label, a cycle-separator and an edge list, we define $['G]_\mathbf{S}$ by

$$['G]_\mathbf{S} = \left[\cdots \left[['G]^{b_1}_{S_1,T_1} \right]^{b_2}_{S_2,T_2} \cdots \right]^{b_t}_{S_t,T_t},$$

which we call a *depth t subarea* of G. We should note here that it is easy to
identify a depth t subarea by using $\mathbf{S}$; for a given $\mathbf{S}$, we can determine whether
$v \in V$ is in the subarea $['G]_\mathbf{S}$ by using only $O(\log n)$-space.

Armed with this method of constructing/specifying subareas we now implement our algorithm idea discussed above as a recursive procedure ExtendReach
(see Algorithm 1). First we explain what it computes. We use global variables
to keep the input graph G, the triangulated planar embedding of its underlying graph 'G, the start vertex v_0, and the goal vertex v_*. As we will see in the
next section, the triangulated planar embedding is $O(\log n)$-space computable.
Hence, we can compute it whenever needed; thus, for simplicity we assume here
that the embedding is given also as a part of the input. Note that for the space
complexity this additional input data is not counted. On the other hand, we
define global variables A and R that are kept in the work space. The variable A
is for keeping grid-separator vertices that are currently considered; the vertices
v_0 and v_* are also kept in A. The array R captures the reachability information
for vertices in A; for any $v \in$ A, R$[v] = \mathbf{true}$ iff the reachability of v from v_0 has
been confirmed. Besides these data in the global variables, the procedure takes
arguments $\mathbf{S}$ and ℓ, where $\mathbf{S}$ specifies the current subarea of 'G and ℓ is a bound
on the length of path extensions. Our task is to update the reachability from v_0
for all vertices in A by using paths of length $\leq \ell$ in the current subarea. More
precisely, the procedure ExtendReach($\mathbf{S}, \ell$) does the following: for each vertex
$v \in$ A, it sets R$[v] = \mathbf{true}$ if and only if there is a path to v in $G[\text{A} \cup V_\mathbf{S}]$ of
length $\leq \ell$ from some vertex $u \in$ A whose value R$[u]$ before the execution equals
$\mathbf{true}$, where $V_\mathbf{S}$ is the set of vertices of the current subarea of 'G specified by
$\mathbf{S}$. Since any vertex in G that is reachable from v_0 is reachable by a path of
length at most $2^{\lceil \log n \rceil}$, the procedure ExtendReach can be used to determine

---

[2] The algorithm CycleSep is defined based on the separator algorithm of Lemma 4
that assumes a triangulated graph as input. Thus, in order to apply CycleSep to
divide '$G[V^b]$ further, we need to get it triangulated.

the reachability of vertex v_* from v_0 as follows, which is our planar graph reachability algorithm: (1) Set $A \leftarrow \{v_0, v_*\}$, $R[v_0] \leftarrow$ **true**, and $R[v_*] \leftarrow$ **false**; and (2) Execute ExtendReach($\langle\rangle, 2^{\lceil \log n \rceil}$), and then output $R[v_*]$.

Next we give some additional explanation concerning Algorithm 1. Consider any execution of ExtendReach for given arguments S and ℓ (together with data kept in its global variables). Let V_S be the set of vertices of the subarea $['G]_S$ specified by S. There are two cases. If V_S has less than $144c_{\text{sep}}^2$ vertices, then the procedure updates the value of R in a straightforward way. As we will see later $G[A \cup V_S]$ has at most $O(\sqrt{n})$ vertices; hence, we can use any standard linear-space and polynomial-time algorithm (e.g., breadth-first search) to do this task. Otherwise, ExtendReach divides the current subarea $['G]_S$ into two smaller subareas with new separator vertex set S'_{t+1} that is added to A. It then explore two subareas by using numbers in the universal sequence σ_s to control the length of paths in recursive calls.

The correctness of Algorithm 1 is demonstrated in Lemma 3 below. From this, as summarized in Theorem 1, it is clear that our algorithm correctly determines the reachability of vertex v_* from vertex v_0 in the input graph G.

Lemma 3. *For any input instance G, v_0, and v_* of the planar graph reachability problem, consider any execution of ExtendReach(S, ℓ) for some $S = \langle (b_1, S_1, T_1), \ldots, (b_t, S_t, T_t) \rangle$ and $\ell = 2^s$. Let V_S denote the set of vertices of $['G]_S$. For each vertex v that is in A before the execution, $R[v]$ is set to true during the execution if and only if there is a path to v in $G[A \cup V_S]$ of length at most 2^s, from some vertex $u \in A$ whose value $R[u]$ before the execution equals true.*

Proof. Suppose that there is a path $p = u_0, u_1, \ldots, u_h$ in $G[A \cup V_S]$, where (i) u_0 and u_h both belong to A, (ii) $h \leq 2^s$, and (iii) $R[u_0] =$ true before executing the procedure. For the lemma, it suffices to show that $R[u_h]$ is set true during the execution of the procedure.

We prove our assertion by induction on the size of V_S. If $|V_S| \leq 144c_{\text{sep}}^2$, then it is clear from the description of the procedure. Consider the case where $|V_S| > 144c_{\text{sep}}^2$. Then the part from line 7 of the procedure is executed. Let S_{t+1}, T_{t+1}^0, and T_{t+1}^1 be the separator and the edge lists computed there. For any $b \in \{0, 1\}$, let S^b denote $\langle (b_1, S_1, T_1), \ldots, (b_t, S_t, T_t), (b, S_{t+1}, T_{t+1}^b) \rangle$; also let $'G^b = ['G]_{S^b} (= [['G]_S]_{(b, S_{t+1}, T_{t+1}^b)})$ and $V^b = V('G^b) \setminus S_{t+1}$.

We observe that p can be decomposed into some number $x \leq |A \cup S_{t+1}| - 1$ of subpaths $p_1, p_2, \ldots, p_x$, such that (i) both start(p_j) and end(p_j) belong to $A \cup S_{t+1}$ for each $j \in [x]$, (ii) the internal vertices of p_j belong either to V^0 or V^1 for each $j \in [x]$, and (iii) end(p_j) = start(p_{j+1}) for each $j \in [x-1]$, where by start(p_j) and end(p_j) we mean the start and end vertices of p_j respectively. Let h_j denote the number of edges in path p_j. By construction (i) $h_j \geq 1$ for all $j \in [x]$, and (ii) $\sum_{j \in [x]} h_j = h \leq 2^s$. Then by Lemma 1(a), the sequence $\langle h_1, \ldots, h_x \rangle$ is dominated by the universal sequence $\sigma_s = \langle c_1, \ldots, c_{2^s+1} \rangle$. That is, there exists a subsequence $\langle c_{k_1}, c_{k_2}, \ldots c_{k_x} \rangle$ of σ_s such that $h_j \leq c_{k_j}$ for all $j \in [x]$. Thus, for any $j \in [x]$, if $R[\text{start}(p_j)] =$ true before the execution of ExtendReach(S^0, c_{k_j}),

---

Algorithm 1. ExtendReach($\mathbf{S}, \ell$)

Given: (as arguments) A sequence $\mathbf{S} = \langle (b_1, S_1, T_1), \ldots, (b_t, S_t, T_t) \rangle$ of triples of a binary label, a cycle-separator, and an additional triangulation edge list, and a bound $\ell = 2^s$ on the length of path.
// In this description we use $V_{\mathbf{S}}$ to denote the set of vertices of $['G]_{\mathbf{S}}$.
(as global variables) The input graph G, its triangulated planar embedding, the source vertex v_0, the goal vertex v_*, a set $\mathbf{A}$ of the currently considered vertices, and a Boolean array $\mathbf{R}$ specifying known reachability from v_0, for all $v \in \mathbf{A}$.

Task: For each vertex $v \in \mathbf{A}$, set $\mathbf{R}[v] = \mathbf{true}$ if there is a path to v in $G[\mathbf{A} \cup V_{\mathbf{S}}]$ of length at most 2^s from some vertex $u \in \mathbf{A}$ whose value $\mathbf{R}[u]$ before the current procedure execution equals $\mathbf{true}$.
// Invariant: $\mathbf{A} = \{v_0, v_*\} \cup \bigcup_{i \in [t]} S_i$. $\mathbf{R}[v] = \mathbf{true} \Rightarrow v$ is reachable from v_0 in G.

1: **if** the number of vertices of V_t is less than $144 c_{\mathrm{sep}}^2$ **then**
2:     $R_t \leftarrow \{u \in \mathbf{A} : \mathbf{R}[u] = \mathbf{true}\}$;
3:     **for** each vertex $v \in \mathbf{A}$ **do**
4:         $\mathbf{R}[v] \leftarrow \mathbf{true}$ iff v is reachable from some $u \in R_t$ in $G[\mathbf{A} \cup V_{\mathbf{S}}]$ by a path of length $\leq \ell$; // Use any linear space and polynomial-time algorithm here.
5:     **end for**
6: **else**
7:     Use CycleSep and AddTri to create a new cycle separator S_{t+1} of $['G]_{\mathbf{S}}$ and its additional triangulation edge lists T_{t+1}^0 and T_{t+1}^1;
8:     $S'_{t+1} \leftarrow S_{t+1} \setminus \mathbf{A}$; $\mathbf{A} \leftarrow \mathbf{A} \cup S'_{t+1}$;
9:     $\mathbf{R}[v] \leftarrow \mathbf{false}$ for each vertex $v \in S'_{t+1}$;
10:     **for** each $c_i = 2^{s_i}$ in the universal sequence σ_s (where $i \in [2^{s+1} - 1]$) **do**
11:         ExtendReach($\langle (b_1, S_1, T_1), \ldots, (b_2, S_t, T_t), (0, S_{t+1}, T_{t+1}^0) \rangle, c_i$);
12:         ExtendReach($\langle (b_1, S_1, T_1), \ldots, (b_2, S_t, T_t), (1, S_{t+1}, T_{t+1}^1) \rangle, c_i$);
13:     **end for**
14:     $\mathbf{A} \leftarrow \mathbf{A} \setminus S'_{t+1}$;
15: **end if**

---

then we have $\mathbf{R}[\mathrm{end}(p_j)] = \mathbf{true}$ after the execution because of the induction hypothesis. Hence, by executing the fragment:

$$\texttt{ExtendReach}(\mathbf{S}^0, c_{k_1}); \quad \texttt{ExtendReach}(\mathbf{S}^1, c_{k_1}); \quad \cdots \quad \texttt{ExtendReach}(\mathbf{S}^0, c_{k_{x-1}});$$
$$\texttt{ExtendReach}(\mathbf{S}^1, c_{k_{x-1}}); \quad \texttt{ExtendReach}(\mathbf{S}^0, c_{k_x}); \quad \texttt{ExtendReach}(\mathbf{S}^1, c_{k_x});$$

we have $\mathbf{R}[\mathrm{end}(p_x)] = \mathbf{true}$ since $\mathbf{R}[\mathrm{start}(p_1)] = \mathbf{true}$ by our assumption. Therefore, $\mathbf{R}[u_h]$ (where $u_h = \mathrm{end}(p_x)$) is set $\mathbf{true}$ as desired since the above fragment must be executed as a part of the execution of line 10–13 of the procedure. □

By analyzing the time and space complexity of our algorithm, we conclude as follows.

Theorem 1. *For any input instance G, v_0, and v_* of the planar directed graph reachability problem (where n is the number of vertices of G), our planar graph reachability algorithm determines whether there is a path from v_0 to v_* in G in $\widetilde{O}(\sqrt{n})$ space and polynomial-time.*

Proof. The correctness of the algorithm follows immediately from Lemma 3. For the complexity analysis, we consider the essential part, that is, the execution of $\texttt{ExtendReach}(\langle\,\rangle, 2^{\lceil \log n \rceil})$.

As a key step for estimating the space and time bounds, we show here a bound on the depth of recursion during the execution. Let $t_{\max}$ denote the maximum depth of recursive calls in the execution, for which we would like to show that $t_{\max} \leq 2.5 \log n$ holds. Consider any depth t recursive call of $\texttt{ExtendReach}$; in other words, the execution of $\texttt{ExtendReach}(\mathbf{S}, \ell)$ with a sequence $\mathbf{S}$ of length t. (Thus, the inital call of $\texttt{ExtendReach}$ is regarded as depth 0 recursive call.) Here some depth t subarea $['G]_{\mathbf{S}}$ is examined; let n_t be the number of vertices of this subarea. Assume that $n_t \geq 144 c_{\text{sep}}^2$. Then two smaller subareas of $['G]_{\mathbf{S}}$ are created and $\texttt{ExtendReach}$ is recursively executed on them. Let n_{t+1} denote the number of vertices of a lager one of these two smaller subareas. Then by Lemma 2 we have $n_{t+1} \leq 2n_t/3 + c_{\text{sep}}\sqrt{n_t} \leq 3n_t/4$ since $n_t \geq 144 c_{\text{sep}}^2$. Hence, $t_{\max}$ is bounded by $2.5 \log n$ as desired because $n(3/4)^{2.5 \log n} < 144 c_{\text{sep}}^2$.

We bound the memory space used in the execution $\texttt{ExtendReach}(\langle\,\rangle, 2^{\lceil \log n \rceil})$. For this, it is enough to bound the number of vertices in $\mathbf{A}$ because the number of words needed to keep in the work memory space during the execution is proportional to $|\mathbf{A}|$. Note further that $\mathbf{A} = \{v_0, v_*\} \cup \bigcup_{i \in [t]} S_i$ at any depth t recursive call of $\texttt{ExtendReach}$. On the other hand, by using the above notation, it follows from the above and Lemma 2 we have

$$|\mathbf{A} \setminus \{v_0, v_*\}| \leq \sum_{i \in [t_{\max}]} |S_i| \leq \sum_{i \in [t_{\max}]} c_{\text{sep}}\sqrt{n_i}$$

$$\leq \sum_{i \in [t_{\max}]} c_{\text{sep}}\sqrt{\left(\frac{3}{4}\right)^{i-1} n} \leq \left(c_{\text{sep}}\sqrt{n}\right)\left(\sum_{i \geq 0}\left(\frac{3}{4}\right)^{i/2}\right) = O(\sqrt{n}),$$

which gives us the desired space bound.

For bounding the time complexity by some polynomial, it suffices to show that the total number of calls of $\texttt{ExtendReach}$ is polynomially bounded. To see this, we estimate $N(t, 2^s)$, the max. number of calls of $\texttt{ExtendReach}$ during any depth t recursive call of $\texttt{ExtendReach}(\mathbf{S}, 2^s)$ that occurs in the execution of $\texttt{ExtendReach}(\langle\,\rangle, 2^{\lceil \log n \rceil})$. (Precisely speaking, $N(t, 2^s) = 0$ if no call of type $\texttt{ExtendReach}(\mathbf{S}, 2^s)$ occurs.) Clearly, $N(t_{\max}, 2^s) = 0$ for any s. Also it is easy to see that $N(t, 2^0) = 2 + 2N(t+1, 2^0)$ for any $t < t_{\max}$; hence, we have $N(t, 2^0) \leq 2 \cdot (2^{t_{\max}-t}-1) \leq 2^{t_{\max}-t+1}$. Consider any $t < t_{\max}$ and $s \geq 1$. From the description of $\texttt{ExtendReach}$ and the property of the universal sequence σ_s (Lemma 1(b)), we have

$$N(t, 2^s) = 2\sum_{i \in [2^{s+1}]} (1 + N(t+1, c_i)) = 2^{s+2} + \sum_{0 \leq j \leq s} 2^{s-j} N(t+1, 2^j),$$

from which we can derive $N(t, 2^s) = 2N(t+1, 2^s) + 2N(t, 2^{s-1})$. Then by induction we can show

$$N(t, 2^s) \leq 2^{t_{\max}-t+s+1}\binom{t_{\max}-t+s}{s}.$$

Thus, $N(0, 2^{\lceil \log n \rceil})$, the total number of calls of `ExtendReach` is polynomially bounded. $\square$

4   Cycle-Separators

We expand here the notion of a cycle-separator and Lemma 2 used in the previous section. Throughout this section, we consider only undirected graphs. In particular, we fix any sufficiently large planar undirected graph $G = (V, E)$ and discuss a cycle-separator for G; all symbols using G are used to denote some graph related to G.

Roughly speaking, a cycle-separator is a separator S consisting of cycles. In this paper, we assume some orientation for each cycle, and our cycle-separators is required to separate G into two subgraphs by considering a part located left (resp., right) of cycles (*cf.* Figure 2).

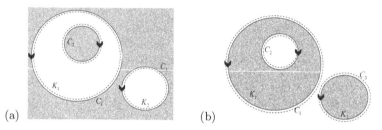

(a)   (b)

S consists of cycles C_1, C_2, C_3, each of which has an orientation specified as in the figure. Dashed lines indicate the cuts corresponding to edges located in the (a) left and (b) right of each cycle. These cuts are used to identify $G[V^0]$ and $G[V^1]$

Fig. 2. An example of a cycle-separator

Recall that we do not assume that our input graph comes equipped with a planar embedding. This is unnecessary for our purposes since Allender and Mahajan [1] showed that the problem of computing a planar embedding can be reduced to the undirected graph reachability problem. Hence, by using the algorithm `UReach` of Reingold, we can compute a planar embedding of G by using $O(\log n)$-space. Also it is also easy to to obtain some triangulation w.r.t. this embedding, and thus we may assume an $O(\log n)$-space algorithm that computes a triangulated planar embedding for a give planar graph. In our reachability algorithm, this $O(\log n)$-space algorithm is used (implicitly) before starting the actual computation and so in the description that follows we proceed as though our input graph G is given with some triangulated planar embedding.

The Planar Separator Theorem guarantees that every planar graph has a separator of size $O(\sqrt{n})$ that disconnects a graph into two subgraphs each of which has at most $2n/3$ vertices, which we call a *2/3-separator*. Imai et al. has shown an algorithm that computes a *2/3-separator* by using $O(\sqrt{n})$-space and in polynomial-time. Though we use such a separator algorithm as a blackbox, we introduce some modification so that we can specify *two* subgraphs disconnected by a separator in order to use them in the context of sublinear-space computation. A *labeled-separator* of G is a pair of a separator S and a set $\tau = \{v_1, \ldots, v_k\}$ of vertices of G (which we simply denote by $S\tau$) such that no two vertices of τ belong to the same connected component of $G[V \setminus S]$. Graphs $G^0_{S\tau}$ and $G^1_{S\tau}$ are two disconnected subgraphs of $G[V \setminus S]$ defined by $S\tau$; $G^0_{S\tau} = \bigcup_{i=1}^{k} K_i$ where each K_i is the connected component of $G[V \setminus S]$ containing v_i, and $G^1_{S\tau}$ is a subgraph of G consisting of all the other connected components of $G[V \setminus S]$. By the planarity, we can show that $G[V \setminus S]$ has at most $2|S| - 4$ connected components. (Recall that we assumed G is triangulated and hence connected.) Thus, each labeled-separator can be stored in $\tilde{O}(|S|)$-space. Furthermore, by using UReach, we can identify, for each $v_i \in \tau$, the connected component K_i containing v_i in $O(\log n)$-space. Since counting is also possible in $O(\log n)$-space, for a given 2/3-separator, we can in fact collect connected components $K_1, \ldots, K_k$ of $G[V \setminus S]$ (and their representative vertices $v_1, \ldots, v_k$) so that $|V(G^0_{S\tau})| \leq 2|V|/3$ and $|V(G^1_{S\tau})| \leq 2|V|/3$ hold with $\tau = \{v_1, \ldots, v_k\}$. In summary, we have the following separator algorithm that is the basis of our cycle-separator algorithm.

Lemma 4. *There exists an $\tilde{O}(\sqrt{n})$-space and polynomial-time algorithm that yields a 2/3-labeled-separator of size $\leq c_{\mathrm{sep}}\sqrt{n}$ for a given planar graph, where c_{sep} is some constant, which has been used in the previous section.*

Recall that we assume some planar embedding of G; the following notions are defined with respect to this embedding. For any cycle C of G, we use a sequence $\langle u_1, \ldots, u_r \rangle$ of vertices of G in the order of appearing in C under one direction. We call such a sequence as a *cycle representation*. With this orientation, we define the left and the right of the cycle C. Our main technical lemma (see a technical report version [ECCC, TR14-071] for the full proof) is to show a way to compute a set S' of cycles from a given separator S that can be used as a separator in $\tilde{O}(|S|)$-space and polynomial-time. More specifically, by using cycles in S', we define two subsets V^0 and V^1 of V as sets of vertices respectively located left and right of the cycles. Then they satisfy Lemma 2; that is, $G[V^0]$ and $G[V^1]$ are subgraphs covering G and sharing only vertices in S', which corresponds to 'G^0, 'G^1, and the separator S in the grid case. Furthermore, their size is (approximately) bounded by $2n/3$, and since $S' \subseteq S$ as a set, we have $|S'| \leq c_{\mathrm{sep}}\sqrt{n}$. This S' is called a *cycle-separator* in this paper. We also provide a way to identify graphs $G[V^0]$ and $G[V^1]$, which is used as a basis of an algorithm identifying ['G]$_\mathbf{s}$.

5 Conclusion

It should be noted that, though restricted to grid graphs, the problem studied by Asano et al. in [2,3] is the shortest path problem, a natural generalization of the reachability problem. In order to keep the discussion in this paper as simple as possible, and focus on the key ideas, we have restricted our attention here to the graph reachability problem. However, it is not hard to see that our algorithm for reachability can be modified to the shortest path problem (with a modest increase in the polynomial time bound).

Similarly, the focus in Asano et al. in [2,3] is on space efficient and yet practically useful algorithms, including time-space tradeoffs. In this paper, on the other hand, our motivation has been in extending a graph class that is solvable in $\widetilde{O}(\sqrt{n})$-space and polynomial-time, and the specific time complexity of algorithms is not so important so long as it is within some polynomial. In fact, since the algorithm of Reingold for the undirected reachability is used heavily, we need very large polynomial to bound our algorithm's running time.

Since we use Reingold's undirected reachability algorithm, our algorithm (and also the one by Imai et al.) have no natural implementation in the NNJAG model. While the worst-case instances for NNJAG given in [5] are non-planar, it is an interesting question whether we have similar worst-case instances based on some planar directed graphs. A more important and challenging question is to define some model in which our algorithm can be naturally implemented and show some limitation of space efficient computation.

References

1. Allender, E., Mahajan, M.: The complexity of planarity testing. Information and Computation 189(1), 117–134 (2004)
2. Asano, T., Doerr, B.: Memory-constrained algorithms for shortest path problem. In: Proc. of the 23th Canadian Conf. on Comp. Geometry, CCCG 1993 (2011)
3. Asano, T., Kirkpatrick, D.: A $O(\sqrt{n})$-space algorithm for reporting a shortest path on a grid graph (in preparation)
4. Barnes, G., Buss, J.F., Ruzzo, W.L., Schieber, B.: A sublinear space, polynomial time algorithm for directed s-t connectivity. In: Proc. Structure in Complexity Theory Conference, pp. 27–33. IEEE (1992)
5. Edmonds, J., Poon, C.K., Achlioptas, D.: Tight lower bounds for st-connectivity on the NNJAG model. SIAM J. Comput. 28(6), 2257–2284 (1999)
6. Gazit, H., Miller, G.L.: A parallel algorithm for finding a separator in planer graphs. In: Proc. of the 28th Ann. Sympos. on Foundations of Comp. Sci. (FOCS 1987), pp. 238–248 (1987)
7. Henzinger, M.R., Klein, P., Rao, S., Subramanian, S.: Faster shortest-path algorithms for planar graphs. Journal of Comput. Syst. Sci. 55, 3–23 (1997)
8. Imai, T., Nakagawa, K., Pavan, A., Vinodchandran, N.V., Watanabe, O.: An $O(n^{\frac{1}{2}+\epsilon})$-space and polynomial-time algorithm for directed planar reachability. In: Proc. of the 28th Conf. on Comput. Complexity, pp. 277–286 (2013)
9. Miller, G.L.: Finding small simple cycle separators for 2-connected planar graphs. J. Comput. Syst. Sci. 32(3), 265–279 (1986)
10. Lipton, R.J., Tarjan, R.E.: A separator theorem for planar graphs. SIAM Journal on Applied Mathematics 36(2), 177–189 (1979)
11. Pippenger, N., Fischer, M.J.: Relations among complexity measures. J. ACM 26(2), 361–381 (1979)
12. Reingold, O.: Undirected connectivity in log-space. J. ACM 55(4) (2008)
13. Wigderson, A.: The complexity of graph connectivity. In: Havel, I.M., Koubek, V. (eds.) MFCS 1992. LNCS, vol. 629, pp. 112–132. Springer, Heidelberg (1992)

Forbidden Induced Subgraphs and the Price of Connectivity for Feedback Vertex Set

Rémy Belmonte[1,*], Pim van 't Hof[2,**],
Marcin Kamiński[3,***], and Daniël Paulusma[4,†]

[1] Dept. of Architecture and Architectural Engineering, Kyoto University, Japan
remybelmonte@gmail.com
[2] Department of Informatics, University of Bergen, Norway
pim.vanthof@ii.uib.no
[3] Institute of Computer Science, University of Warsaw, Poland
mjk@mimuw.edu.pl
[4] School of Engineering and Computing Sciences, Durham University, UK
daniel.paulusma@durham.ac.uk

Abstract. Let $\mathrm{fvs}(G)$ and $\mathrm{cfvs}(G)$ denote the cardinalities of a minimum feedback vertex set and a minimum connected feedback vertex set of a graph G, respectively. For a graph class $\mathcal{G}$, the price of connectivity for feedback vertex set (poc-fvs) for $\mathcal{G}$ is defined as the maximum ratio $\mathrm{cfvs}(G)/\mathrm{fvs}(G)$ over all connected graphs G in $\mathcal{G}$. It is known that the poc-fvs for general graphs is unbounded. We study the poc-fvs for graph classes defined by a finite family $\mathcal{H}$ of forbidden induced subgraphs. We characterize exactly those finite families $\mathcal{H}$ for which the poc-fvs for $\mathcal{H}$-free graphs is bounded by a constant. Prior to our work, such a result was only known for the case where $|\mathcal{H}| = 1$.

1 Introduction

A *feedback vertex set* of a graph is a subset of its vertices whose removal yields an acyclic graph, and a feedback vertex set is connected if it induces a connected graph. We write $\mathrm{fvs}(G)$ and $\mathrm{cfvs}(G)$ to denote the cardinalities of a minimum feedback vertex set and a minimum connected feedback vertex set of a graph G, respectively. Let $\mathcal{G}$ be a class of graphs. The *price of connectivity for feedback vertex set* (poc-fvs) for $\mathcal{G}$ is defined to be the maximum ratio $\mathrm{cfvs}(G)/\mathrm{fvs}(G)$ over all connected graphs G in $\mathcal{G}$. Graphs consisting of two disjoint cycles that are connected to each other by an arbitrarily long path show that the poc-fvs for general graphs is not upper bounded by a constant, and the same clearly holds for planar graphs. Interestingly, Grigoriev and Sitters [6] showed that the

* Supported by the ELC project (Grant-in-Aid for Scientific Research on Innovative Areas, MEXT Japan).
** Supported by the Research Council of Norway (197548/F20).
*** Supported by Foundation for Polish Science (HOMING PLUS/2011-4/8) and National Science Center (SONATA 2012/07/D/ST6/02432).
† Supported by EPSRC (EP/G043434/1) and Royal Society (JP100692).

E. Csuhaj-Varjú et al. (Eds.): MFCS 2014, Part II, LNCS 8635, pp. 57–68, 2014.
© Springer-Verlag Berlin Heidelberg 2014

poc-fvs for planar graphs of minimum degree at least 3 is at most 11. Schweitzer and Schweitzer [7] later improved this upper bound from 11 to 5, and showed the upper bound of 5 to be tight.

In a previous paper [1], we studied the poc-fvs for graph classes characterized by a single forbidden induced subgraph. We proved that the poc-fvs for H-free graphs is bounded by a constant c_H if and only if H is a linear forest, i.e., a disjoint union of paths. In fact, we obtained a more refined tetrachotomy result that determines, for every graph H, which of the following cases holds: (i) $\mathrm{cfvs}(G) = \mathrm{fvs}(G)$ for every connected H-free graph G; (ii) there exists a constant c_H such that $\mathrm{cfvs}(G) \leq \mathrm{fvs}(G) + c_H$ for every connected H-free graph G; (iii) there exists a constant c_H such that $\mathrm{cfvs}(G) \leq c_H \cdot \mathrm{fvs}(G)$ for every connected H-free graph G; (iv) there does not exist a constant c_H such that $\mathrm{cfvs}(G) \leq c_H \cdot \mathrm{fvs}(G)$ for every connected H-free graph G.

The concept of "price of connectivity", introduced by Cardinal and Levy [4], has been studied for other parameters as well. One such parameter is the vertex cover number of a graph. Let $\tau(G)$ and $\tau_c(G)$ denote the cardinalities of a minimum vertex cover and a minimum connected vertex cover of a graphs G, respectively. For a graph class $\mathcal{G}$, the price of connectivity for vertex cover for $\mathcal{G}$ is defined as the worst-case ratio $\tau_c(G)/\tau(G)$ over all connected graphs G in $\mathcal{G}$. It is known that for general graphs, the price of connectivity for vertex cover is upper bounded by 2, and this bound is sharp [2]. Cardinal and Levy [4] showed that for n-vertex graphs with average degree ϵn, this bound can be improved to $2/(1 + \epsilon)$. Camby et al. [2] provided forbidden induced subgraph characterizations of graph classes for which the price of connectivity for vertex cover is upper bounded by 1, 4/3, and 3/2, respectively.

The price of connectivity for dominating set (poc-ds) for a graph class $\mathcal{G}$ is defined as the maximum ration $\gamma_c(G)/\gamma(G)$ over all connected graphs G in $\mathcal{G}$, where $\gamma_c(G)$ and $\gamma(G)$ denote the domination number and the connected domination number of G, respectively. It is easy to prove that the poc-ds for general graphs is upper bounded by 3 [5]. Motivated by the work of Zverovich [8], Camby and Schaudt [3] studied the poc-ds for (P_k, C_k)-free graphs for several values of k. Their results show that the poc-ds for (P_8, P_9)-free graphs is upper bounded by 2, while the general upper bound of 3 is asymptotically sharp for (P_9, C_9)-free graphs.

Our Contribution. We continue the line of research on the price of connectivity for feedback vertex set we initiated in [1]. For a family of graphs $\mathcal{H}$, a graph G is called $\mathcal{H}$-free if G does not contain an induced subgraph isomorphic to any graph $H \in \mathcal{H}$. The vast majority of well-studied graph classes have forbidden induced subgraphs characterizations, and such characterizations can often be exploited when proving structural or algorithmic properties of these graph classes. In fact, for every hereditary graph class $\mathcal{G}$, that is, for every graph class $\mathcal{G}$ that is closed under taking induced subgraphs, there exists a family $\mathcal{H}$ of graphs such that $\mathcal{G}$ is exactly the class of $\mathcal{H}$-free graphs. Notable examples of graphs classes that can be characterized using a *finite* family of forbidden induced subgraphs include claw-free graphs, line graphs, proper interval graphs, split graphs and cographs.

Our main result establishes a dichotomy between the finite families $\mathcal{H}$ for which the price of connectivity for feedback vertex set for $\mathcal{H}$-free graphs is upper bounded by a constant $c_{\mathcal{H}}$ and the families $\mathcal{H}$ for which such a constant $c_{\mathcal{H}}$ does not exist. This can be seen as an extension of the case (iii) from [1] (mentioned above) from monogenic to finitely defined classes of graphs. In order to formally state our main result, we need to introduce some terminology.

For two graphs H_1 and H_2, we write $H_1 + H_2$ to denote the disjoint union of H_1 and H_2. We write sH to denote the disjoint union of s copies of H. For any $r \geq 3$, we write C_r to denote the cycle on r vertices. For any three integers i, j, k with $i, j \geq 3$ and $k \geq 1$, we define $B_{i,j,k}$ to be the graph obtained from $C_i + C_j$ by choosing a vertex x in C_i and a vertex y in C_j, and adding a path of length k between x and y.

It is clear that the price of connectivity for feedback vertex set for the class of all butterflies is not bounded by a constant, since $\mathrm{fvs}(B_{i,j,k}) = 2$ and $\mathrm{cfvs}(B_{i,j,k}) = k + 1$ for every $i, j \geq 3$ and $k \geq 1$. Roughly speaking, our main result states that the price of connectivity for feedback vertex set for the class of $\mathcal{H}$-free graphs is bounded by a constant $c_{\mathcal{H}}$ if and only if the forbidden induced subgraphs in $\mathcal{H}$ prevent arbitrarily large butterflies from appearing as induced subgraphs. To make this statement concrete, we need the following definition.

Definition 1. *Let $i, j \geq 3$ be two integers, let $\mathcal{H}$ be a family of graphs, and let $N = 2 \cdot \max_{H \in \mathcal{H}} |V(H)| + 1$. The family $\mathcal{H}$ covers the pair (i, j) if $\mathcal{H}$ contains an induced subgraph of $B_{i,j,N}$. A graph H covers the pair (i, j) if $\{H\}$ covers (i, j).*

The following theorem provides a sufficient and necessary condition for a finite family $\mathcal{H}$ to have the property that the poc-fvs for $\mathcal{H}$-free graphs is upper bounded by a constant.

Theorem 1. *Let $\mathcal{H}$ be a finite family of graphs. Then the poc-fvs for $\mathcal{H}$-free graphs is upper bounded by a constant $c_{\mathcal{H}}$ if and only if $\mathcal{H}$ covers the pair (i, j) for every $i, j \geq 3$.*

Section 2 is devoted to the proof of Theorem 1. In Section 3, we prove a sequence of lemmata that show exactly which graphs H cover which pairs (i, j). In Section 4, we present some applications of the results in Sections 2 and 3. In particular, we describe a procedure that, given a positive integer k, yields an explicit description of all the minimal graph families $\mathcal{H}$ with $|\mathcal{H}| = k$ for which the poc-fvs for $\mathcal{H}$-free graphs is upper bounded by a constant. For $k = 1$, this immediately yields the aforementioned result from [1], stating that the poc-fvs for H-free graphs is upper bounded by a constant if and only if H is a linear forest (Corollary 1). We also demonstrate the procedure for the case $k = 2$, and obtain an explicit description of exactly those families $\{H_1, H_2\}$ for which the poc-fvs for $\{H_1, H_2\}$-free graphs is upper bounded by a constant (Corollary 2). Section 5 contains some concluding remarks.

We end this section by defining some additional terminology that will be used throughout the paper. For any $k, p, q \geq 1$, let P_k denote the path on k vertices, and let $T_k^{p,q}$ denote the graph obtained from $P_k + P_p + P_q$ by making a new vertex adjacent to one end-vertex of each path. For any $k \geq 0$ and $r \geq 3$, let D_k^r denote the graph obtained from $P_k + C_r$ by adding an edge between a vertex of the cycle and an end-vertex of the path; in particular, D_0^r is isomorphic to C_r.

2   Proof of Theorem 1

In this section, we prove the dichotomy result given in Theorem 1. We will make use of the following simple observation.

Observation 1. *Let i, j, k, ℓ be integers such that $i, j \geq 3$ and $\ell \geq k \geq 1$. A graph on at most k vertices is an induced subgraph of $B_{i,j,k}$ if and only if it is an induced subgraph of $B_{i,j,\ell}$.*

Proof (of Theorem 1). First suppose there exists a pair (i, j) with $i, j \geq 3$ such that $\mathcal{H}$ does not cover (i, j). For contradiction, suppose there exists a constant $c_{\mathcal{H}}$ as in the statement of the theorem. By Definition 1, $\mathcal{H}$ does not contain an induced subgraph of $B_{i,j,N}$, and hence $B_{i,j,N}$ is $\mathcal{H}$-free. As a result of Observation 1, $B_{i,j,k}$ is $\mathcal{H}$-free for every $k \geq N$. In particular, the graph $B_{i,j,N+2c_{\mathcal{H}}}$ is $\mathcal{H}$-free. Note that $\mathrm{fvs}(B_{i,j,N+2c_{\mathcal{H}}}) = 2$ and $\mathrm{cfvs}(B_{i,j,N+2c_{\mathcal{H}}}) = N + 2c_{\mathcal{H}} + 1$. This implies that $\mathrm{cfvs}(B_{i,j,N+2c_{\mathcal{H}}}) > c_{\mathcal{H}} \cdot \mathrm{fvs}(B_{i,j,N+2c_{\mathcal{H}}})$, yielding the desired contradiction.

For the converse direction, suppose $\mathcal{H}$ covers the pair (i, j) for every $i, j \geq 3$. Let G be a connected $\mathcal{H}$-free graph. Observe that $\mathrm{cfvs}(G) = \mathrm{fvs}(G)$ if G is a cycle or a tree, so we assume that G is neither a cycle nor a tree. Let F be a minimum feedback vertex set of G, and without loss of generality assume that each vertex in F lies on a cycle and has degree at least 3 in G. Below, we will prove that the distance in G between any two vertices of F is at most $5N$. To see why this suffices to prove the theorem, observe that we can transform F into a connected feedback vertex set of G of size at most $5N \cdot |F| = 5N \cdot \mathrm{fvs}(G)$ by choosing an arbitrary vertex $x \in F$ and adding, for each $y \in F \setminus \{x\}$, all the internal vertices of a shortest path between x and y.

Let $x, y \in F$, and let P be a shortest path from x to y. For contradiction, suppose P has length at least $5N + 1$. Recall that by the definition of F, there exist cycles C_x and C_y that contain x and y, respectively; assume, without loss of generality, that C_x and C_y are induced cycles in G. Let $X = \{v \in V(C_x) \mid d_{G[V(C_x)]}(v, x) \leq N\}$. Note that X induces the cycle C_x in case $|V(C_x)| \leq 2N$, and X induces a path of length at most $2N$ otherwise. We also define $Y = \{v \in V(C_y) \mid d_{G[V(C_y)]}(v, y) \leq N\}$. We partition the vertex set of P into three sets: $L = \{v \in V(P) \mid d_G(v, x) \leq 2N + 1\}$, $M = \{v \in V(P) \mid d_G(v, x) \geq 2N + 2 \text{ and } d_G(v, y) \geq 2N + 2\}$, and $R = \{v \in V(P) \mid d_G(v, y) \leq 2N + 1\}$. For any two distinct vertices u and v on the path P, we say that u is to the left of v (and, equivalently, v is to the right of u) if the subpath of P from x to u does not contain v.

Claim 1. $G[X \cup L]$ *contains a graph in* $\{D_N^i \mid i \geq 3\} \cup \{T_N^{N,N}\}$ *as an induced subgraph.*

We prove Claim 1 as follows. Let x' be the vertex of P closest to y that has a neighbor $x_1 \in X \setminus \{x\}$; possibly $x' = x$. Let P' be the subpath of P from x to x'. By the definition of X, the distance between x_1 and x is at most N, implying that $d_G(x, x') \leq N + 1$. Since P is a shortest path from x to y, we find that the length of P' is at most $N + 1$. Let x'' be the unique vertex of P such that x'' is to the right of x' and $d_G(x'', x') = N$, and let P'' be the subpath of P from x' to x''. Since $|L| = 2N + 2$, path P' has length at most $N + 1$, and path P'' has length N, it follows that $V(P'') \subseteq L$. Observe that x' is the only vertex of P'' that has a neighbor in $X \setminus \{x\}$.

Suppose $x = x'$. Then $X \cap V(P) = \{x\}$, and hence $G[X \cup V(P'')]$ is isomorphic to either $D_N^{|V(C_x)|}$ or $T_N^{N,N}$, implying that the claim holds in this case. From now on, we assume that $x' \neq x$. We distinguish two cases, depending on how many neighbors x' has in X.

If x' has at least two neighbors in X, then x' has two neighbors x_1, x_2 in X such that there is a path in X from x_1 to x_2 whose internal vertices are not adjacent to x'. This path, together with the edges $x_1 x'$ and $x_2 x'$, forms an induced cycle C in G. Then $G[V(C) \cup V(P'')]$ is isomorphic to $D_N^{|V(C)|}$, so the claim holds.

Now suppose x' has exactly one neighbor $x_1 \in X$. If X induces a cycle in G, then the cycle $G[X]$, the path P'', and the edge $x' x_1$ together form a graph that is isomorphic to $D_N^{|X|}$, so the claim holds. Suppose X induces a path in G; recall that this path has exactly $2N + 1$ vertices, and x is the middle vertex of this path. If $x_1 = x$, then $G[X \cup V(P'')]$ is isomorphic to $T_N^{N,N}$. Suppose $x_1 \neq x$. Let P_X be the unique path in $G[X]$ from x_1 to x. Then the graph $G[V(P_X) \cup V(P')]$ contains an induced cycle C such that x' lies on C, and the graph $G[V(C) \cup V(P'')]$ is isomorphic to $D_N^{|V(C)|}$. This completes the proof of Claim 1.

Let G_x be an induced subgraph of $G[X \cup L]$ that is isomorphic to a graph in $\{D_N^i \mid i \geq 3\} \cup \{T_N^{N,N}\}$ and that is constructed from the cycle C_x in the way described in the proof of Claim 1. In particular, let x'' be the vertex of G_x that is closest to y in G. Recall that x'' is a vertex of P and has degree 1 in G_x. It is clear from the construction of G_x that every vertex in G_x has distance at most $2N + 1$ to x. By symmetry, we can define an induced subgraph G_y of $G[Y \cup R]$ and a vertex y'' in G_y in an analogous way, that is, G_y is isomorphic to a graph in $\{D_N^i \mid i \geq 3\} \cup \{T_N^{N,N}\}$, and y'' is the vertex of G_y that is closest to x in G.

Let P^* be the subpath of P from x'' to y''. The fact that P is a shortest path from x to y implies that x'' and y'' are the only two vertices of G_x and G_y that are adjacent to internal vertices of P^*. Moreover, there are no edges between G_x and G_y, as otherwise there would be a path from x to y of length at most $4N + 2$, contradicting the fact that P is a shortest path from x to y. Let G^* denote the induced subgraph of G obtained from $G_x + G_y$ by connecting x'' and y'' using the path P^*. We distinguish four cases, and obtain a contradiction in each case. We will repeatedly use the fact that in each case, G^* can be obtained from a "large" butterfly by deleting at most two vertices.

Case 1. G_x *is isomorphic to* D_N^i *and* G_y *is isomorphic to* D_N^j *for some* $i, j \geq 3$.

In this case, G^* is isomorphic to $B_{i,j,k}$ for some $k \geq 2N$. Since $\mathcal{H}$ covers the pair (i, j), there exists a graph $H \in \mathcal{H}$ such that H is an induced subgraph of $B_{i,j,N}$ by Definition 1. Due to Observation 1, H is also an induced subgraph of G^* and hence also of G. This contradicts the assumption that G is $\mathcal{H}$-free.

Case 2. G_x *is isomorphic to* D_N^i *for some* $i \geq 3$ *and* G_y *is isomorphic to* $T_N^{N,N}$.

Since $\mathcal{H}$ covers the pair $(i, 2N)$, there exists a graph $H \in \mathcal{H}$ such that H is an induced subgraph of $B_{i,2N,N}$. Since $|V(H)| \leq N$, the graph H contains at most one cycle, and this cycle, if it exists, is of length i. Hence it is clear that H is also an induced subgraph of G^*. This contradicts the assumption that G and thus G^* is $\mathcal{H}$-free.

Case 3. G_x *is isomorphic to* $T_N^{N,N}$ *and* G_y *is isomorphic to* D_N^i *for some* $i \geq 3$.

By symmetry, we obtain a contradiction in the same way as in Case 2.

Case 4. Both G_x *and* G_y *are isomorphic to* $T_N^{N,N}$.

Since $\mathcal{H}$ covers the pair $(2N, 2N)$, there exists a graph $H \in \mathcal{H}$ such that H is an induced subgraph of $B_{2N,2N,N}$. This graph H has at most N vertices, which implies that H has no cycle. But then H is an induced subgraph of G^*, again yielding the desired contradiction. This completes the proof of Theorem 1. $\qquad \square$

3   Which Graphs H Cover Which Pairs (i, j)?

Recall that by Definition 1, a graph H covers a pair (i, j) if and only if H is an induced subgraph of $B_{i,j,N}$, where $N = 2 \cdot |V(H)| + 1$. In particular, if a graph H is not an induced subgraph of a butterfly, then H does not cover any pair (i, j). However, it is important to note that some induced subgraphs of $B_{i,j,N}$ cover more pairs than others. For example, as we will see in Lemma 6, a linear forest covers all pairs (i, j) with $i, j \geq 3$, but this is not the case for any induced subgraph of $B_{i,j,N}$ that is not a linear forest.

In this section, we will prove exactly which pairs (i, j) are covered by which graphs H. For convenience, we first describe all the possible induced subgraphs of $B_{i,j,N}$ in the following observation.

Observation 2. *Let* H *be a graph, let* $N = 2 \cdot |V(H)| + 1$, *and let* $i, j \geq 3$ *be two integers. Then* H *is an induced subgraph of* $B_{i,j,N}$ *if and only if* H *is isomorphic to the disjoint union of a linear forest (possibly on zero vertices) and at most one of the following graphs:*

(i) D_ℓ^i *for some* $\ell \geq 0$;
(ii) D_ℓ^j *for some* $\ell \geq 0$;
(iii) $D_\ell^i + D_{\ell'}^j$ *for some* $\ell, \ell' \geq 0$;
(iv) $T_k^{p,q}$ *for some* $k, p, q \geq 1$ *such that* $p + q + 2 \leq \max\{i, j\}$;
(v) $T_k^{p,q} + T_{k'}^{p',q'}$ *for some* $k, p, q, k', p', q' \geq 1$ *such that* $p + q + 2 \leq i$ *and* $p' + q' + 2 \leq j$;

(vi) $D_\ell^i + T_k^{p,q}$ *for some* $\ell \geq 0$ *and* $k, p, q \geq 1$ *such that* $p + q + 2 \leq j$;

(vii) $D_\ell^j + T_k^{p,q}$ *for some* $\ell \geq 0$ *and* $k, p, q \geq 1$ *such that* $p + q + 2 \leq i$.

The lemmata below show, for each of the induced subgraphs described in Observation 2, exactly which pairs (i, j) they cover. In the statement of each of the lemmata, we refer to a table in which the set of covered pairs is depicted. This will be helpful in the applications presented in Section 4. The rather straightforward proofs of Lemmata 2–5 have been omitted due to page restrictions.

Lemma 1. *Let H be a graph, let $p \geq 3$, and let $\mathcal{X}$ be the set consisting of the pairs (i, j) with $i, j \geq 3$ and $p \in \{i, j\}$; see the left table in Figure 1 for an illustration of the pairs in $\mathcal{X}$.*

(i) If H is an induced subgraph of D_k^p for some $k \geq 0$, then H covers all the pairs in $\mathcal{X}$.

(ii) If D_k^p is an induced subgraph of H for some $k \geq 0$, then H covers only pairs in $\mathcal{X}$.

Proof. Let $N = 2 \cdot |V(H)| + 1$. Suppose H is an induced subgraph of D_k^p for some $k \geq 0$. Then H is also an induced subgraph of $B_{i,j,N}$ for every $i, j \geq 3$ such that $p \in \{i, j\}$. Hence, by Definition 1, H covers the pairs (p, j) and (i, p) for every $i, j \geq 3$.

Now suppose D_k^p is an induced subgraph of H for some $k \geq 0$. Then H contains a cycle of length p. Hence it is clear that if H is an induced subgraph of a butterfly $B_{i,j,N}$, then we must have $p \in \{i, j\}$. This shows that H only covers pairs that belong to $\mathcal{X}$. $\qquad\square$

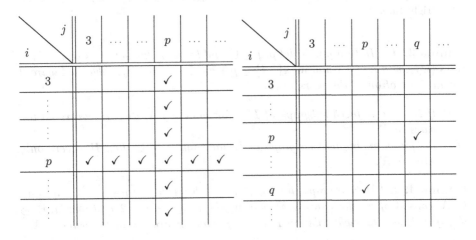

Fig. 1. The ticked cells represent the pairs (i, j) covered by H when H is isomorphic to D_k^p for some $k \geq 0$ (left table) and when H is isomorphic to $D_k^p + D_k^q$ for some $k \geq 0$ (right table)

Lemma 2. *Let H be a graph, let $p, q \geq 3$, and let $\mathcal{X} = \{(p,q),(q,p)\}$; see the right table in Figure 1 for an illustration of the pairs in $\mathcal{X}$.*

(i) If H is an induced subgraph of $D_k^p + D_k^q$ for some $k \geq 0$, then H covers all the pairs in $\mathcal{X}$.
(ii) If $D_k^p + D_k^q$ is an induced subgraph of H for some $k \geq 0$, then H covers only pairs in $\mathcal{X}$.

i \ j	3	...	...	$p+q+2$	...	...
3				✓	✓	✓
⋮				✓	✓	✓
⋮				✓	✓	✓
$p+q+2$	✓	✓	✓	✓	✓	✓
⋮	✓	✓	✓	✓	✓	✓
⋮	✓	✓	✓	✓	✓	✓

i \ j	3	...	$p+q+2$	...	$p'+q'+2$	...
3						
⋮						
$p+q+2$					✓	✓
⋮					✓	✓
$p'+q'+2$			✓	✓	✓	✓
⋮			✓	✓	✓	✓

Fig. 2. The ticked cells represent the pairs (i,j) covered by H when H is isomorphic to $T_r^{p,q}$ for some $r \geq 1$ (left table) and when H is isomorphic to $T_r^{p,q} + T_r^{p',q'}$ for some $r \geq 1$ (right table)

Lemma 3. *Let H be a graph, let $p, q \geq 1$, and let $\mathcal{X}$ be the set consisting of the pairs (i,j) with $i,j \geq 3$ and $\max\{i,j\} \geq p+q+2$; see the left table in Figure 2 for an illustration of the pairs in $\mathcal{X}$.*

(i) If H is an induced subgraph of $T_r^{p,q}$ for some $r \geq 1$, then H covers all the pairs in $\mathcal{X}$.
(ii) If $T_r^{p,q}$ is an induced subgraph of H for some $r \geq 1$, then H covers only pairs in $\mathcal{X}$.

Lemma 4. *Let H be a graph, let $p, q, p', q' \geq 1$ be such that $p+q \leq p'+q'$, and let $\mathcal{X}$ consist of all the pairs (i,j) with $\min\{i,j\} \geq p+q+2$ and $\max\{i,j\} \geq p'+q'+2$; see the right table in Figure 2 for an illustration of the pairs in $\mathcal{X}$.*

(i) If H is an induced subgraph of $T_r^{p,q} + T_r^{p',q'}$ for some $r \geq 1$, then H covers all the pairs in $\mathcal{X}$.
(ii) If $T_r^{p,q} + T_r^{p',q'}$ is an induced subgraph of H for some $r \geq 1$, then H covers only pairs in $\mathcal{X}$.

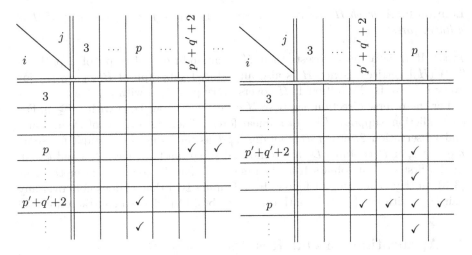

Fig. 3. The ticked cells represent the pairs (i,j) covered by H when H is isomorphic to $D_k^p + T_r^{p',q'}$ for some $k \geq 0$ and $r \geq 1$ in the case where $p < p' + q' + 2$ (left table) and in the case where $p > p' + q' + 2$ (right table)

Fig. 4. The ticked cells represent the pairs (i,j) covered by H when H is isomorphic to $D_k^p + T_r^{p',q'}$ for some $k \geq 0$ and $r \geq 1$ in the case where $p = p' + q' + 2$

Lemma 5. *Let H be a graph, let $p \geq 3$ and $p',q' \geq 1$, and let $\mathcal{X}$ be the set consisting of the pairs (i,j) with either $i = p$ and $j \geq p' + q' + 2$ or $i \geq p' + q' + 2$ and $j = p$; see the left and right tables in Figure 3 and the table in Figure 4 for an illustration of the pairs in $\mathcal{X}$ in the cases where $p < p' + q' + 2$, $p > p' + q' + 2$, and $p = p' + q' + 2$, respectively.*

(i) If H is an induced subgraph of $D_k^p + T_r^{p',q'}$ for some $k \geq 0$ and $r \geq 1$, then H covers all the pairs in $\mathcal{X}$.

(ii) If $D_k^p + T_r^{p',q'}$ is an induced subgraph of H for some $k \geq 0$ and $r \geq 1$, then H covers only pairs in $\mathcal{X}$.

Lemma 6. *A graph H covers every pair (i,j) with $i,j \geq 3$ if and only if H is a linear forest.*

Proof. If H is a linear forest, then H is an induced subgraph of a path on $2 \cdot |V(H)|$ vertices. Hence H is also an induced subgraph of $B_{i,j,2|V(H)|+1}$ for every $i,j \geq 3$. By Definition 1, H covers every pair (i,j) with $i,j \geq 3$.

For the reverse direction, suppose H covers every pair (i,j) with $i,j \geq 3$. For contradiction, suppose H is not a linear forest. Then, as a result of Definition 1 and Observation 2, either H contains $T_r^{p,q}$ as an induced subgraph for some $p,q,r \geq 1$, or H contains D_k^p as an induced subgraph for some $p \geq 3$ and $k \geq 0$. In the first case, it follows from Lemma 3(ii) that H does not cover the pair $(3,3)$. In the second case, it follows from Lemma 1(ii) that H does not cover any pair (i,j) with $r \notin \{i,j\}$. In both cases, we obtain the desired contradiction. □

4   Applications of Our Results

In this section, we show how we can apply Theorem 1 and the lemmata from Section 3 in order to obtain some concrete characterizations. Let us first remark that the following result, previously obtained in [1], immediately follows from Theorem 1 and Lemma 6.

Corollary 1 ([1]). *Let H be a graph. Then the poc-fvs for H-free graphs is upper bounded by a constant c_H if and only if H is a linear forest.*

Obtaining similar characterizations for finite families $\mathcal{H}$ with $|\mathcal{H}| \geq 2$ is more involved, but can be done using the procedure we informally describe below. We then illustrate the procedure in Corollary 2 below for the case where $|\mathcal{H}| = 2$.

Let $k \geq 2$. Suppose we want to characterize the families of graphs $\mathcal{H}$ with $|\mathcal{H}| = k$ for which the poc-fvs for $\mathcal{H}$-free graphs is upper bounded by a constant. It follows from Theorem 1 and Lemma 6 that the poc-fvs for $\mathcal{H}$-free graphs is bounded whenever $\mathcal{H}$ contains a linear forest. What about families $\mathcal{H}$ that do not contain a linear forest?

Consider the infinite table containing all the pairs (i,j) with $i,j \geq 3$. From Lemmata 3–5 and Figures 1–4, we can observe two important things. First, the only graphs H that cover the pair $(3,3)$ are induced subgraphs of $2D_\ell^3$ for some $\ell \geq 0$. Second, the only graphs H that cover infinitely many rows and columns of this table are induced subgraphs of $T_r^{p,q} + T_r^{p',q'}$ for some $r,p,q,p',q' \geq 1$. Hence, any finite family $\mathcal{H}$ that covers all pairs (i,j) must contain at least one graph of both types. Formally, we have the following observation (observe that every linear forest is an induced subgraph of $2D_\ell^3$ for some $\ell \geq 0$ and of $T_r^{p,q} + T_r^{p',q'}$ for some $r,p,q,p',q' \geq 1$):

Observation 3. *Let $\mathcal{H}$ be a finite family of graphs such that $|\mathcal{H}| \geq 2$. If the poc-fvs for $\mathcal{H}$-free graphs is upper bounded by a constant $c_\mathcal{H}$, then $\mathcal{H}$ contains an induced subgraph of $2D_\ell^3$ for some $\ell \geq 0$ and an induced subgraph of $T_r^{p,q} + T_r^{p',q'}$ for some $r,p,q,p',q' \geq 1$.*

Suppose $\mathcal{H}$ is a family of k graphs such that the poc-fvs for $\mathcal{H}$-free graphs is bounded by a constant. By Observation 3, $\mathcal{H}$ contains a graph H_1 that is an induced subgraph of $T_r^{p,q} + T_r^{p',q'}$ for some $r, p, q, p', q' \geq 1$.

If H_1 is also an induced subgraph of $T_r^{p,q}$ for some $r, p, q \geq 1$, or if $\mathcal{H}$ contains another graph that is of this form, then Lemma 3 and Figure 2 show that there are only finitely many pairs (i, j) that are not covered by H_1. These cells need to be covered by the remaining graphs in $\mathcal{H}$. Using Lemmata 3–5, we can determine exactly which combination of graphs covers exactly those remaining pairs.

Suppose $\mathcal{H}$ does not contain induced subgraph of $T_r^{p,q}$ for any $r, p, q \geq 1$. Then, by Lemma 4, there are finitely many rows and columns in which no pair is covered by H_1. In particular, since $p, q, p', q' \geq 1$, the pairs $(i, 3)$ and $(3, j)$ are not covered for any $i, j \geq 3$. From the lemmata in Section 3 and the corresponding tables, it it clear that the only graphs H that cover infinitely many pairs of this type are induced subgraphs of $T_r^{p,q}$ for some $r, p, q \geq 1$ or of $D_{r'}^3 + T_r^{p,q}$ for some $r' \geq 0$ and $p, q \geq 1$. Hence, $\mathcal{H}$ must contain a graph H_2 that is isomorphic to such an induced subgraph. Similarly, if the pairs $(i, 4)$ and $(4, j)$ are not covered for any $i, j \geq 3$, then $\mathcal{H}$ must contain an induced subgraph of $T_r^{p,q}$ for some $r, p, q \geq 1$ or of $D_{r'}^4 + T_r^{p,q}$ for some $r' \geq 0$ and $p, q \geq 1$, etcetera. Once all rows and columns contain only finitely many pairs that are not covered yet, we can determine all possible combinations of graphs that cover those last pairs.

To illustrate the above procedure, we now give an explicit description of exactly those families $\{H_1, H_2\}$ for which the poc-fvs for $\{H_1, H_2\}$-free graphs is upper bounded by a constant.

Corollary 2. *Let H_1 and H_2 be two graphs, and let $\mathcal{H} = \{H_1, H_2\}$. Then the poc-fvs for $\mathcal{H}$-free graphs is upper bounded by a constant $c_{\mathcal{H}}$ if only if there exist integers $\ell \geq 0$ and $r \geq 1$ such that one of the following conditions holds:*

- *H_1 or H_2 is a linear forest;*
- *H_1 and H_2 are induced subgraphs of D_ℓ^3 and $2T_r^{1,1}$, respectively;*
- *H_1 and H_2 are induced subgraphs of $2D_\ell^3$ and $T_r^{1,1}$, respectively.*

Proof. First suppose that the price of connectivity for feedback vertex set for $\mathcal{H}$-free graphs is bounded by some constant $c_{\mathcal{H}}$, and suppose that neither H_1 nor H_2 is a linear forest. Due to Observation 3, we may without loss of generality assume that H_1 is an induced subgraph of $2D_\ell^3$ for some $\ell \geq 0$ and H_2 is an induced subgraph of $T_r^{p,q} + T_r^{p',q'}$ for some $r, p, q, p', q' \geq 1$. From Lemmata 1 and 2 and the assumption that H_1 is not a linear forest, it follows that H_1 does not cover the pair $(4, 4)$. Hence H_2 must cover this pair. This, together with Lemma 4, implies that $p = q = p' = q' = 1$, i.e., H_2 is an induced subgraph of $2T_r^{1,1}$ for some $r \geq 1$.

If H_1 is an induced subgraph of $D_{\ell'}^3$ for some $\ell' \geq 0$, then the second condition holds and we are done. Suppose this is not the case. Then H_1 covers only the pair $(3, 3)$ due to Lemma 2. This means that all the pairs (i, j) with $i, j \geq 3$ and $3 \in \{i, j\}$, apart from $(3, 3)$, must be covered by H_2. From Lemma 3 and 4 it is clear that this only holds if H_2 is an induced subgraph of $T_{r'}^{1,1}$ for some $r' \geq 1$. Hence the third condition holds.

The converse direction follows by combining Theorem 1 with Lemma 6, Lemmata 1 and 4, and Lemmata 2 and 3, respectively.   □

5 Conclusion

Recall that in [1], we proved for every graph H which of the following cases holds: (i) $\mathrm{cfvs}(G) = \mathrm{fvs}(G)$ for every connected H-free graph G; (ii) there exists a constant c_H such that $\mathrm{cfvs}(G) \leq \mathrm{fvs}(G) + c_H$ for every connected H-free graph G; (iii) there exists a constant c_H such that $\mathrm{cfvs}(G) \leq c_H \cdot \mathrm{fvs}(G)$ for every connected H-free graph G; (iv) there does not exist a constant c_H such that $\mathrm{cfvs}(G) \leq c_H \cdot \mathrm{fvs}(G)$ for every connected H-free graph G. Theorem 1 extends the case of (iii) to all finite families $\mathcal{H}$. A natural question to ask is to characterize all finite families $\mathcal{H}$ for (i) and (ii) as well.

Another natural question to ask is whether Theorem 1 can be extended to families $\mathcal{H}$ which are not finite, i.e., to all hereditary classes of graphs. Definition 1 and Theorem 1 show that for any finite family $\mathcal{H}$, the poc-fvs for $\mathcal{H}$-free graphs is bounded essentially when the graphs in this class do not contain arbitrarily large induced butterflies. The following example shows that when $\mathcal{H}$ is infinite, it is no longer only butterflies that can cause the poc-fvs to be unbounded. Let G be a graph obtained from K_3 by first duplicating every edge once, and then subdividing every edge arbitrarily many times. Let $\mathcal{G}$ be the class of all graphs that can be constructed this way. In order to make $\mathcal{G}$ hereditary, we take its closure under the induced subgraph relation. Let $\mathcal{G}'$ be the resulting graph class. Observe that graphs in this class have arbitrarily large minimum connected feedback vertex sets, while $\mathrm{fvs}(G) \leq 2$ for every graph $G \in \mathcal{G}'$. Hence, the poc-fvs for $\mathcal{G}'$ is not bounded. However, no graph in this family contains a butterfly as an induced subgraph.

References

1. Belmonte, R., van 't Hof, P., Kamiński, M., Paulusma, D.: The price of connectivity for feedback vertex set. In: Eurocomb 2013. CRM Series, vol. 16, pp. 123–128 (2013)
2. Camby, E., Cardinal, J., Fiorini, S., Schaudt, O.: The price of connectivity for vertex cover. Discrete Mathematics & Theoretical Computer Science 16(1), 207–224 (2014)
3. Camby, E., Schaudt, O.: A note on connected dominating set in graphs without long paths and cycles. Manuscript, arXiv:1303.2868 (2013)
4. Cardinal, J., Levy, E.: Connected vertex covers in dense graphs. Theor. Comput. Sci. 411(26-28), 2581–2590 (2010)
5. Duchet, P., Meyniel, H.: On Hadwiger's number and the stability number. Ann. Discrete Math. 13, 71–74 (1982)
6. Grigoriev, A., Sitters, R.: Connected feedback vertex set in planar graphs. In: Paul, C., Habib, M. (eds.) WG 2009. LNCS, vol. 5911, pp. 143–153. Springer, Heidelberg (2010)
7. Schweitzer, P., Schweitzer, P.: Connecting face hitting sets in planar graphs. Inf. Process. Lett. 111(1), 11–15 (2010)
8. Zverovich, I.E.: Perfect connected-dominant graphs. Discuss. Math. Graph Theory 23, 159–162 (2003)

Network-Based Dissolution

René van Bevern[1], Robert Bredereck[1], Jiehua Chen[1], Vincent Froese[1],
Rolf Niedermeier[1], and Gerhard J. Woeginger[2]

[1] Institut für Softwaretechnik und Theoretische Informatik, TU Berlin, Germany
[2] Department of Mathematics and Computer Science,
TU Eindhoven, The Netherlands

Abstract. We introduce a graph-theoretic dissolution model that applies to a number of redistribution scenarios such as gerrymandering in political districting or work balancing in an online situation. The central aspect of our model is the deletion of certain vertices and the redistribution of their loads to neighboring vertices in a perfectly balanced way.

We investigate how the underlying graph structure, the pre-knowledge of which vertices should be deleted, and the relation between old and new vertex loads influence the computational complexity of the underlying graph problems. Our results establish a clear borderline between tractable and intractable cases.

1 Introduction

Motivated by applications in areas like political redistricting, economization, and distributed systems, we introduce a class of graph modification problems that we call *network-based dissolution*. We are given an undirected graph where each vertex carries a load consisting of discrete entities (e.g. voters, tasks, data). These loads are *balanced*: all vertices carry the same load. Now a certain number of vertices has to be *dissolved*, that is, they are to be deleted from the graph and their loads are to be redistributed among their neighbors such that afterwards all loads are balanced again.

Indeed, our dissolution problem comes in two flavors called DISSOLUTION and BIASED DISSOLUTION. DISSOLUTION is the basic version, as described in the preceding paragraph. BIASED DISSOLUTION is a variant that is motivated by gerrymandering in the context of political districting. It is centered around a bipartisan scenario with two types A and B of discrete entities. The goal is to find a redistribution that maximizes the number of vertices in which the A-entities form a majority. See Section 2 for a formal definition of these models.

Our main focus lies on analyzing the computational complexity of network-based dissolution problems, and in getting a good understanding of polynomial-time solvable and NP-hard cases.

Three Application Scenarios. We discuss three example scenarios for dissolution applications in some detail. The first and third example relate to BIASED DISSOLUTION, while the second example is closer to DISSOLUTION.

E. Csuhaj-Varjú et al. (Eds.): MFCS 2014, Part II, LNCS 8635, pp. 69–80, 2014.
© Springer-Verlag Berlin Heidelberg 2014

Our first example comes from political districting, the process of setting electoral districts. Let us consider a situation with two political parties (A and B) and an electorate of voters that each support either A or B. The electorate is currently divided into n districts, each consisting of precisely s individual voters. A district is won by the party that receives the majority of votes in this district. The local government performs an electoral reform that reduces the number of districts, and the local governor (from party A) is in charge of the redistricting process. His goal is of course to let party A win as many districts as possible while dissolving some districts and moving their voters to adjacent districts. All resulting new districts should have equal sizes s_{new} (where $s_{new} > s$). In the *network-based dissolution* model, the districts and their neighborhoods are represented by an undirected graph: vertices represent districts and edges indicate that two districts are adjacent.

Our second example concerns economization in a fairly general form. Let us consider a company with n employees, each producing s units of a desirable good during an eight-hour working day; for concreteness, let us say that each employee proves s theorems per working day. Now, due to the increasing support of automatic theorem provers, each employee is able to prove s_{new} theorems per day ($s_{new} > s$). Hence, without lowering the total number of proved theorems per day, some employees may be moved to a special task force for improving automatic theorem provers: this will secure the company's future competitiveness in proving theorems, without decreasing the overall theorem output. By company regulations, all theorem proving employees have to be treated equally and should have identical workloads. In the *network-based dissolution* model, employees correspond to vertices. Employees in the special task force are dissolved and disappear from the scene of action; their workload is to be taken over by neighboring employees who are comparable in qualification and research interests.

Our third and last example concerns storage updates in parallel or distributed systems. Let us consider a distributed storage array consisting of n storage nodes, each having a capacity of s storage units, of which some space is free. As the prices on cheap hard disk space are rapidly decreasing, the operators want to upgrade the storage capacity of some nodes and to deactivate other nodes for saving energy and cost. As their distributed storage concept takes full advantage only in case all nodes have equal capacity, they want to upgrade all (non-deactivated) nodes to the same capacity s_{new} and move capacities from deactivated nodes to non-deactivated neighboring nodes. In the resulting system, every non-deactivated node should only use half of its storage capacity.

Related Work. We are not aware of any previous work on our network-based dissolution problem. Our main inspiration comes from the area of political districting, and in particular from gerrymandering [8, 11, 12] and from supervised regionalization methods [5]. Of course, graph-theoretic models have been employed before for political districting; for instance Mehrota et al. [10] draws a connection to graph partitioning, and Duque [4] and Maravalle and Simeone [9] use graphs to model geographic information in the regionalization problem. These models are tailored towards very specific applications and are mainly used for the purpose of developing efficient heuristic algorithms, often relying on mathematical programming techniques. The computational hardness of districting problems has been known for many years [1].

Remark on Nomenclature. For the ease of presentation, throughout the paper we will adopt a political districting point of view on network-based dissolution: the words districts and vertices are used interchangeably, and the entities in districts are referred to as voters or supporters.

Contributions and Organization of This Paper. We propose two simple models DISSOLUTION and BIASED DISSOLUTION for network-based dissolution (Section 2). In the main body of the paper, we provide a variety of computational tractability and intractability results for both models. Furthermore, we investigate how the structure of the underlying graphs or an in-advance fixing of vertices to be dissolved influence the computational complexity (mainly in terms of polynomial-time solvability versus NP-hard cases).

- In Section 3, with network flow techniques we show that BIASED DISSOLUTION is polynomial-time solvable if the set of districts to be dissolved and the set of districts to be won are both specified as part of the input. The general version is NP-hard for every fixed $s \geq 3$.
- Section 4 presents a complexity dichotomy for DISSOLUTION and BIASED DISSOLUTION with respect to the old district size s and the increase Δ_s in district size (= difference between new and old district size). DISSOLUTION is polynomial-time solvable for $s = \Delta_s$, and BIASED DISSOLUTION is polynomial-time solvable for $s = \Delta_s = 1$; all other cases are NP-hard.
- Section 5 analyzes the complexity of DISSOLUTION and BIASED DISSOLUTION for various specially structured graphs, including planar graphs (NP-hard), cliques (polynomial-time solvable), and graphs of bounded treewidth (linear-time solvable if s and Δ_s are constant).

Due to the lack of space many proofs are only contained in the full version of the paper, which is available on arXiv (arXiv:1402.2664 [cs.DM]).

2 Formal Setting

Let $G = (V, E)$ be an undirected graph representing n districts. Let $s, \Delta_s \in \mathbb{N}^+$ be the *district size* and *district size increase*, respectively. For a subset $V' \subseteq V$ of districts, let $Z(V', G) = \{(x, y) \mid x \in V' \wedge y \in V(G) \setminus V' \wedge \{x, y\} \in E(G)\}$ be the set of pairs of districts in V' and their neighbors that are not in V'. The central notion for our studies is that of a *dissolution* , which basically describes a valid movement of voters from dissolved districts into remaining districts. The formal definition is the following:

Definition 1 (Dissolution). *Let $G = (V, E)$ be an undirected graph and let $D \subset V$ be a subset of districts to dissolve and $z \colon Z(D, G) \to \{0, \ldots, s\}$ be a function that describes how many voters shall be moved from one district to its non-dissolved neighbors. Then, (D, z) is called an (s, Δ_s)-dissolution for G if*

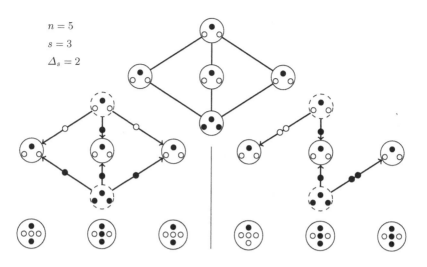

Fig. 1. An illustration of a 1-biased $(3, 2)$-dissolution (left) and a 2-biased $(3, 2)$-dissolution (right). Black circles represent A-supporters while white circles represent B-supporters. The graph on the top shows a neighborhood graph of five districts, each district consisting of three voters. The task is to dissolve two districts such that each remaining district contains five voters. The graphs in the middle show two possible realizations of dissolutions. The graphs on the bottom show the two corresponding outcomes. The arrows point from the districts to be dissolved to the "goal districts" and the black/white circle labels on the arrows indicate which kind of voters are moved along the arrows.

a) *no voter remains in any dissolved district:*

$$\forall v' \in D : \sum_{(v',v)\in Z(D,G)} z(v',v) = s, \text{ and}$$

b) *the size of all remaining (non-dissolved) districts increases by Δ_s:*

$$\forall v \in V \setminus D : \sum_{(v',v)\in Z(D,G)} z(v',v) = \Delta_s.$$

Throughout this work, we use $s_{new} := s + \Delta_s$ to denote the new district size, $d := |D| = |V| \cdot \Delta_s/s_{new}$ to denote the number of dissolved districts, and $r := |V| - d$ to denote the number of remaining, non-dissolved districts.

We write *dissolution* instead of (s, Δ_s)-dissolution when s and Δ_s are clear from the context. By definition, a dissolution only ensures that the numbers of voters moving between districts fulfill the given constraints on the district sizes, that is, the size of each remaining district increases by Δ_s. Figure 1 gives an example illustrating two possible $(3, 2)$-dissolutions for a 5-vertex graph.

Motivated from social choice application scenarios, we additionally assume that each voter supports one of two parties A and B. We then seek a dissolution such that the number of remaining districts won by party A is maximized. Here, a district is won by the party that is supported by a strict majority of the voters inside the district. This yields the notion of a *biased dissolution*, which is defined as follows:

Definition 2 (Biased dissolution). *Let G be an undirected graph and let $\alpha \colon V(G) \to \{0, \ldots, s\}$ be an A-supporter distribution, where $\alpha(v)$ denotes the number of A-supporters in district $v \in V$. Let (D, z) be an (s, Δ_s)-dissolution for G. Let $r_\alpha \in \mathbb{N}$ be the minimum number of districts that party A shall win after the dissolution and $z_\alpha \colon Z(D, G) \to \{0, \ldots, s\}$ be an A-supporter movement, where $z_\alpha(v', v)$ denotes the number of A-supporters moving from district v' to district v. Finally, let $R_\alpha \subseteq V(G) \setminus D$ be a size-r_α subset of districts. Then, $(D, z, z_\alpha, R_\alpha)$ is called an r_α-biased (s, Δ_s)-dissolution for (G, α) if and only if*

c) *a district cannot receive more A-supporters from a dissolved district than the total number of voters it receives from that district:*

$$\forall (v', v) \in Z(D, G) : z_\alpha(v', v) \le z(v', v),$$

d) *no A-supporters remain in any dissolved district:*

$$\forall v' \in D : \sum_{(v', v) \in Z(D, G)} z_\alpha(v', v) = \alpha(v'), \text{ and}$$

e) *each district in R_α has a strict majority of A-supporters:*

$$\forall v \in R_\alpha : \alpha(v) + \sum_{(v', v) \in Z(D, G)} z_\alpha(v', v) > \frac{s + \Delta_s}{2}.$$

We also say that a district wins *if it has a strict majority of A-supporters and* loses *otherwise.*

Figure 1 shows two biased dissolutions: one with $r_\alpha = 1$ and the other one with $r_\alpha = 2$. We are now ready to formally state the definitions of the two dissolution problems that we discuss in this work:

DISSOLUTION
Input: An undirected graph $G = (V, E)$ and positive integers s and Δ_s.
Question: Is there an (s, Δ_s)-dissolution for G?

BIASED DISSOLUTION
Input: An undirected graph $G = (V, E)$, positive integers s, Δ_s, r_α, and an A-supporter distribution $\alpha : V \to \{0, \ldots, s\}$.
Question: Is there an r_α-biased (s, Δ_s)-dissolution for (G, α)?

Note that DISSOLUTION is equivalent to BIASED DISSOLUTION with $r_\alpha = 0$. As we will see later, both DISSOLUTION and BIASED DISSOLUTION are NP-hard in general. In this work, we additionally look into special cases of our dissolution problems and investigate where the causes of intractability lie.

3 Complexity for Partially Known Dissolutions

In this section, we discuss some relevant special cases of our (in general) NP-hard dissolution problems. These include situations where the districts to be dissolved or to win are fixed in advance. We see that BIASED DISSOLUTION is only polynomial-time solvable if both are fixed, and NP-hard otherwise.

Sometimes, the districts to be dissolved and the districts to win are already determined beforehand. For this case, we show that BIASED DISSOLUTION can be modeled as a network flow problem which can be solved in polynomial time.

Theorem 1. *Let* $I = (G = (V, E), s, \Delta_s, r_\alpha, \alpha)$ *be a* BIASED DISSOLUTION *instance, and let* $D, R_\alpha \subset V$ *be two disjoint subsets of districts. The problem of deciding whether* (G, α) *admits an* r_α*-biased* (s, Δ_s)*-dissolution in which* D *is the set of dissolved districts and in which all districts in* R_α *are won can be reduced in linear time to a maximum flow problem with* $2|V| + 2$ *nodes,* $2|V| + 3|E|$ *arcs, and maximum arc capacity* $\max(s, \Delta_s)$.

With the above flow network construction we can design a polynomial-time algorithm for BIASED DISSOLUTION when the number of districts is a constant.

Corollary 1. *Any instance* $((V, E), s, \Delta_s, \alpha)$ *of* BIASED DISSOLUTION *can be solved in time* $O(3^{|V|} \cdot (\max(s, \Delta_s) \cdot |V| \cdot |E| + |V|^3))$.

On the contrary, we obtain NP-hardness for BIASED DISSOLUTION once one of the two sets D and R_α is unknown. For the case that only the set D of dissolved districts is given beforehand, the remaining task is to decide how many A-supporters are moved to a certain non-dissolved district. However, we will see in Section 4.2 that in the hardness construction for Theorem 2 it is already determined which districts are to be dissolved. Furthermore, DISSOLUTION is the special case of BIASED DISSOLUTION with $r_\alpha = 0$ (which implies $R_\alpha = \emptyset$) and DISSOLUTION is NP-hard for the case of $s \neq \Delta_s$ (Theorem 2).

4 Complexity Dichotomy with Respect to District Sizes

In this section, we study the computational complexity of DISSOLUTION and BIASED DISSOLUTION with respect to the ratio of the two integers: old district size s and district size increase Δ_s. We start by showing some useful structural observations for dissolutions in Section 4.1 before we come to the results for DISSOLUTION in Section 4.2 and for BIASED DISSOLUTION in Section 4.3.

4.1 Structural Properties

Using the flow construction from Theorem 1, we can show the equivalence of (s, Δ_s)-dissolutions and star partitions for the cases where s is any multiple of Δ_s.

Lemma 1. *There exists a* $(t \cdot \Delta_s, \Delta_s)$*-dissolution for an undirected graph* G *if and only if* G *has a* t*-star partition.*

We observe a symmetry concerning the district size s and the district size increase Δ_s in the sense that exchanging their values yields an equivalent instance of DISSOLUTION. Intuitively, the idea behind the following lemma is that the roles of dissolved and non-dissolved districts in a given (s, Δ_s)-dissolution can in fact be exchanged by "reversing" the movement of voters to obtain a (Δ_s, s)-dissolution.

Lemma 2. *There exists an (s, Δ_s)-dissolution for an undirected graph G if and only if there exists a (Δ_s, s)-dissolution for G.*

4.2 Complexity Dichotomy for Dissolution

In this subsection, we show a P vs. NP dichotomy of DISSOLUTION with respect to the district size s and the size increase Δ_s. Using Lemma 1, we can show that finding an (s, s)-dissolution essentially corresponds to finding a perfect matching and can thus be done in polynomial time. If $s \neq \Delta_s$, then DISSOLUTION becomes NP-hard. We use Bézout's identity to encode the NP-complete EXACT COVER BY t-SETS problem into our dissolution problem.

Theorem 2. *If $s = \Delta_s$, DISSOLUTION is solvable in $O(n^\omega)$ time (where ω is the matrix multiplication exponent); otherwise the problem is NP-complete.*

Proof. Let $I = (G, s, \Delta_s)$ be a DISSOLUTION instance with $\Delta_s = s$. Set $t := s/\Delta_s = 1$. Lemma 1 implies that I is a yes-instance if and only if G has a t-star partition. A t-star partition with $t = 1$ is indeed a perfect matching, which can be computed in $O(n^\omega)$ time, where ω is the smallest exponent such that matrix multiplication can be computed in $O(n^\omega)$ time. Currently, the smallest known upper bound of ω is 2.3727 [13].

For the case $s \neq \Delta_s$, we show that DISSOLUTION is NP-complete if $s > \Delta_s$. Due to Lemma 2, this also transfers to the cases where $s < \Delta_s$. First, given a DISSOLUTION instance (G, s, Δ_s) and a function $z : Z(D, G) \to \{0, \ldots, s\}$ where $D \subset V(G)$, one can check in polynomial time whether (D, z) is an (s, Δ_s)-dissolution. Thus, DISSOLUTION is in NP.

To show the NP-hardness result, we give a reduction from the NP-complete EXACT COVER BY t-SETS problem [6] for $t := (s + \Delta_s)/g > 2$, where $g := \gcd(s, \Delta_s) \leq \Delta_s$ is the greatest common divisor of s and Δ_s. Given a finite set X and a collection $\mathcal{C}$ of subsets of X of size t, EXACT COVER BY t-SETS asks whether there is a subcollection $\mathcal{C}' \subseteq \mathcal{C}$ that partitions X, that is, each element of X is contained in exactly one subset in $\mathcal{C}'$.

Let $(X, \mathcal{C})$ be an EXACT COVER BY t-SETS instance. We construct a DISSOLUTION instance (G, s, Δ_s) with a neighborhood graph $G = (V, E)$ defined as follows: For each element $u \in X$, add a clique C_u of properly chosen size q to G and let v_u denote an arbitrary fixed vertex in C_u. For each subset $S \in \mathcal{C}$, add a clique C_S of properly chosen size $r \geq t$ to G and connect each v_u for $u \in S$ to a unique vertex in C_S. Figure 2 shows an example of the constructed neighborhood graph for $t = 3$.

Next, we explain how to choose the values of q and r. We set $q := x_q + y_q$, where $x_q \geq 0$ and $y_q \geq 0$ are integers satisfying $x_q s - y_q \Delta_s = g$. Such integers exist

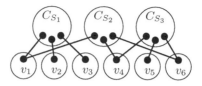

Fig. 2. The constructed instance for $t = 3$

by Bézout's identity. The intuition behind is as follows: Dissolving x_q districts in C_u and moving the voters to y_q districts in C_u creates an overflow of exactly g voters that have to move out of C_u. Notice that the only way to move voters into or out of C_u is via district v_u. Moreover, in any dissolution, exactly x_q districts in C_u are dissolved because dissolving more districts leads to an overflow of at least $g + s + \Delta_s > s$ voters, which is more than v_u can move, whereas dissolving less districts yields a demand of at least $s + \Delta_s - g > \Delta_s$ voters, which is more than v_u can receive. Thus, v_u must be dissolved since there is an overflow of g voters to move out of C_u and this can only be done via district v_u.

The value of $r \geq t$ is chosen in such a way that, for each $S \in \mathcal{C}$ and each $u \in S$, it is possible to move g voters from v_u to C_S (recall that v_u must be dissolved). In other words, we require C_S to be able to receive in total $t \cdot g = s + \Delta_s$ voters in at least t non-dissolved districts. Thus, we set $r := x_r + y_r$, where $x_r \geq 0$ and $y_r \geq t$ are integers satisfying $x_r s - y_r \Delta_s = -(s + \Delta_s)$. Again, since $-(s + \Delta_s)$ is divisible by g, such integers exist by our preliminary discussion. It is thus possible to dissolve x_r districts in C_S moving the voters to the remaining y_r districts in C_S such that we end up with a demand of $s + \Delta_s$ voters in C_S. Note that the only other possibility is to dissolve $x_r + 1$ districts in C_S in order to end up with a demand of zero voters. In this case, no voters of any other districts connected to C_S can move to C_S. By the construction of C_u above, it is clear that it is also not possible to move any voters out of C_S because no v_u can receive voters in any dissolution. Thus, for any dissolution, it holds that either all or none of the v_u connected to some C_S move g voters to C_S.

The proof of correctness is as follows. Suppose $(X, \mathcal{C})$ is a yes-instance, that is, there exists a partition $\mathcal{C}' \subseteq \mathcal{C}$ of X. We can thus dissolve x_q districts in each C_u (including v_u) and move the voters such that all y_q non-dissolved districts receive exactly Δ_s voters. This is always possible since C_u is a clique. If we do so, then, by construction, g voters have to move out of each v_u. Since $\mathcal{C}'$ partitions X, each $u \in X$ is contained in exactly one subset $S \in \mathcal{C}'$. We can thus move the g voters from each v_u to C_S. Now, for each $S \in \mathcal{C}'$, we dissolve any x_r districts that are not adjacent to any v_u and for the subsets in $\mathcal{C} \setminus \mathcal{C}'$, we simply dissolve $x_r + 1$ arbitrary districts in the corresponding cliques. By the above discussion of the construction, we know that this in fact yields an (s, Δ_s)-dissolution. Hence, (G, s, Δ_s) is a yes-instance.

Now assume that there exists an (s, Δ_s)-dissolution for (G, s, Δ_s). As we have already seen in the above discussion, any (s, Δ_s)-dissolution generates an overflow of g voters in each C_u that has to be moved over v_u to some district in C_S.

Furthermore, each C_S either receives g voters from all its adjacent v_u or no voters at all. Therefore, the subsets S corresponding to cliques C_S that receive $t \cdot g$ voters form a partition of X, showing that $(X, \mathcal{C})$ is a yes-instance.   □

4.3   Complexity of Biased Dissolution

Since DISSOLUTION is a special case of BIASED DISSOLUTION, the NP-hardness results for $s \neq \Delta_s$ transfer to BIASED DISSOLUTION. It remains to see whether BIASED DISSOLUTION remains polynomial-time solvable when $s = \Delta_s$. Interestingly, this is true for $s = \Delta_s = 1$.

We introduce a notion called "edge set" for a given dissolution (D, z) of a given graph G. Let $E_z \subseteq E(G)$ contain all edges $\{x, y\}$ with $(x, y) \in Z(D, G)$ and $z(x, y) > 0$. Then, we call E_z the *edge set used* by the dissolution (D, z).

The following lemma shows that finding an r_α-biased $(1, 1)$-dissolution essentially corresponds to finding a maximum-weight perfect matching.

Lemma 3. *Let $(G = (V, E), s = 1, \Delta_s = 1, r_\alpha, \alpha)$ be a* BIASED DISSOLUTION *instance. There exists an r_α-biased $(1, 1)$-dissolution for (G, α) if and only if there exists a perfect matching of weight at least r_α in (G, w) with $w(\{x, y\}) := 1$ if $\alpha(x) = \alpha(y) = 1$ and $w(\{x, y\}) := 0$ otherwise.*

As shown in the proof of Theorem 2, the edge set used by a $(1, 1)$-dissolution is a perfect matching. By appropriately setting α and r_α we can enforce that the edge set used by any r_α-biased $(2, 2)$-dissolution only induces cycles of lengths divisible by four. We end up with a restricted two-factor problem which was already studied in the literature [7] and can be used to show NP-hardness (Theorem 3).

Lemma 4. *Let $G = (V, E)$ be an undirected graph with $4q$ vertices ($q \in \mathbb{N}$). Then G has a two-factor E' whose cycle lengths are all multiples of four, if and only if (G, α) admits a q-biased $(2, 2)$-dissolution where $\alpha(v) = 1$ for all $v \in V$.*

Theorem 3. BIASED DISSOLUTION *on graphs $G = (V, E)$ can be solved in $O(|V|(|E| + |V| \log |V|))$ time if $s = \Delta_s = 1$; otherwise it is NP-complete for any constant value $s = \Delta_s \geq 2$.*

5   Complexity on Special Graph Classes

In a companion paper [3], we have shown that computing star partitions—and hence by Lemma 1 also DISSOLUTION—remains NP-hard even on subcubic grid graphs and split graphs. In this section, we discuss the complexity of BIASED DISSOLUTION on special graph classes.

An interesting special case of BIASED DISSOLUTION occurs if voters can move from any district to any other district, that is, the neighborhood graph is a clique. Then, the existence of an (s, Δ_s)-dissolution depends only on the number $|V|$ of districts, the district size s, and the size increase Δ_s. Clearly, a DISSOLUTION instance is a yes-instance if and only if $d := |V| \cdot \Delta_s / (s + \Delta_s)$ is an integer.

We show that BIASED DISSOLUTION can likewise be solved in polynomial time if the neighborhood graph is a clique. Yuster [14, Theorem 2.3] showed that the H-FACTOR problem is solvable in linear time on graphs of bounded treewidth when the size of H is constant. This includes the case of finding x-star partitions, that is, $(x, 1)$-dissolutions resp. $(1, x)$-dissolutions when x is constant. We can show that the more general problem BIASED DISSOLUTION is solvable in linear time on graphs of bounded treewidth when s and Δ_s are constants. By a polynomial-time reduction from the NP-hard PERFECT PLANAR H-MATCHING problem [2], we get NP-hardness for DISSOLUTION on planar graphs.

Theorem 4

(1) BIASED DISSOLUTION *is solvable in* $O(|V|^2)$ *time on cliques.*
(2) BIASED DISSOLUTION *is solvable in linear time on graphs of bounded treewidth when s and Δ_s are constant.*
(3) DISSOLUTION *on planar graphs is NP-complete for all $s \neq \Delta_s$ such that Δ_s divides s or s divides Δ_s. It is polynomial-time solvable for $s = \Delta_s$.*

Proof (Sketch for (1)). In fact, we show how to solve the optimization version of BIASED DISSOLUTION, where we maximize the number r_α of winning districts. Intuitively, it appears to be a reasonable approach to dissolve districts pursuing the following two objectives: Any losing district should contain as few A-supporters as possible and any winning district should contain exactly the amount that is required to have a majority. Dissolving districts this way minimizes the number of "wasted" A-supporters. We now show that this greedy strategy is indeed optimal.

Let $G = (V, \binom{V}{2})$ be a clique, let α be an A-supporter distribution over V, and let s and Δ_s be the district size and the district size increase. With G being complete, we are free to move voters from any dissolved district to any non-dissolved district. Let $\mu := \lfloor (s + \Delta_s)/2 \rfloor + 1$ be the minimum number of A-supporters required to win a district. Thus, a district with less than $(\mu - \Delta_s)$ A-supporters can never win. Denote by $\mathcal{L} := \{v \in V \mid \alpha(v) < \mu - \Delta_s\}$ the set of *non-winnable* districts.

Our first claim corresponds to the first objective above, that is, the losing districts should contain a minimal number of A-supporters.

Claim 1. *Let $v, w \in V$ be two districts with $\alpha(v) \leq \alpha(w)$. If there exists an r_α-biased dissolution where v is winning and w is losing, then there also exists an r_α-biased dissolution where v is losing and w is winning.*

The next claim basically corresponds to the second objective, in the sense that districts with a large number of A-supporters (possibly more than the required) should be dissolved in order to distribute the voters more efficiently.

Claim 2. *Let $v, w \in V$ be two districts with $\alpha(v) \leq \alpha(w)$. Assume that there exists an r_α-biased dissolution where r_α is optimal. If v is dissolved, then the following holds: (i) If w is losing, then there also exists an r_α-biased dissolution where w is dissolved and v is losing. (ii) If w is winning and v is winnable, that is, $v \notin \mathcal{L}$, then there exists an r_α-biased dissolution where w is dissolved and v is winning.*

Using the two claims above, we now show how to compute an optimal biased dissolution. In order to find a biased dissolution with the maximum number of winning districts, we seek a dissolution which loses a minimum number of remaining districts. Thus, for each $\ell \in \{0, \ldots, r\}$, we check whether it is possible to dissolve d districts such that at most ℓ of the remaining r districts lose. To this end, assume that the districts $v_1, \ldots, v_n$ are ordered by increasing number of A-supporters, that is, $\alpha(v_1) \leq \alpha(v_2) \leq \ldots \leq \alpha(v_n)$ and let $V_\ell := \{v_1, \ldots, v_\ell\}$. Now, if there exists an $(r-\ell)$-biased dissolution, then there also exists an $(r-\ell)$-biased dissolution where the losing districts are exactly V_ℓ. This follows by repeated application of the exchange arguments of Claim 1 and Claim 2(i). Hence, given ℓ, we have to check whether there is a set $D \subseteq V \setminus V_\ell$ of d districts that can be dissolved in such a way that all non-dissolved districts in $V \setminus (V_\ell \cup D)$ win and the districts in V_ℓ lose.

First, note that in order to achieve this, all districts in $\mathcal{L} \setminus V_\ell$ have to be dissolved because they cannot win in any way. Clearly, if $|\mathcal{L} \setminus V_\ell| > d$, then it is simply not possible to lose only ℓ districts and we can immediately go to the next iteration with $\ell := \ell + 1$. Therefore, we assume that $|\mathcal{L} \setminus V_\ell| \leq d$ and let $d' := d - |\mathcal{L} \setminus V_\ell|$ be the number of additional districts to dissolve in $V \setminus (\mathcal{L} \cup V_\ell)$. By Claim 2(ii), it follows that we can assume that the d' districts with the maximum number of A-supporters are dissolved, that is, $V^{d'} := \{v_{n-d'+1}, \ldots, v_n\}$. Thus, we set $D := \mathcal{L} \setminus V_\ell \cup V^{d'}$ and check whether there are enough A-supporters in D to let all $r - \ell$ remaining districts in $V \setminus (V_\ell \cup D)$ win.

Sorting the districts by the number of A-supporters (as preprocessing) requires $O(n \log n)$ arithmetic operations. For up to n values of ℓ, to check whether the remaining districts in $V \setminus (V_\ell \cup D)$ can win requires $O(n)$ arithmetic operations each. Thus, assuming constant-time arithmetics, we end up with $O(n^2)$ time.   □

6   Conclusion

We initiated a graph-theoretic combinatorial approach to concrete redistribution problems occurring in various application domains. Obviously, the two basic problems DISSOLUTION and BIASED DISSOLUTION concern highly simplified situations and will not be able to model all interesting aspects of redistribution scenarios. For instance, our constraint that before and after the dissolution all vertex loads are perfectly balanced may be too restrictive for many applications. All in all, we consider our simple (and yet fairly realistic) models as a first step into a fruitful research direction that might yield a stronger linking of graph-theoretic concepts with districting methods and other application scenarios.

We end with a few specific challenges for future research. We have left open whether the P vs. NP dichotomy for general graphs fully carries over to the planar case: it might be possible that planar graphs allow for some further tractable cases with respect to the relation between old and new district sizes. Moreover, with redistricting applications in mind it might be of interest to study special cases of planar graphs (such as grid-like structures) in quest of finding polynomial-time solvable special cases of network-based dissolution problems.

Having identified several NP-hard special cases of DISSOLUTION and BIASED DISSOLUTION, it is a natural endeavor to investigate their polynomial-time approximability and their parameterized complexity; in the latter case one also needs to identify fruitful parameterizations.

Acknowledgments. René van Bevern was supported by the DFG, project DAPA (NI 369/12), Robert Bredereck by the DFG, project PAWS (NI 369/10), Jiehua Chen by the Studienstiftung des Deutschen Volkes, Vincent Froese by the DFG, project DAMM (NI 369/13), and Gerhard J. Woeginger while visiting TU Berlin by the Alexander von Humboldt Foundation, Bonn, Germany.

References

[1] Altman, M.: Districting Principles and Democratic Representation. PhD thesis, California Institute of Technology (1998)

[2] Berman, F., Johnson, D., Leighton, T., Shor, P.W., Snyder, L.: Generalized planar matching. Journal of Algorithms 11(2), 153–184 (1990)

[3] van Bevern, R., Bredereck, R., Bulteau, L., Chen, J., Froese, V., Niedermeier, R., Woeginger, G.J.: Star partitions of perfect graphs. In: Esparza, J., Fraigniaud, P., Husfeldt, T., Koutsoupias, E. (eds.) ICALP 2014. LNCS, vol. 8572, pp. 174–185. Springer, Heidelberg (2014)

[4] Duque, J.C.: Design of Homogeneous Territorial Units: A Methodological Proposal and Applications. PhD thesis, University of Barcelona (2004)

[5] Duque, J.C., Ramos, R., Surinach, J.: Supervised regionalization methods: A survey. International Regional Science Review 30(3), 195–220 (2007)

[6] Garey, M.R., Johnson, D.S.: Computers and Intractability. W. H. Freeman (1979)

[7] Hell, P., Kirkpatrick, D.G., Kratochvíl, J., Kríz, I.: On restricted two-factors. SIAM Journal on Discrete Mathematics 1(4), 472–484 (1988)

[8] Landau, Z., Su, F.: Fair division and redistricting. Social Choice and Welfare 32(3), 479–492 (2009)

[9] Maravalle, M., Simeone, B.: A spanning tree heuristic for regional clustering. Communications in Statistics—Theory and Methods 24(3), 625–639 (1995)

[10] Mehrota, A., Johnson, E.L., Nemhauser, G.L.: An optimization based heuristic for political districting. Management Science 44(8), 1100–1114 (1998)

[11] Puppe, C., Tasnádi, A.: A computational approach to unbiased districting. Mathematical and Computer Modelling 48, 1455–1460 (2008)

[12] Puppe, C., Tasnádi, A.: Optimal redistricting under geographical constraints: Why "pack and crack" does not work. Economics Letters 105(1), 93–96 (2009)

[13] Vassilevska Williams, V.: Multiplying matrices faster than Coppersmith-Winograd. In: Proc. 44th STOC, pp. 887–898. ACM (2012)

[14] Yuster, R.: Combinatorial and computational aspects of graph packing and graph decomposition. Computer Science Review 1(1), 12–26 (2007)

On Unification of QBF Resolution-Based Calculi

Olaf Beyersdorff[1], Leroy Chew[1], and Mikoláš Janota[2]

[1] School of Computing, University of Leeds, United Kingdom
[2] INESC-ID, Lisbon, Portugal

Abstract. Several calculi for quantified Boolean formulas (QBFs) exist, but relations between them are not yet fully understood. This paper defines a novel calculus, which is resolution-based and enables unification of the principal existing resolution-based QBF calculi, namely Q-resolution, long-distance Q-resolution and the expansion-based calculus ∀Exp+Res. All these calculi play an important role in QBF solving. This paper shows simulation results for the new calculus and some of its variants. Further, we demonstrate how to obtain winning strategies for the universal player from proofs in the calculus. We believe that this new proof system provides an underpinning necessary for formal analysis of modern QBF solvers.

1 Introduction

Traditionally, classifying a problem as NP-hard was ultimately understood as evidence for its infeasibility. Sharply contrasting this view, we have today fast algorithms for many important computational tasks with underlying NP-hard problems. One particularly compelling example of tremendous success is the area of SAT solving [25] where fast algorithms are being developed and tested for the classical NP-complete problem of satisfiability of propositional formulas (SAT). Modern SAT-solvers routinely solve industrial instances with even millions of variables. However, from a theoretical perspective, this success of SAT solvers is not well understood. The main theoretical approach to it comes via proof complexity. In particular, resolution and its subsystems have been very successfully analysed in terms of proof complexity and sharp bounds are known on the size and space for many important principles in resolution (cf. [6]). This is very important information as the main algorithmic approaches to SAT such as DPLL and CDCL are known to correspond to (systems of) resolution [2,7,14,26], and therefore bounds on size and space of proofs directly translate into bounds on running time and memory consumption of SAT solvers.

In the last decade, there has been ever-increasing interest to transfer the successful approach of SAT-solving to the more expressive case of *quantified propositional formulas (QBF)*. Due to its PSPACE completeness, QBF is far more expressive than SAT and thus applies to further fields such as formal verification or planning [27,4]. As for SAT, proof complexity provides the main theoretical approach towards understanding the performance and limitations of QBF-solving. However, compared to proof complexity of classical propositional

E. Csuhaj-Varjú et al. (Eds.): MFCS 2014, Part II, LNCS 8635, pp. 81–93, 2014.
© Springer-Verlag Berlin Heidelberg 2014

logic, QBF proof complexity is at a much earlier stage and also poses additional challenges. Currently, a handful of systems exist, and they correspond to different approaches in QBF-solving. In particular, Kleine Büning et al. [20] define a resolution-like calculus called *Q-resolution*. There are several extensions of Q-resolution; notably *long-distance Q-resolution* [1], which is believed to be more powerful than plain Q-resolution [10]. Q-resolution and its extensions are important as they model QBF solving based on CDCL [12]. Apart from CDCL, another main approach to QBF-solving is through expansion of quantifiers [5,3,16]. Recently, a proof system $\forall Exp+Res$ was introduced with the motivation to trace expansion-based QBF solvers [15]. $\forall Exp+Res$ also uses resolution, but is conceptually very different from Q-resolution. The precise relation of $\forall Exp+Res$ to Q-resolution is currently open (cf. [17]), but we conjecture that the two systems are incomparable as it has been shown that expansion-based solving can exponentially outperform DPLL-based solving.

In general, it is fair to say that relations between the different types of QBF systems mentioned above are currently not well understood. The objective of the present paper is to unify these approaches. Towards this aim we define a calculus that is able to capture the existing QBF resolution-based calculi and yet remains amenable to machine manipulation. Our main contributions are as follows. (1) We introduce two novel calculi IR-calc and IRM-calc, which are shown to be sound and complete for QBF. (2) IR-calc p-simulates Q-resolution and $\forall Exp+Res$, i.e., proofs in either Q-resolution or $\forall Exp+Res$ can be efficiently translated into IR-calc. (3) The variant IRM-calc p-simulates long-distance Q-resolution. (4) We show how to extract winning strategies for the universal player from proofs in IR-calc and IRM-calc. Indeed, unified certification of QBF solvers or certification of solvers combining expansion and DPLL is of immense practical importance [13,1,10] and presents one of the main motivations for our research. To the best of our knowledge, constructions of strategies from expansion-based solvers were not known prior to this paper.

The rest of the paper is structured as follows. Section 2 overviews concepts and notation used throughout the paper. Section 3 introduces novel calculi and Section 4 shows how winning strategies for the universal player are constructed; this is used as an argument for soundness. Section 5 shows p-simulation results for the new calculi. Finally, Section 6 concludes the paper with a discussion. Due to space restrictions some proofs are sketched or omitted.

2   Preliminaries

A *literal* is a Boolean variable or its negation; we say that the literal x is *complementary* to the literal $\neg x$ and vice versa. If l is a literal, $\neg l$ denotes the complementary literal, i.e. $\neg\neg x = x$. A *clause* is a disjunction of zero or more literals. The empty clause is denoted by $\bot$, which is semantically equivalent to false. A formula in *conjunctive normal form* (CNF) is a conjunction of clauses. Whenever convenient, a clause is treated as a set of literals and a CNF formula

$$\frac{}{C} \text{ (Axiom)} \qquad \frac{C_1 \cup \{x\} \qquad C_2 \cup \{\neg x\}}{C_1 \cup C_2} \text{ (Res)}$$

C is a clause in the matrix. Variable x is existential. If $z \in C_1$, then $\neg z \notin C_2$.

$$\frac{C \cup \{u\}}{C} \text{ (}\forall\text{-Red)} \qquad \begin{array}{l} \text{Variable } u \text{ is universal. If } x \in C \text{ is} \\ \text{existential, then } \mathsf{lv}(x) < \mathsf{lv}(u). \end{array}$$

Fig. 1. The rules of Q-Res [20]

as a set of clauses. For a literal $l = x$ or $l = \neg x$, we write $\mathsf{var}(l)$ for x and extend this notation to $\mathsf{var}(C)$ for a clause C and $\mathsf{var}(\psi)$ for a CNF ψ.

A *proof system* (Cook, Reckhow [8]) for a language L over alphabet Γ is a polynomial-time computable partial function $f : \Gamma^* \rightarrow \Gamma^*$ with $rng(f) = L$. An f-*proof* of string y is a string x such that $f(x) = y$. In the systems that we consider here, proofs are sequences of clauses; a *refutation* is a proof deriving $\bot$. A proof system f for L *p-simulates* a system g for L if there exists a polynomial-time computable function t that translates g-proofs into f-proofs, i.e., for all $x \in \Gamma^*$ we have $g(x) = f(t(x))$.

Quantified Boolean Formulas (QBFs) [19] extend propositional logic with quantifiers with the standard semantics that $\forall x. \Psi$ is satisfied by the same truth assignments as $\Psi[0/x] \wedge \Psi[1/x]$ and $\exists x. \Psi$ as $\Psi[0/x] \vee \Psi[1/x]$. Unless specified otherwise, we assume that QBFs are in *closed prenex* form with a CNF *matrix*, i.e., we consider the form $\mathcal{Q}_1 X_1 \ldots \mathcal{Q}_k X_k. \phi$, where X_i are pairwise disjoint sets of variables; $\mathcal{Q}_i \in \{\exists, \forall\}$ and $\mathcal{Q}_i \neq \mathcal{Q}_{i+1}$. The formula ϕ is in CNF and is defined only on variables $X_1 \cup \ldots \cup X_k$. The propositional part ϕ of a QBF is called the *matrix* and the rest the *prefix*. If a variable x is in the set X_i, we say that x is at *level* i and write $\mathsf{lv}(x) = i$; we write $\mathsf{lv}(l)$ for $\mathsf{lv}(\mathsf{var}(l))$. A closed QBF is *false* (resp. *true*), iff it is semantically equivalent to the constant 0 (resp. 1).

Often it is useful to think of a QBF $\mathcal{Q}_1 X_1 \ldots \mathcal{Q}_k X_k. \phi$ as a *game* between the *universal* and the *existential player*. In the i-th step of the game, the player $\mathcal{Q}_i$ assigns values to the variables X_i. The existential player wins the game iff the matrix ϕ evaluates to 1 under the assignment constructed in the game. The universal player wins iff the matrix ϕ evaluates to 0. A QBF is false iff there exists a *winning strategy* for the universal player, i.e. if the universal player can win any possible game.

2.1 Resolution-Based Calculi for QBF

This section gives a brief overview of the main existing resolution-based calculi for QBF. *Q-resolution (Q-Res)*, by Kleine Büning et al. [20], is a resolution-like calculus that operates on QBFs in prenex form where the matrix is a CNF. The rules are given in Figure 1. *Long-distance resolution (LD-Q-Res)* appears originally in the work of Zhang and Malik [33] and was formalized into a calculus by Balabanov and Jiang [1]. It merges complementary literals of a universal

$$\frac{}{C}\ (\text{Axiom}) \qquad \frac{D \cup \{u\}}{D}\ (\forall\text{-Red}) \qquad \frac{D \cup \{u^*\}}{D}\ (\forall\text{-Red}^*)$$

C is a clause in the matrix. Literal u is universal and $\mathsf{lv}(u) \geq \mathsf{lv}(l)$ for all $l \in D$.

$$\frac{C_1 \cup U_1 \cup \{x\} \qquad C_2 \cup U_2 \cup \{\neg x\}}{C_1 \cup C_2 \cup U}\ (\text{Res})$$

Variable x is existential. If for $l_1 \in C_1, l_2 \in C_2$, $\mathsf{var}(l_1) = \mathsf{var}(l_2) = z$ then $l_1 = l_2 \neq z^*$. U_1, U_2 contain only universal literals with $\mathsf{var}(U_1) = \mathsf{var}(U_2)$. For each $u \in \mathsf{var}(U_1)$ we require $\mathsf{lv}(x) < \mathsf{lv}(u)$. If for $w_1 \in U_1, w_2 \in U_2$, $\mathsf{var}(w_1) = \mathsf{var}(w_2) = u$ then $w_1 = \neg w_2$, $w_1 = u^*$ or $w_2 = u^*$. U is defined as $\{u^* \mid u \in \mathsf{var}(U_1)\}$.

Fig. 2. The rules of LD-Q-Res [1]

$$\frac{}{\{l^{\tau_l} \mid l \in C, l \text{ is existential}\} \cup \{\tau(l) \mid l \in C, l \text{ is universal}\}}\ (\text{Axiom})$$

C is a clause from the matrix and τ is an assignment to all universal variables. τ_l are partial assignments obtained by restricting τ to variables u with $\mathsf{lv}(u) < \mathsf{lv}(l)$.

$$\frac{C_1 \cup \{x^\tau\} \qquad C_2 \cup \{\neg x^\tau\}}{C_1 \cup C_2}\ (\text{Res})$$

Fig. 3. The rules of ∀Exp+Res (adapted from [18])

variable u into the special literal u^*. These special literals prohibit certain resolution steps. In particular, different literals of a universal variable u may be merged only if $\mathsf{lv}(x) < \mathsf{lv}(u)$, where x is the resolution variable. The rules are given in Figure 2. Note that the rules do not prohibit resolving $w^* \vee x \vee C_1$ and $u^* \vee \neg x \vee C_2$ with $\mathsf{lv}(w) \leq \mathsf{lv}(u) < \mathsf{lv}(x)$ as long as $w \neq u$.

A different calculus ∀Exp+Res based on expansions was introduced in [18]. In Figure 3 we present an adapted version of this calculus so that it is congruent with the other resolution-based calculi (semantically it is the same as in [18]). The ∀Exp+Res calculus operates on clauses that comprise only existential variables from the original QBF; but additionally, each existential variable x is annotated with a substitution to those universal variables that precede x in the quantification order. For instance, the clause $x \vee b^{0/u}$ can be derived from the original clause $x \vee u$ under the prefix $\exists x \forall u \exists b$.

Besides the aforementioned resolution-based calculi, there is a system by Klieber et al. [23,22], which operates on pairs of sets of literals, rather than clauses; this system is in its workings akin to LD-Q-Res. Van Gelder defines an extension of Q-Res, called *QU-resolution*, which additionally supports resolution over universal variables [32]. Another extension of Q-Res are *variable dependencies* [29,30,31] which enable more flexible ∀-reduction than traditional Q-Res.

For proofs of true QBFs *term-resolution* was developed [11] or *models* in the form of Boolean functions [21] but those do not provide polynomially-verifiable proof systems. Some limitations of term-resolution were shown by Janota et al. [15]. A comparison of sequent calculi [24] and Q-Res was done by Egly [9].

3 Instantiation-Based Calculi IR-calc and IRM-calc

We begin by setting up a framework allowing us to define our new calculi. The framework hinges on the concept of annotated clauses. An *extended assignment* is a partial mapping from the boolean variables to $\{0, 1, *\}$. An *annotated clause* is a clause where each literal is annotated by an extended assignment to universal variables. For an extended assignment σ to universal variables we write $l^{[\sigma]}$ to denote an annotated literal where $[\sigma] = \{c/u \in \sigma \mid \mathsf{lv}(u) < \mathsf{lv}(l)\}$. Two (extended) assignments τ and μ are called *contradictory* if there exists a variable $x \in \mathsf{dom}(\tau) \cap \mathsf{dom}(\mu)$ with $\tau(x) \neq \mu(x)$.

Further we define operations that let us modify annotations of a clause by *instantiation*. For (extended) assignments τ and μ, we write $\tau \veebar \mu$ for the assignment σ defined as follows: $\sigma(x) = \tau(x)$ if $x \in \mathsf{dom}(\tau)$, otherwise $\sigma(x) = \mu(x)$ if $x \in \mathsf{dom}(\mu)$. The operation $\tau \veebar \mu$ is referred to as *completion* because μ provides values for variables that are not defined in τ. The operation is associative and therefore we can omit parentheses. In contrast, it is *not* commutative. The following properties hold: (i) For non-contradictory μ and τ, we have $\mu \veebar \tau = \tau \veebar \mu = \mu \cup \tau$. (ii) $\tau \veebar \tau = \tau$.

We consider an auxiliary function $\mathsf{inst}(\tau, C)$, which for an extended assignment τ and an annotated clause C returns $\{l^{[\sigma \veebar \tau]} \mid l^\sigma \in C\}$.

Our first new system IR-calc operates on clauses annotated with usual assignments with range $\{0, 1\}$. The calculus introduces clauses from the matrix and allows to instantiate and resolve clauses; hence the name IR-calc. It comprises the rules in Figure 4.

$$\frac{}{\left\{x^{[\tau]} \mid x \in C, x \text{ is existential}\right\}} \text{ (Axiom)}$$

C is a non-tautological clause from the matrix. $\tau = \{0/u \mid u \text{ is universal in } C\}$, where the notation $0/u$ for literals u is shorthand for $0/x$ if $u = x$ and $1/x$ if $u = \neg x$.

$$\frac{x^\tau \vee C_1 \qquad \neg x^\tau \vee C_2}{C_1 \cup C_2} \text{ (Resolution)} \qquad\qquad \frac{C}{\mathsf{inst}(\tau, C)} \text{ (Instantiation)}$$

τ is an assignment to universal variables with $\mathsf{rng}(\tau) \subseteq \{0, 1\}$.

Fig. 4. The rules of IR-calc

Axiom and instantiation rules as in IR-calc in Figure 4.

$$\frac{x^{\tau \cup \xi} \vee C_1 \qquad \neg x^{\tau \cup \sigma} \vee C_2}{\mathsf{inst}(\sigma, C_1) \cup \mathsf{inst}(\xi, C_2)} \text{ (Resolution)}$$

$\mathsf{dom}(\tau)$, $\mathsf{dom}(\xi)$ and $\mathsf{dom}(\sigma)$ are mutually disjoint. $\mathsf{rng}(\tau) = \{0, 1\}$

$$\frac{C \vee b^{\mu} \vee b^{\sigma}}{C \vee b^{\xi}} \text{ (Merging)}$$

$\mathsf{dom}(\mu) = \mathsf{dom}(\sigma)$. $\xi = \{c/u \mid c/u \in \mu, c/u \in \sigma\} \cup \{*/u \mid c/u \in \mu, d/u \in \sigma, c \neq d\}$

Fig. 5. The rules of IRM-calc

Our second system IRM-calc is an extension of IR-calc where we allow extended assignments with range $\{0, 1, *\}$. To introduce $*$ we include a new rule called *merging*. IRM-calc is defined in Figure 5. The resolution rule can now deal with $*$, but when $\sigma = \xi = \emptyset$ we have exactly the resolution rule from Figure 4.

Example 1. Consider the (true) QBF $\exists x \forall u w \exists b. (x \vee u \vee b) \wedge (\neg x \vee \neg u \vee b) \wedge (u \vee w \vee \neg b)$. In both calculi axioms yield $x \vee b^{0/u}$, $\neg x \vee b^{1/u}$, and $\neg b^{0/w, 0/u}$. In IR-calc we resolve to get $b^{0/u} \vee b^{1/u}$. IRM-calc further derives $b^{*/u}$ by merging. Intuitively, $b^{0/u} \vee b^{1/u}$ means that the existential player must play so that for any assignment to w either $b = 1$ if $u = 0$, or $b = 0$ if $u = 1$. So for instance, the player might choose to play $b = 1$ if $w = 0$ and $u = 1$, and if $w = 1$ and $u = 0$. The clause $b^{*/u}$ can be seen as a shorthand for the clause $b^{0/u} \vee b^{1/u}$. Note that it would be *unsound* to derive the clause b (with no annotation). This would mean that b must be 1 regardless of the moves of the universal player. However, b needs to be 0 when $u = w = 0$ due to the third axiom. ▲

Note that in $\forall$Exp+Res, propositional variables are introduced so that their annotations assign *all* relevant variables. Like so each literal corresponds to a value of a Skolem function in a specific point. In contrast, in IR-calc, variables are annotated "lazily", i.e. it enables us to reason about multiple points of Skolem functions at the same time. This is analogous to *specializaion* of free variables by constants in first-order logic (FOL). Similarly, resolution in IR-calc is analogous to resolution in Robinson's FOL resolution [28]. IRM-calc additionally enables "compressing" literals with contradictory annotations.

4   Soundness and Extraction of Winning Strategies

The purpose of this section is twofold: show how to obtain a *winning strategy* for the universal player given an IRM-calc proof, and, to show that IRM-calc is *sound* (and therefore also IR-calc). First we show how to obtain a winning strategy for the universal player from a proof. From this, the soundness of the calculus follows because a QBF is false if and only if such strategy exists.

The approach we follow is similar to the one used for Q-Res [13] or LD-Q-Res [10]. Consider a QBF $\Gamma = \exists E \forall U. \Phi$, where E and U are sets of variables and Φ is a QBF (potentially with further quantification). Let π be an IRM-calc refutation of Γ, and let ϵ be a total assignment to E. The assignment ϵ represents a move of the existential player. Reduce π to a refutation π_ϵ of $\forall U. \Phi|_\epsilon$. To obtain a response of the universal player, we construct an assignment μ to the variables U such that reducing π_ϵ by μ gives a refutation of $\Phi|_{\epsilon \cup \mu}$.

Let $\pi_{\epsilon,\mu}$ be the proof resulting from reducing π_ϵ by μ. The game continues with $\phi|_{\epsilon \cup \mu}$ and $\pi_{\epsilon,\mu}$. In each of these steps, two quantifier levels are removed from the given QBF and a refutation for each of the intermediate formulas is produced. This guarantees a winning strategy for the universal player because in the end the existential player will be faced with an unsatisfiable formula without universal variables. We follow this notation for the rest of the section.

To reduce a refutation π by the existential assignment ϵ, we reduce the leaves of π by ϵ and repeat the steps of π with certain modifications. Instantiation steps are repeated with no discrimination. Merging is repeated in the reduced proof unless either of the merged literals is not in the reduced clause and then the clause is left as it is. Whenever a resolution step is possible, repeat it in the reduced proof. If it is not possible, the resolvent in the reduced proof is obtained from the antecedent that is not $\top$ and does *not* contain the pivot literal. If such does not exist, the resolvent is marked as $\top$ (effectively removing it from the proof). When producing a resolvent from a single antecedent, additional instantiation is required. This instantiation is the same one as done by the original resolution step but any $*$ is replaced by 0 (indeed, we can choose the constant arbitrarily). Like so, domains of annotations are preserved. In the end, any clauses marked as $\top$ are removed.

To obtain an assignment to the variables U, collect all the assignments μ to U appearing in annotations in π_ϵ; any variable not appearing in π_ϵ is given an arbitrary value. To obtain $\pi_{\epsilon,\mu}$, remove occurrences of U-variables from the annotation in the proof π_ϵ. This will leave us with a valid refutation because we will show in Theorem 3 that for each variable in U only a single value constant annotation can appear in the entire proof π_ϵ.

To show that this procedure is correct, we need to argue that the reduction returns a valid IRM-calc refutation π_ϵ, and that π_ϵ does not contain annotations giving contradictory values to variables in U. We start with the first claim.

Lemma 2. *The above reduction yields a valid IRM-calc refutation π_ϵ of $\forall U. \Phi|_\epsilon$.*

We omit the proof, which proceeds by induction on the derivation depth.

Lemma 3. *Let π be an IRM-calc refutation of a QBF formula starting with a block of universally quantified variables U. Consider the set of annotations μ on variables U that appear anywhere in π. Then μ is non-contradictory and does not contain instances of $*$.*

Proof. The proof proceeds by induction on the derivation depth. Let μ_C denote the set of annotations to variables in U appearing anywhere in the derivation of

C (i.e., we only consider the connected component of the proof dag with sink C). The induction hypothesis states:

(i) The set μ_C is non-contradictory.
(ii) For every literal $l^\sigma \in C$, it holds that $\mu_C \subseteq \sigma$.
(iii) $*/u \notin \mu_C$, for any $u \in U$.

Base Case. Condition (i) is satisfied by the axioms because we are assuming there are no complementary literals in clauses in the matrix. Condition (ii) is satisfied because all existential literals are at a higher level than the variables of U. Condition (iii) holds because we do not instantiate by $*/u$ in the axiom rule.

Instantiation. Let $u \in U$ and $C = \mathsf{inst}(c/u, C')$ in the proof π. By induction hypothesis, u either appears in the annotations of all the literals l^ξ in C' or it does not appear in any of them. In the first case, the instantiation step is ineffective. In the second case, c/u is added to all literals in C. By induction hypothesis u does not appear in any annotation of any clause in the sub-proof deriving C', and hence C is the first clause containing u.

Resolution. Let C be derived by resolving $x^{\tau \cup \xi} \vee C_1$ and $\neg x^{\tau \cup \sigma} \vee C_2$. Let $u \in U$, consider the following cases.

Case 1. For some $c \in \{0, 1\}$, $c/u \in \sigma$ and $u \notin \mathsf{dom}(\xi)$. By induction hypothesis, u does not appear in the annotations of C_1. Hence $\mathsf{inst}(\sigma, C_1)$ adds c/u to all the annotations in C_1.

Case 2. $c/u \in \tau$. By induction hypothesis, c/u appears in all annotations of C_1, C_2 and hence in all annotations of the resolvent.

Case 3. $u \notin \mathsf{dom}(\tau) \cup \mathsf{dom}(\sigma) \cup \mathsf{dom}(\xi)$. Then u does not appear as annotation anywhere in the derivation of either of the antecedents and neither it will appear in the resolvent.

Merging. Because of (i) we do not obtain $*$ for variables in U.   □

Therefore we obtain winning strategies:

Theorem 4. *The construction above yields a winning strategy for the universal player.*

The soundness of IRM-calc follows directly from Theorem 4.

Corollary 5. *The calculi IR-calc and IRM-calc are sound.*

5   Completeness and Simulations of Known QBF Systems

In this section we prove that our calculi simulate the main existing resolution-based QBF proof systems. As a by-product, this also shows completeness of our proof systems IR-calc and IRM-calc. We start by simulating Q-resolution, which is even possible with our simpler calculus IR-calc.

Theorem 6. *IR-calc p-simulates Q-Res.*

Proof (Sketch). Let $C_1, \ldots, C_k$ be a Q-Res proof. We translate the clauses into $D_1, \ldots, D_k$, which will form the skeleton of a proof in IR-calc.

- For an axiom C_i in Q-Res we introduce the same clause D_i by the axiom rule of IR-calc, i.e., we remove all universal variables and add annotations.
- If C_i is obtained via $\forall$-reduction from C_j, then $D_i = D_j$.
- Consider now the case that C_i is derived by resolving C_j and C_k with pivot variable x. Then $D_j = x^\tau \vee K_j$ and $D_k = x^\sigma \vee K_k$. We instantiate to get $D'_j = \mathsf{inst}(\sigma, D_j)$ and $D'_k = \mathsf{inst}(\tau, D_k)$. Define D'_i as the resolvent of D'_j and D'_k. In order to obtain D_i we must ensure that there are no identical literals with different annotations. For this consider the set $\zeta = \{c/u \mid c/u \in t, l^t \in D'_i\}$ and define $D_i = \mathsf{inst}(\zeta, D'_i)$. This guarantees that we will always have fewer literals in D_i than in C_i, and we get a refutation.

We have to prove that the resolution steps are valid, by showing that τ and σ are not contradictory and ζ does not contain contradictory annotations. This follows from the next claim, which can be proven by induction (omitted here).

Claim. For all existential literals l we have $l \in C_i$ iff $l^t \in D_i$ for some annotation t. Additionally, if $0/u \in t$ for a literal u, then $u \in C_i$ (where for a variable x, we equivalently denote the annotation $1/x$ by $0/\neg x$). $\quad\square$

Despite its simplicity, IR-calc is powerful enough to also simulate the expansion based proof system $\forall$Exp+Res from [18].

Theorem 7. *IR-calc p-simulates $\forall$Exp+Res.*

Proof. Let $C_1, \ldots, C_k$ be an $\forall$Exp+Res proof. We transform it into an IR-calc proof $D_1, \ldots, D_k$ as follows. If C_i is an axiom from clause C and assignment τ we construct D_i by taking the axiom in IR-calc of C and then instantiating with $\mathsf{inst}(\tau, C)$. If C_i is derived by resolving C_j, C_k over variable x^τ, then D_i is derived by resolving D_j, D_k over variable x^τ. This yields a valid IR-calc proof because $l^t \in D_i$ iff $l^t \in C_i$, which is preserved under applications of both rules. $\quad\square$

We now come to the simulation of a more powerful system than Q-resolution, namely LD-Q-Res from [1]. We show that this system is simulated by IRM-calc. The proof uses a similar, but more involved technique as in Theorem 6.

Theorem 8. *IRM-calc p-simulates LD-Q-Res.*

Proof (Sketch). Consider an LD-Q-Res refutation $C_1, \ldots, C_n$. We construct clauses $D_1, \ldots, D_n$, which will form the skeleton of the IRM-calc proof. The construction will preserve the following four invariants for $i = 1, \ldots, n$.

(1) For an existential literal l, it holds that $l \in C_i$ iff $l^t \in D_i$ for some t.
(2) The clause D_i has no literals l^{t_1} and l^{t_2} such that $t_1 \neq t_2$.
(3) If $l^t \in D_i$ with $0/u \in t$, then $u \in C_i$ or $u^* \in C_i$, likewise if $l^t \in D_i$ with $1/u \in t$, then $\neg u \in C_i$ or $u^* \in C_i$.
(4) If $l^t \in D_i$ with $*/u \in t$, then $u^* \in C_i$.

The actual construction proceeds as follows. If C_i is an axiom, D_i is constructed by the axiom rule from the same clause. If C_i is a $\forall$-reduction of C_j with $j < i$, then we set D_i equal to D_j. If C_i is obtained by a resolution step from C_j and C_k with $j < k < i$, the clause D_i is obtained by a resolution step from D_j and D_k, yielding clause K, and by performing some additional steps on K. Firstly, we let $\theta = \{c/u \mid c \in \{0, 1\}, c/u \in t, l^t \in K\} \cup \{0/u \mid */u \in t, l^t \in K\}$ and perform instantiation on K by substitutions in θ, in any order, to derive K'. Like so, all annotations in K' have the same domain. We merge all pairs of literals $l^\sigma, l^\tau \in K'$ with $\tau \neq \sigma$ (in any order) to derive D_i.

To show that this construction yields a valid IRM-calc refutation, we first need to prove the invariants above. This proceeds by induction on i. We omit the base case and the $\forall$-reduction and just sketch the case of a resolution step.

For this consider C_j, C_k being resolved in LD-Q-Res to obtain C_i. As only the resolved variable is removed, which is removed completely due to condition (2), D_i fulfills (1). By induction hypothesis we know that there can be at most two copies of each variable when we derive K. Their annotations have the same domain in K', because instantiation by θ applies the entire domain of all annotations in the clause to all its literals. It then follows that all copies of identical literals are merged into one literal in D_i. Therefore (2) holds for D_i.

To prove (3) consider the case where $l^t \in D_i$ with $0/u \in t$. The case with $1/u \in t$ is analogous. We know that $0/u$ appearing in D_i means that $0/u$ must appear in K' as merging cannot produce a new annotation $0/u$. Existence of $0/u$ in K' means that either $*/u$ appears in K or $0/u$ appears in K. No new annotations are created in a resolution step, so either $*/u$ or $0/u$ must appear in one or more of D_j, D_k. By induction hypothesis this means that u or u^* appears in $C_j \cup C_k$, hence also in C_i.

To show condition (4), let $l^t \in D_i$ with $*/u \in t$. Then either $*/u$ is present in K', or $0/u$ and $1/u$ are present in K' and will be merged. In the first case it is clear that some $*/u$ annotation appears in K and thus in D_j or in D_k, in which case from (4) of the induction hypothesis u^* must appear in C_i. In the second case it is possible that $0/u$ in K' was obtained from $*/u$ in K. Thus as already argued, u^* must appear in C_i. If instead $1/u, 0/u$ are both present in K then they must come from the original clauses D_j, D_k. If they both appear in the same clause D_j, then by condition (3) it must be the case that u^* appears in C_j and thus in C_i. If, however, they appear in different clauses, then by (3) either of the clauses C_j, C_k contains u^* or they contain literals over u of opposite polarity. Both situations merge the literals to $u^* \in C_i$.

We now show that these invariants imply that we indeed obtain a valid IRM-calc proof. We only need to consider the resolution steps. Suppose $x^{t_1} \in D_j$ and $\neg x^{t_2} \in D_k$ where C_j and C_k are resolved on x to get C_i in the LD-Q-Res proof. To perform the resolution step between D_j and D_k we need to ensure that we do not have $c/u \in t_1, d/u \in t_2$ where $c \neq d$ or $c = d = *$. Assume on the contrary that $*/u \in t_1$ and $c/u \in t_2$. By (4) we have $u^* \in C_j$, and by (3) some literal of u is in C_k. But as $\mathsf{lv}(u) < \mathsf{lv}(x)$ the LD-resolution of C_j and C_k on variable x is forbidden, giving a contradiction. Similarly, if there is $0/u \in t_1$ and $1/u \in t_2$, then either we get the same situation or we have two opposite literals of u in the different clauses C_j, C_k. In either case the resolution of C_j, C_k is forbidden. Hence the IRM-calc proof is correct.

It is not difficult to see that the IRM-calc proof is indeed a refutation and all steps of the construction can be performed in polynomial time, thus we obtain a p-simulation. □

6 Conclusion

This paper introduces two novel calculi for quantified Boolean formulas. Both of these calculi are anchored in a common framework of *annotated clauses*. The first calculus, IR-calc, provides the rules of resolution and instantiation of clauses. The second calculus, IRM-calc, additionally enables *merging* literals with contradictory annotations. The paper demonstrates that the simple calculus IR-calc already p-simulates Q-resolution and the expansion-based system ∀Exp+Res. The extended version IRM-calc additionally p-simulates long-distance Q-resolution. The paper further demonstrates that refutations in the introduced calculi enable generation of winning strategies of the universal player—a favorable property from a practical perspective [1].

The contribution of the paper is both practical and theoretical. From a practical perspective, a calculus unifying the existing calculi for QBF enables a uniform certification of off-the-shelf QBF solvers. From a theoretical perspective, a unifying calculus provides an underpinning necessary for complexity characterizations of existing solvers as well as for furthering our understanding of the strengths of the underlying proof systems.

Acknowledgments. This work was supported by FCT grants ATTEST (CMU-PT-/ELE/0009/2009), POLARIS (PTDC/EIA-CCO/123051/2010), INESC-ID's multiannual PIDDAC funding PEst-OE/EEI/LA0021/2013, grant no. 48138 from the John Templeton Foundation, and a Doctoral Training Grant from EPSRC (2nd author).

References

1. Balabanov, V., Jiang, J.H.R.: Unified QBF certification and its applications. Formal Methods in System Design 41(1), 45–65 (2012)
2. Beame, P., Kautz, H.A., Sabharwal, A.: Towards understanding and harnessing the potential of clause learning. J. Artif. Intell. Res. (JAIR) 22, 319–351 (2004)
3. Benedetti, M.: Evaluating QBFs via symbolic Skolemization. In: Baader, F., Voronkov, A. (eds.) LPAR 2004. LNCS (LNAI), vol. 3452, pp. 285–300. Springer, Heidelberg (2005)
4. Benedetti, M., Mangassarian, H.: QBF-based formal verification: Experience and perspectives. JSAT 5(1-4), 133–191 (2008)
5. Biere, A.: Resolve and expand. In: Hoos, H.H., Mitchell, D.G. (eds.) SAT 2004. LNCS, vol. 3542, pp. 59–70. Springer, Heidelberg (2005)
6. Buss, S.R.: Towards NP-P via proof complexity and search. Ann. Pure Appl. Logic 163(7), 906–917 (2012)

7. Buss, S.R., Hoffmann, J., Johannsen, J.: Resolution trees with lemmas: Resolution refinements that characterize DLL algorithms with clause learning. Logical Methods in Computer Science 4(4) (2008)
8. Cook, S.A., Reckhow, R.A.: The relative efficiency of propositional proof systems. J. Symb. Log. 44(1), 36–50 (1979)
9. Egly, U.: On sequent systems and resolution for QBFs. In: Cimatti, A., Sebastiani, R. (eds.) SAT 2012. LNCS, vol. 7317, pp. 100–113. Springer, Heidelberg (2012)
10. Egly, U., Lonsing, F., Widl, M.: Long-distance resolution: Proof generation and strategy extraction in search-based QBF solving. In: McMillan, K., Middeldorp, A., Voronkov, A. (eds.) LPAR-19 2013. LNCS, vol. 8312, pp. 291–308. Springer, Heidelberg (2013)
11. Giunchiglia, E., Narizzano, M., Tacchella, A.: Clause/term resolution and learning in the evaluation of quantified Boolean formulas. JAIR 26(1), 371–416 (2006)
12. Giunchiglia, E., Marin, P., Narizzano, M.: Reasoning with quantified boolean formulas. In: Handbook of Satisfiability, pp. 761–780. IOS Press (2009)
13. Goultiaeva, A., Van Gelder, A., Bacchus, F.: A uniform approach for generating proofs and strategies for both true and false QBF formulas. In: IJCAI (2011)
14. Hertel, P., Bacchus, F., Pitassi, T., Van Gelder, A.: Clause learning can effectively p-simulate general propositional resolution. In: AAAI (2008)
15. Janota, M., Grigore, R., Marques-Silva, J.: On QBF proofs and preprocessing. In: McMillan, K., Middeldorp, A., Voronkov, A. (eds.) LPAR-19 2013. LNCS, vol. 8312, pp. 473–489. Springer, Heidelberg (2013)
16. Janota, M., Klieber, W., Marques-Silva, J., Clarke, E.: Solving QBF with counterexample guided refinement. In: Cimatti, A., Sebastiani, R. (eds.) SAT 2012. LNCS, vol. 7317, pp. 114–128. Springer, Heidelberg (2012)
17. Janota, M., Marques-Silva, J.: ∀Exp+Res does not P-Simulate Q-resolution. In: International Workshop on Quantified Boolean Formulas (2013)
18. Janota, M., Marques-Silva, J.: On propositional QBF expansions and Q-resolution. In: Järvisalo, M., Van Gelder, A. (eds.) SAT 2013. LNCS, vol. 7962, pp. 67–82. Springer, Heidelberg (2013)
19. Kleine Büning, H., Bubeck, U.: Theory of quantified boolean formulas. In: Handbook of Satisfiability, pp. 735–760. IOS Press (2009)
20. Kleine Büning, H., Karpinski, M., Flögel, A.: Resolution for quantified Boolean formulas. Inf. Comput. 117(1), 12–18 (1995)
21. Kleine Büning, H., Subramani, K., Zhao, X.: Boolean functions as models for quantified boolean formulas. J. Autom. Reasoning 39(1), 49–75 (2007)
22. Klieber, W., Janota, M., Marques-Silva, J., Clarke, E.: Solving QBF with free variables. In: Schulte, C. (ed.) CP 2013. LNCS, vol. 8124, pp. 415–431. Springer, Heidelberg (2013)
23. Klieber, W., Sapra, S., Gao, S., Clarke, E.: A non-prenex, non-clausal QBF solver with game-state learning. In: Strichman, O., Szeider, S. (eds.) SAT 2010. LNCS, vol. 6175, pp. 128–142. Springer, Heidelberg (2010)
24. Krajíček, J., Pudlák, P.: Quantified propositional calculi and fragments of bounded arithmetic. Mathematical Logic Quarterly 36(1), 29–46 (1990)
25. Marques Silva, J.P., Lynce, I., Malik, S.: Conflict-driven clause learning SAT solvers. In: Handbook of Satisfiability, IOS Press (2009)
26. Pipatsrisawat, K., Darwiche, A.: On the power of clause-learning SAT solvers as resolution engines. Artif. Intell. 175(2), 512–525 (2011)
27. Rintanen, J.: Asymptotically optimal encodings of conformant planning in QBF. In: AAAI, pp. 1045–1050. AAAI Press (2007)

28. Robinson, J.A.: A machine-oriented logic based on the resolution principle. J. ACM 12(1), 23–41 (1965)
29. Samer, M., Szeider, S.: Backdoor sets of quantified Boolean formulas. J. Autom. Reasoning 42(1), 77–97 (2009)
30. Slivovsky, F., Szeider, S.: Variable dependencies and Q-Resolution. In: International Workshop on Quantified Boolean Formulas (2013)
31. Van Gelder, A.: Variable independence and resolution paths for quantified Boolean formulas. In: Lee, J. (ed.) CP 2011. LNCS, vol. 6876, pp. 789–803. Springer, Heidelberg (2011)
32. Van Gelder, A.: Contributions to the theory of practical quantified Boolean formula solving. In: Milano, M. (ed.) CP 2012. LNCS, vol. 7514, pp. 647–663. Springer, Heidelberg (2012)
33. Zhang, L., Malik, S.: Conflict driven learning in a quantified Boolean satisfiability solver. In: ICCAD, pp. 442–449 (2002)

Minimum Planar Multi-sink Cuts
with Connectivity Priors*

Ivona Bezáková and Zachary Langley

Rochester Institute of Technology, Rochester, NY, USA
{ib,zbl9222}@cs.rit.edu

Abstract. Given is a connected positively weighted undirected planar graph G embedded in the plane, a source vertex s, and a set of sink vertices T. An (s,T)-cut in G corresponds to a cycle or a collection of edge-disjoint cycles in the planar dual graph G^* that define a planar region containing s but not T. A cut with a connectivity prior does not separate the vertices in T from each other: we focus on the most natural prior where the cut corresponds to a (simple, i.e., no repeated vertices) cycle in G^*. We present an algorithm that finds a minimum simple (s,T)-cut in $O(n^4)$ time for n vertices. To the best of our knowledge, this is the first polynomial-time algorithm for minimum cuts with connectivity priors. Such cuts have applications in computer vision and medical imaging.

1 Introduction

We address the problem of finding a minimum simple single-source-multi-sink cut in a positively weighted undirected planar graph $G = (V, E, w)$ embedded in the plane. In particular, given a source vertex s, and a set of sink vertices T, a cut $S \subseteq V$ is said to be *a simple (s,T)-cut* if S contains s and does not contain any vertex in T and the dual edges of the cut edges $\{(u,v) \mid u \in S, v \notin S\}$ form a simple cycle, i.e., no repeated vertices, in the dual graph. We present an $O(n^4)$ algorithm that finds a simple (s,T)-cut of the smallest weight in a positively weighted planar graph with n vertices. For a small example, see Figure 1(b).

Graph cuts are an important algorithmic tool in computer vision, see, e.g., [7,6,13]. For example, in the simplest form of image segmentation, a user is asked to identify a point (seed) inside an object and in the background. Viewing the input image as a graph, typically the 2D grid of pixels with edge weights representing (dis)similarity of neighboring pixels, a natural segmentation approach isolates the object by identifying a minimum cut between the two seeds. However, if the object contains thin parts (for example, if trying to isolate a vein on an ultrasound image), it is likely that a minimum cut will opt to sever the thin parts

---

* This material is based upon work supported by the National Science Foundation, Award No. CCF-1319987. Part of the work was done while the first author visited the Simons Institute for the Theory of Computing at the University of California, Berkeley.

E. Csuhaj-Varjú et al. (Eds.): MFCS 2014, Part II, LNCS 8635, pp. 94–105, 2014.
© Springer-Verlag Berlin Heidelberg 2014

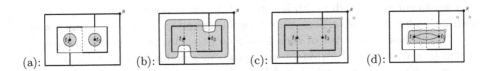

Fig. 1. (a) Minimum (s, T)-cut, weight 4ε: thick edges are of weight ∞, the five dashed edges of weight ε, and the other four edges of weight 1. (b) Minimum simple (s, T)-cut, weight $4 + 2\varepsilon$. (c)-(d) Common pitfalls involving shortest paths between sinks in the dual graph: (c) Merge all involved faces into a single face f and compute minimum (s, f)-cut; here results in weight 2∞. (d) Cut the plane along the shortest paths, find a shortest cycle that separates s from the "cut-out" face; here results in a non-simple cut (the middle two faces are visited twice). Small circles denote dual vertices.

from the object, see, e.g., [13]. A typical solution is to ask the user to identify multiple seeds inside the object. This might still result in the cut set containing several disconnected regions, as seen in the example in Figure 1(a). The problem can be remedied by enforcing a connectivity prior on the cut [13,6,14]. In other words, we do not look for a minimum cut but instead look for a smallest cut that somehow "connects" the seeds inside the object. Arguably the most natural connectivity prior is to require both the cut set as well as its complement to induce a connected graph. For connected planar inputs, this corresponds to finding a simple dual cycle that separates the seeds inside the object from the seed(s) outside. We also briefly discuss a connectivity prior where the cut corresponds to a non-self-crossing tour. While we admit that our running time is prohibitive for large inputs, we note that several applications first preprocess the input by contracting subcomponents, obtaining a much smaller graph. Our algorithms are to the best of our knowledge the first provably polynomial algorithms for minimum (s, T)-cut problems with connectivity priors.

The algorithms are based on a dynamic programming approach along a dual shortest path tree connecting the sinks. We first observe that there is a minimum simple (s, T)-cut such that its corresponding dual cycle does not cross this tree. However, the cycle might touch the tree arbitrarily many times, even along the same tree branch from both sides. In such case we form the cycle by concatenating paths that connect pairs of vertices on the tree. The tricky part is to ensure that the paths separate the source from the sinks and that the concatenation results in a simple cycle.

Instead of exactly computing the minimum length of such separating paths, we merely *bound* it to speed up the algorithm. In particular, we either get the shortest possible length, or the quantity we compute is smaller than the shortest length yet larger than the value of a minimum (s, T)-cut in which case the quantity will be eventually eliminated from the minimum cut computation. To guarantee that the cycle does not cross the tree, we "cut" the plane along the tree, thus preventing paths from crossing it. This involves duplicating each tree vertex as many times as is its degree in the tree, with no edges between the copies of the same vertex. The most challenging aspect of the computation is to ensure that we do not go through multiple copies of the same vertex – this is a problem with forbidden pairs of vertices, searching for a shortest cycle separating s and T, using at most one vertex of each forbidden pair. Our polynomial-time result contrasts with a related forbidden pair problem searching for a shortest cycle through a given vertex s, which is NP-hard even if the graph is planar and all forbidden pairs are on the outerface [8].

We remark that two natural, fast, and seemingly working approaches based on finding shortest sink-sink paths in the dual graph and contracting/merging components along the paths do not yield correct algorithms for this problem; see Figure 1(c)-(d).

We note that the single-source-multi-sinks problem in directed planar graphs remains open as our techniques are specific to undirected graphs since we expect to be able to traverse a path in both directions.

Related Works. The case of a single sink t has been extensively studied as any minimum (s, t)-cut is simple. The latest algorithms run in time $O(n \log \log n)$ for undirected graphs due to Łącki and Sankowski [12], and in $O(n \log n)$ time for directed graphs, see, e. g., Borradaile and Klein [4] and the references within. Chalermsook, Fakcharoenphol, and Nanongkai [9] gave an $O(n \log^2 n)$ algorithm that finds an overall minimum cut (no pre-specified vertices to separate).

For multiple sinks, Chambers, de Verdière, Erickson, Lazarus, and Whittlesey [11] showed that the problem of finding a minimum simple multi-source-multi-sink cut in a planar graph is NP-hard (see the proof of Theorem 3.1). Bienstock and Monma [3] designed a polynomial-time algorithm that finds a shortest circuit separating a set of vertices from the outerface.

Several recent papers address various "multi" cut and flow problems in planar graphs. Bateni, Hajiaghayi, Klein, and Mathieu [1] designed a PTAS to approximate the weight of a minimum multiway cut where a set of terminals needs to be separated from each other. Borradaile, Klein, Mozes, Nussbaum, and Wulff-Nilsen [5] gave a near-linear algorithm for the multi-source-multi-sink flow problem.

For surfaces with genus g, Chambers, Erickson, and Nayyeri [10] designed $g^{O(g)} n \log n$ algorithms for finding minimum single-source-single-sink cuts in surface-embedded graphs. Chambers et al. [11] showed that the problem of finding a shortest splitting (i. e., simple and not homeomorphic to a disk) cycle is NP-hard but fixed parameter tractable with respect to g. Cabello [8] studied shortest contractible (i. e., homotopic to a constant function) and shortest separating (i. e., that split the surface into two connected components) cycles in embedded graphs, showing that a shortest contractible cycle can be found in polynomial time and the shortest separating cycle problem is NP-hard.

In a recent work [2], we investigated the existence and counting of "contiguous" cuts among all minimum single-source-multi-sink cuts in planar graphs. Unfortunately, the results heavily rely on the max-flow min-cut duality and do not extend to problems that go beyond minimum cuts.

2 Preliminaries

Let $G = (V, E, w)$ be a connected positively weighted undirected planar graph embedded in the plane. Its dual (multi-)graph $G^* = (V^*, E^*, w^*)$ is defined as follows. V^* is the set of faces of G. For every edge $e \in E$ bordering faces f_1, f_2, the set E^* contains the edge $e^* = (f_1, f_2)$ of the same weight as e, i.e., $w^*(e^*) = w(e)$. The planar embedding of G yields the corresponding planar embedding of G^*. For a given vertex $s \in V$ and a set of vertices $T \subseteq V$, $s \notin T$, an (s, T)-cut is a set of vertices $S \subseteq V$ such that $s \in S$ and $S \cap T = \emptyset$. The edges cut by the cut, i.e., $\{(u, v) \in E \mid u \in S, v \notin S\}$, correspond to dual edges $C = \{(u, v)^* \mid u \in S, v \notin S\}$. If C forms a simple cycle (no repeated vertices in the dual graph), then we say that the (s, T)-cut is *simple* or a *bond*. We allow C to be of length 1 (a dual self-loop), as well as of length 2 (formed by two different dual edges). The *weight* of an (s, T)-cut is the sum of the edge weights cut by the cut, i.e., $\sum_{(u,v) \in E : u \in S, v \notin S} w(u, v)$. For simple (s, T)-cuts this directly corresponds to the length of the corresponding dual cycle.

Observation 1. *A simple (s, T)-cut exists if and only if, after removing s from G, the sinks are still connected.*

Therefore, we assume that a simple (s, T)-cut exists. We also assume that $|T| > 1$ since for $|T| = 1$ any minimum (s, T)-cut is simple. Moreover, we assume that the degree of every vertex is > 1, as vertices of degree 1 can be merged with their neighbor. By a "cycle" or a "path" we mean a simple cycle or a simple path (no repeated vertices). If p is a path or a cycle, we denote by $|p|$ its length, i.e., the sum of the weights of edges on p. As we work with embedded graphs, we abuse terminology and identify a path or a cycle with the corresponding curve.

Let α and β be (possibly closed) non-self-intersecting curves in the plane. We say that α and β *cross* if there is a maximal curve χ (possibly just a point x) in $\alpha \cap \beta$ such that β continues on different sides of α at the end-points of χ; we refer to such χ as a cross segment. If α and β do not cross at χ, they *touch* at χ. If $\alpha \cap \beta \neq \emptyset$ and they share more than just their end-points, we say that α and β *intersect*.

Let G_0 be a graph obtained from G by enlarging each $u \in T \cup \{s\}$ into a small new face u^* with ∞-weighted edges bounding it. Notice that there is a straightforward bijection between simple (s, T)-cuts in G and simple cycles in $G_0^* - \{u^* \mid u \in T \cup \{s\}\}$ that define a planar region containing s^* but no $t^* \in T^*$, where $T^* := \{t^* \mid t \in T\}$. We refer to such cycles in G_0^* as (s^*, T^*)-*separating cycles*. To simplify our language, the remainder of this text takes place in the dual graph G_0^*. In particular, unless otherwise specified, by vertices and edges we mean vertices and edges of G_0^* and by the source and sinks we mean s^* and T^*.

Let c_{opt} be a minimum (s^*, T^*)-separating cycle in G_0^*. In this extended abstract we focus on the computation of the weight $|c_{opt}|$; the algorithm can be extended to obtain the corresponding cut within the same running time.

3   Graph H: Cutting along a Shortest Path Tree

Choose an arbitrary $t_1^* \in T^*$ and build a shortest[1] path tree τ from t_1^* to every $t_2^* \in T^*$, see Figure 2(a). Notice that every leaf of τ is a sink and that no sink nor s^* is an internal vertex of τ due to the ∞-weighted edges. Due to space constraints we omit the proof of the following lemma.

Lemma 1. *There exists a minimum (s^*, T^*)-separating cycle c such that c does not cross τ.*

Remark 1. We could have considered shortest paths between every pair of sinks which, in most cases, would have sped up the algorithm in practice. In the worst case, however, the union of all such paths forms a tree; using a tree simplifies the exposition in the paper.

Suppose we do a clockwise depth-first traversal of τ from t_1^* until we process the entire tree and return back to t_1^*. In a clockwise traversal, the neighbors of every vertex are considered in a clockwise cyclic manner (that is, if we use edge (u_1, v) to get to v, then we continue with edge (v, u_2) where u_2 is the clockwise next neighbor of v after u_1). Let $a_0, a_1, \dots, a_{\ell-1}$ be the sequence of vertices encountered by the traversal, in this order and with repetitions – each vertex appears in the sequence as many times as is its degree in τ. See Figure 2(a).

We associate certain edges of G_0^* with each a_x: let $v_{x,0}, v_{x,1}, \dots, v_{x,d_x+1}$ be the neighbors of a_x listed in the clockwise order from a_{x-1} to a_{x+1}, where $a_{-1} := a_{\ell-1}$ and $a_\ell := a_0$.

We construct a graph H analogous to G_0^* that will prevent us from crossing τ. Intuitively, the graph H corresponds to "cutting" the plane along the tree τ. We remove all vertices in τ and add a new vertex b_x for each a_x; we include edges of G_0^* that do not involve τ, plus edges (b_x, b_{x+1}) for each x, and edges from b_x to each $v_{x,i}$ for $1 \le i \le d_x$. (If $v_{x,i} = a_k$, then let $v_{x,i} := b_k$.) See Figure 2(b). Formally,

$$V(H) = V_0^* - V(\tau) \cup \{b_x \mid x \in \{0, 1, \dots, \ell-1\}\},$$
$$E(H) = \{(u, v) \in E_0^* \mid u, v \in V_0^* - V(\tau)\} \cup \{(b_x, b_x + 1) \mid 0 \le x < \ell\} \cup$$
$$\{e_{x,i} := (b_x, v_{x,i}) \mid 1 \le i \le d_x\}.$$

For convenience we define $e_{x,0} := (b_{x-1}, b_x)$ and $e_{x,d_x+1} := (b_x, b_{x+1})$. To simplify our expressions, we define $b_{x+y} := b_{(x+y)} \mod \ell$ for $0 \le x < \ell$ and $y \in \mathbf{Z}$ such that $x+y \notin \{0, 1, \dots, \ell-1\}$. Also, let $B = \{b_0, \dots, b_{\ell-1}\}$. Notice that every sink $t^* \in T^*$ corresponds to a unique a_{x_t} and therefore there is a unique corresponding vertex b_{x_t} in H. We abuse notation and use T^* and s^* to refer to the sinks and the source in H. Also, $b_0, \dots, b_{\ell-1}$ is a cycle in H that bounds a single face f_B.

From now on the entire discussion takes place in H.

By *clockwise distance* from b_x to b_y, $x, y \in \{0, 1, \dots, \ell-1\}$, we mean $y - x$ if $x \le y$ and $\ell + y - x$ if $x > y$. Intuitively, this is the number of vertices,

---

[1] We allow the paths to use two ∞-weighted edges, at the start and at the end.

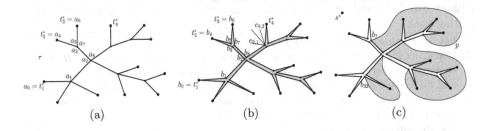

Fig. 2. (a) Tree τ, vertices a_x (notice that $a_2 = a_8$, $a_3 = a_5 = a_7$, etc.). (b) Graph H: vertices b_x and edges $e_{x,0}, \ldots, e_{x,d_x+1}$ (depicted are $e_{9,0} = (b_8, b_9)$, $e_{9,1}$, $e_{9,2}$, $e_{9,3}$, $e_{9,4} = (b_9, b_{10})$); face f_B is shaded. (c) Region R_p, here shown for a b_7-b_{33} source-sinks separating path p.

with repetition, we encounter during the traversal of τ from a_x to a_y. We write $b_x \prec b_y \prec b_z$ for $x, y, z \in \{0, 1, \ldots, \ell - 1\}$ if $x < y < z$, or, if $x > z$, then either $x < y$ or $y < z$. We say that such b_y is *clockwise between* b_x and b_z.

Lemma 2. *Let $b_{i_1}, b_{i_2}, b_{i_3}, b_{i_4}$ be such that $b_{i_1} \prec b_{i_2} \prec b_{i_3} \prec b_{i_4} \prec b_{i_1}$. Then, $a_{i_1} = a_{i_3}$ and $a_{i_2} = a_{i_4}$ cannot be both true.*

Proof. Suppose $a_{i_1} = a_{i_3}$ and suppose that as we traverse τ from a_{i_1}, we encounter a_{i_2} before returning back to $a_{i_1} = a_{i_3}$. Then, by the nature of the depth-first traversal, we must have processed all copies of a_{i_2} before returning back to a_{i_1}. Therefore, $a_{i_2} \neq a_{i_4}$. □

Notice that any (s^*, T^*)-separating cycle in G_0^* that does not cross τ corresponds to a cycle in H that separates s^* from all T^*. The converse is not always true since a cycle in H may visit b_x, b_y where $x \neq y$ and hence $b_x \neq b_y$, yet $a_x = a_y$, leading to a non-simple cycle in G_0^*. We say that a cycle in H is (s^*, T^*)-*separating* if it yields a (simple) (s^*, T^*)-separating cycle in G_0^*.

4 Source-Sinks Separating Paths

If a minimum (s^*, T^*)-separating cycle in G_0^* corresponds to a cycle in H that touches f_B, we will form it by concatenating paths between pairs of vertices in B. We need to be careful about concatenation of intersecting paths, as well as about separating s^* from T^*.

Definition 1. *Let p be a b_x-b_y path in H that does not go through the source or any of the sinks, and, if it goes through a vertex $b_z \in B$, then $b_x \preceq b_z \preceq b_y$. We define the* path-region(s) *R_p as the planar region(s) on the right of p, bounded by p and the clockwise b_x-b_y part of the boundary of f_B. We say that p is source-sinks separating if $s^* \notin R_p$ and that p is no-B if, except for b_x and b_y, it does not go through any vertices in B.*

Algorithm 1. Bounding the length of a shortest $e_{x,i}$-$e_{y,j}$ no-B source-sink separating path

1. Let q be a shortest $e_{y,j}$-s^* path in H.
2. Construct H_q by "cutting" the plane open along q and removing vertices in $B \cup \{s^*\}$. In particular, for every vertex u on $q - \{b_y, s^*\}$, replace it by two new vertices u_1, u_2, and for every v such that $(u, v) \in E(H)$, add edge (u_1, v) if v is on the left of q, or (u_2, v) if it is on the right of q. Additionally, add edges (u_1, u'_1) and (u_2, u'_2) for every (u, u') on q and (v_1, v_2) for every $v_1, v_2 \in V(H) - B - q$ such that $(v_1, v_2) \in E(H)$. Use the same edge weights as in H.
3. For every $u \in q$, compute the shortest distance $\mathrm{dist}_{H_q}[v_{x,i}, u_1]$ from $v_{x,i}$ to u_1 where $e_{x,i} = (b_x, v_{x,i})$. Let $\mathrm{dist}_q[u, b_y]$ be the distance from u to b_y along q.
4. **return** $\beta[e_{x,i}, e_{y,j}] := w(e_{x,i}) + \min_{u \in q - \{b_y, s^*\}} \mathrm{dist}_{H_q}[v_{x,i}, u_1] + \mathrm{dist}_q[u, b_y]$.

The definition is depicted in Figure 2(c). Notice that all the sinks that lie clockwise between b_x and b_y are strictly inside $R_p \cup f_B$.

In this section we bound the length of a shortest b_x-b_y no-B source-sink separating path p that starts with the edge $e_{x,i}$ and ends with the edge $e_{y,j}$. We refer to such paths as $e_{x,i}$-$e_{y,j}$ no-B source-sink separating paths. Algorithm 1 summarizes the computation. It relies on the following lemmas; we omit their proofs due to space constraints. The proof of Lemma 4 is of a similar flavor as the forthcoming proof of Lemma 7. We note that the algorithm works even if $x = y$ (when searching for a no-B cycle c through b_x, the only vertex in B on c).

Lemma 3. Let q be a shortest $e_{y,j}$-s^* path in $H - B \cup \{b_y\}$. There exists a shortest $e_{x,i}$-$e_{y,j}$ no-B source-sink separating path that does not cross q.

Lemma 4. Suppose there exists an $e_{x,i}$-$e_{y,j}$ no-B source-sink separating path in H; let p be a shortest such path. Then, the quantity computed by Algorithm 1 satisfies $\beta[e_{x,i}, e_{y,j}] \leq |p|$. Moreover, if $\beta[e_{x,i}, e_{y,j}]$ holds a numerical value, then $\beta[e_{x,i}, e_{y,j}] = |p|$, or $\beta[e_{x,i}, e_{y,j}] > |c_{opt}|$.

Lemma 5. Algorithm 1 can be implemented in time $O(n)$. Moreover, across different $e_{x,i}$, $e_{y,j}$ pairs, the computation of the length of the shortest $e_{x,i}$-$e_{y,j}$ no-B source-sink separating path can be done in overall time $O(n^2 \log n)$.

5   Minimum (s^*, T^*)-Separating Cycle

If a minimum (s^*, T^*)-separating cycle goes through B, we decompose it into source-sinks separating paths of the following type.

Definition 2. For $b_x, b_y \in B$, $b_x \neq b_y$, and i, j such that $0 \leq i \leq d_x$ and $1 < j \leq d_y + 1$, let $P[x, i, y, j]$ be the set of all b_x-b_y source-sinks-separating paths p in H such that

1. p leaves b_x by an edge $e_{x,i'}$ where $i < i'$,
2. p enters b_y by an edge $e_{y,j'}$ where $j' < j$, and
3. $a_{z_1} \neq a_{z_2}$ for every b_{z_1}, b_{z_2} on p where $b_{z_1} \neq b_{z_2}$.

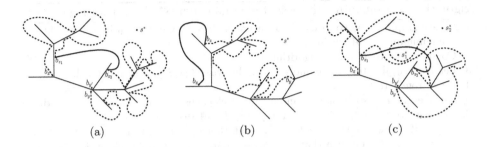

Fig. 3. (a)-(b): Possibilities for a shortest $e_{x,i}$-$e_{y,j}$ source-sinks separating path, schematic view: (a) Case 1: there exists $b_{y'}$ between b_x and b_y such that $a_{y'} = a_y$; the path is split into three subpaths p_1, p_2 (non-dashed, no-B), and p_3. (b) Case 3: there is no such $b_{y'}$ or $b_{x'}$. (c): Proof of Lemma 8, paths found by Algorithm 2 might intersect: Region R'_{s*}. We show two possible locations for s^*: if $s^* = s_1^*$, then s^* is in a region bounded by subpaths of p'_1, p'_2, and p'_3 and $L[x,i,y,j] > |c_{opt}|$; if $s^* = s_2^*$, we get a contradiction with the selection of p'_1, p'_2, and p'_3.

For $b_x = b_y$, $i < j$, let $P[x,i,y,j]$ be the set containing a single path, b_x, of length 0.

In other words, $p \in P[x,i,y,j]$ is a b_x-b_y source-sinks-separating path that leaves b_x by an edge that comes after $e_{x,i}$, enters b_y by an edge that comes before $e_{y,j}$, and it gives rise to a simple path in G_0^* (i. e., repeated vertices are not allowed). Recall also that a b_x-b_y source-sink separating path visits only those vertices in B that are clockwise between b_x and b_y.

We bound the length of a shortest path $p \in P[x,i,y,j]$ using dynamic programming. In particular, we compute $L[x,i,y,j]$ such that $L[x,i,y,j] = |p|$, or $|p| \geq L[x,i,y,j] > |c_{opt}|$. The algorithm, summarized in Algorithm 2, proceeds by gradually increasing the clockwise distance of b_x and b_y. Step 3 deals with the base case, $b_x = b_y$. For the inductive case, we distinguish three possibilities based on the relative position of b_x and b_y, as shown in Figure 3(a)-(b). Consider the $b_x, b_{x+1}, \ldots, b_y$ path in H and the corresponding walk $a_x, a_{x+1}, \ldots, a_y$ in G_0^*. Steps 7-9 (**case 1**) deal with the case when a_y is visited multiple times by the walk, steps 10-12 (**case 2**) with the case when a_x is visited multiple times, and steps 13-14 (**case 3**) with the case when both a_x and a_y are visited exactly once.

The following lemma analyzes the structure of a shortest path p in $P[x,i,y,j]$ and provides rationale for steps 7-9 of the algorithm; we defer analogous lemmas for steps 13-14 and 15 to the full version of the paper. The two subsequent lemmas provide bounds on $L[x,i,y,j]$.

Lemma 6. *Let p be a shortest path in $P[x,i,y,j]$. If b_x, b_y follow case 1, then there exist z_1, k_1, z_2, k_2 satisfying the conditions in step 9 such that p is a concatenation of paths p_1, p_2, and p_3, where $p_1 \in P[x,i,z_1,k_1]$, p_2 is an e_{z_1,k_1}-e_{z_2,k_2} no-B source-sinks-separating path, and $p_3 \in P[z_2,k_2,y,j]$.*

Algorithm 2. Computing the weight of a minimum (s^*, T^*)-separating cycle

---

1. Create graph H_b by contracting B into a single vertex b. Let L_0 be the weight of the minimum (s^*, b)-cut in H_b.
2. Compute $\beta[]$ for H using Algorithm 1.
3. Let $L[x, i, y, j] = 0$ if $x = y$ and $i < j$. Otherwise, $L[x, i, y, j]$ is undefined.
4. **for** d from 1 to $\ell - 1$ **do**
5.   **for** every x, y such that b_x and b_y are at clockwise distance d and $a_x \neq a_y$ **do**
6.   **for** every $i, j, 0 \leq i \leq d_x + 1, 0 \leq j \leq d_y + 1$ **do**
7.   **if** there is a $b_{y'}$, $b_x \prec b_{y'} \prec b_y$, such that $a_y = a_{y'}$ **then**
8.   Let $b_{y'}$ be such that $b_x \prec b_{y'} \prec b_y$, $a_y = a_{y'}$, and the clockwise distance from b_x to $b_{y'}$ is smallest possible.
9.   Let

$$L[x, i, y, j] := \min_{z_1, k_1, z_2, k_2} \{L[x, i, z_1, k_1] + \beta[e_{z_1, k_1}, e_{z_2, k_2}] + L[z_2, k_2, y, j]\},$$

         where z_1, k_1, z_2, k_2 range over all possibilities such that

-   $b_x \preceq b_{z_1} \prec b_{y'} \prec b_{z_2} \preceq b_y$,
-   $a_x \neq a_{z_1}$ or $b_x = b_{z_1}$, and $a_{z_2} \neq a_y$ or $b_{z_2} = b_y$, and
-   if $b_x = b_{z_1}$ then $i < k_1$, and if $b_{z_2} = b_y$ then $k_2 < j$.

10.   **else**
11.   **if** there is a $b_{x'}$, $b_x \prec b_{x'} \prec b_y$, such that $a_x = a_{x'}$ **then**
12.   Computation of $L[x, i, y, j]$ is analogous to the computation above.
13.   **else**
14.   Let

$$L[x, i, y, j] := \min_{i', z, k} \{\beta[e_{x, i'}, e_{z, k}] + L[z, k, y, j]\},$$

         where i', z, k range over all possibilities where $i < i'$ and $b_x \prec b_z \preceq b_y$.
15. **return** $L^* := \min\{L_0, \min_{x, i, y, j}\{\beta[e_{y, j}, e_{x, i}] + L[x, i, y, j]\}\}$, where x, i, y, j range over all possibilities such that either $b_x = b_y$ and $i < j$, or $a_x \neq a_y$, $i > 0$ and $j \leq d_y$.

Proof. Since $a_y = a_{y'}$, path p needs to "jump" over $b_{y'}$ because of condition 3 of Definition 2. Let $b_{z_1} \in p$, $b_x \preceq b_{z_1} \prec b_{y'}$, be such that the clockwise distance from b_{z_1} to $b_{y'}$ is smallest possible. Such z_1 must exist since p starts in B but it leaves it before it reaches $b_{y'}$. Let b_{z_2} be the next vertex in B, after b_{z_1}, encountered when traversing p from b_{z_1}. Note that b_{z_2} exists as p eventually gets to $b_y \in B$. Thus, p can be decomposed into several segments: a b_x-b_{z_1} path p_1, a b_{z_1}-b_{z_2} no-B path p_2, and a b_{z_2}-b_y path p_3.

Notice that $b_{y'} \prec b_{z_2} \preceq b_y$. This is because p enters vertices in B only between b_x and b_y and the p_1 segment blocks off access to B between b_x and b_{z_1}, and b_{z_1} is the clockwise closest vertex to $b_{y'}$ such that $b_{z_1} \in p$ and $b_x \preceq b_{z_1} \prec b_{y'}$. Also notice that for every $b_{z'}$ on p_3, it must be that $b_{z_2} \preceq b_{z'} \preceq b_y$, as access to vertices in B between b_x and b_{z_2} is blocked off by p_1 and p_2. Similarly, for every $b_{z'}$ on p_1, it must be that $b_x \preceq b_{z'} \preceq b_{z_1}$. Therefore, all the p_1, p_2, p_3 segments are source-sinks separating, since $s^* \notin R_p$.

Next we argue that $p_1 \in P[x, i, z_1, k_1]$, where k_1 is such that p leaves b_{z_1} by the edge e_{z_1, k_1}. If $b_x = b_{z_1}$, then, since $p \in P[x, i, y, j]$, condition 1 of Definition

2 implies that $i < k_1$. Therefore, we have $p_1 = b_x$ and $p_1 \in P[x, i, z_1, k_1]$. If $b_x \neq b_{z_1}$, then condition 1 of Definition 2 holds for $p_1 \in P[x, i, z_1, k_1]$ because $p \in P[x, i, y, j]$ and p and p_1 share the starting edge. Condition 3 holds for p_1 since it holds for p. Condition 2 holds because if p_1 entered b_{z_1} by an edge $e_{z_1, k'}$, $k_1 \leq k'$, yet p leaves b_{z_1} by the edge e_{z_1, k_1} and then it continues to b_{z_2}, $b_x \preceq b_{z_1} \prec b_{z_2}$, we would get a loop, a contradiction with p being a path. Thus, $p_1 \in P[x, i, z_1, k_1]$.

By analogous reasons we have that $p_3 \in P[z_2, k_2, y, j]$, where k_2 is such that p enters b_{z_2} by the edge e_{z_2, k_2}.   □

Lemma 7. *If $P[x, i, y, j] \neq \emptyset$, then $L[x, i, y, j] \leq |p|$, where p is a shortest path in $P[x, i, y, j]$.*

Proof. The proof proceeds by induction on the clockwise distance from b_x to b_y. The base case, $b_x = b_y$, follows from step 3 of Algorithm 2. For the inductive case, we distinguish the three cases for positions of b_x and b_y.

If b_x, b_y follow case 1, then p can be decomposed into p_1, p_2, and p_3 as described in Lemma 6. Let z_1, k_1, z_2, k_2 be the corresponding values. Then, $L[x, i, y, j] \leq L[x, i, z_1, k_1] + \beta[e_{z_1, k_1}, e_{z_2, k_2}] + L[z_2, k_2, y, j]$, since $L[]$ is computed as a minimization that considers z_1, k_1, z_2, k_2 as one of the options. By Lemma 4, $\beta[e_{z_1, k_1}, e_{z_2, k_2}] \leq |p_2|$. By the inductive hypothesis, $L[x, i, z_1, k_1] \leq |p'_1| \leq |p_1|$, where p'_1 is a shortest path in $P[x, i, z_1, k_1]$. Similarly, $L[z_2, k_2, y, j] \leq |p_2|$. Therefore, $L[x, i, y, j] \leq |p_1| + |p_2| + |p_3| = |p|$. Cases 2 and 3 are analogous.   □

Lemma 8. *If $L[x, i, y, j]$ holds a numerical value, then $L[x, i, y, j] = |p|$, where p is a shortest path in $P[x, i, y, j]$, or $L[x, i, y, j] > |c_{opt}|$.*

Proof. We proceed by induction on the clockwise distance from b_x to b_y. For the base case, we have $b_x = b_y$, and step 3 computes $L[x, i, y, j]$ correctly.

For the inductive case, suppose that b_x and b_y fall under case 1. Let z'_1, k'_1, z'_2, k'_2 be the values that minimize the expression in step 9. Then, $L[x, i, y, j] = L[x, i, z'_1, k'_1] + \beta[e_{z'_1, k'_1}, e_{z'_2, k'_2}] + L[z'_2, k'_2, y, j]$. By the inductive hypothesis, $L[x, i, z'_1, k'_1] > |c_{opt}|$ or $L[x, i, z'_1, k'_1] = |p'_1|$, and $\beta[e_{z'_1, k'_1}, e_{z'_2, k'_2}] > |c_{opt}|$ or $\beta[e_{z'_1, k'_1}, e_{z'_2, k'_2}] = |p'_2|$, where p'_1 and p'_2 are shortest paths in $P[x, i, z'_1, k'_1]$ and $P[z'_2, k'_2, y, j]$, respectively. By Lemma 4, $\beta[e_{z'_1, k'_1}, e_{z'_2, k'_2}] > |c_{opt}|$ or $\beta[e_{z'_1, k'_1}, e_{z'_2, k'_2}] = |p'_2|$, where p'_2 is a shortest $e_{z'_1, k'_1}$-$e_{z'_2, k'_2}$ no-B source-sinks separating path. If $L[x, i, z'_1, k'_1] > |c_{opt}|$ or $L[z'_2, k'_2, y, j] > |c_{opt}|$ or $\beta[e_{z'_1, k'_1}, e_{z'_2, k'_2}] > |c_{opt}|$, we have $L[x, i, y, j] > |c_{opt}|$ since $L[x, i, z'_1, k_1]$, $\beta[e_{z'_1, k'_1}, e_{z'_2, k'_2}]$, and $L[z'_2, k'_2, y, j]$ are nonnegative.

It remains to deal with the case when $L[x, i, z'_1, k'_1] = |p'_1|$, $\beta[e_{z'_1, k'_1}, e_{z'_2, k'_2}] = |p'_2|$, and $L[z'_2, k'_2, y, j] = |p'_3|$. By Lemma 7, we have $L[x, i, y, j] = |p'_1| + |p'_2| + |p'_3| \leq |p|$. Let p' be the concatenation of p'_1, p'_2, and p'_3. We will show that either p'_1, p'_2, and p'_3 do not intersect, in which case $p' \in P[x, i, y, j]$ and, therefore, $L[x, i, y, j] = |p'| = |p|$; or they do intersect, in which case $L[x, i, y, j] > |c_{opt}|$.

Claim. For every b_u, b_v, $b_u \neq b_v$, on p', we have $a_u \neq a_v$.

PROOF: Suppose, by contradiction, that there are b_u, b_v, $b_x \preceq b_u \prec b_v \preceq b_y$, on p' such that $a_u = a_v$. Since $p'_1 \in P[x, i, z'_1, k'_1]$, it cannot be that both b_u and b_v are on p'_1 due to condition 3 in Definition 2. Similarly, b_u and b_v cannot both be on p'_3. And, as p'_2 is a no-B path and $a_{z'_1} \neq a_{z'_2}$ due to Lemma 2 applied to z'_1, y', z'_2, y, vertices b_u and b_v cannot both be on p'_2. Recall also that p'_1 does not contain a vertex $b_{x'}$ with $a_{x'} = a_y$, as $b_{y'}$ has the smallest clockwise distance from b_x and it comes after $b_{z'_1}$. Thus, b_u is on p'_1 and b_v on p'_3. Since $a_u = a_v \neq a_y$, we get $b_u \prec b_{y'} \prec b_v \prec b_y$, a contradiction with Lemma 2.   ◇

If p'_1, p'_2, and p'_3 do not intersect, then $p' \in P[x, i, y, j]$; this is due to the above claim and the p'_k's being from their respective $P[]$. Thus, $L[x, i, y, j] = |p'| = |p|$.

If p'_1, p'_2, and p'_3 intersect, their concatenation results in a walk with one or more loops, not a path. Let us look at the path-regions $R_{p'_1}$, $R_{p'_2}$, and $R_{p'_3}$. The union of these regions and f_B contains all the sinks between b_x and b_y strictly inside. Its complement contains the source. Let R' be the complement of $R_{p'_1} \cup R_{p'_2} \cup R_{p'_3} \cup f_B$. Let R'_{s*} be the maximal simply connected planar region in R' that contains s^*. If R'_{s*} borders no b_v, $b_y \preceq b_v \preceq b_x$, then R'_{s*} is bounded by sub-paths of p'_1, p'_2 and p'_3, see Figure 3(c) where we assume $s^* = s^*_1$. Let c be the cycle in G^*_0 corresponding to the boundary of R'_{s*}; notice that c is simple. Since R'_{s*} does not contain any sinks, c is an (s^*, T^*)-separating cycle. Then, $L[x, i, y, j] = |p'_1| + |p'_2| + |p'_3| > |c| \geq |c_{opt}|$.

Finally, if R'_{s*} borders some b_v for $b_y \preceq b_v \preceq b_x$, see Figure 3(c) (assume $s^* = s^*_2$), then R'_{s*} is enclosed by the clockwise b_y-b_x part of the boundary of f_B and by a b_x-b_y path p'' that is formed by concatenating segments of p'_1, p'_2, and p'_3. Observe that $p'' \in P[x, i, y, j]$ and $|p''| < |p'_1| + |p'_2| + |p'_3| \leq |p|$. This is a contradiction with p being a shortest path in $P[x, i, y, j]$.   □

Corollary 1. *If $P[x, i, y, j] \neq \emptyset$, then $L[x, i, y, j] = |p|$ where p is a shortest path in $P[x, i, y, j]$, or $L[x, i, y, j] > |c_{opt}|$. If $P[x, i, y, j] = \emptyset$, then either $L[x, i, y, j]$ is undefined, or $L[x, i, y, j] > |c_{opt}|$.*

The corollary follows from observing that if $P[x, i, y, j] \neq \emptyset$, then $L[x, i, y, j]$ holds a numerical value. Now we are ready for the main theorem and the sketch of its proof; we defer the complete proof to the full version of the paper.

Theorem 1. *Algorithm 2 computes the weight of a minimum (s^*, T^*)-separating cycle in G^*_0 (and a minimum simple (s, T)-cut in G). It runs in time $O(n^4)$.*

Proof (sketch). Similarly as in Lemma 7, we get $L^* \leq |c_{opt}|$. Suppose $L^* < |c_{opt}|$. Since the value of L_0 corresponds to an (s^*, T^*)-separating cycle, we get $|c_{opt}| \leq L_0$. Thus, $L^* = \beta[e_{y', j'}, e_{x', i'}] + L[x', i', y', j']$ for some x', i', y', j'. If either quantity is $> |c_{opt}|$, we get $|c_{opt}| < |c_{opt}|$, a contradiction. If both quantities are computed correctly, we get a shorter (s^*, T^*)-separating cycle than c_{opt}, a contradiction. Therefore, $L^* = |c_{opt}|$.

The running time is $O(n^4)$ because there are $O(n^2)$ possibilities for $(x, i), (y, j)$ since they correspond to a pair of edges; and the computation $L[x, i, y, j]$ considers another $O(n^2)$ pairs of $(z_1, k_1), (z_2, k_2)$.   □

*Remark 2 (**Extensions**).* The presented approach can be used for other connectivity priors. For example, a cut is contiguous if the dual cut-edges form a non-crossing tour that separates s^* from T^* (we allow repeated vertices but not edges as long as the tour can be drawn in a non-self-crossing manner in the infinitesimal neighborhood of each vertex). In a prior work [2], we computed how many of the minimum (s, T)-cuts are contiguous; however, the earlier approach did not extend to finding, among all contiguous cuts, the one with the smallest weight. Algorithm 2 can be modified to allow $a_x = a_y$, but one has to be careful not to use the edges of τ more than once. We leave the details for the journal version of this paper.

References

1. Bateni, M., Hajiaghayi, M., Klein, P.N., Mathieu, C.: A polynomial-time approximation scheme for planar multiway cut. In: Proceedings of the 23rd Annual ACM-SIAM Symposium on Discrete Algorithms (SODA). pp. 639–655 (2012)
2. Bezáková, I., Langley, Z.: Contiguous minimum single-source-multi-sink cuts in weighted planar graphs. In: Gudmundsson, J., Mestre, J., Viglas, T. (eds.) COCOON 2012. LNCS, vol. 7434, pp. 49–60. Springer, Heidelberg (2012)
3. Bienstock, D., Monma, C.L.: On the complexity of embedding planar graphs to minimize certain distance measures. Algorithmica 5(1), 93–109 (1990)
4. Borradaile, G., Klein, P.N.: An $O(n \log n)$ algorithm for maximum st-flow in a directed planar graph. J. ACM 56(2) (2009)
5. Borradaile, G., Klein, P.N., Mozes, S., Nussbaum, Y., Wulff-Nilsen, C.: Multiple-source multiple-sink maximum flow in directed planar graphs in near-linear time. In: Proceedings of the IEEE 52nd Annual Symposium on Foundations of Computer Science (FOCS), pp. 170–179 (2011)
6. Boykov, Y., Veksler, O.: Graph cuts in vision and graphics: Theories and applications. In: Handbook of Mathematical Models in Computer Vision. Springer (2006)
7. Boykov, Y., Veksler, O., Zabih, R.: Fast approximate energy minimization via graph cuts. IEEE Trans. Pattern Anal. Mach. Intell. 23(11), 1222–1239 (2001)
8. Cabello, S.: Finding shortest contractible and shortest separating cycles in embedded graphs. ACM Trans. on Algorithms 6(2) (2010); Ext. abstr. in SODA 2009
9. Chalermsook, P., Fakcharoenphol, J., Nanongkai, D.: A deterministic near-linear time algorithm for finding minimum cuts in planar graphs. In: Proceedings of the 15th Annual ACM-SIAM Symp. on Discr. Algorithms (SODA), pp. 828–829 (2004)
10. Chambers, E.W., Erickson, J., Nayyeri, A.: Minimum cuts and shortest homologous cycles. In: Proceedings of the 25th Annual ACM Symposium on Computational Geometry (SCG), pp. 377–385 (2009)
11. Chambers, E.W., de Verdière, É.C., Erickson, J., Lazarus, F., Whittlesey, K.: Splitting (complicated) surfaces is hard. Comput. Geom. 41(1-2), 94–110 (2008)
12. Łącki, J., Sankowski, P.: Min-cuts and shortest cycles in planar graphs in $O(n \log \log n)$ time. In: Demetrescu, C., Halldórsson, M.M. (eds.) ESA 2011. LNCS, vol. 6942, pp. 155–166. Springer, Heidelberg (2011)
13. Vicente, S., Kolmogorov, V., Rother, C.: Graph cut based image segmentation with connectivity priors. In: IEEE Computer Society Conference on Computer Vision and Pattern Recognition, CVPR (2008)
14. Zeng, Y., Samaras, D., Chen, W., Peng, Q.: Topology cuts: A novel min-cut/max-flow algorithm for topology preserving segmentation in N-D images. Computer Vision Image Understanding 112, 81–90 (2008)

The Price of Envy-Freeness
in Machine Scheduling[*]

Vittorio Bilò[1], Angelo Fanelli[2], Michele Flammini[3],
Gianpiero Monaco[3], and Luca Moscardelli[4]

[1] University of Salento, Italy
vittorio.bilo@unisalento.it
[2] CNRS, (UMR-6211), France
angelo.fanelli@gmail.com
[3] Gran Sasso Science Institute, L'Aquila, Italy
{michele.flammini,gianpiero.monaco}@univaq.it
[4] University of Chieti-Pescara, Italy
luca.moscardelli@unich.it

Abstract. We consider k-envy-free assignments for scheduling problems in which the completion time of each machine is not k times larger than the one she could achieve by getting the jobs of another machine, for a given factor $k \geq 1$. We introduce and investigate the notion of price of k-envy-freeness, defined as the ratio between the makespan of the best k-envy-free assignment and that of an optimal allocation achievable without envy-freeness constraints. We provide exact or asymptotically tight bounds on the price of k-envy-freeness for all the basic scheduling models, that is unrelated, related and identical machines. Moreover, we show how to efficiently compute such allocations with a worsening multiplicative factor being at most the best approximation ratio for the minimum makespan problem guaranteed by a polynomial time algorithm for each specific model. Finally, we extend our results to the case of restricted assignments and to the objective of minimizing the sum of the completion times of all the machines.

1 Introduction

The evolution of scheduling closely tracked the development of computers. Given m machines that have to process n jobs, minimizing the makespan of an assignment of the jobs to the machines is one of the most well-studied problem in the Theory of Algorithms [12,16,17,19]. In more details, assuming that the processing of job i on machine j requires time $p_{ij} > 0$, the completion time of machine j (under a certain assignment) is given by the sum of the processing times of all the jobs allocated to j. The makespan of an assignment is the maximum completion time among all the machines (we stress that an assignment is not forced to use all the available machines) and the objective of the scheduling problem is to find an assignment of minimum makespan.

---

[*] This research was partially supported by the PRIN 2010–2011 research project ARS TechnoMedia (Algorithmics for Social Technological Networks), funded by the Italian Ministry of University and Research.

E. Csuhaj-Varjú et al. (Eds.): MFCS 2014, Part II, LNCS 8635, pp. 106–117, 2014.
© Springer-Verlag Berlin Heidelberg 2014

In the literature, three different models of machines have been adopted. The general setting illustrated above is called scheduling problem with *unrelated machines* [19]. An interesting particular scenario is the case with *related machines* [17], where each job i has a load $l_i > 0$ and each machine j has a speed of processing $s_j > 0$, and thus the processing time of job i on machine j is given by $p_{ij} = l_i/s_j$. Finally, the even more specific setting in which the speed of each machine is 1 is referred to as the scheduling problem with *identical machines* [12,16]. Even this latter problem is NP-hard [16].

The approximability of the scheduling problem has been well understood for all the three models described above. However, all the proposed solutions do not envisage fair allocations in which no machine prefers (or envies) the set of the tasks assigned to another machine, i.e., for which her completion time would be strictly smaller. In the literature, such fairness property is referred to as "envy-freeness" [8,9]. Specifically, consider a scenario in which a set of tasks (jobs) has to be allocated among employees (machines) in such a way that the last task finishes as soon as possible. It is natural to consider fair allocations, that is allocations where no employee prefers (or envies) the set of tasks assigned to some other employee, i.e., a set of tasks for which her completion time would be strictly smaller than her actual one.

It is possible to consider two different variants of this model, depending on the fact that an employee (i) can envy the set of tasks assigned to any other employee or (ii) can only envy the set of tasks of other employees getting at least one job: in the latter case, employees not getting any job do not create envy. In the following, we provide some scenarios motivating both variants.

For the first variant, consider a company that receives an order of tasks that must be assigned among its m employees. For equity reasons, in order to make the workers satisfied with their task assignment so that they are as productive as they can, the tasks should be assigned in such a way that no envy is induced among the employees.

For the second variant, consider a scenario in which a company, in order to fulfill a complex job composed by several tasks, has to engage a set of employees that, for law or trade union reasons have to be all paid out the same wage. Again, for making the workers as productive as they can, it is required that no envy is induced, but in this case we are interested only in the envy among the *engaged* employees, i.e. the ones receiving at least a task to perform.

We notice that the existence of envy-free schedules is not guaranteed in the first variant of the model. For instance, consider a scenario where the number of machines is strictly greater than the number of jobs. Clearly at least one machine would not get any job and all the machines getting at least one job would be envious. Therefore, in the following of this paper we focus on the second varant of the model, in which envy-freeness is required only among machines getting at least one job.

We adopt a more general definition of envy-free allocations, namely the k-*envy-freeness* (for any $k \geq 1$): Given an assignment and two machines j, j' (where both j and j' get jobs), we say that j k-envies j' if the completion time of j

is at least k times the completion time she would have when getting the set of jobs assigned to j'. In other words, an assignment is k-envy-free if no machine would decrease her completion time by a factor at least k by being assigned all the jobs allocated to another machine. Notice that a k-envy-free assignment always exists: a trivial one can be obtained by allocating all the jobs to a single machine, even if it might have a dramatically high makespan.

We are interested in analyzing the loss of performance due to the adoption of envy-free allocations. Our study has an optimistic nature and, then, aims at quantifying the efficiency loss in the best k-envy-free assignment. Therefore, we introduce the **price of k-envy-freeness**, defined as the ratio between the makespan of the best k-envy-free assignment and that of an optimal assignment. In the literature, other papers performed similar optimistic studies, see, for instance, [1,6]. The price of k-envy-freeness represents an ideal limitation to the efficiency achievable by any k-envy-free assignment. In our work, we also show how to efficiently compute k-envy-free assignments which nicely compare with the performance of the best possible ones. We point out that the computation of non-trivial k-envy-free assignments is necessary to achieve good quality solutions, since the ratio between the makespan of the worst k-envy-free assignment and that of an optimal assignment can be very high. In particular, it is unbounded for unrelated machines, $n\frac{s_{max}}{s_{min}}$ for related ones, where s_{max} (resp. s_{min}) is the maximum (resp. minimum) speed among all the machines, and n for identical machines.

Related Work. The scheduling problem with unrelated machines has been studied in [19]. The authors provide a 2-approximation polynomial time algorithm and show that the problem cannot be approximated in polynomial time within a factor less than $3/2$. Polynomial time approximation schemes for related and identical machines have been presented in [17] and [16], respectively.

The problem of fair allocation is a longstanding issue, thus, the literature on this topic includes hundreds of references. For a nice review, we refer the reader to the book [4]. One common notion of fairness, recurring in many papers and therefore adopted for central problems, is that of envy-freeness. For instance, the classical Vickrey auctions [23], as well as some optimal Bayesian auctions [2,20], generate envy-free outcomes. An interesting paper explicitly dealing with envy-free auctions is [13]. Studies on envy-free divisions, typically referred to as envy-free cake cutting, can be found in [3,8,9]. Furthermore, [10,14] consider algorithmic issues related to the envy-free pricing problem, that is a scenario in which a seller has to set (envy-free) prices and allocations of items to buyers in order to maximize the total revenue.

Concerning scheduling problems, an important stream of research is the one focusing on envy-free algorithmic mechanism design. Roughly speaking, algorithmic mechanism design is the attempt of motivating the machines, through payments or incentives, to follow desired behaviors (*truthful mechanisms*). Upper and lower bounds on the approximation ratio achieved by truthful mechanisms have been given in [7,18,22]. However, such papers are not concerned with fair allocations. To the best of our knowledge, envy-free mechanisms for the scheduling problem with unrelated machines have been first considered in [15]. the authors prove a lower bound of $2 - 1/m$ and an upper bound of $(m + 1)/2$ on the performance guarantee of envy-free truthful mechanisms. Such upper and lower bounds have been improved in [5] to $O(\log m)$ and $\Omega\left(\frac{\log m}{\log \log m}\right)$, respectively. Recently, [11] shows that no truthful mechanism can guarantee an envy-free allocation with a makespan less than a factor of $O(\log m)$ the optimal one, thus closing the gap. It is worth noticing that, for $k = 1$, our model can be seen as a special case of the one considered in [5,11,15] when the same payment is provided to all the machines receiving at least a job, while no payment is given to the other machines.

The work most closely related to our study is [6]. The authors consider the envy-free scheduling problem with unrelated machines with some substantial differences with respect to our setting. Specifically, i) they only consider 1-envy-free assignments (while we consider k-envy-free assignments, for any $k \geq 1$); ii) the objective in their work is that of minimizing the sum of the completion times of all jobs (while we mainly consider the makespan); iii) in their setting all the machines contribute to create envy (while in our setting only machines getting at least one job are considered for the envy-freeness). Not surprisingly, the authors prove that, in their setting, the price of envy-freeness is unbounded.

Our Results. We consider the price of k-envy-freeness in the scheduling problem, that is, the ratio between the makespan of the best k-envy-free assignment and that of an optimal assignment. We investigate the cases of unrelated, related and identical machines and provide exact or asymptotically tight bounds on the price of k-envy-freeness. We stress that low values of k implies a greater attitude to envy, which tremendously reduces the set of k-envy-free assignments. A natural threshold that arose in our analysis of the cases with related and identical machines is the value $k = 2$, as it can be appreciated in the following table where we summarize our main results. They are fully described in Section 3.

		Identical	Related	Unrelated
$k = 1$	UB and LB	$\min\{n, m\}$	$\min\{n, m\}$	$2^{\min\{n,m\}-1}$
$k \in (1, 2)$	UB	$\frac{2k}{k-1}$	$2k\sqrt{\frac{m}{k-1}}$	$\left(1 + \frac{1}{k}\right)^{\min\{n,m\}-1}$
	LB	$\Omega\left(\frac{2k}{k-1}\right)$	$\Omega\left(\sqrt{\frac{m}{k-1}}\right)$	$\left(1 + \frac{1}{k}\right)^{\min\{n,m\}-1}$
$k \geq 2$	UB	$1 + \frac{1}{k}$	$2 + \max\left\{1, \sqrt{\frac{m}{k}}\right\}$	$\left(1 + \frac{1}{k}\right)^{\min\{n,m\}-1}$
	LB	$1 + \frac{1}{k}$	$\max\left\{1, \sqrt{\frac{m}{k}}\right\}$	$\left(1 + \frac{1}{k}\right)^{\min\{n,m\}-1}$

A further result derives from the fact that our upper bound proofs are constructive and, therefore, they de facto provide polynomial time algorithms able to calculate good k-envy-free assignments. Such an extension is discussed in Section 3.4. Furthermore, in Subsection 4.1 we also consider the *restricted* scheduling problem, where each job can be assigned only to a subset of machines. Moreover, besides considering the problem of minimizing the makespan, we consider in Subsection 4.2 the problem of minimizing the sum of the completion times of all the machines.

Due to space constraints, some proofs are omitted.

2   Preliminaries

In the scheduling problem, there are $m \geq 2$ *machines* and n indivisible *jobs* to be assigned to the machines. In the *unrelated* case, the time of running job i on machine j is given by $p_{ij} > 0$. In the *related* setting, each job i has a load $l_i > 0$, each machine has a speed of processing $s_j > 0$, and the processing time of job i on machine j is given by $p_{ij} = l_i/s_j$. We refer to the specific setting in which the speed of each machine is 1 as *identical*, where $p_{ij} = l_i$.

For an integer $h > 0$, define $[h] := \{1, \ldots, h\}$. In the related and identical setting, we denote with $L = \sum_{i \in [n]} l_i$ the total load of all the jobs and with $l_{max} = \max_{i \in [n]} l_i$ the maximum load of a job.

An *assignment* or *solution* $\mathbf{N}$ is specified by a partition of the set of jobs into m components, i.e., $(N_j)_{j \in [m]}$, where N_j denotes the set of jobs assigned to machine j. Let Q be a set of jobs, we use the notation $C_j(Q)$ to denote the completion time of machine j on the set Q, i.e., $C_j(Q) = \sum_{i \in Q} p_{ij}$. Thus $C_j(N_j)$ denotes the completion time of machine j under the assignment $\mathbf{N}$. For the related and identical settings, let $L_j(\mathbf{N})$ be the total load of the jobs assigned by $\mathbf{N}$ to machine $j \in [m]$, i.e., $L_j(\mathbf{N}) = \sum_{i \in N_j} l_i$ and $L_{min}(\mathbf{N}) = \min\{L_j(\mathbf{N}) : j \in [m] \land N_j \neq \emptyset\}$ (resp. $L_{max}(\mathbf{N}) = \max\{L_j(\mathbf{N}) : j \in [m] \land N_j \neq \emptyset\}$) the minimum (resp. maximum) load of the non-empty machines in $\mathbf{N}$. Notice that, in the related setting, we have $C_j(N_j) = L_j(\mathbf{N})/s_j$ and, for the identical one, $C_j(N_j) = L_j(\mathbf{N})$. The *makespan* of assignment $\mathbf{N}$ is defined as $\mathcal{M}(\mathbf{N}) = \max_{j \in [m]} C_j(N_j)$, that is the maximum processing time among all the machines. An *optimal* assignment is one minimizing the makespan. We denote by $\mathbf{O}$ an optimal assignment.

Given an assignment $\mathbf{N}$, a real value $k \geq 1$, and two machines j, j' such that $N_j \neq \emptyset$ and $N_{j'} \neq \emptyset$, we say that j k-**envies** j' if $C_j(N_j) > kC_j(N_{j'})$. An assignment $\mathbf{N}$ is k-**envy-free** if $C_j(N_j) \leq kC_j(N_{j'})$ for every pair of machines (j, j') such that $N_j \neq \emptyset$ and $N_{j'} \neq \emptyset$. Notice that a k-envy-free assignment can always be obtained by assigning all jobs to a single machine. The **price of k-envy-freeness** (PoEF$_k$) is defined as the ratio between the makespan of the best k-envy-free assignment and the makespan of an optimal assignment. More formally, let $\mathcal{F}_k$ be the set of the k-envy-free assignments, then PoEF$_k = \min_{\mathbf{N} \in \mathcal{F}_k} \frac{\mathcal{M}(\mathbf{N})}{\mathcal{M}(\mathbf{O})}$.

We conclude this section with some preliminary general results.

Proposition 1. *For the scheduling problem with related machines,* PoEF$_k \leq \min\{n, m\}$ *for any* $k \geq 1$.

Proof. Assume that machine 1 is the fastest one, i.e., $s_1 \geq s_j$ for each $j \in [m]$. Clearly, the solution $\mathbf{N}$ assigning all jobs to machine 1 is k-envy-free for any $k \geq 1$ and has $\mathcal{M}(\mathbf{N}) = \frac{L}{s_1} \leq \frac{n l_{max}}{s_1}$. By $\mathcal{M}(\mathbf{O}) \geq \frac{l_{max}}{s_1}$ and $\mathcal{M}(\mathbf{O}) \geq \frac{L}{ms_1}$, we obtain the claim.   □

Such a simple upper bound on the price of k-envy-freeness proves to be tight when $k = 1$ even for the setting of identical machines.

Proposition 2. *For the scheduling problem with identical machines, there exists an instance for which* PoEF$_k = \min\{n, m\}$ *when* $k = 1$.

We now show that, for finite values of k, a price of k-envy-freeness equal to 1 cannot be achieved even in the setting of identical machines.

Proposition 3. *For the scheduling problem with identical machines, no value of k (possibly depending on n and m) can guarantee* $\text{PoEF}_k = 1$.

In the next lemma, we give an important result which helps to characterize the performance of k-envy-free solutions in the case of related machines.

Lemma 1. *For a value $k \geq 1$, an instance of the scheduling problem with related machines, and an integer $2 \leq h \leq \min\left\{m, \left\lfloor \frac{L(k-1)}{kl_{max}} \right\rfloor\right\}$, there always exists a k-envy-free solution $\mathbf{N}$ using exactly h machines and such that $\mathcal{M}(\mathbf{N}) \leq \frac{L/h+l_{max}}{s_h}$, where s_h is the speed of the h-th fastest machine.*

3   Results

3.1   Identical Machines

In this subsection, we consider the scheduling problem with identical machines. For the case of $k \geq 2$, we can prove a constant upper bound on the price of k-envy freeness.

Theorem 1. *For the scheduling problem with identical machines,* $\text{PoEF}_k \leq 1 + 1/k$ *for any $k \geq 2$.*

Proof. We argue that applying Algorithm 1 to any initial assignment $\mathbf{S}$, we get a k-envy free assignment $\mathbf{N}$ with makespan at most $\mathcal{M}(\mathbf{S})(1 + 1/k)$. The claim follows by choosing as the starting assignment $\mathbf{S}$ an optimal solution $\mathbf{O}$.

Initially Algorithm 1 manipulates the starting assignment in such a way that it becomes an assignment with makespan 1 with the minimal number of non-empty machines, and such that the machines are numbered so that to a smaller index corresponds a larger or equal load. After the first phase we assume that the jobs are assigned to machines in $[\overline{m}]$.

Since machine $\overline{m}$ is the least loaded one, if $L_{\overline{m}}(\mathbf{S}) \geq 1/k$, then $\mathbf{S}$ is k-envy-free and the claim follows. On the other side, if $L_{\overline{m}}(\mathbf{S}) < 1/k$, we move all the jobs in $\mathbf{S}$ from machine $\overline{m}$ to machine $\overline{m} - 1$ obtaining a new assignment $\mathbf{N}$ which is k-envy-free. In fact, in the new assignment $\mathbf{N}$, machine $\overline{m} - 1$ gets a load larger than 1, thus becoming the most loaded machine, whereas any other machine has a load smaller than 1. Machine $\overline{m} - 1$ does not envy any other machine, since $L_{\overline{m}-1}(\mathbf{N}) = L_{\overline{m}-1}(\mathbf{S}) + L_{\overline{m}}(\mathbf{S}) \leq 2L_{\overline{m}-1}(\mathbf{S}) \leq kL_{\overline{m}-1}(\mathbf{S}) \leq kL_j(\mathbf{N})$, for each $j \leq \overline{m} - 1$ and $k \geq 2$. Thus, we can conclude that the new assignment is k-envy-free. Finally we see that the makespan of $\mathbf{N}$ is at most $L_{\overline{m}-1}(\mathbf{N}) = L_{\overline{m}-1}(\mathbf{S}) + L_{\overline{m}}(\mathbf{S}) \leq L_{\overline{m}-1}(\mathbf{S}) + 1/k \leq (1 + 1/k)L_{\overline{m}-1}(\mathbf{N}) \leq \mathcal{M}(\mathbf{S})(1 + 1/k)$. The claim follows.   □

The next result shows that the above upper bound is tight for any $k \geq 2$.

Algorithm 1.

1: **Input:** assignment $\mathbf{S}$
2: Rescale the loads in such a way that $\mathcal{M}(\mathbf{S}) = 1$
3: **while** there exists a pair of machines (j, j') s.t. $L_j(\mathbf{S}) + L_{j'}(\mathbf{S}) \le 1$ **do**
4:      $S_j \leftarrow S_j \cup S_{j'}$
5:      $S_{j'} \leftarrow \emptyset$
6: **end while**
7: Renumber the machines in non-increasing order of loads
        $L_j(\mathbf{S}) \ge L_{j+1}(\mathbf{S})$ for each $j \in [m-1]$
8: Let $[\overline{m}]$ be the set of machines with at least one job assigned
9: Create a new assignment $\mathbf{N}$ defined as follows
10: **if** $L_{\overline{m}}(\mathbf{S}) < 1/k$ **then**
11:      $N_j \leftarrow S_j$ for each $j < \overline{m} - 1$
12:      $N_{\overline{m}-1} \leftarrow S_{\overline{m}-1} \cup S_{\overline{m}}$
13:      $N_j \leftarrow \emptyset$ for each $j > \overline{m} - 1$
14: **else**
15:      $N_j \leftarrow S_j$ for each $j \in [m]$
16: **end if**
17: **return N**

Proposition 4. *For the scheduling problem with identical machines, given any $k \ge 2$, there exists an instance for which $\mathrm{PoEF}_k \ge 1 + 1/k - \epsilon$, for any $\epsilon > 0$.*

For the remaining case of $k \in (1, 2)$, the following bounds hold.

Theorem 2. *For the scheduling problem with identical machines, $\mathrm{PoEF}_k \le \min\left\{\frac{2k}{k-1}, n, m\right\}$ for any $k \in (1, 2)$.*

Theorem 3. *For the scheduling problem with identical machines, given any $k \in (1, 2)$, there exists an instance for which $\mathrm{PoEF}_k = \Omega\left(\min\left\{\frac{2k}{k-1}, n, m\right\}\right)$.*

3.2   Related Machines

In this subsection, we consider the scheduling problem with related machines.

Theorem 4. *For the scheduling problem with related machines, $\mathrm{PoEF}_k \le 2 + \max\left\{1, \sqrt{\frac{m}{k}}\right\}$ for any $k \ge 2$.*

Proof. Given an instance of the scheduling problem with related machines, consider any assignment $\mathbf{S}$. Let us normalize the machine speeds and the loads of the jobs so that the fastest machine has speed 1 and the makespan of solution $\mathbf{S}$ is 1, i.e., $\mathcal{M}(\mathbf{S}) = 1$. Let us rename the machines in such a way that $s_j \ge s_{j+1}$ for any $j = 1, \ldots, m-1$; notice that $L_1(\mathbf{S}) \le 1$ and we can assume that $L_j(\mathbf{S}) \ge L_{j+1}(\mathbf{S})$ for any $j = 1, \ldots, m - 1$ (otherwise by swapping S_j and S_{j+1} a solution having equal or better makespan could be obtained).

Denote by $M_1 = \{1, \ldots, |M_1|\}$ the set of machines having load at least $1/k$ in $\mathbf{S}$, i.e., $L_j(\mathbf{S}) \ge 1/k$ for any $j \in M_1$, and by M_2 the set of the remaining machines. Note that it also holds $s_j \geq 1/k$ for any $j \in M_1$. Moreover, it is easy to check that no pair (j, j') of machines in M_1 is such that j k-envies j'.

In the following we build a new allocation $\mathbf{N}$ starting from allocation $\mathbf{S}$.

Let L_j^i be the load of each machine j at the moment in which the job i is considered for allocation by Algorithm 2. The new assignment $\mathbf{N}$ is obtained as described in Algorithm 2.

Algorithm 2.

1: **Input:** assignment $\mathbf{S}$
2: $\mathbf{N} \leftarrow \mathbf{S}$
3: $M' \leftarrow \emptyset$
4: $j' \leftarrow |M_1|$
5: **while** $j' < m$ **do**
6:     $j \leftarrow j' + 1$
7:     **if** $k > m$ or $\sum_{p \geq j} L_p(\mathbf{S}) \leq \sqrt{\frac{m}{k}}$ **then**
8:         **for each job in** $i \in \bigcup_{p \geq j} S_p$ **do**
9:             Let $j'' \in M_1 \cup M'$ be the machine with the current smallest load
10:            **if** $L_1^i + l_i \leq k L_{j''}^i$ **then**
11:                $N_1 \leftarrow N_1 \cup \{i\}$ ▷ Assign job i to machine 1
12:            **else**
13:                $N_{j''} \leftarrow N_{j''} \cup \{i\}$ ▷ Assign job i to machine j''
14:            **end if**
15:        **end for**
16:        $j' \leftarrow m$
17:    **else**
18:        $M' \leftarrow M' \cup \{j\}$
19:        Let j' such that $\frac{1}{k} - L_j(\mathbf{S}) \leq \sum_{p=j+1}^{j'} L_p(\mathbf{S}) \leq \frac{2}{k} - L_j(\mathbf{S})$
20:        $N_j \leftarrow \bigcup_{p=j}^{j'} S_p$
21:    **end if**
22: **end while**
23: **return** $\mathbf{N}$

Assignment $\mathbf{N}$ is initially set equal to assignment $\mathbf{S}$. When lines 18–20 are executed, it means that $k \leq m$ and $\sum_{p \geq j} L_p(\mathbf{S}) > \sqrt{\frac{m}{k}}$. It follows that $L_j(\mathbf{S}) > \frac{1}{\sqrt{mk}}$ (and therefore also $s_j > \frac{1}{\sqrt{mk}}$). In this case, the only machine receiving some new jobs is machine j. Since the load of any machine in M_2 is less than $1/k$, we can gather all the jobs of machines $j+1, j+2, \ldots, j'$ of total load between $\frac{1}{k} - L_j(\mathbf{S})$ and $\frac{2}{k} - L_j(\mathbf{S})$, and add in $\mathbf{N}$ all such jobs to machine j. We obtain $C_j(N_j) = \frac{L_j(\mathbf{N})}{s_j} \leq \frac{\frac{2}{k}}{\frac{1}{\sqrt{m}\sqrt{k}}} = 2\sqrt{\frac{m}{k}}$. Notice that machine j cannot be k-envied by any other machine, and (since $k \geq 2$) cannot k-envy other machines with load at least $1/k$ (all the machines in $M_1 \cup M'$ have load at least $1/k$ in assignment $\mathbf{N}$).

Note that lines 8–16 can be executed only once. When they are executed, it means that $k > m$ or $\sum_{p \geq j} L_p(\mathbf{S}) \leq \sqrt{\frac{m}{k}}$. When $k > m$, the load in $\mathbf{S}$ of each machine in M_2 is at most $1/m$ and the total load of all machines in M_2 is at most 1. Therefore, in any case, the total load of all machines $p \geq j$ is at most $\max\left\{1, \frac{\sqrt{m}}{\sqrt{k}}\right\}$. Notice that, since (i) $k \geq 2$, (ii) the load of each machine in M_1 is at least $1/k$ already in allocation **S** and (iii) the load of each job to be assigned is at most $1/k$, it is always possible to maintain k-envy-free an allocation by assigning each job either to machine 1 or (in case the assignment to machine 1 would result in a state non being k-envy-free) to the machine of $M_1 \cup M'$ having the smallest load at that moment. In fact, consider any job i belonging in **S** to some machine $p \geq j$, and, for any $j \in M_1 \cup M'$, let L_j^i be the load of machine j at the moment in which the job i is considered for assignation by Algorithm 2. Assigning job i to machine j'' results in a k-envy-free state because $L_{j''}^i + l_i \leq kL_{j''}^i$ as conditions (i), (ii) and (iii) hold.

Let us now compute the makespan of assignment **N**, by considering only the machines receiving some new jobs in lines 8–16 of Algorithm 2:

The total load added to machine 1 is at most $\max\left\{1, \frac{\sqrt{m}}{\sqrt{k}}\right\}$ and therefore the total load $C_1(N_1)$ of machine 1 at the end of the process is at most $1 + \max\left\{1, \frac{\sqrt{m}}{\sqrt{k}}\right\}$.

For any machine $j \in M_1 \setminus \{1\}$, let $last(j)$ be the last job assigned to machine j and $\ell_j = l_{last(j)}$ its load. Since $last(j)$ has not been assigned to machine 1, it must hold that $L_1^{last(j)} + \ell_j > kL_{j'}^{last(j)}$ for some $j' \in M_1 \setminus \{1\}$. In particular, $L_1^{last(j)} + \ell_j > kL_j^{last(j)}$ because $last(j)$ has been assigned to the machine with minimum load at that moment. Since the total load that can be given to machine 1 is at most $1 + \max\left\{1, \sqrt{\frac{m}{k}}\right\}$, it follows that $L_j^{last(j)} < \frac{L_1^{last(j)} + \ell_j}{k} \leq \frac{1 + \max\left\{1, \sqrt{\frac{m}{k}}\right\}}{k}$.

Finally, since $last(j)$ is the last job assigned to machine j, $L_j(\mathbf{N}) = L_j^{last(j)} + \ell_j \leq L_j^{last(j)} + 1/k$ and the completion time of machine j is $C_j(N_j) = \frac{L_j(\mathbf{N})}{s_j} \leq$

$$\frac{L_j^{last(j)} + 1/k}{1/k} \leq k\left(\frac{1 + \max\left\{1, \sqrt{\frac{m}{k}}\right\}}{k} + \frac{1}{k}\right) = 2 + \max\left\{1, \sqrt{\frac{m}{k}}\right\}.$$

The claim follows by choosing $\mathbf{O} = \mathbf{S}$.   □

For the case of $k \in (1, 2)$, the following upper bound holds.

Theorem 5. *For the scheduling problem with related machines,* $\mathrm{PoEF}_k \leq \min\left\{n, m, 2k\sqrt{\frac{m}{k-1}}\right\}$ *for any* $k \in (1, 2)$.

We now show that the two upper bounds proved in Theorems 4 and 5 are asymptotically tight.

Proposition 5. *For the scheduling problem with related machines, given any* $k \geq 1$*, there exists an instance for which* $\mathrm{PoEF}_k \geq \max\left\{1, \sqrt{\frac{m}{k}}\right\}$*. Moreover, for any* $k \in \left(1, \frac{3+\sqrt{11}}{6}\right)$*, there exists an instance for which* $\mathrm{PoEF}_k = \Omega\left(\sqrt{\frac{m}{k-1}}\right)$*.*

Note that, for each $k \in \left[\frac{3+\sqrt{11}}{6}, 2\right)$, it holds $\mathrm{PoEF}_k \leq 2k\sqrt{\frac{m}{k-1}} = O(\sqrt{m})$ by Theorem 5, while, by Proposition 5, we have $\mathrm{PoEF}_k \geq \max\left\{1, \sqrt{\frac{m}{k}}\right\} = \Omega(\sqrt{m})$.

This shows that all the bounds on the PoEF_k presented in this subsection are asymptotically tight.

3.3   Unrelated Machines

In this subsection, we consider the scheduling problem with unrelated machines. In this case we are able to give an exact characterization of the price of k-envy-freeness as witnessed by the upper and lower bounds given in the following.

Theorem 6. *For the scheduling problem with unrelated machines,* $\text{PoEF}_k \leq \left(1 + \frac{1}{k}\right)^{\min\{n,m\}-1}$ *for any* $k \geq 1$.

Proposition 6. *For the scheduling problem with unrelated machines, given any* $k \geq 1$ *and* $\epsilon > 0$, *there exists an instance for which* $\text{PoEF}_k = \left(1 + \frac{1}{k+\epsilon}\right)^{\min\{n,m\}-1}$.

3.4   Complexity

An important feature of the proofs we used to upper bound the PoEF_k in the various cases is that they rely on polynomial time algorithms constructing k-envy-free assignments of reasonable low makespan. In particular, for identical machines with $k \in (1, 2)$, the algorithm used in the proof of Theorem 2 does not require any information to be executed; hence, it indeed constructs a k-envy-free assignment whose performance guarantee coincides with the upper bound on the PoEF_k.

For all the other cases, given an input solution **S**, all the designed algorithms rearrange the allocations defined by **S** so as to obtain in polynomial time a k-envy-free assignment **N** such that $\mathcal{M}(\mathbf{N}) \leq \text{PoEF}_k \cdot \mathcal{M}(\mathbf{S})$. This means that, when given as input a solution **S** such that $\mathcal{M}(\mathbf{S}) \leq \alpha \cdot \mathcal{M}(\mathbf{O})$, each algorithm computes in polynomial time a k-envy-free assignment **N** such that $\mathcal{M}(\mathbf{N}) \leq \alpha \cdot \text{PoEF}_k \cdot \mathcal{M}(\mathbf{O})$. By recalling that there exists a PTAS for the scheduling problem with related and identical machines and a 2-approximation algorithm for the case of unrelated ones and by the fact that our upper bounds of PoEF_k are tight or asymptotically tight, it follows that we are able to compute in polynomial time k-envy-free assignments of best possible quality, when dealing with related and identical machines, and of at least half the best possible quality, when dealing with unrelated ones.

4   Extensions

4.1   Restricted Scheduling

In this subsection, we focus on the case in which a job cannot be assigned to every machine: for any job i there is a set $M_i \subseteq \{1, \ldots, m\}$ containing the machines being admissible for job i. We have to clarify the definition of k-envy-freeness in this setting: in assignment $\mathbf{N}$, machine j k-envies machine j' if $C_j(N_j) > kC_j(N_{j'})$ and for each job $i \in N_{j'}$, $j \in M_i$. It can be easily verified that also in the case of restricted scheduling an envy–free solution always exists. In fact, starting from any feasible assignment, an envy–free solution can be obtained as follows: while there exist two machines j and j' such that j k-envies j', assign to machine j also all the jobs of machine j'.

As already remarked in the introduction, the setting of unrelated machines studied in Subsection 3.3 includes the case of restricted (unrelated) machines, as it is possible to assign a very large value to p_{ij} whenever machine $j \notin M_i$, so that neither an optimal solution, nor a k-envy-free one minimizing the makespan can assign a job to a machine not being admissible for it. Therefore, for the restricted case, it remains to analyze the related and identical settings. In the related case, for which the upper bound provided in Theorem 6 clearly holds, it is possible to modify the instance exploited in Proposition 6 so that it becomes a restricted instance of the related setting, and the following theorem holds.

Proposition 7. *For the restricted scheduling problem with related machines, given any $k \geq 1$ and $\epsilon > 0$, there exists an instance for which* $\text{PoEF}_k = \left(1 + \frac{1}{k+\epsilon}\right)^{\min\{n,m\}-1}$.

Finally, for the case of identical machines, a trivial upper bound equal to $\min\{n, m\}$ holds as any solution (k-envy–free or not) approximates an optimal one by at most $\min\{n, m\}$, and the following lower bound holds.

Proposition 8. *For the restricted scheduling problem with identical machines, given any $k \geq 1$, there exists an instance for which* $\text{PoEF}_k = \Omega(\min\{n, m\})$.

4.2   Sum of Completion Times

In this subsection, we extend our study to the case where the objective is that of minimizing the sum of the completion times of all the machines. We refer to such a case as scheduling SUM problem. Formally, given an assignment $\mathbf{N}$ where $C_j(N_j)$ denotes the completion time of machine j under the assignment $\mathbf{N}$, an optimal assignment minimizes the sum $\sum_{j=1}^{m} C_j(N_j)$. We notice that an optimal solution can be trivially determined by assigning each job to the machine providing it the minimum possible processing time.

By exploiting ideas used for the minimum makespan we are able to show that $\text{PoEF}_k = 1$ for related (and then identical) machines, and that $\text{PoEF}_k = \left(1 + \frac{1}{k}\right)^{\min\{n,m\}-1}$ for unrelated machines. All the details will be given in the full version of the paper.

References

1. Anshelevich, E., Dasgupta, A., Kleinberg, J.M., Tardos, E., Wexler, T., Roughgarden, T.: The Price of Stability for Network Design with Fair Cost Allocation. SIAM Journal on Computing 38(4), 1602–1623 (2008)

2. Bulow, J., Roberts, J.: The Simple Economics of Optimal Auctions. The Journal of Political Economy 97(5), 1060–1090 (1989)
3. Brams, S.J., Taylor, A.D.: An Envy-Free Cake Division Protocol. The American Mathematical Monthly 102(1), 9–18 (1995)
4. Brams, S.J., Taylor, A.D.: Fair Division: From Cake-Cutting to Dispute Resolution. Cambridge University Press (1996)
5. Cohen, E., Feldman, M., Fiat, A., Kaplan, H., Olonetsky, S.: Envy-Free Makespan Approximation. SIAM Journal on Computing 41(1), 12–25 (2012)
6. Caragiannis, I., Kaklamanis, C., Kanellopoulos, P., Kyropoulou, M.: The Efficiency of Fair Division. Theory of Computing Systems 50(4), 589–610 (2012)
7. Christodoulou, G., Koutsoupias, E., Vidali, A.: A Lower Bound for Scheduling Mechanisms. Algorithmica 55(4), 729–740 (2009)
8. Dubins, L.E., Spanier, E.H.: How to cut a cake fairly. American Mathematical Monthly 68, 1–17 (1961)
9. Foley, D.: Resource allocation and the public sector. Yale Economics Essays 7, 45–98 (1967)
10. Feldman, M., Fiat, A., Leonardi, S., Sankowski, P.: Revenue maximizing envy-free multi-unit auctions with budgets. In: Proceedings of the ACM Conference on Electronic Commerce (EC), pp. 532–549 (2012)
11. Fiat, A., Levavi, A.: Tight Lower Bounds on Envy-Free Makespan Approximation. In: Goldberg, P.W. (ed.) WINE 2012. LNCS, vol. 7695, pp. 553–558. Springer, Heidelberg (2012)
12. Graham, R.L.: Bounds for Certain Multiprocessing Anomalies. Bell System Technical Journal 45, 1563–1581 (1966)
13. Goldberg, A.V., Hartline, J.D.: Envy-free auctions for digital goods. In: Proceedings of the ACM Conference on Electronic Commerce (EC), pp. 29–35 (2003)
14. Guruswami, V., Hartline, J.D., Karlin, A.R., Kempe, D., Kenyon, C., McSherry, F.: On profit-maximizing envy-free pricing. In: Proceedings of the ACM-SIAM Symposium on Discrete Algorithms (SODA), pp. 1164–1173 (2005)
15. Hartline, J., Ieong, S., Mualem, A., Schapira, M., Zohar, A.: Multi-dimensional envy-free scheduling mechanisms. Technical Report 1144, The Hebrew Univ (2008)
16. Hochbaum, D.S., Shmoys, D.B.: Using Dual Approximation Algorithms for Scheduling Problems: Theoretical and Practical Results. Journal of ACM 34(1), 144–162 (1987)
17. Hochbaum, D.S., Shmoys, D.B.: A Polynomial Approximation Scheme for Scheduling on Uniform Processors: Using the Dual Approximation Approach. SIAM Journal on Computing 17(3), 539–551 (1988)
18. Koutsoupias, E., Vidali, A.: A Lower Bound of $1 + \varphi$ for Truthful Scheduling Mechanisms. Algorithmica 66(1), 211–223 (2013)
19. Lenstra, J.K., Shmoys, D.B.: E Tardos. Approximation algorithms for scheduling unrelated parallel machines. Mathematical Programing 46, 259–271 (1990)
20. Myerson, R.B.: Optimal Auction Design. Mathematics of Operations Research 6, 58–73 (1981)
21. Nagura, J.: On the interval containing at least one prime number. Proceedings of the Japan Academy, Series A 28, 177–181 (1952)
22. Nisan, N., Ronen, A.: Algorithmic Mechanism Design. Games and Economic Behavior 35(1-2), 166–196 (2001)
23. Vickrey, W.: Counterspeculation, Auctions, and Competitive Sealed Tenders. Journal of Finance 16, 8–37 (1961)

On the Complexity of Some Ordering Problems

Beate Bollig[*]

TU Dortmund, LS2 Informatik, Germany

Abstract. Two different ordering problems are investigated. Ordered binary decision diagrams (OBDDs) are a popular data structure for Boolean functions. Some applications work with a restricted variant called complete OBDDs. This model has also been investigated in complexity theory, e.g., in property testing. It is well-known that the size of an OBDD for the representation of a given function may depend significantly on the chosen variable ordering but the computation of an optimal ordering is NP-hard. Since optimal variable orderings for OBDDs are not necessarily optimal for the complete model, the complexity to find an optimal variable ordering for complete OBDDs is investigated. Here, using a new reduction idea it is shown that the problem is NP-hard. Among the many areas of applications OBDDs have been used in the design and analysis of implicit graph algorithms where the choice of a good vertex encoding is of additional importance to represent a given input graph in small size. The computational complexity of the vertex encoding problem is unknown but in the paper a first step is done to determine its complexity by showing that a restricted case is NP-hard.

1 Introduction

Ordered binary decision diagrams (OBDDs), introduced by Bryant in his seminal paper in 1986 [9], are a popular data structure for Boolean functions and among the many areas of applications are verification, model checking, computer aided design, and also the design and analysis of implicit graph algorithms (for a survey see, e.g., [25]). Complete OBDDs are a restricted variant of OBDDs which are closely related to deterministic finite automata for so-called Boolean languages L, where $L \subseteq \{0,1\}^n$, $n \in \mathbb{N}$. Already in [19] the model has been applied to design parallel algorithms for fundamental Boolean operations by a breadth first search approach. Furthermore, complete OBDDs have been used in the analysis of implicit graph algorithms [21,27] and to represent simple games and to solve computational problems on them [1]. Moreover, also in complexity theory complete OBDDs have been investigated, e.g., in property testing (see, e.g., [8,14,17,20]). Therefore, complete OBDDs seem to be a fundamental model. Maybe the most important issue of OBDDs is the possibility to choose the variable ordering and in applications it is an important problem to pick a good one since the size of an OBDD representing a function f, defined on n Boolean

---

[*] The author is supported by DFG project BO 2755/1-2.

E. Csuhaj-Varjú et al. (Eds.): MFCS 2014, Part II, LNCS 8635, pp. 118–129, 2014.

variables and essentially dependent on all of them, heavily depends on the chosen variable ordering and may vary between linear and exponential size with respect to n. An example for such a function is the most significant bit of binary addition. Therefore, the choice of good variable orderings is a key problem for the usability of OBDDs. It is well-known that we cannot expect efficient algorithms for the computation of optimal variable orderings for a function given by an OBDD representation since the corresponding optimization problem is NP-hard [6]. Even more we cannot hope to design efficient approximation algorithms unless NP = P [22]. Optimal variable orderings for OBDDs do not necessarily carry over to optimal variable orderings for the complete model representing the same function. An example is the multiplexer, also called direct storage access function, (for a formal definition see Section 4). Hence, it seems to be a natural question to ask for the computational complexity of the variable ordering problem for complete OBDDs. The NP-completeness or nonapproximability proofs for OBDDs do not work for the complete model because one key idea in the reductions is the property that in OBDDs not all variables have to be tested on a path from the source to one of the sinks. Here, we prove that the decision variant of the problem to compute an optimal variable ordering for a function given by a complete OBDD is NP-complete. The problems for OBDDs and complete OBDDs look quite similar but it is unclear how to relate the complexity of the problems directly. Therefore, for the reduction we choose the same NP-complete problem as in the NP-completeness proof for OBDDs but our construction is different and we use a new reduction idea. Although many exponential lower bounds on the size of (complete) OBDDs for Boolean functions are known and the method how to obtain such bounds is simple, there are only few functions where the size of (complete) OBDDs is asymptotically known exactly (see, e.g., [2,5,7].) Therefore, the proof may also be interesting on its own right in order to strengthen the ability to prove tight lower bounds. Moreover, knowledge on the relation of good variable orderings and Boolean functions is fundamental for the design of heuristics to compute good orderings.

Given the rapid growth of application-based networks, an heuristic approach to deal with very large structured graphs are implicit OBDD-based graph algorithms (see, e.g., [12,15,21,27]). Vertices of an input graph are binary encoded and the edge set of the input graph is represented by its characteristic function. Since OBDDs are able to take advantage over the presence of regular substructures, this approach leads sometimes to sublinear graph representations [13,16,18]. Here, the vertex encoding problem is of additional importance in order to deal with OBDDs of reasonable size (for a simple example see Section 4). The computational complexity of the vertex encoding problem is unknown but in the paper a first step is done to determine the complexity.

The rest of the paper is organized as follows. In Section 2 we recall the main definitions concerning OBDDs and complete OBDDs. Furthermore, we give a short summary on the known results how the sizes of the two models are related. Section 3 is devoted to the NP-completeness proof of the variable ordering problem for complete OBDDs. We describe carefully why the NP-completeness proof for OBDDs does not work for the complete variant. In the last section we start to investigate the computational complexity of the problem to improve a given vertex encoding with respect to the corresponding OBDD size in the implicit setting. For a restricted class of vertex encodings we can prove that the corresponding optimization problem is NP-hard. As a by-product there is a nice observation about optimal variable orderings and the conjunction of Boolean functions.

2   Preliminaries

We briefly recall the main notions concerning OBDDs and discuss the relation between OBDDs and complete OBDDs.

On (Complete) Ordered Binary Decision Diagrams. OBDDs are a popular dynamic data structure in areas working with Boolean functions, like circuit verification or model checking. (For a history of results on binary decision diagrams see, e.g., [25]).

Definition 1. Let $X_n = \{x_1, \ldots, x_n\}$ be a set of Boolean variables. A variable ordering π on X_n is given by a permutation on $\{1, \ldots, n\}$ leading to the ordered list $x_{\pi(1)}, \ldots, x_{\pi(n)}$ of the variables. A π-OBDD on X_n is a directed acyclic graph $G = (V, E)$ whose sinks are labeled by the Boolean constants 0 and 1 and whose non-sink (or decision) nodes are labeled by Boolean variables from X_n. Each decision node has two outgoing edges, one labeled by 0 and the other by 1. The edges between decision nodes have to respect the variable ordering π, i.e., if an edge leads from an x_i-node to an x_j-node, then $\pi^{-1}(i) < \pi^{-1}(j)$ (x_i precedes x_j in $x_{\pi(1)}, \ldots, x_{\pi(n)}$). Each node v represents a Boolean function $f_v \in B_n$, i.e., $f_v : \{0,1\}^n \to \{0,1\}$, defined in the following way. In order to evaluate $f_v(b)$, $b \in \{0,1\}^n$, start at v. After reaching an x_i-node choose the outgoing edge with label b_i until a sink is reached. The label of this sink defines $f_v(b)$. The size of a π-OBDD G is equal to the number of its decision nodes. A π-OBDD of minimal size for a given function f and a fixed variable ordering π is unique up to isomorphism. A π-OBDD for a function f is called reduced if it is the minimal π-OBDD for f. The π-OBDD size or complexity of a function f, denoted by π-OBDD(f), is the size of the reduced π-OBDD representing f. An OBDD is a π-OBDD for an arbitrary variable ordering π. The OBDD size of f is the minimum of all π-OBDD(f).

A variable ordering is called a *natural variable ordering* if π is the identity $1, 2, \ldots, n$. Obviously a variable ordering π can be identified with the corresponding ordering $x_{\pi(1)}, \ldots, x_{\pi(n)}$ of the variables if the meaning is clear from the context. A *k-interleaved variable ordering* on k variable vectors $x^{(i)} = (x_1^{(i)}, \ldots, x_n^{(i)})$, $1 \leq i \leq k$, is defined in the following way:

$$x_{\tau(1)}^{(1)}, \ldots, x_{\tau(1)}^{(k)}, x_{\tau(2)}^{(1)}, \ldots, x_{\tau(2)}^{(k)}, \ldots, x_{\tau(n)}^{(k)},$$

where τ is a permutation on $\{1, 2, \ldots, n\}$.

Complete OBDDs are closely related to nonuniform finite automata for Boolean languages L, where $L \subseteq \{0,1\}^n$ (see, e.g., Section 3.2 in [25]). Here, there are only edges between nodes labeled by neighboring variables, i.e., if an edge leads from an x_i-node to an x_j-node, then $\pi^{-1}(i) = \pi^{-1}(j) - 1$.

Definition 2. *An OBDD on X_n is complete if all paths from the source to one of the sinks have length n. The width of a complete OBDD is the maximal number of nodes labeled by the same variable. A complete π-OBDD of minimal size for a given function f and a fixed variable ordering π is unique up to isomorphism. A π-OBDD for a function f is called* quasi-reduced *if it is the minimal complete π-OBDD for f. The complete π-OBDD size of a function f, denoted by π-QOBDD(f), is the size of the quasi-reduced π-OBDD representing f. A complete OBDD, or QOBDD for short, is a complete π-OBDD for an arbitrary variable ordering π. The complete OBDD size or QOBDD size of f is the minimum of all π-QOBDD(f).*

Complete OBDDs with respect to natural variable orderings differ from deterministic finite automata only in the minor aspect that there can also be nodes that represent the constant function 0.

Let f be a Boolean function on the variables $x_1, \ldots, x_n$. The *subfunction* $f_{|x_i = c}$, $1 \leq i \leq n$ and $c \in \{0,1\}$, is defined as $f(x_1, \ldots, x_{i-1}, c, x_{i+1}, \ldots, x_n)$. A function f *depends essentially* on a Boolean variable z if $f_{|z=0} \neq f_{|z=1}$. The size of the (quasi-)reduced π-OBDD representing f is described by the following result.

Proposition 1 ([23]). *Let $a_1, \ldots, a_{i-1} \in \{0,1\}$. The number of $x_{\pi(i)}$-nodes in the quasi-reduced (reduced) π-OBDD for f is equal to the number of different subfunctions $f_{|x_{\pi(1)}=a_1, \ldots, x_{\pi(i-1)}=a_{i-1}}$ (that essentially depend on $x_{\pi(i)}$).*

In a reduced OBDD each node encodes a different function, whereas in a quasi-reduced OBDD each node labeled by the same variable represents a different function.

On the Relation between OBDDs and QOBDDs. Wegener has compared the size of quasi-reduced OBDDs with the size of reduced OBDDs for functions defined on n Boolean variables. For the natural variable ordering he has proved that the quotient is at most $1 + \mathcal{O}(2^{-n/3} \cdot n)$ for almost all Boolean functions, i.e., all but a fraction of $\mathcal{O}(2^{-n/3})$. This result does not rule out the possibility that for many functions there exists some ordering of the variables where the difference is significantly larger but it has also been shown that the maximal quotient of the quasi-reduced OBDD and the size of the reduced OBDD with respect to the same variable ordering for a function f where the maximum is taken over all variable orderings is at most $1 + \mathcal{O}(2^{-n/3} \cdot n)$ for almost all Boolean functions (Theorem 3 and 4 in [26]). Nevertheless, one may ask for the quotient of the representation sizes not for almost all but for some important Boolean functions. It is obvious that π-QOBDD(f) $\leq (n+1)\pi$-OBDD(f) for all Boolean functions $f \in B_n$. Furthermore, it is not difficult to see that π-QOBDD(f) = $\Theta(n \cdot (\pi\text{-OBDD}(f)))$ for some function f that depends essentially on all n variables. The multiplexer (see Section 4 for the formal definition) is an example for such a function and a variable ordering where the address variables are tested before the data variables. In [7] the question whether there exists a Boolean function $f \in B_n$ that depend essentially on all variables and QOBDD(f_n) = $\Theta(n \cdot \text{OBDD}(f_n))$ has been answered in the affirmative. Therefore, we can conclude that the OBDD size of a function may be a size factor of $\Theta(n)$ smaller that its QOBDD size. The multiplexer has been a good candidate for a function with the largest possible gap between the OBDD and the QOBDD size but it has turned out that the multiplexer only leads to a size gap of $\Theta(n/\log n)$. Despite all these results we hardly know anything about the relation between the variable orderings that lead to minimal OBDDs and variable orderings for minimal QOBDDs for a given function f.

3   On the Variable Ordering Problem for Complete OBDDs

The complexity of the problem OPTIMAL QOBDD is investigated.

Definition 3 (Optimal QOBDD). *Given a QOBDD G and a size bound s, the answer to the problem OPTIMAL QOBDD is yes iff the function represented by G can be represented by a QOBDD (respecting an arbitrary variable ordering) with at most s nodes.*

Theorem 1. *The problem OPTIMAL QOBDD is NP-complete.*

Sketch of Proof
The problem OPTIMAL QOBDD is in NP. A QOBDD can be guessed. The equivalence of QOBDDs with respect to different variable orderings can be verified similarly to the case for OBDDs in deterministic polynomial time [11]. As in [6,24] our NP-hardness proof uses a polynomial time reduction from the well-known NP-complete problem Optimal Linear Arrangement (OLA for short).

Definition 4 (Optimal Linear Arrangement). *Given an undirected graph $H = (V = \{1, 2, \ldots, n\}, E)$ and a bound b, the answer to the problem Optimal Linear Arrangement (OLA) is yes iff there is a permutation τ on $\{1, 2, \ldots, n\}$ such that*

$$cost(\tau) := \sum_{\{u,v\} \in E} |\tau^{-1}(u) - \tau^{-1}(v)| \leq b.$$

The cost of τ measures the length of all edges if the vertices of H are arranged in linear order with respect to τ.

In the following we present a polynomial time reduction from OLA to OPTIMAL QOBDD. Let $H = (V, E)$ and b be given and $m := |E|$. W.l.o.g. we assume that the degree of each vertex, i.e., the number of its neighbors, is at least 2. For the polynomial reduction we have to transform the input (H, b) for OLA into an input (G, s) for OPTIMAL QOBDD such that the QOBDD size of the function represented by G is at most s iff the cost of an optimal linear arrangement for H is at most b. For the ith edge $\{j, k\}$, $1 \le i \le m$ and $j, k \in \{1, 2, \ldots, n\}$, we introduce an edge-function $f_i(v_1, v_2, \ldots, v_n) = (v_j \vee v_k)$. It is easy to see that the size of a quasi-reduced QOBDD representing f_i with respect to a variable ordering $v_{\pi(1)}, v_{\pi(2)}, \ldots, v_{\pi(n)}$ is $n + |\pi^{-1}(j) - \pi^{-1}(k)|$ plus additional nodes representing the constant function 0.

Now, we are faced with two problems. First, in order to obtain a single Boolean function the edge functions have to be combined to one function. Second, we have to make sure that representations for different edge functions do not share nodes. Also in the NP-hardness proof for OBDDs for each edge of the input graph for OLA a corresponding Boolean function has been defined (which is different from our edge function). In order to avoid the sharing of nodes for different edge functions, in case of OBDDs different edge functions have been defined on disjoint sets of variables. To ensure that the orderings for the variables of the different edge functions are not completely independent but correspond to an ordering of the vertices of the input graph for OLA, another function has been added. Its representation size is very large and and its OBDD size is optimal if all variables that correspond to the same vertex in the input graph for OLA are tested one after another. Unfortunately, this idea does not work for QOBDDs. Hence, we use another construction to combine the edge function and to prevent the unwanted sharing of BDD nodes. The idea is to use two simple functions to frame the edge functions. For the framing we use counting functions. We will show that there exists an optimal variable ordering where the variables that represent the vertices of the input graph for OLA are tested in the middle. Therefore, in the corresponding QOBDD the representations for the edge functions cannot share BDD nodes.

For a variable vector $z = (z_1, \ldots, z_n)$, $n \in \mathbb{N}$, let $\| z \|$ be $\sum_{i=1}^{n} z_i$. The function $F \in B^{2m+n}$ is defined on the variable vectors $u = (u_1, u_2, \ldots, u_m)$, $v = (v_1, v_2, \ldots, v_n)$, and $w = (w_1, w_2, \ldots, w_m)$, and

$$F(u, v, w) := \bigvee_{i=1}^{m} (\| u \| = i) \wedge f_i(v) \wedge (\| w \| = i).$$

The u- and the w-variables are called weight variables and the v-variables are called vertex variables since they represent the vertices of the input graph H for OLA. We call a vertex variable v_j essential for an edge function f_i iff j is incident to the ith edge in H. F is symmetric on the u-variables and on the w-variables, respectively. Here, a function is symmetric on two variables x_i and x_j if the function does not change when exchanging the variables x_i and x_j, and symmetry is an equivalent relation on the set of variables a function is defined on. If $P = \{i_1, \ldots, i_m\}$ is the set of positions of the u-variables (or w-variables, respectively) in a variable ordering π, it does not matter which u-variable is tested on which position in P for the corresponding size of a QOBDD for F. Moreover, the roles of the u-and the w-variables are exchangeable, therefore, in the remaining part of the section we assume w.l.o.g. that the u-variables are tested in the ordering $u_1, u_2, \ldots, u_m$ and the w-variables in $w_1, w_2, \ldots, w_m$ and u_1 is the first variable of all u- and w-variables.

Our transformation computes the (quasi-reduced) OBDD representing F with respect to the ordering $u_1, u_2, \ldots, u_m, v_1, v_2, \ldots, v_n, w_1, w_2, \ldots, w_m$ in polynomial time.

A *sandwich variable ordering* is a variable ordering where the v-variables are tested between the u- and the w-variables, and all u-variables as well as all w-variables are tested one after another.

The following lemma is not difficult to prove.

Lemma 1. *Let π be a sandwich variable ordering and let π' be the subordering of π on the v-variables. Then the π-QOBDD size of $F(u, v, w)$ is*

$$m \cdot (m+1)/2 + n \cdot m + cost(\pi') + m \cdot (m+1)/2 + (m-1) + (n+m).$$

We are now able to define the size bound s in our reduction:

$$s := m \cdot (m+1)/2 + n \cdot m + b + m \cdot (m+1)/2 + (m-1) + (n+m).$$

In order to prove the correctness of our reduction we have to show that the input graph H for OLA has a linear arrangement whose cost is bounded by b iff F can represented by a QOBDD with at most s nodes. Using Lemma 1 the only-if-part is easy. The if-part of the correctness proof is more involved. By our considerations above it remains to prove that some optimal variable ordering of $F = (u, v, w)$ is a sandwich variable ordering.

The idea is to change a given (optimal) variable ordering π in three phases until it is a sandwich variable ordering. If each phase does not increase the size of the BDD representation, we are done. In all phases we do not change the ordering among the u-variables, among the v-variables, and among the w-variables, respectively. Remember that the roles of the u- and the w-variables are symmetric, therefore, we assume w.l.o.g. that the first weight variable is a u-variable. First, we ensure that all u-variables are tested before all w-variables. We do this by exchanging the positions of the first w-variable in the variable ordering and the following u-variable without increasing the size of the corresponding QOBDD. Since the procedure can be iterated, we are done.

Lemma 2. *Let π be a variable ordering on the u-, v-, and w-variables and let i_k be the position of the variable u_k and j_k be the position of the variable w_k, $1 \leq j \leq m$. Furthermore, let j_1 be between i_l and i_{l+1}, $l \in \{1, \ldots, m-1\}$. Let π' be the variable ordering where the variable w_1 is at position i_{l+1} and u_{l+1} is at position j_1 and all other variables are ordered according to π. Then π'-QOBDD(F) is not larger than π-QOBDD(F).*

Next, we change the variable ordering in such a way that the u-variables are tested in the beginning.

Lemma 3. *Let π be a variable ordering on the u-, v-, and w-variables where all u-variables are before the w-variable. Let π' be the variable ordering that starts with the u-variables followed by the remaining variables in the same order as in π. Then π'-QOBDD(F) is not larger than π-QOBDD(F).*

Finally, we modify the variable ordering such that the w-variables are tested in the end. This is the most laborious part of the correctness proof and we have to count the QOBDD nodes very carefully using a sophisticated accounting method solving some combinatorial problems which may be interesting on their own right. The w-variables are divided into blocks of maximal length that consist only of w-variables. If the number of vertex variables between the u-variables and the first w-variable in the given variable ordering and the length of the first block of w-variables, is not too large, the first vertex variable after the first block of w-variables jumps in the ordering at the position just before the first block of w-variables. Otherwise, all vertex variables jump at once before the w-variables. In the first case the procedure is iterated until the w-variables are the last variables in the variable ordering.

Lemma 4. *Let π be a variable ordering on the u-, v-, and w-variables that starts with the u-variables. Let π' be the variable ordering where the u-variables are in the beginning of the ordering, the v-variables are ordered in the same suborder as in π, and the w-variables are the last variables in the ordering. Then π'-QOBDD(F) is not larger than π-QOBDD(F).*

By Lemma 2-4 we have shown the following result.

Corollary 1. *There exists a sandwich variable ordering that is optimal for the function $F(u, v, w)$.*

4   On the Optimal Vertex Encoding Problem

This section is devoted to the vertex encoding problem in the implicit setting. It seems to be a natural idea to represent highly regular graphs by means of data structures smaller than adjacency matrices or adjacency lists. Boolean encodings for the vertices can be used to characterize sets of vertices or edges by their characteristic Boolean functions, and data structures like OBDDs can be used to represent and manipulate the input graphs. Let $G = (V, E)$ be a graph with N vertices $v_0, \ldots v_{N-1}$ and $|z|_2 := \sum_{i=0}^{n-1} z_i 2^i$, where $z = (z_0, \ldots, z_{n-1}) \in \{0,1\}^n$ and $n = \lceil \log N \rceil$. Now, E can be represented by its characteristic function, where $x, y \in \{0,1\}^n$ and $\chi_E(x, y) = 1 \Leftrightarrow (|x|_2, |y|_2 < N) \wedge (v_{|x|_2}, v_{|y|_2}) \in E$. Undirected edges can be represented by symmetric directed ones.

Several very natural and large graph classes have OBDD representations which yield a much better space behavior than that of explicit representations (see, [13,16,18]). Obviously, besides the choice of the variable ordering, the encoding of the vertices can influence the size of the OBDD-based graph representations. A simple example is the following graph $G_n = (V, E)$ defined on 2^n vertices v_i, $0 \leq i \leq 2^n - 1$, and 2^{n-2} edges (see Figure 1). To present a bad vertex encoding, we start with the definition of the hidden weighted bit function (HWB) introduced by Bryant [10].

Definition 5. *The hidden weighted bit function* $\text{HWB}_n : \{0,1\}^n \to \{0,1\}$ *computes the bit* b_{sum} *on the input* $b = (b_1, \ldots, b_n)$, *where* $sum := \sum_{i=0}^{n} b_i$ *and* $b_0 := 0$.

The OBDD complexity HWB_n is exponential [10], and the best lower bound is $\Omega(2^{n/5})$ [4]. Another interesting property of HWB_n is that exactly half of the inputs are mapped to the function value 1. Now, let $b^i = (b_0^i, \ldots, b_{n-1}^i)$ be the binary representation of an integer $i \in \{0, \ldots, 2^n - 1\}$. There exists an edge between a vertex v_i and a vertex v_j iff $i = 2^{n-1} + j$ and $b_R^i = (b_0^i, \ldots, b_{n-2}^i)$ is in $\text{HWB}_{n-1}^{-1}(1)$.

Fig. 1. The graph G_n used as an example for a good and a bad vertex encoding. G_n has 2^n vertices and 2^{n-2} edges.

It is not difficult to show that in this case the complexity of the OBDD representation for G_n is exponential. To be more precise, it can be proved that the size of an OBDD for the representation of the characteristic function of E is $\Omega(n^{-1/2} 2^{(n-1)/5})$ (for a similar proof see [3]). The isomorphic graph $G_n' = (V, E')$ has an edge between a vertex v_i and a vertex v_j iff $b_{n-1}^i = b_{n-2}^i = 1$, $b_{n-1}^j = b_{n-2}^j = 0$ and $b_k^i = b_k^j$ for $j \in \{0, \ldots, n-3\}$ (or vice versa). Using Proposition 1 it is easy to see that the π-OBDD size for the characteristic function of the edge set E' is $4 + 3(n - 2)$, where π is the 2-interleaved variable ordering $x_{n-1}, y_{n-1}, x_{n-2}, \ldots, x_0, y_0$. There are four nodes to check whether $x_{n-1}, y_{n-1}, x_{n-2}, y_{n-2}$ are set to 1010 and an equality check for $x_R = (x_{n-3}, \ldots, x_0)$ and $y_R = (y_{n-3}, \ldots, y_0)$. Summarizing, we have seen that a re-encoding of the vertices of a given graph may have a large impact on the OBDD representation size. Therefore, we define the following problem.

Definition 6 (Optimal Vertex Encoding). *Let $B_{n,n}$ be the set of multiple output Boolean functions that map inputs from $\{0, 1\}^n$ to the set $\{0, 1\}^n$, and let g be a bijective function in $B_{n,n}$. Furthermore, let $\mathcal{X}_E$ be the characteristic Boolean function of an implicitly defined graph $G = (V, E)$ and $\mathcal{X}_{E_g}$ be the characteristic Boolean function of the graph $G_g = (V, E_g)$ that is isomorphic to G with respect to g.*

Let π be an arbitrary variable ordering on $2n$ variables. Given a π-OBDD H representing $\mathcal{X}_E$ and a size bound s, the answer to the problem Optimal Vertex Encoding (OVE) is yes iff there exists a bijective function h, $h \in B_{n,n}$, such that the π-OBDD size for $\mathcal{X}_{E_h}$ is at most s.

The computational complexity of the problem OVE is unknown. We start to investigate a restricted class of vertex encodings, where the encoding function h, $h : \{0, 1\}^n \to \{0, 1\}^n$, can be described by a permutation of the variables h is defined on, i.e., $h(x_0, \ldots, x_{n-1}) = x_{\pi(0)}, \ldots, x_{\pi(n-1)}$ for a given permutation π on the set $\{0, \ldots, n-1\}$. Let ROVE (Restricted Optimal Vertex Encoding) be the corresponding optimization problem.

Theorem 2. *The problem* ROVE *is* NP-*hard.*

The problem ROVE looks similar to the classical NP-complete problem Optimal OBDD: given an OBDD for a function f_n and a size bound s in $\mathbb{N}$, we ask whether there exists an OBDD with respect to an arbitrary variable ordering representing f_n with at most s nodes [6]. Therefore, it seems to be obvious to choose the problem Optimal OBDD for a polynomial time reduction to ROVE. We have to transform the input (F, s) for Optimal OBDD into an input (H, s') for ROVE in polynomial time, where H is an OBDD with respect to a fixed variable ordering π representing the characteristic function of the edge relation of an input graph G. The transformation has to guarantee that there exists an OBDD with at most s nodes representing the same function as the OBDD F iff there exists a restricted vertex encoding h such that the π-OBDD size for the implicit representation of G_h is at most s'.

First, we discuss an approach which does not work. The motivation for the investigation is to learn more about the properties of variable orderings for (complete) OBDDs hoping that more insights lead to better heuristics for the corresponding problems. The idea is the following. We use a similar construction as presented above for the bad vertex encoding. Let F be a π-OBDD for the function f_n defined on the variables $x_0, \ldots, x_{n-1}$. We construct a bipartite graph G on 2^{n+1} vertices v_i, $0 \leq i \leq 2^{n+1} - 1$, and $|f_n^{-1}(1)|$ edges. Again let $b^i = (b_0^i, \ldots, b_n^i)$ be the binary representation of an integer $i \in \{0, \ldots, 2^{n+1} - 1\}$. There exists an edge between a vertex v_i and a vertex v_j iff $i = 2^n + j$ and $b_R^i = (b_0^i, \ldots, b_{n-1}^i)$ is an input in $f_n^{-1}(1)$. If F is a π-OBDD for f_n, we can easily construct in polynomial time an OBDD with respect to a 2-interleaved variable ordering according to π on the variables $x = (x_0, \ldots, x_{n-1})$ and $y = (y_0, \ldots, y_{n-1})$ representing the function $f_n'(x, y) := f_n(x) \wedge \mathrm{EQ}_n(x, y)$. Next, we extend the 2-interleaved variable ordering by starting with two new variables x_n, y_n. Afterwards, we compute in polynomial time another OBDD H with respect to the new variable ordering for the function $(x_n \oplus y_n) \wedge f_n'(x, y)$. It is not difficult to see that H represents the characteristic function of G's edge relation. The intuition is that an optimal 2-interleaved variable ordering for the representation of G's edge relation leads to an optimal variable ordering for the representation of f_n by deleting the variable x_n and all y-variables. The following proposition gives a hint why this polynomial transformation fails.

Proposition 2. *Let f_n and g_n be Boolean functions in B_n, and let π be an optimal variable ordering for f_n, i.e., π-OBDD$(f_n) = $ OBDD(f_n). Furthermore the function g_n has the property that the size of a reduced OBDD representing g_n is independent of the choice of the variable ordering, i.e., π'-OBDD$(g_n) = $ OBDD(g_n) for every variable ordering π'. Let h_n be the conjunction of f_n and g_n. Then π is not necessarily an optimal variable ordering for the OBDD representation of h_n.*

Proof. Our counterexample to prove Proposition 2 is the following one. The function g_n is the parity function and f_n is the multiplexer, often also called direct storage access function, which is defined in the following way.

Definition 7. *Let $n = 2^k$. The multiplexer* MUX_n *is defined on $n + k$ variables* $a_{k-1}, \ldots, a_0, x_0, \ldots, x_{n-1}$. *The output of* $\mathrm{MUX}_n(a, x)$ *is* $x_{|a|_2}$.

The variable ordering π is $a_0, a_1, \ldots, a_k, x_0, x_1, \ldots, x_{n-1}$ and π' is the variable ordering $a_{k-1}, a_{k-2}, \ldots, a_{k-m}, x_0, x_1, \ldots, x_{n-1}, a_{k-m-1}, \ldots, a_0$. The variable ordering π is optimal for the multiplexer [5]. Using Proposition 1 it is not difficult to see that π-OBDD(h) is $\Theta(n^2)$ but π'-OBDD(h) is $\mathcal{O}(n^2/\log n)$. The proof does also work for the combination of f_n and g_n by disjunction.   □

Proposition 2 justifies why in the transformation above the size bound s' for the input of ROVE cannot easily be adapted. In the remaining part of the section we present the idea how to prove Theorem 2. Again the aim is to construct a polynomial reduction from Optimal OBDD to ROVE. Let F be a π-OBDD for the function f_n defined on the variables $x_0, \ldots, x_{n-1}$. We construct a bipartite graph G on 2^{n+1} vertices v_i, $0 \le i \le 2^{n+1} - 1$, and $2^n \cdot |f^{-1}(1)|$ edges. Again let $b^i = (b_0^i, \ldots, b_n^i)$ be the binary representation of an integer $i \in \{0, \ldots, 2^{n+1} - 1\}$. There exists an edge between a vertex v_i and a vertex v_j iff $i \ge 2^n$, $j \le 2^n$, and $b_R^i = (b_0^i, \ldots, b_{n-1}^i)$ is in $f_n^{-1}(1)$.

If F is a π-OBDD for f_n, we can easily construct in linear time two OBDDs H_x and H_y with respect to a 2-interleaved variable ordering according to π on the variables $x = (x_0, \ldots, x_{n-1})$ and $y = (y_0, \ldots, y_{n-1})$ representing the functions $f_n(x)$ and $f_n(y)$. Next, we extend the 2-interleaved variable ordering adding two new variables x_n and y_n at the beginning. Afterwards, we compute in linear time two OBDDs H_x' and H_y' for the functions $(x_n \overline{y_n}) \wedge f_n(x)$ and $(\overline{x_n} y_n) \wedge f_n(y)$. In the last step we construct an OBDD H for the disjunction of these two functions. It is not difficult to see that H represent the characteristic function of G's edge relation. Let s be the bound of the size in the input for Optimal OBDD, then we set the size bound s' in the input for ROVE to $2s + 3$. It remains to prove the correctness of the polynomial reduction.

References

1. Berghammer, R., Bolus, S.: On the use of binary decision diagrams for solving problems on simple games. European Journal of Operational Research 222(3), 529–541 (2012)
2. Bollig, B.: On the size of (generalized) OBDDs for threshold functions. Inf. Process. Lett. 109(10), 499–503 (2009)
3. Bollig, B.: On symbolic representations of maximum matchings and (un)directed graphs. In: Calude, C.S., Sassone, V. (eds.) TCS 2010. IFIP AICT, vol. 323, pp. 286–300. Springer, Heidelberg (2010)
4. Bollig, B., Löbbing, M., Sauerhoff, M., Wegener, I.: On the complexity of the hidden weighted bit function for various BDD models. Theoretical Informatics and Applications 33, 103–115 (1999)
5. Bollig, B., Range, N., Wegener, I.: Exact OBDD bounds for some fundamental functions. Theory of Computing Systems 47(2), 593–609 (2010)
6. Bollig, B., Wegener, I.: Improving the variable ordering of OBDDs is NP-complete. IEEE Trans. Computers 45(9), 993–1002 (1996)

7. Bollig, B., Wegener, I.: Asymptotically optimal bounds for OBDDs and the solution of some basic OBDD problems. Journal of Computing and System Science 61(3), 558–579 (2000)
8. Brody, J., Matulef, K., Wu, C.: Lower bounds for testing computability by small width OBDDs. In: Ogihara, M., Tarui, J. (eds.) TAMC 2011. LNCS, vol. 6648, pp. 320–331. Springer, Heidelberg (2011)
9. Bryant, R.: Graph-based algorithms for boolean function manipulation. IEEE Trans. Computers 35(8), 677–691 (1986)
10. Bryant, R.: On the complexity of VLSI implementations and graph representations of boolean functions with application to integer multiplication. IEEE Trans. Computers 40, 205–213 (1991)
11. Fortune, F., Hopcroft, J., Schmidt, E.: The complexity of equivalence and containment for free single variable program schemes. In: Ausiello, G., Böhm, C. (eds.) ICALP 1978. LNCS, vol. 62, pp. 227–240. Springer, Heidelberg (1978)
12. Gentilini, R., Piazza, C., Policriti, A.: Symbolic graphs: linear solutions to connectivity related problems. Algorithmica 50(1), 120–158 (2008)
13. Gillé, M.: OBDD-based representation of interval graphs. In: Brandstädt, A., Jansen, K., Reischuk, R. (eds.) WG 2013. LNCS, vol. 8165, pp. 286–297. Springer, Heidelberg (2013)
14. Goldreich, O.: On testing computability by small width OBDDs. In: Serna, M., Shaltiel, R., Jansen, K., Rolim, J. (eds.) APPROX and RANDOM 2010. LNCS, vol. 6302, pp. 574–587. Springer, Heidelberg (2010)
15. Hachtel, G., Somenzi, F.: A symbolic algorithm for maximum flow in 0-1 networks. Formal Methods in System Design 10, 207–219 (1997)
16. Meer, K., Rautenbach, D.: On the OBDD size for graphs of bounded tree- and clique-width. Discrete Mathematics 309(4), 843–851 (2009)
17. Newman, I.: Testing membership in languages that have small width branching programs. SIAM J. Comput. 31(5), 1557–1570 (2002)
18. Nunkesser, R., Woelfel, P.: Representations of graphs by OBDDs. Discrete Applied Mathematics 157(2), 247–261 (2009)
19. Ochi, H., Yasuoka, K., Yajima, S.: Breadth-first manipulation of very large binary-decision diagrams. In: ICCAD, pp. 48–55 (1993)
20. Ron, D., Tsur, G.: Testing computability by width-two OBDDs. Theor. Comput. Sci. 420, 64–79 (2012)
21. Sawitzki, D.: The complexity of problems on implicitly represented inputs. In: Wiedermann, J., Tel, G., Pokorný, J., Bieliková, M., Štuller, J. (eds.) SOFSEM 2006. LNCS, vol. 3831, pp. 471–482. Springer, Heidelberg (2006)
22. Sieling, D.: The nonapproximability of OBDD minimization. Information and Computation 172(2), 103–138 (2002)
23. Sieling, D., Wegener, I.: NC-algorithms for operations on binary decision diagrams. Parallel Processing Letters 3, 3–12 (1993)
24. Tani, S., Hamagushi, K., Yajima, S.: The complexity of the optimal variable ordering problems of a shared binary decision diagram. In: Ng, K.W., Balasubramanian, N.V., Raghavan, P., Chin, F.Y.L. (eds.) ISAAC 1993. LNCS, vol. 762, pp. 389–396. Springer, Heidelberg (1993)
25. Wegener, I.: Branching programs and binary decision diagrams: theory and applications. SIAM (2000)
26. Wegener, I.: The size of reduced OBDDs and optimal read-once branching programs for almost all boolean functions. IEEE Trans. Computers 43(11), 1262–1269 (1994)
27. Woelfel, P.: Symbolic topological sorting with OBDDs. J. Discrete Algorithms 4(1), 51–71 (2006)

The Relationship between Multiplicative Complexity and Nonlinearity

Joan Boyar and Magnus Gausdal Find

Department of Mathematics and Computer Science,
University of Southern Denmark

Abstract. We consider the relationship between nonlinearity and multiplicative complexity for Boolean functions with multiple outputs, studying how large a multiplicative complexity is necessary and sufficient to provide a desired nonlinearity. For quadratic circuits, we show that there is a tight connection between error correcting codes and circuits computing functions with high nonlinearity. Using known coding theory results, the lower bound proven here, for quadratic circuits for functions with n inputs and n outputs and high nonlinearity, shows that at least $2.32n$ AND gates are necessary. We further show that one cannot prove stronger lower bounds by only appealing to the nonlinearity of a function; we show a bilinear circuit computing a function with almost optimal nonlinearity with the number of AND gates being exactly the length of such a shortest code. For general circuits, we exhibit a concrete function with multiplicative complexity at least $2n - 3$.

1 Definitions and Preliminaries

Let $\mathbb{F}_2$ be the finite field of order 2 and $\mathbb{F}_2^n$ the n-dimensional vector space over $\mathbb{F}_2$. We denote by $[n]$ the set $\{1, \ldots, n\}$. An (n, m)-function is a mapping from $\mathbb{F}_2^n$ to $\mathbb{F}_2^m$ and we refer to these as the *Boolean functions*.

It is well known that every $(n, 1)$-function f can be written uniquely as a multilinear polynomial over $\mathbb{F}_2$

$$f(\mathbf{x}_1, \ldots, \mathbf{x}_n) = \sum_{X \subseteq [n]} \alpha_X \prod_{i \in X} \mathbf{x}_i.$$

This polynomial is called the *Zhegalkin polynomial* or the *algebraic normal form* of f. For the rest of this paper most, but not all, arithmetic will be in $\mathbb{F}_2$. We trust that the reader will find it clear whether arithmetic is in $\mathbb{F}_2$, $\mathbb{F}_{2^n}$, or $\mathbb{R}$ when not explicitly stated, and will not address it further.

The *degree* of f is the largest $|X|$ such that $\alpha_X = 1$. For an (n, m)-function f, we let f_i be the $(n, 1)$-function defined by the ith output bit of f, and say that the degree of f is the largest degree of f_i for $i \in [m]$. A function is *affine* if it has degree 1, and *quadratic* if it has degree 2. For $T \subseteq [m]$ we let

$$f_T = \sum_{i \in T} f_i,$$

E. Csuhaj-Varjú et al. (Eds.): MFCS 2014, Part II, LNCS 8635, pp. 130–140, 2014.
© Springer-Verlag Berlin Heidelberg 2014

and for $\mathbf{v} \in \mathbb{F}_2^n$ we let $|\mathbf{v}|$ denote the *Hamming weight* of $\mathbf{v}$, that is, the number of nonzero entries in $\mathbf{v}$, and let $|\mathbf{u} + \mathbf{v}|$ be the *Hamming distance* between the two vectors $\mathbf{u}$ and $\mathbf{v}$.

We will use several facts on the nonlinearity of Boolean functions. We refer to the two chapters [6,7] by Carlet for proofs and references. The *nonlinearity* of an $(n,1)$-function f is the Hamming distance to the closest affine function, more precisely

$$NL(f) = 2^n - \max_{\mathbf{a} \in \mathbb{F}_2^n, b \in \mathbb{F}_2} |\{\mathbf{x} \in \mathbb{F}_2^n | \langle \mathbf{a}, \mathbf{x} \rangle + b = f(\mathbf{x})\}|,$$

where $\langle \mathbf{a}, \mathbf{x} \rangle = \sum_{i=1}^{n} \mathbf{a}_i \mathbf{x}_i$. For an (n,m)-function f, the nonlinearity is defined as

$$NL(f) = \min_{T \subseteq [m], T \neq \emptyset} \{NL(f_T)\}.$$

The nonlinearity of an (n,m)-function is always between 0 and $2^{n-1} - 2^{\frac{n}{2}-1}$. The (n,m)-functions meeting this bound are called *bent functions*. Bent $(n,1)$ functions exist if and only if n is even. A standard example of a bent $(n,1)$-function is the *inner product*, on $n = 2k$ variables, defined as:

$$IP_{2k}(\mathbf{x}_1, \ldots, \mathbf{x}_k, \mathbf{y}_1, \ldots, \mathbf{y}_k) = \langle \mathbf{x}, \mathbf{y} \rangle .$$

This function is clearly quadratic. If we identify $\mathbb{F}_2^n$ with $\mathbb{F}_{2^n}$, a standard example of a bent $(2n,n)$-function is the *finite field multiplication* function:

$$f(\mathbf{x}, \mathbf{y}) = \mathbf{x} \cdot \mathbf{y} \tag{1}$$

where multiplication is in $\mathbb{F}_{2^n}$.

If $n = m$, $NL(f)$ is between 0 and $2^{n-1} - 2^{\frac{n-1}{2}}$ [8], and functions meeting this bound are called *almost bent*. These exist only for odd n. As remarked by Carlet, this name is a bit misleading since the name indicates that they are suboptimal, which they are not. Again, if we identify $\mathbb{F}_2^n$ and $\mathbb{F}_{2^n}$, for $1 \leq i \leq \frac{n-1}{2}$ and $gcd(i,n) = 1$, the so called *Gold functions* defined as

$$G(\mathbf{x}) = \mathbf{x}^{2^i+1} = \mathbf{x} \cdot \left(\mathbf{x}^{2^i}\right) \tag{2}$$

are almost bent. This function is quadratic since the mapping $\mathbf{x} \mapsto \mathbf{x}^2$ is affine in $\mathbb{F}_2^n$, and each output bit of finite field multiplication is quadratic in the inputs, see also [7].

An *XOR-AND circuit* is a Boolean circuit where each of the gates is either $\oplus$ (XOR, addition in $\mathbb{F}_2$), $\wedge$ (AND, multiplication in $\mathbb{F}_2$) or the constant 1. In this paper we are mainly concerned with the number of $\wedge$ gates, so we allow $\oplus$-gates to have unbounded fan-in while $\wedge$-gates have fan-in 2. A circuit is *quadratic* if every AND gate computes a quadratic function. A quadratic circuit is *bilinear* if the input partitioned into two sets, and each input to an AND gate is a linear combination of variables from one of these two sets, with the other input using the opposite set of the partition.

The *multiplicative complexity* of an (n, m)-function, f, is the smallest number of AND gates in any XOR-AND circuit computing f. Some relations between nonlinearity and multiplicative complexity are known. In particular, if an $(n, 1)$-function is to have a certain nonlinearity, it is known exactly how many AND gates are necessary and sufficient.

Corollary 1 ([1]). *If the $(n, 1)$-function, f, has multiplicative complexity M, it has nonlinearity at most $2^{n-1} - 2^{n-M-1}$. Furthermore this is tight: for $M \leq \frac{n}{2}$, there exists a simple quadratic function with this nonlinearity.*

The upper bound holds for all M, but gives something nontrivial only when $M < \frac{n}{2}$.

1.1   Linear Codes

Most bounds in this paper will come from coding theory. In this subsection, we briefly review the necessary facts. For more information, see chapter 17 in [10] or the older but comprehensive [20].

A *linear (error correcting) code* of *length* s is a linear subspace, $\mathcal{C}$ of $\mathbb{F}_2^s$. The *dimension* of a code is the dimension of the subspace, $\mathcal{C}$, and the elements of $\mathcal{C}$ are called *codewords*. The (minimum) *distance* d of $\mathcal{C}$ is defined as

$$d = \min_{\mathbf{x} \neq \mathbf{y} \in \mathcal{C},} |\mathbf{x} + \mathbf{y}|.$$

The following fact is well known

Proposition 1. *For every linear code, $\mathcal{C}$, the distance is exactly the minimum weight among non-zero codewords.*

Let $L(m, d)$ be the length of the shortest linear m-dimensional code over $\mathbb{F}_2$ with distance d. We will use lower and upper bounds on $L(m, d)$. One lower bound is the following [16], see also [20], page 563.

Theorem 1 (McEliece, Rodemich, Rumsey, Welch). *For $0 < \delta < 1/2$, let $\mathcal{C} \subseteq \{0, 1\}^s$ be a linear code with dimension m and distance δs. Then the rate $R = \frac{m}{s}$ of the code satisfies $R \leq \min_{0 \leq u \leq 1 - 2\delta} B(u, \delta)$, where $B(u, \delta) = 1 + h(u^2) - h(u^2 + 2\delta u + 2\delta)$, $h(x) = H_2\left(\frac{1 - \sqrt{1-x}}{2}\right)$, and $H_2(x) = -x \log x - (1 - x) \log(1 - x)$.*

An upper bound is the following, see [10].

Theorem 2 (Gilbert-Varshamov). *A linear code $\mathcal{C} \subseteq \{0, 1\}^s$ of dimension m and distance d exists provided that $\sum_{i=0}^{d-2} \binom{s-1}{i} < 2^{s-m}$.*

2   Introduction

In several practical settings, such as homomorphic encryption and secure multiparty computation (see e.g. [23] and [13]), the number of AND gates is significantly more important than the number of XOR gates, hence one is interested in (n, m)-functions with as few AND gates as possible.

Encryption functions should have high nonlinearity to be resistant against linear and differential attacks (see again [7] and the references therein). This is an explicit design criteria for modern cryptographic systems, such as AES, [9], which has been used has a benchmark for several implementations of homomorphic encryption. A natural question to ask is how these nonlinearity and multiplicative complexity are related to each other: how large does one measure need to be in order for the other to have at least a certain value? As stated in Section 1, for every desired nonlinearity, it is known exactly how many AND gates are necessary and sufficient for an $(n, 1)$-function to achieve this. We study this same question for functions with multiple bits of output.

Our Contributions. Let f be an (n, m)-function with nonlinearity $2^{n-1}-2^{n-M-1}$. We show that any quadratic circuit with s AND gates computing f defines an m-dimensional linear code $\mathbb{F}_2^s$ with distance M, so lower bounds on the size of such codes show lower bounds on the number of AND gates in such a circuit. In particular this implies that any quadratic circuit computing an almost bent function must have at least $L(n, \frac{n-1}{2})$ AND gates, and that any quadratic function from $2n$ bits to n bits with optimal nonlinearity requires quadratic circuits with $L(n, n)$ AND gates. Since the finite field multiplication function is bent, the $L(n, n)$ lower bound applies, so the well known result in [3,15], described in the section 2.1, follows immediately as a corollary.

On the other hand, we show that appealing only to the nonlinearity of a function cannot lead to much stronger lower bounds on the multiplicative complexity, by showing the existence of *quadratic* (in fact, *bilinear*) *circuits* with $L\left(n, \frac{n}{2}\right)$ AND gates computing a function from n bits to n bits with nonlinearity at least $2^{n-1} - 2^{\frac{n}{2}+3\sqrt{n}}$ which is close to the optimum.

Although almost all Boolean functions with n inputs and one output have multiplicative complexity at least $2^{n/2} - O(n)$ [2], no concrete function of this type has been shown to have multiplicative complexity more than $n - 1$. We give a concrete function with n inputs and n outputs with multiplicative complexity at least $2n - 3$.

Using known coding theory bounds, the lower bound proven here, for quadratic circuits for functions with n inputs and n outputs and high nonlinearity, shows that at least $2.32n$ AND gates are necessary. Using a known upper bound on $L\left(n, \frac{n}{2}\right)$ gives that circuits for (n, n)-functions with nonlinearity at least $2^{n-1} - 2^{\frac{n}{2}+3\sqrt{n}}$ can be designed using at most $2.95n$ AND gates. This is a factor less than 6 times larger than the multiplicative complexity of $(n, 1)$-functions with similar nonlinearity.

2.1 Related Results

To the best of our knowledge, our lower bound of $2.32n$ AND gates is the largest lower bound on the number of AND gates for quadratic circuits. Previous results showing relations between error correcting codes and bilinear and quadratic circuits include the work of [3,15] where it is shown that a bilinear or quadratic circuit computing finite field multiplication of two $\mathbb{F}_{q^n}$ elements induces an error correcting code over $\mathbb{F}_q$ of dimension n and distance n. For $q = 2$, Theorem 1 implies that such a circuit must have at least $3.52n$ multiplications (AND gates). If n is the number of input bits, this corresponds to a lower bound of $1.76n$. For $q > 2$, the gates (or lines in a straight-line program) have field elements as inputs, and the total number of multiplications and divisions is counted. Kaminski and Bshouty show a lower bound of $3n - o(n)$ for bilinear circuits [12] and extend it to general circuits [4]. This proof is not based on coding theoretic techniques, but rather the study of Hankel matrices related to the bilinear transformation.

Suppose some (n, m)-function f has a certain nonlinearity D. If we identify $f_1, \ldots, f_m, x_1, \ldots, x_n$ and the constant 1 with their truth tables as vectors in $\mathbb{F}_2^{2^n}$, then $\mathcal{C} = span\{f_1, \ldots, f_m, x_1, \ldots, x_n, 1\}$ is a code in $\mathbb{F}_2^{2^n}$ with dimension $n + m + 1$ and distance D, and limitations and possibilities for codes transfer to results on nonlinearity (see the survey [7] and the references therein). However this says nothing about the multiplicative complexity of the function f.

The structure of quadratic circuits has itself been studied by Mirwald and Schnorr [17]. Among other things they show that for quadratic $(n, 1)$- and $(n, 2)$-functions, quadratic circuits are optimal. It is still not known whether this is true for (n, m)-functions in general.

3 Lower Bounds on Multiplicative Complexity

The multiplicative complexity of an $(n, 1)$-function is between 0 and $(1 + o(1)) 2^{n/2}$ [18] (see also [11]), and almost all such functions have multiplicative complexity at least $2^{n/2} - O(n)$ [2]. However, there is no value of n where a concrete $(n, 1)$-function has been exhibited with a proof that more than $n - 1$ AND gates are necessary to compute it. A lower bound of $n - 1$ follows by the simple *degree bound*[1]: a function with degree d has multiplicative complexity at least $d - 1$ [19]. Here we show that repeated use of the degree bound gives a concrete (n, n)-function, exhibiting a lower bound of $2n - 3$. To the best of our knowledge this is the first example of lower bound on the multiplicative complexity for (n, n)-functions.

Theorem 3. *The (n, n)-function f defined as $f_i(\mathbf{x}) = \prod_{j \in [n] \setminus \{i\}} \mathbf{x}_j$, has multiplicative complexity at least $2n - 3$.*

Proof. Consider the first AND gate, A, with degree at least $n - 1$. Such a gate exists since the outputs have degree $n - 1$. By the degree bound, A must have at least $p \geq n - 3$ AND gates with degree at most $n - 2$ in its subcircuit. Call these

---
[1] Notice that despite the name, this is not the same as Strassen's degree bound as described in [21] and Chapter 8 of [5].

AND gates $A_1, \ldots, A_p$. None of these AND gates can be an output gate. Suppose there are q additional AND gates (including A), where some of these must have degree at least $n - 1$. Call these AND gates $B_1, \ldots, B_q$. Then, for every $i \in [n]$, there exist $P_i \subseteq [p]$ and $Q_i \subseteq [q]$ such that $f_i = \sum_{j \in P_i} A_j + \sum_{j \in Q_i} B_j$. We can think of each B_j as a vector in $\mathbb{F}_2^n$, where the ith coordinate is 1 if the term $\prod_{k \in [n] \setminus \{i\}} x_k$ is present in the Zhegalkin polynomial of the function computed by B_j. Since each A_j has degree at most $n - 2$, all the A_j are zero vectors in this representation, so $span(A_1, \ldots, A_p, B_1, \ldots, B_q) = span(B_1, \ldots, B_q)$. It follows that $\{f_1, \ldots, f_n\} \subseteq span(B_1, \ldots, B_q)$. Since

$$n = dim(\{f_1, \ldots, f_n\}) \leq dim(span(B_1, \ldots, B_q)) \leq q,$$

we conclude that the circuit has at least $q + p \geq 2n - 3$ AND gates. □

4   Nonlinearity and Multiplicative Complexity

This section is devoted to showing a relation between the nonlinearity and the multiplicative complexity of quadratic circuits. We first show a connection between nonlinearity, multiplicative complexity and certain linear codes. Applying this connection, Theorem 1 gives a bound on any quadratic (n, m)-function.

Theorem 4. *Let the (n, m)-function, f, have $NL(f) \geq 2^{n-1} - 2^{n-M-1}$, where $M \leq \frac{n}{2}$. Then a quadratic circuit with s AND gates computing f exhibits an m-dimensional linear code over $\mathbb{F}_2^s$ with distance M.*

Proof. Let C be a quadratic circuit with s AND gates computing f, and let $A_1, \ldots, A_s$, be the AND gates. Since C is quadratic, for each $i \in [m]$ there exist $S_i \subseteq [s]$ and $X_i \subseteq [n]$ such that f_i can be written as

$$f_i = \sum_{j \in S_i} A_j + \sum_{j \in X_i} \mathbf{x}_j.$$

Without loss of generality, we can assume that $X_i = \emptyset$ for all i, since both nonlinearity and multiplicative complexity are invariant under the addition of affine terms. For each $i \in [m]$, we define the vector $\mathbf{v}_i \in \mathbb{F}_2^s$, where $\mathbf{v}_{i,j} = 1$ if and only if there is a directed path from A_j to the ith output. By the nonlinearity of f, we have that for each $i \in [m]$,

$$NL(f_i) \geq 2^{n-1} - 2^{n-M-1}.$$

Applying Corollary 1, the multiplicative complexity of f_i is at least M, hence $|\mathbf{v}_i| \geq M$. Similarly, for any nonempty $T \subseteq [m]$ we can associate a vector $\mathbf{v}_T$ by setting

$$\mathbf{v}_T = \sum_{i \in T} \mathbf{v}_i.$$

Since the circuit is quadratic, it holds that if $|\mathbf{v}_T| \leq p$, the multiplicative complexity of $f_T = \sum_{i \in T} f_i$ is at most p. Applying the definition of nonlinearity to f_T, $NL(f_T) \geq 2^{n-1} - 2^{n-M-1}$. Corollary 1 implies that the multiplicative complexity of f_T is at least M, so we have that $|\mathbf{v}_T| \geq M$ when $T \neq \emptyset$.

In conclusion, every nonzero vector in the m dimensional vector space $\mathcal{C} = span_{\mathbb{F}_2}\{\mathbf{v}_1, \dots, \mathbf{v}_m\}$ has Hamming weight at least M. By Proposition 1, $\mathcal{C}$ is a linear code with dimension m and distance at least M.   □

Applying this theorem to quadratic almost bent functions, we have that a quadratic circuit computing such a function has at least $L(n, \frac{n-1}{2})$ AND gates. Combining this with Theorem 1, calculations show:

Corollary 2. *Any quadratic circuit computing an almost bent (n,n)-function has at least $L(n, \frac{n-1}{2})$ AND gates. For sufficiently large n, $L(n, \frac{n-1}{2}) > 2.32n$.*

The corollary above applies to e.g. the almost bent Gold functions G defined in Eqn. (2). For bent $(2n, n)$-functions, using Theorem 4 with $M = n$ and applying Theorem 1, calculations show:

Corollary 3. *A quadratic circuit computing any bent $(2n, n)$-function has at least $L(n, n)$ AND gates. For sufficiently large n, $L(n, n) > 3.52n$.*

This applies to e.g. the finite field multiplication function as defined in Eqn. (1), reproving the known result on multiplicative complexity for quadratic circuits for field multiplication mentioned in Section 2.1.

For both Corollaries 2 and 3, any improved lower bounds on codes lengths would give an improved lower bound on the multiplicative complexity. For Corollary 2 this technique cannot prove better lower bounds than $L(n, \frac{n-1}{2})$. Theorem 2 implies that $L(n, \frac{n-1}{2}) \leq 2.95n$. Below we show that this is not merely a limitation of the proof strategy; there exist quadratic circuits with $L(n, \frac{n-1}{2})$ AND gates with nonlinearity close to the optimal. To the best of our knowledge this is the first example of highly nonlinear (n,n)-functions with linear multiplicative complexity, and therefore it might be a useful building block for cryptographic purposes.

Before proving the next theorem, we need a technical lemma on the probability that a random matrix has small rank. A simple proof of this can be found in e.g. [14].

Lemma 1 (Komargodski, Raz, Tal). *A random $k \times k$ matrix has rank at most d with probability at most $2^{k-(k-d)^2}$.*

Theorem 5. *There exist (n,n)-functions with multiplicative complexity at most $L(n, \frac{n-1}{2})$ and nonlinearity at least $2^{n-1} - 2^{\frac{n}{2}+3\sqrt{n}-1}$.*

Proof. For simplicity we show the upper bound for $L(n, \frac{n}{2})$ AND gates. It is elementary to verify that it holds for $L(n, \frac{n-1}{2})$ AND gates as well. We give a probabilistic construction of a quadratic (in fact, bilinear) circuit with $s = L(n, \frac{n}{2})$ AND gates, then we show that with high probability, the function computed by this circuit has the desired nonlinearity.

For the construction of the circuit, we first define the value computed by the ith AND gate as $A_i(\mathbf{x}) = L_i(\mathbf{x})R_i(\mathbf{x})$ where L_i is a random sum over $\mathbf{x}_1, \ldots, \mathbf{x}_{n/2}$ and R_i is a random sum over $\mathbf{x}_{n/2+1}, \ldots, \mathbf{x}_n$. In the following, we will identify sums over $\mathbf{x}_1, \ldots, \mathbf{x}_n$ with vectors in $\mathbb{F}_2^n$ and sums over $A_1, \ldots, A_s$ with vectors in $\mathbb{F}_2^s$.

Let $\mathcal{C}$ be an n-dimensional code of length $L(n, \frac{n}{2})$ with distance $\frac{n}{2}$ and let $\mathbf{y}_1, \ldots, \mathbf{y}_n \in \mathbb{F}_2^s$ be a basis for $\mathcal{C}$. Now we define the corresponding sums over $A_1, \ldots, A_s$ to be the outputs computed by the circuit. This completes the construction of the circuit. Now fix $r(\mathbf{x}) \in span_{\mathbb{F}_2}\{\mathbf{y}_1, \ldots, \mathbf{y}_n\}$, $r \neq \mathbf{0}$. We want to show that r has the desired nonlinearity with high probability. By an appropriate relabeling of the AND gates, we can write r as

$$r(\mathbf{x}) = \sum_{i=1}^{q} A_i(\mathbf{x}) = \sum_{i=1}^{q} L_i(\mathbf{x})R_i(\mathbf{x}) \tag{3}$$

for some $q \geq \frac{n}{2}$. We now assume that

$$t = \mathrm{rk}\{R_1, \ldots, R_q\} \geq \frac{n}{2} - \frac{3\sqrt{n}}{2}. \tag{4}$$

At the end of the proof, we will show that this is true with high probability. Again by an appropriate relabeling, we let $\{R_1, \ldots, R_t\}$ be a basis of $span\{R_1, \ldots, R_q\}$. If $q > t$, for $j > t$, we can write $R_j = \sum_{i=1}^{t} \alpha_{j,i} R_i$. In particular for $j = q$, we can substitute this into (3) and obtain

$$r(\mathbf{x}) = \sum_{i}^{q-1} \left(L_i(\mathbf{x}) + \alpha_{q,i} L_q(\mathbf{x})\right) R_i(\mathbf{x})$$

where we let $\alpha_{q,i} = 0$ for $i > t$. If $\{L_1, \ldots, L_q\}$ are independently, uniformly randomly distributed, then so are $\{L_1 + \alpha_{q,1} L_q, \ldots, L_{q-1} + \alpha_{q,q-1} L_q\}$. Continuing this process, we get that for $\frac{n}{2} \geq t \geq \frac{n}{2} - \frac{3\sqrt{n}}{2}$, there are sums $L'_1, \ldots, L'_t, R'_1, \ldots, R'_t$ such that

$$r(\mathbf{x}) = \sum_{i=1}^{t} L'_i(\mathbf{x})R'_i(\mathbf{x})$$

where the $\{L'_1, \ldots, L'_t\}$ are independently, uniformly random and the $\{R'_1, \ldots, R'_t\}$ are linearly independent. We now further assume that

$$u = \mathrm{rk}(L'_1, \ldots, L'_t) \geq t - \frac{3\sqrt{n}}{2}. \tag{5}$$

Again, we will show at the end of this proof that this is true with high probability. Applying a similar procedure as above, we get that for some

$$u \geq t - \frac{3\sqrt{n}}{2} \geq \frac{n}{2} - 3\sqrt{n}$$

there exist sums $\tilde{L}_1, \ldots, \tilde{L}_u$ and $\tilde{R}_1, \ldots, \tilde{R}_u$, such that

$$r(\mathbf{x}) = \sum_{i=1}^{u} \tilde{L}_i(\mathbf{x})\tilde{R}_i(\mathbf{x}),$$

where all $\tilde{L}_1, \ldots, \tilde{L}_u$ and all $\tilde{R}_1, \ldots \tilde{R}_u$ are linearly independent. Thus, there exists a linear bijection $(\mathbf{x}_1, \ldots, \mathbf{x}_n) \mapsto (\mathbf{z}_1, \ldots, \mathbf{z}_n)$ with $\mathbf{z}_1 = \tilde{L}_1, \ldots, \mathbf{z}_u = \tilde{L}_u, \mathbf{z}_{u+1} = \tilde{R}_1, \ldots, \mathbf{z}_{2u} = \tilde{R}_u$, such that

$$\tilde{r}(\mathbf{z}) = \mathbf{z}_1\mathbf{z}_{u+1} + \ldots, \mathbf{z}_u\mathbf{z}_{2u}$$

where r and $\tilde{r}$ are equivalent up to a linear bijection on the inputs. Since non-linearity is invariant under linear bijections, we just need to determine the non-linearity of $\tilde{r}$. Given the high nonlinearity of IP_n, it is elementary to verify that

$$NL(\tilde{r}) = 2^{n-2u}\left(2^{2u-1} - 2^{u-1}\right) = 2^{n-1} - 2^{n-u-1}.$$

If $u \geq \frac{n}{2} - 3\sqrt{n}$, this is at least $2^{n-1} - 2^{\frac{n}{2}+3\sqrt{n}-1}$.

Now it remains to show that the probability of either (4) or (5) occurring is so small that a union bound over all the $2^n - 1$ choices of r gives that with high probability, *every* $r \in span\{\mathbf{y}_1, \ldots, \mathbf{y}_n\}$ has at least the desired nonlinearity.

For (4), we can think of the $q \geq \frac{n}{2}$ vectors $R_1, \ldots, R_q$ as rows in a $q \times \frac{n}{2}$ matrix. We will consider the upper left $\frac{n}{2} \times \frac{n}{2}$ submatrix. By Lemma 1 this has rank at most $\frac{n}{2} - \frac{3\sqrt{n}}{2}$ with probability at most

$$2^{\frac{n}{2}-\left(\frac{n}{2}-\left(\frac{n}{2}-\frac{3\sqrt{n}}{2}\right)\right)^2} = 2^{\frac{n}{2}-\frac{9n}{4}} = 2^{-\frac{7n}{4}}$$

Similarly for (5) we can consider the $\frac{n}{2} \geq t \geq \frac{n}{2} - \frac{3\sqrt{n}}{2}$ vectors $L'_1, \ldots, L'_t$ as the rows in a $t \times \frac{n}{2}$ matrix. Consider the top left $t \times t$ submatrix. Again, by Lemma 1, the probability of this matrix having rank at most $t - \frac{3\sqrt{n}}{2}$ is at most

$$2^{t-\left(t-\left(t-\frac{3\sqrt{n}}{2}\right)\right)^2} \leq 2^{\frac{n}{2}-\frac{9n}{4}} = 2^{-\frac{7n}{4}}$$

There are $2^n - 1$ choices of r, so by the union bound, the total probability of at least one of (4) or (5) failing for a least one choice is at most $2 \cdot (2^n - 1) \cdot 2^{-\frac{7n}{4}}$, which tends to zero, so in fact the described construction will have the desired nonlinearity with high probability. □

We should note that it is not hard to improve in the constants in the proof and show that in fact the described function has nonlinearity at least $2^{n-1} - 2^{\frac{n}{2}+c\sqrt{n}}$ for some constant $c < 3$. However, the proof given does not allow improvement to e.g. $c = 2$.

5   Open Problems

Strassen [22] (see also [5], Proposition 14.1, p. 351) proved that for an *infinite field*, $\mathbb{K}$, if the quadratic function $F: \mathbb{K}^n \to \mathbb{K}^m$ can be computed with M multi-plications/divisions, then it can be computed in M multiplications by a quadratic circuit. However, it is unknown whether a similar result holds for *finite fields* in particular for $\mathbb{F}_2$. Mirwald and Schnorr [17] showed that for quadratic $(n, 1)$- and $(n, 2)$-functions, quadratic circuits are optimal. It is still not known whether this is true for (n, m)-functions in general. It would be very interesting to determine if the bounds proven here for quadratic circuits also hold for general circuits.

When inspecting the proof of Theorem 4, one can make a weaker assumption on the circuit than it being quadratic. For example, it is sufficient if it holds that for every AND gate, A, there is a unique AND gate, A' (which might be equal to A), such that every path from A to an output goes through A'. Can one find a larger, interesting class of circuits where the proof holds?

The function defined in Theorem 3 has multiplicative complexity at least $2n - 3$ and at most $3n - 6$. What is the exact value?

References

1. Boyar, J., Find, M., Peralta, R.: Four measures of nonlinearity. In: Spirakis, P.G., Serna, M. (eds.) CIAC 2013. LNCS, vol. 7878, pp. 61–72. Springer, Heidelberg (2013), eprint with correction available at the Cryptology ePrint Archive, Report 2013/633 (2013), http://eprint.iacr.org/
2. Boyar, J., Peralta, R., Pochuev, D.: On the multiplicative complexity of Boolean functions over the basis $(\wedge, \oplus, 1)$. Theor. Comput. Sci. 235(1), 43–57 (2000)
3. Brown, M.R., Dobkin, D.P.: An improved lower bound on polynomial multiplication. IEEE Trans. Computers 29(5), 337–340 (1980)
4. Bshouty, N.H., Kaminski, M.: Polynomial multiplication over finite fields: from quadratic to straight-line complexity. Computational Complexity 15(3), 252–262 (2006)
5. Bürgisser, P., Clausen, M., Shokrollahi, M.A.: Algebraic Complexity Theory. Grundlehren der mathematischen Wissenschaften, vol. 315. Springer (1997)
6. Carlet, C.: Boolean functions for cryptography and error correcting codes. In: Crama, Y., Hammer, P.L. (eds.) Boolean Models and Methods in Mathematics, Computer Science, and Engineering, ch. 8, pp. 257–397. Cambridge University Press, Cambridge (2010)
7. Carlet, C.: Vectorial Boolean functions for cryptography. In: Crama, Y., Hammer, P.L. (eds.) Boolean Models and Methods in Mathematics, Computer Science, and Engineering, ch. 9, pp. 398–469. Cambridge Univ. Press, Cambridge (2010)
8. Chabaud, F., Vaudenay, S.: Links between differential and linear cryptanalysis. In: De Santis, A. (ed.) EUROCRYPT 1994. LNCS, vol. 950, pp. 356–365. Springer, Heidelberg (1995)
9. Daemen, J., Rijmen, V.: The Design of Rijndael: AES-The Advanced Encryption Standard. Security and Cryptology. Springer (2002)
10. Jukna, S.: Extremal Combinatorics: with Applications in Computer Science, 2nd edn. Texts in Theoretical Computer Science. Springer (2011)
11. Jukna, S.: Boolean Function Complexity: Advances and Frontiers. Springer, Heidelberg (2012)
12. Kaminski, M., Bshouty, N.H.: Multiplicative complexity of polynomial multiplication over finite fields. J. ACM 36(1), 150–170 (1989)

13. Kolesnikov, V., Schneider, T.: Improved garbled circuit: Free XOR gates and applications. In: Aceto, L., Damgård, I., Goldberg, L.A., Halldórsson, M.M., Ingólfsdóttir, A., Walukiewicz, I. (eds.) ICALP 2008, Part II. LNCS, vol. 5126, pp. 486–498. Springer, Heidelberg (2008)

14. Komargodski, I., Raz, R., Tal, A.: Improved average-case lower bounds for demorgan formula size. In: FOCS, pp. 588–597 (2013)

15. Lempel, A., Seroussi, G., Winograd, S.: On the complexity of multiplication in finite fields. Theor. Comput. Sci. 22, 285–296 (1983)

16. McEliece, R.J., Rodemich, E.R., Rumsey Jr., H., Welch, L.R.: New upper bounds on the rate of a code via the Delsarte-MacWilliams inequalities. IEEE Trans. Inform. Theory 23(2), 157–166 (1977)

17. Mirwald, R., Schnorr, C.P.: The multiplicative complexity of quadratic Boolean forms. Theor. Comput. Sci. 102(2), 307–328 (1992)

18. Nechiporuk, E.I.: On the complexity of schemes in some bases containing nontrivial elements with zero weights. Problemy Kibernetiki 8, 123–160 (1962) (in Russian)

19. Schnorr, C.P.: The multiplicative complexity of Boolean functions. In: Mora, T. (ed.) AAECC 1988. LNCS, vol. 357, pp. 45–58. Springer, Heidelberg (1989)

20. Sloane, N., MacWilliams, F.: The Theory of Error Correcting Codes. North-Holland Math. Library 16 (1977)

21. Strassen, V.: Die berechnungskomplexität von elementarsymmetrischen funktionen und von interpolationskoeffizienten. Numerische Mathematik 20(3), 238–251 (1973)

22. Strassen, V.: Vermeidung von Divisionen. Journal für die reine und angewandte Mathematik 264, 184–202 (1973)

23. Vaikuntanathan, V.: Computing blindfolded: New developments in fully homomorphic encryption. In: Ostrovsky, R. (ed.) FOCS, pp. 5–16. IEEE (2011)

Dual Connectedness of Edge-Bicolored Graphs and Beyond

Leizhen Cai* and Junjie Ye

Department of Computer Science and Engineering,
The Chinese University of Hong Kong, Shatin, Hong Kong SAR, China
{lcai,jjye}@cse.cuhk.edu.hk

Abstract. Let G be an edge-bicolored graph where each edge is colored either red or blue. We study problems of obtaining an induced subgraph H from G that simultaneously satisfies given properties for H's red graph and blue graph. In particular, we consider DUALLY CONNECTED INDUCED SUBGRAPH problem — find from G a k-vertex induced subgraph whose red and blue graphs are both connected, and DUAL SEPARATOR problem — delete at most k vertices to simultaneously disconnect red and blue graphs of G.

We will discuss various algorithmic and complexity issues for DU-ALLY CONNECTED INDUCED SUBGRAPH and DUAL SEPARATOR problems: NP-completeness, polynomial-time algorithms, W[1]-hardness, and FPT algorithms. As by-products, we deduce that it is NP-complete and W[1]-hard to find k-vertex (resp., $(n-k)$-vertex) strongly connected induced subgraphs from n-vertex digraphs. We will also give a complete characterization of the complexity of the problem of obtaining a k-vertex induced subgraph H from G that simultaneously satisfies given heredi-tary properties for H's red and blue graphs.

Keywords: Edge-bicolored graph, dually connected, dual separator.

1 Introduction

Edge-colored graphs are fundamental in graph theory and have been extensively studied in the literature, especially for alternating cycles, monochromatic sub-graphs, heterchromatic subgraphs and partitions [1,12]. In this paper, we focus on edge-bicolored graphs — simple undirected graphs G where each edge is uniquely colored by either blue or red, and we use G_b and G_r to denote the red and blue graphs of G respectively. We are interested in finding an induced subgraph from G that simultaneously satisfies specified properties for its red and blue graphs. In particular, we study the following three closely related prob-lems concerning the fundamental property of being connected for edge-bicolored graphs G.

* Partially supported by GRF grant CUHK410212 of the Research Grants Council of Hong Kong.

- DUALLY CONNECTED INDUCED SUBGRAPH: Does G contain exactly k vertices V' such that both $G_b[V']$ and $G_r[V']$ are connected?
- DUALLY CONNECTED DELETION: Does G contain exactly k vertices V' such that both $G_b - V'$ and $G_r - V'$ are connected?
- DUAL SEPARATOR: Does G contain at most k vertices V' such that both $G_b - V'$ and $G_r - V'$ are disconnected?

Related Work: In connection with our dually connected subgraph problems, Gai et al. [7] defined a *common connected component* of two graphs G_1 and G_2 on the same vertex set V as a maximal subset $V' \in V$ such that induced subgraphs $G_1[V']$ and $G_2[V']$ are both connected, and they also mentioned three typical applications in computational biology. Using partition refinement to maintain connectivity dynamically, they obtained an algorithm for finding all common connected components in $O(n \log n + m \log^2 n)$ time. For the same problem, Bin-Xuan et al. [2] used their technique of competitive graph search to produce an algorithm with running time $O(n + m \log^2 n)$. We also note that when both G_1 and G_2 are paths, the problem of finding all common connected subgraphs coincides with the well studied problem of finding all *common intervals* of two permutations [18], a problem with many applications.

On the other hand, despite an enormous amount of work on induced subgraph and vertex deletion problems on uncolored graphs, we are unaware of any systematic investigation of the type of problems we study in this paper.

Our Contributions: We study both traditional and parameterized complexities of the above three problems, which has further inspired general induced subgraph problems on edge-bicolored graphs. The following list summarizes our results.

1. DUALLY CONNECTED INDUCED SUBGRAPH is NP-complete and W[1]-hard even when both G_b and G_r are trees, but is solvable in $O(n^2 \alpha(n^2, n))$ time when G is a complete graph, where $\alpha(n^2, n)$ is inverse of Ackermann's function.
2. DUALLY CONNECTED DELETION is NP-complete and W[1]-hard but admits an FPT algorithm when both G_b and G_r are trees.
3. DUAL SEPARATOR is NP-complete.
4. It is NP-complete and W[1]-hard to obtain k-vertex (resp. $(n-k)$-vertex) strongly connected induced subgraphs from n-vertex digraphs.
5. We give a complete characterization of both classical and parameterized complexities of the INDUCED (Π_b, Π_r)-SUBGRAPH problem for hereditary properties Π_b and Π_r: Does an edge-bicolored graph G contain a k-vertex induced subgraph whose blue and red graphs simultaneously satisfy properties Π_b and Π_r respectively?
6. We give FPT algorithms for parametric dual problems of INDUCED (Π_b, Π_r)-SUBGRAPH when properties Π_b and Π_r admit finite forbidden induced subgraph characterizations.

Notation and Definitions: For a graph G, $V(G)$ and $E(G)$ denote its vertex set and edge set respectively, and n and m, respectively, are numbers of vertices and edges of G. For a subset $V' \subseteq V(G)$, $N_G(V')$ denotes the neighbors of V' in $V(G) - V'$ and $G[V']$ the subgraph of G induced by V'. A graph property Π is a collection of graphs, and it is *hereditary* if every induced subgraph of a graph in Π also belongs to Π. It is well-known that Π is hereditary iff it has a forbidden induced subgraph characterization.

For an edge-bicolored graph $G = (V, E_b \cup E_r)$, $G_b = (V, E_b)$ and $G_r = (V, E_r)$, respectively, denote the blue graph and red graph of G. We say that G is *dually connected* if both G_b and G_r are connected, and a *dual tree* if both G_b and G_r are trees. A *dually connected component* of G is a maximal dually connected induced subgraph of G. A subset $V' \subseteq V(G)$ is a *dual separator* of G if both $G_b - V'$ and $G_r - V'$ are disconnected. We use $\alpha(n^2, n)$ for inverse of Ackermann's function.

Remark: In this paper we require monochromatic subgraphs to be spanning subgraphs of G, but for some applications we may disregard isolated vertices in monochromatic subgraphs. For example, we can define $G_b = G[E_b]$ or $G_b = (V_b, E_b)$ with $V_b \subseteq V$. Our results in the paper are also valid for these two alternative definitions of monochromatic subgraphs, except our FPT algorithm for DUALLY CONNECTED DELETION on dual trees (see Problem 1 in Section 5).

2   Dually Connected Induced Subgraphs

Although all dually connected components in an edge-bicolored graph can be found in $O(n + m \log^2 n)$ time [2], it is surprisingly difficult to determine whether an edge-bicolored graph contains a dually connected induced subgraph on exactly k vertices. We will show that DUALLY CONNECTED INDUCED SUBGRAPH is solvable in $O(n^2 \alpha(n^2, n))$ time when G is a complete graph, but NP-complete and W[1]-hard when G is a dual tree, i.e., both blue and red graphs of G are trees, which rules out efficient ways to list all common connected subgraphs of two trees. We begin with a lemma for edge-bicolored complete graphs.

Lemma 1. *A dually connected edge-bicolored complete graph G contains, for every $4 \le k \le n$, a k-vertex dually connected induced subgraph.*

Proof. For a vertex v, if $G - v$ remains dually connected, we can delete v from G and regard the smaller graph as G. Therefore, we need only consider the case that G contains a vertex v such that $G - v$ is not dually connected. W.l.o.g., we may assume that v is a cut vertex of G_b. Since G_r is connected, v is not an isolated vertex of G_r and hence not adjacent to all vertices of G_b. Therefore G_b has a vertex x such that $d_{G_b}(v, x) = 2$. Let y be a vertex of $G_b - v$ not in the component containing x. Then we have $d_{G_b}(x, y) \ge 3$.

We now use a breadth-first search from v to obtain $k \ge 4$ vertices S, including $\{v, x, y\}$, such that $G_b[S]$ is connected. Since $d_{G_b[S]}(x, y) \ge d_{G_b}(x, y) \ge 3$, we see that in $G_b[S]$, no vertex is adjacent to both x and y, and hence every vertex is adjacent to at least one of x and y in the complement of $G_b[S]$, i.e., graph $G_r[S]$. Since $\{x, y\}$ is an edge in $G_r[S]$, any pair of vertices in $G_r[S]$ has distance at most 3 and hence $G_r[S]$ is connected, implying that $G[S]$ is dually connected. ∎

Theorem 1. Dually Connected Induced Subgraph *can be solved in* $O(n^2\alpha(n^2, n))$ *time for edge-bicolored complete graphs* G.

Proof. First we find a largest dully connected component H in G. If G itself is dully connected, then set H to G. Otherwise, one of G_b or G_r is disconnected, and G's dually connected components are equivalent to its *maximal strong modules* [11], which can be found in linear time by *modular decomposition* [11]. If $k \leq 3$ or $|V(H)| < k$, then the answer is "No"; otherwise the answer is "Yes" by Lemma 1.

Now we discuss how to find a k-vertex dully connected subgraph inside H. Order vertices of H as $v_1, v_2, \ldots, v_h$ with $h = |V(H)| > k$, and let $V_i = \{v_1, \ldots, v_i\}$. By Lemma 1, we only need to find the smallest index $i > k$ such that $H[V_i]$ is dually connected but $H[V_{i-1}]$ is not, and then find our required k-vertex subgraph inside $H[V_i]$.

For this purpose, we construct H by adding $v_1, v_2, \ldots, v_h$ one by one in this order and, in the process, we use *disjoint sets* to dynamically maintain components of $H_b[V_i]$ and $H_r[V_i]$. For the blue graph H_b (similar for the red graph H_r), blue sets are components of $H_b[V_k]$ initially. In adding vertex v_i to H ($i > k$), we create a blue singleton set $\{v_i^b\}$ for vertex v_i, and for each blue edge v_iv_j with $j < i$, we merge $\{v_i^b\}$ with the blue set containing v_j^b. The procedure stops once there is only one blue set and one red set, i.e., $H[V_i]$ is dually connected but $H[V_{i-1}]$ is not. Now we can use the proof of Lemma 1 to find a k-vertex dually connected subgraph in $O(n^2)$ time. Using standard Union-Find data structure, we can find the required $H[V_i]$ in $O(n^2\alpha(n^2, n))$ time, which is also an upper bound of our algorithm. ∎

We now introduce a structure called *dual 2t-path* that will be useful in proving the intractability of Dually Connected Induced Subgraph and also Dually Connected Deletion in the next section. For any $t \geq 3$, a dual $2t$-path P^* is the edge-bicolored graph formed by taking the union of a blue path $P_b = v_1v_2\ldots v_{2t}$ and red path $P_r = v_{2t}v_{2t-2}\ldots v_4v_2v_{2t-1}v_{2t-3}\ldots v_3v_1$ (see Figure 1 for an example). We denote the two ends v_1 and v_{2t} of P^* by v^b and v^r respectively.

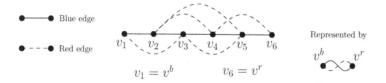

Fig. 1. Dual $2t$-path for $t = 3$

Lemma 2. *In an edge-bicolored graph* $G = (V, E_b \cup E_r)$, *if* $V^* \subseteq V$ *induces a dual 2t-path* P^* *with ends* v^b *and* v^r *such that the only edges between* V^* *and* $V - V^*$ *are blue edges (resp., red edges) between* v^b *(resp.,* v^r*) and* $V - V^*$, *then for any dually connected induced subgraph* G' *of* G *that contains a vertex in* V^* *and a vertex in* $V - V^*$, G' *must contain all vertices of* V^*.

Proof. Deleting some but not all vertices in V^* of P^* will disconnect the blue or red graph of G'. ∎

Theorem 2. DUALLY CONNECTED INDUCED SUBGRAPH *is NP-complete and W[1]-hard for dual trees.*

Proof. The problem is clearly in NP, and we give a polynomial and FPT reduction from the classical NP- and W[1]-complete CLIQUE problem [8] to prove the theorem. For an instance (G, k) of CLIQUE, we construct an edge-bicolored graph G' such that both G'_b and G'_r are trees (see Figure 2 for an example):

1. Set $p = k(k-1)$ and create a new vertex v^*.
2. Replace every vertex v of G by a dual p-path P_v^* with end vertices v^b and v^r, and refer to vertices in P_v^* as *path-vertices*. Add blue edge $v^b v^*$ and red edge $v^r v^*$.
3. For each edge $e = uv$ of G, create *edge-vertex* $\tilde{e}$, and replace e by blue edge $u^b \tilde{e}$ and red edge $v^r \tilde{e}$.

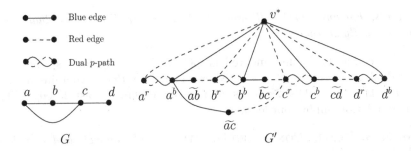

Fig. 2. Construction of G' from G

It is easy to see that the construction of G' takes polynomial time, and that G'_b and G'_r are both trees. We claim that G has a k-clique iff G' has $k' = 1 + kp + p/2$ vertices S such that $G'[S]$ is dually connected.

Assume that G has a k-clique $\{v_1, v_2, \ldots, v_k\}$. Let S be the union of $\{v^*\}$, path-vertices of all v_i and edge-vertices of all $v_i v_j$. The size of $|S|$ is $1 + kp + k(k-1)/2 = k'$. Since edge-vertex of each $v_i v_j$ is dually connected to v^* through v_i^b and v_j^r, $G'[S]$ is dually connected.

Conversely, suppose that G' contains k' vertices S such that $G'[S]$ is dually connected. Since all dual p-paths are dually connected through v^*, S must contain v^*. Also by Lemma 2, S contains either all or no vertices of any dual p-path P_v^*. Therefore S contains path-vertices of at most k dual p-paths as $|S| = 1 + kp + p/2$. Furthermore, since an edge-vertex $\widetilde{xy}$ is dually connected to v^* through both vertices x^b and y^r, S must contain both x^b and y^r when S contains an edge-vertex $\widetilde{xy}$. Thus S contains at most $k(k-1)/2$ edge-vertices. It follows that S contains path-vertices of exactly k dual p-paths, and their corresponding vertices in G form a k-clique of G. ∎

We can regard the complement graph of G' in the above proof as a graph with the third color, and obtain the following result to complement Theorem 1.

Corollary 1. *Given an edge-tricolored complete graph, it is NP-complete and W[1]-hard to find an induced subgraph on exactly k vertices that is connected in every monochromatic graph.*

3   Dual Connectedness by Vertex Deletion

The intractability of DUALLY CONNECTED INDUCED SUBGRAPH calls for an investigation of the parameterized complexity of its dual problem DUALLY CONNECTED DELETION: *Can we delete exactly k vertices from an edge-bicolored graph so that the resulting graph is dually connected?*

We show that DUALLY CONNECTED DELETION is also W[1]-hard but becomes FPT for dual trees, which is in contrast to the W[1]-hardness of DUALLY CONNECTED INDUCED SUBGRAPH on dual trees. Our FPT algorithm uses the following connection with a vertex cover problem that is solvable by the random separation method of Cai, Chan and Chan [4].

Lemma 3. *For any dual tree T and k vertices S of T, $T - S$ is a dual tree iff S covers exactly $2k$ edges.*

Proof. Note that both blue and red graphs of $T - S$ are forests, and an n-vertex dual tree contains $2(n - 1)$ edges. Therefore $T - S$ contains at most $2(n - k - 1) = 2(n - 1) - 2k$ edges, and thus S covers at least $2k$ edges. This min-max relation implies our lemma. ∎

Theorem 3. DUALLY CONNECTED DELETION *is NP-complete and W[1]-hard, but FPT on dual trees.*

Proof. We start with an FPT algorithm for the problem on dual trees T. By Lemma 3, it suffices to find k vertices in T that cover exactly $2k$ edges. We use a modification of the random separation algorithm of Cai, Chan and Chan [4] for finding a subset of vertices to cover exactly k edges.

First, we regard T as an uncolored graph and produce a random black-white coloring for the vertices of T. We begin by using black to color all vertices with degree more than $2k$, and then we randomly and independently color each uncolored vertex by black or white with probability $\frac{1}{2}$. Given a black-white coloring of vertices of T, a set S of k vertices is a *well-colored solution* if

1. S covers exactly $2k$ edges, and
2. all vertices in S are white and all vertices in $N_T(S)$ are black.

Let V_w denote the set of white vertices, and refer to connected components of $T[V_w]$ as *white components*. For a white component H_i, let n_i be the number of vertices in H_i and e_i the number of edges covered by vertices of H_i. Then a well-colored solution consists of a collection $\mathcal{H}'$ of white components satisfying

$$\sum_{H_i \in \mathcal{H}'} n_i = k \quad \text{and} \quad \sum_{H_i \in \mathcal{H}'} e_i = 2k.$$

Therefore we can easily formulate the problem of finding a well-colored solution as a 0-1 knapsack problem, and solve it in $O(kn)$ time using the standard dynamic programming algorithm for the 0-1 knapsack problem. Note that it takes $O(n)$ time to compute all n_i and e_i.

Since a well-colored solution S satisfies $|S \cup N_T(S)| \leq 3k$, our random black-white coloring has probability at least 2^{-3k} to produce a well-colored solution. Therefore when T has a solution, we can find it with probability at least 2^{-3k} in $O(kn)$ time. We can derandomize the algorithm by a family of $(n, 3k)$-universal sets of size $8^k k^{O(\log k)} \log n$ [16], and thus obtain a deterministic FPT algorithm running in time $8^k k^{O(\log k)} n \log n$.

For our problem on general edge-bicolored graphs, we give an FPT reduction from the classical W[1]-complete INDEPENDENT SET problem [6] to show W[1]-hardness. For an arbitrary instance (G, k) of INDEPENDENT SET, we construct an edge-bicolored graph G' from G as follows (see Figure 3 for an example):

1. Replace each edge uv of G by the *replacement gadget* H_{uv} in Figure 3.
2. Create a dual $2k$-path P^* with end vertices v^b and v^r, and connect every vertex of G to v^b by a blue edge and to v^r by a red edge.

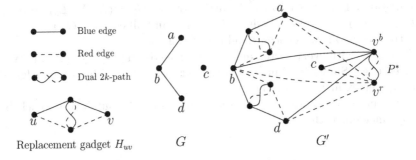

Fig. 3. Construction of G' by using the replacement gadget H_{uv}

The construction clearly takes polynomial time, and we show that G has an independent k-set iff we can deleting k vertices from G' to obtain a dually connected graph.

If G contains an independent set S with k vertices, then for each edge e of G, at least one end-vertex, say v, of e remains in $G' - S$. It is easy to verify that $G' - S$ is dually connected as all vertices of H_e in $G' - S$ are dually connected to v, which is dully connected to the dual path P^* in Step 2.

Conversely, suppose that G' contains k vertices S such that $G' - S$ is dually connected. By the property of dual paths (Lemma 2), neither dual $2k$-path P^* nor dual $2k$-path in any H_e contains any vertex from S. Therefore all vertices in S are vertices of G, and we show that S is an independent set of G. For any two vertices $u, v \in S$, if uv is an edge of G, then the dual $2k$-path in H_{uv} is disconnected from G' after deleting S, contrary to the assumption that $G' - S$ is dually connected. Therefore no two vertices in S are adjacent in G, and thus S is an independent k-set of G.   ∎

4   Dual Separators

Complimenting dual connectedness, we now consider problems of disconnecting blue and red graphs simultaneously by vertex deletion. In particular, we study DUAL SEPARATOR and DUAL SEPARATOR FOR TWO TERMINALS: *Is it possible to disconnect two given vertices s and t in both G_b and G_r of an edge-bicolored graph G by removing $\leq k$ vertices?*

Although minimum separators in uncolored graphs can be found in polynomial time, it is intractable to find minimum-size dual separators in edge-bicolored graphs as we will show that both dual separator problems are NP-complete. However, the parameterized complexity of these two problems remain open.

Theorem 4. DUAL SEPARATOR FOR TWO TERMINALS *is NP-complete.*

Proof. The problem is clearly in NP, and we prove the theorem by a reduction from VERTEX COVER on cubic graphs, whose NP-completeness was established by Garey, Johnson and Stockmeyer [9]. Given a cubic graph $G = (V, E)$, we construct an edge-bicolored graph G' as follows (see Figure 4 for an example):

1. Partition edges of G into two bipartite graphs $G_b = (X_b, Y_b; E_b)$ and $G_r = (X_r, Y_r; E_r)$, and color all edges of G_b blue and all edges of G_r red.
2. Introduce two new vertices s and t as terminals.
3. Connect s with every vertex in X_b by a blue edge and every vertex in X_r by a red edge. Similarly, connect t with every vertex in Y_b by a blue edge and every vertex in Y_r by a red edge.
4. Turn the above multigraph into a simple graph by subdividing each blue edge incident with s or t.

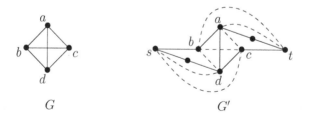

$$G \qquad\qquad\qquad\qquad G'$$

Fig. 4. In graph G', blue edges are solid and red edges are dashed

The above construction takes polynomial time since we can partition edges of any cubic graph G into two bipartite graphs G_b and G_r in polynomial time by using, for instance, a proper edge 4-coloring of G.

In G', it is easy to see that every monochromatic (s, t)-path goes through some edge of G, and every edge of G is contained in some monochromatic (s, t)-path of G'. This clearly implies that for any set S of vertices of G, S is a vertex cover of G iff it is a dual (s,t)-separator of G'. Furthermore, we notice that if an (s,t)-separator S' of G' involves some vertices V^* used for subdividing blue edges, we can always replace V^* by its neighbors in $X_b \cup Y_b$ to get an (s,t)-separator S with $|S| \leq |S'|$. Therefore we can conclude that G admits a vertex cover with $\leq k$ vertices iff G' has an (s,t)-separator with $\leq k$ vertices, and hence we have a required polynomial reduction. ∎

The above theorem enables us to show the hardness of DUAL SEPARATOR.

Theorem 5. DUAL SEPARATOR *is NP-complete.*

Proof. We give a polynomial reduction from DUAL SEPARATOR FOR TWO TER-MINALS by constructing a new graph G'' from the graph G' in the proof of Theorem 4. Let S'_b (resp., T'_b) denote vertices that subdivide blue edges incident with s and t in G'. To *duplicate* a vertex v, we create a new vertex v' and add all blue (resp., red) edges between v and vertices corresponding to $N_{G'_b}(v)$ (resp., $N_{G'_r}(v)$). We construct G'' as follows:

1. Take graph G', and duplicate k copies of s and t. Denote by S vertex s and its k duplicates, and denote by T vertex t and its k duplicates.
2. Make k duplicates of each vertex in S'_b (resp., T'_b). Denote by S''_b (resp., T''_b) vertices S'_b (resp., T'_b) and all their duplicates.
3. Between S''_b (resp., T''_b) and T (resp., S), add all possible edges and color them red.

It is easy to see that G' has a dual (s,t)-separator of size k iff G'' has a dual (S,T)-separator of size k. We claim that G'' has a dual (S,T)-separator of size k iff it has a dual separator of size k, which will prove the theorem.

A dual (S,T)-separator is certainly a dual separator of G''. Conversely, suppose that G'' does not have a dual (S,T)-separator of size k. Then after deleting k vertices V^*, we will get a graph G^* such that S and T are connected in G^*_b or G^*_r. Note that each vertex in $W = S \cup T \cup S''_b \cup T''_b$ has $k+1$ copies, and thus it is useless to delete any vertices in W. Hence we can assume that $V^* \cap W = \emptyset$. Suppose that S and T are connected in G^*_r. Since $V(G) = X_r \cup Y_r$, each vertex in $V(G) \cap V(G^*)$ is adjacent to a vertex in S or T with red edges. Furthermore all vertices in $S''_b \cup T''_b$ are adjacent to S or T with red edges by Step 3 in constructing G'', implying that G^*_r is connected. By a similar argument, we can deduce that G^*_b is connected when S and T are connected in G^*_b. Thus we have proved this theorem. ∎

5   Concluding Remarks

Results on connectedness and separators for edge-bicolored graphs have shown a rich diversity of the complexity of induced subgraph problems on edge-bicolored graphs, which extends an invitation for studying various induced subgraph problems on edge-bicolored graphs. In fact we have obtained a complete characterization of INDUCED (Π_b, Π_r)-SUBGRAPH on edge-bicolored graphs for hereditary properties Π_b and Π_r, and also obtained FPT algorithms for its parametric dual problems for properties with finite forbidden induced subgraph characterizations. Furthermore, the work in the paper also enables us to obtain some results for digraphs. Due to space limit, proofs for the following theorems are omitted and will appear in the full paper.

Building on a characterization of Khot and Raman [13] for induced subgraph problems on uncolored graphs and Ramsey's theorem, we completely character-ize the complexity of INDUCED (Π_b, Π_r)-SUBGRAPH on edge-bicolored graphs for hereditary properties Π_b and Π_r, which depends on whether Π_b and Π_r include all complete graphs K_i or trivial graphs $\overline{K_i}$ (see Figure 5 for an illustration).

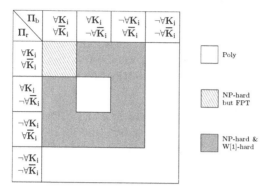

Fig. 5. For a property Π, $\forall K_i = $ "Π includes all complete graphs" and $\neg\forall K_i = $ "Π excludes some complete graphs". Similar for trivial graphs $\overline{K_i}$.

Theorem 6. *For hereditary properties Π_b and Π_r, the complexity of INDUCED (Π_b, Π_r)-SUBGRAPH is completely determined as follows:*

1. *NP-hard and W[1]-hard if one of Π_b and Π_r includes all trivial graphs, and the other excludes some complete graphs but includes all trivial graphs or vice versa.*
2. *NP-hard but FPT if both Π_b and Π_r include all complete graphs and all trivial graphs.*
3. *Polynomial-time solvable if both Π_b and Π_r exclude some trivial graphs, or one of Π_b and Π_r excludes some complete graphs and some trivial graphs.*

We remark that the above theorem implies a complete characterization of the DUAL Π-SUBGRAPH problem, i.e., INDUCED (Π, Π)-SUBGRAPH, for hereditary Π (see the main diagonal of Figure 5).

For the parametric dual problem of INDUCED (Π_b, Π_r)-SUBGRAPH, i.e., delet-ing k vertices to obtain an induced (Π_b, Π_r)-graph, we can easily deduce the fol-lowing general result on edge-colored multigraphs as a corollary of a well-known result of the first author regarding graph modification problems [3].

Theorem 7. *Let $G_1, \ldots, G_t$ be graphs, and $\Pi_1, \ldots, \Pi_t$ graph properties charac-terizable by finite forbidden induced subgraphs. It is FPT to determine whether there are k vertices in $V = \bigcup_{i=1}^t V(G_i)$ such that $G[V(G_i) \cap V]$ is a Π_i-graph for every $1 \leq i \leq t$.*

Edge-bicolored graphs also have close connections with digraphs, and dually connected graphs resemble strongly connected digraphs. We may use ideas in this paper to study subgraph problems on digraphs. In particular, we can easily modify proofs of Theorem 2 and Theorem 3 to obtain the following results for strongly connected subgraphs.

Theorem 8. *It is NP-complete and W[1]-hard to determine whether a digraph G contain exactly k vertices V' such that $G[V']$ (resp., $G - V'$) is strongly connected.*

We hope that our work will stimulate further research on simultaneous subgraph problems for edge-bicolored graphs and edge-bicolored multigraphs in general. Indeed, many fundamental and interesting problems are awaiting to be investigated, and we list some open problems here.

Problem 1. Determine whether DUALLY CONNECTED DELETION is FPT on "dual trees" when blue and red graphs are defined by $G[E_b]$ and $G[E_r]$, instead of (V, E_b) and (V, E_r).

Problem 2. Determine the parameterized complexity of DUAL SEPARATOR and DUAL SEPARATOR FOR TWO TERMINALS.

For hereditary properties Π not covered by Theorem 7, the parameterized complexity of DUAL Π-GRAPH DELETION is open for various fundamental properties Π. We note that for every property Π in the following problem, FPT algorithm exists for turning an uncolored graph into a Π-graph by deleting k vertices [10,5,17,14,15].

Problem 3. Determine parameterized complexities of DUAL Π-GRAPH DELETION for Π being acyclic, bipartite, chordal, and planar graphs, respectively.

For a different flavor, we may also consider modifying an edge-bicolored graph into a required graph by edge recoloring. We have obtained some interesting results in connection with dual connectedness and separators, and we will report our findings in a separate paper.

Problem 4. Determine complexities of turning an edge-bicolored graph into a dual Π-graph for Π being acyclic, bipartite, chordal, and planar graphs, respectively, by edge recoloring.

We expect many exciting results concerning simultaneous subgraphs in edge-bicolored graphs and edge-colored multigraphs in general, which may also shed light on other graph problems such as problems on digraphs.

Acknowledgement. We are grateful to Michel Habib for bringing our attention to the work on common connected components and common intervals.

References

1. Bang-Jensen, J., Gutin, G.: Alternating cycles and paths in edge-coloured multi-graphs: a survey. Discrete Mathematics 165, 39–60 (1997)
2. Bui-Xuan, B.M., Habib, M., Paul, C.: Competitive graph searches. Theoretical Computer Science 393(1), 72–80 (2008)
3. Cai, L.: Fixed-parameter tractability of graph modification problems for hereditary properties. Information Processing Letters 58(4), 171–176 (1996)
4. Cai, L., Chan, S.M., Chan, S.O.: Random separation: A new method for solving fixed-cardinality optimization problems. In: Bodlaender, H.L., Langston, M.A. (eds.) IWPEC 2006. LNCS, vol. 4169, pp. 239–250. Springer, Heidelberg (2006)
5. Dehne, F., Fellows, M., Langston, M., Rosamond, F., Stevens, K.: An $O(2^{O(k)}n^3)$ FPT algorithm for the undirected feedback vertex set problem. Theory of Computing Systems 41(3), 479–492 (2007)
6. Downey, R.G., Fellows, M.R.: Parameterized Complexity. Springer, New York (1999)
7. Gai, A.T., Habib, M., Paul, C., Raffinot, M.: Identifying common connected components of graphs. Technical Report RR-LIRMM-03016, LIRMM, Université de Montpellier II (2003)
8. Garey, M.R., Johnson, D.S.: Computers and Intractability: A Guide to the Theory of NP-Completeness. W.H. Freeman, New York (1979)
9. Garey, M.R., Johnson, D.S., Stockmeyer, L.: Some simplified NP-complete graph problems. Theoretical Computer Science 1(3), 237–267 (1976)
10. Guo, J., Gramm, J., Hüffner, F., Niedermeier, R., Wernicke, S.: Compression-based fixed-parameter algorithms for feedback vertex set and edge bipartization. Journal of Computer and System Sciences 72, 1386–1396 (2006)
11. Habib, M., Paul, C.: A survey of the algorithmic aspects of modular decomposition. Computer Science Review 4(1), 41–59 (2010)
12. Kano, M., Li, X.: Monochromatic and heterochromatic subgraphs in edge-colored graphs-a survey. Graphs and Combinatorics 24(4), 237–263 (2008)
13. Khot, S., Raman, V.: Parameterized complexity of finding subgraphs with hereditary properties. Theoretical Computer Science 289(2), 997–1008 (2002)
14. Marx, D.: Chordal deletion is fixed-parameter tractable. Algorithmica 57(4), 747–768 (2010)
15. Marx, D., Schlotter, I.: Obtaining a planar graph by vertex deletion. Algorithmica 62(3-4), 807–822 (2012)
16. Naor, M., Schulman, L.J., Srinivasan, A.: Splitters and near-optimal derandomization. In: Proceedings of the 36th Annual Symposium of Foundations of Computer Science, pp. 182–191 (1995)
17. Reed, B., Smith, K., Vetta, A.: Finding odd cycle transversals. Operations Research Letters 32(4), 299–301 (2004)
18. Uno, T., Yagiura, M.: Fast algorithms to enumerate all common intervals of two permutations. Algorithmica 26(2), 290–309 (2000)

Combinatorial Voter Control in Elections[*]

Jiehua Chen[1], Piotr Faliszewski[2], Rolf Niedermeier[1], and Nimrod Talmon[1]

[1] Institut für Softwaretechnik und Theoretische Informatik, TU Berlin, Germany
{jiehua.chen@tu-berlin.de,rolf.niedermeier}@tu-berlin.de,
nimrodtalmon77@gmail.com
[2] AGH University of Science and Technology, Krakow, Poland
faliszew@agh.edu.pl

Abstract. Voter control problems model situations such as an external agent trying to affect the result of an election by adding voters, for example by convincing some voters to vote who would otherwise not attend the election. Traditionally, voters are added one at a time, with the goal of making a distinguished alternative win by adding a minimum number of voters. In this paper, we initiate the study of combinatorial variants of control by adding voters: In our setting, when we choose to add a voter v, we also have to add a whole bundle $\kappa(v)$ of voters associated with v. We study the computational complexity of this problem for two of the most basic voting rules, namely the Plurality rule and the Condorcet rule.

1 Introduction

We study the computational complexity of control by adding voters [2, 18], investigating the case where the sets of voters that we can add have some combinatorial structure. The problem of election control by adding voters models situations where some agent (e.g., a campaign manager for one of the alternatives) tries to ensure a given alternative's victory by convincing some undecided voters to vote. Traditionally, in this problem we are given a description of an election (that is, a set C of alternatives and a set V of voters who decided to vote), and also a set W of undecided voters (for each voter in $V \cup W$ we assume to know how this voter intends to vote which is given by a linear order of the set C; we might have good approximation of this knowledge from preelection polls). Our goal is to ensure that our preferred alternative p becomes a winner, by convincing as few voters from W to vote as possible (provided that it is at all possible to ensure p's victory in this way).

Control by adding voters corresponds, for example, to situations where supporters of a given alternative make direct appeals to other supporters of the alternative to vote (for example, they may stress the importance of voting, or

---

[*] JC was supported by the Studienstiftung des Deutschen Volkes. PF was supported by the DFG project PAWS (NI 369/10). NT was supported by the DFG Research Training Group "Methods for Discrete Structures" (GRK 1408). This work has been partly supported by COST Action IC1205 on Computational Social Choice.

E. Csuhaj-Varjú et al. (Eds.): MFCS 2014, Part II, LNCS 8635, pp. 153–164, 2014.
© Springer-Verlag Berlin Heidelberg 2014

help with the voting process by offering rides to the voting locations, etc.). Unfortunately, in its traditional phrasing, control by adding voters does not model larger-scale attempts at convincing people to vote. For example, a campaign manager might be interested in airing a TV advertisement that would motivate supporters of a given alternative to vote (though, of course, it might also motivate some of this alternative's enemies), or maybe launch viral campaigns, where friends convince their own friends to vote. It is clear that the sets of voters that we can add should have some sort of a combinatorial structure (e.g., a TV advertisement appeals to a particular group of voters and we can add all of them at the unit cost of airing the advertisement).

The goal of our work is to formally define an appropriate computational problem modeling a combinatorial variant of control by adding voters and to study its computational complexity. Specifically, we focus on the Plurality rule and the Condorcet rule, mainly because the Plurality rule is the most widely used rule in practice, and it is one of the few rules for which the standard variant of control by adding voters is solvable in polynomial time [2], whereas for the Condorcet rule the problem is polynomial-time solvable for the case of single-peaked elections [14]. For the case of single-peaked elections, in essence, all our hardness results for the Condorcet rule directly translate to all Condorcet-consistent voting rules, a large and important family of voting rules. We defer the formal details, definitions, and concrete results to the following sections. Instead, we state the high-level, main messages of our work:

- Many typical variants of combinatorial control by adding voters are intractable, but there is also a rich landscape of tractable cases.
- Assuming that voters have single-peaked preferences does not lower the complexity of the problem (even though it does so in many election problems [6, 9, 14]). On the contrary, assuming single-crossing preferences does lower the complexity of the problem.

We believe that our setting of combinatorial control, and—more generally—combinatorial voting, offers a very fertile ground for future research and we intend the current paper as an initial step.

Related Work. In all previous work on election control, the authors always assumed that one could affect each entity of the election at unit cost only (e.g., one could add a voter at a unit cost; adding two voters always was twice as expensive as adding a single voter). Only the paper of Faliszewski et al. [15], where the authors study control in weighted elections, could be seen as an exception: One could think of adding a voter of weight w as adding a group of w voters of unit weight. On the one hand, the weighted election model does not allow one to express rich combinatorial structures as those that we study here, and on the other hand, in our study we consider unweighted elections only (though adding weights to our model would be seamless).

The specific combinatorial flavor of our model has been inspired by the seminal work of Rothkopf et al. [23] on *combinatorial auctions* (see, e.g., Sandholm [24] for additional information). There, bidders can place bids on combinations of items. While in combinatorial auctions one "bundles" items to bid on, in our scenario one bundles voters.

In the computational social choice literature, combinatorial voting is typically associated with scenarios where voters express opinions over a set of items that themselves have a specific combinatorial structure (typically, one uses CP-nets to model preferences over such alternative sets [5]). For example, Conitzer et al. [10] studied a form of control in this setting. In contrast, we use the standard model of elections (where all alternatives and preference orders are given explicitly), but we have a combinatorial structure of the sets of voters that can be added.

2   Preliminaries

We assume familiarity with standard notions regarding algorithms and complexity theory. For each nonnegative integer z, we write $[z]$ to mean $\{1, \ldots, z\}$.

Elections. An election $E := (C, V)$ consists of a set C of m alternatives and a set V of $|V|$ voters $v_1, v_2, \ldots, v_{|V|}$. Each voter v has a linear order $\succ_v$ over the set C, which we call a *preference order*. We call a voter $v \in V$ a *c-voter* if she prefers $c \in C$ the best. Given a set C of alternatives, if not stated explicitly, we write $\langle C \rangle$ to denote an arbitrary but fixed preference order over C.

Voting Rules. A voting rule $\mathcal{R}$ is a function that given an election E outputs a (possibly empty) set $\mathcal{R}(E) \subseteq C$ of the (tied) election winners. We study the Plurality rule and the Condorcet rule. Given an election, the *Plurality score* of an alternative c is the number of voters that have c at the first position in their preference orders; an alternative is a Plurality winner if it has the maximum Plurality score. An alternative c is a *Condorcet winner* if it beats all other alternatives in head-to-head contests. That is, c is a *Condorcet winner* in election $E = (C, V)$ if for each alternative $c' \in C \setminus \{c\}$ it holds that $|\{v \in V \mid c \succ_v c'\}| > |\{v \in V \mid c' \succ_v c\}|$. Condorcet's rule elects the (unique) Condorcet winner if it exists, and returns an empty set otherwise. A voting rule is *Condorcet-consistent* if it elects a Condorcet winner when there is one (however, if there is no Condorcet winner, then a Condorcet-consistent rule is free to provide any set of winners).

Domain Restrictions. Intuitively, an election is *single-peaked* [4] if it is possible to order the alternatives on a line in such a way that for each voter v the following holds: If c is v's most preferred alternative, then for each two alternatives c_i and c_j that both are on the same side of c (with respect to the ordering of the alternatives on the line), among c_i and c_j, v prefers the one closer to c. For example, single-peaked elections arise when we view the alternatives on the standard political left-right spectrum and voters form their preferences based solely on alternatives' positions on this spectrum. There are polynomial-time algorithms that given an election decide if it is single-peaked and, if so, provide a societal axis for it [1, 13]. *Single-crossing* elections, introduced by Roberts [22], capture a similar idea as single-peaked ones, but from a different perspective. This time we assume that it is possible to order the voters so that for each two alternatives a and b either all voters rank a and b identically, or there is a single point along this order where voters switch from preferring one of the alternatives to preferring the other one. There are polynomial-time algorithms that decide if an election is single-crossing and, if so, produce the voter order witnessing this fact [12, 7].

Combinatorial Bundling Functions. Given a voter set X, a combinatorial bundling function $\kappa : X \to 2^X$ (abbreviated as *bundling function*) is a function assigning to each voter a subset of voters. For convenience, for each subset $X' \subseteq X$, we let $\kappa(X') = \bigcup_{x \in X'} \kappa(x)$. For $x \in X$, $\kappa(x)$ is called x's *bundle* (and for this bundle, x is called its *leader*). We assume that $x \in \kappa(x)$ and so $\kappa(x)$ is never empty. We typically write b to denote the maximum bundle size under a given κ (which will always be clear from context). Intuitively, we use combinatorial bundling functions to describe the sets of voters that we can add to an election at a unit cost. For example, one can think of $\kappa(x)$ as the group of voters that join the election under x's influence.

We are interested in various special cases of bundling functions. We say that κ is *leader-anonymous* if for each two voters x and y with the same preference order $\kappa(x) = \kappa(y)$ holds. Furthermore, κ is *follower-anonymous* if for each two voters x and y with the same preference orders, and each voter z, it holds that $x \in \kappa(z)$ if and only if $y \in \kappa(z)$. We call κ *anonymous* if it is both leader-anonymous and follower-anonymous. The swap distance between two voters v_i and v_j is the minimum number of swaps of consecutive alternatives that transform v_i's preference order into that of v_j. Given a number $d \in \mathbb{N}$, we call κ a *full-d bundling function* if for each $x \in X$, $\kappa(x)$ is exactly the set of all $y \in X$ such that the swap distance between the preference orders of x and y is at most d.

Central Problem. We consider the following problem for a given voting rule $\mathcal{R}$:

> $\mathcal{R}$ COMBINATORIAL CONSTRUCTIVE CONTROL BY ADDING VOTERS ($\mathcal{R}$-C-CC-AV)
> **Input:** An election $E = (C, V)$, a set W of (unregistered) voters with $V \cap W = \emptyset$, a bundling function $\kappa : W \to 2^W$, a preferred alternative $p \in C$, and a bound $k \in \mathbb{N}$.
> **Question:** Is there a subset of voters $W' \subseteq W$ of size at most k such that $p \in \mathcal{R}(C, V \cup \kappa(W'))$, where $\mathcal{R}(C, X)$ is the set of winners of the election (C, X) under the rule $\mathcal{R}$?

We note that we use here a so-called nonunique-winner model. For a control action to be successful, it suffices for p to be one of the tied winners. Throughout this work, we refer to the set W' of voters such that p wins election $(C, V \cup \kappa(W'))$ as the solution and denote k as the solution size.

$\mathcal{R}$-C-CC-AV is a generalization of the well-studied problem $\mathcal{R}$ CONSTRUCTIVE CONTROL BY ADDING VOTERS ($\mathcal{R}$-CC-AV) (in which κ is fixed so that for each $w \in W$ we have $\kappa(w) = \{w\}$). The non-combinatorial problem CC-AV is polynomial-time solvable for the Plurality rule [2], but is NP-complete for the Condorcet rule [20].

Table 1. Computational complexity classification of Plurality-C-CC-AV (since the non-combinatorial problem CC-AV is already NP-hard for Condorcet's rule, we concentrate here on the Plurality rule). Each row and column in the table corresponds to a parameter such that each cell contains results for the two corresponding parameters combined. Due to symmetry, there is no need to consider the cells under the main diagonal, therefore they are painted in gray. ILP-FPT means FPT based on a formulation as an integer linear program.

	m	n	k	b	d
# alternatives (m)				Non-anonymous: W[2]-h wrt. k even if $m = 2$ [Thm. 2] Anonymous: ILP-FPT wrt. m [Thm. 3]	
# unreg. voters (n)				FPT wrt. n	
solution size (k)				XP [Obs. 1] Anonymous & $b = 3$: W[1]-h wrt. k [Thm. 1]	Single-peaked & full-1 κ: W[1]-h wrt. k [Thm. 8]
max. bundle size (b)				$b = 2$: NP-h [Thm. 4] and P for full-d κ [Thm. 5] $b = 3$: NP-h even for full-d κ [Thm. 6] $b \geq 4$: NP-h even for full-1 κ [Thm. 7]	
max. swap dist. (d)					$d = 1$: W[1]-h wrt. k [Thm. 8] Single-crossing & full-d κ: P [Thm. 9]

Parameterized Complexity. An instance (I, k) of a parameterized problem consists of the actual instance I and an integer k being the *parameter* [11]. A parameterized problem is called *fixed-parameter tractable* (is in FPT) if there is an algorithm solving it in $f(k) \cdot |I|^{O(1)}$ time, for an arbitrary computable function f only depending on parameter k, whereas an algorithm with running-time $|I|^{f(k)}$ only shows membership in the class XP (clearly, FPT $\subseteq$ XP). One can show that a parameterized problem L is (presumably) not fixed-parameter tractable by devising a *parameterized reduction* from a W[1]-hard or a W[2]-hard problem to L. A parameterized reduction from a parameterized problem L to another parameterized problem L' is a function that, given an instance (I, k), computes in $f(k) \cdot |I|^{O(1)}$ time an instance (I', k'), such that $k' \leq g(k)$ and $(I, k) \in L \Leftrightarrow (I', k') \in L'$. Betzler et al. [3] survey parameterized complexity investigations in voting.

Our Contributions. As $\mathcal{R}$-C-CC-AV is generally NP-hard even for $\mathcal{R}$ being the Plurality rule, we show several fixed-parameter tractability results for some of the natural parameterizations of $\mathcal{R}$-C-CC-AV; we almost completely resolve the complexity of C-CC-AV, for the Plurality rule and the Condorcet rule, as a function of the maximum bundle size b and the maximum distance d from a voter v to the farthest element of her bundle. Further, we show that the problem remains hard even when restricting the elections to be single-peaked, but that it is polynomial-time solvable when we focus on single-crossing elections. Our results for Plurality elections are summarized in Table 1.

3   Complexity for Unrestricted Elections

In this section we provide our results for the case of unrestricted elections, where voters may have arbitrary preference orders. In the next section we will consider single-peaked and single-crossing elections that only allow "reasonable" preference orders.

Number of Voters, Number of Alternatives, and Solution Size. We start our discussion by considering parameters "the number m of alternatives", "the number n of unregistered voters", and "the solution size k". A simple brute-force algorithm, checking all possible combinations of k bundles, proves that both PLURALITY-C-CC-AV and CONDORCET-C-CC-AV are in XP for parameter k, and in FPT for parameter n (the latter holds because $k \leq n$). Indeed, the same result holds for all voting rules that are XP/FPT-time computable for the respective parameters.

Observation 1. *Both* PLURALITY-C-CC-AV *and* CONDORCET-C-CC-AV *are solvable in* $O(n^k \cdot n \cdot m \cdot \text{winner})$ *time, where* winner *is the complexity of determining Plurality/Condorcet winners.*

The XP result for PLURALITY-C-CC-AV with respect to the parameter k probably cannot be improved to fixed-parameter tractability. Indeed, for parameter k we show that the problem is W[1]-hard, even for anonymous bundling functions and for maximum bundle size three.

Theorem 1. PLURALITY-C-CC-AV *is* NP-*hard and* W[1]-*hard when parameterized by the solution size* k, *even when the maximum bundle size* b *is three and the bundling function is anonymous.*

Proof (Sketch). We provide a parameterized reduction from the W[1]-hard problem CLIQUE parameterized by the parameter h [11], which asks for the existence of a complete subgraph with h vertices in an input graph G.

Let (G, h) be a CLIQUE instance. Without loss of generality, we assume that G is connected, that $h \geq 3$, and that each vertex in G has degree at least $h - 1$. We construct an election $E = (C, V)$ with $C := \{p, w, g\} \cup \{c_e \mid e \in E(G)\}$, and set p to be the preferred alternative. The registered voter set V consists of $\binom{h}{2} + h$ voters each with preference order $w \succ \langle C \setminus \{w\} \rangle$, another $\binom{h}{2}$ voters each with preference order $g \succ \langle C \setminus \{g\} \rangle$, and another h voters each with preference order $p \succ \langle C \setminus \{p\} \rangle$. For each vertex $u \in V(G)$, we define $C(u) := \{c_e \mid e \in E(G) \wedge u \in e\}$, and construct the set W of unregistered voters as follows:

(i) For each vertex $u \in V(G)$, we add an unregistered g-voter w_u with preference order $g \succ \langle C(u) \rangle \succ \langle C \setminus (\{g\} \cup C(u)) \rangle$ and we set $\kappa(w_u) = \{w_u\}$.

(ii) For each edge $e = \{u, u'\} \in E(G)$, we add an unregistered p-voter w_e with preference order $p \succ c_e \succ \langle C \setminus \{p, c_e\} \rangle$ and we set $\kappa(w_e) = \{w_u, w_{u'}, w_e\}$.

Since all the unregistered voters have different preference orders (this is so because G is connected, $h \geq 3$, and each vertex has degree at least $h - 1$), every bundling function for our instance is anonymous. Finally, we set $k := \binom{h}{2}$. $\square$

If we drop the anonymity requirement for the bundling function, then we obtain a stronger intractability result. For parameter k, the problem becomes

W[2]-hard, even for two alternatives. This is quite remarkable because typically election problems with a small number of alternatives are easy (they can be solved either through brute-force attacks or through integer linear programming attacks employing the famous FPT algorithm of Lenstra [19]; see the survey of Betzler et al. [3] for examples, but note that there are also known examples of problems where a small number of alternatives does not seem to help [8]). Further, since our proof uses only two alternatives, it applies to almost all natural voting rules: For two alternatives almost all of them (including the Condorcet rule) are equivalent to the Plurality rule. The reduction is from the W[2]-complete problem SET COVER parameterized by the solution size [11].

Theorem 2. *Both* PLURALITY-C-CC-AV *and* CONDORCET-C-CC-AV *parameterized by the solution size k are* W[2]-*hard, even for two alternatives.*

If we require the bundling function to be anonymous, then C-CC-AV can be formulated as an integer linear program where the number of variables and the number of constraints are bounded by some function in the number m of alternatives. Hence, C-CC-AV is fixed-parameter tractable due to Lenstra [19].

Theorem 3. *For anonymous bundling functions, both* PLURALITY-C-CC-AV *and* CONDORCET-C-CC-AV *parameterized by the number m of alternatives are fixed-parameter tractable.*

Combinatorial Parameters. We focus now on the complexity of PLURALITY-C-CC-AV as a function of two combinatorial parameters: (a) the maximum swap distance d between the leader and his followers in one bundle, and (b) the maximum size b of each voter's bundle.

First, if $b = 1$, then C-CC-AV reduces to CC-AV and, thus, can be solved by a greedy algorithm in polynomial time [2]. However, for arbitrary bundling functions, PLURALITY-C-CC-AV becomes intractable as soon as $b = 2$.

Theorem 4. PLURALITY-C-CC-AV *is* NP-*hard even if the maximum bundle size b is two.*

Proof (Sketch). We reduce from a restricted variant of 3SAT, where each clause has either two or three literals, each variable occurs exactly four times, twice as a positive literal, and twice as a negative literal. This variant is still NP-hard (the proof is analogous to the one shown for [25, Theorem 2.1]). Given such a restricted 3SAT instance $(\mathcal{C}, \mathcal{X})$, where $\mathcal{C}$ is the set of clauses over the set of variables $\mathcal{X}$, we construct an election (C, V) with $C := \{p, w\} \cup \{c_i \mid C_i \in \mathcal{C}\}$; we call c_i the *clause* alternatives. We set $k := 4|\mathcal{X}|$. We construct the set V such that the initial score of w is $4|\mathcal{X}|$, the initial score of each clause alternative c_i is $4|\mathcal{X}| - |C_i| + 1$, and the initial score of p is zero. We construct the set W of unregistered voters as follows (we will often write ℓ_j to refer to a literal that contains variable x_j; depending on the context, ℓ_j will mean either x_j or $\neg x_j$ and the exact meaning will always be clear):

1. for each variable $x_j \in \mathcal{X}$, we construct four p-voters, denoted by $p_1^j, p_2^j, p_3^j, p_4^j$;
2. for each clause $C_i \in \mathcal{C}$ and each literal ℓ contained in C_i, we construct a c_i-voter, denoted by c_i^ℓ; we call such voter a *clause* voter.

We define the assignment function κ as follows: For each variable $x_j \in \mathcal{X}$ that occurs as a negative literal $(\neg x_j)$ in clauses C_i and C_s, and as a positive literal (x_j) in clauses C_r and C_t, we set $\kappa(p_1^j) = \{p_1^j, c_i^{\neg x_j}\}$, $\kappa(c_i^{\neg x_j}) = \{c_i^{\neg x_j}, p_2^j\}$, $\kappa(p_2^j) = \{p_2^j, c_r^{x_j}\}$, $\kappa(c_r^{x_j}) = \{c_r^{x_j}, p_3^j\}$, $\kappa(p_3^j) = \{p_3^j, c_s^{\neg x_j})\}$, $\kappa(c_s^{\neg x_j}) = \{c_s^{\neg x_j}, p_4^j\}$, $\kappa(p_4^j) = \{p_4^j, c_t^{x_j}\}$, $\kappa(c_t^{x_j}) = \{c_t^{x_j}, p_1^j\}$.

The general idea is that in order to let p win, all p-voters must be in $\kappa(W')$ and no clause alternative should gain more than $(|C_i| - 1)$ points. We now show that if $(\mathcal{C}, \mathcal{X})$ has a satisfying truth assignment, then there is a size-k subset $W' \subseteq W$ such that p wins the election $(C, V \cup \kappa(W'))$ (recall that $k = 4|\mathcal{X}|$). The proof for the reverse direction is omitted.

Let $\beta : \mathcal{X} \to \{T, F\}$ be a satisfying truth assignment function for $(\mathcal{C}, \mathcal{X})$. Intuitively, β will guide us through constructing the set W' in the following way: First, for each variable x_j, we put into W' those voters $c_i^{\ell_j}$ for whom β sets ℓ_j to false (this way in $\kappa(W')$ we include $2|\mathcal{X}|$ p-voters and, for each clause c_i, at most $(|C_i|-1)$ c_i-voters). Then, for each clause voter $c_i^{\ell_j}$ already in W', we also add the voter p_a^j, $1 \le a \le 4$, that contains $c_i^{\ell_j}$ in his or her bundle (this way we include in $\kappa(W')$ additional $2|\mathcal{X}|$ p-voters without increasing the number of clause voters included). Formally, we define W' as follows: $W' := \{c_i^{\neg x_j}, p_a^j \mid \neg x_j \in C_i \wedge \beta(x_j) = T \wedge c_i^{\neg x_j} \in \kappa(p_a^j)\} \cup \{c_i^{x_j}, p_a^j \mid x_j \in C_i \wedge \beta(x_j) = F \wedge c_i^{\neg x_j} \in \kappa(p_a^j)\}$. As per our intuitive argument, one can verify that all p-voters are contained in $\kappa(W')$ and each clause alternative c_i gains at most $(|C_i| - 1)$ points.   □

The situation is different for full-d bundling functions, because we can extend the greedy algorithm by Bartholdi et al. [2] to bundles of size two.

Theorem 5. *If κ is a full-d bundling function and the maximum bundle size b is two, then* PLURALITY-C-CC-AV *is polynomial-time solvable.*

However, as soon as $b = 3$, we obtain NP-hardness, by modifying the reduction used in Theorem 4.

Theorem 6. *If κ is a full-d bundling function, then* PLURALITY-C-CC-AV *is NP-hard even if the maximum bundle size b is three.*

Taking also the swap distance d into account, we find out that both PLURALITY-C-CC-AV and CONDORCET-C-CC-AV are NP-hard, even if $d = 1$. This stands in contrast to the case where $d = 0$, where $\mathcal{R}$-C-CC-AV reduces to the CC-AV problem (perhaps for the weighted voters [15]), which, for Plurality voting, is polynomial-time solvable by a simple greedy algorithm.

Theorem 7. PLURALITY-C-CC-AV *is NP-hard even for full-1 bundling functions and even if the maximum bundle size b is four.*

4   Single-Peaked and Single-Crossing Elections

In this section, we focus on instances with full-d bundling functions, and we do so because without this restriction the hardness results from previous sections easily translate to our restricted domains (at least for the case of the Plurality rule). We find that the results for the combinatorial variant of control by adding voters for single-peaked and single-crossing elections are quite different than those for the non-combinatorial case. Indeed, both for Plurality and for Condorcet, the voter control problems for single-peaked elections and for single-crossing elections are solvable in polynomial time for the non-combinatorial case [6, 14, 21]. For the combinatorial case, we show hardness for both PLURALITY-C-CC-AV and CONDORCET-C-CC-AV for single-peaked elections, but give polynomial-time algorithms for single-crossing elections. We mention that the intractability results can also be seen as regarding anonymous bundling functions because all full-d bundling functions are leader-anonymous and follower-anonymous.

Theorem 8. *Both* PLURALITY-C-CC-AV *and* CONDORCET-C-CC-AV *parameterized by the solution size k are* W[1]*-hard for single-peaked elections, even for full-1 bundling functions.*

Proof (Sketch). We provide a parameterized reduction from the W[1]-complete PARTIAL VERTEX COVER (PVC) parameterized by "solution size" h [17], which asks for a set of at most h vertices in a graph G, which intersects with at least ℓ edges. Given a PVC instance (G, h, ℓ), we set $k := h$, construct an election $E = (C, V)$ with $C := \{p, w\} \cup \{a_i, \overline{a}_i, b_i, \overline{b}_i \mid u_i \in V(G)\}$, and set p to be the preferred alternative, such that the initial score of w is $h + \ell$, and is zero for all other alternatives. We do so by creating $h + \ell$ registered voters who all have the same preference order $\succ$ such that it differs from the following *canonical preference order*: $p \succ w \succ a_1 \succ \overline{a}_1 \succ \ldots \succ a_{|V(G)|} \succ \overline{a}_{|V(G)|} \succ b_1 \succ \overline{b}_1 \succ \ldots \succ b_{|V(G)|} \succ \overline{b}_{|V(G)|}$ by only the first pair $\{p, w\}$, which is swapped.

For each set P of disjoint pairs of alternatives, neighboring with respect to the canonical preference order, we define the preference order diff-order(P) to be identical to the canonical preference order, except that all the pairs of alternatives in P are swapped. The unregistered voter set W consists of the following three types of voters:

(i) for each edge $e = \{u_i, u_j\} \in E(G)$ we have an *edge voter* w_e with preference order diff-order($\{\{a_i, \overline{a}_i\}, \{a_j, \overline{a}_j\}\}$),

(ii) for each edge $e = \{u_i, u_j\} \in E(G)$ we have a *dummy voter* d_e with preference order diff-order($\{\{p, w\}, \{a_i, \overline{a}_i\}, \{a_j, \overline{a}_j\}\}$), and

(iii) for each vertex $u_i \in V(G)$ we have a *vertex voter* w_i^u with preference order diff-order($\{\{a_i, \overline{a}_i\}\}$).

The preference orders of the voters in $V \cup W$ are single-peaked with respect to the axis $\langle \overline{B} \rangle \succ \langle \overline{A} \rangle \succ p \succ w \succ \langle A \rangle \succ \langle B \rangle$, where $\langle \overline{B} \rangle := \overline{b}_{|V(G)|} \succ \overline{b}_{|V(G)|-1} \succ \ldots \succ \overline{b}_1$, $\langle \overline{A} \rangle := \overline{a}_{|V(G)|} \succ \overline{a}_{|V(G)|-1} \succ \ldots \succ \overline{a}_1$, $\langle A \rangle := a_1 \succ a_2 \succ \ldots \succ a_{|V(G)|}$, and $\langle B \rangle := b_1 \succ b_2 \succ \ldots \succ b_{|V(G)|}$. Finally, we define the function κ such that it is a full-1 bundling function. $\qquad \square$

We now present some tractability results for single-crossing elections. Consider an $\mathcal{R}$-C-CC-AV instance $((C, V), W, d, \kappa, p \in C, k)$ such that $(C, V \cup W)$ is single-crossing. This has a crucial consequence for full-d bundling functions: For each unregistered voter $w \in W$, the voters in bundle $\kappa(w)$ appear consecutively along the single-crossing order restricted to only the voters in W.[1] Using the following lemmas, we can show that PLURALITY-C-CC-AV and CONDORCET-C-CC-AV are polynomial-time solvable in some cases.

Lemma 1. *Let $I = ((C, V), W, d, \kappa, p \in C, k)$ be a PLURALITY-C-CC-AV instance such that $(C, V \cup W)$ is single-crossing and κ is a full-d bundling function. Then, the following statements hold:*

(i) *The p-voters are ordered consecutively along the single-crossing order.*
(ii) *If I is a yes instance, then there is a subset $W' \subseteq W$ of size at most k such that all bundles of voters $w \in W'$ contain only p-voters, except at most two bundles which may contain some non-p-voters.*

Lemma 2. *Let $(C, V \cup \kappa(W'))$ be a single-crossing election with single-crossing voter order $\langle x_1, x_2, \ldots, x_z \rangle$ and set $X_{\text{median}} := \{x_{\lceil z/2 \rceil}\} \cup \{x_{z/2+1} \text{ if } z \text{ is even}\}$, where $z = |V| + |\kappa(W')|$. Alternative p is a (unique) Condorcet winner in $(C, V \cup \kappa(W'))$ if and only if every voter in X_{median} is a p-voter.*

Theorem 9. *Both PLURALITY-C-CC-AV and CONDORCET-C-CC-AV are polynomial-time solvable for the single-crossing case with full-d bundling functions.*

Proof. First, we find a (unique) single-crossing voter order for $(C, V \cup W)$ in quadratic time [12, 7]. Due to Lemma 1 and Lemma 2, we only need to store the most preferred alternative of each voter to find the solution set W'. Thus, the running-time from now on only depends on the number of voters. We start with the Plurality rule and let $\alpha := \langle w_1, w_2, \ldots, w_{|W|} \rangle$ be a single-crossing voter order.

Due to Lemma 1 (ii), the two bundles in $\kappa(W')$ which may contain non-p-voters appear at the beginning and at the end of the p-voter block, along the single-crossing order. We first guess these two bundles, and after this initial guess, all remaining bundles in the solution contain only p-voters (Lemma 1 (i)). Thus, the remaining task is to find the maximum score that p can gain by selecting k' bundles containing only p-voters. This problem is equivalent to the MAXIMUM INTERVAL COVER problem, which is solvable in $O(|W|^2)$ time (Golab et al. [16, Section 3.2]).

For the Condorcet rule, we propose a slightly different algorithm. The goal is to find a minimum-size subset $W' \subseteq W$ such that p is the (unique) Condorcet winner in $(C, V \cup \kappa(W'))$. Let $\beta := \langle x_1, x_2, \ldots, x_z \rangle$ be a single-crossing voter order for $(C, V \cup W)$. Considering Lemma 2, we begin by guessing at most two voters in $V \cup W$ whose bundles may contain the median p-voter (or, possibly, several p-voters) along the single-crossing order of voters restricted to the final

---

[1] Note that for each single-crossing election, the order of the voters possessing the single-crossing property is, in essence, unique.

election (for simplicity, we define the bundle of each registered voter to be its singleton). The voters in the union of these two bundles must be consecutively ordered. Let those voters be $x_i, x_{i+1}, \ldots, x_{i+j}$ (where $i \geq 1$ and $j \geq 0$), let $W_1 := \{x_s \in W \mid s < i\}$, and let $W_2 := \{x_s \in W \mid s > i + j\}$. We guess two integers $z_1 \leq |W_1|$ and $z_2 \leq |W_1|$ with the property that there are two subsets $B_1 \subseteq W_1$ and $B_2 \subseteq W_2$ with $|B_1| = z_1$ and $|B_2| = z_2$ such that the median voter(s) in $V \cup B_1 \cup \{x_i, x_{i+1}, \ldots, x_{i+j}\} \cup B_2$ are indeed p-voters (for now, only the sizes z_1 and z_2 matter, not the actual sets). These four guesses cost $O(|V \cup W|^2 \cdot |W|^2)$ time. The remaining task is to find two minimum-size subsets W_1' and W_2' such that $\kappa(W_1') \subseteq W_1$, $\kappa(W_2') \subseteq W_2$, $|\kappa(W_1')| = z_1$, and $|\kappa(W_2')| = z_2$. As already discussed, this can be done in $O(|W|^2)$ time [16]. We conclude that one can find a minimum-size subset $W' \subseteq W$ such that p is the (unique) Condorcet winner in $(C, V \cup \kappa(W'))$ in $O(|V \cup W|^2 \cdot |W|^4)$ time. $\quad\square$

5 Conclusion

We provide opportunities for future research. First, we did not discuss destructive control and the related problem of combinatorial deletion of voters. For Plurality, we conjecture that combinatorial addition of voters for destructive control, and combinatorial deletion of voters for either constructive or destructive control behave similarly to combinatorial addition of voters for constructive control.

Another, even wider field of future research is to study other combinatorial voting models—this may include controlling the swap distance, "probabilistic bundling", "reverse bundling", or using other distance measures than the swap distance. Naturally, it would also be interesting to consider other problems than election control (with bribery being perhaps the most natural candidate).

Finally, instead of studying a "leader-follower model" as we did, one might also be interested in an "enemy model" referring to control by adding alternatives: The alternatives of an election "hate" each other such that if one alternative is added to the election, then all of its enemies are also added to the election.

References

[1] Bartholdi III, J.J., Trick, M.: Stable matching with preferences derived from a psychological model. Oper. Res. Lett. 5(4), 165–169 (1986)
[2] Bartholdi III, J.J., Tovey, C.A., Trick, M.A.: How hard is it to control an election. Math. Comput. Model. 16(8-9), 27–40 (1992)
[3] Betzler, N., Bredereck, R., Chen, J., Niedermeier, R.: Studies in computational aspects of voting. In: Bodlaender, H.L., Downey, R., Fomin, F.V., Marx, D. (eds.) Fellows Festschrift 2012. LNCS, vol. 7370, pp. 318–363. Springer, Heidelberg (2012)
[4] Black, D.: On the rationale of group decision making. J. Polit. Econ. 56(1), 23–34 (1948)
[5] Boutilier, C., Brafman, R.I., Domshlak, C., Hoos, H.H., Poole, D.: CP-nets: A tool for representing and reasoning with conditional *ceteris paribus* preference statements. J. Artificial Intelligence Res. 21, 135–191 (2004)

[6] Brandt, F., Brill, M., Hemaspaandra, E., Hemaspaandra, L.A.: Bypassing combinatorial protections: Polynomial-time algorithms for single-peaked electorates. In: Proc. 24th AAAI, pp. 715–722 (2010)

[7] Bredereck, R., Chen, J., Woeginger, G.: A characterization of the single-crossing domain. Soc. Choice Welf. 41(4), 989–998 (2013)

[8] Bredereck, R., Chen, J., Faliszewski, P., Nichterlein, A., Niedermeier, R.: Prices matter for the parameterized complexity of shift bribery. In: Proc. 28th AAAI (to appear, 2014)

[9] Conitzer, V.: Eliciting single-peaked preferences using comparison queries. J. Artificial Intelligence Res. 35, 161–191 (2009)

[10] Conitzer, V., Lang, J., Xia, L.: How hard is it to control sequential elections via the agenda? In: Proc. 21st IJCAI, pp. 103–108 (July 2009)

[11] Downey, R.G., Fellows, M.R.: Fundamentals of Parameterized Complexity. Springer (2013)

[12] Elkind, E., Faliszewski, P., Slinko, A.: Clone structures in voters' preferences. In: Proc. 13th EC, pp. 496–513 (2012)

[13] Escoffier, B., Lang, J., Öztürk, M.: Single-peaked consistency and its complexity. In: Proc. 18th ECAI, pp. 366–370 (2008)

[14] Faliszewski, P., Hemaspaandra, E., Hemaspaandra, L.A., Rothe, J.: The shield that never was: Societies with single-peaked preferences are more open to manipulation and control. Inform. and Comput. 209(2), 89–107 (2011)

[15] Faliszewski, P., Hemaspaandra, E., Hemaspaandra, L.A.: Weighted electoral control. In: Proc. 12th AAMAS, pp. 367–374 (2013)

[16] Golab, L., Karloff, H., Korn, F., Saha, A., Srivastava, D.: Sequential dependencies. In: In 35th PVLDB, vol. 2(1), pp. 574–585 (2009)

[17] Guo, J., Niedermeier, R., Wernicke, S.: Parameterized complexity of Vertex Cover variants. Theory Comput. Syst. 41(3), 501–520 (2007)

[18] Hemaspaandra, E., Hemaspaandra, L.A., Rothe, J.: Anyone but him: The complexity of precluding an alternative. Artif. Intell. 171(5-6), 255–285 (2007)

[19] Lenstra, H.W.: Integer programming with a fixed number of variables. Math. Oper. Res. 8(4), 538–548 (1983)

[20] Liu, H., Feng, H., Zhu, D., Luan, J.: Parameterized computational complexity of control problems in voting systems. Theor. Comput. Sci. 410, 2746–2753 (2009)

[21] Magiera, K., Faliszewski, P.: How hard is control in single-crossing elections? In: Proc. 21st ECAI (to appear, 2014)

[22] Roberts, K.W.: Voting over income tax schedules. J. Public Econ. 8, 329–340 (1977)

[23] Rothkopf, M.H., Pekeč, A., Harstad, R.M.: Computationally manageable combinational auctions. Manage. Sci. 44(8), 1131–1147 (1998)

[24] Sandholm, T.: Optimal winner determination algorithms. In: Cramton, Shoham, Steinberg (eds.) Combinatorial Auctions. ch. 14. MIT Press (2006)

[25] Tovey, C.A.: A simplified NP-complete satisfiability problem. Discrete Appl. Math. 8(1), 85–89 (1984)

An Improved Deterministic #SAT Algorithm for Small De Morgan Formulas

Ruiwen Chen[1], Valentine Kabanets[1], and Nitin Saurabh[2]

[1] Simon Fraser University, Burnaby, Canada
ruiwenc@sfu.ca, kabanets@cs.sfu.ca
[2] Institute of Mathematical Sciences, Chennai, India
nitin@imsc.res.in

Abstract. We give a deterministic #SAT algorithm for de Morgan formulas of size up to $n^{2.63}$, which runs in time $2^{n-n^{\Omega(1)}}$. This improves upon the deterministic #SAT algorithm of [3], which has similar running time but works only for formulas of size less than $n^{2.5}$.

Our new algorithm is based on the shrinkage of de Morgan formulas under random restrictions, shown by Paterson and Zwick [12]. We prove a *concentrated* and *constructive* version of their shrinkage result. Namely, we give a deterministic polynomial-time algorithm that selects variables in a given de Morgan formula so that, *with high probability* over the random assignments to the chosen variables, the original formula shrinks in size, when simplified using a *deterministic polynomial-time* formula-simplification algorithm.

Keywords: de Morgan formulas, random restrictions, shrinkage, SAT algorithms.

1 Introduction

Subbotovskaya [16] introduced the method of *random restrictions* to prove that PARITY requires de Morgan formulas of size $\Omega(n^{1.5})$, where a de Morgan formula is a boolean formula over the basis $\{\vee, \wedge, \neg\}$. She showed that a random restriction of all but a fraction p of the input variables yields a new formula whose size is expected to reduce by at least the factor $p^{1.5}$. That is, the *shrinkage exponent* Γ for de Morgan formulas is at least 1.5, where the shrinkage exponent is defined as the least upper bound on γ such that the expected formula size shrinks by the factor p^{γ} under a random restriction leaving p fraction of variables free.

Impagliazzo and Nisan [9] argued that Subbotovskaya's bound $\Gamma \geqslant 1.5$ is not optimal, by showing that $\Gamma \geqslant 1.556$. Paterson and Zwick [12] improved upon [9], getting $\Gamma \geqslant (5-\sqrt{3})/2 \approx 1.63$. Finally, Håstad [6] proved the tight bound $\Gamma = 2$; combined with Andreev's construction [1], this yields a function in P requiring de Morgan formulas of size $\Omega(n^{3-o(1)})$.

While the original motivation to study shrinkage in [16,9,12,6] was to prove formula lower bounds, the same results turn out to be useful also for designing nontrivial SAT algorithms for small de Morgan formulas. Santhanam [14] strengthened

E. Csuhaj-Varjú et al. (Eds.): MFCS 2014, Part II, LNCS 8635, pp. 165–176, 2014.
© Springer-Verlag Berlin Heidelberg 2014

Subbotovskaya's *expected* shrinkage result to *concentrated* shrinkage, i.e., shrinkage with high probability, and used this to get a deterministic #SAT algorithm (counting the number of satisfying assignments) for linear-size de Morgan formulas, with the running time $2^{n-\Omega(n)}$. Santhanam's algorithm deterministically selects a most frequent variable in the current formula, and recurses on the two subformulas obtained by restricting the chosen variable to 0 and 1; after $n - \Omega(n)$ recursive calls, almost all obtained formulas depend on fewer than the actual number of free variables remaining, which leads to nontrivial savings over the brute-force SAT algorithm for the original formula. A similar algorithm works also for formulas of size less than $n^{2.5}$, with the running time $2^{n-n^{\Omega(1)}}$ [3].

Motivated by average-case formula lower bounds, Komargodksi et al. [11] (building upon [8]) showed a concentrated-shrinkage version of Håstad's optimal result for the shrinkage exponent $\Gamma = 2$. Combined with the aforementioned algorithm of Chen et al. [3], this yields a nontrivial *randomized* zero-error #SAT algorithm for de Morgan formulas of size $n^{3-o(1)}$, running in time $2^{n-n^{\Omega(1)}}$.

The main question addressed by our paper is whether there is a *deterministic* #SAT algorithm, with similar running time, for formulas of size close to n^3. This question is interesting since getting a deterministic algorithm often yields deeper understanding of the problem by revealing additional structural properties. It also provides better understanding of the role of randomness in efficient algorithms, as part of research on *derandomization*.

We give a deterministic #SAT algorithm for formulas of size up to $n^{2.63}$. In the process, we refine the results of Paterson and Zwick [12] on shrinkage of de Morgan formulas by making their results *constructive* in a certain precise sense. We provide more details next.

1.1   Our Main Results and Techniques

Our main result is a *deterministic* #SAT algorithm for de Morgan formulas of size up to $n^{2.63}$, running in time $2^{n-n^{\Omega(1)}}$.

Theorem 1 (Main). *There is a deterministic algorithm for counting the number of satisfying assignments in a given de Morgan formula on n variables of size at most $n^{2.63}$ which runs in time at most $2^{n-n^{\delta}}$, for some constant $0 < \delta < 1$.*

As in [14,3], we use a deterministic algorithm to choose a next variable to restrict, and then recurse on the two resulting restrictions of this variable to 0 and 1. Instead of Subbotovskaya-inspired selection procedure (choosing the most frequent variable), we use the weight function introduced by Paterson and Zwick [12], which measures the potential savings for each one-variable restriction, and selects a variable with the biggest savings. Since [12] gives the shrinkage exponent $\Gamma \approx 1.63$, rather than Subbotovskaya's 1.5, this could potentially lead to an improved #SAT algorithm for larger de Morgan formulas.

However, computing the savings, as defined by [12], is NP-hard, as it requires computing the size of a smallest logical formula equivalent to a given one-variable restriction. In fact, the shrinkage result of [12] is *nonconstructive* in the following sense: the expected shrinkage in size is proved for the minimal logical formula computing the restricted boolean function, rather than for the formula obtained from the original formula using efficiently computable simplification rules. In contrast, the shrinkage results of [16,6] are constructive: the restricted formula is expected to shrink in size when simplified using a certain explicit set of logical rules, so that the new, simplified formula is computable in polynomial time from the original restricted formula.

While the constructiveness of shrinkage is unimportant for proving formula lower bounds, it is crucial for designing shrinkage-based #SAT algorithms for de Morgan formulas, such as those in [14,3,11]. Our main technical contribution is a proof of the *constructive* version of the result in [12]: we give deterministic polynomial-time algorithms for formula simplification and extend the analysis of [12] to show expected shrinkage of formulas with respect to this efficiently computable simplification procedure. The same simplification procedure allows us to choose, in deterministic polynomial-time, which variable should be restricted next. The merit of deterministic variable selection and concentrated and constructive shrinkage, for a shrinkage exponent Γ, is that they yield a deterministic satisfiability algorithm for de Morgan formulas up to size $n^{\Gamma+1-o(1)}$, using an approach of [3].

Namely, once we have this constructive shrinkage result, based on restricting one variable at a time, we apply the martingale-based analysis of [10,3] to derive a *concentrated* version of constructive shrinkage, showing that almost all random settings of the selected variables yield restricted formulas of reduced size, where the restricted formulas are simplified by our efficient procedure. The shrinkage exponent $\Gamma = (5 - \sqrt{3})/2 \approx 1.63$ is the same as in [12]. Using [3], we then get a deterministic #SAT algorithm, running in time $2^{n-n^{\Omega(1)}}$, that works for de Morgan formulas of size up to $n^{\Gamma+1-o(1)} \approx n^{2.63}$.

1.2 Related Work

The deep interplay between lower bounds and satisfiability algorithms has been witnessed in several circuit models. For example, Paturi, Pudlak and Zane [13] give a randomized algorithm for k-SAT running in time $O(n^2 s 2^{n-n/k})$, where n is the number of variables and s is the formula size; they also show that PARITY requires depth-3 circuits of size $\Omega(n^{1/4}2^{\sqrt{n}})$. More generally, Williams [18] shows that a "better-than-trivial" algorithm for Circuit Satisfiability, for a class $\mathcal{C}$ of circuits, implies a super-polynomial lower bounds against the circuit class $\mathcal{C}$ for some language in NEXP; using this approach, Williams [19] obtains a super-polynomial lower bound against ACC^0 circuits[1] by designing a nontrivial SAT algorithm for ACC^0 circuits.

Following [14], Seto and Tamaki [15] get a nontrivial #SAT algorithm for general linear-size formulas (over an arbitrary basis). Impagliazzo et al. [7] use a generalization of Håstad's Switching Lemma [5], an analogue of shrinkage for

---

[1] constant-depth, unbounded fanin circuits, using AND, OR, NOT, and (MOD m) gates, for any integer m.

AC^0 circuits[2], to give a nontrivial randomized zero-error #SAT algorithm for depth-d AC^0 circuits on n inputs of size up to $2^{n^{1/(d-1)}}$. Beame et al. [2] give a nontrivial deterministic #SAT algorithm for AC^0 circuits, however, only for circuits of much smaller size than that of [7].

Recently, the method of (pseudo) random restrictions has also been used to get pseudorandom generators (yielding additive-approximation #SAT algorithms) for small de Morgan formulas [8] and AC^0 circuits [17].

Remainder of the Paper. We give basic definitions in Section 2. Section 3 contains our efficient formula-simplification procedures. We use these procedures in Section 4 to prove a constructive and concentrated shrinkage result for de Morgan formulas. This is then used in Section 5 to describe and analyze our #SAT algorithm from Theorem 1. Section 6 contains some open questions. Some proofs had to be omitted from this extended abstract due to space limitations; for a more complete version, please see [4].

2 Preliminaries

A *(de Morgan) formula* is a binary tree where each leaf is labeled by a literal (a variable x or its negation $\bar{x}$) or a constant (0 or 1), and each internal node is labeled by $\wedge$ or $\vee$. A formula naturally computes a boolean function on its input variables.

Let F be a formula with no constant leaves. We define the *size* of F, denoted by $L(F)$, as number of leaves in F. Following [12], we define a *twig* to be a subtree with exactly two leaves. Let $T(F)$ be the number of twigs in F. We define the *weight* of F as $w(F) = L(F) + \alpha \cdot T(F)$, where $\alpha = \sqrt{3} - 1 \approx 0.732$. For convenience, if F is a constant, we define $L(F) = w(F) = 0$. We say F is *trivial* if it is a constant or a literal. Note that we define the size and weight only for formulas which are either constants or with no constant leaves; this is without loss of generality since constants can always be eliminated using a simplification procedure below.

It is easy to see that $L(F) + \alpha \leqslant w(F) \leqslant L(F)(1 + \alpha/2)$, since the number of twigs in a formula is at least one and at most half of the number of leaves.

We denote by $F|_{x=1}$ the formula obtained from F by substituting each appearance of x by 1 and $\bar{x}$ by 0; $F|_{x=0}$ is similar. We say a formula $\vee$-*depends* ($\wedge$-*depends*) on a literal y if there is a path from the root to a leaf labeled by y such that every internal node on the path (including the root) is labeled by $\vee$ (by $\wedge$).

3 Formula Simplification Procedures

3.1 Basic Simplification

We define a procedure **Simplify** to eliminate constants, redundant literals and redundant twigs in a formula. The procedure includes the standard constant

---
[2] constant-depth, unbounded fanin circuits, using AND, OR, and NOT gates.

simplification rules and a natural extension of the one-variable simplification rules from [6].

Simplify(F):
If F is trivial, done. Otherwise, apply the following transformations whenever applicable. We denote by y a literal and G a subformula.

1. **Constant elimination.**
   (a) If a subformula is of the form $0 \wedge G$, replace it by 0.
   (b) If a subformula is of the form $1 \vee G$, replace it by 1.
   (c) If a subformula is of the form $1 \wedge G$ or $0 \vee G$, replace it by G.
2. **One-variable simplification.**
   (a) If a subformula is of the form $y \vee G$ and y or $\bar{y}$ appears in G, replace the subformula by $y \vee G|_{y=0}$.
   (b) If a subformula is of the form $y \wedge G$ and y or $\bar{y}$ appears in G, replace the subformula by $y \wedge G|_{y=1}$.
   (c) If a subformula G is of the form $G_1 \vee G_2$ for non-trivial G_1 and G_2, and G $\vee$-depends on a literal y, then replace G by $y \vee G|_{y=0}$.
   (d) If a subformula G is of the form $G_1 \wedge G_2$ for non-trivial G_1 and G_2, and G $\wedge$-depends on a literal y, then replace G by $y \wedge G|_{y=1}$.

We call a formula *simplified* if it is invariant under **Simplify**. Note that a simplified formula may not be a smallest logically equivalent formula; e.g., $(x \wedge y) \vee (\bar{x} \wedge y)$ is already simplified but it is logically equivalent to y.

The rules 1(a)–(c) and 2(a)–(b) are from [6,14]. Rules 2(c)–(d) are a natural generalization of the one-variable rule of [6], which allow us to eliminate more redundant literals and reduce the formula weight. For example, the formula $(x \vee y) \vee (x \wedge y)$ simplifies to $x \vee y$ under our rules but not the rules in [6,14]. For another example, the formula $(x \vee y) \vee (z \wedge w)$ with weight $4 + 2\alpha$ simplifies to $x \vee (y \vee (z \wedge w))$ with weight $4 + \alpha$.

The next lemma (proof omitted) shows **Simplify** is efficient.

Lemma 1. **Simplify** *runs in polynomial time.*

3.2 Simplification under All One-Variable Restrictions

Here we consider how a formula simplifies when one of its variables is restricted. Let F be a formula. We define a recursive procedure **RestrictSimplify** which produces a collection of formulas for F under all one-variable restrictions. We denote the output of the procedure by $\{F_y\}$, where y ranges over all literals. Note that each F_y is logically equivalent to $F|_{y=1}$.

The idea behind the transformations in **RestrictSimplify** is the following. When a formula simplifies to a literal under some one-variable restriction, then the formula must be logically equivalent to some special form. For example, if we know that $F|_{x=1}$ simplifies to a literal y, then F itself must be logically equivalent to $(x \wedge y) \vee (\bar{x} \wedge G)$ for some G. This logically equivalent form may help to simplify F under other one-variable restrictions.

RestrictSimplify(F):

If F is a constant c, then let $F_y := c$ for all y. If F is a literal, then let $F_y := F|_{y=1}$ for all y.

If F is $G \vee H$ or $G \wedge H$, recursively call **RestrictSimplify** to compute $\{G_y\}$ and $\{H_y\}$, and initialize each $F_y := \mathbf{Simplify}(G_y \vee H_y)$ or $F_y := \mathbf{Simplify}(G_y \wedge H_y)$, respectively. Then apply the following transformations whenever possible. We suppose there are two literals x and y over distinct variables such that $F_x = y$.

1. If $F_{\overline{x}} = y$, then let $F_w := y|_{w=1}$ for every literal w.
2. If $F_{\overline{x}} = z$ for some literal $z \notin \{x, \overline{x}, y\}$, then let $F_w := \mathbf{Simplify}((x \wedge y) \vee (\overline{x} \wedge z)|_{w=1})$ for every literal w.
3. (a) If neither x nor $\overline{x}$ appears in F_y, then let $F_y := 1$; (b) otherwise, let $F_y := \mathbf{Simplify}(x \vee (F_y|_{x=0}))$.
4. (a) If neither x nor $\overline{x}$ appears in $F_{\overline{y}}$, then let $F_{\overline{y}} := 0$; (b) otherwise, let $F_{\overline{y}} := \mathbf{Simplify}(\overline{x} \wedge (F_{\overline{y}}|_{x=0}))$.
5. For $z \notin \{x, \overline{x}, y, \overline{y}\}$, if neither x nor $\overline{x}$ appears in F_z, then let $F_z := y$.

Correctness of RestrictSimplify. The above transformations are based on logical implications. In case 1, $F_x = F_{\overline{x}} = y$ implies that $F \equiv y$. In case 2, $F_x = y$ and $F_{\overline{x}} = z$ implies that $F \equiv (x \wedge y) \vee (\overline{x} \wedge z)$. Note that in this case z might be $\overline{y}$. In case 3, we have $F_y|_{x=1} \equiv F_x|_{y=1} = 1$; if neither x nor $\overline{x}$ appears in F_y then $F_y = F_y|_{x=1} \equiv 1$, otherwise $F_y \equiv x \vee (F_y|_{x=0})$. Case 4 is dual to case 3. In case 5, if neither x nor $\overline{x}$ appears in F_z then $F_z = F_z|_{x=1} \equiv F_x|_{z=1} = y$.

Remark 1. It is possible to introduce more simplifications rules in **Restrict-Simplify**, e.g., when F_x is a constant for some literal x, or when, in case 5, x or $\overline{x}$ appears in $F_z{}^3$. However, such simplifications are not needed for our proof of constructive shrinkage.

It is easy to show that **RestrictSimplify** is efficient.

Lemma 2. **RestrictSimplify** *runs in polynomial time.*

The *solo structure* of a formula F is the relation on literals defined by $x \Rightarrow y$ if $F_x = y$, where the collection of formulas $\{F_x\}$ is produced by the procedure **RestrictSimplify**. The following lemma gives all possible solo structures; it resembles the characterization of solo structures for boolean functions from [12].

Lemma 3. *The solo structure of a non-trivial formula F must be in one of the following forms:*

(i) *the empty relation,*
(ii) *there exists y such that for all literals $x \notin \{y, \overline{y}\}$ we have $x \Rightarrow y$ in the relation,*
(iii) $\{x_1 \Rightarrow y, \ \ldots, \ x_k \Rightarrow y\}$ *for some $k \geqslant 1$ and x_i's are over distinct variables,*
(iv) $\{x \Rightarrow y, \ y \Rightarrow x, \ \overline{x} \Rightarrow \overline{y}, \ \overline{y} \Rightarrow \overline{x}\}$,
(v) $\{x \Rightarrow y, \ \overline{x} \Rightarrow z\}$,
(vi) $\{x \Rightarrow y, \ y \Rightarrow x\}$,
(vii) $\{x \Rightarrow y, \ \overline{y} \Rightarrow \overline{x}\}$.

---

[3] then we could let $F_z := (x \wedge y) \vee (\overline{x} \wedge \mathbf{Simplify}(F_z|_{x=0}))$.

4   Constructive and Concentrated Shrinkage

Here we prove a constructive and concentrated version of the shrinkage result from [12]. For each literal y of a given formula F, we define the savings (reduction in weight of F) when we replace F by the new formula F_y, as computed by the procedure **RestrictSimplify**. We first prove that the lower bound on the average savings (over all variables of F) shown by [12] continues to hold with respect to our efficiently computable one-variable restrictions F_y.

4.1   Average Savings under One-Variable Restrictions

Assume a formula F is simplified; otherwise, let $F := \mathbf{Simplify}(F)$. For a formula F and a literal y, we define $\sigma_y(F) = w(F) - w(F_y)$, where F_y is produced by **RestrictSimplify**. Let $\sigma(F) = \sum_x(\sigma_x(F) + \sigma_{\overline{x}}(F))$, where the summation ranges over all variables of F. The quantity $\sigma(F)$ measures the total *savings* under all one-variable restrictions.

Theorem 2. *For any formula F, it holds that $\sigma(F)/w(F) \geqslant 2\gamma$, where $\gamma = (5 - \sqrt{3})/2 \approx 1.63$.*

The proof is by induction, as in [12]. The difficulty here is that we need to apply the "syntactic simplifications" defined by the procedure **RestrictSimplify**, instead of using the smallest logically equivalent formulas as in [12].

For the base case, the following lemma can be proved by enumerating all possible formulas of size at most 4 (the proof is omitted).

Lemma 4. *For any simplified F of size at most 4, we have $\sigma(F)/w(F) \geqslant 2\gamma$.*

For formulas of size larger than 4, we consider whether one child of the root is trivial. Without loss of generality, we assume the root is labeled by $\vee$; the other case is dual. The following lemma considers if one child of the root is trivial. The proof is omitted here but it is similar to [12].

Lemma 5. *If F is a simplified formula of the form $x \vee G$ for some literal x and subformula G, and $L(F) \geqslant 5$, then $\sigma(F)/w(F) \geqslant 2\gamma$.*

Now we consider formulas where both children of the root are non-trivial.

Lemma 6. *Suppose F is of the form $G \vee H$ with $L(F) \geqslant 5$ and G, H are non-trivial. Then $\sigma(F)/w(F) \geqslant 2\gamma$.*

Intuitively, we need to take care of the cases where both G and H simplify to literals on distinct variables (thereby forming a new twig); otherwise the result holds by the induction hypothesis. Suppose $G_x \vee H_x$ is a twig for some literal x. Then $\sigma_x(F) = \sigma_x(G) + \sigma_x(H) - \alpha$, i.e., we get the savings from restricting x in G and H, but then need to pay the penalty α for the twig created. We will argue that there are "extra savings" from restricting other literals in the formula F that can be used to compensate for the penalty α at x.

Proof. We shall need the following basic property of **RestrictSimplify**.

Claim. For $F = G \vee H$ or $F = G \wedge H$, we have $w(F_y) \leqslant w(G_y) + w(H_y)$, for all literals y except those where G_y and H_y are literals over distinct variables.

Proof (of Claim). Let $F = G \vee H$; the other case is identical. For $F_y :=$ **Simplify**$(G_y \vee H_y)$, the required inequality holds initially. All transformations, except 3(b) and 4(b), produce the smallest logically equivalent formula; rules 3(b) and 4(b) do not increase the weight of the formula.   □

We first prove that, for a literal x, if G_x and H_x are not literals over distinct variables, then $\sigma_x(F) \geqslant \sigma_x(G) + \sigma_x(H)$. Indeed, since $w(F) = w(G) + w(H)$, this follows from $w(F_x) \leqslant w(G_x) + w(H_x)$, which holds by the claim above.

Next, let k be the number of different literals x such that $G_x \vee H_x$ is a twig (i.e., G_x and H_x are literals over distinct variables). Thus there are k twigs created as we consider all possible one-variable restrictions. We will argue that, for different cases of k, the weight $k\alpha$ of these new twigs can be compensated from savings in other restrictions.

Case $k = 0$: We have $\sigma_y(F) \geqslant \sigma_y(G) + \sigma_y(H)$ for all literals y, and thus $\sigma(F) \geqslant \sigma(G) + \sigma(H)$. The result is by the induction hypothesis on G and H.

Case $1 \leqslant k \leqslant 2$: Let x be such that $G_x = y$ and $H_x = z$. Without loss of generality, assume x, y, z are distinct variables. Consider F under the restrictions $y = 1$ and $z = 1$. We will argue that the extra savings from applying **Simplify** on $G_y \vee H_y$ and $G_z \vee H_z$ are at least $2 > k\alpha$.

Since $G_x = y$, transformation 3(a)–(b) in **RestrictSimplify** guarantee that either G_y is constant 1 or it $\vee$-depends on x. Similarly either H_z is constant 1 or it $\vee$-depends on x. Since $H_y|_{x=1} \equiv H_x|_{y=1} = z$, we get that H_y is not a constant (it depends on z), and if it is a literal it must be z. Similarly G_z is not a constant (it depends on y), and if it is a literal it must be y.

We first consider the case that either G_y or H_z is constant 1. If $G_y = H_z = 1$, then there are at least 2 savings from simplifying $G_y \vee H_y$ and $G_z \vee H_z$ by eliminating constants. If $G_y = 1$ and H_y is not a literal, then there are at least 2 savings from simplifying $G_y \vee H_y$. If $G_y = 1$, $H_y = z$ and $H_z \neq 1$, we first have one saving from simplifying $G_y \vee H_y$; then since $H_y = z$ and $H_z \neq 1$, by the transformation 3(b) in **RestrictSimplify** H_z $\vee$-depends on y, and since G_z depends on y, we get another saving from simplifying $G_z \vee H_z$. The cases where $H_z = 1$ are similar.

Next we consider that both G_y and H_z $\vee$-depends on x. In the following we analyze different possibilities for H_y and G_z.

- If x appears in both H_y and G_z, then there are at least 2 savings from simplifying $G_y \vee H_y$ and $G_z \vee H_z$ by eliminating x.
- If x appears in H_y but not G_z, then by the transformation 5 in **RestrictSimplify** we have $G_z = y$, and thus G_y $\vee$-depends on both x and z. Then since H_y depends on both x and z, we have two savings from simplifying $G_y \vee H_y$ by eliminating both x and z from H_y.

- If x appears in G_z but not H_y, this is similar to the previous case.
- If x appears in neither H_y nor G_z, then by the transformation 5 in **RestrictSimplify** we have $G_z = y$ and $H_y = z$. Thus G_y $\vee$-depends on both x and z, and H_z $\vee$-depends on both x and y. Therefore we have at least 2 savings, one from simplifying $G_y \vee H_y$ by eliminating z, and another from simplifying $G_z \vee H_z$ by eliminating y.

Case $k \geqslant 3$: By Lemma 3, the solo structure of G and H must be one of cases (ii), (iii), or (iv).

First assume that either G or H is in case (ii) of Lemma 3. Without loss of generality, suppose G is in case (ii); then G is logically equivalent to a literal y but itself is non-trivial, which implies that $w(G) \geqslant 4 + \alpha$. (The smallest non-trivial, simplified formula equivalent to a literal has size at least 4). We have that $w(G_z) = 1$ for at least k literals $z \notin \{y, \overline{y}\}$, and $w(G_y) = w(G_{\overline{y}}) = 0$. Then by the fact that $w(F) = w(G) + w(H)$ and the induction hypothesis on H, we have

$$\sigma(F) \geqslant k(w(G) - 1) + 2w(G) + \sigma(H) - k\alpha$$
$$\geqslant 2\gamma \cdot w(F) + (2 + k - 2\gamma)w(G) - k(1 + \alpha) \geqslant 2\gamma \cdot w(F).$$

If both G and H are in case (iv), then, under each restriction, they reduce to literals on the same variable. Since in case (iii) all x_i's are over distinct variables, it is not possible that one of G and H is in case (iv) while the other is in case (iii). Thus, we now only need to analyze if both G and H are in case (iii).

Without loss of generality, suppose that $x_1, \ldots, x_k, y, z$ are distinct variables such that $G_{x_i} = y$ and $H_{x_i} = z$ for $i = 1, \ldots, k$. By the transformation 3 in **RestrictSimplify**, either $G_y = 1$ or G_y $\vee$-depends on $x_1, \ldots, x_k$; and H_z is similar.

If every x_i appears in H_y, then there are k savings from simplifying $G_y \vee H_y$ by eliminating x_i's. Similarly, if every x_i appears in G_z, there are also k savings from simplifying $G_z \vee H_z$.

If some x_i does not appear in H_y and some x_i does not appear in G_z. By the transformation 5 in **RestrictSimplify**, we have $H_y = z$ and $G_z = y$. Therefore,

$$\sigma_{x_i}(F) = w(F) - (2 + \alpha), \quad i = 1, \ldots, k$$
$$\sigma_y(F) \geqslant 1 + (w(H) - 1) = w(H)$$
$$\sigma_z(F) \geqslant 1 + (w(G) - 1) = w(G)$$
$$\sum_v \sigma_{\overline{v}}(F) \geqslant L(F) \geqslant w(F)/(1 + \alpha/2), \quad v \text{ ranges over all variables of } F$$

Summing the above cases together yields $\sigma(F) \geqslant 2\gamma \cdot w(F)$.   □

Proof (Theorem 2). The proof is by combining the base case in Lemma 4 and the two inductive cases in Lemma 5 and Lemma 6.   □

4.2   Concentrated Shrinkage

Theorem 2 characterizes the average shrinkage of the weight of a formula when a randomly chosen literal is restricted. Given a formula F on n variables, if we randomly pick one variable and randomly assign it 0 or 1, the weight of the restricted formula (produced by **RestrictSimplify**) reduces by at least $\gamma \cdot w(F)/n$ on average.

The procedure **RestrictSimplify** also allows us to deterministically pick the variable with the best savings in polynomial time. That is, given a formula F, we run **RestrictSimplify** to produce a collection of formulas $\{F_y\}$, and then pick a variable x such that $\sigma_x(F) + \sigma_{\overline{x}}(F)$ is maximized. We show that randomly restricting such a variable significantly reduces the expected weight of the simplified formula.

Lemma 7. *Let F be a formula on n variables. Let x be the variable such that $\sigma_x(F) + \sigma_{\overline{x}}(F)$ is maximized. Let F' be F_x or $F_{\overline{x}}$ with equal probability. Then we have $w(F') \leqslant w(F) - 1$ and $\mathbf{E}[w(F')] \leqslant \left(1 - \frac{1}{n}\right)^{\gamma} \cdot w(F)$.*

Proof. Restricting one variable eliminates at least one leaf; therefore $w(F') \leqslant w(F) - 1$. By Theorem 2, $n(\sigma_x(F) + \sigma_{\overline{x}}(F)) \geqslant \sigma(F) \geqslant 2\gamma \cdot w(F)$. Then we have $\mathbf{E}[w(F')] = w(F) - \frac{1}{2}(\sigma_x(F) + \sigma_{\overline{x}}(F)) \leqslant \left(1 - \frac{\gamma}{n}\right) \cdot w(F) \leqslant \left(1 - \frac{1}{n}\right)^{\gamma} \cdot w(F)$. $\square$

Next we use the martingale-based analysis from [10,3] to derive a "high-probability shrinkage" result from Lemma 7. Let $F_0 = F$ be a formula on n variables. For $1 \leqslant i \leqslant n$, let F_i be the (random) formula obtained from F_{i-1} by assigning the variable with the best savings with a random value $R_i \in \{0,1\}$. The following Lemma shows the weight of a given de Morgan formula reduces with high probability under the restriction process. The proof, which is similar to [3], is omitted here due to space constraints.

Lemma 8 (Concentrated weight shrinkage). *For any de Morgan formula F on n variables and any $k > 10$, $\mathbf{Pr}\left[w(F_{n-k}) \geqslant 2 \cdot w(F) \cdot \left(\frac{k}{n}\right)^{\gamma}\right] < 2^{-k/10}$.*

Finally, by $w(F)/(1 + \alpha/2) \leqslant L(F) \leqslant w(F)$ for all F, we get from Lemma 8 the desired concentrated constructive shrinkage with respect to the restriction process defined above.

Corollary 1 (Concentrated constructive shrinkage). *Let F be an arbitrary de Morgan formula. There exist constants $c, d > 1$ such that, for any $k > 10$, $\mathbf{Pr}\left[L(F_{n-k}) \geqslant c \cdot L(F) \cdot \left(\frac{k}{n}\right)^{\gamma}\right] < 2^{-k/d}$.*

5   #SAT Algorithm for $n^{2.63}$-size de Morgan Formulas

Here we prove our main result.

Theorem 3. *There is a deterministic algorithm for counting the number of satisfying assignments in a given formula on n variables of size at most $n^{2.63}$ which runs in time $t(n) \leqslant 2^{n-n^{\delta}}$, for some constant $0 < \delta < 1$.*

Proof. Suppose we have a formula F on n variables of size $n^{1+\gamma-\epsilon}$ for a small constant $\epsilon > 0$. Let $k = n^\alpha$ such that $\alpha < \epsilon/\gamma$. We build a restriction decision tree with 2^{n-k} branches as follows:

> Starting with F at the root, run **RestrictSimplify** to produce a collection $\{F_y\}$, pick the variable x which will make the largest reduction in the weight of the current formula. Make the two formulas F_x and $F_{\bar{x}}$ the children of the current node. Continue recursively on F_x and $F_{\bar{x}}$ until get a full binary tree of depth exactly $n - k$.

Note that constructing this decision tree takes time $2^{n-k}\text{poly}(n)$, since the procedure **RestrictSimplify** runs in polynomial time. By Corollary 1, all but at most $2^{-k/d}$ fraction of the leaves have the formula size $L(F_{n-k}) < c \cdot L(F) \left(\frac{k}{n}\right)^\gamma = cn^{1-\epsilon+\gamma\alpha}$.

To solve #SAT for all "big" formulas (those that haven't shrunk), we use brute-force enumeration over all possible assignments to the k free variables left. The running time is at most $2^{n-k} \cdot 2^{-k/d} \cdot 2^k \cdot \text{poly}(n) \leqslant 2^{n-k/d} \cdot \text{poly}(n)$.

For "small" formulas (those that shrunk to the size less than $cn^{1-\epsilon+\gamma\alpha}$), we use memoization. First, we enumerate all formulas of such size, and compute and store the number of satisfying assignments for each of them. Then, as we go over the leaves of the decision tree that correspond to small formulas, we simply look up the stored answers for these formulas.

There are at most $2^{O(n^{1-\epsilon+\gamma\alpha})}\text{poly}(n)$ such formulas, and counting the satisfying assignments for each one (with k inputs) takes time $2^k\text{poly}(n)$. Including pre-processing, computing #SAT for all small formulas takes time at most $2^{n-k} \cdot \text{poly}(n) + 2^{O(n^{1-\epsilon+\gamma\alpha})} \cdot 2^k \cdot \text{poly}(n) \leqslant 2^{n-k} \cdot \text{poly}(n)$.

The overall running time of our #SAT algorithm is bounded by 2^{n-n^δ} for some $\delta > 0$. $\square$

6   Open Questions

The main open problem is to get a nontrivial deterministic #SAT algorithm for de Morgan formulas of size up to $n^{3-o(1)}$. Can one derandomize the zero-error algorithm of [11] that is based on Håstad's shrinkage result [6]?

Can one improve the analysis of the shrinkage result of [12] (by considering more general patterns than just twigs), getting a better shrinkage exponent? If so, this could lead to a deterministic #SAT algorithm for larger de Morgan formulas.

References

1. Andreev, A.E.: On a method of obtaining more than quadratic effective lower bounds for the complexity of π-schemes. Vestnik Moskovskogo Universiteta. Matematika 42(1), 70–73 (1987); english translation in Moscow University Mathematics Bulletin

2. Beame, P., Impagliazzo, R., Srinivasan, S.: Approximating AC^0 by small height decision trees and a deterministic algorithm for $\#AC^0SAT$. In: Proceedings of the Twenty-Seventh Annual IEEE Conference on Computational Complexity, pp. 117–125 (2012)
3. Chen, R., Kabanets, V., Kolokolova, A., Shaltiel, R., Zuckerman, D.: Mining circuit lower bound proofs for meta-algorithms. In: Proceedings of the Twenty-Ninth Annual IEEE Conference on Computational Complexity (2014)
4. Chen, R., Kabanets, V., Saurabh, N.: An improved deterministic $\#SAT$ algorithm for small De Morgan formulas. Electronic Colloquium on Computational Complexity (ECCC) 20, 150 (2013)
5. Håstad, J.: Almost optimal lower bounds for small depth circuits. In: Proceedings of the Eighteenth Annual ACM Symposium on Theory of Computing, pp. 6–20 (1986)
6. Håstad, J.: The shrinkage exponent of de Morgan formulae is 2. SIAM Journal on Computing 27, 48–64 (1998)
7. Impagliazzo, R., Matthews, W., Paturi, R.: A satisfiability algorithm for AC^0. In: Proceedings of the Twenty-Third Annual ACM-SIAM Symposium on Discrete Algorithms, pp. 961–972 (2012)
8. Impagliazzo, R., Meka, R., Zuckerman, D.: Pseudorandomness from shrinkage. In: Proceedings of the Fifty-Third Annual IEEE Symposium on Foundations of Computer Science, pp. 111–119 (2012)
9. Impagliazzo, R., Nisan, N.: The effect of random restrictions on formula size. Random Structures and Algorithms 4(2), 121–134 (1993)
10. Komargodski, I., Raz, R.: Average-case lower bounds for formula size. In: Proceedings of the Forty-Fifth Annual ACM Symposium on Theory of Computing, pp. 171–180 (2013)
11. Komargodski, I., Raz, R., Tal, A.: Improved average-case lower bounds for DeMorgan formula size. In: Proceedings of the Fifty-Fourth Annual IEEE Symposium on Foundations of Computer Science, pp. 588–597 (2013)
12. Paterson, M., Zwick, U.: Shrinkage of de Morgan formulae under restriction. Random Structures and Algorithms 4(2), 135–150 (1993)
13. Paturi, R., Pudlák, P., Zane, F.: Satisfiability coding lemma. Chicago Journal of Theoretical Computer Science (1999)
14. Santhanam, R.: Fighting perebor: New and improved algorithms for formula and QBF satisfiability. In: Proceedings of the Fifty-First Annual IEEE Symposium on Foundations of Computer Science, pp. 183–192 (2010)
15. Seto, K., Tamaki, S.: A satisfiability algorithm and average-case hardness for formulas over the full binary basis. In: Proceedings of the Twenty-Seventh Annual IEEE Conference on Computational Complexity, pp. 107–116 (2012)
16. Subbotovskaya, B.: Realizations of linear function by formulas using $\vee$, &, $^-$. Doklady Akademii Nauk SSSR 136(3), 553–555 (1961); english translation in Soviet Mathematics Doklady
17. Trevisan, L., Xue, T.: A derandomized switching lemma and an improved derandomization of AC^0. In: Proceedings of the Twenty-Eighth Annual IEEE Conference on Computational Complexity, pp. 242–247 (2013)
18. Williams, R.: Improving exhaustive search implies superpolynomial lower bounds. In: Proceedings of the Forty-Second Annual ACM Symposium on Theory of Computing, pp. 231–240 (2010)
19. Williams, R.: Non-uniform ACC circuit lower bounds. In: Proceedings of the Twenty-Sixth Annual IEEE Conference on Computational Complexity, pp. 115–125 (2011)

On the Limits of Depth Reduction at Depth 3 Over Small Finite Fields

Suryajith Chillara and Partha Mukhopadhyay

Chennai Mathematical Institute, Chennai, India
{suryajith,partham}@cmi.ac.in

Abstract. In a surprising recent result, Gupta et al. [GKKS13b] have proved that over $\mathbb{Q}$ any $n^{O(1)}$-variate and n-degree polynomial in VP can also be computed by a depth three $\Sigma\Pi\Sigma$ circuit of size $2^{O(\sqrt{n}\log^{3/2} n)}$ [1]. Over fixed-size finite fields, Grigoriev and Karpinski proved that any $\Sigma\Pi\Sigma$ circuit that computes the determinant (or the permanent) polynomial of a $n \times n$ matrix must be of size $2^{\Omega(n)}$. In this paper, for an explicit polynomial in VP (over fixed-size finite fields), we prove that any $\Sigma\Pi\Sigma$ circuit computing it must be of size $2^{\Omega(n\log n)}$. The explicit polynomial that we consider is the iterated matrix multiplication polynomial of n generic matrices of size $n \times n$. The importance of this result is that over fixed-size fields there is *no depth reduction technique* that can be used to compute all the $n^{O(1)}$-variate and n-degree polynomials in VP by depth 3 circuits of size $2^{o(n\log n)}$. The result of [GK98] can only rule out such a possibility for $\Sigma\Pi\Sigma$ circuits of size $2^{o(n)}$.

We also give an example of an explicit polynomial ($\mathrm{NW}_{n,\epsilon}(X)$) in VNP (which is not known to be in VP), for which any $\Sigma\Pi\Sigma$ circuit computing it (over fixed-size fields) must be of size $2^{\Omega(n\log n)}$. The polynomial we consider is constructed from the combinatorial design of Nisan and Wigderson [NW94], and is closely related to the polynomials considered in many recent papers where strong depth 4 circuit size lower bounds were shown [KSS13, KLSS14, KS13b, KS14].

1 Introduction

In a recent breakthrough, Gupta et al. [GKKS13b] have proved that over $\mathbb{Q}$, if an $n^{O(1)}$-variate polynomial of degree d is computable by an arithmetic circuit of size s, then it can also be computed by a depth three $\Sigma\Pi\Sigma$ circuit of size $2^{O(\sqrt{d}\log d\log n\log s)}$. Using this result, they get a $\Sigma\Pi\Sigma$ circuit of size $2^{O(\sqrt{n}\log n)}$ computing the determinant polynomial of a $n \times n$ matrix (over $\mathbb{Q}$). Before this result, no depth 3 circuit for Determinant of size smaller than $2^{O(n\log n)}$ was known (over any field of characteristic $\neq 2$).

The situation is very different over *fixed-size finite fields*. Grigoriev and Karpinski proved that over fixed-size finite fields, any depth 3 circuit for the determinant polynomial of a $n \times n$ matrix must be of size $2^{\Omega(n)}$ [GK98]. Although

---

[1] In a nice follow-up work, Tavenas has improved the upper bound to $2^{O(\sqrt{n}\log n)}$. The main ingredient in his proof is an improved depth 4 reduction [Tav13].

E. Csuhaj-Varjú et al. (Eds.): MFCS 2014, Part II, LNCS 8635, pp. 177–188, 2014.

Grigoriev and Karpinski proved the lower bound result only for the determinant polynomial, it is a folklore result that some modification of their argument can show a similar depth 3 circuit size lower bound for the permanent polynomial as well [2]. Over any field, Ryser's formula for Permanent gives a $\Sigma\Pi\Sigma$ circuit of size $2^{O(n)}$ (for an exposition of this result, see [Fei09]). Thus, for the permanent polynomial the depth 3 complexity (over fixed-size finite fields) is essentially $2^{\Theta(n)}$.

The result of [GKKS13b] is obtained through an ingenious depth reduction technique but their technique is tailored to the fields of zero characteristic. In particular, the main technical ingredients of their proof are the well-known monomial formula of Fischer [Fis94] and the duality trick of Saxena [Sax08]. These techniques do not work over finite fields. In contrast to the situation over $\mathbb{Q}$, over the fixed-size finite fields a natural question is to ask whether one can find a new depth reduction technique over fixed-size finite fields such that any $n^{O(1)}$-variate and degree n polynomial in VP can also be computed by a $\Sigma\Pi\Sigma$ circuit of size $2^{o(n\log n)}$.

Question 1. Over any fixed-size finite field $\mathbb{F}_q$, is it possible to compute any $n^{O(1)}$-variate and n-degree polynomial in VP by a $\Sigma\Pi\Sigma$ circuit of size $2^{o(n\ln n)}$?

Note that any $n^{O(1)}$-variate and n-degree polynomial can be trivially computed by a $\Sigma\Pi\Sigma$ circuit of size $2^{O(n\log n)}$ by writing it explicitly as a sum of all $n^{O(n)}$ possible monomials.

We give a negative answer to the aforementioned question by showing that over fixed-size finite fields, any $\Sigma\Pi\Sigma$ circuit computing the iterated matrix multiplication polynomial (which is in VP for any field) must be of size $2^{\Omega(n\log n)}$ (See Subsection 2.1, for the definition of the polynomial). More precisely, we prove that any $\Sigma\Pi\Sigma$ circuit computing the iterated matrix multiplication polynomial of n generic $n \times n$ matrices (denoted by $\mathrm{IMM}_{n,n}(X)$), must be of size $2^{\Omega(n\log n)}$.

Previously, Nisan and Wigderson [NW97] proved a size lower bound of $\Omega(n^{d-1}/d!)$ for any homogeneous $\Sigma\Pi\Sigma$ circuit computing the iterated matrix multiplication polynomial over d generic $n \times n$ matrices. Kumar et al. [KMN13] improved the bound to $\Omega(n^{d-1}/2^d)$. These results work over any field. Over fields of zero characteristic, Shpilka and Wigderson proved a near quadratic lower bound for the size of depth 3 circuits computing the trace of the iterated matrix multiplication polynomial [SW01].

Recently Tavenas [Tav13], by improving upon the previous works of Agrawal and Vinay [AV08], and Koiran [Koi12] proved that any $n^{O(1)}$-variate, n-degree polynomial in VP has a depth four $\Sigma\Pi^{[O(\sqrt{n})]}\Sigma\Pi^{[\sqrt{n}]}$ circuit of size $2^{O(\sqrt{n}\log n)}$. Subsequently, Kayal et al. [KSS13] proved a size lower bound of $2^{\Omega(\sqrt{n}\log n)}$ for a polynomial in VNP which is constructed from the combinatorial design of Nisan and Wigderson [NW94]. In a beautiful follow up result, Fournier et al. [FLMS13] proved that a similar lower bound of $2^{\Omega(\sqrt{n}\log n)}$ is also attainable by

---

[2] Saptharishi gives a nice exposition of this result in his unpublished survey and he attributes it to Koutis and Srinivasan [Sap13].

the iterated matrix multiplication polynomial (see [CM14], for a unified analysis of the depth 4 lower bounds of [KSS13] and [FLMS13]). The main technique used was *the method of shifted partial derivatives* which was also used to prove $2^{\Omega(\sqrt{n})}$ size lower bound for $\Sigma\Pi^{[O(\sqrt{n})]}\Sigma\Pi^{[\sqrt{n}]}$ circuits computing Determinant or Permanent polynomial [GKKS13a]. Recent work of Kumar and Saraf [KS13a] shows that the depth reduction as shown by Tavenas [Tav13] is optimal even for the homogeneous formulas. This strengthens the result of [FLMS13] who proved the optimality of depth reduction for the circuits. Very recently, a series of papers show strong depth 4 lower bounds even for homogeneous depth 4 formulas with no bottom fan-in restriction [KLSS14, KS13b, KS14].

Similar to the situation at depth 4, we also give an example of an explicit n^2-variate and n-degree polynomial in VNP (which is not known to be in VP) such that over fixed-size finite fields, any depth three $\Sigma\Pi\Sigma$ circuit computing it must be of size $2^{\Omega(n \log n)}$. This polynomial family, denoted by $\mathrm{NW}_{n,\epsilon}(X)$ (see Subsection 2.1, for the definition of the polynomial) is closely related to the polynomial family introduced by Kayal et al. [KSS13]. In fact, from our proof idea it will be clear that the strong depth 3 size lower bound results that we show for $\mathrm{NW}_{n,\epsilon}(X)$ and $\mathrm{IMM}_{n,n}(X)$ polynomials are not really influenced by the fact that the polynomials are either in VNP or VP. Rather, the bounds are determined by a combinatorial property of the subspaces generated by a set of carefully chosen derivatives. Our main theorem is the following.

Theorem 2. *Over any fixed-size finite field $\mathbb{F}_q$, any depth three $\Sigma\Pi\Sigma$ circuit computing the polynomials $\mathrm{NW}_{n,\epsilon}(X)$ or $\mathrm{IMM}_{n,n}(X)$ must be of size at least $2^{\delta n \log n}$, where the parameters δ and $\epsilon(< 1/2)$ are in $(0,1)$ and depend only on q.*

As an important consequence of the above theorem, we have the following corollary.

Corollary 3. *Over any fixed-size finite field $\mathbb{F}_q$, there is no depth reduction technique that can be used to compute all the $n^{O(1)}$-variate and n-degree polynomials in VP by depth 3 circuits of size $2^{o(n \log n)}$.*

The result of [GK98] only says that over fixed-size finite fields, not all the $n^{O(1)}$-variate and n-degree polynomials in VP can be computed by $\Sigma\Pi\Sigma$ circuits of size $2^{o(n)}$. Our main theorem (Theorem 2) can also be viewed as the first quantitative improvement over the result of [GK98].

Proof Idea

Our proof technique is quite simple and it borrows ideas mostly from the proof technique of Grigoriev and Karpinski [GK98]. Recall that a $\Sigma\Pi\Sigma$ circuit (over a field $\mathbb{F}$) with s multiplication gates computes a polynomial of the form $\sum_{i=1}^{s}\prod_{j=1}^{d_i} L_{i,j}(x_1,\ldots,x_n)$ where $L_{i,j}$s are affine linear functions over $\mathbb{F}$ and

$\{x_1, x_2, \ldots, x_n\}$ are the variables appearing in the polynomial. A recurring notion in many papers related to $\Sigma\Pi\Sigma$ circuits is the notion of *rank* of a product gate. Let $T = L_1 L_2 \ldots L_d$ be a product gate such that each L_i is an affine linear form over the underlying field. By rank of T, one simply means the maximum rank of the homogeneous linear system corresponding to set of affine functions $\{L_1, L_2, \ldots, L_d\}$.

Over fixed-size finite fields, $\Sigma\Pi\Sigma$ circuits enjoy a nice property that the derivatives of the high rank product gates can be eliminated except for a few erroneous points (denoted by E). This property was first observed by Grigoriev and Karpinski in [GK98]. The intuition is simple. If a product gate has many linearly independent functions, then it is likely that a large number of linear functions will be set to zero if we randomly substitute the variables with elements from the field. Then the derivatives (of relatively low order) of the polynomial obtained from the product gate will disappear on a random point with very high probability. To quantify the notion of *high rank*, Grigoriev and Karpinski fixed a threshold for the rank of the product gates to $\Theta(n)$. Since they were looking for a $2^{\Omega(n)}$ lower bound for the Determinant of a $n \times n$ matrix and the rank of the entire derivative space of of the determinant polynomial is $2^{O(n)}$, it was natural for them to fix the threshold to be $\Theta(n)$. Since the dimension of the derivative spaces of the polynomial families $\{NW_{n,\epsilon}(X)\}_{n>0}$ and $\{IMM_{n,n}(X)\}_{n>0}$ is $2^{\Omega(n\log n)}$, it is possible for us to choose the threshold for the rank of the product gates to be $\Theta(n \log n)$. This allows us to bound the size of the error set meaningfully. We formalize this in Lemma 4.

We now give a high level description of the proof technique in [GK98] to motivate our proof strategy. Roughly speaking, they consider the space H spanned by $\Theta(n)$ order derivatives of the determinant. This makes the dimension of H to be of the order of $2^{\Theta(n)}$. For a point $a \in \mathbb{F}_q^N$, the subspace H_a is the space of functions in H that evaluate to zero at a. From the rank analysis on the circuit side, they get that the dimension of the space of functions that may not be zero outside the error set E is bounded. More precisely, $\mathrm{codim}(\cap_{a \notin E} H_a)$ is small. Grigoriev and Karpinski then considered the group of invertible matrices G of order $n \times n$ over $\mathbb{F}_q$. For any $g \in G$, they define a $\mathbb{F}_q$-linear operator $T_g : H \to H$ by the formula $(T_g(f))(a) = f(ga)$. The fact that the derivative space of the determinant polynomial of a $n \times n$ matrix is invariant under $GL_n(\mathbb{F}_q)$ action was crucially used in defining the map. Then they consider the plane $P = \cap_{a \in G \setminus E} H_a \subset H$. Since $\mathrm{codim}(\cap_{a \notin E} H_a)$ is bounded, the same bound applies for $\mathrm{codim}(P)$ as well. The most remarkable idea in their work was to prove that $\mathrm{codim}(\cap_{b \in G} H_b)$ in H is small given that $\mathrm{codim}(P)$ is small. Notice that the plane P is defined only on $G \setminus E$ and not on the entire group G. To achieve this, they prove that the full invertible group G can be covered by taking only a few translates of $G \setminus E$ from G. This was done by appealing to a graph theoretic lemma of Lovász [Lov75]. Now, it is not hard to see that we can bound $\mathrm{codim}(\cap_{b \in G} H_b)$ to a quantity smaller than $\dim(H)$. This shows us that there exists a nonzero function in H that evaluates to zero on the entire group G. Since the elements in H are only multilinear polynomials, they finally prove that it is impossible to have such a function in H by showing that no nonzero multilinear polynomial can vanish over the entire group G.

The group symmetry based argument of [GK98] is tailored to the determinant polynomial and it can not be directly applied to the polynomials that we consider. The main technical contribution of this work is to replace the group symmetry based argument by a new argument that makes the proof strategy robust enough to handle the family of polynomials that we consider. We carefully choose a subspace H (of sufficiently large dimension) of the derivative spaces of these polynomials which has an additional structure. The subspace H is spanned by a *downward closed* set of monomials (see, Definition 8). Let $\mathbb{F}_q$ be the finite field and N be the number of the variables in the polynomial under consideration. The basic idea is to prove that the dimension of the space H of the polynomial being considered is more than the dimension of the set of functions in H which do not evaluate to zero over the entire space $\mathbb{F}_q^N$. Since the subspace H contains only multilinear polynomials, we can then conclude that a nonzero multilinear polynomial in H will evaluate to zero on entire $\mathbb{F}_q^N$, which is not possible by combinatorial nullstellensatz [Alo99].

To implement this, we define a linear map $T_u : H \to H$ by $T_u(f(X)) = f(X-u)$ for any function $f : \mathbb{F}_q^N \to \mathbb{F}_q$ and $u \in \mathbb{F}_q^N$. The map is well-defined by the downward closed structure of the generating set for H. Also the map T_u is one to one for any $u \in \mathbb{F}_q^N$. As before, for a point $a \in \mathbb{F}_q^N$, the subspace H_a is the space of functions in H that evaluates to zero on a. Let $P = \cap_{a \in \mathbb{F}_q^N \setminus E} H_a$. Then by Lemma 4, we get that codim $P = \text{codim}(\cap_{a \in \mathbb{F}_q^N \setminus E} H_a)$ is small. Notice that the plane P is defined over $\mathbb{F}_q^N \setminus E$ and not over the entire space $\mathbb{F}_q^N$. Similar to the argument in [GK98], we also use the graph theoretic lemma of Lovász to prove that the entire space $\mathbb{F}_q^N$ can be covered by only a few translates of $\mathbb{F}_q^N \setminus E$. Then it is simple to observe that $\text{codim}(\cap_{b \in \mathbb{F}_q^N} H_b)$ is small compared to the dimension of H. As a consequence we get that a nonzero multilinear polynomial in H must evaluate to zero over $\mathbb{F}_q^N$, which is not possible by the combinatorial nullstellensatz.

1.1 Organization

We define the polynomial families $\{NW_{n,\epsilon}(X)\}_{n>0}$ and $\{IMM_{n,n}(X)\}_{n>0}$ in Section 2. We recall known results related to the derivative space of $\Sigma\Pi\Sigma$ circuits in Section 3. In section 4, we study the derivative spaces of our polynomial families. We prove Theorem 2 in Section 5. The omitted proofs (Lemma 4, Lemma 5, and Proposition 9) can be found in the full version of the paper [Chi14].

2 Preliminaries

2.1 The Polynomial Families

A multivariate polynomial family $\{f_n(X) \in \mathbb{F}[x_1, x_2, \ldots, x_n] : n \geq 1\}$ is in the class VP if f_n has degree at most poly(n) and can be computed by an arithmetic circuit of size poly(n). It is in VNP if it can be expressed as: $f_n(X) = \sum_{Y \in \{0,1\}^m} g_{n+m}(X, Y)$, where $m = |Y| = \text{poly}(n)$ and g_{n+m} is a polynomial in VP.

The Polynomial Family from the Combinatorial Design. Let $\mathbb{F}$ be any field[3]. For integers $n > 0$ ranging over prime powers and $0 < \epsilon < 1$, we define a polynomial family $\{NW_{n,\epsilon}(X)\}_{n>0}$ in $\mathbb{F}_q[X]$ as follows.

$$NW_{n,\epsilon}(X) = \sum_{a(z) \in \mathbb{F}_n[z]} x_{1a(1)} x_{2a(2)} \cdots x_{na(n)}$$

where $a(z)$ runs over all univariate polynomials of degree $< \epsilon n$. The finite field $\mathbb{F}_n$ is naturally identified with the numbers $\{1, 2, \ldots, n\}$. Notice that the number of monomials in $NW_{n,\epsilon}(X)$ is $n^{\epsilon n}$. From the explicitness of the polynomial, it is clear that $\{NW_{n,\epsilon}(X)\}_{n>0}$ is in VNP for any $\epsilon \in (0, 1)$. In [KSS13], a very similar family of polynomials was introduced where the degree of the univariate polynomial was bounded by $\epsilon \sqrt{n}$.

The Iterated Matrix Multiplication Polynomial. The iterated matrix multiplication polynomial of n generic $n \times n$ matrices $X^{(1)}, X^{(2)}, \ldots, X^{(n)}$ is the $(1,1)$th entry of the product of the matrices. More formally, let $X^{(1)}, X^{(2)}, \ldots, X^{(n)}$ be n generic $n \times n$ matrices with disjoint sets of variables and $x_{ij}^{(k)}$ be the variable in $X^{(k)}$ indexed by $(i, j) \in [n] \times [n]$. Then the iterated matrix multiplication polynomial (denoted by the family $\{IMM_{n,n}(X)\}_{n>0}$) is defined as follows.

$$IMM_{n,n}(X) = \sum_{i_1, i_2, \ldots, i_{n-1} \in [n]} x_{1i_1}^{(1)} x_{i_1 i_2}^{(2)} \cdots x_{i_{(n-2)} i_{(n-1)}}^{(n-1)} x_{i_{(n-1)} 1}^{(n)}$$

Notice that $IMM_{n,n}(X)$ is a $n^2(n-2) + 2n$-variate polynomial of degree n. For our application, we consider $n = 2m$ where m ranges over the positive integers. Over any field $\mathbb{F}$, the polynomial family $\{IMM_{n,n}(X)\}_{n>0}$ can be computed in VP. This can be seen by observing that $IMM_{n,n}(X)$ can be computed by a poly(n) sized algebraic branching program.

3   The Derivative Space of $\Sigma\Pi\Sigma$ Circuits Over Small Fields

In this section we fix the field $\mathbb{F}$ to be a fixed-size finite field $\mathbb{F}_q$. Let C be a $\Sigma\Pi\Sigma$ circuit of top fan-in s computing a $N = n^{O(1)}$-variate polynomial of degree n. Consider a Π gate $T = L_1 L_2 \ldots L_d$ in C. Let r be the rank of the (homogeneous)-linear system corresponding to $\{L_1, L_2, \ldots, L_d\}$ by viewing each L_i as a vector in

---

[3] In the lower bound proof for $NW_{n,\epsilon}(X)$, we will consider $\mathbb{F}$ to be any fixed finite field $\mathbb{F}_q$.

$\mathbb{F}_q^{N+1}$. Fix a threshold for the rank of the system of linear functions $r_0 = \beta n \ln n$, where $\beta > 0$ is a constant to be fixed later in the analysis. In our application, the parameter N is at least n^2, so the threshold for the rank is meaningful. W.l.o.g, let $\{L_1, L_2, \ldots, L_r\}$ be a set of affine linear forms in $\{L_1, L_2, \ldots, L_d\}$ whose homogeneous system forms a maximal independent set of linear functions.

Low Rank Gates : $r \leq r_0$ Over the finite field $\mathbb{F}_q$, we have $x^q = x$. We express $T : \mathbb{F}_q^N \to \mathbb{F}_q$ as a linear combination of $\{L_1^{e_1} L_2^{e_2} \ldots L_r^{e_r} : e_i < q \text{ for all } i \in [r]\}$. Since, the derivatives of all orders lie in the same space, the dimension of the set of partial derivatives of T of all orders is bounded by $q^r \leq q^{r_0}$.

High Rank Gates : $r > r_0$ Let the rank of a high rank gate T be $y\beta n \ln n$ where $y \geq 1$. We assign values to the variables uniformly at random from $\mathbb{F}_q$ and compute the probability that at most n linearly independent functions evaluate to zero.

Using the threshold for the rank of the product gates, the following lemma can be easily proved. Essentially, the proof can be reworked from [GK98] for suitable parameters. It shows that the derivative space of a $\Sigma\Pi\Sigma$ circuit can be approximated by just the derivative space of the low rank product gates of the circuit over a large subset of $\mathbb{F}_q^N$.

Lemma 4. *Let $\mathbb{F}_q$ be a fixed-size finite field. Then there exist constants $0 < \delta(q), \beta(q), \mu(q) < 1$ such that the following is true. Let C be a $\Sigma\Pi\Sigma$ circuit of top fan-in $s \leq e^{\delta n \ln n}$ computing a $N = n^{O(1)}$-variate and n-degree polynomial $f(\mathbf{X})$ over the finite field $\mathbb{F}_q$. Then, there exists a set $E \subset \mathbb{F}_q^N$ of size at most $q^N \mu^{n \ln n}$ such that the dimension of the space spanned by the derivatives of order $\leq n$ of C restricted to $\mathbb{F}_q^N \setminus E$ is $\leq s\, q^{\beta n \ln n}$.*

In the proof, we need $\delta n \ln n$ to be strictly less than $\frac{y\beta n}{q} \ln n - n \ln y$. That is,

$$\delta < \frac{y\beta}{q} - \frac{\ln y}{\ln n} . \tag{1}$$

It is worth emphasizing that, when we consider the derivatives, what we really mean is the formal derivatives of C as polynomials. In the above lemma we view the derivatives as functions from $\mathbb{F}_q^N \to \mathbb{F}_q$. Then it follows from the above analysis that the dimension of the space spanned by the functions corresponding to the derivatives of order $\leq n$ of C restricted to $\mathbb{F}_q^N \setminus E$ is $\leq s\, q^{\beta n \ln n}$. This way of viewing derivatives either as *formal polynomials* or as *functions* is implicit in the work of [GK98]. In Section 5, we show how to fix the parameters δ, β, and μ which depend only on the field size q.

4   Derivative Spaces of the Polynomial Families

In this section, we study the derivative spaces of $\mathrm{NW}_{n,\epsilon}(\mathbf{X})$ and $\mathrm{IMM}_{n,n}(\mathbf{X})$ polynomials. Instead of considering the full derivative spaces, we focus on a set of carefully chosen derivatives and consider the subspaces spanned by them.

The Derivative Space of $\{\mathrm{NW}_{n,\epsilon}(\mathrm{X})\}_{n>0}$ Polynomial Family

A set of variables $D = \{x_{i_1 j_1}, x_{i_2 j_2}, \ldots, x_{i_t j_t}\}$ is called an admissible set if i_ks (for $1 \leq k \leq t$) are all distinct and $\epsilon n \leq t \leq n$. Let H be the subspace spanned by the set of the partial derivatives of the polynomial $\mathrm{NW}_{n,\epsilon}(\mathrm{X})$ with respect to the admissible sets of variables. More formally, $H :=$ $\mathbb{F}_q$-span $\left\{ \frac{\partial \mathrm{NW}_{n,\epsilon}(\mathrm{X})}{\partial D} : D \text{ is an admissible set of variables}\right\}$. Since the monomials of the $\mathrm{NW}_{n,\epsilon}(\mathrm{X})$ polynomial are defined by the univariate polynomials of degree $< \epsilon n$, each partial derivative with respect to such a set D yields a multilinear monomial. If we choose ϵ such that $n - \epsilon n > \epsilon n$ (i.e. $\epsilon < 1/2$), then after the differentiation, all the monomials of length $n - \epsilon n$ are distinct. This follows from the fact that the monomials are generated from the image of the univariate polynomials of degree $< \epsilon n$.

Let us treat these monomials as functions from $\mathbb{F}_q^{n^2} \to \mathbb{F}_q$. The following simple lemma says that the functions corresponding to any set of distinct monomials are linearly independent.

Lemma 5. *Let $m_1(\mathrm{X}), m_2(\mathrm{X}), \ldots, m_k(\mathrm{X})$ be any set of k distinct multilinear monomials in $\mathbb{F}_q[x_1, x_2, \ldots, x_N]$. For $1 \leq i \leq k$, let $f_i : \mathbb{F}_q^N \to \mathbb{F}_q$ be the function corresponding to the monomial $m_i(\mathrm{X})$, i.e. $f_i(\mathrm{X}) = m_i(\mathrm{X})$. Then, f_is are linearly independent in the q^N dimensional vector space over $\mathbb{F}_q$.*

Consider the derivatives of $\mathrm{NW}_{n,\epsilon}(\mathrm{X})$ corresponding to the sets $\{x_{1a(1)}, x_{2a(2)}, \ldots, x_{\epsilon n a(\epsilon n)}\}$ for all univariate polynomials a of degree $< \epsilon n$. From Lemma 5, it follows that $\dim(H) \geq n^{\epsilon n} = e^{\epsilon n \ln n}$. W.l.o.g, we can assume that the constant function $\mathbf{1} : \mathbb{F}_q^{n^2} \to \mathbb{F}_q$ given by $\forall x, \mathbf{1}(x) = 1$ is also in H. This corresponds to the derivatives of order n.

The Derivative Space of $\{\mathrm{IMM}_{n,n}(\mathrm{X})\}_{n>0}$ Polynomial Family

For our application, we consider $n = 2m$ where m ranges over the positive integers. Consider the set of matrices $\mathrm{X}^{(1)}, \mathrm{X}^{(3)}, \ldots, \mathrm{X}^{(2m-1)}$ corresponding to the odd places. Let S be any set of m variables chosen as follows. Choose any variable from the first row of $\mathrm{X}^{(1)}$ and choose any one variable from each of the matrices $\mathrm{X}^{(3)}, \ldots, \mathrm{X}^{(2m-1)}$. We call such a set S an admissible set.

If we differentiate $\mathrm{IMM}_{n,n}(\mathrm{X})$ with respect to two different admissible sets of variables S and S', then we get two different monomials of length m each. This follows from the structure of the monomials in the $\mathrm{IMM}_{n,n}(\mathrm{X})$ polynomial, whenever we fix two variables from $\mathrm{X}^{(i-1)}$ and $\mathrm{X}^{(i+1)}$, the variable from $\mathrm{X}^{(i)}$ gets fixed. So the number of such monomials after differentiation is exactly $n^{2m-1} = e^{(n-1)\ln n}$.

Let m_S be the monomial obtained after differentiating $\mathrm{IMM}_{n,n}(\mathrm{X})$ by the set of variables in S and $\mathrm{var}(m_S)$ be the set of variables in m_S. Consider the derivatives of $\mathrm{IMM}_{n,n}(\mathrm{X})$ with respect to the following sets of variables: $\{S \cup T : T \subseteq \mathrm{var}(m_S)\}$ where S ranges over all admissible sets.

Let H be the subspace spanned by these derivatives. More formally, $H :=$ $\mathbb{F}_q$-span $\left\{ \frac{\partial \mathrm{IMM}_{n,n}(X)}{\partial D} : D = S \cup T \text{ where } T \subseteq \mathrm{var}(m_S); S \text{ is an admissible set} \right\}$. As before, we can assume that the constant function 1 is in H. From Lemma 5, we know that $\dim(H) \geq e^{(n-1)\ln n}$. Now to unify the arguments for $\mathrm{NW}_{n,\epsilon}(X)$ and $\mathrm{IMM}_{n,n}(X)$ polynomials, we introduce the following notion.

Downward Closed Property

Definition 6. *A set of multilinear monomials $\mathcal{M}$ is said to be downward closed if the following property holds. If $m(X) \in \mathcal{M}$ and multilinear monomial $m'(X)$ is such that $\mathrm{var}(m'(X)) \subseteq \mathrm{var}(m(X))$, then $m'(X) \in \mathcal{M}$.*

Now we consider a downward closed set of monomials $\mathcal{M}$ over N variables. These monomials can be viewed as functions from $\mathbb{F}_q^N$ to $\mathbb{F}_q$. W.l.o.g, we assume that the constant function 1 is also in $\mathcal{M}$ (constant function 1 corresponds to a monomial with an empty set of variables). Let H be the subspace spanned by these functions in $\mathcal{M}$.

For any $u \in \mathbb{F}_q^N$, define an operator T_u such that $(T_u(f))(X) = f(X - u)$ for any function $f : \mathbb{F}_q^N \to \mathbb{F}_q$. The following proposition is simple to prove.

Proposition 7. *Let H be the subspace spanned by a downward closed set of monomials $\mathcal{M}$ over the set of variables $\{x_1, x_2, \ldots, x_N\}$. Then for any $u \in \mathbb{F}_q^N$, T_u is a linear map from H to H. Moreover, the map T_u is one-one for any $u \in \mathbb{F}_q^N$.*

Proof. Let $g(X)$ be an arbitrary function in H which can be expressed as follows: $g(X) = \sum_{i \geq 1} c_i m_i(X)$ where $m_i(X) \in \mathcal{M}$, and $c_i \in \mathbb{F}_q$ for all $i \geq 1$. Observe that, $(T_u(g))(X) = g(X - u) = \sum_{i \geq 1} c_i m_i(X - u)$. It is sufficient to prove that $m(X - u) \in H$ for all $m(X) \in \mathcal{M}$. We can express $m(X - u)$ as follows: $m(X - u) = \sum_{S \subseteq \mathrm{var}(m(X))} c_S \prod_{x_r \in S} x_r$. where $c_S \in \mathbb{F}_q$. For every $S \subseteq \mathrm{var}(m(X))$, $\prod_{x_r \in S} x_r \in \mathcal{M}$ because $\mathcal{M}$ is downward closed. Since the choice of S was arbitrary, $m(X - u) \in H$. It is obvious that T_u is a linear map.

To see that T_u is a one-one map, consider any nonzero function $g(X)$ in H which can be defined as $g(X) = \sum_{i \geq 1} c_i m_i(X)$ where $m_i(X) \in \mathcal{M}$, and $c_i \in \mathbb{F}_q$ for all $i \geq 1$. Now, we notice that the highest degree monomials always survive in $T_u(g(X))$. Hence, the kernel of the map T_u is trivial for all $u \in \mathbb{F}_q^N$. $\qquad\square$

It is not difficult to observe that the derivative spaces that we select for $\mathrm{NW}_{n,\epsilon}(X)$ and $\mathrm{IMM}_{n,n}(X)$ are spanned by downward closed sets of monomials.

Lemma 8. *The generator sets for the derivative subspaces H for $\mathrm{NW}_{n,\epsilon}(X)$ and $\mathrm{IMM}_{n,n}(X)$ polynomials are downward closed.*

Proof. Let us consider the $\mathrm{NW}_{n,\epsilon}(X)$ polynomial first. Let $m \in H$ be any monomial and D be the admissible set such that $m = \frac{\partial \mathrm{NW}_{n,\epsilon}(X)}{\partial D}$. Let m' be any monomial such that $\mathrm{var}(m') \subseteq \mathrm{var}(m)$. Then $m' = \frac{\partial \mathrm{NW}_{n,\epsilon}(X)}{\partial D'}$ where $D' = D \cup (\mathrm{var}(m) \setminus \mathrm{var}(m'))$.

Similarly for the $\mathrm{IMM}_{n,n}(\mathrm{X})$ polynomial, consider any $m \in H$. Then $m = \frac{\partial \mathrm{IMM}_{n,n}(\mathrm{X})}{\partial D}$ and $D = S \cup T$ for an admissible set S and $T \subseteq \mathrm{var}(m_S)$. If m' is any monomial such that $\mathrm{var}(m') \subseteq \mathrm{var}(m)$, then $m' = \frac{\partial \mathrm{IMM}_{n,n}(\mathrm{X})}{\partial D'}$ where $D' = S \cup (T \cup (\mathrm{var}(m) \setminus \mathrm{var}(m')))$. Clearly $T \cup (\mathrm{var}(m) \setminus \mathrm{var}(m')) \subseteq \mathrm{var}(m_S)$. $\square$

5   A Covering Argument

In this section, we adapt the covering argument of [GK98] to prove the lower bound results. In [GK98], the covering argument was given over the set of invertible matrices. Here we adapt their argument suitably over the entire space $\mathbb{F}_q^N$. As defined in the section 4, the subspace H represents the chosen derivative subspace of either the $\mathrm{NW}_{n,\epsilon}(\mathrm{X})$ polynomial or the $\mathrm{IMM}_{n,n}(\mathrm{X})$ polynomial.

Define the subspace $H_a := \{f \in H : f(a) = 0\}$ for $a \in \mathbb{F}_q^N$. Let us recall that E is the set of points over which some of the product gates with large rank may not evaluate to zero. Let the set of points $\mathbb{F}_q^N \setminus E$ be denoted by A. Then Lemma 4 says that in H, we get that $\mathrm{codim}(\bigcap_{a \in A} H_a) < s\, q^{r_0}$. We note the following simple observation.

Proposition 9. *For any $u, a \in \mathbb{F}_q^N$, we have that $T_u(H_a) = H_{u+a}$.*

Let $P = \bigcap_{a \in A} H_a$. Let $S \subset \mathbb{F}_q^N$ be a set such that we can cover the entire space $\mathbb{F}_q^N$ by the shifts of A with the elements from S: $\bigcup_{u \in S} u + A = \mathbb{F}_q^N$. Now by applying the map T_u to P which is one-one, we get the following: $T_u(P) = \bigcap_{a \in A} T_u(H_a) = \bigcap_{b \in u+A} H_b$. By a further intersection over S, we get the following.

$$\bigcap_{u \in S} T_u(P) = \bigcap_{u \in S} \bigcap_{b \in u+A} H_b = \bigcap_{b \in \mathbb{F}_q^N} H_b \tag{2}$$

From Equation 2, we get the following estimate.

$$\mathrm{codim}\left(\bigcap_{b \in \mathbb{F}_q^N} H_b\right) = \mathrm{codim}\left(\bigcap_{u \in S} T_u(P)\right) \leq |S|\ \mathrm{codim}(P) \leq |S|\, s\, q^{r_0} \tag{3}$$

The $\mathrm{codim}\left(\bigcap_{b \in \mathbb{F}_q^N} H_b\right)$ refers to the dimension of the set of functions in H which do not evaluate to zero over all the points in $\mathbb{F}_q^N$.

Next, we show an upper-bound estimate for the size of the set S. This follows from a simple adaptation of the dominating set based argument given in [GK98]. Consider the directed graph $G = (V, R)$ defined as follows. The points in $\mathbb{F}_q^N$ are the vertices of the graph. For $u_1, u_2 \in \mathbb{F}_q^N$, the edge $u_1 \to u_2$ is in R iff $u_2 = u_1 + b$ for any $b \in A$. Clearly the in-degree and out-degree of any vertex are equal to $|A|$. Now, we recall Lemma 2 of [GK98] to estimate the size of S.

Lemma 10 (Lovász, [Lov75]). *Let (V, R) be a directed (regular) graph with $|V| = m$ vertices and with the in-degree and the out-degree of each vertex both equal to d. Then there exists a subset $U \subset V$ of size $O(\frac{m}{d} \log(d+1))$ such that for any vertex $v \in V$ there is a vertex $u \in U$ forming an edge $(u, v) \in R$.*

Let c_0 be the constant fixed by the lemma in its $O()$ notation. By Lemma 10, we get the following estimate: $|S| \leq c_0 \frac{|\mathbb{F}_q^N|}{|A|} \log(|A|+1) \leq c_0 \frac{q^N}{q^N - |E|} \log(q^N - |E| + 1) \leq c_0 (\log q) N \frac{q^N}{q^N - |E|} = O(N)$ (for fixed q). The last equation follows from the estimate for $|E|$ from the section 3.

Fixing the Parameters. Consider the inequality 1 which is $\delta < \frac{y\beta}{q} - \frac{\ln y}{\ln n}$. Fix the values for β, δ, and μ in Lemma 4 as follows. Set $\beta = \frac{1}{10 \ln q}, \delta = \frac{1}{20q \ln q}, \nu = \frac{\delta}{2}$, and $\mu = e^{-\nu}$. Consider the function $g(y) = y - \frac{10q \ln q}{\ln n} \ln y - 0.50$. Since $g(y)$ is a monotonically increasing function (for n appropriately larger than a threshold value depending on q) which takes the value of 0.50 at $y = 1$, $g(y) > 0$ for $y \geq 1$ and thus $\delta < \frac{y\beta}{q} - \frac{\ln y}{\ln n}$ for the chosen values of β and δ. Also, $\frac{y\beta}{q} - \frac{\ln y}{\ln n} - \delta > \nu$ and thus $|E| \leq q^N \mu^{n \ln n}$.

From Section 4, we know that $\dim(H)$ for $\mathrm{NW}_{n,\epsilon}(\mathrm{X})$ is at least $e^{\epsilon n \ln n}$. Consider the upper bound on $\mathrm{codim}\left(\bigcap_{b \in \mathbb{F}_q^N} H_b\right)$ given by the inequality 3. If we choose ϵ in such a way that $e^{\epsilon n \ln n} > |S| \, s \, q^{r_0}$, then there will be a multilinear polynomial f in H such that f will evaluate to zero over all points in $\mathbb{F}_q^N$. Then, $\dim(H) > n^{\epsilon n} = e^{\epsilon n \ln n} \implies e^{\epsilon n \ln n} > |S| \, s \, q^{r_0} = e^{\delta n \ln n + (\beta \ln q) n \ln n + \ln N}$. Considering the terms of the order of $n \ln n$ in the exponent, it is enough to choose $\epsilon (< 1/2)$ such that the following holds: $\epsilon > \delta + \beta \ln q = \frac{1}{20q \ln q} + \frac{1}{10}$.

Since the $\dim(H)$ for $\mathrm{IMM}_{n,n}(\mathrm{X})$ is $\geq e^{(n-1)\ln n}$, the chosen values of β and δ clearly suffice. Finally, we recall from the combinatorial nullstellensatz [Alo99] that no non-zero multilinear polynomial can be zero over $\mathbb{F}_q^N$. Thus, we get the main theorem (restated from Section 1).

Theorem 11. *For any fixed-size finite field $\mathbb{F}_q$, any depth three $\Sigma\Pi\Sigma$ circuit computing the polynomials $\mathrm{NW}_{n,\epsilon}(\mathrm{X})$ or $\mathrm{IMM}_{n,n}(\mathrm{X})$ must be of size at least $2^{\delta n \log n}$ where the parameters δ and $\epsilon (< 1/2)$ are in $(0,1)$ and depend only on q.*

The main interesting open problem that remains after our work, is to prove that over the fixed-size fields, any $\Sigma\Pi\Sigma$ circuit computing the determinant polynomial for a $n \times n$ matrix must be of size $2^{\Omega(n \log n)}$.

References

[Alo99]   Alon, N.: Combinatorial nullstellensatz. Combinatorics, Probability and Computing 8 (1999)

[AV08]   Agrawal, M.: V Vinay. Arithmetic circuits: A chasm at depth four. In: Proceedings-Annual Symposium on Foundations of Computer Science, pp. 67–75. IEEE (2008)

[Chi14]   Chillara, S.: (2014), http://www.cmi.ac.in/~suryajith/Depth3.pdf

[CM14]   Chillara, S., Mukhopadhyay, P.: Depth-4 lower bounds, determinantal complexity: A unified approach. In: STACS, pp. 239–250 (2014)

[Fei09]   Feige, U.: The permanent and the determinant (2009)

[Fis94]   Fischer, I.: Sums of like powers of multivariate linear forms. Mathematics Magazine 67(1), 59–61 (1994)

[FLMS13]   Fournier, H., Limaye, N., Malod, G., Srinivasan, S.: Lower bounds for depth 4 formulas computing iterated matrix multiplication. In: To Appear in the proceedings of STOC 2014. Electronic Colloquium on Computational Complexity (ECCC), vol. 20, p. 100 (2013)

[GK98]   Grigoriev, D., Karpinski, M.: An exponential lower bound for depth 3 arithmetic circuits. In: STOC, pp. 577–582 (1998)

[GKKS13a] Gupta, A., Kamath, P., Kayal, N., Saptharishi, R.: Approaching the chasm at depth four. In: IEEE Conference on Computational Complexity, pp. 65–73 (2013)

[GKKS13b] Gupta, A., Kamath, P., Kayal, N., Saptharishi, R.: Arithmetic circuits: A chasm at depth three. In: FOCS, pp. 578–587 (2013)

[KLSS14]   Kayal, N., Limaye, N., Saha, C., Srinivasan, S.: An exponential lower bound for homogeneous depth four arithmetic formulas. In: To appear in the Proceedings of STOC 2014. Electronic Colloquium on Computational Complexity (ECCC), vol. 21, p. 5 (2014)

[KMN13]   Kumar, M., Maheshwari, G., Sarma M.N., J.: Arithmetic circuit lower bounds via maxRank. In: Fomin, F.V., Freivalds, R., Kwiatkowska, M., Peleg, D. (eds.) ICALP 2013, Part I. LNCS, vol. 7965, pp. 661–672. Springer, Heidelberg (2013)

[Koi12]   Koiran, P.: Arithmetic circuits: The chasm at depth four gets wider. Theor. Comput. Sci. 448, 56–65 (2012)

[KS13a]   Kumar, M., Saraf, S.: The limits of depth reduction for arithmetic formulas: It's all about the top fan-in. In: To appear in the Proceedings of STOC 2014. Electronic Colloquium on Computational Complexity (ECCC), vol. 20, p. 153 (2013)

[KS13b]   Kumar, M., Saraf, S.: Superpolynomial lower bounds for general homogeneous depth 4 arithmetic circuits. In: Esparza, J., Fraigniaud, P., Husfeldt, T., Koutsoupias, E. (eds.) ICALP 2014. LNCS, vol. 8572, pp. 751–762. Springer, Heidelberg (2014)

[KS14]   Kumar, M., Saraf, S.: On the power of homogeneous depth 4 arithmetic circuits. Electronic Colloquium on Computational Complexity (ECCC) 21, 45 (2014)

[KSS13]   Kayal, N., Saha, C., Saptharishi, R.: A super-polynomial lower bound for regular arithmetic formulas. In: To appear in the Proceedings of STOC 2014. Electronic Colloquium on Computational Complexity (ECCC), vol. 20, p. 91 (2013)

[Lov75]   Lovász, L.: On the ratio of optimal integral and fractional covers. Discrete mathematics 13(4), 383–390 (1975)

[NW94]   Nisan, N., Wigderson, A.: Hardness vs randomness. J. Comput. Syst. Sci. 49(2), 149–167 (1994)

[NW97]   Nisan, N., Wigderson, A.: Lower bounds on arithmetic circuits via partial derivatives. Computational Complexity 6(3), 217–234 (1997)

[Sap13]   Saptharishi, R.: Personal communication (2013)

[Sax08]   Saxena, N.: Diagonal circuit identity testing and lower bounds. In: Aceto, L., Damgård, I., Goldberg, L.A., Halldórsson, M.M., Ingólfsdóttir, A., Walukiewicz, I. (eds.) ICALP 2008, Part I. LNCS, vol. 5125, pp. 60–71. Springer, Heidelberg (2008)

[SW01]   Shpilka, A., Wigderson, A.: Depth-3 arithmetic circuits over fields of characteristic zero. Computational Complexity 10(1), 1–27 (2001)

[Tav13]   Tavenas, S.: Improved bounds for reduction to depth 4 and depth 3. In: Chatterjee, K., Sgall, J. (eds.) MFCS 2013. LNCS, vol. 8087, pp. 813–824. Springer, Heidelberg (2013)

Hitting Forbidden Subgraphs in Graphs of Bounded Treewidth[*]

Marek Cygan[1], Dániel Marx[2], Marcin Pilipczuk[3], and Michał Pilipczuk[3]

[1] Institute of Informatics, University of Warsaw, Poland
cygan@mimuw.edu.pl
[2] Institute for Computer Science and Control,
Hungarian Academy of Sciences (MTA SZTAKI), Hungary
dmarx@cs.bme.hu
[3] Department of Informatics, University of Bergen, Norway
{Marcin.Pilipczuk,Michal.Pilipczuk}@ii.uib.no

Abstract. We study the complexity of a generic hitting problem H-SUBGRAPH HITTING, where given a fixed pattern graph H and an input graph G, we seek for the minimum size of a set $X \subseteq V(G)$ that hits all subgraphs of G isomorphic to H. In the colorful variant of the problem, each vertex of G is precolored with some color from $V(H)$ and we require to hit only H-subgraphs with matching colors. Standard techniques (e.g., Courcelle's theorem) show that, for every fixed H and the problem is fixed-parameter tractable parameterized by the treewidth of G; however, it is not clear how exactly the running time should depend on treewidth. For the colorful variant, we demonstrate matching upper and lower bounds showing that the dependence of the running time on treewidth of G is tightly governed by $\mu(H)$, the maximum size of a minimal vertex separator in H. That is, we show for every fixed H that, on a graph of treewidth t, the colorful problem can be solved in time $2^{O(t^{\mu(H)})} \cdot |V(G)|$, but cannot be solved in time $2^{o(t^{\mu(H)})} \cdot |V(G)|^{O(1)}$, assuming the Exponential Time Hypothesis (ETH). Furthermore, we give some preliminary results showing that, in the absence of colors, the parameterized complexity landscape of H-SUBGRAPH HITTING is much richer.

1 Introduction

The "optimality programme" is a thriving trend within parameterized complexity, which focuses on pursuing tight bounds on the time complexity of parameterized problems. Instead of just determining whether the problem is fixed-parameter tractable, that is, whether the problem with a certain parameter k can be solved in time $f(k) \cdot n^{O(1)}$ for some computable function $f(k)$, the

[*] The research leading to these results has received funding from the European Research Council under the European Union's Seventh Framework Programme (FP/2007-2013) / ERC Grant Agreement n. 267959, and n. 280152, as well as OTKA grant NK10564 and Polish National Science Centre grant DEC-2012/05/D/ST6/03214.

goal is to determine the best possible dependence $f(k)$ on the parameter k. For several problems, matching upper and lower bounds have been obtained for the function $f(k)$. The lower bounds are under the complexity assumption Exponential Time Hypothesis (ETH), which roughly states than n-variable 3SAT cannot be solved in time $2^{o(n)}$; see, e.g., the survey of Lokshtanov et al. [11].

One area where this line of research was particularly successful is the study of fixed-parameter algorithms parameterized by the treewidth of the input graph and understanding how the running time has to depend on the treewidth. Classic results on model checking monadic second-order logic on graphs of bounded treewidth, such as Courcelle's Theorem, provide a unified and generic way of proving fixed-parameter tractability of most of the tractable cases of this parameterization [1,5]. While these results show that certain problems are solvable in time $f(t) \cdot n$ on graphs of treewidth t for some function f, the exact function $f(t)$ resulting from this approach is usually hard to determine and far from optimal. To get reasonable upper bounds on $f(t)$, one typically resorts to constructing a dynamic programming algorithm, which often is straightforward, but tedious.

The question whether the straightforward dynamic programming algorithms for bounded treewidth graphs are optimal received particular attention in 2011. On the hardness side, Lokshtanov, Marx and Saurabh proved that many natural algorithms are probably optimal [10,12]. In particular, they showed that there are problems for which the $2^{\mathcal{O}(t \log t)} n$ time algorithms are best possible, assuming ETH. On the algorithmic side, Cygan et al. [6] presented a new technique, called *Cut&Count*, that improved the running time of the previously known (natural) algorithms for many connectivity problems. For example, previously only $2^{\mathcal{O}(t \log t)} \cdot n^{\mathcal{O}(1)}$ algorithms were known for HAMILTONIAN CYCLE and FEEDBACK VERTEX SET, which was improved to $2^{\mathcal{O}(t)} \cdot n^{\mathcal{O}(1)}$ by Cut&Count. These results indicated that not only proving tight bounds for algorithms on tree decompositions is within our reach, but such a research may lead to surprising algorithmic developments. Further work includes derandomization of Cut&Count in [3,8], an attempt to provide a meta-theorem to describe problems solvable in single-exponential time [13], and a new algorithm for PLANARIZATION [9].

We continue here this line of research by investigating a family of subgraph-hitting problems parameterized by treewidth and find surprisingly tight bounds for a number of problems. An interesting conceptual message of our results is that, for every integer $c \geq 1$, there are fairly natural problems where the best possible dependence on treewidth is of the form $2^{\mathcal{O}(t^c)}$.

Studied Problems and Motivation. In our paper we focus on the following generic H-SUBGRAPH HITTING problem: for a pattern graph H and an input graph G, what is the minimum size of a set $X \subseteq V(G)$ that hits all subgraphs of G that are isomorphic to H? (Henceforth we call them H-*subgraphs* for brevity.) This problem generalizes a few ones studied in the literature, for example VERTEX COVER (for $H = P_2$), where a tight $2^t \cdot t^{\mathcal{O}(1)} \cdot |V(G)|$ time bound is known [10], or finding largest induced subgraph of maximum degree at most Δ (for $H = K_{1,\Delta+1}$), which is $W[1]$-hard for treewidth parameter if Δ is a part of the input [2], but, to the best of our knowledge, no detailed study of treewidth parameterization for constant Δ has been done before. We also study the following *colorful* variant COLORFUL H-SUBGRAPH HITTING, where the input graph G is additionally equipped with a coloring $\sigma : V(G) \to V(H)$, and we are only interested in hitting H-subgraphs whose all vertices match their colors.

A direct source of motivation for our study is the work of Pilipczuk [13], which attempted to describe graph problems admitting fixed-parameter algorithms with running time of the form $2^{\mathcal{O}(t)} \cdot |V(G)|^{\mathcal{O}(1)}$, where t is the treewidth of G. The proposed description is a logical formalism where one can quantify existence of some vertex/edge sets, whose properties can be verified "locally" by requesting satisfaction of a formula of modal logic in every vertex. In particular, Pilipczuk argued that the language for expressing local properties needs to be somehow modal, as it cannot be able to discover cycles in a constant-radius neighborhood of a vertex. This claim was supported by a lower bound: unless ETH fails, for any constant $\ell \geq 5$, the problem of finding the minimum size of a set that hits all the cycles C_ℓ in a graph of treewidth t cannot be solved in time $2^{o(t^2)} \cdot |V(G)|^{\mathcal{O}(1)}$. Motivated by this result, we think that it is natural to investigate the complexity of hitting subgraphs for more general patterns H, instead of just cycles.

We may see the colorful variant as an intermediate step towards full understanding of the complexity of H-SUBGRAPH HITTING, but it is also an interesting problem on its own. It often turns out that the colorful variants of problems are easier to investigate, while their study reveals useful insights; a remarkable example is the kernelization lower bound for SET COVER and related problems [7]. In our case, if we allow colors, a major combinatorial difficulty vanishes: when the algorithm keeps track of different parts of the pattern H that appear in the graph G, and combines a few parts into a larger one, the coloring σ ensures that the parts are vertex-disjoint. Hence, the colorful variant is easier to study, whereas at the same time it reveals interesting insight into the standard variant.

Our Results and Techniques. In the case of COLORFUL H-SUBGRAPH HITTING, we obtain a tight bounds for the complexity of the treewidth parameterization. First, note that, in the presence of colors, one actually can solve COLORFUL H-SUBGRAPH HITTING for each connected component of H independently; hence, we may focus only on connected patterns H. Second, we observe that there are two special cases. If H is a path then COLORFUL H-SUBGRAPH HITTING reduces to a maximum flow/minimum cut problem, and hence is polynomial-time solvable. If H is a clique, then any H-subgraph of G needs to be contained in a single bag of any tree decomposition, and there is a simple $2^{\mathcal{O}(t)}|V(G)|$-time algorithm, where t is the treewidth of G. Finally, for the remaining cases we show that the dependence on treewidth is tightly connected to the value of $\mu(H)$, the maximum size of a minimal vertex separator in H (a separator S is minimal if there are two vertices x, y such that S is an xy-separator, but no proper subset of S is). We prove the following matching upper and lower bounds.

Theorem 1. *A* COLORFUL H-SUBGRAPH HITTING *instance* (G, σ) *can be solved in time* $2^{\mathcal{O}(t^{\mu(H)})}|V(G)|$ *in the case when* H *is connected and is not a clique, where* t *is the treewidth of* G.

Theorem 2. *Let H be a graph that contains a connected component that is neither a path nor a clique. Then, unless ETH fails, there does not exist an algorithm that, given a* COLORFUL H-SUBGRAPH HITTING *instance (G, σ) and a tree decomposition of G of width t, resolves (G, σ) in time $2^{o(t^{\mu(H)})}|V(G)|^{\mathcal{O}(1)}$.*

In all the theorems in this work we treat H as a fixed graph of constant size, and hence the factors hidden in the $\mathcal{O}$-notation may depend on the size of H.

In the absence of colors, we give preliminary results showing that the parameterized complexity of the treewidth parameterization of H-SUBGRAPH HITTING is more involved than the one of the colorful counterpart. In this setting, we are able to relate the dependence on treewidth only to a larger parameter of the graph H. Let $\mu^\star(H)$ be the maximum size of $N_H(A)$, where A iterates over connected subsets of $V(H)$ such that $N_H(N_H[A]) \neq \emptyset$, i.e., $N_H[A]$ is not a whole connected component of H. Observe that $\mu(H) \leq \mu^\star(H)$ for any H. First, we were able to construct a counterpart of Theorem 1 only with the exponent $\mu^\star(H)$.

Theorem 3. *Assume H contains a connected component that is not a clique. Then, given a graph G of treewidth t, one can solve H-SUBGRAPH HITTING on G in time $2^{\mathcal{O}(t^{\mu^\star(H)} \log t)}|V(G)|$.*

We remark that for COLORFUL H-SUBGRAPH HITTING, an algorithm with running time $2^{\mathcal{O}(t^{\mu^\star(H)})}|V(G)|$ (as opposed to $\mu(H)$ in the exponent in Theorem 1) is rather straightforward: in the state of dynamic programming one needs to remember, for every subset X of the bag of size at most $\mu^\star(G)$, all forgotten connected parts of H that are attached to X and not hit by the constructed solution. To decrease the exponent to $\mu(H)$, we introduce a "prediction-like" definition of a state of the dynamic programming, leading to highly involved proof of correctness. For the problem without colors, however, even an algorithm with the exponent $\mu^\star(H)$ (Theorem 3) is far from trivial. We cannot limit ourselves to keeping track of forgotten connected parts of the graph H independently of each other, since in the absence of colors these parts may not be vertex-disjoint and, hence, we would not be able to reason about their union in latter bags of the tree decomposition. To cope with this issue, we show that the set of forgotten (not necessarily connected) parts of the graph H that are subgraphs of G can be represented as a *witness graph* with $\mathcal{O}(t^{\mu^\star(H)})$ vertices and edges. As there are only $2^{\mathcal{O}(t^{\mu^\star(H)} \log t)}$ possible graphs of this size, the running time bound follows.

We also observe that the bound of $\mathcal{O}(t^{\mu^\star(H)})$ on the size of a witness graph is not tight for many patterns H. For example, if H is a path, then we are able to find a witness graph with $\mathcal{O}(t)$ vertices and edges, and the algorithm of Theorem 3 runs in $2^{\mathcal{O}(t \log t)}|V(G)|$ time.

From the lower bound perspective, we were not able to prove an analog of Theorem 2 in the absence of colors. However, there is a good reason for that: we show that for any fixed $h \geq 2$ and $H = K_{2,h}$, the H-SUBGRAPH HITTING problem is solvable in time $2^{\mathcal{O}(t^2 \log t)}|V(G)|$ for a graph G of treewidth t. This should be put in contrast with $\mu^\star(K_{2,h}) = \mu(K_{2,h}) = h$. Moreover, the lower bound of $2^{o(t^h)}$ can be proven if we break the symmetry of $K_{2,h}$ by attaching a triangle to each of the two degree-h vertices of $K_{2,h}$. This indicates that the optimal dependency on t in an algorithm for H-SUBGRAPH HITTING may heavily rely on the symmetries of H, and may be more difficult to pinpoint.

2   Preliminaries

Graph Notation. In most cases, we use standard graph notation. A *t-boundaried graph* is a graph G with a prescribed (possibly empty) *boundary* $\partial G \subseteq V(G)$ with $|\partial G| \leq t$, and an injective function $\lambda_G : \partial G \to \{1, 2, \ldots, t\}$. For a vertex $v \in \partial G$ the value $\lambda_G(v)$ is called the *label* of v.

A *colored graph* is a graph G with a function $\sigma : V(G) \to \mathbb{L}$, where $\mathbb{L}$ is some finite set of colors. A graph G is *H-colored*, for some other graph H, if $\mathbb{L} = V(H)$. We also say in this case that σ is an *H-coloring* of G.

A *homomorphism* of graphs H and G is a function $\pi : V(H) \to V(G)$ such that $ab \in E(H)$ implies $\pi(a)\pi(b) \in E(G)$. In the H-colored setting, i.e., when G is H-colored, we also require that $\sigma(\pi(a)) = a$ for any $a \in V(H)$ (every vertex of H is mapped onto appropriate color). The notion extends also to t-boundaried graphs: if both H and G are t-boundaried, we require that whenever $a \in \partial H$ then $\pi(a) \in \partial G$ and $\lambda_G(\pi(a)) = \lambda_H(a)$. Note, however, that we allow that a vertex of int H is mapped onto a vertex of ∂G.

An *H-subgraph of G* is any injective homomorphism $\pi : V(H) \to V(G)$. Recall that in the t-boundaried setting, we require that the labels are preserved, whereas in the colored setting, we require that the homomorphism respect colors. In the latter case, we call it a *σ-H-subgraph of G* for clarity.

We say that a set $X \subseteq V(G)$ *hits* a $(\sigma$-$)H$-subgraph π if $X \cap \pi(V(H)) \neq \emptyset$. The (COLORFUL) H-SUBGRAPH HITTING problem asks for a minimum possible size of a set that hits all $(\sigma$-$)H$-subgraphs of G.

Tree Decompositions. In this work, we view tree decompositions as rooted: in a tree decomposition $(\mathtt{T}, \beta)$, $\mathtt{T}$ is a rooted tree and $\beta(w)$ is a bag at node $w \in V(\mathtt{T})$. We moreover define $\gamma(w) = \bigcup_{w' \preceq w} \beta(w')$, where the union iterates over all descendants w' of w in $\mathtt{T}$, and $\alpha(w) = \gamma(w) \setminus \beta(w)$. In all our algorithms, by using a recent 5-approximation algorithm for treewidth [4], we assume that we are given a tree decomposition $(\mathtt{T}, \beta)$ where each bag is of size at most t; this linear shift in the value of t is irrelevant for the complexity bounds, but makes the notation much cleaner. Moreover, we assume that we are additionally equipped with a labeling $\Lambda : V(G) \to \{1, 2, \ldots, t\}$ that is injective on each bag $\beta(w)$; observe that it is straightforward to compute such a labeling in a top-down manner on $\mathtt{T}$. Consequently, we may treat each graph $G[\gamma(w)]$ as a t-boundaried graph with $\partial G[\gamma(w)] = \beta(w)$ and labeling $\Lambda|_{\beta(w)}$.

Important Graph Invariants, Chunks And Slices. For two vertices $a, b \in V(H)$, a set $S \subseteq V(H) \setminus \{a, b\}$ is an *ab-separator* if a and b are not in the same connected component of $H \setminus S$. The set S is additionally a *minimal ab-separator* if no proper subset of S is an ab-separator. A set S is a *minimal separator* if it is a

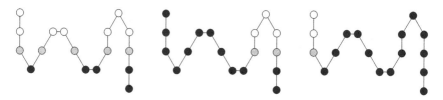

Fig. 1. White and gray vertices denote a slice (left), chunk (centre) and separator chunk (right) in a graph H being a path. The gray vertices belong to the boundary.

minimal ab-separator for some $a, b \in V(H)$. For a graph H, by $\mu(H)$ we denote the maximum size of a minimal separator in H.

For an induced subgraph $H' = H[D]$, $D \subseteq V(H)$, we define the boundary $\partial H' = N_H(V(H) \setminus D)$ and the interior $\operatorname{int} H' = D \setminus \partial H[D]$; thus $V(H') = \partial H' \uplus \operatorname{int} H'$. Observe that $N_H(\operatorname{int} H') \subseteq \partial H'$. An induced subgraph H' of H is a *slice* if $N_H(\operatorname{int} H') = \partial H'$, and a *chunk* if additionally $H[\operatorname{int} H']$ is connected. For a set $A \subseteq V(H)$, we use $\mathbf{p}[A]$ ($\mathbf{c}[A]$) to denote the unique slice (chunk) with interior A (if it exists). The intuition behind this definition is that, when we consider some bag $\beta(w)$ in a tree decomposition, a slice is a part of H that may already be present in $G[\gamma(w)]$ and we want to keep track of it. If a slice (chunk) $\mathbf{p}$ is additionally equipped with a injective labeling $\lambda_{\mathbf{p}} : \partial \mathbf{p} \rightarrow \{1, 2, \ldots, t\}$, then we call the resulting t-boundaried graph a *t-slice* (*t-chunk*, respectively).

By $\mu^\star(H)$ we denote the maximum size of $\partial \mathbf{c}$, where $\mathbf{c}$ iterates over all chunks of H. We remark here that both $\mu(H)$ and $\mu^\star(H)$ are positive only for graphs H that contain at least one connected component that is not a clique, as otherwise there are no chunks with nonempty boundary nor minimal separators in H.

Observe that if S is a minimal ab-separator in H, and A is the connected component of $H \setminus S$ that contains a, then $N_H(A) = S$ and $\mathbf{c}[A]$ is a chunk in H with boundary S. Consequently, $\mu(H) \leq \mu^\star(H)$ for any graph H. A chunk $\mathbf{c}$ for which $\partial \mathbf{c}$ is a minimal separator in H is henceforth called a *separator chunk*. See also Figure 1 for an illustration.

3   General Algorithm for H-Subgraph Hitting

In this section we sketch an algorithm for H-SUBGRAPH HITTING running in time $2^{\mathcal{O}(t^{\mu^\star(H)} \log t)} |V(G)|$, where t is the width of the tree decomposition we are working on. The general idea is the natural one: for each node w of the tree decomposition, for each set $\widehat{X} \subseteq \beta(w)$ and for each family $\mathbb{P}$ of t-slices, we would like to find the minimum size of a set $X \subseteq \alpha(w)$ such that, if we treat $G[\gamma(w)]$ as a t-boundaried graph with $\partial G[\gamma(w)] = \beta(w)$ and labeling $\Lambda|_{\beta(w)}$, then any slice that is a subgraph of $G[\gamma(w) \setminus (X \cup \widehat{X})]$ belongs to $\mathbb{P}$. However, as there can be as many as $t^{|H|}$ t-slices, we have too many choices for the family $\mathbb{P}$.

The essence of the proof, encapsulated in the next lemma, is to show that each "reasonable" choice of $\mathbb{P}$ can be encoded as a *witness graph* of essentially size $\mathcal{O}(t^{\mu^\star(H)})$. Such a claim would give a $2^{\mathcal{O}(t^{\mu^\star(H)} \log t)}$ bound on the number of possible witness graphs, and provide a good bound on the size of state space.

Lemma 1. *Assume H contains a connected component that is not a clique. Then, for any t-boundaried graph (G, λ) there exists a t-boundaried graph $(\widehat{G}, \lambda)$ that (a) is a subgraph of (G, λ), (b) $\partial G = \partial \widehat{G}$ and $G[\partial G] = \widehat{G}[\partial \widehat{G}]$, (c) $\widehat{G} \setminus E(\widehat{G}[\partial \widehat{G}])$ contains $\mathcal{O}(t^{\mu^*(H)})$ vertices and edges, and, (d) for any t-slice $\mathbf{p}$ and any set $Y \subseteq V(G)$ such that $|Y| + |V(\mathbf{p})| \leq |V(H)|$, there exists a $\mathbf{p}$-subgraph in $(G \setminus Y, \lambda)$ if and only if there exists one in $(\widehat{G} \setminus Y, \lambda)$.*

Proof. We define $\widehat{G}$ by a recursive procedure. We start with $\widehat{G} = G[\partial G]$. Then, for every t-chunk $\mathbf{c} = (H', \lambda')$, we invoke a procedure $\mathtt{enhance}(\mathbf{c}, \emptyset)$. The procedure $\mathtt{enhance}(\mathbf{c}, X)$, for $X \subseteq V(G)$, first tries to find a $\mathbf{c}$-subgraph π in $(G \backslash X, \lambda)$. If there is none, the procedure terminates. Otherwise, it first adds all edges and vertices of $\pi(\mathbf{c})$ to $\widehat{G}$ that are not yet present there. Second, if $|X| < |V(H)|$, then it recursively invokes $\mathtt{enhance}(\mathbf{c}, X \cup \{v\})$ for each $v \in \pi(\mathbf{c})$.

We first bound the size of the constructed graph $\widehat{G}$. There are at most $2^{|V(H)|} t^{\mu^*(H)}$ choices for the chunk, since a chunk $\mathbf{c}$ is defined by its vertex set, and there are at most $t^{\mu^*(H)}$ labellings of its boundary. The procedure $\mathtt{enhance}(\mathbf{c}, X)$ at each step adds at most one copy of H to G, and branches into at most $|V(H)|$ directions. The depth of the recursion is bounded by $|V(H)|$. Hence, in total at most $2^{|V(H)|} t^{\mu^*(H)} \cdot (|V(H)| + |E(H)|) \cdot |V(H)|^{|V(H)|}$ edges and vertices are added to $\widehat{G}$, except for the initial graph $G[\partial G]$.

It remains to argue that $\widehat{G}$ satisfies property (d). Clearly, since $(\widehat{G}, \lambda)$ is a subgraph of (G, λ), the implication in one direction is trivial. In the other direction, we start with the following claim.

Claim 4. *For any set $Z \subseteq V(G)$ of size at most $|V(H)|$, and for any t-chunk $\mathbf{c}$, if there exists a $\mathbf{c}$-subgraph in $(G \setminus Z, \lambda)$ then there exists also one in $(\widehat{G} \setminus Z, \lambda)$.*

Proof. Let π be a $\mathbf{c}$-subgraph in $(G \backslash Z, \lambda)$. Define $X_0 = \emptyset$. We will construct sets $X_0 \subsetneq X_1 \subsetneq \ldots$, where $X_i \subseteq Z$ for every i, and analyse the calls to the procedure $\mathtt{enhance}(\mathbf{c}, X_i)$ in the process of constructing $\widehat{G}$.

Assume that $\mathtt{enhance}(\mathbf{c}, X_i)$ has been invoked at some point during the construction; clearly this is true for $X_0 = \emptyset$. Since we assume $X_i \subseteq Z$, there exists a $\mathbf{c}$-subgraph in $(G \setminus X_i, \lambda)$ — π is one such example. Hence, $\mathtt{enhance}(\mathbf{c}, X_i)$ has found a $\mathbf{c}$-subgraph π_i, and added its image to $\widehat{G}$. If π_i is a $\mathbf{c}$-subgraph also in $(\widehat{G} \setminus Z, \lambda)$, then we are done. Otherwise, there exists $v_i \in Z \setminus X_i$ that is also present in the image of π_i. In particular, since $|Z| \leq |V(H)|$, we have $|X_i| < |V(H)|$ and the call $\mathtt{enhance}(\mathbf{c}, X_i \cup \{v_i\})$ has been invoked. We define $X_{i+1} := X_i \cup \{v_i\}$.

Since the sizes of sets X_i grow at each step, for some X_i, $i \leq |Z|$, we reach the conclusion that π_i is a $\mathbf{c}$-subgraph of $(\widehat{G} \setminus Z, \lambda)$, and the claim is proven.   ⌙

Fix now a set $Y \subseteq V(G)$ and a t-slice $\mathbf{p}$ with labeling $\lambda_{\mathbf{p}}$ and with $|Y| + |V(\mathbf{p})| \leq |V(H)|$. Let π be a $\mathbf{p}$-subgraph of $(G \setminus Y, \lambda)$. Let $A_1, A_2, \ldots, A_r$ be the connected components of $H[\text{int } \mathbf{p}]$. Define $H_i = N_H[A_i]$, and observe that each H_i is a chunk with $\partial H_i = N_H(A_i) \subseteq \partial \mathbf{p}$. We define $\lambda_i = \lambda_{\mathbf{p}}|_{\partial H_i}$ to obtain a t-chunk $\mathbf{c}_i = (H_i, \lambda_i)$. By the properties of a t-slice, each vertex of $\mathbf{p}$ is present in at least one graph $\mathbf{c}_i$, and vertices of $\partial \mathbf{p}$ may be present in more than one.

We now inductively define injective homomorphisms $\pi_0, \pi_1, \ldots, \pi_r$ such that of π_i maps the subgraph of $\mathbf{p}$ induced by $\partial\mathbf{p} \cup \bigcup_{j \leq i} A_j$ to $(\widehat{G} \setminus Y, \lambda)$, and does not use any vertex of $\bigcup_{j>i} \pi(A_j)$. Observe that π_r is a $\mathbf{p}$-subgraph of $(\widehat{G} \setminus Y, \lambda)$. Hence, this construction will conclude the proof of the lemma.

For base case, recall that $\pi(\partial\mathbf{p}) \subseteq \partial G = \partial\widehat{G}$ and define $\pi_0 = \pi|_{\partial\mathbf{p}}$. For the inductive case, assume that π_{i-1} has been constructed for some $1 \leq i \leq r$. Define

$$Z_i = Y \cup \pi(\partial\mathbf{p} \setminus \partial H_i) \cup \bigcup_{j<i} \pi_{i-1}(A_j) \cup \bigcup_{j>i} \pi(A_j).$$

Note that since π and π_{i-1} are injective and Y is disjoint with $\pi(\partial\mathbf{p})$, then we have that $Z_i \cap \pi(\partial\mathbf{p}) = \pi(\partial\mathbf{p} \setminus \partial H_i)$. This observation and the inductive assumption on π_{i-1} imply that the mapping $\pi|_{V(H_i)}$ does not use any vertex of Z_i. Thus, $\pi|_{V(H_i)}$ is a $\mathbf{c}_i$-subgraph in $(G \setminus Z_i, \lambda)$. Observe moreover that $|Z_i| \leq |Y| + |V(\mathbf{p})| \leq |V(H)|$. By Claim 4, there exists a $\mathbf{c}_i$-subgraph π'_i in $(\widehat{G} \setminus Z_i, \lambda)$. Observe that, since π'_i and π_{i-1} are required to preserve labellings on boundaries of their preimages, $\pi_i := \pi'_i \cup \pi_{i-1}$ is a function and a homomorphism. Moreover, by the definition of Z_i, π_i is injective and does not use any vertex of $\bigcup_{j>i} \pi(A_j)$. Hence, π_i satisfies all the required conditions, and the inductive construction is completed. This concludes the proof of the lemma. □

Using Lemma 1, we now define states of dynamic programming algorithm on the input tree decomposition $(\mathbf{T}, \beta)$. For every node $w \in V(\mathbf{T})$, a *state* is a pair $\mathbf{s} = (\widehat{X}, \widehat{G})$ where $\widehat{X} \subseteq \beta(w)$ and $\widehat{G}$ is a graph with $\mathcal{O}(t^{\mu^*(H)})$ vertices and edges such that $\beta(w) \setminus \widehat{X} \subseteq V(\widehat{G})$ and $\widehat{G}[\beta(w) \setminus \widehat{X}] = G[\beta(w) \setminus \widehat{X}]$. We treat $\widehat{G}$ as a t-boundaried graph with $\partial\widehat{G} = \beta(w) \setminus \widehat{X}$ and labeling $\Lambda|_{\beta(w) \setminus \widehat{X}}$. We say that a set $X \subseteq \alpha(w)$ is *feasible* for w and $\mathbf{s}$ if for every $Y \subseteq \beta(w) \setminus \widehat{X}$ and for every t-slice $\mathbf{p}$ such that $|Y| + |V(\mathbf{p})| \leq |V(H)|$, if there is a $\mathbf{p}$-subgraph in $(G[\gamma(w) \setminus (X \cup \widehat{X} \cup Y)], \Lambda|_{\beta(w) \setminus (\widehat{X} \cup Y)})$ then there is also one in $(\widehat{G} \setminus Y, \Lambda|_{\beta(w) \setminus (\widehat{X} \cup Y)})$. For every w and every state $\mathbf{s}$, we would like to compute $T[w, \mathbf{s}]$, the minimum possible size of a feasible set X. Note that the answer to the input H-SUBGRAPH HITTING instance is the minimum value of $T[\text{root}(\mathbf{T}), (\emptyset, \widehat{G})]$ where $\widehat{G}$ iterates over all graphs of with $\mathcal{O}(t^{\mu^*(H)})$ vertices and edges that do not contain the t-slice $(H, \emptyset)$ as a subgraph. Hence, it remains to show how to compute the values $T[w, \mathbf{s}]$ in a bottom-up manner in the tree decomposition, which is relatively standard.

4   Discussion on Special Cases of H-Subgraph Hitting

As announced in the introduction, we now discuss a few special cases of H-SUBGRAPH HITTING. First, let us consider H being a path, $H = P_h$ for some $h \geq 3$. Note that $\mu(P_h) = 1$, while $\mu^*(P_h) = 2$ for $h \geq 5$. Observe that in the dynamic programming algorithm of the previous section we have that $G[\gamma(w) \setminus (X \cup X_w)]$ does not contain an H-subgraph and, hence, the witness graph obtained through Lemma 1 does not contain an H-subgraph as well. However, graphs excluding P_h as a subgraph have very rigid structure: any their depth-first search tree has depth bounded by h. Using this insight, we can derive the following improvement of Lemma 1, that improves the running time of Theorem 3 to $2^{\mathcal{O}(t \log t)}|V(G)|$ for H being a path.[1]

Lemma 2. *Assume H is a path. Then, for any t-boundaried graph (G, λ) that does not contain an H-subgraph, there exists a witness graph as in Lemma 1 with $\mathcal{O}(t)$ vertices and edges.*

Second, let us consider $H = K_{2,h}$ (the complete biclique with 2 vertices on one side, and h on the other), for some $h \geq 2$. Observe that $\mu^\star(K_{2,h}) = \mu(K_{2,h}) = h$. On the other hand, we note the following.

Lemma 3. *Assume $H = K_{2,h}$ for some $h \geq 2$. If the witness graph given by Lemma 1 does not admit an H-subgraph, then it has $\mathcal{O}(t^2)$ vertices and edges.*

Proof. Since the constructed witness graph $\widehat{G}$ does not admit an H-subgraph, each two vertices $v_1, v_2 \in \partial\widehat{G}$ have less than h common neighbours in $\widehat{G}$, as otherwise there is a H-subgraph in $\widehat{G} \setminus \partial\widehat{G}$ on vertices v_1, v_2 and h vertices of $N_{\widehat{G}}(v_1) \cap N_{\widehat{G}}(v_2)$. Hence

$$\sum_{v \in V(\widehat{G}) \setminus \partial\widehat{G}} \binom{|N_{\widehat{G}}(v) \cap \partial\widehat{G}|}{2} \leq (h-1)\binom{|\partial\widehat{G}|}{2} \leq (h-1)\binom{t}{2}. \tag{1}$$

Let $V(H) = \{a_1, a_2, b_1, b_2, \ldots, b_h\}$ where $A := \{a_1, a_2\}$ and $B := \{b_1, b_2, \ldots, b_h\}$ are bipartition classes of H. Note that there are only two types of proper chunks in H: $N_H[a_i]$, $i = 1, 2$ and $N_H[b_j]$, $1 \leq j \leq h$. Hence, one can easily verify that in the construction of the witness graph $\widehat{G}$ of Lemma 1 every vertex $v \in \widehat{G} \setminus \partial\widehat{G}$ has at least two neighbours in $\partial\widehat{G}$, and $\widehat{G} \setminus \partial\widehat{G}$ is edgeless. Then we have $|N_{\widehat{G}}(v) \cap \partial\widehat{G}| \leq 2\binom{|N_{\widehat{G}}(v) \cap \partial\widehat{G}|}{2}$ for each $v \in V(\widehat{G}) \setminus \partial\widehat{G}$. Consequently, by (1) there are at most $2(h-1)\binom{t}{2}$ edges of $\widehat{G}$ with exactly one endpoint in $\partial\widehat{G}$, whereas there are at most $\binom{t}{2}$ edges in $\widehat{G}[\partial\widehat{G}]$. The lemma follows. □

Lemma 3 together with a dynamic programming as in Section 3 imply that $K_{2,h}$-SUBGRAPH HITTING can be solved in $2^{\mathcal{O}(t^2 \log t)}|V(G)|$ time, in spite of the fact that $\mu^\star(K_{2,h}) = \mu(K_{2,h}) = h$.

We now show that a slight modification of $K_{2,h}$ enables us to prove a much higher lower bound. For this, let us consider a graph H_h for $h \geq 2$ defined as $K_{2,h}$ with triangles attached to both degree-h vertices. Note that $\mu(H_h) = \mu^\star(H_h) = h$. One may view H_h as $K_{2,h}$ with some symmetries broken, so that the proof of Lemma 3 does not extend to H_h. We observe that the lower bound proof of Theorem 2 works, with small modifications, also for the case of H_h-SUBGRAPH HITTING.

Theorem 5. *Unless ETH fails, for every $h \geq 2$ there does not exist an algorithm that, given a H_h-SUBGRAPH HITTING instance G and a tree decomposition of G of width t, resolves G in time $2^{o(t^h)}|V(G)|^{\mathcal{O}(1)}$.*

---

[1] The proofs Lemma 2 and Theorem 5 are deferred to the full version of the paper.

Furthermore, the proof of Theorem 5 does not need to assume that h is a constant. Thus, we obtain the following interesting double-exponential lower bound.

Corollary 1. *Unless ETH fails, there does not exist an algorithm that, given a graph G with a tree decomposition of width t, and an integer $h = \mathcal{O}(\log |V(G)|)$, finds in $2^{2^{o(t)}}|V(G)|^{\mathcal{O}(1)}$ time the minimum size of a set that hits all H_h-subgraphs of G.*

5   Overview of the Proof for Colorful Variant

5.1   Proof Sketch of Theorem 1

In this sketch, we focus on the definition of a state that will be used in the dynamic programming algorithm on the input tree decomposition $(\mathtt{T}, \beta)$. A *potential chunk* is a separator t-chunk $\mathbf{c}[A]$. A *state* at node $w \in V(\mathtt{T})$ is a pair $(\widehat{X}, \mathbb{C})$ where $\widehat{X} \subseteq \beta(w)$ and $\mathbb{C}$ is a family of potential chunks, where each chunk $\mathbf{c}$ in $\mathbb{C}$: (i) uses only labels of $\Lambda^{-1}(\beta(w) \setminus \widehat{X})$; and (ii) the mapping $\pi : \partial\mathbf{c} \to \beta(w) \setminus \widehat{X}$ that maps a vertex of $\partial\mathbf{c}$ to a vertex with the same label is a homomorphism from $H[\partial\mathbf{c}]$ to G (in particular, it respects colors). Observe that, as $|\partial\mathbf{c}| \leq \mu(H)$ for any separator chunk $\mathbf{c}$, there are $\mathcal{O}(t^{\mu(H)})$ possible separator t-chunks, and hence $2^{\mathcal{O}(t^{\mu(H)})}$ possible states for a fixed node w.

The intuitive idea behind a state is that, for node $w \in V(\mathtt{T})$ and state $(\widehat{X}, \mathbb{C})$, we investigate the possibility of the following: for a solution X we are looking for, it holds that $\widehat{X} = X \cap \beta(w)$ and the family $\mathbb{C}$ is exactly the set of possible separator chunks of H that are subgraphs of $G \setminus X$, where the subgraph relation is defined as on t-boundaried graphs and $G \setminus X$ is equipped with $\partial G \setminus X = \beta(w) \setminus X$ and labeling $\Lambda_w|_{\beta(w) \setminus X}$. The difficult part of the proof is to show that this information is sufficient, in particular, it suffices to keep track only of the separator chunks, and not all proper chunks of H. We emphasize here that the intended meaning of the set $\mathbb{C}$ is that it represents separator chunks present in the entire $G \setminus X$, not $G[\gamma(w)] \setminus X$. That is, to be able to limit ourselves only to separator chunks, we need to encode in the state some prediction for the future. This makes our dynamic programming algorithm rather non-standard.

Let us proceed to a more formal definition of the dynamic programming table. For a bag w and a state $\mathbf{s} = (\widehat{X}, \mathbb{C})$ at w we define the graph $G(w, \mathbf{s})$ as follows. We first take the graph $G[\gamma(w)] \setminus \widehat{X}$ and then, for each chunk $\mathbf{c} \in \mathbb{C}$ we add a disjoint copy of $\mathbf{c}$ to $G(w, \mathbf{s})$ and identify the pairs of vertices with the same label in $\partial\mathbf{c}$ and in $\beta(w)$. Note that $G[\gamma(w)] \setminus \widehat{X}$ is an induced subgraph of $G(w, \mathbf{s})$: by the properties of elements of $\mathbb{C}$, no new edge has been introduced between two vertices of $\beta(w)$. We make $G(w, \mathbf{s})$ a t-boundaried graph in a natural way: $\partial G(w, \mathbf{s}) = \beta(w) \setminus \widehat{X}$ with labeling $\Lambda|_{\partial G(w,\mathbf{s})}$. Here we exploit the crucial property of the colored version of the problem: no two isomorphic chunks glued to $\beta(w)$ can participate together in any σ-H-subgraph, since their vertices have the same colors. Therefore, attaching undeletable chunks of $\mathbb{C}$ explicitly to $\beta(w)$ is equivalent to just allowing these chunks to be present either in the future, or in the forgotten part of the graph.

For each bag w and for each state $\mathbf{s} = (\widehat{X}, \mathbb{C})$ we say that a set $X \subseteq \alpha(w)$ is *feasible* if $G(w, \mathbf{s}) \setminus X$ does not contain any σ-H-subgraph, and for any separator t-chunk $\mathbf{c}$ of H, if there is a $\mathbf{c}$-subgraph in $G(w, \mathbf{s}) \setminus X$ then $\mathbf{c} \in \mathbb{C}$. We would like to compute the value $T[w, \mathbf{s}]$ that equals the minimum size of a feasible set X, using the standard bottom-up dynamic programming. Observe that $T[\mathtt{root}(\mathtt{T}), \emptyset]$ is the minimum size of a solution for COLORFUL H-SUBGRAPH HITTING.

5.2 Proof Sketch of Theorem 2

The proof of this theorem is inspired by the approach used in [13] for the lower bound for C_ℓ-SUBGRAPH HITTING.

Consider a minimal separator S in H such that $\mu := |S| = \mu(H)$ and A, B are two such distinct connected components of $H \setminus S$ with $N_H(A) = N_H(B) = S$.

With some simple preprocessing, we may assume we are given n-variable 3-CNF formula Φ where each variable appears exactly three times, at least once positively and at least once negatively. Let s be a smallest integer such that $s^\mu \geq 3n$; $s = \mathcal{O}(n^{1/\mu})$. We start by introducing a set M of $s\mu$ vertices $w_{i,c}$, $1 \leq i \leq s$, $c \in S$, with coloring $\sigma(w_{i,c}) = c$. The set M is the central part of the constructed graph G. In particular, each connected component of $G \setminus M$ will be of constant size, immediately implying that G has treewidth $\mathcal{O}(n^{1/\mu})$.

To each clause C of Φ, and to each literal l in C, assign a function $f_{C,l} : S \to \{1, 2, \ldots, s\}$ such that $f_{C,l} \neq f_{C',l'}$ for $(C, l) \neq (C', l')$. Observe that this is possible due to the assumption $s^\mu \geq 3n$ and the preprocessing step.

The main idea is as follows. For each clause C and literal l in C, we attach a copy of $H[N_H[A]]$ and a copy of $H[N_H[B]]$ to $\{w_{f_{C,l}(c),c} \mid c \in S\}$ in a natural way. For each variable x we use constant-size gadgets to we wire up all the copies of $H[N_H[A]]$ that correspond to an occurrence of that variable, so that with minimum budget we may hit all copies corresponding to positive occurrences of x or all copies corresponding to negative occurrences; this choice corresponds to the decision on the value of x. Similarly, for each clause C we use constant-size gadgets to wire up all the copies of $H[N_H[B]]$ that correspond to literals in C, so that with minimum budget we may hit all but one of these copies; this choice corresponds to the decision which literal of C is satisfied by an assignment. Finally, we attach to the construction a large number of copies of $H \setminus (A \cup B \cup S)$, so that a small solution needs to hit any σ-H-subgraph of (G, σ) in a vertex of $A \cup B$. The construction enforces that, whenever a clause C chooses a literal l to satisfy C, it leaves the corresponding copy of $H[B]$ not hit, forcing the solution to hit the corresponding copy of $H[A]$, and therefore forcing the correct assignment of the variable in l.

6 Conclusions and Open Problems

Our preliminary study of the treewidth parameterization of the H-SUBGRAPH HITTING problem revealed that its parameterized complexity is highly involved. Whereas for the more graspable colored version we obtained essentially tight bounds, a large gap between lower and upper bounds remains for the standard version. In particular, the following two questions arise: Can we improve the running time of Theorem 3 to factor $t^{\mu(H)}$ in the exponent? Is there any relatively general symmetry-breaking assumption on H that would allow us to show a $2^{o(t^{\mu(H)})}$ lower bound in the absence of colors?

In a broader view, let us remark that the complexity of the treewidth parameterization of *minor-hitting* problems is also currently highly unclear. Here, for a minor-closed graph class $\mathcal{G}$ and input graph G, we seek for the minimum size of a set $X \subseteq V(G)$ such that $G \setminus X \in \mathcal{G}$, or, equivalently, X hits all minimal forbidden minors of $\mathcal{G}$. A straightforward dynamic programming algorithm has double-exponential dependency on the width of the decomposition. However, it was recently shown that $\mathcal{G}$ being the class of planar graphs, a $2^{\mathcal{O}(t \log t)}|V(G)|$-time algorithm exists [9]. Can this result be generalized to more graph classes?

References

1. Arnborg, S., Lagergren, J., Seese, D.: Easy problems for tree-decomposable graphs. J. Algorithms 12(2), 308–340 (1991)
2. Betzler, N., Bredereck, R., Niedermeier, R., Uhlmann, J.: On bounded-degree vertex deletion parameterized by treewidth. Discrete Appl. Math. 160(1-2), 53–60 (2012)
3. Bodlaender, H.L., Cygan, M., Kratsch, S., Nederlof, J.: Deterministic single exponential time algorithms for connectivity problems parameterized by treewidth. In: Fomin, F.V., Freivalds, R., Kwiatkowska, M., Peleg, D. (eds.) ICALP 2013, Part I. LNCS, vol. 7965, pp. 196–207. Springer, Heidelberg (2013)
4. Bodlaender, H.L., Drange, P.G., Dregi, M.S., Fomin, F.V., Lokshtanov, D., Pilipczuk, M.: An $\mathcal{O}(c^k n)$ 5-approximation algorithm for treewidth. In: FOCS, pp. 499–508 (2013)
5. Courcelle, B.: The monadic second-order logic of graphs I: Recognizable sets of finite graphs. Inf. Comput. 85, 12–75 (1990)
6. Cygan, M., Nederlof, J., Pilipczuk, M., Pilipczuk, M., van Rooij, J.M.M., Wojtaszczyk, J.O.: Solving connectivity problems parameterized by treewidth in single exponential time. In: FOCS, pp. 150–159 (2011)
7. Dom, M., Lokshtanov, D., Saurabh, S.: Incompressibility through colors and iDs. In: Albers, S., Marchetti-Spaccamela, A., Matias, Y., Nikoletseas, S., Thomas, W. (eds.) ICALP 2009, Part I. LNCS, vol. 5555, pp. 378–389. Springer, Heidelberg (2009)
8. Fomin, F.V., Lokshtanov, D., Saurabh, S.: Efficient computation of representative sets with applications in parameterized and exact algorithms. In: SODA, pp. 142–151 (2014)
9. Jansen, B.M.P., Lokshtanov, D., Saurabh, S.: A near-optimal planarization algorithm. In: SODA, pp. 1802–1811 (2014)
10. Lokshtanov, D., Marx, D., Saurabh, S.: Known algorithms on graphs on bounded treewidth are probably optimal. In: SODA, pp. 777–789 (2011)
11. Lokshtanov, D., Marx, D., Saurabh, S.: Lower bounds based on the exponential time hypothesis. Bulletin of the EATCS 105, 41–72 (2011)
12. Lokshtanov, D., Marx, D., Saurabh, S.: Slightly superexponential parameterized problems. In: SODA, pp. 760–776 (2011)
13. Pilipczuk, M.: Problems parameterized by treewidth tractable in single exponential time: A logical approach. In: Murlak, F., Sankowski, P. (eds.) MFCS 2011. LNCS, vol. 6907, pp. 520–531. Springer, Heidelberg (2011)

Probabilistic Analysis of Power Assignments[*]

Maurits de Graaf[1,2] and Bodo Manthey[1]

[1] University of Twente, Department of Applied Mathematics,
Enschede, The Netherlands
{m.degraaf,b.manthey}@utwente.nl
[2] Thales Nederland B.V., Huizen, The Netherlands

Abstract. A fundamental problem for wireless ad hoc networks is the assignment of suitable transmission powers to the wireless devices such that the resulting communication graph is connected. The goal is to minimize the total transmit power in order to maximize the life-time of the network. Our aim is a probabilistic analysis of this power assignment problem. We prove complete convergence for arbitrary combinations of the dimension d and the distance-power gradient p. Furthermore, we prove that the expected approximation ratio of the simple spanning tree heuristic is strictly less than its worst-case ratio of 2.

Our main technical novelties are two-fold: First, we find a way to deal with the unbounded degree that the communication network induced by the optimal power assignment can have. Minimum spanning trees and traveling salesman tours, for which strong concentration results are known in Euclidean space, have bounded degree, which is heavily exploited in their analysis. Second, we apply a recent generalization of Azuma-Hoeffding's inequality to prove complete convergence for the case $p \geq d$ for both power assignments and minimum spanning trees (MSTs). As far as we are aware, complete convergence for $p > d$ has not been proved yet for any Euclidean functional.

1 Introduction

Wireless ad hoc networks have received significant attention due to their many applications in, for instance, environmental monitoring or emergency disaster relief, where wiring is difficult. Unlike wired networks, wireless ad hoc networks lack a backbone infrastructure. Communication takes place either through single-hop transmission or by relaying through intermediate nodes. We consider the case that each node can adjust its transmit power for the purpose of power conservation. In the assignment of transmit powers, two conflicting effects have to be taken into account: if the transmit powers are too low, the resulting network may be disconnected. If the transmit powers are too high, the nodes run out of energy quickly. The goal of the power assignment problem is to assign transmit powers to the transceivers such that the resulting network is connected and the sum of transmit powers is minimized [12].

---

[*] A full version with all proofs is available at http://arxiv.org/abs/1403.5882.

E. Csuhaj-Varjú et al. (Eds.): MFCS 2014, Part II, LNCS 8635, pp. 201–212, 2014.
© Springer-Verlag Berlin Heidelberg 2014

1.1   Problem Statement and Previous Results

We consider a set of vertices $X \subseteq [0,1]^d$, which represent the sensors, $|X| = n$, and assume that $\|u - v\|^p$, for some $p \in \mathbb{R}$ (called the *distance-power gradient* or *path loss exponent*), is the power required to successfully transmit a signal from u to v. This is called the power-attenuation model, where the strength of the signal decreases with $1/r^p$ for distance r, and is a simple yet very common model for power assignments in wireless networks [14]. In practice, we typically have $1 \le p \le 6$ [13].

A power assignment $\mathsf{pa} : X \to [0, \infty)$ is an assignment of transmit powers to the nodes in X. Given pa, we have an edge between two nodes u and v if both $\mathsf{pa}(x), \mathsf{pa}(y) \ge \|x - y\|^p$. If the resulting graph is connected, we call it a *PA graph*. Our goal is to find a PA graph and a corresponding power assignment pa that minimizes $\sum_{v \in X} \mathsf{pa}(v)$. Note that any PA graph $G = (X, E)$ induces a power assignment by $\mathsf{pa}(v) = \max_{u \in X:\{u,v\} \in E} \|u - v\|^p$.

PA graphs can in many aspects be regarded as a tree as we are only interested in connectedness, but it can contain more edges in general. However, we can simply ignore edges and restrict ourselves to a spanning tree of the PA graph.

The minimal connected power assignment problem is NP-hard for $d \ge 2$ and APX-hard for $d \ge 3$ [4]. For $d = 1$, i.e., when the sensors are located on a line, the problem can be solved by dynamic programming [11]. A simple approximation algorithm for minimum power assignments is the minimum spanning tree heuristic (MST heuristic), which achieves a tight worst-case approximation ratio of 2 [11]. This has been improved by Althaus et al. [1], who devised an approximation algorithm that achieves an approximation ratio of 5/3. A first average-case analysis of the MST heuristic was presented by de Graaf et al. [6]: First, they analyzed the expected approximation ratio of the MST heuristic for the (non-geometric, non-metric) case of independent edge lengths. Second, they proved convergence of the total power consumption of the assignment computed by the MST heuristic for the special case of $p = d$, but not of the optimal power assignment. They left as open problems, first, an average-case analysis of the MST heuristic for random geometric instances and, second, the convergence of the value of the optimal power assignment.

1.2   Our Contribution

In this paper, we conduct an average-case analysis of the optimal power assignment problem for Euclidean instances. The points are drawn independently and uniformly from the d-dimensional unit hypercube $[0,1]^d$. We believe that probabilistic analysis is better-suited for performance evaluation in wireless ad hoc networks than worst-case analysis, as the positions of the sensors – in particular if deployed in areas that are difficult to access – are subjected to randomness.

Roughly speaking, our contributions are as follows:

1. We show that the power assignment functional has sufficiently nice properties in order to apply Yukich's general framework for Euclidean functionals [16] to obtain concentration results (Section 3).

2. Combining these insights with a recent generalization of the Azuma-Hoeffding bound [15], we obtain concentration of measure and complete convergence for all combinations of d and $p \geq 1$, even for the case $p \geq d$ (Section 4). In addition, we obtain complete convergence for $p \geq d$ for minimum-weight spanning trees. As far as we are aware, complete convergence for $p \geq d$ has not been proved yet for such functionals. The only exception we are aware of are minimum spanning trees for the case $p = d$ [16, Sect. 6.4].

3. We provide a probabilistic analysis of the MST heuristic for the geometric case. We show that its expected approximation ratio is strictly smaller than its worst-case approximation ratio of 2 [11] for any d and p (Section 5).

Our main technical contributions are two-fold: First, we introduce a transmit power redistribution argument to deal with the unbounded degree that graphs induced by the optimal transmit power assignment can have. The unboundedness of the degree makes the analysis of the power assignment functional PA challenging. The reason is that removing a vertex can cause the graph to fall into a large number of components and it might be costly to connect these components without the removed vertex. In contrast, the degree of any minimum spanning tree, for which strong concentration results are known in Euclidean space for $p \leq d$, is bounded for every fixed d, and this is heavily exploited in the analysis. (The concentration result by de Graaf et al. [6] for the power assignment obtained from the MST heuristic also exploits that MSTs have bounded degree.)

Second, we apply a recent generalization of Azuma-Hoeffding's inequality by Warnke [15] to prove complete convergence for the case $p \geq d$ for both power assignments and minimum spanning trees. We introduce the notion of *typically smooth* Euclidean functionals, prove convergence of such functionals, and show that minimum spanning trees and power assignments are typically smooth. In this sense, our proof of complete convergence provides an alternative and generic way to prove complete convergence, whereas Yukich's proof for minimum spanning trees is tailored to the case $p = d$. In order to prove complete convergence with our approach, one only needs to prove convergence in mean, which is often much simpler than complete convergence, and typically smoothness. Thus, we provide a simple method to prove complete convergence of Euclidean functionals along the lines of Yukich's result that, in the presence of concentration of measure, convergence in mean implies complete convergence [16, Cor. 6.4].

2   Definitions and Notation

Throughout the paper, d (the dimension) and p (the distance-power gradient) are fixed constants. For three points x, y, v, we by $\overline{xv}$ the line through x and v, and we denote by $\angle(x, v, y)$ the angle between $\overline{xv}$ and $\overline{yv}$.

A *Euclidean functional* is a function F^p for $p > 0$ that maps finite sets of points from the unit hypercube $[0, 1]^d$ to some non-negative real number and is translation invariant and homogeneous of order p [16, page 18]. From now on, we omit the superscript p of Euclidean functionals, as p is always fixed and clear from the context.

PA_B is the canonical boundary functional of PA (we refer to Yukich [16] for boundary functionals of other optimization problems): given a hyperrectangle $R \subseteq \mathbb{R}^d$ with $X \subseteq R$, this means that a solution is an assignment $\mathsf{pa}(x)$ of power to the nodes $x \in X$ such that

- x and y are connected if $\mathsf{pa}(x), \mathsf{pa}(y) \geq \|x - y\|^p$,
- x is connected to the boundary of R if the distance of x to the boundary of R is at most $\mathsf{pa}(x)^{1/p}$, and
- the resulting graph, called a *boundary PA graph*, is either connected or consists of connected components that are all connected to the boundary.

Then $\mathsf{PA}_B(X, R)$ is the minimum value for $\sum_{x \in X} \mathsf{pa}(x)$ that can be achieved by a boundary PA graph. Note that in the boundary functional, no power is assigned to the boundary. It is straight-forward to see that PA and PA_B are Euclidean functionals for all $p > 0$ according to Yukich [16, page 18].

For a hyperrectangle $R \subseteq \mathbb{R}^d$, let $\operatorname{diam} R = \max_{x,y \in R} \|x - y\|$ denote the diameter of R. For a Euclidean functional F, let $\mathsf{F}(n) = \mathsf{F}(\{U_1, \ldots, U_n\})$, where $U_1, \ldots, U_n$ are drawn uniformly and independently from $[0, 1]^d$. Let $\gamma_{\mathsf{F}}^{d,p} = \lim_{n \to \infty} \frac{\mathbb{E}(\mathsf{F}(n))}{n^{\frac{d-p}{d}}}$. (In principle, $\gamma_{\mathsf{F}}^{d,p}$ need not exist, but it does exist for all functionals considered in this paper.)

A sequence $(R_n)_{n \in \mathbb{N}}$ of random variables *converges in mean* to a constant γ if $\lim_{n \to \infty} \mathbb{E}(|R_n - \gamma|) = 0$. The sequence $(R_n)_{n \in \mathbb{N}}$ *converges completely to a constant γ* if we have $\sum_{n=1}^{\infty} \mathbb{P}(|R_n - \gamma| > \varepsilon) < \infty$ for all $\varepsilon > 0$ [16, page 33].

Besides PA, we consider two other Euclidean functions: $\mathsf{MST}(X)$ denotes the length of the minimum spanning tree with lengths raised to the power p. $\mathsf{PT}(X)$ denotes the total power consumption of the assignment obtained from the MST heuristic, again with lengths raised to the power p. The MST heuristic proceeds as follows: First, we compute a minimum spanning tree of X. Then let $\mathsf{pa}(x) = \max\{\|x - y\|^p \mid \{x, y\} \text{ is an edge of the MST}\}$. By construction and a simple analysis, we have $\mathsf{MST}(X) \leq \mathsf{PA}(X) \leq \mathsf{PT}(X) \leq 2 \cdot \mathsf{MST}(X)$ [11].

For $n \in \mathbb{N}$, let $[n] = \{1, \ldots, n\}$.

3   Properties of the Power Assignment Functional

After showing that optimal PA graphs can have unbounded degree and providing a lemma that helps solving this problem, we show that the power assignment functional fits into Yukich's framework for Euclidean functionals [16].

3.1   Degrees and Cones

As opposed to minimum spanning trees, whose maximum degree is bounded from above by a constant that depends only on the dimension d, a technical challenge is that the maximum degree in an optimal PA graph cannot be bounded by a constant in the dimension. This holds even for the simplest case of $d = 1$ and $p > 1$. We conjecture that the same holds also for $p = 1$, but proving this seems to be more difficult and not to add much.

Lemma 3.1. *For all $p > 1$, all integers $d \geq 1$, and for infinitely many n, there exist instances of n points in $[0, 1]^d$ such that the unique optimal PA graph is a tree with a maximum degree of $n - 1$.*

The unboundedness of the degree of PA graphs make the analysis of the functional PA challenging. The technical reason is that removing a vertex can cause the PA graph to fall into a non-constant number of components. The following lemma is the crucial ingredient to get over this "degree hurdle".

Lemma 3.2. *Let $x, y \in X$, let $v \in [0, 1]^d$, and assume that x and y have power $\mathsf{pa}(x) \geq \|x - v\|^p$ and $\mathsf{pa}(y) \geq \|y - v\|^p$, respectively. Assume further that $\|x - v\| \leq \|y - v\|$ and that $\angle(x, v, y) \leq \alpha$ with $\alpha \leq \pi/3$. Then the following holds:*

(a) $\mathsf{pa}(y) \geq \|x - y\|^p$, i.e., y has sufficient power to reach x.
(b) If x and y are not connected (i.e., $\mathsf{pa}(x) < \|x - y\|^p$), then $\|y - v\| > \frac{\sin(2\alpha)}{\sin(\alpha)} \cdot \|x - v\|$.

For instance, $\alpha = \pi/6$ results in a factor of $\sqrt{3} = \sin(\pi/3)/\sin(\pi/6)$. In the following, we invoke this lemma always with $\alpha = \pi/6$, but this choice is arbitrary as long as $\alpha < \pi/3$, which causes $\sin(2\alpha)/\sin(\alpha)$ to be strictly larger than 1.

3.2 Deterministic Properties

In this section, we state properties of the power assignment functional. Subadditivity (Lemma 3.3), superadditivity (Lemma 3.4), and growth bound (Lemma 3.5) are straightforward.

Lemma 3.3 (subadditivity). PA *is subadditive [16, (2.2)] for all $p > 0$ and all $d \geq 1$, i.e., for any point sets X and Y and any hyperrectangle $R \subseteq \mathbb{R}^d$ with $X, Y \subseteq R$, we have $\mathsf{PA}(X \cup Y) \leq \mathsf{PA}(X) + \mathsf{PA}(Y) + O((\operatorname{diam} R)^p)$.*

Lemma 3.4 (superadditivity). PA_B *is superadditive for all $p \geq 1$ and $d \geq 1$ [16, (3.3)], i.e., for any X, hyperrectangle $R \subseteq \mathbb{R}^d$ with $X \subseteq R$ and partition of R into hyperrectangles R_1 and R_2, we have $\mathsf{PA}_B^p(X, R) \geq \mathsf{PA}_B^p(X \cap R_1, R_1) + \mathsf{PA}_B^p(X \cap R_2, R_2)$.*

Lemma 3.5 (growth bound). *For any $X \subseteq [0, 1]^d$ and $0 < p$ and $d \geq 1$, we have $\mathsf{PA}_B(X) \leq \mathsf{PA}(X) \leq O\left(\max\left\{n^{\frac{d-p}{d}}, 1\right\}\right)$.*

The following lemma shows that PA is smooth, which roughly means that adding or removing a few points does not have a huge impact on the function value. Its proof requires Lemma 3.2 to deal with the fact that optimal PA graphs can have unbounded degree.

Lemma 3.6. *The power assignment functional* PA *is smooth for all* $0 < p \le d$ *[16, (3.8)], i.e.,* $\left| \mathsf{PA}^p(X \cup Y) - \mathsf{PA}^p(X) \right| = O\left(|Y|^{\frac{d-p}{d}} \right)$ *for all point sets* $X, Y \subseteq [0,1]^d$.

Proof. One direction is straightforward: $\mathsf{PA}(X \cup Y) - \mathsf{PA}(X)$ is bounded by $\Psi = O\left(|Y|^{\frac{d-p}{d}} \right)$, because the optimal PA graph for Y has a value of at most Ψ by Lemma 3.5. Then we can take the PA graph for Y and connect it to the tree for X with a single edge, which costs at most $O(1) \le \Psi$ because $p \le d$.

For the other direction, consider the optimal PA graph T for $X \cup Y$. The problem is that the degrees $\deg_T(v)$ of vertices $v \in Y$ can be unbounded (Lemma 3.1). (If the maximum degree were bounded, then we could argue in the same way as for the MST functional.) The idea is to exploit the fact that removing $v \in Y$ also frees some power. Roughly speaking, we proceed as follows: Let $v \in Y$ be a vertex of possibly large degree. We add the power of v to some vertices close to v. The graph obtained from removing v and distributing its energy has only a constant number of components.

To prove this, Lemma 3.2 is crucial. We consider cones rooted at v with the following properties:

- The cones have a small angle α, meaning that for every cone C and every $x, y \in C$, we have $\angle(x, v, y) \le \alpha$. We choose $\alpha = \pi/6$.
- Every point in $[0,1]^d$ is covered by some cone.
- There is a finite number of cones. (This can be achieved because d is a constant.)

Let $C_1, \ldots, C_m$ be these cones. By abusing notation, let C_i also denote all points $x \in C_i \cap (X \cup Y \setminus \{v\})$ that are adjacent to v in T. For C_i, let x_i be the point in C_i that is closest to v and adjacent to v (breaking ties arbitrarily), and let y_i be the point in C_i that is farthest from v and adjacent to v (again breaking ties arbitrarily). (For completeness, we remark that then C_i can be ignored if $C_i \cap X = \emptyset$.) Let $\ell_i = \|y_i - v\|$ be the maximum distance of any point in C_i to v, and let $\ell = \max_i \ell_i$.

We increase the power of x_i by ℓ^p / m. Since the power of v is at least ℓ^p and we have m cones, we can account for this with v's power because we remove v. Because $\alpha = \pi/6$ and x_i is closest to v, any point in C_i is closer to x_i than to v. According to Lemma 3.2(a), every point in C_i has sufficient power to reach x_i. Thus, if x_i can reach a point $z \in C_i$, then there is an established connection between them.

From this and increasing x_i's power to at least ℓ^p / m, there is an edge between x_i and every point $z \in C_i$ that has a distance of at most $\ell / \sqrt[p]{m}$ from v. We recall that m and p are constants.

Now let $z_1, \ldots, z_k \in C_i$ be the vertices in C_i that are not connected to x_i because x_i has too little power. We assume that they are sorted by increasing distance from v. Thus, $z_k = y_i$. We can assume that no two z_j and $z_{j'}$ are in the same component after removal of v. Otherwise, we can simply ignore one of the edges $\{v, z_j\}$ and $\{v, z_{j'}\}$ without changing the components.

Since z_j and z_{j+1} were connected to v and they are not connected to each other, we can apply Lemma 3.2(b), which implies that $\|z_{j+1} - v\| \geq \sqrt{3} \cdot \|z_j - v\|$. Furthermore, $\|z_1 - v\| \geq \ell/\sqrt[d]{m}$ by assumption. Iterating this argument yields $\ell = \|z_k - v\| \geq \sqrt{3}^{k-1} \|z_1 - v\| \geq \sqrt{3}^{k-1} \cdot \ell/\sqrt[d]{m}$. This implies $k \leq \log_{\sqrt{3}}(\sqrt[d]{m}) + 1$. Thus, removing v and redistributing its energy as described causes the PA graph to fall into at most a constant number of components. Removing $|Y|$ points causes the PA graph to fall into at most $O(|Y|)$ components. These components can be connected with costs $O(|Y|^{\frac{d-p}{d}})$ by choosing one point per component and applying Lemma 3.5. □

Lemma 3.7. PA_B *is smooth for all* $1 \leq p \leq d$ *[16, (3.8)]*.

Crucial for convergence of PA is that PA, which is subadditive, and PA_B, which is superadditive, are close to each other. Then both are approximately both subadditive and superadditive. The following lemma states that indeed PA and PA_B do not differ too much for $1 \leq p < d$.

Lemma 3.8. PA *is point-wise close to* PA_B *for* $1 \leq p < d$ *[16, (3.10)], i.e.,*
$$\left| \mathsf{PA}^p(X) - \mathsf{PA}_B^p(X, [0,1]^d) \right| = o\!\left(n^{\frac{d-p}{d}}\right) \text{ for every set } X \subseteq [0,1]^d \text{ of } n \text{ points.}$$

3.3 Probabilistic Properties

For $p > d$, smoothness is not guaranteed to hold, and for $p \geq d$, point-wise closeness is not guaranteed to hold. But similar properties typically hold for random point sets, namely smoothness in mean (Definition 3.10) and closeness in mean (Definition 3.12). In the following, let $X = \{U_1, \ldots, U_n\}$. Recall that $U_1, \ldots, U_n$ are drawn uniformly and independently from $[0,1]^d$. We need the following bound on the longest edge of an optimal PA graph.

Lemma 3.9 (longest edge). *For every constant* $\beta > 0$, *there exists a constant* $c_{\text{edge}} = c_{\text{edge}}(\beta)$ *such that, with a probability of at least* $1 - n^{-\beta}$, *every edge of an optimal PA graph and an optimal boundary PA graph* PA_B *is of length at most* $r_{\text{edge}} = c_{\text{edge}} \cdot (\log n/n)^{1/d}$.

Yukich gave two different notions of smoothness in mean [16, (4.13) and (4.20) & (4.21)]. We use the stronger notion, which implies the other.

Definition 3.10 (smooth in mean [16, (4.20), (4.21)]). *A Euclidean functional* F *is called* smooth in mean *if, for every constant* $\beta > 0$, *there exists a constant* $c = c(\beta)$ *such that the following holds with a probability of at least* $1 - n^{-\beta}$:
$$\left| \mathsf{F}(n) - \mathsf{F}(n \pm k) \right| \leq ck \cdot \left(\tfrac{\log n}{n}\right)^{p/d} \quad \text{and} \quad \left| \mathsf{F}_B(n) - \mathsf{F}_B(n \pm k) \right| = ck \cdot \left(\tfrac{\log n}{n}\right)^{p/d}.$$
for all $0 \leq k \leq n/2$.

Lemma 3.11. PA_B *and* PA *are smooth in mean for all* $p > 0$ *and all* d.

Definition 3.12 (close in mean [16, (4.11)]). *A Euclidean functional* F *is* close in mean *to its boundary functional* F_B *if* $\mathbb{E}\left(|\mathsf{F}(n) - \mathsf{F}_B(n)|\right) = o(n^{\frac{d-p}{d}})$.

Lemma 3.13. PA *is close in mean to* PA_B *for all* d *and* $p \geq 1$.

4 Convergence

4.1 Standard Convergence

Our findings of Sections 3.2 yield complete convergence of PA for $p < d$ (Theorem 4.1). Together with the probabilistic properties of Section 3.3, we obtain convergence in mean in a straightforward way for all combinations of d and p (Theorem 4.2). In Sections 4.2 and 4.3, we prove complete convergence for $p \geq d$.

Theorem 4.1. *For all d and p with $1 \leq p < d$, there exists a constant $\gamma_{\mathsf{PA}}^{d,p}$ such that $\frac{\mathsf{PA}^p(n)}{n^{\frac{d-p}{d}}}$ converges completely to $\gamma_{\mathsf{PA}}^{d,p}$.*

Theorem 4.2. *For all $p \geq 1$ and $d \geq 1$, there exists a constant $\gamma_{\mathsf{PA}}^{d,p}$ such that $\lim_{n\to\infty} \frac{\mathbb{E}(\mathsf{PA}^p(n))}{n^{\frac{d-p}{d}}} = \lim_{n\to\infty} \frac{\mathbb{E}(\mathsf{PA}_B^p(n))}{n^{\frac{d-p}{d}}} = \gamma_{\mathsf{PA}}^{d,p}$.*

4.2 Concentration with Warnke's Inequality

McDiarmid's or Azuma-Hoeffding's inequality are powerful tools to prove concentration of measure for a function that depends on many independent random variables, all of which have only a bounded influence on the function value. If we consider smoothness in mean (see Lemma 3.11), then we have the situation that the influence of a single variable is typically very small (namely $O((\log n/n)^{p/d})$), but can be quite large in the worst case (namely $O(1)$). Unfortunately, this situation is not covered by McDiarmid's or Azuma-Hoeffding's inequality. Fortunately, Warnke [15] proved a generalization specifically for the case that the influence of single variables is typically bounded and fulfills a weaker bound in the worst case.

The following theorem is a simplified version (personal communication with Lutz Warnke) of Warnke's concentration inequality [15, Theorem 2], tailored to our needs.

Theorem 4.3 (Warnke). *Let $U_1, \ldots, U_n$ be a family of independent random variables with $U_i \in [0,1]^d$ for each i. Suppose that there are numbers $c_{\mathrm{good}} \leq c_{\mathrm{bad}}$ and an event Γ such that the function $\mathsf{F} : ([0,1]^d)^n \to \mathbb{R}$ satisfies*

$$\max_{i \in [n]} \max_{x \in [0,1]^d} |\mathsf{F}(U_1, \ldots, U_n) - \mathsf{F}(U_1, \ldots, U_{i-1}, x, U_{i+1}, \ldots, U_k)|$$

$$\leq \begin{cases} c_{\mathrm{good}} & \text{if } \Gamma \text{ holds and} \\ c_{\mathrm{bad}} & \text{otherwise.} \end{cases} \tag{1}$$

Then, for any $t \geq 0$ and $\gamma \in (0,1]$ and $\eta = \gamma(c_{\mathrm{bad}} - c_{\mathrm{good}})$, we have

$$\mathbb{P}\big(|\mathsf{F}(n) - \mathbb{E}(\mathsf{F}(n))| \geq t\big) \leq 2 \exp\big(-\frac{t^2}{2n(c_{\mathrm{good}}+\eta)^2}\big) + \frac{n}{\gamma} \cdot \mathbb{P}(\neg\Gamma). \tag{2}$$

Next, we introduce *typical smoothness*, which means that, with high probability, a single point does not have a significant influence on the value of F, and we apply Theorem 4.3 for typically smooth functionals F. The bound of $c \cdot (\log n/n)^{p/d}$ in Definition 4.4 below for the typical influence of a single point is somewhat arbitrary, but works for PA and MST. This bound is also essentially the smallest possible, as there can be regions of diameter $c' \cdot (\log n/n)^{1/d}$ for some small constant $c' > 0$ that contain no or only a single point. It might be possible to obtain convergence results for other functionals for weaker notions of typical smoothness.

Definition 4.4 (typically smooth). *A Euclidean functional* F *is typically smooth if, for every $\beta > 0$, there exists a constant $c = c(\beta)$ such that*

$$\max_{x \in [0,1]^d, i \in [n]} \left| F(U_1, \ldots, U_n) - F(U_1, \ldots, U_{i-1}, x, U_{i+1}, \ldots, U_n) \right| \leq c \cdot \left(\frac{\log n}{n} \right)^{p/d}$$

with a probability of at least $1 - n^{-\beta}$.

Theorem 4.5 (concentration of typically smooth functionals). *Let $p, d \geq 1$. Assume that* F *is typically smooth. Then*

$$\mathbb{P}\big(| F(n) - \mathbb{E}(F(n)) | \geq t \big) \leq O(n^{-\beta}) + \exp\big(-\tfrac{t^2 n^{\frac{2p}{d}-1}}{C(\log n)^{2p/d}}\big)$$

for an arbitrarily large constant $\beta > 0$ and another constant $C > 0$ that depends on β.

Choosing $t = n^{\frac{d-p}{d}} / \log n$ yields a nontrivial concentration result that suffices to prove complete convergence of typically smooth Euclidean functionals.

Corollary 4.6. *Let $p, d \geq 1$. Assume that* F *is typically smooth. Then*

$$\mathbb{P}\big(| F(n) - \mathbb{E}(F(n)) | > n^{\frac{d-p}{d}} / \log n \big) \leq O\big(n^{-\beta} + \exp\big(-\tfrac{n}{C(\log n)^{2+\frac{2p}{d}}}\big)\big) \qquad (3)$$

for any constant β and C depending on β as in Theorem 4.5.

4.3 Complete Convergence for $p \geq d$

In this section, we show that typical smoothness (Definition 4.4) suffices for complete convergence. This implies complete convergence of MST and PA by Lemma 4.8 below.

Theorem 4.7. *Let $p, d \geq 1$. Assume that* F *is typically smooth and $F(n)/n^{\frac{d-p}{d}}$ converges in mean to $\gamma_F^{d,p}$. Then $F(n)/n^{\frac{d-p}{d}}$ converges completely to $\gamma_F^{d,p}$.*

Although similar in flavor, smoothness in mean does not immediately imply typical smoothness or vice versa: the latter makes only a statement about *single* points at *worst-case* positions. The former only makes a statement about adding and removing *several* points at *random* positions. However, the proofs of smoothness in mean for MST and PA do not exploit this, and we can adapt them to yield typical smoothness.

Lemma 4.8. PA *and* MST *are typically smooth.*

Corollary 4.9. *For all d and p with $p \geq 1$,* $\mathsf{MST}(n)/n^{\frac{d-p}{d}}$ *and* $\mathsf{PA}(n)/n^{\frac{d-p}{d}}$ *converge completely to constants* $\gamma_{\mathsf{MST}}^{d,p}$ *and* $\gamma_{\mathsf{PA}}^{d,p}$, *respectively.*

5   Average-Case Ratio of the MST Heuristic

In this section, we show that the average-case approximation ratio of the MST heuristic for power assignments is strictly better than its worst-case ratio of 2. First, we prove that the average-case bound is strictly (albeit marginally) better than 2 for any combination of d and p. Second, we show a simple improved bound for the 1-dimensional case.

5.1   The General Case

The idea behind showing that the MST heuristic performs better on average than in the worst case is as follows: the weight of the PA graph obtained from the MST heuristic can not only be upper-bounded by twice the weight of an MST, but it is in fact easy to prove that it can be upper-bounded by twice the weight of the heavier half of the edges of the MST [6]. Thus, we only have to show that the lighter half of the edges of the MST contributes $\Omega(n^{\frac{d-p}{d}})$ to the value of the MST in expectation.

For simplicity, we assume that the number $n = 2m + 1$ of points is odd. The case of even n is similar but slightly more technical. We draw points $X = \{U_1, \ldots, U_n\}$ as described above. Let $\mathsf{PT}(X)$ denote the power required in the power assignment obtained from the MST. Furthermore, let H denote the m heaviest edges of the MST, and let L denote the m lightest edges of the MST. We omit the parameter X since it is clear from the context. Then we have

$$\mathsf{H} + \mathsf{L} = \mathsf{MST} \leq \mathsf{PA} \leq \mathsf{PT} \leq 2\,\mathsf{H} = 2\,\mathsf{MST} - 2\,\mathsf{L} \leq 2\,\mathsf{MST} \qquad (4)$$

since the weight of the PA graph obtained from an MST can not only be upper bounded by twice the weight of a minimum-weight spanning tree, but it is easy to show that the PA graph obtained from the MST is in fact by twice the weight of the heavier half of the edges of a minimum-weight spanning tree [6]. We can show that $\mathbb{E}(\mathsf{L}) = \Omega(n^{\frac{d-p}{d}})$. This yields the following result.

Theorem 5.1. *For any $d \geq 1$ and any $p \geq 1$, we have*

$$\gamma_{\mathsf{MST}}^{d,p} \leq \gamma_{\mathsf{PA}}^{d,p} \leq 2(\gamma_{\mathsf{MST}}^{d,p} - C) < 2\gamma_{\mathsf{MST}}^{d,p}$$

for some constant $C > 0$ that depends only on d and p.

By exploiting that PA converges completely, we can obtain a bound on the expected approximation ratio from the above result.

Corollary 5.2. *For any $d \geq 1$ and $p \geq 1$ and sufficiently large n, the expected approximation ratio of the MST heuristic for power assignments is bounded from above by a constant strictly smaller than 2.*

5.2   An Improved Bound for the One-Dimensional Case

The case $d = 1$ is much simpler than the general case, because the MST is just a Hamiltonian path starting at the left-most and ending at the right-most point. Furthermore, we also know precisely what the MST heuristic does: assume that a point x_i lies between x_{i-1} and x_{i+1}. The MST heuristic assigns power $\mathsf{PA}(x_i) = \max\{|x_i - x_{i-1}|, |x_i - x_{i+1}|\}^p$ to x_i. The example that proves that the MST heuristic is no better than a worst-case 2-approximation shows that it is bad if x_i is very close to either side and good if x_i is approximately in the middle between x_{i-1} and x_{i+1}. By analyzing $\gamma_{\mathsf{MST}}^{1,p}$ and $\gamma_{\mathsf{PA}}^{1,p}$ carefully, we obtain the following theorem.

Theorem 5.3. *For all $p \geq 1$, we have $\gamma_{\mathsf{MST}}^{1,p} \leq \gamma_{\mathsf{PA}}^{1,p} \leq (2 - 2^{-p}) \cdot \gamma_{\mathsf{MST}}^{1,p}$.*

The high probability bounds for the bound of $2 - 2^{-p}$ of the approximation ratio of the power assignment obtained from the spanning tree together with the observation that in case of any "failure" event we can use the worst-case approximation ratio of 2 yields the following corollary.

Corollary 5.4. *The expected approximation ratio of the MST heuristic is at most $2 - 2^{-p} + o(1)$.*

6   Conclusions and Open Problems

We have proved complete convergence of Euclidean functionals that are *typically smooth* (Definition 4.4) for the case that the distance-power gradient p is larger than the dimension d. The case $p > d$ appears naturally in the case of transmission questions for wireless networks. As examples, we have obtained complete convergence for the MST and the PA functional. To prove this, we have used a recent concentration of measure result by Warnke [15]. His concentration inequality might be of independent interest to the algorithms community. As a technical challenge, we have had to deal with the fact that the degree of an optimal power assignment graph can be unbounded.

To conclude this paper, let us mention some problems for further research:

1. Is it possible to prove complete convergence of other functionals for $p \geq d$? The most prominent one would be the traveling salesman problem (TSP).
2. Is it possible to prove improved bounds on the approximation ratio of the MST heuristic?
3. Can our findings about power assignments be generalized to other problems in wireless communication, such as the k-station network coverage problem of Funke et al. [5], where transmit powers are assigned to at most k stations such that X can be reached from at least one sender, or power assignments in the SINR model [7,9]? Interestingly, in the SINR model the MST turns out to be a good solution to schedule all links within a short time [8,10]. More general, can this framework also be exploited to analyze other approximation algorithms for geometric optimization problems? As far as we are aware, besides partitioning heuristics [2,16], the only other algorithm analyzed within this framework is Christofides' algorithm for the TSP [3].

References

1. Althaus, E., Calinescu, G., Mandoiu, I.I., Prasad, S.K., Tchervenski, N., Zelikovsky, A.: Power efficient range assignment for symmetric connectivity in static ad hoc wireless networks. Wireless Networks 12(3), 287–299 (2006)
2. Bläser, M., Manthey, B., Rao, B.V.R.: Smoothed analysis of partitioning algorithms for Euclidean functionals. Algorithmica 66(2), 397–418 (2013)
3. Bläser, M., Panagiotou, K., Rao, B.V.R.: A probabilistic analysis of Christofides' algorithm. In: Fomin, F.V., Kaski, P. (eds.) SWAT 2012. LNCS, vol. 7357, pp. 225–236. Springer, Heidelberg (2012)
4. Clementi, A.E.F., Penna, P., Silvestri, R.: On the power assignment problem in radio networks. Mobile Networks and Applications 9(2), 125–140 (2004)
5. Funke, S., Laue, S., Lotker, Z., Naujoks, R.: Power assignment problems in wireless communication: Covering points by disks, reaching few receivers quickly, and energy-efficient travelling salesman tours. Ad Hoc Networks 9(6), 1028–1035 (2011)
6. de Graaf, M., Boucherie, R.J., Hurink, J.L., van Ommeren, J.K.: An average case analysis of the minimum spanning tree heuristic for the range assignment problem. Memorandum 11259 (revised version), Department of Applied Mathematics, University of Twente (2013)
7. Halldórsson, M.M., Holzer, S., Mitra, P., Wattenhofer, R.: The power of non-uniform wireless power. In: Proc. of the 24th Ann. ACM-SIAM Symp. on Discrete Algorithms (SODA), pp. 1595–1606. SIAM (2013)
8. Halldórsson, M.M., Mitra, P.: Wireless connectivity and capacity. In: Rabani, Y. (ed.) Proc. of the 23rd Ann. ACM-SIAM Symp. on Discrete Algorithms (SODA), pp. 516–526. SIAM (2012)
9. Kesselheim, T.: A constant-factor approximation for wireless capacity maximization with power control in the SINR model. In: Proc. of the 22nd Ann. ACM-SIAM Symp. on Discrete Algorithms (SODA), pp. 1549–1559. SIAM (2011)
10. Khan, M., Pandurangan, G., Pei, G., Vullikanti, A.K.S.: Brief announcement: A fast distributed approximation algorithm for minimum spanning trees in the SINR model. In: Aguilera, M.K. (ed.) DISC 2012. LNCS, vol. 7611, pp. 409–410. Springer, Heidelberg (2012)
11. Kirousis, L.M., Kranakis, E., Krizanc, D., Pelc, A.: Power consumption in packet radio networks. Theoretical Computer Science 243(1-2), 289–305 (2000)
12. Lloyd, E.L., Liu, R., Marathe, M.V., Ramanathan, R., Ravi, S.S.: Algorithmic aspects of topology control problems for ad hoc networks. Mobile Networks and Applications 10(1-2), 19–34 (2005)
13. Pahlavan, K., Levesque, A.H.: Wireless Information Networks. Wiley (1995)
14. Rappaport, T.S.: Wireless Communication. Prentice Hall (2002)
15. Warnke, L.: On the method of typical bounded differences. Computing Research Repository 1212.5796 [math.CO], arXiv (2012)
16. Yukich, J.E.: Probability Theory of Classical Euclidean Optimization Problems. Lecture Notes in Mathematics, vol. 1675. Springer, Heidelberg (1998)

Existence of Secure Equilibrium
in Multi-player Games with Perfect Information

Julie De Pril[1,*], János Flesch[2], Jeroen Kuipers[3],
Gijs Schoenmakers[3], and Koos Vrieze[3]

[1] Département de Mathématique, Université de Mons, Belgium
[2] Department of Quantitative Economics, Maastricht University, The Netherlands
[3] Department of Knowledge Engineering, Maastricht University, The Netherlands

Abstract. A secure equilibrium is a refinement of Nash equilibrium, which provides some security to the players against deviations when a player changes his strategy to another best response strategy. The concept of secure equilibrium is specifically developed for assume-guarantee synthesis and has already been applied in this context. Yet, not much is known about its existence in games with more than two players. In this paper, we establish the existence of secure equilibrium in two classes of multi-player perfect information turn-based games: (1) in games with possibly probabilistic transitions, having countable state and finite action spaces and bounded and continuous payoff functions, and (2) in games with only deterministic transitions, having arbitrary state and action spaces and Borel payoff functions with a finite range (in particular, qualitative Borel payoff functions). We show that these results apply to several types of games studied in the literature.

1 Introduction

The Game: We examine multi-player perfect information turn-based games with possibly probabilistic transitions. In such a game, each state is associated with a player, who controls this state. Play of the game starts at the initial state. At every state that play visits, the player who controls this state has to choose an action from a given action space. Next, play moves to a new state according to a probability measure, which may depend on the current state and the chosen action. This induces an infinite sequence of states and actions, and depending on this play, each player receives a payoff. These payoffs can be fairly general. For example, they could arise as some aggregation of instantaneous rewards that the players receive at the periods of the game. A frequently used aggregation would be taking the discounted sum of the instantaneous rewards. As another example, the payoffs could represent reachability objectives, then a player's payoff would be either 1 or 0 depending on whether a certain set of states is reached.

Two-player zero-sum games with possibly probabilistic transitions have been applied in the model-checking of reactive systems where randomness occurs,

---

* F.R.S.-FNRS postdoctoral researcher. Work partially supported by the European project CASSTING (FP7-ICT-601148).

E. Csuhaj-Varjú et al. (Eds.): MFCS 2014, Part II, LNCS 8635, pp. 213–225, 2014.

because they allow to model the interactions between a system and its environment. However, complex systems are usually made up of several components with objectives that are not necessarily antagonistic, that is why multi-player non zero-sum games are better suited in such cases.

Nash Equilibrium: In these games, Nash equilibrium is a prominent solution concept. A Nash equilibrium is a strategy profile such that no player can improve his payoff by individually deviating to another strategy. In various classes of perfect information games, a Nash equilibrium is known to exist [9,10,13].

Secure Equilibrium: Despite the obvious appeal of Nash equilibrium, certain applications call for additional properties. Chatterjee et al [4] introduced the concept of secure equilibrium, which they specifically designed for assume-guarantee synthesis. They gave a definition of secure equilibrium [4, Definition 8] in qualitative n-player games[1], and then a characterization [4, Proposition 4], which however turns out not to be equivalent. With their definition, such kind of equilibrium may fail to exist even in very simple games (cf. [7, Remark 1] or [6, Example 2.2.34]). That is why we choose to call a *strongly secure equilibrium* an equilibrium according to [4, Definition 8] (see Remark 1), and we choose to call, as it has already been done in [1], a *secure equilibrium* an equilibrium according to the alternative characterization given in [4, Proposition 4], extended to the quantitative framework. Note that, with these definitions, every secure equilibrium is automatically strongly secure if there are only two players.

Thus, a strategy profile is a secure equilibrium if it is a Nash equilibrium and moreover the following security property holds: if any player individually deviates to another best response strategy, then it cannot be the case that all opponents are weakly worse off due to this deviation and at least one opponent is even strictly hurt. For applications of secure equilibrium, we refer to [4,3,5].

Only little is known about the existence of secure equilibrium. To our knowledge, the available existence results are for games with only two players[2] and for only deterministic transitions. Chatterjee et al [4] proved the existence of a secure equilibrium in two-player games in which the payoff function of each player is the indicator function of a Borel subset of plays. Recently, the existence of a secure equilibrium, even a subgame-perfect secure equilibrium, has been shown [1] in two-player games in which each player's goal is to reach a certain set of states and his payoff is determined by the number of moves it takes to get there. Very recently, the existence of a secure equilibrium has been proved [2] in a class of two-player quantitative games which includes payoff functions like sup, inf, lim sup, lim inf, mean-payoff, and discounted sum.

Our Contribution: We address the existence problem of secure equilibrium for multi-player perfect information games. We establish the existence of secure equilibrium in two classes of such games. First, when probabilistic transitions

---

[1] Note that Chatterjee et al [4] gave no existence result in the n-player case.
[2] However, in [3], the existence of secure equilibria is proved in the special case of 3-player qualitative games where the third player can win unconditionally.

are allowed[3], we prove that a secure equilibrium exists, provided that the state space is countable, the action spaces are finite, and the payoff functions are bounded and continuous. To our knowledge, it is the first existence result of secure equilibria in *multi-player quantitative* games. Second, for games with only deterministic transitions, we prove that a secure equilibrium exists if the payoff functions are Borel measurable and have a finite range (in particular, for Borel qualitative objectives). For the latter result, we impose no restriction on the state and action spaces. We show that these results apply to several classes of games studied in the literature. Regarding proof techniques, the proof of the first result relies on an inductive procedure that removes certain actions of the game, while the proof of the second result exploits a transformation of the payoffs.

Structure of the Paper: Section 2 is dedicated to the model. Section 3 presents the main results and mentions some classes of games to which the results apply. Sections 4 and 5 contain the formal proofs of the results (the proofs of the intermediary lemmas can be found in [7]). Finally, Section 6 concludes with some remarks and an algorithmic result for quantitative reachability objectives.

2   The Model

We distinguish two types of perfect information games: games that may contain probabilistic transitions, and games that only has deterministic transitions.

Games with Probabilistic Transitions. A multi-player perfect information game with possibly probabilistic transitions is given by:

1. A finite set of players N, with $|N| \geq 2$.
2. A countable state space S, containing an initial state $\tilde{s}$.
3. A controlling player $i(s) \in N$ for every state $s \in S$.
4. A nonempty and finite action space $A(s)$ for every state $s \in S$.
5. A probability measure $q(s, a)$ for every state $s \in S$ and action $a \in A(s)$, which assigns, to every $z \in S(s, a)$ (the set of possible successor states when choosing action a in state s), the probability $q(s, a)(z)$ of transition from state s to state z under action a.
   Let $\mathbb{N} = \{0, 1, 2, \ldots\}$. Let $\mathcal{H}$ be the set of all sequences of the form $(s^0, a^0, \ldots, s^{n-1}, a^{n-1}, s^n)$, where $n \in \mathbb{N}$, such that $s^0 = \tilde{s}$, and $a^m \in A(s^m)$ and $s^{m+1} \in S(s^m, a^m)$ for all $m = 0, 1, \ldots, n-1$. Let $\mathcal{P}$ be the set of all infinite sequences of the form $(s^m, a^m)_{m \in \mathbb{N}}$ such that $s^0 = \tilde{s}$, and $a^m \in A(s^m)$ and $s^{m+1} \in S(s^m, a^m)$ for all $m \in \mathbb{N}$. The elements of $\mathcal{H}$ are called *histories* and the elements of $\mathcal{P}$ are called *plays*. We endow $\mathcal{P}$ with the topology induced by the cylinder sets $\mathcal{C}(h) = \{p \in \mathcal{P} | \ p \text{ starts with } h\}$ for $h \in \mathcal{H}$. In this topology, a sequence of plays $(p_m)_{m \in \mathbb{N}}$ converges to a play p precisely when for every $k \in \mathbb{N}$ there exists an $N_k \in \mathbb{N}$ such that p_m coincides with p on the first k coordinates for every $m \geq N_k$.
6. A payoff function $u_i : \mathcal{P} \to \mathbb{R}$ for every player $i \in N$, which is bounded and Borel measurable.

---

[3] Such games are also called *turn-based multi-player stochastic games*.

The game is played as follows at periods in $\mathbb{N} = \{0, 1, 2, \ldots\}$. Play starts at period 0 in state $s^0 = \tilde{s}$, where the controlling player $i(s^0)$ chooses an action a^0 from $A(s^0)$. Then, transition occurs according to the probability measure $q(s^0, a^0)$ to a state s^1. At period 1, the controlling player $i(s^1)$ chooses an action a^1 from $A(s^1)$. Then, transition occurs according to the probability measure $q(s^1, a^1)$ to a state s^2, and so on. The realization of this process is a play $p = (s^0, a^0, s^1, a^1, s^2, \ldots)$, and each player $i \in N$ receives payoff $u_i(p)$.

We can assume w.l.o.g. that the sets $S(s, a)$, for $s \in S$ and $a \in A(s)$, are mutually disjoint, and their union is S. This means that each state can be visited in exactly one way from the initial state $\tilde{s}$, so there is a bijection between states and histories. For this reason, we will work with states instead of histories.

Strategies: For every player $i \in N$, let S_i be the set of those states (histories) which are controlled by him. A strategy for a player $i \in N$ is a function σ_i that assigns an action $\sigma_i(s) \in A(s)$ to every state $s \in S_i$. The interpretation is that $\sigma_i(s)$ is the recommended action if state s is reached. A strategy profile is a tuple $(\sigma_1, \ldots, \sigma_{|N|})$ where σ_i is a strategy for every player i. Given a strategy profile $\sigma = (\sigma_1, \ldots, \sigma_{|N|})$ and a player i, we denote by σ_{-i} the profile of strategies of player i's opponents, i.e. $\sigma_{-i} = (\sigma_j)_{j \in N, j \neq i}$. A strategy profile σ induces a unique probability measure on the sigma-algebra of the Borel sets of $\mathcal{P}$. The corresponding expected payoff for player i is denoted by $u_i(\sigma)$.

Nash Equilibrium: A strategy profile σ^* is called a Nash equilibrium, if no player can improve his expected payoff by a unilateral deviation, i.e. $u_i(\tau_i, \sigma^*_{-i}) \leq u_i(\sigma^*)$ for every player $i \in N$ and every strategy τ_i for player i. In other words, every player plays a best response to the strategies of his opponents.

Secure Equilibrium: A strategy profile σ^* is called a secure equilibrium, if it is a Nash equilibrium and if, additionally, no player $i \in N$ has a strategy τ_i with $u_i(\tau_i, \sigma^*_{-i}) = u_i(\sigma^*)$ such that we have for all players $j \in N \setminus \{i\}$ that $u_j(\tau_i, \sigma^*_{-i}) \leq u_j(\sigma^*)$ and for some player $k \in N \setminus \{i\}$ that $u_k(\tau_i, \sigma^*_{-i}) < u_k(\sigma^*)$.

The interpretation of the additional property is the following. Consider a Nash equilibrium σ^* and a strategy τ_i for some player i. By deviating to τ_i, player i either receives a worse expected payoff than with his original strategy σ^*_i, or the same expected payoff at best. In the former case, it is not in player i's interest to deviate to τ_i. In the latter case, however, even though player i is indifferent, his opponents could get hurt. The property thus prevents the case that this deviation is weakly worse for all opponents of player i, and that this deviation is even hurting a player. So, in a certain sense, player is opponents are secure against such deviations by player i.

An equivalent formulation of secure equilibrium is the following: a strategy profile σ^* is called a secure equilibrium, if it is a Nash equilibrium and if, additionally, the following property holds for every player $i \in N$: if τ_i is a strategy for player i such that $u_i(\tau_i, \sigma^*_{-i}) = u_i(\sigma^*)$ and $u_j(\tau_i, \sigma^*_{-i}) < u_j(\sigma^*)$ for some player $j \in N$, then there is a player $k \in N$ such that $u_k(\tau_i, \sigma^*_{-i}) > u_k(\sigma^*)$.

Remark 1. There is a refinement of secure equilibrium, which plays an important role in the proofs of our main results, Theorems 1 and 2. We call a strategy profile

σ^* a *sum-secure equilibrium*, if it is a Nash equilibrium and if additionally the following property holds for every player $i \in N$: if τ_i is a strategy for player i such that $u_i(\tau_i, \sigma^*_{-i}) = u_i(\sigma^*)$ then $\sum_{j \in N \setminus \{i\}} u_j(\tau_i, \sigma^*_{-i}) \geq \sum_{j \in N \setminus \{i\}} u_j(\sigma^*)$. In fact, in Theorems 1 and 2, we prove the existence of sum-secure equilibria.

An even stronger concept is the following. A strategy profile σ^* is called a *strongly secure equilibrium*, if it is a Nash equilibrium and if additionally the following property holds for every player $i \in N$: if τ_i is a strategy for player i such that $u_i(\tau_i, \sigma^*_{-i}) = u_i(\sigma^*)$ then $u_j(\tau_i, \sigma^*_{-i}) \geq u_j(\sigma^*)$ for all $j \neq i$. Note that every secure equilibrium is also strongly secure if there are only two players. The concept of strongly secure equilibrium has the serious drawback though that it fails to exist even in very simple games (cf. [7, Remark 1] or [6, Example 2.2.34]).

Games with Deterministic Transitions. Another type of perfect information games arises when the game has only deterministic transitions, i.e. when the set $S(s, a)$ is a singleton for every state $s \in S$ and every action $a \in A(s)$. In this case, we do not need to take care of measurability conditions for the calculation of expected payoffs. Hence, we can drop the assumptions that the state space is countable and the action spaces are finite, and they can be arbitrary.

3 The Main Results

In this section, we present and discuss our main results for the existence of secure equilibrium. First we examine the case of probabilistic transitions.

Theorem 1. *Take a perfect information game, possibly having probabilistic transitions, with countable state and finite action spaces. If every player's payoff function is bounded and continuous[4], then the game admits a secure equilibrium.*

By assuming that the state space is countable, we avoid measure theoretic complications. Without the hypotheses that the action spaces are finite and the payoff functions are continuous, even a Nash equilibrium may fail to exist (cf. [7, Examples 1 and 2]).

For games with only deterministic transitions, we obtain the following result.

Theorem 2. *Take a perfect information game with deterministic transitions, with arbitrary state and action spaces. If every player's payoff function is Borel measurable and has a finite range, then the game admits a secure equilibrium.*

Theorem 2 assumes that the range of the payoff functions is finite. This assumption is only useful for deterministic transitions, otherwise the range of the expected payoffs may become infinite. Without this assumption, even a Nash equilibrium can fail to exist (cf. [7, Example 1]).

The methods of proving Theorems 1 and 2 are different. The proof of Theorem 1 uses an inductive procedure that eliminates certain actions in certain states. This procedure terminates with a game, in which one can identify an

---

[4] Notice that any continuous function is Borel measurable.

interesting strategy profile that can be enhanced with punishment strategies to be a secure equilibrium in the original game. The proof of Theorem 2 relies on a transformation of the payoffs. In the new game, a Nash equilibrium exists, and it is a secure equilibrium of the original game.

The above two theorems apply to various classes of games that have been studied in the literature. We mention a number of them. We only discuss the assumptions imposed on the payoff functions, as it is clear when a game satisfies the rest of the assumptions:

1. In *games with a finite horizon*, the payoff functions are continuous, so Theorem 1 is applicable to such games if the other hypotheses are satisfied. The same observation holds for *discounted games*, where the players aggregate instantaneous payoffs by taking the discounted sum.
2. All qualitative payoff functions have a finite range, so Theorem 2 directly implies the existence of a secure equilibrium in multi-player games with deterministic transitions and *qualitative Borel objectives* (in particular, objectives like reachability, safety, (co–)Büchi, parity,...).
3. Now consider a game played on a graph with quantitative payoff functions. For *quantitative reachability objectives*, by [1, Remark 2.5], we can use Theorem 1 to find a secure equilibrium as long as the transitions are deterministic (the transformation drastically changes expected payoffs). For *quantitative safety objectives*, continuous payoff functions can also be defined: if $F_i \subseteq S$ is the safety set of player i, let $u_i(p) = 1 - \frac{1}{n+1}$ for any play p if, along p, the set $S \setminus F_i$ is reached at period n for the first time and let $u_i(p) = 1$ if $S \setminus F_i$ is never reached along p. And in a similar way for *quantitative Büchi objectives*: if $B_i \subseteq S$ is the Büchi set of player i, let $M(p)$ denote, for any play p, the set of periods at which the play p is in a state in B_i. Then, define $u_i(p) = \sum_{k \in M(p)} \frac{1}{2^k}$.
4. Theorem 2 implies the existence of a secure equilibrium in games played on *finite weighted*[5] graphs with deterministic transitions, where the payoff functions are computed as the sup, inf, lim sup or lim inf of the weights appearing along plays (as in [2]). Indeed, these functions have finite range in such games.

4   The Proof of Theorem 1

In this section, we provide a formal proof of Theorem 1. Consider a perfect information game G, having possibly probabilistic transitions, with a countable state and finite action spaces. Assume that the payoff function of every player is bounded and continuous.

Preliminaries. We keep the game G fixed. We introduce some notation, define some preliminary notions and state some properties of them.

For every player $i \in N$, we use the notation Σ_i for the set of strategies of player i, so $\Sigma_i = \times_{s \in S_i} A(s)$. We endow Σ_i with the product topology $\mathcal{T}_i$. Since the set

---

[5] Each edge of the graph is labelled by a $|N|$-tuple of real values.

S_i is countable and the action spaces are finite, the topological space $(\Sigma_i, \mathcal{T}_i)$ is compact and metrizable. Hence, the set of strategy profiles $\Sigma = \times_{i \in N} \Sigma_i$, endowed with the product topology $\mathcal{T} = \times_{i \in N} \mathcal{T}_i$, is also compact and metrizable. So, these spaces are sequentially compact, meaning that every sequence in them has a convergent subsequence. This is one of the main consequences of assuming that the action spaces are finite.

We assumed that the payoffs are continuous on the set of plays. The next lemma states that the expected payoffs are continuous as well.

Lemma 1. *For all $i \in N$, the expected payoff function $u_i : \Sigma \to \mathbb{R}$ is continuous.*

Given a state $s \in S$, we define the subgame $G(s)$ as the game that arises when state s is reached (i.e. past play has followed the unique history from the initial state $\tilde{s}$ to s). Every strategy profile σ induces a strategy profile in $G(s)$, as well as an expected payoff for every player i, which we denote by $u_i(\sigma|s)$.

For every player $i \in N$, we derive a zero-sum perfect information game G_i from G by making the following modification. There are two players: player i and an imaginary player $-i$, who replaces the set of opponents $N \setminus \{i\}$ of player i. So, whenever a state is reached controlled by a player from $N \setminus \{i\}$, player $-i$ can choose the action. Player i tries to maximize his expected payoff given by u_i, whereas player $-i$ tries to minimize this expected payoff, so $u_{-i} = -u_i$. We say that this zero-sum game G_i has a value, denoted by v_i, if player i has a strategy σ_i^* and his opponents have a strategy profile σ_{-i}^*, which is thus a strategy for player $-i$, such that $u_i(\sigma_i^*, \tau_{-i}) \geq v_i$ for every strategy profile τ_{-i} and $u_i(\tau_i, \sigma_{-i}^*) \leq v_i$ for every strategy τ_i for player i. This means that σ_i^* guarantees for player i that he receives an expected payoff of at least v_i, and σ_{-i}^* guarantees that player i does not receive an expected payoff of more than v_i. We call the strategy σ_i^* optimal for player i and the strategy profile σ_{-i}^* optimal for player i's opponents. Note that $(\sigma_i^*, \sigma_{-i}^*)$ forms a Nash equilibrium in G_i, and vice versa, if (τ_i, τ_{-i}) is a Nash equilibrium in G_i, then τ_i and τ_{-i} are optimal. In a similar way, we can also speak of the value of the subgame $G_i(s)$ of G_i, for a state $s \in S$, which we denote by $v_i(s)$.

The next lemma states that, for every player $i \in N$ and state $s \in S$, the game $G_i(s)$ admits a value. Moreover, one can find a strategy for player i that induces an optimal strategy in every game $G_i(s)$, for $s \in S$, and player i's opponents have a similar strategy profile. This follows from the existence of a subgame-perfect Nash equilibrium in our setting (where subgame-perfect Nash equilibrium refers to a strategy profile that induces a Nash equilibrium in every subgame), see [11, Theorem 1.2]. It is also an easy extension of [9, Corollary 4.2], with almost the same proof (based on approximations of the game by finite horizon truncations).

Lemma 2. *Take $i \in N$. The value $v_i(s)$ of the game $G_i(s)$ exists for all $s \in S$. Moreover, player i has a strategy σ_i^* such that σ_i^* induces an optimal strategy in $G_i(s)$, for all $s \in S$. Similarly, the opponents of player i have a strategy profile σ_{-i}^* such that σ_{-i}^* induces an optimal strategy profile in $G_i(s)$, for all $s \in S$.*

For every player $i \in N$, every state $s \in S$ and every action $a \in A(s)$, define

$$v_i(s, a) = \sum_{z \in S} q(s, a)(z) \cdot v_i(z).$$

This is in expectation the value for player i in the subgame of G that arises if player $i(s)$ chooses action a in state s. Obviously, for the controlling player $i(s)$ we have

$$v_{i(s)}(s) = \max_{a \in A(s)} v_{i(s)}(s, a). \tag{1}$$

Here, the maximum is attained due to the finiteness of $A(s)$. Let us call an action $a \in A(s)$ optimal in state s if $v_{i(s)}(s) = v_{i(s)}(s, a)$. We have for every $j \in N \setminus \{i(s)\}$ that

$$v_j(s) = \min_{a \in A(s)} v_j(s, a). \tag{2}$$

Towards a Restricted Game with Only Optimal Actions. We now define a procedure that inductively eliminates all actions that are not optimal and terminates with a specific game G^∞, in which all actions are optimal.

Take a nonempty set $A'(s) \subseteq A(s)$ for every state $s \in S$. The sets $A'(s)$, for $s \in S$, induce a game G' that we derive from G as follows: the set of states S' of G' consists of those states $z \in S$ for which the unique history that starts at $\tilde{s}$ and ends at z only uses actions in the sets $A'(s)$, for $s \in S$. These are the states that a play can visit, with possibly probability zero, when the actions are restricted to the sets $A'(s)$, for $s \in S$. The action space of G' in every state $s \in S'$ is then $A'(s)$. Further, the payoff functions of G' are obtained by restricting the payoff functions of G to plays corresponding to these new state and action spaces.

Let G^0 be the game G, let $S^0 = S$ and let $A^0(s) = A(s)$ for every $s \in S$. Then, at every state $s \in S$, we delete all actions that are not optimal. This results in a nonempty action space $A^1(s) \subseteq A^0(s)$ for every $s \in S$. Let G^1 denote the induced game, with state space S^1. In the next step, at every state $s \in S^1$, we delete all actions from $A^1(s)$ that are not optimal in G^1. This gives a nonempty action space $A^2(s) \subseteq A^1(s)$ for every $s \in S^1$. Let G^2 denote the induced game, with state space S^2. By proceeding this way, we obtain for each $k \in \mathbb{N}$ a game G^k with state space S^k and nonempty action spaces $A^k(s)$, for $s \in S^k$. Finally, let $S^\infty = \cap_{k \in \mathbb{N}} S^k$ and $A^\infty(s) = \cap_{k \in \mathbb{N}} A^k(s)$ for every $s \in S^\infty$. Note that the initial state $\tilde{s}$ belongs to S^∞, and also that, due to the finiteness of the action spaces, the sets $A^\infty(s)$, for $s \in S^\infty$, are nonempty. Let G^∞ be the game induced by the sets $A^\infty(s)$, for $s \in S^\infty$. It is clear that the state space of G^∞ is S^∞.

We now define a function $\phi : S \to \mathbb{N} \cup \{\infty\}$. Note that every state $s \in S$ belongs either to $S^k \setminus S^{k+1}$ for a unique $k \in \mathbb{N}$ or to S^∞. In the former case, we define $\phi(s) = k$, whereas in the latter case we define $\phi(s) = \infty$. So, this is the latest iteration in which state s is still included. For notational convenience, we extend the strategy space of player i in G^k, for $k \in \mathbb{N} \cup \{\infty\}$, with strategies σ_i for player i in G with the following property: for every $s \in S$, if $\phi(s) \geq k$ then $\sigma_i(s) \in A^k(s)$, whereas if $\phi(s) < k$ then $\sigma_i(s)$ is an arbitrary action in $A^{\phi(s)}(s)$. By doing so, every strategy in G^k is also a strategy in G^m if $m \leq k$.

Now we consider the subgames of the above defined games. For every $k \in \mathbb{N} \cup \{\infty\}$ and player $i \in N$, by Lemma 2, the game $G_i^k(s)$ has a value $v_i^k(s)$ for every $s \in S$. Moreover, player i has a strategy σ_i^k such that σ_i^k induces an optimal strategy in the subgame $G_i^k(s)$ for every $s \in S^k$. Similarly, the opponents of player i have a strategy profile σ_{-i}^k such that σ_{-i}^k induces an optimal strategy profile in the subgame $G_i^k(s)$ for every $s \in S^k$. Note that σ_i^k can only use optimal actions in G^k, so by construction, σ_i^k is also a strategy for player i in G^{k+1}.

Lemma 3. 1. $v_i^k(s) \leq v_i^{k+1}(s)$ for every $k \in \mathbb{N}$, $i \in N$, and $s \in S^{k+1}$.
2. $\lim_{k \to \infty} v_i^k(s) = v_i^\infty(s)$ for every $i \in N$ and $s \in S^\infty$.

The next lemma shows that there is no need to continue the elimination procedure in a transfinite way.

Lemma 4. Every action is optimal in G^∞, i.e. for all $s \in S^\infty$ and $a \in A^\infty(s)$, we have for the controlling player $i(s)$ that $v_{i(s)}^\infty(s) = v_{i(s)}^\infty(s, a)$. Moreover, if τ is a strategy profile in G^∞, then $u_i(\tau|s) \geq v_i^\infty(s)$ for all $i \in N$ and $s \in S^\infty$.

The Secure Equilibrium. Thanks to the properties of G^∞, we identify an interesting strategy profile in G^∞, which can be enhanced with punishment strategies to become a secure equilibrium in G.

Recall that σ_{-i} and respectively σ_{-i}^∞ are strategy profiles for player i's opponents in G and respectively in G^∞, such that they induce an optimal strategy profile in $G_i(s)$ and respectively $G_i^\infty(s')$, for all $s \in S$ and $s' \in S^\infty$. For each $j \in N \setminus \{i\}$, let $\sigma_{-i,j}$ and respectively $\sigma_{-i,j}^\infty$ be player j's strategy in these strategy profiles. These strategies will play the role of punishing player i if he deviates.

Let ρ^* be a strategy profile in G^∞ which minimizes the sum of the expected payoffs of all the players, i.e. for any strategy profile ρ in G^∞ we have $\sum_{i \in N} u_i(\rho^*) \leq \sum_{i \in N} u_i(\rho)$. Such a strategy profile exists due to the compactness of the set of strategy profiles in G^∞ and due to Lemma 1. The idea is to define a secure equilibrium in G in the following way: follow ρ^*, unless a deviation occurs. If player i deviates, then his opponents should punish player i with σ_{-i}^∞ as long as he chooses actions in G^∞, and they should punish him with σ_{-i} as soon as he chooses an action out of G^∞. Let us specify this strategy profile.

We first define a function L, which assigns to each state the player who has to be punished from this state, or $\perp$ if nobody has to be punished. The idea is to remember the first player who deviated from the strategy profile ρ^*. For the initial state $\tilde{s}$, we have $L(\tilde{s}) = \perp$. For other states, we define it by induction. Suppose that we have defined $L(s)$ for some state $s \in S$. Then, for a state $s' \in S(s, a)$, where $a \in A(s)$, we set:

$$
L(s') := \begin{cases} \perp & \text{if } L(s) = \perp \text{ and } \rho_{i(s)}^*(s) = a, \\ i(s) & \text{if } L(s) = \perp \text{ and } \rho_{i(s)}^*(s) \in A(s) \setminus \{a\}, \\ L(s) & \text{otherwise (i.e. when } L(s) \neq \perp). \end{cases}
$$

Now we define a strategy τ_j for any player $j \in N$ as follows: for any $s \in S_j$, let

$$\tau_j(s) := \begin{cases} \rho_j^*(s) & \text{if } L(s) = \bot, \\ \text{arbitrary action in } A^\infty(s) & \text{if } L(s) = j \text{ and } s \in S^\infty, \\ \text{arbitrary action in } A(s) & \text{if } L(s) = j \text{ and } s \in S \setminus S^\infty, \\ \sigma_{-i,j}^\infty(s) & \text{if } L(s) = i \neq j \text{ and } s \in S^\infty, \\ \sigma_{-i,j}(s) & \text{if } L(s) = i \neq j \text{ and } s \in S \setminus S^\infty. \end{cases}$$

Note that in the second case, when $L(s) = j$ and $s \in S^\infty$, it is not necessary to have any restriction on $\tau_j(s)$, and we only require $\tau_j(s) \in A^\infty(s)$ because it simplifies the arguments of the proof.

We show that $\tau = (\tau_j)_{j \in N}$ is a secure equilibrium in G. Consider a strategy τ_i' for player i. The first part of the following lemma says that, in any state $s \in S^\infty$, it is strictly worse for player i to deviate to an action outside $A^\infty(s)$, unless another player deviated before him (that case is irrelevant for our goal to show that τ is a secure equilibrium). With the help of this, we can handle deviations to actions outside G^∞. The second part of the lemma is about deviations to actions inside G^∞. It claims that player i does not get a better payoff if he deviates to an action inside the game G^∞, given no deviation has occurred before.

Lemma 5. (i) For each state $s \in S^\infty$ such that $L(s) \in \{\bot, i\}$ and $\tau_i'(s) \in A(s) \setminus A^\infty(s)$, we have $u_i(\tau_i', \tau_{-i}|s) < u_i(\tau|s)$.
(ii) For each state $s \in S^\infty$ such that $L(s) = \bot$ and $\tau_i'(s) \in A^\infty(s) \setminus \{\rho_i^*(s)\}$, we have $u_i(\tau_i', \tau_{-i}|s) \leq u_i(\tau|s)$.

It follows from Lemma 5 that $u_i(\tau_i', \tau_{-i}) \leq u_i(\tau)$, and so τ is a Nash equilibrium.

Furthermore, consider the case where $u_i(\tau_i', \tau_{-i}) = u_i(\tau)$. Then, part (i) of Lemma 5 implies that it has probability zero that a state $s \in S^\infty$ with $\tau_i'(s) \in A(s) \setminus A^\infty(s)$ is reached under (τ_i', τ_{-i}). Consequently, when playing according to (τ_i', τ_{-i}) the following properties hold in the game G: (1) only states in S^∞ are visited, and (2) in all states $s \in S^\infty$ that are reached, player i's action $\tau_i'(s)$ belongs to $A^\infty(s)$, and (3) in all states $s \in S_j^\infty$, where $j \in N \setminus \{i\}$, that are reached, player j plays the action given by $\rho_j^*(s)$ if $L(s) = \bot$ and by $\sigma_{-i,j}^\infty(s)$ if $L(s) = i$. Hence, the strategy profile (τ_i', τ_{-i}) does not leave G^∞ when it is played. As $u_i(\tau_i', \tau_{-i}) = u_i(\tau)$, the definition of ρ^* yields

$$\sum_{j \in N \setminus \{i\}} u_j(\tau_i', \tau_{-i}) \geq \sum_{j \in N \setminus \{i\}} u_j(\rho^*) = \sum_{j \in N \setminus \{i\}} u_j(\tau).$$

Thus, τ is a secure equilibrium in G, and the proof of Theorem 1 is complete.

5   The Proof of Theorem 2

In this section, we provide a proof for Theorem 2. Consider a perfect information game G with deterministic transitions, with arbitrary state and action spaces. Assume that the payoff functions are Borel measurable and have a finite range

M, i.e. every player i's payoff function u_i is only taking values in M. Assume also that M contains at least two elements, otherwise the game is trivial.

Let $R = \max_{m \in M} |m|$ and $d = \min_{m,m' \in M, m \neq m'} |m - m'|$, and then choose $\delta = \frac{d}{2|N|R}$. We denote by G^δ the game G with a new payoff function u_i^δ for every player $i \in N$, defined as follows: for every play $p \in \mathcal{P}$, let

$$u_i^\delta(p) = u_i(p) - \delta \cdot \sum_{j \in N, j \neq i} u_j(p).$$

Notice that, for two plays $p, p' \in \mathcal{P}$, if we have $u_i(p) < u_i(p')$ then

$$u_i^\delta(p') - u_i^\delta(p) = (u_i(p') - u_i(p)) - \delta \cdot \sum_{j \in N, j \neq i} (u_j(p') - u_j(p)) \geq d - \delta \cdot (|N| - 1) \cdot 2R > 0,$$

so $u_i^\delta(p) < u_i^\delta(p')$ holds too. Consequently, $u_i^\delta(p) \geq u_i^\delta(p')$ implies $u_i(p) \geq u_i(p')$.

Now suppose that σ^* is a Nash equilibrium in G^δ. Then, due to the previous observation, σ^* is also a Nash equilibrium in the original game G. Now we show that σ^* is a secure equilibrium in G. So, suppose that τ_i is a strategy for some player $i \in N$ such that $u_i(\tau_i, \sigma_{-i}^*) = u_i(\sigma^*)$. Since σ^* is a Nash equilibrium in G^δ, we also have $u_i^\delta(\tau_i, \sigma_{-i}^*) \leq u_i^\delta(\sigma^*)$. Hence

$$\sum_{j \in N, j \neq i} u_j(\tau_i, \sigma_{-i}^*) \geq \sum_{j \in N, j \neq i} u_j(\sigma^*),$$

which proves that σ^* is a secure equilibrium in G indeed.

It remains to prove that G^δ admits a Nash equilibrium. We only provide a sketch, since similar constructions are well known (cf. the result of Mertens and Neyman in [13], and also [16]), and many of these ideas also appeared in the proof of Theorem 1. The important property of G^δ is that the payoff functions u_i^δ, for $i \in N$, are Borel measurable and have a finite range. By applying a corollary of Martin [12], for any $i \in N$, player i has a subgame-perfect optimal strategy σ_i and his opponents have a subgame-perfect optimal strategy profile σ_{-i} in the zero-sum game G_i, in which player i maximizes u_i^δ and the other players are jointly minimizing u_i^δ. It can be checked easily that the following strategy profile is a Nash equilibrium in G^δ: every player i should use the strategy σ_i. As soon as a player deviates, say player i plays another action, then the other players should punish player i in the remaining game by switching to the strategy profile σ_{-i}.

6   Concluding Remarks

Lexicographic Objectives: In the proof of Theorem 1, we in fact showed the existence of a sum-secure equilibrium (see Remark 1). A very similar proof can be given to show that, if $F : \mathbb{R}^{|N|} \to \mathbb{R}$ is a continuous and bounded function, then there exists a Nash equilibrium σ^* such that the following property holds for all $i \in N$: if τ_i is a strategy for player i such that $u_i(\tau_i, \sigma_{-i}^*) = u_i(\sigma^*)$ then $F((u_j(\sigma^*))_{j \in N}) \leq F((u_j(\tau_i, \sigma_{-i}^*))_{j \in N})$. This is closely related to lexicographic preferences: each player's first objective is to maximize his payoff, but in case of a tie between strategies, the secondary objective is to minimize the function F.

Subgame-Perfect Secure Equilibrium: Brihaye et al [1] introduced the concept of subgame-perfect secure equilibrium, and showed its existence in two-player quantitative reachability games. We do not know if Theorem 1 can be extended to subgame-perfect secure equilibrium. Perhaps it is possible to make use of the recently developed techniques for subgame-perfect equilibria in [8,14].

However, Theorem 2 cannot be extended to subgame-perfect secure equilibrium, as even a subgame-perfect equilibrium does not always exist in perfect information games with deterministic transitions and finitely many payoffs [15].

Algorithmic Result: Take a multi-player quantitative reachability game played on a finite graph, with deterministic transitions, where each payoff is determined by the number of moves it takes to get in a particular set of states. As a corollary of Theorem 1 and some results of [1] (the proof of Theorem 4.1, Proposition 4.5 and Remark 4.7), we derive an algorithm to obtain, in EXPSPACE, a secure equilibrium such that finite payoffs are bounded by $2 \cdot |N| \cdot |S|$ in the game. We intend to further investigate algorithmic questions for other classes of objectives.

Acknowledgment. We would like to thank Dario Bauso, Thomas Brihaye, Véronique Bruyère and Guillaume Vigeral for valuable discussions on the concept of secure equilibrium.

References

1. Brihaye, T., Bruyère, V., De Pril, J., Gimbert, H.: On (subgame perfect) secure equilibrium in quantitative reachability games. Logical Methods in Computer Science 9(1) (2013)
2. Bruyère, V., Meunier, N., Raskin, J.-F.: Secure equilibria in weighted games. In: CSL-LICS (2014)
3. Chatterjee, K., Henzinger, T.A.: Assume-guarantee synthesis. In: Grumberg, O., Huth, M. (eds.) TACAS 2007. LNCS, vol. 4424, pp. 261–275. Springer, Heidelberg (2007)
4. Chatterjee, K., Henzinger, T.A., Jurdziński, M.: Games with secure equilibria. Theoretical Computer Science 365(1-2), 67–82 (2006)
5. Chatterjee, K., Raman, V.: Assume-guarantee synthesis for digital contract signing. In: Formal Aspects of Computing, pp. 1–35 (2010)
6. De Pril, J.: Equilibria in multiplayer cost games. Ph.D. Thesis, Université de Mons, Belgium (2013)
7. De Pril, J., Flesch, J., Kuipers, J., Schoenmakers, G., Vrieze, K.: Existence of secure equilibrium in multi-player games with perfect information. CoRR, abs/1405.1615 (2014)
8. Flesch, J., Kuipers, J., Mashiah-Yaakovi, A., Schoenmakers, G., Solan, E., Vrieze, K.: Perfect-information games with lower-semicontinuous payoffs. Mathematics of Operations Research 35(4), 742–755 (2010)
9. Fudenberg, D., Levine, D.: Subgame-perfect equilibria of finite- and infinite-horizon games. Journal of Economic Theory 31(2), 251–268 (1983)

10. Harris, C.J.: Existence and characterization of perfect equilibrium in games of perfect information. Econometrica 53(3), 613–628 (1985)
11. Maitra, A.P., Sudderth, W.D.: Subgame-perfect equilibria for stochastic games. Mathematics of Operations Research 32(3), 711–722 (2007)
12. Martin, D.A.: Borel determinacy. Annals of Mathematics 102, 363–371 (1975)
13. Mertens, J.-F.: Repeated games. In: Proceedings of the International Congress of Mathematicians, pp. 1528–1577. American Mathematical Society (1987)
14. Purves, R.A., Sudderth, W.D.: Perfect information games with upper semicontinuous payoffs. Mathematics of Operations Research 36(3), 468–473 (2011)
15. Solan, E., Vieille, N.: Deterministic multi-player dynkin games. Journal of Mathematical Economics 39(8), 911–929 (2003)
16. Thuijsman, F., Raghavan, T.E.: Perfect information stochastic games and related classes. International Journal of Game Theory 26(3), 403–408 (1997)

An Efficient Quantum Algorithm for Finding Hidden Parabolic Subgroups in the General Linear Group

Thomas Decker[5], Gábor Ivanyos[1], Raghav Kulkarni[2],
Youming Qiao[2,3], and Miklos Santha[2,4]

[1] Institute for Computer Science and Control,
Hungarian Academy of Sciences, Budapest, Hungary
Gabor.Ivanyos@sztaki.mta.hu
[2] Centre for Quantum Technologies, National University of Singapore
kulraghav@gmail.com
[3] Centre for Quantum Computation and Intelligent Systems,
University of Technology, Sydney
jimmyqiao86@gmail.com
[4] LIAFA, Univ. Paris Diderot, CNRS, 75205 Paris, France
miklos.santha@liafa.univ-paris-diderot.fr
[5] EXASOL, Nuremberg, Germany
t.d3ck3r@gmail.com

Abstract. In the theory of algebraic groups, parabolic subgroups form a crucial building block in the structural studies. In the case of general linear groups over a finite field $\mathbb{F}_q$, given a sequence of positive integers $n_1, \ldots, n_k$, where $n = n_1 + \cdots + n_k$, a parabolic subgroup of parameter $(n_1, \ldots, n_k)$ in $\mathrm{GL}_n(\mathbb{F}_q)$ is a conjugate of the subgroup consisting of block lower triangular matrices where the ith block is of size n_i. Our main result is a quantum algorithm of time polynomial in $\log q$ and n for solving the hidden subgroup problem in $\mathrm{GL}_n(\mathbb{F}_q)$, when the hidden subgroup is promised to be a parabolic subgroup. Our algorithm works with no prior knowledge of the parameter of the hidden parabolic subgroup. Prior to this work, such an efficient quantum algorithm was only known for minimal parabolic subgroups (Borel subgroups), for the case when q is not much smaller than n (G. Ivanyos: Quantum Inf. Comput., Vol. 12, pp. 661-669).

1 Introduction

Background. The hidden subgroup problem (HSP for short) is defined as follows. A function f on a group G is said to hide a subgroup $H \leq G$, if f satisfies the following: $f(x) = f(y)$ if and only if x and y are in the same left coset of H (that is, $x^{-1}y \in H$). When such an f is given as a black box, the HSP asks to determine the hidden subgroup H. Note that the problem when the level sets of the hiding f are demanded to be right cosets of H – that is, $f(x) = f(y)$ if and only if $yx^{-1} \in H$ – is equivalent: composing f with taking inverses maps a hiding function via right cosets to a hiding function via left cosets, and vice versa. When we explicitly want to refer to this variant of the problem, we speak about HSP via right cosets.

E. Csuhaj-Varjú et al. (Eds.): MFCS 2014, Part II, LNCS 8635, pp. 226–238, 2014.
© Springer-Verlag Berlin Heidelberg 2014

The complexity of a hidden subgroup algorithm is measured in terms of the number of bits representing the elements of the group G, which is usually $O(\log|G|)$. On classical computers, the problem has exponential query complexity even for abelian groups. In contrast, the quantum query complexity of HSP for any group is polynomial [10], and the HSP for abelian groups can be solved in polynomial time with a quantum computer [5,21]. The latter algorithms are generalizations of Shor's result on order finding and computing discrete logarithms [24]. These algorithms can be further generalized to compute the structure of finite commutative black-box groups [7].

To go beyond the abelian groups is well-motivated by its connection with the graph isomorphism problem. Despite considerable attention, the groups for which the HSP is tractable remain close to being abelian. For example, we know polynomial-time algorithms for the following cases: groups whose derived subgroups are of constant derived length and constant exponent [11], Heisenberg groups [2,1] and more generally two-step nilpotent groups [19], "almost Hamiltonian" groups [12], and groups with a large abelian subgroup and reducible to the abelian case [16]. The limited success in going beyond the abelian case indicates that the nonabelian HSP may be hard, and [23] shows some evidence for this by providing a connection between the HSP in dihedral groups and some supposedly difficult lattice problem.

Instead of considering various ambient groups, another direction is to pose restrictions on the possible hidden subgroups. This can result in efficient algorithms, even over fairly nonabelian ambient groups. For example, if the hidden subgroup is assumed to be normal, then HSP can be solved in quantum polynomial time in groups for which there are efficient quantum Fourier transforms [14,15], and even in a large class of groups, including solvable groups [18]. The methods of [22,13] are able to find sufficiently large non-normal hidden subgroups in certain semidirect products efficiently.

Some restricted subgroups of the general linear groups were also considered in this context. The result by Denney, Moore and Russell in [8] is an efficient quantum algorithm that solves the HSP in the group of 2 by 2 invertible matrices (and related groups) where the hidden subgroup is promised to be a so-called Borel subgroup. In [17], Ivanyos considered finding Borel subgroups in general linear groups of higher degree, and presented an efficient algorithm when the size of the underlying field is not much smaller than the degree.

A well-known superclass of the family of Borel subgroups is the family of parabolic subgroups, whose definition is given below. In this work, we follow the line of research in [8,17], and consider the problem of finding parabolic subgroups in general linear groups. Our main result will be a polynomial-time quantum algorithm for this case, without restrictions on field size.

Parabolic Subgroups of the General Linear Group. Let q be a power of a prime p. The field with q elements is denoted by $\mathbb{F}_q$. The vector space $\mathbb{F}_q^n$ consists of column vectors of length n over $\mathbb{F}_q$. $\mathrm{GL}_n(\mathbb{F}_q)$ stands for the general linear group of degree n over $\mathbb{F}_q$. The elements of $\mathrm{GL}_n(\mathbb{F}_q)$ are the invertible $n \times n$ matrices with entries from $\mathbb{F}_q$. We also use $\mathrm{GL}(V)$ to denote the group of linear automorphisms of the $\mathbb{F}_q$-space V. With this notation, we have $\mathrm{GL}_n(\mathbb{F}_q) \cong \mathrm{GL}(\mathbb{F}_q^n)$ and throughout the paper we will identify these two groups. As a matrix is represented by an array of n^2 elements from $\mathbb{F}_q$, an algorithm is considered efficient if its complexity is polynomial in n and $\log q$.

We now present the definition of parabolic subgroups (see [25]). For a positive integer k, and a sequence of positive integers $n_1, \ldots, n_k$ with $n_1 + \cdots + n_k = n$, the *standard parabolic subgroup* of $\mathrm{GL}_n(\mathbb{F}_q)$ with parameter $(n_1, \ldots, n_k)$ is the subgroup consisting of the invertible lower block triangular matrices of diagonal block sizes $n_1, \ldots, n_k$. Any conjugate of the standard parabolic subgroup is called a *parabolic subgroup*.

To see the geometric meaning of parabolic subgroups, we review the concept of flags of vector spaces. Let 0 also denote the zero vector space. For $\mathbb{F}_q^n$ and $k \geq 1$, a flag F with the parameter $(n_1, \ldots, n_k)$ is a nested sequence of subspaces of $\mathbb{F}_q^n$, that is $\mathbb{F}_q^n = U_0 > U_1 > U_2 > \cdots > U_{k-1} > U_k = 0$, such that for $0 \leq i \leq k-1$, $\dim(U_i) = n_{i+1} + \cdots + n_k$. k is called the length of F. For $g \in \mathrm{GL}_n(\mathbb{F}_q)$, g stabilizes the flag F if for every $i \in [k]$, $g(U_i) = U_i$. Then all group elements in $\mathrm{GL}_n(\mathbb{F}_q)$ stabilizing F form a parabolic subgroup. On the other hand, any parabolic subgroup corresponds to some flag F, namely it consists of the elements in $\mathrm{GL}_n(\mathbb{F}_q)$ stabilizing F.

For example, the standard parablic subgroup B in $\mathrm{GL}_5(\mathbb{F}_q)$ with parameter $(2, 2, 1)$ consists of invertible matrices of the form $\begin{pmatrix} * & * & 0 & 0 & 0 \\ * & * & 0 & 0 & 0 \\ * & * & * & * & 0 \\ * & * & * & * & 0 \\ * & * & * & * & * \end{pmatrix}$. Let $\{e_1, \ldots, e_5\}$ be the standard basis of $\mathbb{F}_q^5$. The flag stabilized by B is $\mathbb{F}_q^5 > \langle e_3, e_4, e_5 \rangle > \langle e_5 \rangle > 0$.

A parabolic subgroup is maximal if there are no parabolic subgroups properly containing it. It is minimal if it does not properly contain any parabolic subgroup. A parabolic subgroup B in $\mathrm{GL}_n(\mathbb{F}_q)$ is maximal if and only if it is the stabilizer of a flag of length 2, that is, it is the stabilizer of some nontrivial subspace. On the other hand, B is minimal if it stabilizes a flag of length n. *Borel subgroups* in $\mathrm{GL}_n(\mathbb{F}_q)$ are just minimal parabolic subgroups. They are conjugates of the subgroup of invertible lower triangular matrices.

Our Results. The main result of this paper is a polynomial-time quantum algorithm for finding parabolic subgroups in general linear groups.

Theorem 1. *Any hidden parabolic subgroup in $\mathrm{GL}_n(\mathbb{F}_q)$ can be found in quantum polynomial time (i.e., in time $\mathrm{poly}(\log q, n)$).*

Note that this algorithm does not require one to know the parameter of the hidden parabolic subgroup in advance. Neither does it pose any restriction on the underlying field size, while the algorithm in [17] for finding Borel subgroups requires the field size to be large enough. The basic idea behind the algorithm is that in certain cases the superposition of the elements in a coset of the subgroup is close to a superposition of the elements of a linear space of matrices. The latter perspective allows the use of standard algorithms for abelian HSPs. Another crucial idea is to make use of the subgroup of common stabilizers of all the vectors on a random hyperplane, and examine its intersection with the hidden parabolic subgroup.

We state without proof the following result: consider certain subgroups of Borel subgroups, namely the *full unipotent subgroups*. They are conjugates of the subgroup of lower triangular matrices with 1's on the diagonal. Following a variant of the idea for Theorem 1, there exists an algorithm for finding full unipotent subgroups whose complexity is polynomial in n and the field size.

The Structure of the Paper. In Section 2 we collect certain preliminaries for the paper. In particular, in Section 2.2 we adapt the standard algorithm for abelian HSP to linear subspaces, which forms the basis of our algorithms. We then present an efficient quantum algorithm for finding maximal parabolic subgroups in Section 3. Section 4 describes a main technical tool, a generalization of the result of [22,8] for finding complements in affine groups. In Section 5 we present the algorithm for finding parabolic subgroups, proving Theorem 1.

2 Preliminary

2.1 Notations and Facts

Throughout the article, q is a prime power. For $n \in \mathbb{N}$, $[n] = \{1, \ldots, n\}$. $\mathcal{M}_n(\mathbb{F}_q)$ is the set of $n \times n$ matrices over $\mathbb{F}_q$. For a finite group G, we will be concerned with finding a subgroup H in G, when it is promised that H is from a fixed family of subgroups $\mathcal{H}$. We use $\mathrm{HSP}(G, \mathcal{H})$ to denote the HSP problem with this promise, and $\mathrm{rHSP}(G, \mathcal{H})$ to denote the HSP via right cosets of $H \in \mathcal{H}$. Let V be a vector space. For a subspace $U \leq V$ and $G = \mathrm{GL}(V)$, let G_U be the subgroup in G consisting of elements that act as *pointwise stabilizers* on U. That is, $G_U = \{X \in \mathrm{GL}(V) : \forall u \in U, Xu = u\}$. Let $G_{\{U\}}$ be the subgroup in G consisting of elements that act as *setwise stabilizers* on U. That is, $G_{\{U\}} = \{X \in \mathrm{GL}(V) : XU = U\}$. Note that $\{G_{\{U\}} : 0 < U < V\}$ is just the set of maximal parabolic subgroups.

Fact 1. *For every prime power q, and for every positive integers $n \geq m$, the probability for a random $n \times m$ matrix M over $\mathbb{F}_q$ to have rank m is no less than what we have in the case of $q = 2$, that is $\frac{1}{2} \cdot \frac{3}{4} \cdot \frac{7}{8} \cdots \approx 0.288788 > 1/4$.*

2.2 The Quantum Fourier Transform of Linear Spaces

In this part we briefly discuss slight generalizations of the Fourier transform of linear spaces over $\mathbb{F}_q$ introduced in [17] and a version useful for certain linear spaces of matrices. Let $V \cong \mathbb{F}_q^m$ be a linear space over the field $\mathbb{F}_q$ and assume that we are given a nonsingular symmetric bilinear function $\phi : V \times V \to \mathbb{F}_q$. By $\mathbb{C}^V$ we denote the Hilbert space of dimension q^m having a designated orthonormal basis consisting of the vectors $|v\rangle$ indexed by the elements $v \in \mathbb{F}_q^m$.

Let $q = p^r$ where p is a prime and let ω be the primitive pth root $e^{\frac{2\pi i}{p}}$ of unity. We define the quantum Fourier transform with respect to ϕ as the linear transformation QFT_ϕ of $\mathbb{C}^V$ which maps

$$|v\rangle \quad \text{to} \quad \frac{1}{\sqrt{|V|}} \sum_{u \in V} \omega^{\mathrm{Tr}(\phi(u,v))} |u\rangle,$$

where $v \in V$ and Tr is the trace map from $\mathbb{F}_q$ to $\mathbb{F}_p$ defined as $\mathrm{Tr}(x) = \sum_{i=0}^{r-1} x^{p^i}$. It turns out that QFT_ϕ is a unitary map and, if the vectors from V are represented by arrays of elements from $\mathbb{F}_q$ that are coordinates in terms of an orthonormal basis of V with respect to ϕ (that is, ϕ is the standard inner product of $\mathbb{F}_q^m$) then QFT_ϕ is just the mth tensor power of the QFT defined in [9] for $\mathbb{F}_q$. (This is the linear transformation of $\mathbb{C}^{\mathbb{F}_q}$ that maps $|x\rangle$ $(x \in \mathbb{F}_q)$ to $\frac{1}{\sqrt{q}} \sum_{y \in \mathbb{F}_q} \omega^{\mathrm{Tr}(xy)} |y\rangle$.) Therefore, in this case, by Lemma 2.2 of [9], QFT_ϕ has a polynomial time approximate implementation on a quantum computer. In the general case, where elements of V are represented by coordinates in terms of a not necessarily orthonormal basis w.r.t. ϕ, the map QFT_ϕ can be efficiently implemented by composing the above transform with linear transformations of $\mathbb{C}^V$ corresponding to appropriate basis changes for V.

For a subset $A \subseteq V$ we adopt the standard notation $|A\rangle$ for the uniform superposition of the elements of A, that is $|A\rangle = \frac{1}{\sqrt{|A|}} \sum_{a \in A} |a\rangle$. Assume that we receive the uniform superposition $|v_0 + W\rangle = \frac{1}{\sqrt{|W|}} \sum_{v \in W} |v_0 + v\rangle$ over the a coset $v_0 + W$ of the $\mathbb{F}_q$-linear subspace W of V and for some $v_0 \in V$. Let $W^\perp$ stand for the subspace of V consisting of the vectors u from $\mathbb{F}_q^m$ such that $\phi(u, v) = 0$ for every $v \in W$. By results from [17], if we measure the state after the Fourier transform, we obtain a uniformly random element of $W^\perp$. If instead of the uniform superposition over the coset $v_0 + W$ we apply the QFT to the superposition $|v_0 + W'\rangle = \frac{1}{\sqrt{|W'|}} \sum_{v \in W'} |v_0 + v\rangle$ over a subset $v_0 + W'$ for $\emptyset \neq W' \subseteq W$, the resulting state is $\sum_{u \in V} c'_u |u\rangle$, where

$$c'_u = \langle u | QFT_\phi | v_0 + W'\rangle = \frac{\omega^{\mathrm{Tr}\phi(v_0, u)}}{\sqrt{|W'||V|}} \sum_{v \in W'} \omega^{\mathrm{Tr}\phi(v, u)}.$$

For $u \in W^\perp$ we have

$$|c'_u| = \frac{|W'|}{\sqrt{|W'||V|}} = \frac{\sqrt{|W'|}}{\sqrt{|W|}} \cdot \frac{1}{\sqrt{|W^\perp|}}, \tag{1}$$

whence, after measurement the chance of obtaining a particular $u \in W^\perp$ is $\frac{|W'|}{|W|}$ times as much as if we had in the case of the uniform distribution over $W^\perp$.

In this paper we consider subspaces and certain subsets of the linear space $\mathcal{M}_n(\mathbb{F}_q)$. If we take the inner product $\phi_0(A, B) = \mathrm{tr}(AB^T)$ the elementary matrices form an orthonormal basis. It follows that QFT_{ϕ_0}, being just the n^2th tensor power of the QFT of $\mathbb{F}_q$, can be efficiently approximated. However, for the purposes of this paper it turns out to be more convenient using the inner product $\phi(AB) = \mathrm{tr}(AB)$. The map QFT_ϕ is the composition of QFT_{ϕ_0} with taking transpose (the latter is just a permutation of the matrix entries). The main advantage of considering QFT_ϕ is that it is invariant in the following sense: we always obtain the same QFT_ϕ even if we write matrices of linear transformations of the space $V = \mathbb{F}_q^n$ in terms of various bases. In particular, in our hidden subgroup algorithms we can think of our matrices in terms of a basis a priori unknown to us in which the hidden subgroup has a natural form, for example lower block triangular.

2.3   A Common Procedure for HSP Algorithms

Suppose we want to find some hidden subgroup H in $G = \mathrm{GL}_n(\mathbb{F}_q)$. Let $V = \mathbb{F}_q^n$. We present the standard procedure that produce a uniform superposition over a coset of the hidden subgroup. This part will be common in (most of) the hidden subgroup algorithms presented in this paper. First we show how to produce the uniform superposition over $\mathrm{GL}(V)$. The uniform superposition $\frac{1}{q^{n^2}} \sum_{X \in \mathcal{M}_n(\mathbb{F}_q)} |X\rangle$ over $\mathcal{M}_n(\mathbb{F}_q)$ can be produced using the QFT for $\mathbb{F}_q^{n^2}$. Then, in an additional qubit we compute a Boolean variable according to whether or not the determinant of X is zero. We measure this qubit, and abort if it indicates that the matrix X has determinant zero. This procedure gives the uniform superposition over $\mathrm{GL}(V)$ with success probability more than $\frac{1}{4}$.

Next we assume that we have the uniform superposition $\frac{1}{\sqrt{|\mathrm{GL}(V)|}} \sum_X |X\rangle|0\rangle$, summing over $X \in \mathrm{GL}(V)$. Recall that f is the function hiding the subgroup. We appended a new quantum register, initialized to zero, for holding the value of f. We compute $f(X)$ in this second register, measure and discard it. The result is $|AH\rangle = \frac{1}{\sqrt{|H|}} \sum_{X \in H} |AX\rangle$ for some unknown $A \in \mathrm{GL}(V)$. A is actually uniformly random, but in this paper we will not make use of this fact.

3   Maximal Parabolic Subgroups

In this section, we settle the HSP when the hidden subgroup is a maximal parabolic subgroup, which will be used in the main algorithm in Section 5. It also helps to illustrate the idea of the idea of approximating a subgroup in the general linear group by a subspace in the linear space of matrices.

Recall that a parabolic subgroup H is maximal if it stabilizes some subspace $0 < U < \mathbb{F}_q^n$. We mentioned in Section 2.1 that they are just setwise stabilizers of subspaces. Determining H is equivalent to finding U. Set $V = \mathbb{F}_q^n$.

Proposition 1. Let $G = \mathrm{GL}_n(\mathbb{F}_q)$, and $\mathcal{H} = \{G_{\{U\}} : 0 < U < V\}$. HSP$(G, \mathcal{H})$ can be solved in quantum polynomial time.

Proof. Let H be the hidden maximal parabolic subgroup, stabilizing some $(n - d)$-dimensional subspace $U \le \mathbb{F}^n$. Note that d is unknown to us. Before describing the algorithm, we observe the following: checking correctness of a guess for U, and hence for H, can be done by applying the oracle to generators of the stabilizer of U, as there are no inclusions between maximal parabolic subgroups.

Now we present the algorithm. First produce a coset superposition $|AH\rangle$ for unknown $A \in \mathrm{GL}(V)$, as described in Section 2.3. Let $W = \{X \in \mathcal{M}_n(\mathbb{F}_q) : XU \le U\}$. In a basis whose last $n - d$ elements are from U, W is the subspace of the matrices of the form $\begin{pmatrix} B & \\ C & D \end{pmatrix}$, where B and C are not necessarily invertible, and the empty space in the upper right corner means a $d \times (n - d)$ block of zeros. Noting that such a matrix is invertiable if and only if B and C are both invertible, we have $H \subset W$ and $\frac{|AH|}{|AW|} = \frac{|H|}{|W|} > \frac{1}{4 \times 4}$. Also, viewing in a basis in which W is block triangular, $(AW)^\perp A$ consists of the matrices of the form $\begin{pmatrix} \\ * \end{pmatrix}$, where $*$ stands for an arbitrary $(n-d)$ times d matrix. This implies that $(AW)^\perp = \{X \in \mathcal{M}_n(\mathbb{F}_q) : XV \leq U \text{ and } XU = 0\}A^{-1}$.

If $d \geq n/2$, we apply QFT to the *left* coset superposition $|A\mathcal{H}\rangle$ and perform a measurement, for any element X in $(AW)^\perp$, the measurement will produce X with probability no less than $\frac{1}{16|(AW)^\perp|}$. It follows that XA will be a particular matrix from $(AW)^\perp A$ with at least $\frac{1}{16|(AW)^\perp|}$. Then more than $\frac{1}{4}$ of the $(n-d) \times d$ matrices have rank $n-d$. It follows that with probability at least $\frac{1}{64}$, the matrix XA will be a matrix from $(AW)^\perp A$ whose image is U. As $XV = XAV$, we can conclude that $XV = U$ with probability more than $\frac{1}{64}$.

For the case $d < n/2$ we consider the HSP via *right* cosets of H, and let act matrices on row vectors from the right. Via the same procedure as above, it will reveal the dual subspace stabilized by H, which determines H uniquely as well.

Finally, though d is not known to us, depending on whether $d \geq n/2$, one of these two procedures with produce U correctly with high probability. So we perform the two procedures alternatively, and use the checking procedure to determine which produces the correct result. This concludes the algorithm.

4   A Tool: Finding Complements in Small Stabilizers

In this section, we introduce and settle a new instance of the hidden subgroup problem. This will be an important technical tool for the main algorithm.

Consider the hidden subgroup problem in the following setting. The ambient group $G \leq \mathrm{GL}_n(\mathbb{F}_q)$ consists of the invertible matrices of the form $\begin{pmatrix} b & \\ v & I \end{pmatrix}$, where $b \in \mathbb{F}_q$, v is a column vector from $\mathbb{F}_q^{n-1}$ and I is the $(n-1) \times (n-1)$ identity matrix. The family of hidden subgroups $\mathcal{H}$ consists of all conjugates of H_0, where H_0 is the subgroup of diagonal matrices in G: $H_0 = \left\{ \begin{pmatrix} b & \\ & I \end{pmatrix} : b \in \mathbb{F}_q^* \right\}$. Note that any conjugate of H_0 is $H_v = \left\{ \begin{pmatrix} b & \\ (b-1)v & I \end{pmatrix} : b \in \mathbb{F}_q^* \right\}$, for some $v \in \mathbb{F}_q^{n-1}$. We will consider the HSP via right cosets in this setting.

The group G has an abelian normal subgroup N consisting of the matrices of the form $\begin{pmatrix} 1 & \\ v & I \end{pmatrix}$ isomorphic to $\mathbb{F}_q^{n-1}$, and the subgroups H_v are the semidirect complements of N. For $n = 2$, G is the affine group $\mathrm{AGL}_1(\mathbb{F}_q)$. The HSP in $\mathrm{AGL}_1(\mathbb{F}_q)$ is solved in quantum polynomial time in [22] over prime fields and in [8] in the general case using the non-commutative Fourier transform of the group $\mathrm{AGL}_1(\mathbb{F}_q)$. The algorithm served as the main technical ingredient in [8] for finding Borel subgroups in $\mathrm{GL}(\mathbb{F}_q^2)$. A generalization for certain similar semidirect product groups is given in [2]. To our knowledge, the first occurrence of the idea of comparing with a coset state in a related abelian group is in [2]. Here, due to the "nice" representation of the group elements, we can apply the same idea in a simpler way, while in [2] it was needed to be combined with a discrete logarithm algorithm which is not necessary here.

Proposition 2. *Let G and $\mathcal{H}$ be as above, and suppose $q = \Omega(n/\log n)$. Then* rHSP$(G, \mathcal{H})$ *can be solved in quantum polynomial time.*

Proof. Assume that the hidden subgroup is $H = H_v$ for some $v \in \mathbb{F}_q^{n-1}$. As right cosets of H are being considered, we have superpositions over right cosets HA for some unknown $A \in G$. The actual information of each matrix X from G is contained in $X - I$, a matrix from the n-dimensional space L of matrices whose last $n - 1$ columns are zero. We will work in L. Set

$$\widetilde{W}' = \{X - I : X \in H\} = \left\{ \begin{pmatrix} b \\ bv \end{pmatrix} : -1 \neq b \in \mathbb{F}_q \right\} \text{ and } W = \left\{ \begin{pmatrix} b \\ bv \end{pmatrix} : b \in \mathbb{F}_q \right\}.$$

Then W is a one-dimensional subspace of L. It turns out that $W = WA$ for every matrix $A \in G$ (that is why it is convenient to consider the HSP via right cosets). It follows that $\{(Y + I)A - I : Y \in W\} = \{YA + (A - I) : Y \in W\} = W + A - I$, whence the set $\{XA - I : X \in H\} = W' + A - I$ for $W' = \widetilde{W}'A$.

Therefore, after an application of the QFT of L to the state $|HA - I\rangle = |W' + A - I\rangle$ and a measurement, we obtain every specific element of $W^\perp$ with probability at least $\frac{q-1}{q} \frac{1}{|W^\perp|}$. More generally, if we do the procedure for a product of $n - 1$ superpositions over right cosets of H we obtain each specific $(n - 1)$-tuple of vectors from $W^\perp$ with probability at least $(\frac{q-1}{q})^{n-1} \frac{1}{|W^\perp|^{n-1}}$. Since the probability that $n - 1$ random elements from a space of dimension $n - 1$ over $\mathbb{F}_q$ span the space is at least $\frac{1}{4}$, therefore, the probability of getting a basis of $W^\perp$ is $\Omega((\frac{q-1}{q})^{n-1})$. Using this basis, we obtain a guess for W and H as H is the set of invertible matrices from $W + I$. A correct guess will be obtained expectedly with $O((\frac{q}{q-1})^{n-1})$ repetitions. This is polynomial if q is $\Omega(n/\log n)$.

Finally we note that for constant q, or more generally for constant characteristic, [11] can be used to obtain a polynomial time algorithm. On the other hand, it is intriguing to study the case of "intermediate" values of q.

5   The Main Algorithm

5.1   The Structure of the Algorithm

In this subsection, we describe the structure of an algorithm for finding parabolic subgroups in general linear groups, proving Theorem 1. Let $G = \mathrm{GL}_n(\mathbb{F}_q)$, $V = \mathbb{F}_q^n$, and the hidden parabolic subgroup H be the stabilizer of the flag $V > U_1 > U_2 > \cdots > U_{k-1} > 0$. Note that the parameter of the flag, including k, is unknown to us. The algorithm will output the hidden flag, from which a generating set of the parabolic subgroup can be constructed easily.

Let $T = U_{k-1}$ denote the smallest subspace in the flag. The algorithm relies on the following subroutines crucially. These two subroutines are described in Section 5.2 and Section 5.3, respectively.

Proposition 3. *Let G, H and T be as above. There exists a quantum polynomial-time algorithm, that given access to an oracle hiding H in G, produces three subspaces W_1, W_2 and W_3, s.t. one of W_i is a nonzero subspace contained in T with high probability.*

Proposition 4. *Let G, H and T be as above. There exists a classical polynomial-time algorithm, that given access to an oracle hiding H in G, and some $0 < W \leq V$, determines whether $W \leq T$, and in the case of $W \leq T$, whether $W = T$.*

Given these two subroutines, the algorithm proceeds as follows. It starts with checking whether $k = 1$, that is whether $H = G$. This can be done easily: produce a set of generators of G, and check whether the oracle returns the same on all of them. If $k = 1$, return the trivial flag $V > 0$.

Otherwise, it repeatedly calls the subroutine in Proposition 3 until that subroutine produces subspaces W_1, W_2 and W_3, such that for some $i \in [3]$, we have $0 < W_i \leq T$. This can be verified by Proposition 4. Let W be this subspace. The second subroutine then also tells whether $W = T$.

After getting $0 < W \leq T$, the algorithm fixes a subspace W' to be any direct complement of W in V, and makes a recursive call to the HSP with a new ambient group G', and a new hidden subgroup H', as follows. G' is $\{X \in \mathrm{GL}(V) : XW' \leq W' \text{ and } (X - I)W = 0\}$, which is isomorphic to $\mathrm{GL}(W') \cong \mathrm{GL}(V/W)$. H' is the stabilizer of the flag $W' > W' \cap U_1 > \cdots > W' \cap U_{k-1} \geq 0$. Note that the oracle restricted to G' realizes a hiding function for H'.

The recursive call then returns a flag in W' as $W' > U_1' > U_2' > \cdots > U_{k'}' > 0$. Let $U_i = \langle U_i' \cup W \rangle$, $i \in [k']$. If $W = T$, then the algorithm outputs the flag $V > U_1 > U_2 > \cdots > U_{k'} > W > 0$. If $W < T$, return $V > U_1 > U_2 > \cdots > U_{k'} > 0$.

It is clear that at most n recursive calls will be made, and the algorithm runs in polynomial time given that the two subroutines run in polynomial time too. We now prove Proposition 3 and 4 in the next two subsections.

5.2   Guessing a Part of the Flag

In this subsection we prove Proposition 3. Recall that $G = \mathrm{GL}_n(\mathbb{F}_q)$, the hidden subgroup H stabilizing of the flag $V > U_1 > \cdots > U_{k-1} > 0$, and $T = U_{k-1}$. The algorithm of [8] for finding hidden Borel subgroups in 2 by 2 matrix groups was based on computing the intersection with the stabilizer of a nonzero vector. Here we follow an extension of the idea to arbitrary dimension n. We consider the *common* stabilizer of $n - 1$ linearly independent vectors.

Pick a random subspace $U' \leq V$ of dimension $n - 1$. Recall that $G_{U'}$ denotes the group of pointwise stabilizers of U'. We also consider the group consisting of the unipotent elements of $G_{U'}$, $N = \{X \in \mathrm{GL}(V) : (X - I)V \leq U' \text{ and } X \in G_{U'}\}$. Note that N is an abelian normal subgroup of $G_{U'}$ of size q^{n-1}. Here we illustrate the form of $G_{U'}$ and N when U' is put in an appropriate basis:

$$\begin{pmatrix} 1 & & & & * \\ & 1 & & & * \\ & & 1 & & * \\ & & & 1 & * \\ & & & & * \end{pmatrix}, \begin{pmatrix} 1 & & & & * \\ & 1 & & & * \\ & & 1 & & * \\ & & & 1 & * \\ & & & & 1 \end{pmatrix}.$$

$$G_{U'} \qquad\qquad\qquad N$$

We will describe three procedures, whose success on producing some $0 < W \le T$ depend on $d := \dim(T)$ and the field size q. Each of these procedures only works for a certain range of d and q, but together they cover all possible cases. Thus, the algorithm needs to run each of these procedures, and return the three results from them. The general idea behind these procedures is to examine the intersection of the random hyperplane U' with T. As $d = \dim(T)$, the probability that U' contains T is $\frac{q^{n-d}-1}{q^n-1} \sim \frac{1}{q^d}$.

Assume first that U' does not contain T. We claim that in this case

$$\sum_{X \in H \cap G_{U'}} (X - I)V = T \tag{2}$$

and

$$\sum_{X \in H \cap N} (X - I)V = U' \cap T. \tag{3}$$

To see this, pick $v_n \in T \setminus U'$, and let $v_1, \ldots, v_{n-1}$ be a basis for U' such that for every $0 < j < k$, the system $v_{n-\dim(U_j)+1}, \ldots, v_{n-\dim(U_{j+1})}$ is a basis for U_j. In the basis $v_1, \ldots, v_n$, the matrices of the elements of N are the matrices with ones in the diagonal, arbitrary elements in the last column except the lowest one, and zero elsewhere. Among these the matrices of the elements of intersection with H are those whose first $n - d$ entries in the last column are also zero:

$$\begin{pmatrix} * & & & & \\ * & * & & & \\ * & * & * & & \\ * & * & * & * & * \\ * & * & * & * & * \end{pmatrix}, \begin{pmatrix} 1 & & & & \\ & 1 & & & \\ & & 1 & & \\ & & & 1 & * \\ & & & & * \end{pmatrix}, \begin{pmatrix} 1 & & & & \\ & 1 & & & \\ & & 1 & & \\ & & & 1 & * \\ & & & & 1 \end{pmatrix}.$$

$$H \qquad\qquad H \cap G_{U'} \qquad\qquad H \cap N$$

Based on the above analysis, the three procedures are as follows.

- If $d > 1$, then $H \cap N$ is nontrivial. As N is abelian, we can efficiently compute $H \cap N$ by the abelian hidden subgroup algorithm. Thus by Equation 3, we can use it to compute W_1 as a guess for a nontrivial subspace of T.
- If $d = 1$ and $q \ge n$, we can compute $H \cap G_{U'}$ in $G_{U'}$ by the algorithm in Proposition 2, and use it to compute W_2 as a guess for T by Equation 2.
- If $d = 1$ and $q < n$, with probability at least $\frac{1}{q} - \frac{1}{q^2} = \Omega(\frac{1}{q}) = \Omega(\frac{1}{n})$, we have that $U' \ge T$ but U' does not contain U_{k-2}. Then we have

$$\sum_{X \in H \cap N} (X - I)V = U' \cap U_{k-2}. \tag{4}$$

To see this, pick $v_n \in U_{k-1} \setminus \{0\}$, $v_{n-1} \in U_{k-2} \setminus U'$, and $v_1, \ldots, v_{n-2}$ s.t. $v_1, \ldots, v_{n-2}, v_n$ is a basis for U' and for every $0 < j < k$, the system $v_{n-\dim(U_j)+1}, \ldots, v_{n-\dim(U_{j+1})}$ is a basis for U_j. In this basis the matrices for the elements of $N \cap H$ are those whose entries are zero except the ones in the diagonal and except the other lowest $\dim U_{k-2}$ entries in the next to last column:

$$
\underbrace{\begin{pmatrix} * & & & & \\ * & * & & & \\ * & * & * & * & \\ * & * & * & * & \\ * & * & * & * & * \end{pmatrix}}_{H}, \quad
\underbrace{\begin{pmatrix} 1 & & & * & \\ & 1 & & * & \\ & & 1 & * & \\ & & & 1 & \\ & & & * & 1 \end{pmatrix}}_{N}, \quad
\underbrace{\begin{pmatrix} 1 & & & & \\ & 1 & & & \\ & & 1 & * & \\ & & & 1 & \\ & & & * & 1 \end{pmatrix}}_{H \cap N}.
$$

Again, we can find $H \cap N$ by the abelian hidden subgroup algorithm and use Equation 4 to compute $V' = U' \cap U_{k-2}$. If $\dim V' = 1$ then return $W_3 = V'$ as the guess for T. Otherwise we take a direct complement V'' of V' and restrict the HSP to the subgroup of the tranformations X such that $(X - I)V'' = 0$ and $XV' \leq V'$ (which is isomorphic to $\mathrm{GL}(V')$) and apply the method in Proposition 1 to compute a subspace W_3 as the guess for T.

5.3   Checking and Recursion

In this subsection we prove Proposition 4. Recall that the goal is to determine whether some subspace $0 < W \leq V$ is contained in $T = U_{k-1}$, the last member of the flag $V > U_1 > \cdots > U_{k-1} > 0$ stabilized by the hidden parabolic subgroup H. If $W \leq V$, we'd like to know whether $W = T$. This can be achieved with the help of the following lemma, whose proof is omitted here.

Lemma 1. *Let H be the stabilizer in $\mathrm{GL}(V)$ of the flag $V > U_1 > U_2 > \ldots > U_{k-1} > 0$, and let $0 < W < V$. Let W' be any direct complement of W in V. Then $U_{k-1} \geq W$ if and only if $H \geq \{X \in \mathrm{GL}(V) : (X - I)V \leq W\}$. Furthermore, if $U_{k-1} \geq W$ then $U_{k-1} = W$ if and only if*
$$H \cap \{X \in \mathrm{GL}(V) : (X - I)V \leq W' \text{ and } (X - I)W' = 0\} = \{I\}.$$

It is clear that this allows us to determine whether $U_{k-1} \geq W$: form a generating set of $\{X \in \mathrm{GL}(V) : (X - I)V \leq W\}$, and query the oracle to see whether all element in the generating set evaluate the same. Similarly if $U_{k-1} \geq W$, we can test whether $U_{k-1} = W$.

Acknowledgements. The research is partially funded by the Singapore Ministry of Education and the National Research Foundation, also through the Tier 3 Grant "Random numbers from quantum processes," MOE2012-T3-1-009. Research partially supported by the European Commission IST STREP project Quantum Algorithms (QALGO) 600700, by the French ANR Blanc program under contract ANR-12-BS02-005 (RDAM project), and by the Hungarian Scientific Research Fund (OTKA), Grant NK105645.

References

1. Bacon, D.: How a Clebsch-Gordan transform helps to solve the Heisenberg hidden subgroup problem. Quantum Inf. Comput. 8, 438–467 (2008)
2. Bacon, D., Childs, A., van Dam, W.: From optimal measurement to efficient quantum algorithms for the hidden subgroup problem over semidirect product groups. In: Proc. 46th IEEE FOCS, pp. 469–478 (2005)
3. Berlekamp, E.R.: Algebraic coding theory. McGraw-Hill, New York (1968)
4. Berlekamp, E.R.: Factoring polynomials over large finite fields. Math. Comput. 24, 713–735 (1970)
5. Boneh, D., Lipton, R.J.: Quantum cryptanalysis of hidden linear functions. In: Coppersmith, D. (ed.) CRYPTO 1995. LNCS, vol. 963, pp. 424–437. Springer, Heidelberg (1995)
6. Cantor, D.G., Zassenhaus, H.: A New Algorithm for Factoring Polynomials Over Finite Field. Math. Comput. 36, 587–592 (1981)
7. Cheung, K., Mosca, M.: Decomposing finite abelian groups. Quantum Inf. Comput. 1, 26–32 (2001)
8. Denney, A., Moore, C., Russell, A.: Finding conjugate stabilizer subgroups in PSL(2;q) and related problems. Quantum Inf. Comput. 10, 282–291 (2010)
9. van Dam, W., Hallgren, S., Ip, L.: Quantum algorithms for some hidden shift problems. SIAM J. Comput. 36, 763–778 (2006)
10. Ettinger, M., Hoyer, P., Knill, E.: The quantum query complexity of the hidden subgroup problem is polynomial. Inform. Proc. Lett. 91, 43–48 (2004)
11. Friedl, K., Ivanyos, G., Magniez, F., Santha, M., Sen, P.: Hidden translation and orbit coset in quantum computing. In: Proc. 35th STOC, pp. 1–9 (2003)
12. Gavinsky, D.: Quantum solution to the hidden subgroup problem for poly-near-Hamiltonian groups. Quantum Inf. Comput. 4, 229–235 (2004)
13. Gonçalves, D.N., Portugal, R., Cosme, C.M.M.: Solutions to the hidden subgroup problem on some metacyclic groups. In: Childs, A., Mosca, M. (eds.) TQC 2009. LNCS, vol. 5906, pp. 1–9. Springer, Heidelberg (2009)
14. Grigni, M., Schulman, L., Vazirani, M., Vazirani, U.: Quantum mechanical algorithms for the nonabelian Hidden Subgroup Problem. In: Proc. 33rd ACM STOC, pp. 68–74 (2001)
15. Hallgren, S., Russell, A., Ta-Shma, A.: Normal subgroup reconstruction and quantum computation using group representations. SIAM J. Comp. 32, 916–934 (2003)
16. Inui, Y., Le Gall, F.: Efficient quantum algorithms for the hidden subgroup problem over semi-direct product groups. Quantum Inf. Comput. 7, 559–570 (2007)
17. Ivanyos, G.: Finding hidden Borel subgroups of the general linear group. Quantum Inf. Comput. 12, 661–669 (2012)
18. Ivanyos, G., Magniez, F., Santha, M.: Efficient quantum algorithms for some instances of the non-Abelian hidden subgroup problem. Int. J. Found. Comp. Sci. 15, 723–739 (2003)
19. Ivanyos, G., Sanselme, L., Santha, M.: An efficient quantum algorithm for the hidden subgroup problem in nil-2 groups. Algorithmica 63(1-2), 91–116 (2012)
20. Jozsa, R.: Quantum factoring, discrete logarithms, and the hidden subgroup problem. Computing in Science and Engineering 3, 34–43 (2001)
21. Yu. Kitaev, A.: Quantum measurements and the Abelian Stabilizer Problem, Technical report arXiv:quant-ph/9511026 (1995)
22. Moore, C., Rockmore, D., Russell, A., Schulman, L.: The power of basis selection in Fourier sampling: Hidden subgroup problems in affine groups. In: Proc. 15th ACM-SIAM SODA, pp. 1106–1115 (2004)

23. Regev, O.: Quantum computation and lattice problems. SIAM J. Comput. 33, 738–760 (2004)
24. Shor, P.: Algorithms for quantum computation: Discrete logarithm and factoring. SIAM J. Comput. 26, 1484–1509 (1997)
25. Springer, T.A.: Linear Algebraic groups, 2nd edn. Progress in mathematics, vol. 9. Birkhäuser, Basel (1998)
26. Watrous, J.: Quantum algorithms for solvable groups. In: Proc. 33rd ACM STOC, pp. 60–67 (2001)

A Note on the Minimum Distance
of Quantum LDPC Codes

Nicolas Delfosse[1], Zhentao Li[2], and Stéphan Thomassé[3]

[1] Département de Physique, Université de Sherbrooke,
Sherbrooke, Québec, J1K2R1, Canada
nicolas.delfosse@usherbrooke.ca
[2] Département d'Informatique UMR CNRS 8548, École Normale Supérieure, France
zhentao.li@ens.fr
[3] LIP, UMR 5668, École Normale Supérieure de Lyon - CNRS - UCBL - INRIA,
Université de Lyon, France
stephan.thomasse@ens-lyon.fr

Abstract. We provide a new lower bound on the minimum distance of a family of quantum LDPC codes based on Cayley graphs proposed by MacKay, Mitchison and Shokrollahi [14]. Our bound is exponential, improving on the quadratic bound of Couvreur, Delfosse and Zémor [3]. This result is obtained by examining a family of subsets of the hypercube which locally satisfy some parity conditions.

1 Introduction

A striking difference between classical and quantum computing is the unavoidable presence of perturbations when we manipulate a quantum system, which induces errors at every step of the computation. This makes essential the use of quantum error correcting codes. Their role is to avoid the accumulation of errors throughout the computation by rapidly identifying the errors which occur.

One of the most satisfying construction of classical error correcting codes capable of a rapid determination of the errors which corrupt the data is the family of Low Density Parity–Check codes (LDPC codes) [9]. It is therefore natural to investigate their quantum generalization. Moreover, Gottesman remarked recently that this family of codes can significantly reduce the overhead due to the use of error correcting codes during a quantum computation [10]. Quantum LDPC codes may therefore become an essential building block for quantum computing.

Quantum LDPC codes have been proposed by MacKay, Mitchison, and Mac-Fadden in [15]. One of the first difficulty which arises is that most of the families of quantum LDPC codes derived from classical constructions lead to a bounded minimum distance, see [20] and references therein. Such a distance is generally not sufficient and it induces a poor error-correction performance.

Only a rare number of constructions of quantum LDPC codes are equipped with an unbounded minimum distance. Most of them are inspired by Kitaev toric codes constructed from the a tiling of the torus [12] such as, color codes

E. Csuhaj-Varjú et al. (Eds.): MFCS 2014, Part II, LNCS 8635, pp. 239–250, 2014.
© Springer-Verlag Berlin Heidelberg 2014

which are based on 3-colored tilings of surfaces [1], hyperbolic codes which are defined from hyperbolic tilings [8,21], or other constructions based on tilings of higher dimensional manifolds [8,11]. These constructions are based on tilings of surfaces or manifolds and their minimum distance depends on the homology of this tiling. The determination of the distance of these codes is thus based on homological properties and general bounds on the minimum distance can be derived from sophisticated homological inequalities [6,4].

In this article, we study a construction of quantum LDPC codes proposed by MacKay, Mitchison and Shokrollahi [15] based on Cayley graphs and studied in [3]. This family does not rely on homological properties and thus homological method cited earlier seems impossible to apply. We relate their minimum distance to a combinatorial property of the hypercube. Then, using an idea of Gromov, we derive a lower bound on the minimum distance of these quantum codes which clearly improves the results of Couvreur, Delfosse and Zémor [3].

The remainder of this article is organized as follows. In Section 2, we recall the definition of linear codes and a construction of quantum codes based on classical codes. Section 3 introduces the quantum codes of MacKay, Mitchison and Shokrollahi [14]. In order to describe the minimum distance of these quantum codes based on Cayley graphs, we introduced two families of subsets of these graphs that we call borders and pseudo-borders in Section 4. We are then interested in the size of pseudo-borders of Cayley graphs. In Section 5, we reduce this problem to the study of t-pseudo-borders of the hypercube, which are a local version of pseudo-borders. Theorem 1, proved in Section 6, establishes a lower bound on the size of t-pseudo-border. As a corollary, we derive a lower bound on the minimum distance of Cayley graphs quantum codes.

2   Minimum Distance of Quantum Codes

A *code of length* n is defined to be a subspace of $\mathbb{F}_2^n$. It contains 2^k elements, called *codewords*, where k is the dimension of the code. The *minimum distance* d of a code is the minimum Hamming distance between two codewords. By linearity, it is also the minimum Hamming weight of a non-zero codeword. This parameter plays an important role in the error correction capability of the code. Indeed, assume we start with a codeword c and that t of its bits are flipped. Denote by c' the resulting vector. If t is smaller than $(d-1)/2$ then we can recover c by looking for the closest codeword to c'. Therefore, we can theoretically correct up to $(d-1)/2$ bit-flip errors. The parameters of a code are denoted $[n, k, d]$.

Every code can be defined as the kernel of a binary matrix H, called a *parity–check matrix* of the code. Alternatively, a code can be given as the space generated by the rows of a matrix called a *generator matrix* of the code. For instance, the following parity–check matrix defines a code of parameter $[7, 4, 3]$.

$$H = \begin{pmatrix} 1\,0\,1\,0\,1\,0\,1 \\ 0\,1\,1\,0\,0\,1\,1 \\ 0\,0\,0\,1\,1\,1\,1 \end{pmatrix}. \tag{1}$$

The code which admits H as a generator matrix has parameters $[7, 3, 4]$.

The space $\mathbb{F}_2^n$ is equipped with the inner product $(x, y) = \sum_{i=1}^{n} x_i y_i$, where $x = (x_1, x_2, \ldots, x_n)$ and $y = (y_1, y_2, \ldots, y_n)$. The orthogonal of a code C of length n is called the *dual code* of C. A code is said to be *self-orthogonal* if it is included in its dual. For example, we can easily check that two rows of the matrix H given in Eq.(1) are orthogonal which means the code generated by the rows of H is self-orthogonal.

Quantum information theory studies the generalization of error correcting codes to the protection of information written in a quantum mechanical system. By analogy with the classical setting, a quantum error correcting code is defined as the embedding of K qubits into N qubits. The CSS construction allows us to define a quantum code from a classical self-orthogonal code [2,19]. As in the classical setting, the minimum distance D of a quantum code is an important parameter which measures the code's performance. The following proposition gives a combinatorial description of the parameters of these quantum codes.

Proposition 1. *Let C be a classical code of parameters $[n, k, d]$. If C is self-orthogonal, then we can associate with C a quantum code of parameters $[[N, K, D]]$, where $N = n$, $K = \dim C^\perp / C = n - 2k$ and when $K \neq 0$, D is the minimum weight of codeword of $C^\perp$ which is not in C: $D = \min\{w(x) \mid x \in C^\perp \backslash C\}$.*

Throughout this article, we only consider this combinatorial definition of the parameters of quantum codes. A complete description of quantum error correcting codes, starting from the postulates of quantum mechanics, can be found for example in [18].

The minimum distance of the classical code $C^\perp$ is $d^\perp = \min\{w(x) \mid x \in C^\perp \backslash \{0\}\}$. If there exists a vector $x \in C^\perp \backslash \{0\}$ of minimum weight which is not in C, then the quantum minimum distance is $D = d^\perp$. In that case, the computation of the minimum distance corresponds to the computation of the minimum distance of the classical code $C^\perp$.

When, D is strictly larger than the classical minimum distance $d^\perp$, the quantum code is called *degenerate*. Then, we do not consider the codewords of C in the computation of D. This essential feature can improve the performance of the quantum code but also makes the determining the minimum distance strikingly more difficult than in the classical setting. In the present work, we obtain a lower bound on the minimum distance of a family of degenerate quantum codes.

3   A Family of Quantum Codes Based on Cayley Graphs

We consider a family of quantum codes constructed from Cayley graphs, which we now define.

Definition 1. *Let G be a group and S be a set of elements of G such that $s \in S$ implies $s^{-1} \in S$. The Cayley graph $\Gamma(G, S)$ is the graph with vertex set G such that two vertices are adjacent if they differ by an element in S.*

In our case, the group G is always $\mathbb{F}_2^r$ and $S = \{c_1, c_2, \ldots, c_n\}$ is a generating set of $\mathbb{F}_2^r$. Thus $\Gamma(G, S)$ is a regular graph of degree n with 2^r vertices.

This graph is connected since S is a generating set. To simplify notation, we assume that the vectors c_i are the n columns of a matrix $H \in M_{r,n}(\mathbb{F}_2)$. We denote by $G(H)$ this graph and $A(H)$ the adjacency matrix of this graph.

For instance, the Cayley graph $G(I_n)$, associated with the identity matrix of size n, is the hypercube of dimension n. Indeed, its vertex set is $\mathbb{F}_2^n$ and two vertices x and y are adjacent if and only the vectors x and y differ in exactly one component.

The following proposition proves that we can associate a quantum code with these graphs [3].

Proposition 2. *Let $H \in M_{r,n}(\mathbb{F}_2)$ be a binary matrix. If n is an even integer, then the adjacency matrix $A(H)$ of the graph $G(H)$ is the generating matrix of a classical self-orthogonal code. We denote by $C(H)$ this self-orthogonal code and by $Q(H)$ the corresponding quantum code.*

We want to determine the minimum distance of these quantum codes $Q(H)$.

Recall that a family of quantum codes associated with a family of self-orthogonal codes (C_i) defines quantum LDPC codes if C_i admits a parity–check matrix H_i which is sparse. In our case the generating matrix of the code $C(H)$ is the adjacency matrix $A(H)$, which is typically sparse since each row is a vector of length 2^r and weight n. When n and r are proportional, each row of $A(H)$ has weight in $O(\log N)$, where $N = 2^r$ is the number of columns of $A(H)$. Some authors consider LDPC codes defined by a parity-check matrix with bounded row weight. However, an unbouded row weight is needed for example to achieve the capacity of the quantum erasure channel [5].

It turns out that regardless of the choice of generators, the resulting graph $G(H)$ is always *locally isomorphic* to the hypercube [3]. That is, the set of vertices within some distance t, depending on H, of a vertex in $G(H)$, is isomorphic to the subgraph of the hypercube induced by all vertices within distance t of a vertex in the hypercube.

We close this subsection with a sketch of the proof of this local isomorphism. By *ball* of radius t centered at a vertex v in a graph G, we mean the subgraph of G induced by all vertices at distance at most t from v. The neighbourhood $N(v)$ of a vertex v of the graph $G(H)$ is the set of vertices incident to v.

The radius of the isomorphism depends on the shortest length d of a relation $\sum_{i=1}^{d} c_i = 0$ between columns of H, which is, by definition, the minimum distance of the code of parity-check matrix H.

Proposition 3. *Let $H \in M_{r,n}(\mathbb{F}_2)$ and let d be the minimum distance of the code of parity–check matrix H. Then, there is a graph isomorphism between any ball of radius $(d-1)/2$ of $G(H)$ and any ball of same radius in $G(I_n)$ where I_n is the identity matrix of size n.*

Proof (sketch). The generating set S is the set of columns of H. The neighbourhood of a vertex x is $\{x + c_1 | c_1 \in S\}$ and the second neighbourhood is $\{x + c_1 + c_2 | c_1 \neq c_2 \in S\}$. If there is no relation between set of 4 different generators (i.e. there is no relation where the sum of four generators is 0) then except for $x + c_1 + c_2$ and $x + c_2 + c_1$ being equal, all these vertices are distinct. So by mapping x to 0, c_i to e_i and $c_i + c_j$ to $e_i + e_j$, we see that vertices at distance at most two from x is isomorphic to the hypercube.

In the general case, we can also map each generator of $G(H)$ to a generator of $G(I_n)$.   □

This result gives more information about the local structure of the graph when we start with a parity–check matrix H defining a code of large distance d. For instance, when $H \in M_{r,n}(\mathbb{F}_2)$ is a random matrix chosen uniformly among the matrices of maximal rank, the distance d is linear in n. We can also choose H as the parity-check matrix of a known code equipped with a large minimum distance.

Our result will be proved by only looking at vertices within distance $(d-1)/2$ of some central vertex and hence we may assume that we are in the hypercube of dimension n.

4   Borders and Pseudo-borders of Cayley Graphs

The aim of this section is to provide a graphical description of the minimum distance of the quantum codes $Q(H)$ introduced in the previous section. This quantum code is associated with the classical self-orthogonal code $C(H)$ generated by the rows of the matrix $A(H)$. By Proposition 1, the minimum distance D of this quantum code is given by

$$D = \min\{w(x) \mid x \in C(H)^{\perp}\backslash C(H)\}.$$

By definition, the codes $C(H)$ and its dual $C(H)^{\perp}$ are subspaces of $\mathbb{F}_2^N$ where $N = 2^r$ is the size of the matrix $A(H)$. Since the columns of $A(H)$ are indexed by the N vertices of the graph $G(H)$, a vector $x \in \mathbb{F}_2^N$ can be regarded as the indicator vector of a subset of the vertex set of $G(H)$. We can then replace the two conditions $x \in C(H)^{\perp}$ and $x \notin C(H)$ by conditions on the set of vertices corresponding to x. In order to describe the vectors x of $C(H)$ and $C(H)^{\perp}$, we introduced two families of subsets of the vertex set of $G(H)$: the borders and the pseudo-borders.

Definition 2. *Let S be a subset of the vertex set of $G(H)$. The* border $B(S)$ *of S in the graph $G(H)$ is the set of vertices of $G(H)$ which belong to an odd number of neighbourhoods $N(v)$ for $v \in S$.*

Equivalently, the border of a subset S is the symmetric difference of all the neigborhoods $N(v)$ for $v \in S$.

Definition 3. *A* pseudo-border *in the graph $G(H)$ is a family $\mathcal{P}$ of vertices of $G(H)$ such that the cardinality of $N(v) \cap \mathcal{P}$ is even for every vertex v of $G(H)$.*

These borders and pseudo-borders correspond to the vectors of the classical code $C(H)$ and its dual.

Proposition 4. *Let x be a vector of $\mathbb{F}_2^N$ where $N = 2^r$ is the number of vertices of $G(H)$. Then x is the indicator vector of a subset $\mathcal{S}_x$ of the vertex set of $G(H)$. Moreover, we have*

- *$x \in C(H)$ if and only if $\mathcal{S}_x$ is border,*
- *$x \in C(H)^\perp$ if and only if $\mathcal{S}_x$ is pseudo-border.*

The idea of the proof is to check the vector properties of $A(H)$ (needed for $x \in C(H)$ or $x \in C(H)^\perp$) translate to properties of neighbourhoods in $G(H)$.

This proposition, combined with Proposition 2, shows that, when n is even, every border is a pseudo-border. In some special cases, every pseudo-border is a border. However, this is generally false and in these cases,, the minimum distance of the quantum code associated with H is equal to

$$D = \min\{|\mathcal{S}| \mid \mathcal{S} \text{ is a pseudo-border which is not a border }\}.$$

Since the graph $G(H)$ is locally isomorphic to the hypercube of dimension n, (Proposition 3), it is natural to investigate the borders and pseudo-borders of the hypercube.

5   Borders and Pseudo-borders of the Hypercube

In this section, we focus on the hypercube and we introduce a local version of the pseudo-borders which is preserved by the local isomorphism of Proposition 3.

We gave the Cayley graph definion of a hypercube as $G(I_n)$ in Section 3. Here we give an alternate definition based on sets. We denote by $[n]$ the set $\{1, 2, \ldots, n\}$. The vertex set of the hypercube is $2^{[n]}$, all subsets of $[n]$ (see Figure 1). Inspired by this graphical structure, for a subset v of $[n]$, we write $N(v)$ for the *neighbourhood* centered at v, i.e. the family of all subsets of $[n]$ that differ by one element from v.

To motivate our new definitions, recall a simple lower bound on the cardinality of a pseudo-border $\mathcal{S}$ which is not a border in $G(H)$. Assume that the parameter d is larger than 7, so that every ball of radius $(d-1)/2 = 3$ of the graph $G(H)$ is isomorphic to a ball of the hypercube of dimension n by Proposition 3. The pseudo-border $\mathcal{S}$ is not empty, otherwise it is a border. Therefore, it contains a vertex u of $G(H)$. We use the following arguments:

- $u \in \mathcal{S}$,
- The cardinality of $\mathcal{S} \cap N(v)$ is even for every neighbour v of the vertex u,
- The ball $B(u, 3)$ of the graph $G(H)$ is isomorphic to a ball of the hypercube of dimension n.

Then, each of the n neighbourhoods $N(v)$ centered at $v \in N(u)$, contains the vertex u, thus it must contain at least another vertex of $\mathcal{S}$. This provides n other vertices of the set $\mathcal{S}$. Since these vertices can appear in at most two different sets $N(v)$, the set $\mathcal{S}$ contains at least $1 + n/2$ distinct vertices.

In order to extend this argument, we introduce t-pseudo-borders.

Definition 4. *Let t and n be two positive integers such that $t < n$. A t-pseudo-border $\mathcal{S}$ of the hypercube $2^{[n]}$ is a family of subsets of $[n]$ such that*

- *$\emptyset \in \mathcal{S}$,*
- *The cardinality of $\mathcal{S} \cap N(v)$ is even for every $v \subset [n]$ of size $|v| \le t - 1$,*
- *$\mathcal{S}$ is included in the ball of radius t centered in $\emptyset$ of the hypercube $2^{[n]}$.*

In other words, a t-pseudo-border is a subset of vertices of a ball of the hypercube satisfying the conditions of the definition of a pseudo-border in this ball. Starting from a pseudo-border of a Cayley graph $G(H)$ and applying the local isomorphism of Proposition 3, we obtain a t-pseudo-border of the hypercube. We are interested in a lower bound on the size of t-pseudo-borders. Thus, we aim to answer the following refinement of the question of the determination of the minimum distance of the quantum codes $Q(H)$.

Question 1. What is the minimum cardinality of a t-pseudo-border of the hypercube $2^{[n]}$ which is not a border?

The results of [3] provide a polynomial lower bound in $O(tn^2)$. To our knowledge, this is the best known lower bound on the cardinality of the t-pseudo-border of the hypercube. Our main result is an exponential lower bound.

Theorem 1. *The minimum cardinality of a t-pseudo-border of the hypercube $2^{[n]}$ which is not a border is at least*

$$\sum_{i=0}^{i \le M} \frac{(n/2)^{i/2}}{i!},$$

where $M = \min\{t - 1, \sqrt{n/2}\}$. When t is larger than $\sqrt{n/2}$, this lower bound is at least $e^{\sqrt{n/2}}$.

This Theorem is proved in Section 6. As an application, we obtain a lower bound on the minimum distance of the Cayley graph quantum codes $Q(H)$.

Corollary 1. *Let $H \in M_{r,n}(\mathbb{F}_2)$, with n an even integer, and let d be the minimum distance of the code of parity–check matrix H. The quantum code $Q(H)$ encodes K qubits into $N = 2^r$ qubits. If $K \ne 0$, then the minimum distance D of $Q(H)$ is at least*

$$D \ge \sum_{i=0}^{i \le M} \frac{(n/2)^{i/2}}{i!},$$

where $M = \min\{(d - 3)/2, \sqrt{n/2}\}$. When d is larger than $\sqrt{n/2}$, this lower bound is at least $e^{\sqrt{n/2}}$.

Proof. We want to bound the minimum distance D of $Q(H)$, which, by Proposition 1, is the minimum weight of a vector x of $C(H)^{\perp} \setminus C(H)$. Such a vector x corresponds to a subset $\mathcal{S}_x$ of vertices of the graph $G(H)$ which is a pseudo-border and which is not a border by Proposition 4. By this bijection $x \mapsto \mathcal{S}_x$, the weight of x corresponds to the cardinality of the set $\mathcal{S}_x$. Therefore, D is the minimum cardinality of a pseudo-border of $G(H)$ which is not a border.

First, let us prove that such a pseudo-border exists. By Proposition 1, the number of encoded qubits K, which is assumed to be positive, is the dimension of the quotient space $K = \dim C(H)^{\perp}/C(H)$. We know, from Proposition 4, that $C(H)^{\perp}$ and $C(H)$ are in one-to-one correspondence with the sets of pseudo-borders and borders respectively. Therefore the positivity of K implies the existence of a pseudo-border $\mathcal{S}$ of $G(H)$ which is not a border.

Let $\mathcal{S}$ be a pseudo-border of $G(H)$ which is not a border of minimum cardinality. Since $\mathcal{S}$ is not a border it is not empty. Let u be a vertex of $\mathcal{S}$ in $G(H)$. Applying the local isomorphism of Proposition 3, we can map the ball of radius $(d-1)/2$ centered at u of $G(H)$ to the ball of same radius centered at $\emptyset$ of the hypercube $2^{[n]}$. By this transformation, the restriction of $\mathcal{S}$ to the ball $B(u, (d-1)/2)$ is sent onto a $(d-1)/2$-pseudo-border $\mathcal{S}_u$ of the hypercube $2^{[n]}$. The set $\mathcal{S}_u$ is not a border, otherwise the set $\mathcal{S}\backslash\mathcal{S}_u$ is a pseudo-border of $G(H)$ which is not a border of size strictly smaller than $|\mathcal{S}|$. This graph isomorphism cannot increase the size of $\mathcal{S}$ thus D is lower bounded by the minimum size of a $(d-1)/2$-pseudo-border of $2^{[n]}$: $D \geq |\mathcal{S}| \geq |\mathcal{S}_u|$. Finally, Theorem 1 provides a lower bound on $|\mathcal{S}_u|$.   □

6   Bound on the Size of Local Pseudo-borders of the Hypercube

This section is devoted to the proof of Theorem 1. So our goal is to derive a lower bound on the size of t-pseudo-borders of the hypercube which are not borders. Here it is more convenient to use the language of sets. We work in the hypercube $2^{[n]}$ defined in Section 5, whose vertices correspond to subsets of $[n] = \{1, 2, \ldots, n\}$.

To show our bound, we consider a t-pseudo-border $\mathcal{S}$ of minimum size. In order to exploit the minimality of this set, we introduce the following operation.

Definition 5. *Let $\mathcal{S}$ be a family of subsets of $[n]$ and v a subset of $[n]$. By flipping $\mathcal{S}$ along v, we mean to swap elements and non-elements of $\mathcal{S}$ in the border $N(v)$.*

Put differently, flipping $\mathcal{S}$ along v gives the symmetric difference $\mathcal{S}\Delta N(v)$. We now show that flipping a minimum t-pseudo-border cannot decrease its size, a key ingredient for our main result.

Lemma 1. *Let $\mathcal{S}$ be a t-pseudo-border of $2^{[n]}$ which is not a border, of minimum cardinality. Then, flipping $\mathcal{S}$ along any number of subsets v such that $2 \leq |v| \leq t-1$ leads to a family of greater or equal cardinality.*

Proof (sketch). By minimality of $\mathcal{S}$, it suffices to check that this transformation conserves the set of t-pseudo-borders which are not borders.   □

6.1   Lower Bounds for 2-Subsets and 4-Subsets

In Section 5, we showed a first lower bound on the size of pseudo-borders. The basic idea is to use the facts that $\emptyset \in S$ and there is an even number of elements of S in $N(v)$ when $|v| = 1$. This shows S has elements at distance 2 from $\emptyset$. It is reasonable to expect that these subsets of S at distance 2 from $\emptyset$ will imply the existence of subsets of S at distance 4 from $\emptyset$ and so on. However, as the distance to $\emptyset$ increases, the local structure of the graph becomes more and more complicated making this problem more complex.

In this section, as an example, we give a lower bound on the number of 4-subsets of a minimal t-pseudo-borders which is not a border. The same tools are then used in Section 6.2 to bound the number of k-subsets of a minimal t-pseudo-border. A k-set (or k-subset) of a set $[n]$ is a subset of $[n]$ of size k. k-sets are used to decompose the hypercube into layers (see Fig. 1).

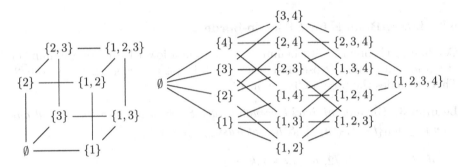

Fig. 1. Left: Hypercube on the set [3]. Right: k-sets of the hypercube on [4].

Definition 6. *An odd k-set with respect to S is a k-set (not necessarily in S) containing an odd number of $(k-1)$-subsets of S.*

Stated differently, odd k-sets are the sets v such that the constraint $|S \cap N(v)|$ is even is not satisfied when we restrict S to the ball of radius $k - 1$ centered in $\emptyset$. Therefore, they can be used to deduce the existence of $(k+1)$-subsets of S as proved in the following lemma.

Lemma 2. *Let $k < t - 1$. If there are o_k odd k-sets with respect to a minimal t-pseudo-border S which is not a border then there are at least $\frac{o_k}{k+1}$ sets of size $k + 1$ in S.*

Proof. The ball centered at each of these odd k-sets contains an even number of elements of S and therefore contains at least one element of S of size $k + 1$.

On the other hand, each $(k+1)$-set contains $k+1$ k-sets and therefore at most $k + 1$ odd k-sets of S. So we need at least $\frac{o_k}{k+1}$ such sets to satisfy the parity condition for all odd k-sets.   □

We now need to lower bound the number of odd $(k+1)$-sets in terms of the number of k-sets. This bound is adapted from a result of Gromov, independently proven by Linial, Meshulam [13], and Wallach [17] (and maybe by others). (See also Lemma 3 of [16] and [7]). The following result is proved in Section 6.3.

Theorem 2. *For every $1 \le k \le t - 2$, if there are s_k k-sets in a minimal t-pseudo-border $\mathcal{S}$ which is not a border then there are at least $\frac{n-(k-1)k}{k+1} s_k$ odd $(k+1)$-sets with respect to $\mathcal{S}$.*

For example, let us consider the first layers of a minimal t-pseudo-border which is not a border $\mathcal{S}$. Assume that $t \ge 3$. We can see that all 1-sets are odd with respect to $\mathcal{S}$. Since there are exactly n 1-sets, this remark immediately gives us at least $\frac{n}{2}$ 2-sets with respect to $\mathcal{S}$. If $t \ge 5$, applying Theorem 2 with $k = 2$, we get at least $\frac{n}{2}\frac{n-2}{3} = \frac{n(n-2)}{2 \cdot 3}$ odd 3-sets with respect to $\mathcal{S}$. Then, we obtain at least $\frac{n(n-2)}{2 \cdot 3 \cdot 4}$ 4-sets in $\mathcal{S}$.

6.2   Lower Bounds for t-Pseudo-borders

Combining Theorem 2 with Lemma 2, we obtain a lower bound on the number of k-sets in a minimal t-pseudo-border which is not a border when $k \le \sqrt{n/2}$. This concludes the proof of Theorem 1.

Lemma 3. *For any minimal t-pseudo-border $\mathcal{S}$ which is not a border and any even $k \le \min\{t-1, \sqrt{n/2}\}$, $\mathcal{S}$ has at least $\frac{n^{k/2}}{2^{k/2}k!}$ k-sets.*

Proof. Since $k \le \sqrt{n/2}$, $n - (k-1)k \ge \frac{n}{2}$.

We prove this by induction on k. Since $\emptyset \in \mathcal{S}$, it is true for $k = 0$. Suppose this is true for $k - 2$. Then by Theorem 2, there are at least $\frac{n}{2(k-1)} \frac{n^{(k-2)/2}}{2^{(k-2)/2}(k-2)!}$ odd $(k+1)$-sets with respect to $\mathcal{S}$. By Lemma 2, $\mathcal{S}$ contains at least

$$\frac{1}{k} \frac{n}{2(k-1)} \frac{n^{(k-2)/2}}{2^{(k-2)/2}(k-2)!} = \frac{n^{k/2}}{2^{k/2}k!}$$

k-sets, as required.   □

The bound in Lemma 3 is maximized at $k = \sqrt{n/2}$ and by Stirling's formula is at least $e^{\sqrt{n/2}}$.

To obtain the lower bound on the size of t-pseudo-borders of $2^{[n]}$ stated in Theorem 1, we simply apply Lemma 3 to all the k-sets of a t-pseudo-border with $k < t$ and $k \le \sqrt{n/2}$.

6.3   Lower Bounds for Odd Sets

Proof (Proof of Theorem 2). Let $\mathcal{S}$ be a minimal t-pseudo-border. We denote by $\mathcal{S}_k$ the set of k-sets of $\mathcal{S}$ and by $\mathcal{O}_{k+1}$ the odd $(k+1)$-sets with respect to $\mathcal{S}$. For an element $i \in [n]$, we write $\partial_i(\mathcal{O}_{k+1})$ for the sets of $\mathcal{O}_{k+1}$ containing i, each with i removed (so $\partial_i(\mathcal{O}_{k+1})$ is a set of k-sets) and $\partial_i(\mathcal{S}_k)$ for the set of $(k-1)$-sets consisting of all elements of $\mathcal{S}_k$ containing i with i itself removed from each k-set.

Let $i \in [n]$ be the index minimizing

$$|\partial_i(\mathcal{O}_{k+1})| + (k-1)|\partial_i(\mathcal{S}_k)| \leq \frac{1}{n}\sum_i (|\partial_i(\mathcal{O}_{k+1})| + (k-1)|\partial_i(\mathcal{S}_k)|)$$

$$= \frac{1}{n}\sum_i |\partial_i(\mathcal{O}_{k+1})| + \frac{k-1}{n}\sum_i |\partial_i(\mathcal{S}_k)|$$

$$= \frac{k+1}{n}|\mathcal{O}_{k+1}| + \frac{(k-1)k}{n}|\mathcal{S}_k|.$$

Since $\mathcal{S}$ is minimal, we may flip on $\partial_i(\mathcal{S}_k)$ and apply Lemma 1. We claim this flip yields a family $\mathcal{S}'$ whose k-sets is exactly $\partial_i(\mathcal{O}_{k+1})$. Indeed, if $f \notin \partial_i(\mathcal{O}_{k+1})$ then $f \cup \{i\} \notin \mathcal{O}_{k+1}$ which means $f \cup \{i\}$ contains an even number of elements of $\mathcal{S}_k$. But except for f itself, these elements of $\mathcal{S}_k$ all contain i. If $f \notin \mathcal{S}_k$, f contains an even number of elements of $\partial_i(\mathcal{S}_k)$ and is therefore not added (flipped) to $\mathcal{S}'$. If $f \in \mathcal{S}_k$, f contains an odd number of elements of $\partial_i(\mathcal{S}_k)$ and is therefore removed (flipped) when building $\mathcal{S}'$. The reverse inclusion is proved similarly.

The only other sets affected by flipping on $\partial_i(\mathcal{S}_k)$ are $(k-2)$-sets and this flips (gains) at most $(k-1)|\partial_i(\mathcal{S}_k)|$ elements of size $k-2$ (since each element of $\partial_i(\mathcal{S}_k)$ has size $k-1$ and contains at most $k-1$ elements (of $\mathcal{S}$) of size $k-2$).

Therefore, by minimality (Lemma 1)

$$|\mathcal{S}_k| \leq |\partial_i(\mathcal{O}_{k+1})| + (k-1)|\partial_i(\mathcal{S}_k)| \leq \frac{k+1}{n}|\mathcal{O}_{k+1}| + \frac{(k-1)k}{n}|\mathcal{S}_k|.$$

Rearranging gives $\frac{n-(k-1)k}{k+1}|\mathcal{S}_k| \leq |\mathcal{O}_{k+1}|$ and the theorem follows.   □

Acknowledgements. Nicolas Delfosse was supported by the Lockheed Martin Corporation. Nicolas Delfosse acknowledges the hospitality of Robert Raussendorf and the University of British Columbia where part of this article was written. The authors wish to thank Benjamin Audoux, Alain Couvreur, Anthony Leverrier, Jean-Pierre Tillich and Gilles Zémor for their comments and thank Jean-Sébastien Séréni for fruitful discussions about Gromov result.

References

1. Bombin, H., Martin-Delgado, M.: Topological quantum distillation. Physical Review Letters 97, 180501 (2006)
2. Calderbank, A., Shor, P.: Good quantum error-correcting codes exist. Physical Review A 54(2), 1098 (1996)
3. Couvreur, A., Delfosse, N., Zémor, G.: A construction of quantum LDPC codes from Cayley graphs. IEEE Transactions on Information Theory 59(9), 6087–6098 (2013)

4. Delfosse, N.: Tradeoffs for reliable quantum information storage in surface codes and color codes. In: Proc. of IEEE International Symposium on Information Theory, ISIT 2013, pp. 917–921 (2013)

5. Delfosse, N., Zémor, G.: Upper bounds on the rate of low density stabilizer codes for the quantum erasure channel. Quantum Information & Computation 13(9-10), 793–826 (2013)

6. Fetaya, E.: Bounding the distance of quantum surface codes. Journal of Mathematical Physics 53, 062202 (2012)

7. Fox, J., Gromov, M., Lafforgue, V., Naor, A., Pach, J.: Overlap properties of geometric expanders. Journal für die reine und angewandte Mathematik (Crelles Journal) 2012(671), 49–83 (2012)

8. Freedman, M., Meyer, D., Luo, F.: Z2-systolic freedom and quantum codes. Mathematics of Quantum Computation, pp. 287–320. Chapman & Hall/CRC (2002)

9. Gallager, R.: Low Density Parity-Check Codes. Ph.D. thesis, Massachusetts Institute of Technology (1963)

10. Gottesman, D.: What is the overhead required for fault-tolerant quantum computation? arXiv preprint arXiv:1310.2984 (2013)

11. Guth, L., Lubotzky, A.: Quantum error-correcting codes and 4-dimensional arithmetic hyperbolic manifolds. arXiv preprint arXiv:1310.5555 (2013)

12. Kitaev, A.: Fault-tolerant quantum computation by anyons. Annals of Physics 303(1), 27 (2003)

13. Linial, N., Meshulam, R.: Homological connectivity of random 2-complexes. Combinatorica 26(4), 475–487 (2006)

14. MacKay, D., Mitchison, G., Shokrollahi, A.: More sparse-graph codes for quantum error-correction (2007),
http://www.inference.phy.cam.ac.uk/mackay/cayley.pdf

15. MacKay, D.J.C., Mitchison, G., McFadden, P.L.: Sparse-graph codes for quantum error correction. IEEE Transaction on Information Theory 50(10), 2315–2330 (2004)

16. Matousek, J., Wagner, U.: On Gromov's method of selecting heavily covered points. arXiv preprint arXiv:1102.3515 (2011)

17. Meshulam, R., Wallach, N.: Homological connectivity of random k-dimensional complexes. Random Structures & Algorithms 34(3), 408–417 (2009)

18. Nielsen, M., Chuang, I.: Quantum Computation and Quantum Information, 1st edn. Cambridge University Press (2000)

19. Steane, A.: Multiple-particle interference and quantum error correction. Proc. of the Royal Society of London. Series A: Mathematical, Physical and Engineering Sciences 452(1996), 2551–2577 (1954)

20. Tillich, J.P., Zémor, G.: Quantum LDPC codes with positive rate and minimum distance proportional to $n^{1/2}$. In: Proc. of IEEE International Symposium on Information Theory, ISIT 2009, pp. 799–803 (2009)

21. Zémor, G.: On cayley graphs, surface codes, and the limits of homological coding for quantum error correction. In: Chee, Y.M., Li, C., Ling, S., Wang, H., Xing, C. (eds.) IWCC 2009. LNCS, vol. 5557, pp. 259–273. Springer, Heidelberg (2009)

Minimum Bisection Is NP-hard
on Unit Disk Graphs

Josep Díaz[1] and George B. Mertzios[2,*]

[1] Departament de Llenguatges i Sistemes Informátics,
Universitat Politécnica de Catalunya, Spain
[2] School of Engineering and Computing Sciences, Durham University, UK
diaz@lsi.upc.edu, george.mertzios@durham.ac.uk

Abstract. In this paper we prove that the MIN-BISECTION problem is
NP-hard on *unit disk graphs*, thus solving a longstanding open question.

Keywords: Minimum bisection problem, unit disk graphs, planar
graphs, NP-hardness.

1 Introduction

The problem of appropriately partitioning the vertices of a given graph into
subsets, such that certain conditions are fulfilled, is a fundamental algorithmic
problem. Apart from their evident theoretical interest, graph partitioning problems have great practical relevance in a wide spectrum of applications, such as
in computer vision, image processing, and VLSI layout design, among others,
as they appear in many divide-and-conquer algorithms (for an overview see [2]).
In particular, the problem of partitioning a graph into equal sized components,
while minimizing the number of edges among the components turns out to be
very important in parallel computing. For instance, to parallelize applications
we usually need to evenly distribute the computational load to processors, while
minimizing the communication between processors.

Given a simple graph $G = (V, E)$ and $k \geq 2$, a *balanced k-partition* of $G =
(V, E)$ is a partition of V into k vertex sets $V_1, V_2, \ldots, V_k$ such that $|V_i| \leq \left\lceil \frac{|V|}{k} \right\rceil$
for every $i = 1, 2, \ldots, k$. The *cut size* (or simply, the *size*) of a balanced k-
partition is the number of edges of G with one endpoint in a set V_i and the other
endpoint in a set V_j, where $i \neq j$. In particular, for $k = 2$, a balanced 2-partition
of G is also termed a *bisection* of G. The *minimum bisection* problem (or simply,
MIN-BISECTION) is the problem, given a graph G, to compute a bisection of G
with the minimum possible size, also known as the *bisection width* of G.

Due to the practical importance of MIN-BISECTION, several heuristics and
exact algorithms have been developed, which are quite efficient in practice [2],
from the first ones in the 70's [16] up to the very efficient one described in [7].
However, from the theoretical viewpoint, MIN-BISECTION has been one of the

---

* Partially supported by the EPSRC Grant EP/K022660/1.

E. Csuhaj-Varjú et al. (Eds.): MFCS 2014, Part II, LNCS 8635, pp. 251–262, 2014.
© Springer-Verlag Berlin Heidelberg 2014

most intriguing problems in algorithmic graph theory so far. This problem is well known to be NP-hard for general graphs [11], while it remains NP-hard when restricted to the class of everywhere dense graphs [18] (i.e. graphs with minimum degree $\Omega(n)$), to the class of bounded maximum degree graphs [18], or to the class of d-regular graphs [5]. On the positive side, very recently it has been proved that MIN-BISECTION is fixed parameter tractable [6], while the currently best known approximation ratio is $O(\log n)$ [20]. Furthermore, it is known that MIN-BISECTION can be solved in polynomial time on trees and hypercubes [9, 18], on graphs with bounded treewidth [13], as well as on grid graphs with a constant number of holes [10, 19].

In spite of this, the complexity status of MIN-BISECTION on planar graphs, on grid graphs with an arbitrary number of holes, and on unit disk graphs have remained longstanding open problems so far [8, 10, 14, 15]. The first two of these problems are equivalent, as there exists a polynomial time reduction from planar graphs to grid graphs with holes [19]. Furthermore, there exists a polynomial time reduction from planar graphs with maximum degree 4 to unit disk graphs [8]. Therefore, since grid graphs with holes are planar graphs of maximum degree 4, there exists a polynomial reduction of MIN-BISECTION from planar graphs to unit disk graphs. Another motivation for studying MIN-BISECTION on unit disk graphs comes from the area of wireless communication networks [1, 3], as the bisection width determines the communication bandwidth of the network [12].

Our Contribution. In this paper we resolve the complexity of MIN-BISECTION on unit disk graphs. In particular, we prove that this problem is NP-hard by providing a polynomial reduction from a variant of the maximum satisfiability problem, namely from the monotone Max-XOR(3) problem (also known as the monotone Max-2-XOR(3) problem). Consider a monotone XOR-boolean formula ϕ with variables $x_1, x_2, \ldots, x_n$, i.e. a boolean formula that is the conjunction of XOR-clauses of the form $(x_i \oplus x_k)$, where no variable is negated. If, in addition, every variable x_i appears in exactly k XOR-clauses in ϕ, then ϕ is called a *monotone XOR(k)* formula. The *monotone Max-XOR(k)* problem is, given a monotone XOR(k) formula ϕ, to compute a truth assignment of the variables $x_1, x_2, \ldots, x_n$ that XOR-satisfies the largest possible number of clauses of ϕ. Recall here that the clause $(x_i \oplus x_k)$ is XOR-satisfied by a truth assignment τ if and only if $x_i \neq x_k$ in τ. Given a monotone XOR(k) formula ϕ, we construct a unit disk graph H_ϕ such that the truth assignments that XOR-satisfy the maximum number of clauses in ϕ correspond bijectively to the minimum bisections in H_ϕ, thus proving that MIN-BISECTION is NP-hard on unit disk graphs.

Organization of the Paper. Necessary definitions and notation are given in Section 2. In Section 3, given a monotone XOR(3)-formula ϕ with n variables, we construct an auxiliary unit disk graph G_n, which depends only on the size n of ϕ (and not on ϕ itself). In Section 4 we present our reduction from the monotone Max-XOR(3) problem to MIN-BISECTION on unit disk graphs, by modifying the graph G_n to a unit disk graph H_ϕ which also depends on the formula ϕ

itself. Finally we discuss the presented results and remaining open problems in Section 5.

2   Preliminaries and Notation

We consider in this article simple undirected graphs with no loops or multiple edges. In an undirected graph $G = (V, E)$, the edge between vertices u and v is denoted by uv, and in this case u and v are said to be *adjacent* in G. For every vertex $u \in V$ the *neighborhood* of u is the set $N(u) = \{v \in V \mid uv \in E\}$ of its adjacent vertices and its *closed neighborhood* is $N[u] = N(u) \cup \{u\}$. The subgraph of G that is *induced* by the vertex subset $S \subseteq V$ is denoted $G[S]$. Furthermore a vertex subset $S \subseteq V$ induces a *clique* in G if $uv \in E$ for every pair $u, v \in S$.

A graph $G = (V, E)$ with n vertices is the *intersection graph* of a family $F = \{S_1, \ldots, S_n\}$ of subsets of a set S if there exists a bijection $\mu : V \to F$ such that for any two distinct vertices $u, v \in V$, $uv \in E$ if and only if $\mu(u) \cap \mu(v) \neq \emptyset$. Then, F is called an *intersection model* of G. A graph G is a *disk* graph if G is the intersection graph of a set of disks (i.e. circles together with their internal area) in the plane. A disk graph G is a *unit disk* graph if there exists a disk intersection model for G where all disks have equal radius (without loss of generality, all their radii are equal to 1). Given a disk (resp. unit disk) graph G, an intersection model of G with disks (resp. unit disks) in the plane is called a *disk* (resp. *unit disk*) *representation* of G. Alternatively, unit disk graphs can be defined as the graphs that can be represented by a set of points on the plane (where every point corresponds to a vertex) such that two vertices intersect if and only if the corresponding points lie at a distance at most some fixed constant c (for example $c = 1$). Although these two definitions of unit disk graphs are equivalent, in this paper we use the representation with the unit disks instead of the representation with the points.

Note that any unit disk representation R of a unit disk graph $G = (V, E)$ can be completely described by specifying the centers c_v of the unit disks D_v, where $v \in V$, while for any disk representation we also need to specify the radius r_v of every disk D_v, $v \in V$. Given a graph G, it is NP-hard to decide whether G is a disk (resp. unit disk) graph [4, 17]. Given a unit disk representation R of a unit disk graph G, in the remainder of the paper we may not distinguish for simplicity between a vertex of G and the corresponding unit disk in R, whenever it is clear from the context. It is well known that the Max-XOR problem is NP-hard. Furthermore, it remains NP-hard even if the given formula ϕ is restricted to be a monotone XOR(3) formula. For the sake of completeness we provide in the next lemma a proof of this fact.

Lemma 1. *Monotone Max-XOR(3) is NP-hard.*

3   Construction of the Unit Disk Graph G_n

In this section we present the construction of the auxiliary unit disk graph G_n, given a monotone XOR(3)-formula ϕ with n variables. Note that G_n depends only on the size of the formula ϕ and not on ϕ itself. Using this auxiliary graph G_n we will then construct in Section 4 the unit disk graph H_ϕ, which depends also on ϕ itself, completing thus the NP-hardness reduction from monotone Max-XOR(3) to the minimum bisection problem on unit disk graphs.

We define G_n by providing a unit disk representation R_n for it. For simplicity of the presentation of this construction, we first define a set of halflines on the plane, on which all centers of the disks are located in the representation R_n.

3.1   The Half-lines Containing the Disk Centers

Denote the variables of the formula ϕ by $\{x_1, x_2, \ldots, x_n\}$. Define for simplicity the values $d_1 = 5.6$ and $d_2 = 7.2$. For every variable x_i, where $i \in \{1, 2, \ldots, n\}$, we define the following four points in the plane:

- $p_{i,0} = (2i \cdot d_1, 2(i-1) \cdot d_2)$ and $p_{i,1} = ((2i-1) \cdot d_1, (2i-1) \cdot d_2)$, which are called the *bend points* for variable x_i, and
- $q_{i,0} = ((2i-1) \cdot d_1, 2(i-1) \cdot d_2)$ and $r_{i,0} = (2i \cdot d_1, 2i \cdot d_2)$, which is called the *auxiliary points* for variable x_i.

Then, starting from point $p_{i,j}$, where $i \in \{1, 2, \ldots, n\}$ and $j \in \{0, 1\}$, we draw in the plane one halfline parallel to the x-axis pointing to the left and one halfline parallel to the y-axis pointing upwards. The union of these two halflines on the plane is called the *track* $T_{i,j}$ of point $p_{i,j}$. Note that, by definition of the points $p_{i,j}$, the tracks $T_{i,0}$ and $T_{i,1}$ do not have any common point, and that, whenever $i \neq k$, the tracks $T_{i,j}$ and $T_{k,\ell}$ have exactly one common point. Furthermore note that, for every $i \in \{1, 2, \ldots, n\}$, both auxiliary points $q_{i,0}$ and $r_{i,0}$ belong to the track $T_{i,0}$.

We will construct the unit disk representation R_n of the graph G_n in such a way that the union of all tracks $T_{i,j}$ will contain the centers of all disks in R_n. The construction of R_n is done by repeatedly placing on the tracks $T_{i,j}$ multiple copies of three particular unit disk representations $Q_1(p)$, $Q_2(p)$, and $Q_3(p)$ (each of them including $2n^6 + 2$ unit disks), which we use as gadgets in our construction. Before we define these gadgets we need to define first the notion of a (t, p)-crowd.

Definition 1. *Let $\varepsilon > 0$ be infinitesimally small. Let $t \geq 1$ and $p = (x_p, y_p)$ be a point in the plane. Then, the* horizontal (t, p)-crowd *(resp. the* vertical (t, p)-crowd*) is a set of t unit disks whose centers are equally distributed between the points $(x_p - \varepsilon, y_p)$ and $(x_p + \varepsilon, y_p)$ (resp. between the points $(x_p^\bullet, y_p - \varepsilon)$ and $(x_p, y_p + \varepsilon)$).*

Note that, by Definition 1, both the horizontal and the vertical (t, p)-crowds represent a clique of t vertices. Furthermore note that both the horizontal and the vertical $(1, p)$-crowds consist of a single unit disk centered at point p. For simplicity of the presentation, we will graphically depict in the following a (t, p)-crowd just by a disk with a *dashed contour* centered at point p, and having the number t written next to it. Furthermore, whenever the point p lies on the horizontal (resp. vertical) halfline of a track $T_{i,j}$, then any (t, p)-crowd will be meant to be a horizontal (resp. vertical) (t, p)-crowd.

3.2 Three Useful Gadgets

Let $p = (p_x, p_y)$ be a point on a track $T_{i,j}$. Whenever p lies on the horizontal halfline of $T_{i,j}$, we define for any $\delta > 0$ (with a slight abuse of notation) the points $p - \delta = (p_x - \delta, p_y)$ and $p + \delta = (p_x + \delta, p_y)$. Similarly, whenever p lies on the vertical halfline of $T_{i,j}$, we define for any $\delta > 0$ the points $p - \delta = (p_x, p_y - \delta)$ and $p + \delta = (p_x, p_y + \delta)$. Assume first that p lies on the *horizontal* halfline of $T_{i,j}$. Then we define the unit disk representation $Q_1(p)$ as follows:

- $Q_1(p)$ consists of the horizontal $(n^3, p + 0.9)$-crowd, the horizontal $(2n^6 - 2n^3 + 2, p + 2.8)$-crowd, and the horizontal $(n^3, p + 4.7)$-crowd, as it is illustrated in Figure 1(a).

Assume now that p lies on the *vertical* halfline of $T_{i,j}$, we define the unit disk representations $Q_2(p)$ and $Q_3(p)$ as follows:

- $Q_2(p)$ consists of a single unit disk centered at point p, the vertical $(n^6, p + 1.8)$-crowd, a single unit disk centered at point $p + 3.6$, and the vertical $(n^6, p + 5.4)$-crowd, as it is illustrated in Figure 1(b).
- $Q_3(p)$ consists of a single unit disk centered at point p, the vertical $(n^6, p + 1.7)$-crowd, a single unit disk centered at point $p + 3.6$, and the vertical $(n^6, p + 5.4)$-crowd, as it is illustrated in Figure 1(c).

In the above definition of the unit disk representation $Q_k(p)$, where $k \in \{1, 2, 3\}$, the point p is called the *origin* of $Q_k(p)$. Note that the origin p of the representation $Q_2(p)$ (resp. $Q_3(p)$) is a center of a unit disk in $Q_2(p)$ (resp. $Q_3(p)$). In contrast, the origin p of the representation $Q_1(p)$ is not a center of any unit disk of $Q_1(p)$, however p lies in $Q_1(p)$ within the area of each of the n^3 unit disks of the horizontal $(n^3, p + 0.9)$-crowd of $Q_1(p)$. For every point p, each of $Q_1(p)$, $Q_2(p)$, and $Q_3(p)$ has in total $2n^6 + 2$ unit disks (cf. Figure 1).

Furthermore, for any $i \in \{1, 2, 3\}$ and any two points p and p' in the plane, the unit disk representation $Q_i(p')$ is an isomorphic copy of the representation $Q_i(p)$, which is placed at the origin p' instead of the origin p. Moreover, for any point p in the vertical halfline of a track $T_{i,j}$, the unit disk representations $Q_2(p)$ and $Q_3(p)$ are almost identical: their only difference is that the vertical $(n^6, p + 1.8)$-crowd in $Q_2(p)$ is replaced by the vertical $(n^6, p + 1.7)$-crowd in $Q_3(p)$, i.e. this whole crowd is just moved downwards by 0.1 in $Q_3(p)$.

Observation 1. *Let $k \in \{1, 2, 3\}$ and $p \in T_{i,j}$, where $i \in \{1, 2, \ldots, n\}$ and $j \in \{0, 1\}$. For every two adjacent vertices u, v in the unit disk graph defined by $Q_k(p)$, u and v belong to a clique of size at least $n^6 + 1$.*

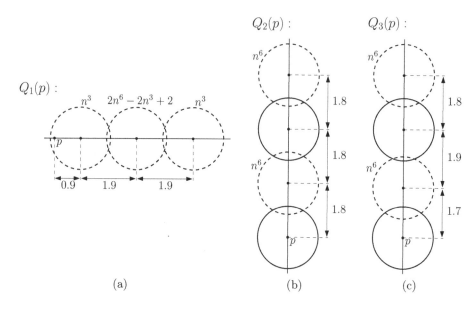

Fig. 1. The unit disk representations $Q_1(p)$, $Q_2(p)$, and $Q_3(p)$, where p is a point on one of the tracks $T_{i,j}$, where $1 \le i \le n$ and $j \in \{0,1\}$

3.3 The Unit Disk Representation R_n of G_n

We are now ready to iteratively construct the unit disk representation R_n of the graph G_n, using the above gadgets $Q_1(p)$, $Q_2(p)$, and $Q_3(p)$, as follows:

(a) for every $i \in \{1, 2, \ldots, n\}$ and for every $j \in \{0, 1\}$, add to R_n:
  - the gadget $Q_1(p)$, with its origin at the point $p = (0, (2(i-1) + j) \cdot d_2)$,
(b) for every $i \in \{1, 2, \ldots, n\}$, add to R_n:
  - the gadgets $Q_1(q_{i,0})$, $Q_2(r_{i,0})$, $Q_3(p_{i,0})$, and $Q_3(p_{i,1})$,
  - the gadgets $Q_1(p)$ and $Q_1(p')$, with their origin at the points $p = (-d_1, (2i-1) \cdot d_2)$ and $p' = (-2d_1, (2i-1) \cdot d_2)$ of the track $T_{i,1}$, respectively,
(c) for every $i, k \in \{1, 2, \ldots, n\}$ and for every $j, \ell \in \{0, 1\}$, where $i \neq k$, add to R_n:
  - the gadgets $Q_1(p)$ and $Q_2(p)$, with their origin at the (unique) point p that lies on the intersection of the tracks $T_{i,j}$ and $T_{k,\ell}$.

This completes the construction of the unit disk representation R_n of the graph $G_n = (V_n, E_n)$, in which the centers of all unit disks lie on some track $T_{i,j}$, where $i \in \{1, 2, \ldots, n\}$ and $j \in \{0, 1\}$.

Definition 2. *Let $i \in \{1, 2, \ldots, n\}$ and $j \in \{0, 1\}$. The vertex set $S_{i,j} \subseteq V_n$ consists of all vertices of those copies of the gadgets $Q_1(p)$, $Q_2(p)$, and $Q_3(p)$, whose origin p belongs to the track $T_{i,j}$.*

For every $v \in V_n$ let c_v be the center of its unit disk in the representation R_n. Note that, by Definition 2, the unique vertex $v \in V_n$, for which $c_v \in T_{i,j} \cap T_{k,\ell}$, where $i < k$ (i.e. c_v lies on the intersection of the vertical halfline of $T_{i,j}$ with the horizontal halfline of $T_{k,\ell}$), we have that $v \in S_{i,j}$. Furthermore note that $\{S_{i,j} : 1 \le i \le n, j \in \{0,1\}\}$ is a partition of the vertex set V_n of G_n. In the next lemma we show that this is also a balanced $2n$-partition of G_n, i.e. $|S_{i,j}| = |S_{k,\ell}|$ for every $i, k \in \{1, 2, \ldots, n\}$ and $j, \ell \in \{0, 1\}$.

Lemma 2. *For every $i \in \{1, 2, \ldots, n\}$ and $j \in \{0, 1\}$, we have that $|S_{i,j}| = 4(n+1)(n^6 + 1)$.*

Consider the intersection point p of two tracks $T_{i,j}$ and $T_{k,\ell}$, where $i \ne k$. Assume without loss of generality that $i < k$, i.e. p belongs to the vertical halfline of $T_{i,j}$ and on the horizontal halfline of $T_{k,\ell}$, cf. Figure 2(a). Then p is the origin of the gadget $Q_2(p)$ in the representation R_n (cf. part (c) of the construction of R_n). Therefore p is the center of a unit disk in R_n, i.e. $p = c_v$ for some $v \in S_{i,j} \subseteq V_n$. All unit disks of R_n that intersect with the disk centered at point p is shown in Figure 2(a). Furthermore, the induced subgraph $G_n[\{v\} \cup N(v)]$ on the vertices of G_n, which correspond to these disks of Figure 2(a), is shown in Figure 2(c). In Figure 2(c) we denote by K_{n^6} and K_{n^3} the cliques with n^6 and with n^3 vertices, respectively, and the thick edge connecting the two K_{n^3}'s depicts the fact that all vertices of the two K_{n^3}'s are adjacent to each other.

Now consider a bend point $p_{i,j}$ of a variable x_i, where $j \in \{0, 1\}$. Then $p_{i,j}$ is the origin of the gadget $Q_3(p_{i,j})$ in the representation R_n (cf. the first bullet of part (b) of the construction of R_n). Therefore $p_{i,j}$ is the center of a unit disk in R_n, i.e. $p = c_v$ for some $v \in S_{i,j} \subseteq V_n$. All unit disks of R_n that intersect with the disk centered at point $p_{i,j}$ are shown in Figure 2(b). Furthermore, the induced subgraph $G_n[\{v\} \cup N(v)]$ of G_n that corresponds to the disks of Figure 2(b), is shown in Figure 2(d). In both Figures 2(a) and 2(b), the area of the intersection of two crowds (i.e. disks with dashed contour) is shaded gray for better visibility.

Lemma 3. *Consider an arbitrary bisection $\mathcal{B}$ of G_n with size strictly less than n^6. Then for every set $S_{i,j}$, $i \in \{1, 2, \ldots, n\}$ and $j \in \{0, 1\}$, all vertices of $S_{i,j}$ belong to the same color class of $\mathcal{B}$.*

4 Minimum Bisection on Unit Disk Graphs

In this section we provide our polynomial-time reduction from the monotone Max-XOR(3) problem to the minimum bisection problem on unit disk graphs. To this end, given a monotone XOR(3) formula ϕ with n variables and $m = \frac{3n}{2}$ clauses, we appropriately modify the auxiliary unit disk graph G_n of Section 3 to obtain the unit disk graph H_ϕ. Then we prove that the truth assignments that satisfy the maximum number of clauses in ϕ correspond bijectively to the minimum bisections in H_ϕ.

We construct the unit disk graph $H_\phi = (V_\phi, E_\phi)$ from $G_n = (V_n, E_n)$ as follows. Let $(x_i \oplus x_k)$ be a clause of ϕ, where $i < k$. Let p_0 (resp. p_1) be the

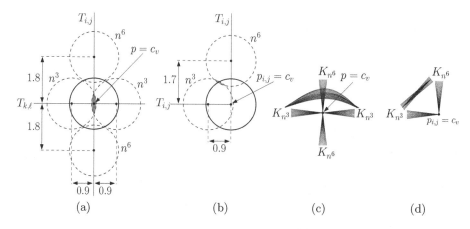

Fig. 2. The disks in R_n (a) around the intersection point $p = c_v$ of two tracks $T_{i,j}$ and $T_{k,\ell}$, where $i < k$, and (b) around the bend point $p_{i,j} = c_v$ of a variable x_i, where $j \in \{0,1\}$. (c) The induced subgraph of G_n on the vertices of part (a), and (d) the induced subgraph of G_n for part (b).

unique point in the unit disk representation R_n that lies on the intersection of the tracks $T_{i,0}$ and $T_{k,1}$ (resp. on the intersection of the tracks $T_{i,1}$ and $T_{k,0}$). For every point $p \in \{p_0, p_1\}$, where we denote $p = (p_x, p_y)$, we modify the gadgets $Q_1(p)$ and $Q_2(p)$ in the representation R_n as follows:

(a) replace the horizontal $(n^3, p + 0.9)$-crowd of $Q_1(p)$ by the horizontal $(n^3 - 1, p + 0.9)$-crowd and a single unit disk centered at $(p_x + 0.9, p_y + 0.02)$,
(b) replace the vertical $(n^6, p + 1.8)$-crowd of $Q_2(p)$ by the vertical $(n^6 - 1, p + 1.8)$-crowd and a single unit disk centered at $(p_x + 0.02, p_y + 1.8)$.

That is, for every point $p \in \{p_0, p_1\}$, we first move one (arbitrary) unit disk of the horizontal $(n^3, p + 0.9)$-crowd of $Q_1(p)$ upwards by 0.02, and then we move one (arbitrary) unit disk of the vertical $(n^6, p + 1.8)$-crowd of $Q_2(p)$ to the right by 0.02. In the resulting unit disk representation these two unit disks intersect, whereas they do not intersect in the representation R_n. Furthermore it is easy to check that for any other pair of unit disks, these disks intersect in the resulting representation if and only if they intersect in R_n.

Denote by R_ϕ the unit disk representation that is obtained from R_n by performing the above modifications for all clauses of the formula ϕ. Then H_ϕ is the unit disk graph induced by R_ϕ. Note that, by construction, the graphs H_ϕ and G_n have exactly the same vertex set, i.e. $V_\phi = V_n$, and that $E_n \subset E_\phi$. In particular, note that the sets $S_{i,j}$ (cf. Definition 2) induce the same subgraphs in both H_ϕ and G_n, and thus the next corollary follows directly by Lemma 3.

Corollary 1. *Consider an arbitrary bisection $\mathcal{B}$ of H_ϕ with size strictly less than n^6. Then for every set $S_{i,j}$, $i \in \{1, 2, \ldots, n\}$ and $j \in \{0,1\}$, all vertices of $S_{i,j}$ belong to the same color class of $\mathcal{B}$.*

Theorem 1. *There exists a truth assignment τ of the formula ϕ that satisfies at least k clauses if and only if the unit disk graph H_ϕ has a bisection with value at most $2n^4(n-1) + 3n - 2k$.*

Proof (sketch). The ($\Rightarrow$) part of the proof is omitted due to lack of space.

($\Leftarrow$) Assume that H_ϕ has a minimum bisection $\mathcal{B}$ with value at most $2n^4(n-1) + 3n - 2k$. Denote the two color classes of $\mathcal{B}$ by blue and red, respectively. Since the size of $\mathcal{B}$ is strictly less than n^6, Corollary 1 implies that for every $i \in \{1, 2, \ldots, n\}$ and $j \in \{0, 1\}$, all vertices of the set $S_{i,j}$ belong to the same color class of $\mathcal{B}$. Therefore, all cut edges of $\mathcal{B}$ have one endpoint in a set $S_{i,j}$ and the other endpoint in a set $S_{k,\ell}$, where $(i,j) \neq (k,\ell)$. Furthermore, since $\mathcal{B}$ is a bisection of H_ϕ, Lemma 2 implies that exactly n of the sets $\{S_{i,j} : 1 \leq i \leq n, j \in \{0,1\}\}$ are colored blue and the other n ones are colored red in $\mathcal{B}$.

First we will prove that, for every $i \in \{1, 2, \ldots, n\}$, the sets $S_{i,0}$ and $S_{i,1}$ belong to different color classes in $\mathcal{B}$. To this end, let $t \geq 0$ be the number of variables x_i, $1 \leq i \leq n$, for which both sets $S_{i,0}$ and $S_{i,1}$ are colored blue (such variables x_i are called *blue*). Then, since $\mathcal{B}$ is a bisection of H_ϕ, there must be also t variables x_i, $1 \leq i \leq n$, for which both sets $S_{i,0}$ and $S_{i,1}$ are colored red (such variables x_i are called *red*), whereas $n - 2t$ variables x_i, for which one of the sets $\{S_{i,0}, S_{i,1}\}$ is colored blue and the other one red (such variables x_i are called *balanced*). Using the minimality of the bisection $\mathcal{B}$, we will prove that $t = 0$.

Every cut edge of $\mathcal{B}$ occurs at the intersection of the tracks of two variables x_i, x_k, where either both x_i, x_k are balanced variables, or one of them is a balanced and the other one is a blue or red variable, or one of them is a blue and the other one is a red variable. Furthermore recall by the construction of the graph H_ϕ from the graph G_n that every clause $(x_i \oplus x_k)$ of the formula ϕ corresponds to an intersection of the tracks of the variables x_i and x_k. Among the m clauses of ϕ, let m_1 of them correspond to intersections of tracks of two balanced variables, m_2 of them correspond to intersections of tracks of a balanced variable and a blue or red variable, and m_3 of them correspond to intersections of tracks of a blue variable and a red variable. Note that $m_1 + m_2 + m_3 \leq m$.

Let $1 \leq i < k \leq n$. In the following we distinguish the three cases of the variables x_i, x_k that can cause a cut edge in the bisection $\mathcal{B}$.

- x_i **and** x_k **are both balanced variables:** in total there are $\frac{(n-2t)(n-2t-1)}{2}$ such pairs of variables, where exactly m_1 of them correspond to a clause $(x_i \oplus x_k)$ of the formula ϕ. It is easy to check that, for every such pair x_i, x_k that does not correspond to a clause of ϕ, the intersection of the tracks of x_i and x_k contributes exactly $2n^3 + 2n^3 = 4n^3$ edges to the value of $\mathcal{B}$. Furthermore, for each of the m_1 other pairs x_i, x_k that correspond to a clause of ϕ, the intersection of the tracks of x_i and x_k contributes either $4n^3$ or $4n^3 + 2$ edges to the value of $\mathcal{B}$. In particular, if the vertices of the sets $S_{i,0}$ and $S_{k,1}$ have the same color in $\mathcal{B}$ then the pair x_i, x_k contributes $4n^3$ edges to the value of $\mathcal{B}$, otherwise it contributes $4n^3 + 2$ edges. Among these m_1 clauses, let m_1^* of them contribute $4n^3$ edges each and the remaining $m_1 - m_1^*$ of them contribute $4n^3 + 2$ edges each.

- **one of x_i, x_k is a balanced variable and the other one is a blue or red variable:** in total there are $(n - 2t)2t$ such pairs of variables, where exactly m_2 of them correspond to a clause $(x_i \oplus x_k)$ of the formula ϕ. It is easy to check that, for every such pair x_i, x_k that does not correspond to a clause of ϕ, the intersection of the tracks of x_i and x_k contributes exactly $2n^3 + 2n^3 = 4n^3$ edges to the value of $\mathcal{B}$. Furthermore, for each of the m_2 other pairs x_i, x_k that correspond to a clause of ϕ, the intersection of the tracks of x_i and x_k contributes $4n^3 + 1$ edges to the value of $\mathcal{B}$.
- **one of x_i, x_k is a blue variable and the other one is a red variable:** in total there are t^2 such pairs of variables, where exactly m_3 of them correspond to a clause $(x_i \oplus x_k)$ of the formula ϕ. It is easy to check that, for every such pair x_i, x_k that does not correspond to a clause of ϕ, the intersection of the tracks of x_i and x_k contributes exactly $4 \cdot 2n^3 = 8n^3$ edges to the value of $\mathcal{B}$. Furthermore, for each of the m_3 other pairs x_i, x_k that correspond to a clause of ϕ, the intersection of the tracks of x_i and x_k contributes $8n^3 + 2$ edges to the value of $\mathcal{B}$.

Therefore, the value of $\mathcal{B}$ can be computed (the exact details are omitted due to lack of space) as $2n^4(n-1) + 4n^3 t + 2(m_1 - m_1^*) + m_2 + 2m_3$. Note now that $0 \leq 2(m_1 - m_1^*) + m_2 + 2m_3 \leq 2m = 3n < 4n^3$. Therefore, since the value of the bisection $\mathcal{B}$ is minimum by assumption, it follows that $t = 0$. Thus for every $i \in \{1, 2, \ldots, n\}$ the variable x_i of ϕ is balanced in the bisection $\mathcal{B}$, i.e. the sets $S_{i,0}$ and $S_{i,1}$ belong to different color classes in $\mathcal{B}$. That is, $m_1 = m$ and $m_2 = m_3 = 0$, and thus the value of $\mathcal{B}$ is equal to $2n^4(n-1) + 2(m - m_1^*)$. On the other hand, since the value of $\mathcal{B}$ is at most $2n^4(n-1) + 3n - 2k$ by assumption, it follows that $2(m - m_1^*) \leq 3n - 2k$. Therefore, since $m = \frac{3n}{2}$, it follows that $m_1^* \geq k$.

We define now from $\mathcal{B}$ the truth assignment τ of ϕ as follows. For every $i \in \{1, 2, \ldots, n\}$, if the vertices of the set $S_{i,0}$ are blue and the vertices of the set $S_{i,1}$ are red in $\mathcal{B}$, then we set $x_i = 0$ in τ. Otherwise, if the vertices of the set $S_{i,0}$ are red and the vertices of the set $S_{i,1}$ are blue in $\mathcal{B}$, then we set $x_i = 1$ in τ. Recall that m_1^* is the number of clauses of ϕ that contribute $4n^3$ edges each to the value of $\mathcal{B}$, while the remaining $m - m_1^*$ clauses of ϕ contribute $4n^3 + 2$ edges each to the value of $\mathcal{B}$. Thus, by the construction of H_ϕ from G_n, for every clause $(x_i \oplus x_k)$ of ϕ that contributes $4n^3$ (resp. $4n^3 + 2$) to the value of $\mathcal{B}$, the vertices of the sets $S_{i,0}$ and $S_{k,1}$ have the same color (resp. $S_{i,0}$ and $S_{k,1}$ have different colors) in $\mathcal{B}$. Therefore, by definition of the truth assignment τ, there are exactly m_1^* clauses $(x_i \oplus x_k)$ of ϕ where $x_i \neq x_k$ in τ, and there are exactly $m - m_1^*$ clauses $(x_i \oplus x_k)$ of ϕ where $x_i = x_k$ in τ. That is, τ satisfies exactly $m_1^* \geq k$ of the m clauses of ϕ. This completes the proof of the theorem. □

We can now state our main result, which follows by Theorem 1 and Lemma 1.

Theorem 2. MIN-BISECTION *is NP-hard on unit disk graphs.*

5   Concluding Remarks

In this paper we proved that MIN-BISECTION is NP-hard on unit disk graphs by providing a polynomial time reduction from the monotone Max-XOR(3) problem, thus solving a longstanding open question. As pointed out in the Introduction, our results indicate that MIN-BISECTION is probably also NP-hard on planar graphs, or equivalently on grid graphs with an arbitrary number of holes, which remains yet to be proved.

References

1. Akyildiz, I., Su, W., Sankarasubramaniam, Y., Cayirci, E.: Wireless sensor networks: A survey. Computer Networks 38, 393–422 (2002)
2. Bichot, C.-E., Siarry, P. (eds.): Graph Partitioning. Wiley (2011)
3. Bradonjic, M., Elsässer, R., Friedrich, T., Sauerwald, T., Stauffer, A.: Efficient broadcast on random geometric graphs. In: Proceedings of the 21st Annual ACM-SIAM Symposium on Discrete Algorithms (SODA), pp. 1412–1421 (2010)
4. Breu, H., Kirkpatrick, D.G.: Unit disk graph recognition is NP-hard. Computational Geometry 9(1-2), 3–24 (1998)
5. Bui, T., Chaudhuri, S., Leighton, T., Sipser, M.: Graph bisection algorithms with good average case behavior. Combinatorica 7, 171–191 (1987)
6. Cygan, M., Lokshtanov, D., Pilipczuk, M., Pilipczuk, M., Saurabh, S.: Minimum bisection is fixed parameter tractable. In: Proceedings of the 46th Annual Symposium on the Theory of Computing, STOC (to appear, 2014)
7. Delling, D., Goldberg, A.V., Razenshteyn, I., Werneck, R.F.: Exact combinatorial branch-and-bound for graph bisection. In: Proceedings of the 14th Meeting on Algorithm Engineering & Experiments (ALENEX), pp. 30–44 (2012)
8. Díaz, J., Penrose, M.D., Petit, J., Serna, M.J.: Approximating layout problems on random geometric graphs. Journal of Algorithms 39(1), 78–116 (2001)
9. Díaz, J., Petit, J., Serna, M.: A survey on graph layout problems. ACM Computing Surveys 34, 313–356 (2002)
10. Feldmann, A.E., Widmayer, P.: An $\mathcal{O}(n^4)$ time algorithm to compute the bisection width of solid grid graphs. In: Demetrescu, C., Halldórsson, M.M. (eds.) ESA 2011. LNCS, vol. 6942, pp. 143–154. Springer, Heidelberg (2011)
11. Garey, M.R., Johnson, D.S.: Computers and intractability: A guide to the theory of NP-completeness. W. H. Freeman & Co. (1979)
12. Hromkovic, J., Klasing, R., Pelc, A., Ruzicka, P., Unger, W.: Dissemination of Information in Communication Networks - Broadcasting, Gossiping, Leader Election, and Fault-Tolerance. In: Texts in Theoretical Computer Science. An EATCS Series, Springer, Heidelberg (2005)
13. Jansen, K., Karpinski, M., Lingas, A., Seidel, E.: Polynomial time approximation schemes for Max-Bisection on planar and geometric graphs. SIAM Journal on Computing 35(1), 110–119 (2005)
14. Kahruman-Anderoglu, S.: Optimization in geometric graphs: Complexity and approximation. PhD thesis, Texas A & M University (2009)
15. Karpinski, M.: Approximability of the minimum bisection problem: An algorithmic challenge. In: Diks, K., Rytter, W. (eds.) MFCS 2002. LNCS, vol. 2420, pp. 59–67. Springer, Heidelberg (2002)

16. Kernighan, B., Lin, S.: An efficient heuristic procedure for partitioning graphs. Bell System Technical Journal 49(2), 291–307 (1970)
17. Kratochvíl, J.: Intersection graphs of noncrossing arc-connected sets in the plane. In: Proceedings of the 4th Int. Symp. on Graph Drawing (GD), pp. 257–270 (1996)
18. MacGregor, R.: On partitioning a graph: A theoretical and empirical study. PhD thesis, University of California, Berkeley (1978)
19. Papadimitriou, C.H., Sideri, M.: The bisection width of grid graphs. Mathematical Systems Theory 29(2), 97–110 (1996)
20. Räcke, H.: Optimal hierarchical decompositions for congestion minimization in networks. In: Proceedings of the 40th Annual ACM Symposium on Theory of Computing (STOC), pp. 255–264 (2008)

Query-Competitive Algorithms for Cheapest Set Problems under Uncertainty

Thomas Erlebach[1], Michael Hoffmann[1], and Frank Kammer[2]

[1] Department of Computer Science, University of Leicester, England
{te17,mh55}@mcs.le.ac.uk
[2] Institut für Informatik, Universität Augsburg, Germany
kammer@informatik.uni-augsburg.de

Abstract. Considering the model of computing under uncertainty where element weights are uncertain but can be obtained at a cost by query operations, we study the problem of identifying a cheapest (minimum-weight) set among a given collection of feasible sets using a minimum number of queries of element weights. For the general case we present an algorithm that makes at most $d \cdot OPT + d$ queries, where d is the maximum cardinality of any given set and OPT is the optimal number of queries needed to identify a cheapest set. For the minimum multi-cut problem in trees with d terminal pairs, we give an algorithm that makes at most $d \cdot OPT + 1$ queries. For the problem of computing a minimum-weight base of a given matroid, we give an algorithm that makes at most $2 \cdot OPT$ queries, generalizing a known result for the minimum spanning tree problem. For each of our algorithms we give matching lower bounds.

1 Introduction

Motivated by applications where exact input data is not always easily available, we consider cheapest set problems under uncertainty: We are given a set E of elements, and a collection $\mathcal{S}$ of *feasible* subsets of E, where $\mathcal{S}$ may be specified explicitly or implicitly. Each element $e \in E$ has an exact weight (or cost) w_e, but initially only an *uncertainty area* A_e, which is a set that contains w_e, is known. We assume that each uncertainty area A_e is either trivial (i.e., a singleton set containing only w_e) or an open set with finite lower limit L_e and finite upper limit U_e (for example, an open interval (L_e, U_e)). The task is to find a cheapest set in $\mathcal{S}$, i.e., a set $S \in \mathcal{S}$ such that $\sum_{e \in S} w_e$ is minimized. It may not be possible to identify a cheapest set based on just the given uncertainty areas. We assume that it is possible to obtain the exact weight w_e of an element $e \in E$ using a *query* operation, but we wish to minimize the number of queries needed.

An algorithm solving the cheapest set problem under uncertainty may make more queries than absolutely necessary. To assess the quality of an algorithm, we use competitive analysis, i.e., for the given instance of the cheapest set problem we compare the number of queries the algorithm makes with the best possible number of queries, which we denote by OPT. An algorithm for a problem under uncertainty that is measured competitively with respect to the number of queries is also called a *query-competitive* algorithm. We restrict the uncertainty areas to

E. Csuhaj-Varjú et al. (Eds.): MFCS 2014, Part II, LNCS 8635, pp. 263–274, 2014.
© Springer-Verlag Berlin Heidelberg 2014

be open sets or singleton sets because it is easy to see (as shown in [5] for the minimum spanning tree problem) that there are no query-competitive algorithms with non-trivial competitive ratio for the uncertainty problems that we consider if closed intervals are allowed as uncertainty areas.

We consider the cheapest set problem under uncertainty both in the general case, where the feasible sets can be arbitrary and are specified explicitly as part of the input, and in special cases that arise when the feasible sets have a certain structure. In the multi-cut problem for trees, the feasible sets are the sets of edges that separate the given terminal pairs. In the minimum matroid base problem, the feasible sets are the bases of a matroid. The minimum spanning tree problem is a special case of the minimum matroid base problem where the independent sets of the matroid are the spanning forests of the given graph.

Motivation for studying the cheapest set problem under uncertainty can be found in numerous application areas. Many optimization problems can be viewed as the problem of selecting a minimum-weight set among all feasible sets. Especially in distributed networks or mobile computing, it is often the case that the exact weight of an element is known only approximately (e.g., an estimate for the cost of a remote service or congestion of a remote link), but it may be possible to obtain the exact weight at an extra cost (e.g., a negotiation with a service provider, or a query message and response exchanged over the network). If the cost for obtaining the exact weight of an input element by a query is not negligible, the objective of minimizing the number of queries needed to solve the problem becomes natural. For example, consider the problem of installing monitoring equipment on links of a tree network so as to monitor all traffic between d given terminal pairs. The cost of the installation on a specific link depends on the total traffic on that link (generated by the terminal pairs and by background traffic), as all packets traversing the link need to be processed. The exact cost of a link can be determined by conducting traffic measurements, but this may be costly. The problem of identifying a set of edges of minimum total cost for solving the monitoring problem, while making a minimum number of traffic measurements on different links, is the multi-cut problem in trees under uncertainty.

Our Results. For the cheapest set problem under uncertainty, we give an algorithm that makes at most $d \cdot OPT + d$ queries, where d is the maximum cardinality of a feasible set in the given instance. We also give a matching lower bound, showing that the algorithm is best possible among deterministic algorithms. For the minimum multi-cut problem for d terminal pairs in trees under uncertainty, we give an algorithm that makes at most $d \cdot OPT + 1$ queries, and we prove a matching lower bound. For the minimum matroid base problem under uncertainty, we give an algorithm that makes at most $2 \cdot OPT$ queries, generalizing a known 2-competitive algorithm U-RED for minimum spanning trees under uncertainty [5]. We remark that the generalisation is not straightforward since in [5] properties of connected components are considered while the matroid setting requires set oriented proofs. The known lower bound for minimum spanning trees under uncertainty [5] implies that our algorithm for minimum matroid base is best possible. Some proofs from Sections 4 to 6 are omitted.

Related Work. The first study of query-competitive algorithms for problems under uncertainty that we are aware of is the work by Kahan [9], who gives query-competitive algorithms with optimal competitive ratio for the problems of computing the maximum, the median and the minimum gap of n real values that are constrained to fall into given real intervals. Bruce et al. [2] consider geometric problems where input points are not known exactly but lie in given uncertainty areas. They propose the concept of witness set algorithms that, in each step, query a set S of elements with the property that any query solution must query at least one element of S. Sets with this property are called witness sets. The competitive ratio of a witness set algorithm is bounded by the maximum size of a witness set. Bruce et al. present 3-competitive algorithms for computing maximal points or the points on the convex hull of a given set of uncertain points in two-dimensional space [2]. Erlebach et al. [5] give a 2-competitive algorithm for computing minimum spanning trees in graphs with uncertain edge weights and show that this is optimal for deterministic algorithms.

Feder et al. study the problem of minimizing the total cost of queries for the problem of computing an approximation of the value of the median of n uncertain values [7] or of the length of a shortest path in a graph with uncertain edge weights [6]. They also consider algorithms that must specify the whole set of queries in advance, rather than querying uncertain elements one by one as in our model. Other related work exploring trade-offs between query cost and solution accuracy includes [13] and [10]. Another line of work considers the problem of computing, for an input with uncertain elements, the minimum and maximum possible cost of an optimal solution, over all possible precise values of the input [12,4]. In these problems, there is no concept of queries. There are also numerous other models of optimization under uncertainty, e.g., min-max regret versions of standard optimization problems [1].

For the standard version of the multi-cut problem in trees, Garg et al. [8] show that the problem is $\mathcal{NP}$-hard and MAX SNP-hard and admits a 2-approximation (with respect to the cost of the multi-cut). The standard version of the minimum matroid base problem can be solved by a greedy algorithm, see e.g. [3]. We conclude in Section 7.

2   Preliminaries

Based on the terminology of Section 1 we now formally define the cheapest set problem under uncertainty (CSU). An instance of the CSU problem is represented by a quadruple $(E, \mathcal{S}, w, A)$ where E is the finite set of elements, $\mathcal{S} \subseteq \mathcal{P}(E)$ is a family of subsets of E, w is a real-valued weight function that maps each element $e \in E$ to its precise weight $w(e)$, and A is a function that maps each element $e \in E$ to its uncertainty area $A(e)$. From here on, we write w_e and A_e for $w(e)$ and $A(e)$, respectively. Note that for the CSU problem we allow w_e to be smaller than 0. The goal of the CSU problem is to find a *cheapest set* S, i.e., a set $S \in \mathcal{S}$ such that $\sum_{e \in S} w_e \leq \sum_{e \in S'} w_e$ for all $S' \in \mathcal{S}$, using a minimum number of queries. Let w' be a function that maps elements to weights and let A' be a function that maps elements to uncertainty areas. We say that w' is *consistent with A'* if $w'_e \in A'_e$ for all $e \in E$. Note that in any instance $(E, \mathcal{S}, w, A)$ of the CSU problem w is consistent with A by definition. For $e \in E$, we refer to all elements of A_e as *potential weights* of e. An instance $(E, \mathcal{S}, w, A)$ is *solved* if there exists a set $S \in \mathcal{S}$ such that for every weight function w' that is consistent with A, S is a cheapest set in the instance $(E, \mathcal{S}, w', A)$. If an instance is solved and S satisfies the condition stated in the previous sentence, we also say that S can be *identified* as cheapest set. An instance that is not solved is also called *unsolved*. A *query* of an element e changes an instance $(E, \mathcal{S}, w, A)$ to $(E, \mathcal{S}, w, A')$ where A' is the same as A apart from A_e being set to $\{w_e\}$.

We say that an element whose area of uncertainty is trivial (i.e., a singleton set) is a *certain* element while an element whose area of uncertainty is non-trivial is an *uncertain* element. For $X \subseteq E$, we denote by X^U the set of its uncertain elements and by X^C the set of its certain elements.

For every instance $(E, \mathcal{S}, w, A)$, a *query solution* is a set of elements that when queried results in a solved instance. A query solution of minimum cardinality is called an *optimal query solution*, and the size of an optimal query solution is denoted by OPT. Throughout this paper, we consider only instances where each area of uncertainty is trivial or a (bounded) open set. For an element $e \in E$, we write L_e for the lower limit and U_e for the upper limit of the uncertainty area of e. If e has a trivial uncertainty area, then $L_e = U_e = w_e$. For any set $T \subseteq E$, we let $T_{min} = \sum_{e \in T} L_e$. Note that T_{min} is a lower bound on the exact weight of the set T. T_{max} is defined analogously via U_e instead of L_e. Furthermore, we sometimes write T_w for $\sum_{e \in T} w_e$. We say that $T \in \mathcal{S}$ *lies within the uncertainty bounds* of $S \in \mathcal{S}$ if for any potential weights of the elements in T there exist potential weights of the elements in $S \setminus T$ such that S is cheaper than T and (other) potential weights of the elements in $S \setminus T$ such that T is cheaper than S. Note explicitly that the definition implies $T \neq S$.

For an instance $(E, \mathcal{S}, w, A)$ the input of an algorithm is $(E, \mathcal{S}, A)$. The aim is to make queries until the resulting instance is solved. The quality of an algorithm is measured by the number of queries it makes compared to OPT of the initial instance. As introduced in [2], for a given instance a set of elements is a *witness set* if the instance cannot be solved without querying at least one element of the set. Let I' be an instance that has resulted from querying some elements in an instance I. Then a witness set of I' is also a witness set of I. Hence, we have the following observation, which can be shown using similar arguments as in [2]:

Observation 1. *If an instance I is solved by an algorithm that queries all elements of all sets $U_1, U_2, \ldots, U_l$ where U_i is a witness set of the instance resulting from I after querying $U_1, \ldots U_{i-1}$, then the value OPT for the instance I is at least l. If only k of the l sets U_i are witness sets, then the value OPT for the instance I is at least k.*

Definition 2. *For an instance $I = (E, \mathcal{S}, w, A)$, a set $S \in \mathcal{S}$ is called a* Robust Potential Cheapest (RPC) *set, if for any potential weights of elements in $E \setminus S$,* *there exist potential weights for the elements in S such that S is a cheapest set among all sets in $\mathcal{S}$.*

We also refer to an RPC set of an instance $(E, \mathcal{S}, w, A)$ as an *RPC set in $\mathcal{S}$* if the instance is clear from the context. The following observation is a direct consequence of the definition. The next lemma is needed to prove Lemma 5.

Observation 3. Let S be an RPC set of an instance I. Then we have that $S_{min} = \min\{X_{min} \mid X \in \mathcal{S}\}$. Further, if S contains only certain elements, I is solved and S is a cheapest set for the instance I.

Lemma 4. *A set S is an RPC set if, for all $X \in \mathcal{S} \backslash \{S\}$, either 1. or 2. holds:*

1. *$S_{min} < X_{min}$, or*
2. *$S_{min} = X_{min}$ and $X^U \not\subset S^U$, i.e., X^U is not a proper subset of S^U.*

Proof. Let w' denote an arbitrary choice of potential values for $E \setminus S$.

Consider any $X \in \mathcal{S}$ for which Condition 1 holds. As the potential values for S can be chosen arbitrarily close to their lower limits (since the uncertainty areas are open sets), there exist potential values w_e^X for $e \in S$ such that S is not more expensive than X_{min} and thus not more expensive than $X_{w'}$.

Consider any $X \in \mathcal{S}$ for which Condition 2 holds. If $X^U = S^U$, the weights of X and S are equal (and hence S is not more expensive than X) for all potential values for $e \in S$. Therefore, assume that $X^U \neq S^U$. As X^U is not a proper subset of S^U, there must exist an element e^* in $X^U \setminus S^U = (X \setminus S)^U$. Let $S' = S \setminus X$ and $X' = X \setminus S$. Note that $e^* \in (X')^U$. By Condition 2, we also have $S'_{min} + (S \cap X)_{min} = S_{min} = X_{min} = X'_{min} + (S \cap X)_{min}$, and hence $S'_{min} = X'_{min}$. As $w'_{e^*} > L_{e^*}$ and $e^* \in X'$, we can choose values w_e^X for the elements $e \in S'$ such that $S'_{w^X} < X'_{w'}$. For any choice of values w_e^X for the elements of $S \cap X$, we then have $S_{w^X} < X_{w'}$.

Combining the arguments from the two previous paragraphs, we have that for each set $X \in \mathcal{S} \backslash \{S\}$, there exist values w_e^X for $e \in S$ such that S is not more expensive than X. If we now choose $w_e^* = \min_{X \in \mathcal{S} \backslash \{S\}} w_e^X$, the weights w_e^* for $e \in S$ are such that S is not more expensive than any of the sets in $\mathcal{S} \setminus \{S\}$. As the choice of w' was arbitrary, we have shown that S is an RPC set. □

Lemma 5. *Every instance of the CSU problem has at least one RPC set.*

Proof. We find an RPC set as follows: Initially, we choose any S such that $S_{min} \leq X_{min}$ for all $X \in \mathcal{S}$. As long as there exists a set $X \in \mathcal{S} \setminus \{S\}$ with $X_{min} = S_{min}$ and $X^U \subset S^U$, set $S := X$. This process must terminate as $|S^U|$ decreases in each step. In the end, S will be an RPC set by Lemma 4. □

Lemma 6. *An instance is solved if and only if all cheapest sets are identified. Moreover, in such an instance, $S_{min} = X_{min}$ and $S^U = X^U$ for any two cheapest sets S and X.*

Proof. As the instance is solved, there is at least one set S that can be identified as a cheapest set. Let X be another cheapest set. Assume that S contains an uncertain element e that is not in X. As all non-trivial areas of uncertainty are open, there is a potential value for e that is higher than w_e. So S_w has the potential to increase while X_w remains the same, and potentially $X_w < S_w$. A contradiction to S_w being identified as a cheapest set.

Similarly, assume that X contains an uncertain element e that is not in S. As all non-trivial areas of uncertainty are open there is a potential value for e that is lower than w_e. So X_w has the potential to decrease while S_w remains the same, and potentially $X_w < S_w$. A contradiction.

Consequently, $X^U = S^U$. Since $S_w = X_w$, the sum of all certain elements in S and in X must also be the same, and $S_{min} = X_{min}$. Hence, if S is identified as a cheapest set, then also X must be identified as a cheapest set.   □

Lemma 7. *Let $I = (E, \mathcal{S}, w, A)$ be an instance of the CSU problem. If S is an RPC set but not a witness set of I, then S is a cheapest set and $X^U = S^U$ for each cheapest set X of I.*

Proof. Let I' be the instance obtained after querying $E \setminus S$ in I. As S is not a witness set, the instance I' must be solved. Since S was an RPC set in I and no element of S was queried, S is also an RPC set in I' and hence S is potentially a cheapest set in I'. As I' is solved, by Lemma 6 we have that S must be identified either as being a cheapest set or as not being a cheapest set in I'. As S is potentially a cheapest set in I', it can only be the case that S is identified as a cheapest set in I'. Hence, S is a cheapest set of I.

Let X be another cheapest set in I, and therefore also in I'. As I' is solved, Lemma 6 implies that $X_{min} = S_{min}$ and $X^U = S^U$ in I'. Assume that an element of X was queried by the query of $E \setminus S$. Then X_{min} in I must be smaller than X_{min} in I' as all non-trivial areas of uncertainty are open. Hence, we must have $X_{min} < S_{min}$ in I. This contradicts S being an RPC set in I. So no element of X was queried by $E \setminus S$, and we have that $X^U = S^U$ also in I.   □

3 Cheapest Set

We denote the maximum number of elements in any set of $\mathcal{S}$ by d. Let us define algorithm SIMPLE for the CSU problem as an algorithm that, while the instance is not solved, queries all elements of an RPC set. First, we need the next lemma.

Lemma 8. *Let $S \in \mathcal{S}$ be an RPC set in $\mathcal{S}$ and let $T \in \mathcal{S}$ be such that T lies within the uncertainty bounds of S. Then S is a witness set.*

Proof. Assume that S is not a witness set. Then S is a cheapest set by Lemma 7, and $E \setminus S$ is a query solution. Let I' be the instance after querying $E \setminus S$. As no element of S has been queried, T still lies within the uncertainty bounds of S. Hence T is still potentially cheaper than S and S cannot be identified in I' as a cheapest set. So by Lemma 6, I' cannot be a solved instance. This is a contradiction to $E \setminus S$ being a query solution. So, S must be a witness set.   □

Theorem 9. *Algorithm SIMPLE makes at most $d \cdot OPT + d$ queries.*

Proof. Let t be the number of sets queried by the algorithm. For $1 \leq i \leq t$, let S_i be the set queried by the algorithm in the ith iteration of the while-loop. We claim that there is at most one $i \in \{1, \ldots, t\}$ such that S_i is not a witness set in iteration i. Choose i as small as possible such that S_i is not a witness set in iteration i. (If no such i exists, all t sets are witness sets, and the claim holds.) By Lemma 7, S_i is a cheapest set in iteration i. Assume for contradiction that there exists $j > i$ such that S_j is not a witness set. Again by Lemma 7, S_j is also a cheapest set and $S_i^U = S_j^U$ before iteration i. As S_i^U was queried at iteration i, at iteration j the set S_j does not contain uncertain elements. By Observation 3, the instance is solved before iteration j. A contradiction.

The algorithm queries t sets of which at least $t - 1$ are witness sets. Hence, $OPT \geq t - 1$ and the algorithm makes at most $dt \leq d \cdot OPT + d$ queries. $\qquad \square$

In the following, we want to reduce the number of queries for variants of the CSU problem that satisfy a special property. The next lemma allows us to determine a witness set for any unsolved instance of the CSU problem.

Lemma 10. *Let $S, T \in \mathcal{S}$ such that S is an RPC set in $\mathcal{S}$ and T is potentially cheaper than S. Then $S \cup T$ is a witness set.*

Proof. Assume that $S \cup T$ is not a witness set. Then it is possible to solve the instance without querying any element of $S \cup T$. As $S \subseteq S \cup T$, S is not a witness set either. By Lemma 7, S is a cheapest set. If we do not query any element of $S \cup T$, T is potentially cheaper than S, and S cannot be identified as a cheapest set. So there is at least one cheapest set that cannot be identified. By Lemma 6, we cannot solve the instance, a contradiction. $\qquad \square$

We say that a variant (special case) of the CSU problem has the *1-gap property* if, for every unsolved instance $(E, \mathcal{S}, w, A)$ of that problem, there exist $S, T \in \mathcal{S}$ such that (1) S is an RPC set in $\mathcal{S}$, (2) T is potentially cheaper than S, and (3) $|S \cup T| \leq d + 1$. We now show that the following algorithm U-SET for 1-gap CSU problems makes at most $d \cdot OPT + 1$ queries.

Algorithm 1. Algorithm U-SET for 1-gap CSU problems

---

1. **while** instance is not yet solved **do**
2.   **if** there exist $S, T \in \mathcal{S}$ such that S is an RPC set
         and T lies within the uncertainty bounds of S **then** query S
3.   **else** query all uncertain elements of $S \cup T$ where S is an RPC set
         and T is a potentially cheaper set than S such that $|S \cup T| \leq d + 1$

Lemma 11. *When Step 3 in algorithm U-SET is executed any time except the first time, the set S must contain a certain element.*

Proof. The first situation that we consider is when Step 3 is executed for the first time. Let us call the RPC set X and the set that is potentially cheaper Y. The second situation we consider is when Step 3 is executed another time (second, third, and so on). This is after the first, so $X \cup Y$ has been queried and the algorithm now takes an RPC set S. Let us assume that S consists only of uncertain elements. Then X and S must be disjoint. All elements of S had in the first situation the same uncertainty information as in the second situation, and S did not lie within the uncertainty bounds of X in the first situation. As X and S are disjoint and X was an RPC set, X could have been potentially cheaper than S for any weights of the elements of S in the first situation. So as S did not lie within the uncertainty bounds of X, there must exist some potential weight for the elements in S such that S cannot be cheaper than X. Since X and S are disjoint, for any potential weight of elements in X, the set S can be more expensive than X. In the second situation $X \cup Y$ has been queried and the weight of X is known precisely. So even for the queried set X, S can still be more expensive than X. Since S is an RPC set, S can also be cheaper than X. Thus, the queried set X lies with in the uncertainty bounds of S. Thus, in the second situation, the algorithm would execute Step 2 and not Step 3. A contradiction. Hence, our assumption that S does not contain any certain elements is false.   □

Theorem 12. *For any uncertainty set problem with the 1-gap property, there exists an algorithm that makes at most $d \cdot OPT + 1$ queries.*

Proof. By Lemmas 8 and 10, all sets of queries performed by the algorithm are witness sets. By Lemma 11, the algorithm requests a set of queries of size $d+1$ at most once, and all other query sets are of size at most d. Hence, by Observation 1, the algorithm makes at most $d \cdot (OPT - 1) + d + 1$ queries.   □

4   Minimum Multicut in Trees

We now consider the minimum multicut problem in trees under uncertainty (MMCTU). An instance of it is given by a tuple $(E, (G, D), w, A)$, where G is an undirected tree with edge set E, D is a set of d terminal pairs that need to be cut, w maps each edge $e \in E$ to its actual weight $w_e > 0$, and A maps each $e \in E$ to its uncertainty area A_e. The family S of feasible sets is not given explicitly, but is determined by G and D: A set $S \subseteq E$ is feasible if removing the edges in S from the tree G separates all terminal pairs in D. In other words, S is the family of all possible multicuts for the given terminal pairs. Multicuts containing more than d edges can be ignored because they must contain redundant edges, so we only need to consider multicuts consisting of at most d edges.

Let S be a potential minimum multicut. Then each element in S cuts at least one terminal pair that is not cut by any other element of S. By considering the elements of $S = \{s_1, \ldots, s_{|S|}\}$ one by one, a partition $P = \{P_1, P_2, \ldots, P_{|S|}\}$ of D is formed, where P_i is the set of terminal pairs that are cut by s_i and not by any $s_1, \ldots, s_{i-1}$. We say P is a *partition induced by* S. We also say that the element $s_i \in S$ *leads to* the element $P_i \in P$.

Lemma 13. *Let P and Q be two partitions of a finite set K. If, for all proper subsets P' of P and Q' of Q, $\bigcup_{X \in P'} X \neq \bigcup_{X \in Q'} X$, then $|P| + |Q| \leq |K| + 1$.*

Lemma 14. *The MMCTU problem has the 1-gap property.*

Proof. Let $I = (E, (G, D), w, A)$ be an unsolved instance of the MMCTU problem. Let S be an RPC set of I. Let $\mathcal{F}_S$ be the family of multicuts that are potentially cheaper than S. Let $T \in \mathcal{F}_S$ be such that $|T \backslash S| \leq |T' \backslash S|$ for all $T' \in \mathcal{F}_S$. Let D' be the set of pairs that are not cut by $S \cap T$. So, let P be a partition of D' induced by $S \backslash T$ and let Q be a partition of D' induced by $T \backslash S$.

Assume that there exist proper subsets P' of P and Q' of Q with $\bigcup_{X \in P'} X = \bigcup_{X \in Q'} X$. Let S' be the set of elements in S that lead to elements in P' and similarly T' be the set of elements in T that lead to elements in Q'. Note that the pairs of $D \backslash D'$ are cut by $S \backslash S'$ as well as by $T \backslash T'$. So the sets $Mix = S' \cup (T \backslash T')$ and $Mix' = T' \cup (S \backslash S')$ are also cuts of all pairs in D. Since

$$S'_{min} + (S \backslash S')_{min} = S_{min} \overset{\substack{S \text{ is an RPC set}}}{\leq} Mix'_{min} = T'_{min} + (S \backslash S')_{min},$$

we have $S'_{min} \leq T'_{min}$. We now consider two cases.

Case 1. If $S'_{max} \leq T'_{min}$, then Mix is always cheaper than T. As T was potentially cheaper than S, Mix also must be potentially cheaper than S. Moreover, $|Mix \backslash S| = |(T \backslash S) \backslash T'| < |T \backslash S|$, which is a contradiction to our choice of T.

Case 2. If $S'_{max} > T'_{min}$, then S can be more expensive than Mix'. This means that Mix' is potentially cheaper than S, and $|Mix' \backslash S| = |T'| < |T \backslash S|$. The existence of Mix' contradicts our choice of T.

Thus, no proper subsets P' of P and Q' of Q as described above exist. Hence, by Lemma 13, $|S \Delta T| \leq |D'| + 1$. As $|S \cap T| \leq d - |D'|$ we get $|S \cup T| \leq d + 1$. $\square$

Lemma 14 and Theorem 12 give us the following.

Theorem 15. *For the MMCTU problem, there exists an algorithm that makes at most $d \cdot OPT + 1$ queries.*

5   Minimum Matroid Base

We present an algorithm for the minimum matroid base under uncertainty (MMBU) problem using at most $2 \cdot OPT$ queries. We first recall the basic notation of matroids (see, e.g., [3]). A matroid $M = (E, I)$ consists of a set of elements E and a set of independent sets $I \subseteq \mathcal{P}(E)$ such that the following properties are satisfied.

Non-emptiness: $\emptyset \in I$,
Heredity: Every subset of a set in I is also in I,
Exchange: If $S, T \in I$ and $|S| < |T|$, then $\exists e \in T$ such that $S \cup \{e\} \in I$.

Each element has a real-valued weight, which may be negative. A subset of E that is not independent is called *dependent*. A *circuit* of M is a dependent set over E such that all its proper subsets are independent. A set $S \subseteq E$ is called a *base* of M if S is independent and for any $e \in E \setminus S$ the set $S \cup \{e\}$ is dependent. We write *circuit* (or *base*) *of M over $E' \subseteq E$* for a circuit (or base) of $(E', \{S \cap E' | S \in I\})$. When M is clear from the context, we write just *circuit* or *base over E'*. The following observation is well known.

Observation 16. Every base of a matroid M has the same number of elements.

Let M be a matroid with a weight function w that assigns each $e \in E$ a weight w_e. M is then called a *weighted matroid*. A *minimum base* of M is a base such that the sum of the weights of its elements is minimum among all bases of M. An instance of the MMBU problem is given by a tuple (E, I, w, A), where $M = (E, I)$ is a matroid, w maps each $e \in E$ to its (actual) weight w_e, and A maps each $e \in E$ to its uncertainty area A_e.

Lemma 17. *Let C be a circuit of M, and let B be a base of M containing an element $e \in C$. Then there exists an element $f \in C$ such that $(B \setminus \{e\}) \cup \{f\}$ is a base of M.*

It follows from the previous lemma that:

Corollary 18. *Let C be a circuit of M, and let e be an element in C with a highest weight among all elements in C. Then a minimum base of $E \setminus \{e\}$ is also a minimum base of E.*

Before introducing the algorithm we define an order of elements denoted by $<_e$. Let f and g be two elements in E. We say $f <_e g$ if $L_f < L_g$ or ($L_f = L_g$ and $U_f < U_g$). Edges with the same upper and lower weight limit are ordered arbitrarily. We also say E is *indexed by* $<_e$ if $\{e_1, \ldots, e_n\} = E$ where $e_i <_e e_{i+1}$. Based on the order $<_e$ and the resulting indexing, the first circuit created by taking the lowest indexed elements of any set $E' \subseteq E$ is called the *first circuit* of E'.

Algorithm 2. U-RED2

1. fix $<_e$; index E by $<_e$; set $E' := E$
2. repeatedly remove an element b from E' that is a highest element of some circuit C of E', i.e., that satisfies $L_b \geq U_c$ for all $c \in C \setminus \{b\}$
3. **if** E' is independent **then**
4.   stop and output E' as a minimum base of M
5. **else**
6.   set $C :=$ the first circuit over E'
7.   set $h :=$ an element in C with a maximum upper limit
8.   set $f :=$ an element in $C \setminus \{h\}$ that could be potentially higher than h
9.   query the witness set $\{f, h\}$ and restart

Algorithm U-RED2 is shown in Algorithm 2. In Step 2, it repeatedly removes an element that can be identified as a highest element in a circuit, based on just the uncertainty areas that are known to the algorithm at that time. In Step 9, we query $\{f, h\}$. The basic idea to prove that $\{f, h\}$ is a witness set is as follows: By the choice of h and f, on the one hand, h could be a highest element of C and would then be excluded from any minimum base over E. On the other hand, even if all elements except h and f are queried, C being the first circuit implies that for each circuit $C' \neq C$ containing h, C' contains an element higher than L_h, i.e., h could be part of any minimum base over E. So, we must query at least one element in $\{f, h\}$.

Lemma 19. *The set $\{f, h\}$ in the algorithm U-RED2 is a witness set.*

Lemma 20. *Once the algorithm does not restart, E' is a minimum base of M.*

Proof. The set E' is independent and hence does not contain any circuits. Thus, E' is the only, and therefore also minimum, base over E'. Since E' was derived from E by removing only highest elements of a circuit, by a repeated application of Corollary 18, E' is a also a base for E. □

Finally we note that $OPT \geq k$ where k is the number of times the algorithm restarts. Since the number of queries requested by the algorithm is $2k$, U-RED2 makes at most $2 \cdot OPT$ queries.

Theorem 21. *The U-RED2 algorithm solves the minimum matroid base under uncertainty problem with at most $2 \cdot OPT$ queries.*

6 Competitive Lower Bounds

In this section we present lower bounds on the competitive ratio of online algorithms for cheapest set under uncertainty as well as its special cases.

An algorithm is called strongly ρ-competitive if $ALG \leq \rho OPT$ for all instances of the problem, and weakly ρ-competitive if there is a constant c such that $ALG \leq \rho OPT + c$ for all instances. Here, ALG is the number of queries made by the algorithm, and OPT is the size of an optimal query solution. A lower bound on the best possible competitive ratio of weakly competitive algorithms is also a lower bound for strongly competitive algorithms. We can prove lower bounds for weakly competitive algorithms, covering both the multiplicative ratio ρ and the additive constant c. The proofs use only singleton sets and open intervals as areas of uncertainty.

Definition 22. *We say that a variant of the cheapest set problem with uncertainty has (deterministic) lower bound $\lambda OPT + \delta$ if the two conditions hold:*
- Any algorithm that satisfies $ALG \leq \rho OPT + O(1)$ for all instances has $\rho \geq \lambda$.
- Any algorithm that satisfies $ALG \leq \lambda OPT + c$ for all instances has $c \geq \delta$.

Theorem 23. *CSU (MMCTU) has a lower bound of $d \cdot OPT + d$ (of $d \cdot OPT + 1$).*

Finally, we remark that the lower bound proof in [5] implies the following:

Theorem 24. *MMBU has a lower bound of $2 \cdot OPT$.*

7 Conclusion

In this paper, we have studied online and offline variants of cheapest set problems under uncertainty. While our lower bounds are tight for deterministic algorithms, an interesting direction for future research is to determine whether query-competitive algorithms with better ratios are possible using randomization. Another direction would be to identify further variants of cheapest set problems whose structure admits better competitive ratios than the general case.

Acknowledgements. We would like to thank Gerhard Woeginger for pointing out that Lemma 13 can be proved using Theorem 1 in [11]. The first author would also like to thank Anita Maring for helpful discussions about lower bound examples for the general cheapest set problem. The second author would like to thank the University of Leicester to support this research in granting him academic study leave.

References

1. Aissi, H., Bazgan, C., Vanderpooten, D.: Min-max and min-max regret versions of combinatorial optimization problems: A survey. European Journal of Operational Research 197(2), 427–438 (2009)
2. Bruce, R., Hoffmann, M., Krizanc, D., Raman, R.: Efficient update strategies for geometric computing with uncertainty. Theory of Computing Systems 38(4), 411–423 (2005)
3. Cook, W.J., Cunningham, W.H., Pulleyblank, W.R., Schrijver, A.: Combinatorial Optimization. John Wiley and Sons, New York (1998)
4. Dorrigiv, R., Fraser, R., He, M., Kamali, S., Kawamura, A., López-Ortiz, A., Seco, D.: On minimum-and maximum-weight minimum spanning trees with neighborhoods. In: Erlebach, T., Persiano, G. (eds.) WAOA 2012. LNCS, vol. 7846, pp. 93–106. Springer, Heidelberg (2013)
5. Erlebach, T., Hoffmann, M., Krizanc, D., Mihalák, M., Raman, R.: Computing minimum spanning trees with uncertainty. In: Albers, S., Weil, P. (eds.) 25th International Symposium on Theoretical Aspects of Computer Science (STACS 2008). LIPIcs, vol. 1, pp. 277–288. Schloss Dagstuhl - Leibniz-Zentrum für Informatik, Germany (2008)
6. Feder, T., Motwani, R., O'Callaghan, L., Olston, C., Panigrahy, R.: Computing shortest paths with uncertainty. Journal of Algorithms 62(1), 1–18 (2007)
7. Feder, T., Motwani, R., Panigrahy, R., Olston, C., Widom, J.: Computing the median with uncertainty. SIAM Journal on Computing 32(2), 538–547 (2003)
8. Garg, N., Vazirani, V.V., Yannakakis, M.: Primal-dual approximation algorithms for integral flow and multicut in trees. Algorithmica 18(1), 3–20 (1997)
9. Kahan, S.: A model for data in motion. In: 23rd Annual ACM Symposium on Theory of Computing (STOC 1991), pp. 267–277 (1991)
10. Khanna, S., Tan, W.C.: On computing functions with uncertainty. In: 20th Symposium on Principles of Database Systems (PODS 2001), pp. 171–182 (2001)
11. Lindström, B.: A theorem on families of sets. Journal of Combinatorial Theory (A) 13, 274–277 (1970)
12. Löffler, M., van Kreveld, M.: Largest and smallest tours and convex hulls for imprecise points. In: Arge, L., Freivalds, R. (eds.) SWAT 2006. LNCS, vol. 4059, pp. 375–387. Springer, Heidelberg (2006)
13. Olston, C., Widom, J.: Offering a precision-performance tradeoff for aggregation queries over replicated data. In: 26th International Conference on Very Large Data Bases (VLDB 2000), pp. 144–155 (2000)

Streaming Kernelization[*]

Stefan Fafianie and Stefan Kratsch

TU Berlin, Germany
{stefan.fafianie,stefan.kratsch}@tu-berlin.de

Abstract. Kernelization is a formalization of preprocessing for combinatorially hard problems. We modify the standard definition for kernelization, which allows any polynomial-time algorithm for the preprocessing, by requiring instead that the preprocessing runs in a streaming setting and uses $\mathcal{O}(poly(k) \log |x|)$ bits of memory on instances (x, k). We obtain several results in this new setting, depending on the number of passes over the input that such a streaming kernelization is allowed to make. EDGE DOMINATING SET turns out as an interesting example because it has no single-pass kernelization but two passes over the input suffice to match the bounds of the best standard kernelization.

1 Introduction

When faced with an NP-hard problem we do not expect to find an efficient algorithm that solves every instance exactly and in polynomial time (as this would imply P = NP). The study of algorithmic techniques offers various paradigms for coping with this situation if we are willing to compromise on efficiency, exactness, or the generality of being applicable to all instances (or several of those). Before we commit to such a compromise it is natural to see how much closer we can come to a solution by spending only polynomial time, i.e., how much we can simplify and shrink the instance by polynomial-time *preprocessing*. This is usually compatible with any way of solving the simplified instance and it finds wide application in practice (e.g., as a part of ILP solvers like CPLEX), although, typically, the applications are of a heuristic flavor with no guarantees for the size of the simplified instance or the amount of simplification.

The notion of *kernelization* is one way of formally capturing preprocessing. A kernelization algorithm applied to some problem instance takes polynomial time in the input size and always returns an equivalent instance (i.e., the instances will have the same answer) of size bounded by a function of some *problem-specific parameter*. For example, the problem of testing whether a given graph G has a vertex cover of size at most k can be efficiently reduced to an equivalent instance (G', k) where G' has $\mathcal{O}(k)$ vertices and $\mathcal{O}(k^2)$ total bit size. The study of kernelization is a vibrant field that has seen a wealth of new techniques and results over the last decade. (The interested reader is referred to recent surveys by Lokshtanov et al. [10] and Misra et al. [11].) In particular, a wide-range of problems is already classified into admitting or not admitting[1] a *polynomial*

---

[*] Supported by the Emmy Noether-program of the DFG, KR 4286/1.
[1] Unless NP ⊆ coNP/poly and the polynomial hierarchy collapses.

E. Csuhaj-Varjú et al. (Eds.): MFCS 2014, Part II, LNCS 8635, pp. 275–286, 2014.

kernelization, where the guaranteed output size bound is polynomial in the chosen parameter. It is seems fair to say that this shows a substantial *theoretical* success of the notion of kernelization.

From a practical point of view, we might have to do more work to convince a practitioner that our positive kernelization results are also *worth implementing*. This includes choice of parameter, computational complexity, and also conceptual difficulty (e.g., number of black box subroutines, huge hidden constants). Stronger parameterizations already receive substantial interest from a theoretical point of view, see e.g., [6], and there is considerable interest in making kernelizations fast, see e.g., [14,7,13,9]. Conceptual difficulty is of course "in the eye of the beholder" and perhaps hard to quantify.

In this work, we take the perspective that kernelizations that work in a restricted model might, depending on the model, be provably robust and useful/implementable (and hopefully also fast). Concretely, in the spirit of studying restricted models, we ask which kernelizations can be made to work in a *streaming model* where the kernelization has a small local memory and only gets to look at the input once, or a bounded number of times. The idea is that the kernelization should maintain a sufficiently good sketch of the input that in the end will be returned as the reduced instance.

We think that this restricted model for kernelization has several further benefits: First of all, it naturally improves the memory access patterns since the input is read sequentially, which should be beneficial already for medium size inputs. (It also works more naturally for huge inputs, but huge instances of NP-hard problems are probably only really addressable by outright use of heuristics or sampling methods.) Second, it is naturally connected to a dynamic/incremental setting since, due to the streaming setting, the algorithm has not much choice but to essentially maintain a simplified instance of bounded size that is equivalent to the input seen so far (or be able to quickly produce one should the end of the stream be declared). Thus, as further input arrives, the problem kernel is adapted to the now slightly larger instance without having to look at the whole instance again. (In a sense, the kernelization could run in parallel to the *creation* of the actual input.) Third, it appears, at least in our positive results, that one could easily translate this to a parallel setting where, effectively, several copies of the algorithm work on different positions on the stream to simplify the instance (this however would require that an algorithm may delete data from the stream).

Our results. In this work we consider a streaming model where elements of a problem instance are presented to a kernelization algorithm in arbitrary order. The algorithm is required to return an equivalent instance of size polynomial in parameter k after the stream has been processed. Furthermore, it is allowed to use $\mathcal{O}(poly(k)\log n)$ bits of memory, i.e., an overhead factor of $\mathcal{O}(\log n)$ is used in order to distinguish between elements of an instance of size n.

We show that d-Hitting Set(k) and d-Set Matching(k) admit streaming kernels of size $\mathcal{O}(k^d \log k)$ while using $\mathcal{O}(k^d \log |U|)$ bits of memory where U is the universal set of an input instance. We then consider a single pass kernel for Edge Dominating Set(k) and find that it requires at least $m-1$ bits of memory for instances with m edges. This rules out streaming kernels with $c \cdot poly(k) \log n$ bits for instances with n vertices since for any fixed c and $poly(k)$ there exist instances with $m - 1 > c \cdot poly(k) \log n$. Insights obtained from this lower bound allow us to develop a general lower bound for the space complexity of single pass kernels for a class of parameterized graph problems.

Despite the lower bound for single pass kernels, we show that EDGE DOMI-NATING SET(k) admits a streaming kernel if it is allowed to make a pass over the input stream twice. Finally, we use communication complexity games in order to rule out similar results for CLUSTER EDITING(k) and MINIMUM FILL-IN(k) and show that multi-pass streaming kernels for these problems must use $\Omega(n)$ bits of local memory for graphs with n vertices, even when a constant number of passes are allowed.

Related work. The data stream model is formalized by Henzinger et al. [8]. Lower bounds for exact and randomized algorithms with a bounded number of passes over the input stream are given for various graph problems and are proven by means of communication complexity. An overview is given by Babcock et al. [2] in which issues that arise in the data stream model are explored. An introduction and overview of algorithms and applications for data streams is given by Muthukrishnan [12].

Organization. Section 2 contains preliminaries and a formalization of kernelization algorithms in the data streaming setting. Single pass kernels for d-HITTING SET(k) and d-SET MATCHING(k) are presented in Section 3. The lower bounds for single pass kernels are given in Section 4. The 2-pass kernel for EDGE DOMINATING SET(k) is shown in Section 5 while lower bounds for multi-pass kernels are given in Section 6. Finally, Section 7 contains concluding remarks.

2 Preliminaries

We use standard notation from graph theory. For a set of edges E, let $V(E)$ be the set of vertices that are incident with edges in E. For a graph $G = (V, E)$, let $G[V]$ denote the subgraph of G induced by V. Furthermore, let $G[E]$ be the subgraph induced by E, i.e. $G[E] = G(V(E), E)$.

A *parameterized problem* is a language $Q \subseteq \Sigma^* \times \mathbb{N}$; the second component k of instances (x, k) is called the *parameter*. A parameterized problem is *fixed-parameter tractable* if there is an algorithm that decides if $(x, k) \in Q$ in $f(k)|x|^{\mathcal{O}(1)}$ time, where f is any computable function. A *kernelization algorithm* (*kernel*) for a parameterized problem $Q \subseteq \Sigma^* \times N$ is an algorithm that, for input $(x, k) \in \Sigma^* \times \mathbb{N}$ outputs a pair $(x', k') \in \Sigma^* \times \mathbb{N}$ in $(|x| + k)^{\mathcal{O}(1)}$ time such that $|x'|, k' < g(k)$ for some computable function g, called the *size* of the kernel, and $(x, k) \in Q \Leftrightarrow (x', k') \in Q$. A *polynomial kernel* is a kernel with polynomial size.

Kernelization in the Data-Streaming Model

An input stream is a sequence of elements of the input problem. We denote the start of an input stream by $\langle$ and let $\rangle$ denote the end, e.g. $\langle e_1, e_2, \ldots, e_m \rangle$

denotes an input stream for a sequence of m elements. We use $\rangle$ to denote a halt in the stream and $\langle$ to denote its continuation, e.g. $\langle e_1, e_2 \rangle$ and $\langle e_3, \ldots, e_m \rangle$ denote the same input stream broken up in two parts.

A *streaming kernelization algorithm* (*streaming kernel*) is an algorithm that receives input (x, k) for a parameterized problem in the following fashion. The algorithm is presented with an input stream where elements of x are presented in a sequence, i.e. adhering to the cash register model [12]. Finally, the algorithm should return a kernel for the problem upon request. A *t-pass streaming kernel* is a streaming kernel that is allowed t passes over the input stream before a kernel is requested.

If x is a graph, then the sequence of elements of x are its edges in arbitrary ordering. In a natural extension to hypergraphs, if x is a family of subsets on some ground set U, then the sequence of elements of x are the sets of this family in arbitrary ordering. We assume that a streaming kernelization algorithm receives parameter k and the size of the vertex set (resp. ground set) before the input stream. Note that this way isolated vertices are given implicitly.

Furthermore, we require that the algorithm uses a limited amount of space at any time during its execution. In the strict streaming kernelization setting the streaming kernel must use at most $p(k) \log |x|$ space where p is a polynomial. We will refer to a 1-pass streaming kernelization algorithm which upholds these space bounds simply as a *streaming kernelization*.

We assume that words of size $\log |x|$ in memory can be compared in $\mathcal{O}(1)$ operations when considering the running time of the streaming kernelization algorithms in each step.

3   Single Pass Kernelization Algorithms

In this section we will show streaming kernelization algorithms for d-Hitting Set(k) and d-Set Matching(k) in the 1-pass data-stream model. These algorithms make a single pass over the input stream after which they output a kernel. We analyze their efficiency with regard to local space and the worst case processing time for a single element in the input stream.

---

d-Hitting Set(k) **Parameter:** k.

Input: A set U and a family $\mathcal{F}$ of subsets of U each of size at most d, i.e. $\mathcal{F} \subseteq \binom{U}{\leq d}$, and $k \in \mathbb{N}$.

Question: Is there a set S of at most k elements of U that has a nonempty intersection with each set in $\mathcal{F}$?

In the following, we describe a single step of the streaming kernelization. After Step t, the algorithm has seen a family $\mathcal{F}_t \subseteq \mathcal{F}$, where $\mathcal{F}$ denotes the whole family of sets provided in the stream. The memory contains some subfamily $\mathcal{F}'_t \subseteq \mathcal{F}_t$, using for each $F \in \mathcal{F}'_t$ a total of at most $d \log n = \mathcal{O}(\log n)$ bits to denote the up to d elements therein. The algorithm maintains the invariant that the number of sets $F \in \mathcal{F}'_t$ that contain any $C \in \binom{U}{\leq d-1}$ as a subset is at most

$(d - |C|)! \cdot (k+1)^{d-|C|}$. For intuition, let us remark that this strongly relates to the sunflower lemma [4]. Now, let us consider Step $t + 1$. The memory contains some $\mathcal{F}_t' \subseteq \mathcal{F}_t$ and a new set F arrives.

1. Iterate over all subsets C of F, ordered by decreasing size.
2. Count the number of sets in $\mathcal{F}_t'$ that contain C as a subset.
3. If the result equals $(d - |C|)! \cdot (k+1)^{d-|C|}$ then the algorithm decides not to store F and ends the computation for Step $t + 1$, i.e., let $\mathcal{F}_{t+1}' = \mathcal{F}_t'$.
4. Else, continue with the next set C.
5. If no set $C \subseteq F$ gave a total of $(d - |C|)! \cdot (k+1)^{d-|C|}$ sets containing F then the algorithm decides to store F, i.e., $\mathcal{F}_{t+1}' = \mathcal{F}_t' \cup \{F\}$. Note that this preserves the invariant for all $C \in \binom{U}{d-1}$ since only the counts for C with $C \subseteq F$ can increase, but all those were seen to be strictly below the threshold $(d-|C|)! \cdot (k+1)^{d-|C|}$ so they can at most reach equality by adding F.

To avoid confusion, let us point out that at any time the algorithm only has a single set $\mathcal{F}_t'$; the index t is used for easier discussion of the changes over time.

Observation 1 *The algorithm stores at most* $d!(k+1)^d = \mathcal{O}(k^d)$ *sets at any point during the computation. This follows directly from the invariant when considering* $C = \emptyset$.

Theorem 1. (★[2]) d-HITTING SET(k) *admits a streaming kernelization which, using* $\mathcal{O}(k^d \log |U|)$ *bits of local memory and* $\mathcal{O}(k^d)$ *time in each step, returns an equivalent instance of size* $\mathcal{O}(k^d \log k)$.

The time spent in each step can be improved from $\mathcal{O}(|\mathcal{F}_t'|) = \mathcal{O}(k^d)$ to $\mathcal{O}(\log |\mathcal{F}_t'|)$ at the cost of an increase in local space by a constant factor. This can be realized with a tree structure $\mathbb{T}$ in which the algorithm maintains the number of sets in $\mathcal{F}_t'$ that contain a set $C \in \binom{U}{\leq d-1}$ as a subset.

Each $C \subseteq F'$, $F' \in \mathcal{F}_t'$ has a corresponding node in $\mathbb{T}$ and in this node the number of supersets of C in $\mathcal{F}_t'$ are stored. Let the root node represent $C = \emptyset$ with a child for each set C of size 1. In general, a node is assigned an element in $e \in \bigcup \mathcal{F}_t'$ and represents $C = C' \cup \{e\}$ where C' is the set represented by its parent, i.e. $|C| = d$ for nodes with depth d. For each node, let e_i be assigned to child node n_i. Let us assume that there is some arbitrary ordering on elements that are in sets of $\mathcal{F}_t'$, e.g. by their identifier. Then we can treat sets C as if they are ordered and we require that $e_i > e$ for each child.

Furthermore, each node has a dictionary, i.e. a collection of (key, value) pairs (e_i, n_i) in order to facilitate quick lookup of its children. By utilizing the ordering on elements, this dictionary can be implemented as a self-balancing binary search tree. This allows us to find a child node and insert new child node in time $\mathcal{O}(\log h)$ if there are h children.

Corollary 1. (★) d-HITTING SET(k) *admits a streaming kernelization which, using* $\mathcal{O}(k^d \log |U|)$ *bits of local memory and* $\mathcal{O}(\log k)$ *time in each step, returns an equivalent instance of size* $\mathcal{O}(k^d \log k)$.

---

[2] Proofs of statements marked with ★ are in the full paper [5].

---

d-SET MATCHING(k)   **Parameter:** k.
Input: A set U and a family $\mathcal{F}$ of subsets of U each of size at most d, i.e. $\mathcal{F} \subseteq \binom{U}{\leq d}$, and $k \in \mathbb{N}$.
Question: Is there a matching M of at least k sets in $\mathcal{F}$, i.e. are there k sets in $\mathcal{F}$ that are pairwise disjoint?

The streaming kernelization will mostly perform the same operations in a single step as the algorithm described above such that only the invariant differs. In this case it is maintained that the number of sets $F \in \mathcal{F}'_t$ that contain any $C \in \binom{U}{\leq d-1}$ as a subset is at most $(d - |C|)! \cdot (d(k-1)+1)^{d-|C|}$.

Observation 2 *The algorithm stores at most $d!(d(k-1)+1)^d = \mathcal{O}(k^d)$ sets at any point during the computation. This follows directly from the invariant when considering $C = \emptyset$.*

Theorem 2. (★) d-SET MATCHING(k) *admits a streaming kernelization which, using $\mathcal{O}(k^d \log |U|)$ bits of local memory and $\mathcal{O}(k^d)$ time in each step, returns an equivalent instance of size $\mathcal{O}(k^d \log k)$.*

Similar to the algorithm described in the previous section, the running time in each step can be improved at the cost of an increase in local space by a constant factor. We omit an explicit proof.

Corollary 2. d-SET MATCHING(k) *admits a streaming kernelization which, using $\mathcal{O}(k^d \log |U|)$ bits of local memory and $\mathcal{O}(\log k)$ time in each step, returns an equivalent instance of size $\mathcal{O}(k^d \log k)$.*

4   Space Lower Bounds for Single Pass Kernels

We will now present lower bounds on the memory requirements of single pass streaming kernelization algorithms for a variety of graph problems. Before giving a general lower bound we first illustrate the essential obstacle by considering the EDGE DOMINATING SET(k) problem. We show that a single pass kernel for EDGE DOMINATING SET(k) requires at least $m - 1$ bits of memory on instances with m edges.

EDGE DOMINATING SET(k)   **Parameter:** k.
Input: A graph $G = (V, E)$ and $k \in \mathbb{N}$.
Question: Is there a set S of at most k edges such that every edge in $E \setminus S$ is incident with an edge in S?

An obstacle that arises for many problems, such as EDGE DOMINATING SET(k), is that they are not monotone under adding additional edges, i.e., additional edges do not always increase the cost of a minimum edge dominating set but may also decrease it. This decrease, however, may in turn depend on the existence of a particular edge in the input. Thus, on an intuitive level, it may be impossible for a streaming kernelization to "decide" which edges to forget, since worst-case analysis effectively makes additional edges behave adversarial. (Note that our lower bound does not depend on assumptions on what the kernelization decides to store.)

Consider the following type of instance as a concrete example of this issue. The input stream contains the number of vertices (immaterial for the example), the parameter value $k = 1$, and a sequence of edges $\langle \{a, v_1\}, \ldots, \{a, v_n\}, \{b, v\} \rangle$. That is, the first n edges form a star with n leaves and center vertex a. In order to use a relatively small amount of local memory the kernelization algorithm is forced to do some compression such that not every edge belonging to this star is stored in local memory. Now a final edge arrives and the algorithm returns a kernel. Note that the status of the problem instance depends on whether or not this edge is disjoint from the star: If it shares at least one vertex v_i with the star then there is an edge dominating set $\{a, v_i\}$ of size one. Otherwise, if it is disjoint then clearly at least two edges are needed. Thus, from the memory state after the final edge we must be able to extract whether or not v is contained in $\{v_1, \ldots, v_n\}$; in other words, this is equivalent to whether or not the output kernelized instance is YES or NO. (We assume that $a, b \notin \{v_1, \ldots, v_n\}$ for this example.) This, however, is a classic problem for streaming algorithms that is related to the *set reconciliation problem* and it is known to require at least n bits [12]; we give a short self-contained proof for our lower bound.

Theorem 3. ($\bigstar$) *A single pass streaming kernelization algorithm for* EDGE DOMINATING SET(k) *requires at least $m - 1$ bits of local memory for instances with m edges.*

General Lower Bound for a Class of Parameterized Graph Problems

In the following we present space lower bounds for a number of parameterized graph problems. By generalizing the previous argument we find a common property that can be used to quickly rule out single pass kernels with $\mathcal{O}(poly(k) \log |x|)$ memory. We then provide a list of parameterized graph problems for which a single pass streaming kernelization algorithm requires at least $|E| - \mathcal{O}(1)$ bits of local memory.

Definition 1. *Let $Q \in \Sigma^* \times \mathbb{N}$ be a parameterized graph problem and let $c, k \in \mathbb{N}$. Then Q has a c-k-stream obstructing graph $G = (V, E)$ if $\forall e_i \in E$, there is a set of edges $R_i := R(e_i) \subseteq \binom{V}{2} \setminus E$ of size c such that $\forall F \subseteq E$, $(G[F \cup R_i], k) \in Q$ if and only if $e_i \in F$.*

In other words, each edge $e_i \in E$ could equally be critical to decide if $(G', k) \in Q$ for a graph instance G' induced by a subset $F \subseteq E$ and a constant sized remainder of edges R_i, depending on what R_i looks like. Note that G may contain isolated vertices which can also be used to form edge sets R_i. We also consider G to be a c-k-stream obstructing graph in the case that the above definition holds except that $\forall F \subseteq E$, $(G[F \cup R_i], k) \in Q$ if and only if $e_i \notin F$. We omit the proofs for this symmetrical definition in this section.

Lemma 1. (★) *Let $Q \in \Sigma^* \times \mathbb{N}$ be a parameterized graph problem and let $c, k \in \mathbb{N}$. If Q has a c-k-stream obstructing graph $G = (V, E)$ with m edges, then a single pass streaming kernelization algorithm for Q requires at least m bits of local memory for instances with at most $m + c$ edges.*

The following theorem is an easy consequence of Lemma 1 for problems that, essentially, have stream obstructing graphs for all numbers m of edges. Intuitively, of course also having such graphs only for an infinite subset of $\mathbb{N}$ suffices to get a similar bound.

Theorem 4. *Let $Q \in \Sigma^* \times \mathbb{N}$ be a parameterized graph problem. If there exist $c, k \in \mathbb{N}$ such that for every $m \in \mathbb{N}$, Q has a c-k-stream obstructing graph G with m edges, then a single pass streaming kernelization algorithm for Q requires at least $|E| - c$ bits of local memory.*

Proof. Let A be a single pass streaming kernelization algorithm for Q. Assume that there is a stream obstructing graph $G_m = (V_m, E_m)$ for Q with m edges for every $m \in \mathbb{N}$. Then for every m there is a group of instances $\mathcal{G}$ where for each $G_i = (V_i, E_i) \in \mathcal{G}$, $E_i = F \cup R_i$ for some $F \subseteq E_m$ and remainder of edges R_i of size c, i.e. $|E_i| \leq m + c$. Let us consider all graph instances $G = (V, E)$ with exactly $|E| = m + c$ edges. Some of these instances are in $\mathcal{G}$, i.e. $E = E_m \cup R_i$ for some R_i. By Lemma 1, A requires at least $m = |E| - c$ bits of local memory in order to distinguish these instances correctly.   □

The following corollary is a result of Theorem 4 and constructions of stream obstructing graphs of arbitrary size for a variety of parameterized graph problems (definitions for most of these are in [3]). These constructions can be found in the full paper [5], where we also exhibit proofs of correctness for a few of them.

Corollary 3. *For each of the following parameterized graph problems, a single pass streaming kernelization requires at least $|E| - \mathcal{O}(1)$ bits of local memory:* EDGE DOMINATING SET(k), CLUSTER EDITING(k), CLUSTER DELETION(k), CLUSTER VERTEX DELETION(k), COGRAPH VERTEX DELETION(k), MINIMUM FILL-IN(k), EDGE BIPARTIZATION(k), FEEDBACK VERTEX SET(k), ODD CYCLE TRANSVERSAL(k), TRIANGLE EDGE DELETION(k), TRIANGLE VERTEX DELETION(k), TRIANGLE PACKING(k), s-STAR PACKING(k), BIPARTITE COLORFUL NEIGHBORHOOD(k).

5   2-Pass Kernel for Edge Dominating Set

Despite the previously shown lower bound of $m - 1$ bits for a single pass kernel, there is a space efficient streaming kernelization algorithm for EDGE DOMINATING SET(k) if we allow it to make a pass over the input stream twice. We will first describe a single step of the streaming kernelization during the first pass. This is effectively a single pass kernel for finding a $2k$-vertex cover. After Step t the algorithm has seen a set $A_t \subseteq E$. Some subset $A'_t \subseteq A_t$ of edges is stored in memory. Let us consider Step $t + 1$ where a new edge $e = \{u, v\}$ arrives.

1. Count the edges in A'_t that are incident with u; do the same for v.
2. Let $A'_{t+1} = A'_t$ if either of these counts is at least $2k + 1$.
3. Otherwise, let $A'_{t+1} = A'_t \cup \{e\}$.
4. If $|A'_{t+1}| > 4k^2 + 2k$, then return a NO instance.

Lemma 2. ($\star$) *After processing any set A_t of edges on the first pass over the input stream the algorithm has a set $A'_t \subseteq A_t$ such that any set S of at most $2k$ vertices is a vertex cover for $G[A_t]$ if and only if S is a vertex cover for $G[A'_t]$.*

Let A' be the edges stored after the first pass. If there are more than $2k$ vertices with degree $2k + 1$ in $G[A']$ then the algorithm returns a NO instance. We will continue with a description of a single step during the second pass. After Step t the algorithm has revisited a set $B_t \subseteq E$. Some subset $B'_t \subseteq B_t$ of edges is stored along with A'. Now, let us consider Step $t + 1$ where the edge $e = \{u, v\}$ is seen for the second time.

1. Let $B'_{t+1} = B'_t \cup \{e\}$ if $u, v \in V(A')$ and $e \notin A'$.
2. Otherwise, let $B'_{t+1} = B'_t$.

Let B' be the edges stored during the second pass. The algorithm will return $G[A' \cup B']$, which is effectively $G[V(A')]$, after both passes have been processed without returning a NO instance.

Lemma 3. ($\star$) *After processing both passes the algorithm has a set $A' \cup B' \subseteq E$ such that there is an edge dominating set S of size at most k for G if and only if there is an edge dominating set S' of size at most k for $G[A' \cup B']$.*

Theorem 5. ($\star$) EDGE DOMINATING SET(k) *admits a two-pass streaming kernelization algorithm which, using $\mathcal{O}(k^3 \log n)$ bits of local memory and $\mathcal{O}(k^2)$ time in each step, returns an equivalent instance of size $\mathcal{O}(k^3 \log k)$.*

If the algorithm stores a counter for the size of A'_t and a tree structure $\mathbb{T}$ in which it maintains the number of sets (edges) in A'_t that are a superset of $C \subseteq \binom{V}{\leq 2}$ as described in Section 3, then the operations in each step can be performed in $\mathcal{O}(\log k)$ time. We give the following corollary and omit the proof.

Corollary 4. EDGE DOMINATING SET(k) *admits a two-pass streaming kernelization algorithm which, using $\mathcal{O}(k^3 \log n)$ bits of local memory and $\mathcal{O}(\log k)$ time in each step, returns an equivalent instance of size $\mathcal{O}(k^3 \log k)$.*

6 Space Lower Bounds for Multi-pass Streaming Kernels

In this section we will show lower bounds for multi-pass streaming kernels for CLUSTER EDITING(k) and MINIMUM FILL-IN(k). Similar to EDGE DOMINATING SET(k), it is difficult to return a trivial answer for these problems when the local memory exceeds a certain bound at some point during the input stream. Additional edges in the stream may turn a NO instance into a YES instance and vice versa, which makes single pass streaming kernels infeasible. Although there is a 2-pass streaming kernel for EDGE DOMINATING SET(k), we will show that a

t-pass streaming kernel for CLUSTER EDITING(k) requires at least $(n-2)/2t$ bits of local memory for instances with n vertices. As a consequence, $\Omega(n)$ bits are required when a constant number of passes are allowed. Furthermore, $\Omega(n/\log n)$ passes are required when the streaming kernel uses at most $\mathcal{O}(\log n)$ bits of memory. We show a similar result for MINIMUM FILL-IN(k).

CLUSTER EDITING(k)   **Parameter:** k.
Input: A graph $G = (V, E)$ and $k \in \mathbb{N}$.
Question: Can we add and/or delete at most k edges such that G becomes a disjoint union of cliques?

Let us consider the following communication game with two players, P_1 and P_2. Let N be a set of n' vertices and let $u, v \notin N$. The players are given a subset of vertices, $V_1 \subseteq N$ and $V_2 \subseteq N$ respectively. Let $C(V_1)$ denote the edges of a clique on $V_1 \cup \{u\}$. Furthermore, let $S(V_2)$ denote the edges of a star with center vertex v and leaves $V_2 \cup \{u\}$. The object of the game is for the players to determine if $G = (N \cup \{u, v\}, C(V_1) \cup S(V_2))$ is a disjoint union of cliques. The cost of the protocol for this game is the number of bits communicated between the players such that they can provide the answer. We can provide a lower bound for this cost by using the notion of fooling sets as shown in the following lemma.

Lemma 4. ([1]) *A function* $f : \{0,1\}^{n'} \times \{0,1\}^{n'}$ *has a size* M *fooling set if there is an M-sized subset $F \subseteq \{0,1\}^{n'} \times \{0,1\}^{n'}$ and value $b \in \{0,1\}$ such that,*

(1) for every pair $(x, y) \in S$, $f(x, y) = b$
(2) for every distinct $(x, y), (x', y') \in F$, either $f(x, y') \neq b$ of $f(x', y) \neq b$.

If f has a size-M fooling set then $C(f) \geq \log M$ where $C(f)$ is the minimum number of bits communicated in a two-party protocol for f.

Let f be a function modeling our communication game where $f(V_1, V_2) = 1$ if G forms a disjoint union of cliques and $f(V_1, V_2) = 0$ otherwise. We provide a fooling set for f in the following lemma.

Lemma 5. *f has a fooling set* $F = \{(W, W) \mid W \subseteq N\}$.

Proof. For every $W \subseteq N$ we have $G = (N \cup \{u, v\}, C(W) \cup S(W))$ in which there is a clique on vertices $W \cup \{u, v\}$ while the vertices in $N \setminus W$ are completely isolated and thus form cliques of size 1, i.e. $f(W, W) = 1$ for every $(W, W) \in F$. Now let us consider pairs $(W, W), (W', W') \in F$. We must show that either $f(W, W') = 0$ or $f(W', W) = 0$. Clearly $W \neq W'$ since $(W, W) \neq (W', W')$. Let us assume w.l.o.g. that $W \setminus W' \neq \emptyset$, i.e. there is a vertex $w \in W \setminus W'$. Then $\{v, w\}, \{v, u\} \in S(W)$ since $w \in W$. However, $\{u, w\} \notin C(W')$ since $w \notin W'$ and by definition also $\{u, w\} \notin S(W)$ since $S(W)$ is a star with center $v \notin \{u, w\}$. Thus, $G = (N \cup \{u, v\}, C(W') \cup S(W))$ is not a disjoint union of cliques, i.e., $f(W', W) = 0$ and the lemma holds. □

The size of F is $2^{n'}$, implying by Lemma 4 that the protocol for f needs at least n' bits of communication. Intuitively, if we use less than n' bits, then by the pigeonhole principle there must be some pairs $(W, W), (W', W') \in F$ for which the protocol is identical. Then the players cannot distinguish between the cases $(W, W), (W, W'), (W', W), (W', W')$, i.e. for each case the same answer will be given and thus the protocol is incorrect. We prove the following theorem by considering how a multi-pass kernel for CLUSTER EDITING(k) with small local memory can be used to beat the lower bound of the communication game.

Theorem 6. *A streaming kernelization algorithm for* CLUSTER EDITING(k) *requires at least* $(n - 2)/2t$ *bits of local memory for instances with* n *vertices if it is allowed to make* t *passes over the input stream.*

Proof. Let us assume that the players have access to a multi-pass streaming kernelization algorithm A for CLUSTER EDITING(k). They can then use A to solve the communication game for $|N| = n' = n - 2$ by simulating passes over an input stream in the following way. First, P_1 initiates A with budget $k = 0$. To let A make a pass over $C(V_1) \cup S(V_2)$, P_1 feeds A with partial input stream $\langle C(V_1) |$. It then sends the current content of the local memory of A to P_2, which is then able to resume A and feeds it with $| S(V_2) \rangle$. In order to let A make multiple passes, P_2 can send the local memory content back to P_1. Finally, when enough passes have been made an instance can be requested from A for which the answer is YES if and only if $f(V_1, V_2) = 1$.

Now suppose A is a t-pass streaming kernel with less than $(n - 2)/2t$ bits of local memory for instances with n vertices. In each pass the local memory is transmitted between P_1 and P_2 twice. Then in total the players communicate less than $n - 2 = n'$ bits of memory. This is a contradiction to the consequence of Lemmata 4 and 5. Therefore A requires at least $(n - 2)/2t$ bits. $\square$

Note that this argument also holds for CLUSTER DELETION(k) and CLUSTER VERTEX DELETION(k) where we can only delete k edges, respectively vertices.

MINIMUM FILL-IN(k) **Parameter:** k.
Input: A graph $G = (V, E)$ and $k \in \mathbb{N}$.
Question: Can we add at most k edges such that G becomes chordal, i.e. G does not contain an induced cycle of length 4?

Consider the following communication game with two players, P_1 and P_2. Let N be a set of n vertices and let $p, u, v \notin N$. The players are given a subset of vertices, $V_1 \subseteq N$ and $V_2 \subseteq N$ respectively. Let $S_u(V_1)$ denote the edges of a star with center vertex u and leaves $V_1 \cup \{p\}$. Furthermore, let $S_v(V_2)$ denote the edges of a star with center vertex v and leaves $V_2 \cup \{p\}$. The object of the game is to determine if $G = (N \cup \{p, u, v\}, S_u(V_1) \cup S_v(V_2))$ is a chordal graph. Let f be a function modeling this communication game, i.e. $f(V_1, V_2) = 1$ if G is chordal and $f(V_1, V_2) = 0$ otherwise. We provide a fooling set for f.

Lemma 6. *f has a fooling set* $F = \{(W, N \setminus W) \mid W \subseteq N\}$.

The size of F is 2^n, implying by Lemma 4 that the protocol for f needs at least n bits of communication. The following results from a similar argument to that of the proof of Theorem 6. We omit an explicit proof.

Theorem 7. *A streaming kernelization algorithm for* MINIMUM FILL-IN(k) *requires at least* $(n-3)/2t$ *bits of local memory for instances with* n *vertices if it is allowed to make* t *passes.*

7   Conclusion

In this paper we have explored kernelization in a data streaming model. Our positive results include single pass kernels for d-HITTING SET(k) and d-SET MATCHING(k), and a 2-pass kernel for EDGE DOMINATING SET(k). We provide a tool for quick identification of a number of parameterized graph problems for which a single pass kernel requires $m - \mathcal{O}(1)$ bits of local memory for instances with m edges. Furthermore, we have shown lower bounds for the space complexity of multi-pass kernels for CLUSTER EDITING(k) and MINIMUM FILL-IN(k).

References

1. Arora, S., Barak, B.: Computational complexity: a modern approach. Cambridge University Press (2009)
2. Babcock, B., Babu, S., Datar, M., Motwani, R., Widom, J.: Models and issues in data stream systems. In: PODS, pp. 1–16. ACM (2002)
3. Downey, R.G., Fellows, M.R.: Fundamentals of Parameterized Complexity. Springer (2013)
4. Erdös, P., Rado, R.: Intersection theorems for systems of sets. Journal of the London Mathematical Society 1(1), 85–90 (1960)
5. Fafianie, S., Kratsch, S.: Streaming kernelization. arXiv report 1405.1356 (2014)
6. Fellows, M.R., Jansen, B.M.P., Rosamond, F.A.: Towards fully multivariate algorithmics: Parameter ecology and the deconstruction of computational complexity. Eur. J. Comb. 34(3), 541–566 (2013)
7. Hagerup, T.: Simpler Linear-Time Kernelization for Planar Dominating Set. In: Marx, D., Rossmanith, P. (eds.) IPEC 2011. LNCS, vol. 7112, pp. 181–193. Springer, Heidelberg (2012)
8. Henzinger, M.R., Raghavan, P., Rajagopalan, S.: Computing on data streams. In: External Memory Algorithms: DIMACS Workshop External Memory and Visualization, May 20-22, vol. 50, p. 107. AMS (1999)
9. Kammer, F.: A Linear-Time Kernelization for the Rooted k-Leaf Outbranching Problem. In: Brandstädt, A., Jansen, K., Reischuk, R. (eds.) WG 2013. LNCS, vol. 8165, pp. 310–320. Springer, Heidelberg (2013)
10. Lokshtanov, D., Misra, N., Saurabh, S.: Kernelization – Preprocessing with a Guarantee. In: Bodlaender, H.L., Downey, R., Fomin, F.V., Marx, D. (eds.) Fellows Festschrift 2012. LNCS, vol. 7370, pp. 129–161. Springer, Heidelberg (2012)
11. Misra, N., Raman, V., Saurabh, S.: Lower bounds on kernelization. Discrete Optimization 8(1), 110–128 (2011)
12. Muthukrishnan, S.: Data streams: Algorithms and applications. Now Publishers Inc. (2005)
13. van Bevern, R.: Towards Optimal and Expressive Kernelization for d-Hitting Set. In: Gudmundsson, J., Mestre, J., Viglas, T. (eds.) COCOON 2012. LNCS, vol. 7434, pp. 121–132. Springer, Heidelberg (2012)
14. van Bevern, R., Hartung, S., Kammer, F., Niedermeier, R., Weller, M.: Linear-Time Computation of a Linear Problem Kernel for Dominating Set on Planar Graphs. In: Marx, D., Rossmanith, P. (eds.) IPEC 2011. LNCS, vol. 7112, pp. 194–206. Springer, Heidelberg (2012)

A Reconfigurations Analogue
of Brooks' Theorem

Carl Feghali, Matthew Johnson, and Daniël Paulusma*

School of Engineering and Computing Sciences, Durham University,
Science Laboratories, South Road, Durham DH1 3LE, United Kingdom
{carl.feghali,matthew.johnson2,daniel.paulusma}@durham.ac.uk

Abstract. Let G be a simple undirected graph on n vertices with maximum degree Δ. Brooks' Theorem states that G has a Δ-colouring unless G is a complete graph, or a cycle with an odd number of vertices. To *recolour* G is to obtain a new proper colouring by changing the colour of one vertex. We show that from a k-colouring, $k > \Delta$, a Δ-colouring of G can be obtained by a sequence of $O(n^2)$ recolourings using only the original k colours unless
  - G is a complete graph or a cycle with an odd number of vertices, or
  - $k = \Delta + 1$, G is Δ-regular and, for each vertex v in G, no two neighbours of v are coloured alike.

We use this result to study the *reconfiguration graph* $R_k(G)$ of the k-colourings of G. The vertex set of $R_k(G)$ is the set of all possible k-colourings of G and two colourings are adjacent if they differ on exactly one vertex. It is known that
  - if $k \leq \Delta(G)$, then $R_k(G)$ might not be connected and it is possible that its connected components have superpolynomial diameter,
  - if $k \geq \Delta(G) + 2$, then $R_k(G)$ is connected and has diameter $O(n^2)$.

We complete this structural classification by settling the missing case:
  - if $k = \Delta(G) + 1$, then $R_k(G)$ consists of isolated vertices and at most one further component which has diameter $O(n^2)$.

We also describe completely the computational complexity classification of the problem of deciding whether two k-colourings of a graph G of maximum degree Δ belong to the same component of $R_k(G)$ by settling the case $k = \Delta(G) + 1$. The problem is
  - $O(n^2)$ time solvable for $k = 3$,
  - PSPACE-complete for $4 \leq k \leq \Delta(G)$,
  - $O(n)$ time solvable for $k = \Delta(G) + 1$,
  - $O(1)$ time solvable for $k \geq \Delta(G) + 2$ (the answer is always yes).

1   Introduction

Definitions and Background. Let $G = (V, E)$ denote a simple undirected graph and let k be a positive integer. A *k-colouring* of G is a function $\gamma : V \to \{1, 2, \ldots, k\}$ such that if $uv \in E$, $\gamma(u) \neq \gamma(v)$. The *k-colouring reconfiguration graph* of G has as its vertex set all possible k-colourings of G, and

* Author supported by EPSRC (EP/K025090/1).

E. Csuhaj-Varjú et al. (Eds.): MFCS 2014, Part II, LNCS 8635, pp. 287–298, 2014.

two k-colourings γ_1 and γ_2 are joined by an edge if, for some vertex $u \in V$, $\gamma_1(u) \neq \gamma_2(u)$, and, for all $v \in V \setminus \{u\}$, $\gamma_1(v) = \gamma_2(v)$; that is, if γ_1 and γ_2 *disagree* on exactly one vertex. The reconfiguration graph is denoted by $R_k(G)$.

The study of reconfiguration graphs of colourings began in [10,11]. The problem of deciding whether two 3-colourings of a graph G are in the same component of $R_3(G)$ was shown to be solvable in time $O(n^2)$ in [12]; it was also proved that the diameter of any component of $R_3(G)$ is $O(n^2)$. In contrast, in [5] the analagous problem for k-colourings, $k \geq 4$, was shown to be PSPACE-complete, and examples of reconfiguration graphs with components of superpolynomial diameter were given. In [2], reconfiguration graphs of k-colourings of chordal graphs were shown to be connected with diameter $O(n^2)$ whenever k is more than the size of the largest clique (and an infinite class of chordal graphs was described whose reconfiguration graphs have diameter $\Omega(n^2)$). In [1] this was generalized to show that if k is at least two greater than the treewidth $tw(G)$ then, again, $R_k(G)$ is connected with diameter $O(n^2)$. (Notice that if $k = tw(G) + 1$, then $R_k(G)$ might not be connected since, for example, G might be a complete graph on $tw(G) + 1$ vertices and then $R_k(G)$ contains no edges.)

Our Results. We study reconfigurations of colourings for graphs of bounded maximum degree. The celebrated theorem of Brooks [8] states that a graph G with maximum degree Δ has a Δ-colouring unless it is the complete graph on $\Delta + 1$ vertices or a cycle with an odd number n of vertices; we denote these two graphs by $K_{\Delta+1}$ and C_n respectively. The question we address is: given a k-colouring γ of G, is there a path from γ to a Δ-colouring in $R_k(G)$? (Note that we are abusing our terminology. When we are working with $R_k(G)$, by a Δ-colouring we mean a k-colouring in which only Δ colours appear on the vertices.) Our first result provides a complete answer to this question. We require two definitions. A k-colouring γ of a graph is *frozen* if, for every vertex v, every colour except $\gamma(v)$ is used on a neighbour of v. Notice that a frozen colouring is an isolated vertex in $R_k(G)$. The length of a shortest path between colourings α and β in $R_k(G)$ is denoted by $d_k(\alpha, \beta)$. We state our results for connected graphs as other graphs can be considered component-wise.

Theorem 1. *Let G be a connected graph on n vertices with maximum degree Δ, and let $k \geq \Delta + 1$. Let α be a k-colouring of G. If α is not frozen and G is not $K_{\Delta+1}$ or, if n is odd, C_n, then there exists a Δ-colouring γ of G such that $d_k(\alpha, \gamma)$ is $O(n^2)$.*

Note that α can only be frozen if $k = \Delta + 1$, and only if G is Δ-regular. Let us briefly note that such colourings do exist: for example a 3-colouring of C_6 in which each colour appears exactly twice on vertices at distance 3, or a 4-colouring of the cube in which diagonally opposite vertices are coloured alike.

As we will see, the case $k = \Delta + 1$ is the only cause of difficulty in the proof of our first result. Using Theorem 1, however, we can, with the aid of one further lemma, give a characterization of $R_{\Delta+1}(G)$ which is our next result.

Theorem 2. *Let G be a connected graph on n vertices with maximum degree $\Delta \geq 3$. Let α and β be $(\Delta + 1)$-colourings of G. If α and β are not frozen colourings, then $d_{\Delta+1}(\alpha, \beta)$ is $O(n^2)$. Moreover, it is possible to decide in time $O(n)$ whether or not there is a path between α and β in $R_{\Delta+1}(G)$.*

Theorem 2 implies that $R_{\Delta+1}(G)$ contains a number of isolated vertices (representing frozen colourings) plus, possibly, one further component. It is possible that the number of isolated vertices is zero (that is, there are no frozen $(\Delta + 1)$-colourings; for example, consider 4-colourings of $K_{3,3}$), or that there are only isolated vertices (consider $R_4(K_4)$ for instance; and Brooks' theorem tells us that complete graphs are the only graphs for which $R_{\Delta+1}(G)$ is edgeless since other graphs have colourings in which only Δ colours are used and by recolouring any vertex with the unused colour we find a neighbouring colouring). We observe that the requirement that $\Delta \geq 3$ is necessary since, for example $R_3(C_n)$, n odd, has more than one component [10, 11].

Consequences of Our Results. Our theorems complete both structural and algorithmic classifications for reconfigurations of colourings of graphs of bounded maximum degree.

In [9] it was noted that if $k \geq \Delta(G) + 2$, $R_k(G)$ is connected with diameter $O(n^2)$. Combined with the results for general graphs noted above, and Theorem 1, we have the following summary of the structure of reconfiguration graphs:
- if $k \leq \Delta(G)$ then $R_k(G)$ might not be connected and it is possible that its connected components have superpolynomial diameter
- if $k = \Delta(G) + 1$ then $R_k(G)$ consists of zero or more isolated vertices and at most one further component which has diameter $O(n^2)$ (if it exists).
- if $k \geq \Delta(G) + 2$ then $R_k(G)$ is connected and has diameter $O(n^2)$.

And we summarise what is known about the computational complexity of the problem of deciding, given a graph and two k-colourings, whether the two colourings belong to the same connected component of $R_k(G)$ using Theorem 2 for the previously missing third case.
- $O(n^2)$ time solvable for $k = 3$,
- PSPACE-complete for $4 \leq k \leq \Delta(G)$,
- $O(n)$ time solvable for $k = \Delta(G) + 1$,
- $O(1)$ time solvable for $k \geq \Delta(G) + 2$ (the answer is always yes).

Related Work. We note that reconfiguration graphs can be defined for any search problem: vertices correspond to solutions and edges join solutions that are "close" to one another; that is, solutions that differ as little as possible (for a given problem, there might be more than one way to define an edge relation). Reconfiguration graphs have been studied for a number of combinatorial problems; the questions asked are typically (as we have seen for colouring) is the graph connected?, what is the diameter of the graph (or of its connected components)?, how difficult is it to decide whether there is a path between a pair of given solutions? Problems studied include boolean satisfiability [13, 21], clique and vertex cover [16], independent set [6, 20], list edge colouring [17], shortest path [3,4], and subset sum [15] (see also a recent survey [14]). Recent work has included looking at finding the *shortest* path in the reconfiguration graph between given solutions [19], and studying the fixed-parameter-tractability of these problems [7,18,23,24].

Further Preliminaries. The degree of a vertex v is denoted by $\deg(v)$. A graph is k-degenerate if every induced subgraph has a vertex with degree at most k. It is well-known that a graph is k-degenerate if and only if there exists a *degeneracy ordering* $v_1, v_2, \ldots, v_n$ of its vertices such that v_i has at most k neighbours v_j with $j < i$. A graph is r-regular if for every vertex v, $\deg(v) = r$.

A remark about our proofs. A common aim is to find a path between a pair of colourings α and β in a reconfiguration graph. That is, to find a sequence of colourings $\gamma_0, \gamma_1, \ldots, \gamma_t$ with $\alpha = \gamma_0$, $\beta = \gamma_t$ such that adjacent colourings disagree on a single vertex. We think of this sequence as a *recolouring* sequence. If, for $1 \leq i \leq t$, v_i is the vertex on which γ_i and γ_{i-1} disagree, then we can think of β as being obtained from α by recolouring the vertices $v_1, \ldots, v_t$ in order. Therefore, rather than explicitly considering the reconfiguration graph, we will often seek to find a recolouring sequence; that is, to describe a sequence of vertices and to say which colour each vertex should be recoloured with. Then we will need to show that if we apply this recolouring sequence to α, we obtain β, and that all the intermediate colourings obtained are proper.

2   Proofs of Theorems

To prove our theorems, we need a number of lemmas that are mostly concerned with $(\Delta + 1)$-colourings which, as we shall see, present the only real difficulty in proving Theorem 1. Some proofs are omitted for space reasons.

We define a number of terms we will use to describe vertices of G with respect to some $(\Delta+1)$-colouring. A vertex v is *locked* if Δ distinct colours appear on its neighbours. A vertex that is not locked is *free*. Clearly a vertex can be recoloured only if it is free. If v is locked and then one of its neighbour is recoloured and v becomes free, we say that v is *unlocked*. A vertex v is *superfree* if there is a colour $c \neq \Delta + 1$ such that neither v nor any of its neighbours is coloured c. A vertex can only be recoloured with a colour other than $\Delta + 1$ if it is superfree. Note there are $\Delta - 1$ distinct colours that must appear on the Δ neighbours of v if it is not superfree. We say that G is in $(\Delta + 1)$-*reduced form* if for every vertex v coloured with $(\Delta+1)$, v and each of its neighbours are locked. This implies that the distance between any pair of vertices coloured $(\Delta + 1)$ is at least 3 as no vertex can have two neighbours coloured $(\Delta + 1)$.

The key to proving Theorem 1 will be to show that from a $(\Delta + 1)$-colouring one can recolour some of the vertices to arrive at a colouring in which colour $\Delta + 1$ appears on fewer vertices. We begin by considering the case where the colour $\Delta + 1$ appears on only one vertex. The following lemma is inspired by a proof of Brooks' theorem [22].

Lemma 1. *Let $G = (V, E)$ be a connected graph on n vertices with maximum degree $\Delta \geq 3$, and let α be a $(\Delta + 1)$-colouring of G with exactly one vertex v coloured $\Delta + 1$. If G does not contain $K_{\Delta+1}$ as a subgraph, then G can be recoloured to a Δ-colouring in $O(n)$ steps.*

Proof. We can assume that G is in $(\Delta + 1)$-reduced form since if v is not locked then we can immediately recolour it; if a neighbour of v is not locked then it can be recoloured and this will unlock v and allow us to recolour it.

Let us fix a labelling of the neighbours of v: let x_i be the neighbour such that $\alpha(x_i) = i$, $1 \leq i \leq \Delta$. Our aim is to find a recolouring sequence that unlocks v. There is one recolouring sequence that we will use several times. Suppose that C is a connected component of a subgraph of G induced by two colours i and j, $\Delta + 1 \notin \{i, j\}$, and no vertex coloured j in C is adjacent to v. First the vertices coloured j are recoloured with $\Delta + 1$. Then the vertices coloured i are recoloured j, and finally the vertices initially coloured j are recoloured i. It is clear that all colourings are proper and the overall effect is to *swap* the colours i and j on C.

We say that any colouring γ where G is in $(\Delta + 1)$-reduced form, only v is coloured $\Delta + 1$ and $\gamma(x_i) = i$, $1 \leq i \leq \Delta$, is *good*. For any good colouring γ, let G_{ij}^γ be the maximal connected component containing x_i of the subgraph of G induced by the vertices coloured i and j by γ.

We make some claims about good colourings. When we claim that v can be unlocked, it is implicit that colour $\Delta + 1$ is not used on any other vertex in the graph so that unlocking v allows us to reach a colouring where $\Delta + 1$ is not used.

Claim 1: If γ is good and $x_j \notin G_{ij}^\gamma$, then v can be unlocked.

If $x_j \notin G_{ij}^\gamma$, then the only vertex adjacent to v in G_{ij}^γ is x_i. Thus the colours i and j can be swapped on G_{ij}^γ. Then v has two neighbours with colour j and is unlocked.

Claim 2: If γ is good and G_{ij}^γ is not a path from x_i to x_j, then v can be unlocked.

By Claim 1, we can assume that x_i and x_j are in G_{ij}^γ. They must have degree 1 in G_{ij} since, as G is in $(\Delta + 1)$-reduced form, they are locked. Suppose that G_{ij}^γ is not a path and consider the shortest path in G_{ij}^γ from x_i to x_j, and the vertex w nearest to x_i on the path that has degree more than 2. Then w has at least three neighbours coloured alike in G and is superfree and can be recoloured with a colour other than i, j or $\Delta + 1$. Call this new colouring γ' and note that, by the choice of w, $G_{ij}^{\gamma'}$ does not contain x_j. Now Claim 1 implies Claim 2.

As G is $K_{\Delta+1}$-free, v and its neighbours are not a clique so we can assume that x_1 and x_2 are not adjacent. Let u be the unique neighbour of x_1 coloured 2. For a good colouring γ, note that u is in G_{12}^γ, and let H_{23}^γ be the component of the subgraph of G induced by the vertices with colour 2 and 3 that contains u.

Claim 3: If γ is good and u has more than one neighbour in H_{23}^γ, then v can be unlocked.

If G_{12}^γ is not a path, then use Claim 2. Otherwise u has two neighbours coloured 1; if u has two neighbours in H_{23}^γ, then it also has two neighbours coloured 3 and is superfree. Recolour it and apply Claim 1.

Claim 4: If γ is good and H_{23}^γ is a path, then v can be unlocked.

By Claim 2 we can assume G_{23}^γ is a path. If $H_{23}^\gamma = G_{23}^\gamma$, then we can use Claim 3. So we assume $H_{23}^\gamma \neq G_{23}^\gamma$ and so $x_2, x_3 \notin H_{23}$ and H_{23} contains no neighbour of v. Let γ' be the colouring obtained by swapping the colours 2 and 3 on H_{23}^γ.

By Claim 3, u is an endvertex of H_{23}^γ. Let the other endvertex be w. (If $w = u$, then u has no neighbour coloured 3 and can be recoloured. Then use Claim 2.)

If $G_{12}^{\gamma'}$ is not a path from x_1 to x_2, we use Claim 2. If it is such a path, then let the unique neighbour of x_1 in $G_{12}^{\gamma'}$ be y and clearly $y \in H_{23}^\gamma$. From x_2 traverse $G_{12}^{\gamma'}$ until the last vertex z that is also in G_{12}^γ is reached. Let t be the next vertex along from z towards x_1 in $G_{12}^{\gamma'}$. Clearly t is also in H_{23}^γ. In fact, we can assume that $w = y = t$ since if y or t has degree 2 in H_{23} as well as in $G_{12}^{\gamma'}$ it has two neighbours coloured 1 and two neighbours coloured 3 in γ' and is superfree. It can be recoloured and then Claim 2 is used.

So $x_1 w z$ is coloured 131 in γ so is in G_{13}^γ. Then z is in both G_{13}^γ and G_{12}^γ so is superfree and can be recoloured so that Claim 2 can be used. This completes the proof of Claim 4.

To complete the proof: we know that the initial colouring α is good. If none of the four claims can be used, then consider H_{23}^α. We know that u has degree 1 in H_{23} but H_{23} is not a path. So traversing edges away from u in H_{23}^α, let s be the first vertex reached with degree 3. Then s is superfree and can be recoloured so that H_{23} becomes a path, and then Claim 4 can be used.   □

In Lemma 3, we shall see how, for *regular* graphs, the number of vertices coloured $\Delta + 1$ can be reduced when more than one is present. First we need some definitions and a lemma. Let P be a path:
- P is *nearly* $(\Delta + 1)$-*locked* if its endvertices are locked and coloured $\Delta + 1$;
- P is $(\Delta + 1)$-*locked* if it is nearly $(\Delta + 1)$-locked and every vertex on the path is locked.

Lemma 2. *Let G be a graph in $(\Delta + 1)$-reduced form. If G has a $(\Delta + 1)$-locked path P, then each endvertex of P is an endvertex of an $(\Delta + 1)$-locked path of length 3.*

A path is *nice* if it is a nearly $(\Delta + 1)$-locked path, it contains free vertices and the endvertices and their neighbours are the only locked vertices. Notice that a nice path is not necessarily induced and, in particular, may contain a $(\Delta + 1)$-locked subpath.

Lemma 3. *Let G be a connected regular graph on n vertices with degree $\Delta \geq 3$, let α be a $(\Delta + 1)$-colouring of G, and suppose that G is in $(\Delta + 1)$-reduced form. If G has at least two $(\Delta + 1)$-locked vertices and is not frozen, then there exists a $(\Delta + 1)$-colouring γ of G, such that $d_{\Delta + 1}(\alpha, \gamma) = O(n)$ and fewer vertices are coloured $\Delta + 1$ with γ than with α.*

Proof. We consider a number of cases.

Case 1: There exists a free vertex u adjacent to a $(\Delta + 1)$-locked path P.

Let b be the vertex on the path adjacent to u. As b is locked it has a neighbour a coloured $\Delta + 1$. Let c be a neighbour of b on P other than a. As c is locked it has a neighbour d coloured $\Delta + 1$.

Since G is in $(\Delta + 1)$-reduced form, u is not adjacent to a or d but might be adjacent to c. In each case, it is routine to verify that by recolouring u to $\Delta + 1$, b and c can both be recoloured unlocking a and d and allowing them to be recoloured. Thus the number of vertices coloured $\Delta + 1$ is reduced.

Case 2: G has a nice path.

Let P be a shortest nice path. Let the endpoints be v and w with neighbours s and t on P respectively. If s and t are adjacent, then the path $vstw$ is $(\Delta + 1)$-locked and has a free vertex adjacent to s so use Case 1. Thus assume that P is induced since the presence of any other edge would imply either a shorter nice path could be found or that the graph was not in $(\Delta + 1)$-reduced form.

We use induction on the number ℓ of free vertices in P to show that there is a sequence of recolourings that lead to a colouring that has fewer vertices coloured $\Delta + 1$.

If $\ell = 1$, let u be the free vertex in P. Recolour u to $\Delta + 1$. Now s and t have two neighbours coloured $\Delta + 1$ and can be recoloured. Then v and w are unlocked and can both be recoloured, and this leaves one vertex on P coloured $\Delta + 1$ rather than two.

Suppose that $\ell = 2$. Let $P = vsu_1u_2tw$ where u_1 and u_2 are free vertices.

Subcase 2.1: u_1 *and* u_2 *do not share a neighbour.* Let x_1 and x_2 be neighbours of u_1 and u_2 not in P. Clearly $x_1 \neq x_2$ and u_1x_2 and u_2x_1 are not edges.

Subcase 2.1.1: x_1 *is locked.* We know x_1 has a $(\Delta + 1)$-locked neighbour, and this must be v (if it is some other vertex z, then vsu_1x_1z is a nice path that is shorter than P).

Suppose x_1s is not an edge. Recolour u_1 to $\Delta + 1$. This unlocks x_1 which can be recoloured with $\alpha(u_1)$ which, in turn, unlocks v and allows us to recolour it with $\alpha(x_1)$. If u_1 is free, it can be recoloured and the number of vertices coloured $\Delta + 1$ is reduced and we are done. If u_1 is locked, then note that s has been unlocked (as it no longer has a neighbour coloured $\alpha(u_1)$). Thus we can recolour s and then recolour u_1 with $\alpha(s)$ and again we have removed one instance of the colour $\Delta + 1$.

Suppose instead that x_1s is an edge. Notice that $\alpha(s)$, $\alpha(u_1)$ and $\alpha(x_1)$ are distinct as the three vertices form a triangle. Recolour u_1 with $\Delta + 1$ and then s with $\alpha(u_1)$. Now v is unlocked and can be recoloured with $\alpha(s)$. If u_1 is free, then recolour it and we are done. Otherwise this sequence of recolourings leaves u_1 locked (with $\alpha(u_1)$ and $\alpha(x_1)$ as the colours on s and x_1 respectively). So, from α, we do the following instead: again start by recolouring u_1 with $\Delta + 1$, but then recolour x_1 with $\alpha(u_1)$ to unlock v. Now that $\alpha(x_1)$ is not used on a neighbour of u_1, u_1 is free and can be recoloured.

Subcase 2.1.2: x_1 *is free.* If x_2 is locked, we can, by symmetry, use the previous subcase, so we can assume that both x_1 and x_2 are free. Recolour u_2 to $\Delta + 1$. Then t is unlocked and can be recoloured which, in turn, unlocks w allowing us to recolour it too. If u_2 is free, we recolour it and are done. If u_1 is free, we recolour it and unlock u_2 and, again, recolour it.

If u_1 and u_2 are both locked, observe that x_1 is still free as it has no neighbour coloured $\Delta + 1$ since $u_2 x_1$ is not an edge. Recolour x_1 to $\Delta + 1$, and then recolour u_1 to $\alpha(x_1)$. Note that now s has no neighbour coloured $\alpha(u_1)$ and is free and can be recoloured so that v is unlocked and can also be recoloured. By recolouring u_1, we also unlock u_2, so we recolour it and are done.

Subcase 2.2: u_1 *and* u_2 *share a neighbour.* Let x_1 be a neighbour of u_1 and u_2. Since P is induced, x_1 is not in P. If x_1 is locked, then let its neighbour coloured $\Delta + 1$ be y. Then vsu_1x_1y is a shorter nice path unless $y = v$. By an analogous argument we need $y = w$. This contradiction tells us that x_1 must be free.

If x_1 is joined to both s and t, then vsx_1tw is a shorter nice path. So, without loss of generality, assume that x_1t is not an edge. Thus as u_2 has a neighbour that is not adjacent to x_1, x_1 has a neighbour x_3 that is not adjacent to u_2.

Subcase 2.2.1: $x_3 = s$. Recolour u_1 with $\Delta + 1$ and then s with $\alpha(u_1)$. Now v is unlocked and can be recoloured with $\alpha(s)$. If u_1 is free, then recolour it and we are done. If u_2 or x_1 is still free, then recolour one of them to unlock u_1, which in turn can be recoloured and are done. Otherwise this sequence of recolourings leaves u_1, u_2 and x_1 locked so x_1 is the only neighbour of u_2 coloured $\alpha(x_1)$. So, from α, we do the following instead: recolour x_1 with $\Delta + 1$ to unlock s and then v. If x_1 can be recoloured, then we do so and are done. Otherwise notice that $\alpha(x_1)$ is not used on a neighbour of u_2. It is thus free and can be recoloured to unlock x_1 and allow us to recolour it.

Subcase 2.2.2: $x_3 \neq s$, *and* x_3 *is free.* First, suppose x_3s is an edge. Recolour u_2 to $\Delta + 1$, t to $\alpha(u_2)$ and w to $\alpha(t)$. If either u_2 or one of its neighbours is now free, u_2 can be recoloured and we are done. Otherwise u_1, u_2 and x_1 are all locked, but x_3 is still free since it has no neighbour coloured $\Delta + 1$. Recolour x_3 to $\Delta + 1$ to unlock x_1; then recolour x_1 to unlock and recolour u_2. As x_3s is an edge, s has two neighbours coloured $\Delta + 1$. Thus we recolour s to unlock v.

If x_3t is an edge we can use a similar argument. So suppose x_3s and x_3t are not edges. Recolour u_2 to $\Delta + 1$, to unlock and recolour first t and then w. It is possible to recolour u_2 unless it and all its neighbours are locked. This implies that u_1, x_1 and u_2 are locked. We consider two subcases.

Subcase 2.2.2.1: u_1x_3 *is not an edge.* We recolour x_3 to $\Delta + 1$ to unlock and recolour x_1 and then u_2. Notice that u_1 is now free since it has no neighbour coloured $\Delta + 1$. Recoloured u_1 unlocks s, so we recolour it, which in turn unlocks v. Observe that x_1 now has two neighbours u_1 and x_3 with colour $\Delta + 1$ so is free. If u_1 or u_3 is free, we can recolour at least one of them directly and we are done. Otherwise, we recolour x_1 so that x_3 and u_1 can now be recoloured.

Subcase 2.2.2.2: u_1x_3 *is an edge.* Recolour u_3 to $\Delta + 1$, then recolour u_1, s and v. Observe that x_1 now has two neighbours u_2 and u_3 with colour $\Delta + 1$.

If u_2 or u_3 are free, we are done. Otherwise, recolour x_1, then recolour u_2 and x_3, and we are done.

Subcase 2.2.3: $x_3 \neq s$, *and* x_3 *is locked.* Then x_3 has a $(\Delta + 1)$-locked neighbour y. If $y = v$, the path $H = vx_3x_1u_2tw$ is nice with two free vertices x_1 and u_2. Furthermore, u_1 is free and a neighbour of x_1 and u_2, in which case H satisfies the previous subcase unless x_3 and t are adjacent in which case use Subcase 2.1. A similar argument can be made if $y = w$ or $y \notin \{v, w\}$.

This completes the case $\ell = 2$.

Now suppose that for all $i < \ell$, if there is a nice path containing i free vertices, the number of vertices coloured $\Delta + 1$ can be reduced. Suppose that the shortest such path is $P = vsu_1u_2 \ldots u_\ell tw$ where $\ell \geq 3$. We recolour u_ℓ to $\Delta + 1$, then t and then w. If u_ℓ or one of its neighbours is free, then u_ℓ can be recoloured and we are done. Otherwise, u_ℓ and $u_{\ell-1}$ are locked. Consider the path $P' = vsu_1 \ldots u_{\ell-2}u_{\ell-1}u_\ell$. By our inductive hypothesis, the number of colour $\Delta + 1$ vertices in P' can be reduced. Case 2 is complete.

After Cases 1 and 2 we are left with:

Case 3: There does not exist a free vertex adjacent to a $(\Delta + 1)$-locked path and G has no nice path.

As G contains more than one $(\Delta+1)$-locked vertex, it contains a nearly $(\Delta+1)$-locked path; let P be the shortest and let v and w be its endvertices. As G is in $(\Delta + 1)$-reduced form, v, w and their neighbours are locked. If P contains no other vertices, it is $(\Delta + 1)$-locked. Otherwise, since there are no nice paths, P contains another locked vertex u. Let y be the neighbour of u coloured $\Delta+1$. If y is on P, then we can assume, without loss of generality, that it is not between v and u. Then, whether or not y is on P, the subpath from v to u plus the edge uy is a shorter nearly $(\Delta + 1)$-locked path. This contradiction proves that G must contain a $(\Delta + 1)$-locked path.

As G is not frozen, it contains a free vertex. Let Q be the shortest path in G that joins a free vertex to a $(\Delta + 1)$-locked vertex. Let v be the $(\Delta + 1)$-locked endvertex. So v is an endpoint of a $(\Delta + 1)$-locked path R, and, by Lemma 2, we can assume that R has length 3.

Let u be the endvertex of Q that is free. By the minimality of Q, u is the only free vertex in Q. Let a be the neighbour of u in Q. As a is locked it has a $(\Delta + 1)$-locked neighbour z. Thus we must have $z = v$ and $Q = vau$.

Let $R = wtsv$. Observe that us, ut, uv and uw cannot be edges as no locked path has a free neighbour. Thus the vertices of R and Q other than v are distinct. Consider the (not necessarily induced) path $M = wtsvau$. Notice also that at is not an edge else the free vertex u is adjacent to the $(\Delta + 1)$-locked path $vatw$.

Suppose M is an induced path. Recolour u with $\Delta + 1$ to unlock and recolour a and then v. If u is not locked, then recolour and we are done. Else notice that the vertices v and s are free, and the vertices u, a, t, w are locked. Consequently, we have that M is a nice path, and by Case 2 we are done.

The only edge that might be present among the vertices of M is as so suppose this exists. Recolour u with $\Delta + 1$ to unlock and recolour first a and then v. If u

or any of its neighbours are free, u can be recoloured and we are done. Otherwise note that recoloured v unlocks s. It follows that the path $H = uastw$ is nice, and we can use Case 2. This completes Case 3.

As each vertex is recoloured a constant number of times, the lemma follows. □

We need one final lemma before we prove Theorem 1.

Lemma 4. Let $G = (V, E)$ be a connected graph on n vertices with maximum degree $\Delta \geq 1$ and degeneracy $\Delta - 1$. Let α be a $(\Delta + 1)$-colouring of G. Then there exists a Δ-colouring γ of G such that $d_{\Delta+1}(\alpha, \gamma) \leq n^2$.

Proof (of Theorem 1). If $k > \Delta + 1$, then, by Brooks' Theorem, a Δ-colouring γ exists in $R_k(G)$ unless G is complete or an odd cycle. We know that, in this case, $R_k(G)$ is connected and has diameter $O(n^2)$ so certainly $d_k(\alpha, \gamma)$ is $O(n^2)$.

Suppose that $k = \Delta + 1$. If G is $(\Delta - 1)$-degenerate, the result follows from Lemma 4. We claim that the only graphs with maximum degree Δ that are not $(\Delta - 1)$-degenerate are Δ-regular graphs. To see this, consider a smallest possible counterexample G that has degeneracy and maximum degree Δ and contains a vertex v with $\deg(v) < \Delta$. Suppose $G - v$ has degeneracy Δ. Then, by the minimality of G, we find that $G - v$ is Δ-regular. This would mean that every neighbor of v in G has more than Δ neighbours, which is not possible. Hence, $G - v$ must have degeneracy $\Delta - 1$. But every induced subgraph of G is either an induced subgraph of $G - v$ or contains v, and, in either case, must contain a vertex of degree less than Δ contradicting the claim that G has degeneracy Δ.

So we can suppose now that G is Δ-regular and in $(\Delta + 1)$-reduced form with α: if not, we try to recolour each vertex with colour $\Delta + 1$ either directly or by first recolouring one of its neighbours. Repeatedly applying Lemma 3 starting from α, we obtain a $(\Delta + 1)$-colouring γ' in $O(n^2)$ steps such that at most one vertex is coloured $(\Delta + 1)$ with γ'. Lemma 1 can now be applied to obtain a Δ-colouring γ from γ' in $O(n)$ steps. Consequently $d_{\Delta+1}(\alpha, \gamma) \leq O(n^2)$ as required. □

We finish the section by considering Theorem 2. First we need:

Lemma 5. Let $G = (V, E)$ be a connected graph on n vertices with maximum degree $\Delta \geq 3$. Let γ_1 and γ_2 be Δ-colourings of G. Then $d_{\Delta+1}(\gamma_1, \gamma_2)$ is $O(n^2)$.

The lemma says that there is a path between any pair of Δ-colourings, but, because we are working with $R_{\Delta+1}(G)$, the intermediate colourings might use $\Delta + 1$ colours.

Proof (of Theorem 2). Theorem 1 implies that from each of α and β there is a path in $R_{\Delta+1}$ to a Δ-colouring; Lemma 5 implies that there is a path between these two Δ-colourings that completes the path from α to β. Consequently, it is possible to decide in $O(n)$ time whether or not there is a path between α and β in $R_{\Delta+1}(G)$: it is necessary only to check for each vertex v in G, for each of α and β, whether v and its neighbours use every colour in $\{1, 2, \ldots, \Delta + 1\}$. If they do not, neither colouring is frozen so there is a path between them. □

3   Conclusions

We have completed the study of reconfiguration graphs of graphs of bounded degree by considering the case where the number of colours is one more than the maximum degree. In Theorem 2, we showed that the reconfiguration graph contains isolated vertices and one further component. As it is easy to recognize a frozen colouring, this means that we can decide in polynomial time whether a given pair of colourings belong to the same component. We make two additional observations about when the reconfiguration graph can have isolated vertices.

Corollary 1. *Let G be a connected regular graph on n vertices with maximum degree $\Delta \geq 3$. If $n \not\equiv 0 \mod (\Delta + 1)$ then $R_{\Delta+1}(G)$ has diameter $O(n^2)$.*

Proof. Let γ be a frozen colouring of G. Let $V_1, V_2, \ldots, V_{\Delta+1}$ be the colour classes of γ. Suppose there exist integers i, j such that $|V_i| > |V_j|$. Because γ is a frozen colouring each $v \in V_i$ has a neighbour in V_j. Hence there is a vertex $u \in V_j$ with at least two neighbours in V_i. Since u has Δ neighbours, it follows that u is free and can thus be recoloured, a contradiction. Therefore $|V_1| = \cdots = |V_{\Delta+1}|$. We have proved that whenever G has a frozen colouring, $n \equiv 0 \mod (\Delta + 1)$, and by Theorem 2 if there is no frozen colouring, $R_{\Delta+1}(G)$ is connected.   □

Corollary 2. *Let G be a connected graph with maximum degree $\Delta \geq 3$ and degeneracy $(\Delta - 1)$. Then $R_{\Delta+1}(G)$ is connected with diameter $O(n^2)$.*

Proof. The result follows immediately from Theorem 2 by observing that a $(\Delta - 1)$-degenerate graph has a vertex with at most $\Delta - 1$ neighbours and is thus free in any $(\Delta + 1)$-colouring of G.   □

Cereceda [9] conjectured that the diameter of the reconfiguration graph on $(k+2)$-colourings of a k-degenerate graph on n vertices is $O(n^2)$. This conjecture has been answered in the positive for values of $k \in \{1, \Delta\}$ [9]. By the previous corollary, we further confirm this conjecture for the value $k = \Delta - 1$.

References

1. Bonamy, M., Bousquet, N.: Recoloring bounded treewidth graphs. In: Proc. LA-GOS 2013. Electronic Notes in Discrete Mathematics, vol. 44, pp. 257–262 (2013)
2. Bonamy, M., Johnson, M., Lignos, I.M., Patel, V., Paulusma, D.: Reconfiguration graphs for vertex colourings of chordal and chordal bipartite graphs. Journal of Combinatorial Optimization 27, 132–143 (2014)
3. Bonsma, P.: The complexity of rerouting shortest paths. In: Rovan, B., Sassone, V., Widmayer, P. (eds.) MFCS 2012. LNCS, vol. 7464, pp. 222–233. Springer, Heidelberg (2012)
4. Bonsma, P.: Rerouting shortest paths in planar graphs. In: Proc. FSTTCS 2012. LIPIcs, vol. 18, pp. 337–349 (2012)
5. Bonsma, P., Cereceda, L.: Finding paths between graph colourings: Pspace-completeness and superpolynomial distances. Theoretical Computer Science 410, 5215–5226 (2009)

6. Bonsma, P., Kamiński, M., Wrochna, M.: Reconfiguring independent sets in claw-free graphs. arXiv, 1403.0359 (2014)
7. Bonsma, P., Mouawad, A.: The complexity of bounded length graph recoloring. arXiv, 1404.0337 (2014)
8. Brooks, R.L.: On colouring the nodes of a network. Mathematical Proceedings of the Cambridge Philosophical Society 37, 194–197 (1941)
9. Cereceda, L.: Mixing graph colourings. PhD thesis, London School of Economics (2007)
10. Cereceda, L., van den Heuvel, J., Johnson, M.: Connectedness of the graph of vertex-colourings. Discrete Mathematics 308, 913–919 (2008)
11. Cereceda, L., van den Heuvel, J., Johnson, M.: Mixing 3-colourings in bipartite graphs. European Journal of Combinatorics 30(7), 1593–1606 (2009)
12. Cereceda, L., van den Heuvel, J., Johnson, M.: Finding paths between 3-colorings. Journal of Graph Theory 67(1), 69–82 (2011)
13. Gopalan, P., Kolaitis, P.G., Maneva, E.N., Papadimitriou, C.H.: The connectivity of boolean satisfiability: Computational and structural dichotomies. SIAM Journal on Computing 38(6), 2330–2355 (2009)
14. van den Heuvel, J.: The complexity of change. In: Blackburn, S.R., Gerke, S., Wildon, M. (eds.) Surveys in Combinatorics 2013, London. Mathematical Society Lecture Notes Series, vol. 409 (2013)
15. Ito, T., Demaine, E.D.: Approximability of the subset sum reconfiguration problem. In: Ogihara, M., Tarui, J. (eds.) TAMC 2011. LNCS, vol. 6648, pp. 58–69. Springer, Heidelberg (2011)
16. Ito, T., Demaine, E.D., Harvey, N.J.A., Papadimitriou, C.H., Sideri, M., Uehara, R., Uno, Y.: On the complexity of reconfiguration problems. Theoretical Computer Science 412(12-14), 1054–1065 (2011)
17. Ito, T., Kaminski, M., Demaine, E.D.: Reconfiguration of list edge-colorings in a graph. Discrete Applied Mathematics 160(15), 2199–2207 (2012)
18. Johnson, M., Kratsch, D., Kratsch, S., Patel, V., Paulusma, D.: Colouring reconfiguration is fixed-parameter tractable. arXiv, 1403.6347 (2014)
19. Kaminski, M., Medvedev, P., Milanic, M.: Shortest paths between shortest paths. Theoretical Computer Science 412(39), 5205–5210 (2011)
20. Kaminski, M., Medvedev, P., Milanic, M.: Complexity of independent set reconfigurability problems. Theoretical Computer Science 439, 9–15 (2012)
21. Makino, K., Tamaki, S., Yamamoto, M.: On the boolean connectivity problem for horn relations. Discrete Applied Mathematics 158(18), 2024–2030 (2010)
22. Melnikov, L.S., Vizing, V.G.: New proof of brooks' theorem. Journal of Combinatorial Theory 7(4), 289–290 (1969)
23. Mouawad, A.E., Nishimura, N., Raman, V.: Vertex cover reconfiguration and beyond. arXiv, 1402.4926 (2014)
24. Mouawad, A.E., Nishimura, N., Raman, V., Simjour, N., Suzuki, A.: On the parameterized complexity of reconfiguration problems. In: Gutin, G., Szeider, S. (eds.) IPEC 2013. LNCS, vol. 8246, pp. 281–294. Springer, Heidelberg (2013)

Intersection Graphs of L-Shapes and Segments in the Plane*

Stefan Felsner[1], Kolja Knauer[2], George B. Mertzios[3], and Torsten Ueckerdt[4]

[1] Institut für Mathematik, Technische Universität Berlin, Germany
[2] LIRMM, Université Montpellier 2, France
[3] School of Engineering and Computing Sciences, Durham University, UK
[4] Department of Mathematics, Karlsruhe Institute of Technology, Germany
`felsner@math.tu-berlin.de`, `kolja.knauer@math.univ-montp2.fr`,
`george.mertzios@durham.ac.uk`, `torsten.ueckerdt@kit.edu`

Abstract. An L-shape is the union of a horizontal and a vertical segment with a common endpoint. These come in four rotations: L, Γ, ⌐ and ⌐. A k-bend path is a simple path in the plane, whose direction changes k times from horizontal to vertical. If a graph admits an intersection representation in which every vertex is represented by an L, an L or Γ, a k-bend path, or a segment, then this graph is called an {L}-graph, {L, Γ}-graph, B_k-VPG-graph or SEG-graph, respectively. Motivated by a theorem of Middendorf and Pfeiffer [Discrete Mathematics, 108(1):365–372, 1992], stating that every {L, Γ}-graph is a SEG-graph, we investigate several known subclasses of SEG-graphs and show that they are {L}-graphs, or B_k-VPG-graphs for some small constant k. We show that all planar 3-trees, all line graphs of planar graphs, and all full subdivisions of planar graphs are {L}-graphs. Furthermore we show that all complements of planar graphs are B_{19}-VPG-graphs and all complements of full subdivisions are B_2-VPG-graphs. Here a full subdivision is a graph in which each edge is subdivided at least once.

Keywords: Intersection graphs, segment graphs, co-planar graphs, k-bend VPG-graphs, planar 3-trees.

1 Introduction and Motivation

A **segment intersection graph**, SEG-graph for short, is a graph that can be represented as follows. Vertices correspond to straight-line segments in the plane and two vertices are adjacent if and only if the corresponding segments intersect. Such representations are called SEG-*representations* and, for convenience, the class of all SEG-graphs is denoted by SEG. SEG-graphs are an important subject of study strongly motivated from an algorithmic point of view. Indeed, having an intersection representation of a graph (in applications graphs often come

---

* This work was partially supported by (i) the DFG ESF EuroGIGA projects COMPOSE and GraDR, (ii) the EPSRC Grant EP/K022660/1 and (iii) the ANR Project EGOS: ANR-12-JS02-002-01.

E. Csuhaj-Varjú et al. (Eds.): MFCS 2014, Part II, LNCS 8635, pp. 299–310, 2014.

along with such a given representation) may allow for designing better or faster algorithms for optimization problems that are hard for general graphs, such as finding a maximum clique in interval graphs.

More than 20 years ago, Middendorf and Pfeiffer [24], considered intersection graphs of **axis-aligned L-shapes** in the plane, where an axis-aligned L-shape is the union of a horizontal and a vertical segment whose intersection is an endpoint of both. In particular, L-shapes come in four possible rotations: L, Γ, ⌐, and ⊓. For a subset X of these four rotations, e.g., $X = \{L\}$ or $X = \{L, Γ\}$, we call a graph an X-*graph* if it admits an X-*representation*, i.e., vertices can be represented by L-shapes from X in the plane, each with a rotation from X, such that two vertices are adjacent if and only if the corresponding L-shapes intersect. Similarly to SEG, we denote the class of all X-graphs by X. The question if an intersection representation with polygonal paths or pseudo-segments can be *stretched* into a SEG-representation is a classical topic in combinatorial geometry and Oriented Matroid Theory. Middendorf and Pfeiffer prove the following interesting relation between intersection graphs of segments and L-shapes.

Theorem 1 (Middendorf and Pfeiffer [24]). *Every* $\{L, Γ\}$-*representation has a combinatorially equivalent* SEG-*representation.*

This theorem is best-possible in the sense that there are examples of $\{L, ⊓\}$-graphs which are no SEG-graphs [7, 24], i.e., such $\{L, ⊓\}$-representations cannot be stretched. We feel that Theorem 1, which of course implies that $\{L, Γ\} \subseteq$ SEG, did not receive a lot of attention in the active field of SEG-graphs. In particular, one could use Theorem 1 to prove that a certain graph class $\mathcal{G}$ is contained in SEG by showing that $\mathcal{G}$ is contained in $\{L, Γ\}$. For example, very recently Pawlik *et al.* [25] discovered a class of triangle-free SEG-graphs with arbitrarily high chromatic number, disproving a famous conjecture of Erdős [18], and it is in fact easier to see that these graphs are $\{L\}$-graphs than to see that they are SEG-graphs. To the best of our knowledge, the stronger result $\mathcal{G} \subseteq \{L, Γ\}$ has never been shown for any non-trivial graph class $\mathcal{G}$. In this paper we initiate this research direction. We consider several graph classes which are known to be contained in SEG and show that they are actually contained in $\{L\}$, which is a proper subclass of $\{L, Γ\}$ [7].

Whenever a graph is not known (or known not) to be an intersection graph of segments or axis-aligned L-shapes, one often considers natural generalizations of these intersection representations. Asinowski *et al.* [3] introduced **intersection graphs of axis-aligned k-bend paths** in the plane, called B_k-VPG-graphs. An (axis-aligned) k-bend path is a simple path in the plane, whose direction changes k times from horizontal to vertical. Clearly, B_1-VPG-graphs are precisely intersection graphs of all four L-shapes; the union of B_k-VPG-graphs for all $k \geq 0$ is exactly the class STRING of intersection graphs of simple curves in the plane [3]. Now if a graph $G \notin$ SEG is a B_k-VPG-graph for some small k, then one might say that G is "not far from being a SEG-graph".

Our Results and Related Work

Let us denote the class of all planar graphs by PLANAR. A recent celebrated result of Chalopin and Gonçalves [6] states that PLANAR $\subset$ SEG, which was conjectured by Scheinerman [26] in 1984. However, their proof is rather involved and there is not much control over the kind of SEG-representations. Here we give an easy proof for a non-trivial subclass of planar graphs, namely *planar 3-trees*. A *3-tree* is an edge-maximal graph of treewidth 3. Every 3-tree can be built up starting from the clique K_4 and adding new vertices, one at a time, whose neighborhood in the so-far constructed graph is a triangle.

Theorem 2. *Every planar 3-tree is an* $\{L\}$*-graph.*

It remains open to generalize Theorem 2 to planar graphs of treewidth 3 (i.e., subgraphs of planar 3-trees). On the other hand it is easy to see that graphs of treewidth at most 2 are $\{L\}$-graphs [8]. Chaplick and the last author show in [9] that planar graphs are B_2-VPG-graphs, improving on an earlier result of Asinowski *et al.* [3]. In [9] it is also conjectured that PLANAR $\subset \{L\}$, which with Theorem 1 would imply the main result of [6], i.e., PLANAR $\subset$ SEG.

Considering line graphs of planar graphs, one easily sees that these graphs are SEG-graphs. Indeed, a straight-line drawing of a planar graph G can be interpreted as a SEG-representation of the line graph $L(G)$ of G, which has the edges of G as its vertices and pairs of incident edges as its edges. We prove the following strengthening result.

Theorem 3. *The line graph of every planar graph is an* $\{L\}$*-graph.*

Kratochvíl and Kuběna [21] consider the class of all complements of planar graphs (co-planar graphs), CO-PLANAR for short. They show that CO-PLANAR are intersection graphs of convex sets in the plane, and ask whether CO-PLANAR $\subset$ SEG. As the INDEPENDENT SET PROBLEM in planar graphs is known to be NP-complete [15], MAX CLIQUE is NP-complete for any graph class $\mathcal{G} \supseteq$ CO-PLANAR, e.g., intersection graphs of convex sets. Indeed, the longstanding open question whether MAX CLIQUE is NP-complete for SEG [22] has recently been answered affirmatively by Cabello, Cardinal and Langerman [4] by showing that every planar graph has an even subdivision whose complement is a SEG-graph. The subdivision is essential in the proof of [4], as it still remains an open problem whether CO-PLANAR $\subset$ SEG [21]. The largest subclass of CO-PLANAR known to be in SEG is the class of complements of partial 2-trees [14]. Here we show that all co-planar graphs are "not far from being SEG-graphs".

Theorem 4. *Every co-planar graph is a* B_{19}*-VPG graph.*

Theorem 4 implies that MAX CLIQUE is NP-complete for B_k-VPG-graphs with $k \geq 19$. On the other hand, the MAX CLIQUE problem for B_0-VPG-graphs can be solved in polynomial time, while VERTEX COLORABILITY remains NP-complete but allows for a 2-approximation [3]. Middendorf and Pfeiffer [24] show that the complement of any *even subdivision* of any graph, i.e., every edge is subdivided with a non-zero even number of vertices, is an $\{L, \daleth\}$-graph. This implies that MAX CLIQUE is NP-complete even for $\{L, \daleth\}$-graphs.

We consider *full subdivisions* of graphs, that is, a subdivision H of a graph G where each edge of G is subdivided at least once. It is not hard to see that a full subdivision H of G is in STRING if and only if G is planar, and that if G is planar, then H is actually a SEG-graph. Here we show that this can be further strengthened, namely that H is in an $\{L\}$-graph. Moreover, we consider the complement of a full subdivision H of an arbitrary graph G, which is in STRING but not necessarily in SEG. Here, similar to the result of Middendorf and Pfeiffer [24] on even subdivisions we show that such a graph H is "not far from being SEG-graph".

Theorem 5. *Let H be a full subdivision of a graph G.*

(i) If G is planar, then H is an $\{L\}$-graph.
(ii) If G is any graph, then the complement of H is a B_2-VPG-graph.

The graph classes considered in this paper are illustrated in Figure 1. We shall prove Theorems 2, 3, 4 and 5 in Sections 2, 3, 4 and 5, respectively, and conclude with some open questions in Section 6. Due to lack of space, the full proof of Theorem 2 is given in the full version [13].

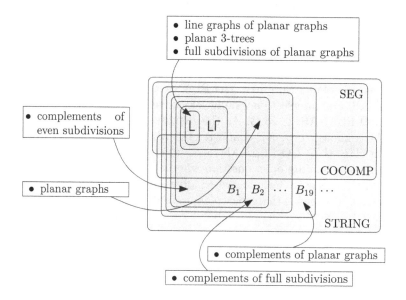

Fig. 1. Graph classes considered in this paper

Related Representations

In the context of *contact representations*, where distinct segments or k-bend paths may not share interior points, it is known that every contact SEG-representation has a combinatorially equivalent contact B_1-VPG-representation, but not vice versa [20]. Contact SEG-graphs are exactly planar Laman graphs and their subgraphs [10], which includes for example all triangle-free planar graphs. Very recently, contact {L}-graphs have been characterized [8]. Necessary and sufficient conditions for stretchability of a contact system of pseudo-segments are known [1, 11].

Let us also mention the closely related concept of *edge*-intersection graphs of paths in a grid (EPG-graphs) introduced by Golumbic *et al.* [16]. There are some notable differences, starting from the fact that *every* graph is an EPG-graph [16]. Nevertheless, analogous questions to the ones posed about VPG-representations of STRING-graphs are posed about EPG-representations of general graphs. In particular, there is a strong interest in finding representations using paths with few bends, see [19] for a recent account.

2   Proof of Theorem 2

Proof (main idea). Let G be a plane 3-tree with a xed plane embedding. We construct an {L}-representation of G satisfying the additional property that for every inner triangular face $\{a, b, c\}$ of G there exists a subset of the plane, called the *private region* of the face, that intersects only the L-paths for a, b and c, and no other L-path. We remark that this technique has also been used by Chalopin *et al.* [5] and refer to Figure 2 for an illustration.   □

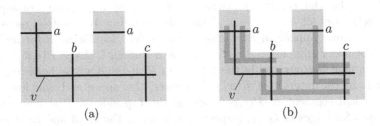

Fig. 2. (a) Introducing an L-shape for vertex v into the private region for the triangle $\{a, b, c\}$. (b) Identifying a pairwise disjoint private regions for the facial triangles $\{a, b, v\}$, $\{a, c, v\}$ and $\{b, c, v\}$.

3   Proof of Theorem 3

Proof. Without loss of generality let G be a maximally planar graph with a fixed plane embedding. (Line graphs of subgraphs of G are induced subgraphs of $L(G)$.) Then G admits a so-called *canonical ordering* –first defined in [12]–, namely an ordering $v_1, \ldots, v_n$ of the vertices of G such that

- Vertices v_1, v_2, v_n form the outer triangle of G in clockwise order. (We draw G such that v_1, v_2 are the highest vertices.)

- For $i = 3, \ldots, n$ vertex v_i lies in the outer face of the induced embedded subgraph $G_{i-1} = G[v_1, \ldots, v_{i-1}]$. Moreover, the neighbors of v_i in G_{i-1} form a path on the outer face of G_{i-1} with at least two vertices.

We shall construct an $\{L\}$-representation of $L(G)$ along a fixed canonical ordering $v_1, \ldots, v_n$ of G. For every $i = 2, \ldots, n$ we shall construct an $\{L\}$-representation of $L(G_i)$ with the following additional properties.

For every outer vertex v of G_i we maintain an auxiliary bottomless rectangle $R(v)$, i.e., an axis-aligned rectangle with bottom-edge at $-\infty$, such that:

- $R(v)$ intersects the horizontal segments of precisely those rectilinear paths for edges in G_i incident to v.
- $R(v)$ does not contain any bends or endpoints of any path for an edge in G_i and does not intersect any $R(w)$ for $w \neq v$.
- the left-to-right order of the bottomless rectangles matches the order of vertices on the counterclockwise outer v_1, v_2-path of G_i.

The bottomless rectangles act as placeholders for the upcoming vertices of $L(G)$. Indeed, all upcoming intersections of paths will be realized inside the corresponding bottomless rectangles. For $i = 2$, the graph G_i consist only of the edge $v_1 v_2$. Hence an $\{L\}$-representation of the one-vertex graph $L(G_2)$ consists of only one L-shape and two disjoint bottomless rectangles $R(v_1)$, $R(v_2)$ intersecting its horizontal segment.

For $i \geq 3$, we shall start with an $\{L\}$-representation of $L(G_{i-1})$. Let $(w_1, \ldots, w_k)$ be the counterclockwise outer path of G_{i-1} that corresponds to the neighbors of v_i in G_{i-1}. The corresponding bottomless rectangles $R(w_1), \ldots, R(w_k)$ appear in this left-to-right order. See Figure 3 for an illustration. For every edge $v_i w_j$, $j = 1, \ldots, k$ we define an L-shape $P(v_i w_j)$ whose vertical segment is contained in the interior of $R(w_j)$ and whose horizontal segment ends in the interior of $R(w_k)$. Moreover, the upper end and lower end of the vertical segment of $P(v_i w_j)$ lies on the top side of $R(w_j)$ and below all L-shapes for edges in G_{i-1}, respectively. Finally, the bend and right end of $P(v_i w_j)$ is placed above the bend of $P(v_i w_{j+1})$ and to the right of the right end of $P(v_i w_{j+1})$ for $j = 1, \ldots, k-1$, see Figure 3.

It is straightforward to check that this way we obtain an $\{L\}$-representation of $L(G_i)$. So it remains to find a set of bottomless rectangles, one for each outer vertex of G_i, satisfying our additional property. We set $R'(v) = R(v)$ for every $v \in V(G_i) \setminus \{v_i, w_1, \ldots, w_k\}$ since these are kept unchanged. Since $R(w_1)$ and $R(w_k)$ are not valid anymore, we define a new bottomless rectangle $R'(w_1) \subset R(w_1)$ such that $R'(w_1)$ is crossed by all horizontal segments that cross $R(w_1)$ and additionally the horizontal segment of $P(v_i w_1)$. Similarly, we define $R'(w_k) \subset R(w_k)$. And finally, we define a new bottomless rectangle $R'(v_i) \subset R(w_k)$ in such a way that it is crossed by the horizontal segments of exactly $P(v_i w_1), \ldots, P(v_i w_k)$. Note that for $1 < j < k$ the outer vertex w_j of G_{i-1} is not an outer vertex of G_i. Then $\{R'(v) \mid v \in v(G_i)\}$ has the desired property. See again Figure 3. $\qquad\square$

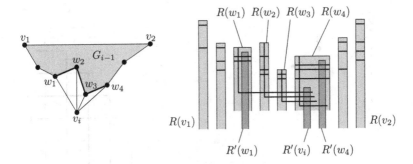

Fig. 3. Along a canonical ordering a vertex v_i is added to G_{i-1}. For each edge between v_i and a vertex in G_{i-1} an L-shape is introduced with its vertical segment in the corresponding bottomless rectangle. The three new bottomless rectangles $R'(w_1), R'(v_i), R'(w_k)$ are highlighted.

4 Proof of Theorem 4

Proof. Let $G = (V, E)$ be any planar graph. We shall construct a B_k-VPG representation of the complement $\bar{G}$ of G for some constant k that is independent of G. Indeed, $k = 19$ is enough. To find the VPG representation we make use of two crucial properties of G: A) G is 4-colorable and B) G is 5-degenerate. Indeed, our construction gives a B_{2d+9}-VPG representation for the complement of any 4-colorable d-degenerate graph. Here a graph is called *d-degenerate* if it admits a vertex ordering such that every vertex has at most d neighbors with smaller index.

Consider any 4-coloring of G with color classes V_1, V_2, V_3, V_4. Further let $\sigma = (v_1, \ldots, v_n)$ be an order of the vertices of V witnessing the degeneracy of G, i.e., for each v_i there are at most 5 neighbors v_j of v_i with $j < i$. We call these neighbors the *back neighbors of v_i*. Consider any ordered pair of color classes, say (V_1, V_2), and denote $W = V_1 \cup V_2$, together with the vertex orders inherited from the order of vertices in V, i.e., $\sigma|_{V_1} = \sigma_1 = (v_1, \ldots, v_{|V_1|})$ and $\sigma|_{V_2} = \sigma_2 = (w_1, \ldots, w_{|V_2|})$. Further consider the axis-aligned rectangle $R = [0, A] \times [0, A]$, where $A = 2(|W| + 2)$. For illustration we divide R into four quarters $[0, A/2] \times [0, A/2]$, $[0, A/2] \times [A/2, A]$, $[A/2, A] \times [0, A/2]$ and $[A/2, A] \times [A/2, A]$. We define a monotone increasing path $Q(v)$ for each $v \in W$ as follows. See Figure 4 for an illustration.

- For $v \in V_1$ let $\{\sigma_2(i_1), \ldots, \sigma_2(i_k)\}$, $i_1 < \cdots < i_k$, be the back neighbors of v in V_2 and $i^* = \max\{0\} \cup \{\sigma_2^{-1}(w) \mid w \in V_2, \sigma^{-1}(w) < \sigma^{-1}(v)\}$ be the largest index with respect to σ_2 of a vertex in V_2 that comes before v in σ or $i^* = 0$ if there is no such vertex. Then we define the path $Q(v)$ so that it starts at $(1, 0)$, uses the horizontal lines at $y = 2i_j - 1$ for $j = 1, \ldots, k$, $y = 2i^* + 1$ and $y = A - 2\sigma_1(v)$ in that order, uses the vertical lines at $x = 1$, $x = 2i_j + 1$ for $j = 1, \ldots, k$ and $x = A - 2\sigma_1(v)$ in that order, and finally ends at $(A, A - 2\sigma_1(v))$.

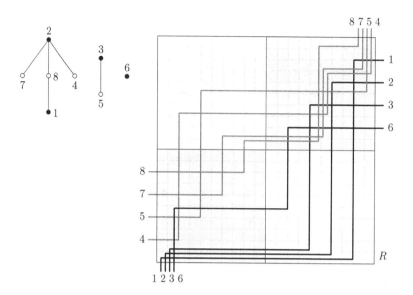

Fig. 4. The induced subgraph $G[W]$ for two color classes $W = V_1 \cup V_2$ of a planar graph G and a VPG representation of its complement $\bar{G}[W]$ in the rectangle $[0, 2(|W|+2)] \times [0, 2(|W|+2)]$

Note that $Q(v)$ avoids the top-left quarter of R, has exactly one bend at $(A - 2\sigma_1(v), A - 2\sigma_1(v))$ in the top-right quarter, and goes above the point $(2i, 2i)$ in the bottom-left quarter if and only if $i \neq i_1, \dots, i_k$ and $i \leq i^*$.

- For $w_i \in V_2$ the path $P(w_i)$ is defined analogous after rotating the rectangle R by 180 degrees and swapping the roles of V_1 and V_2.

It is straightforward to check that $\{Q(v) \mid v \in W\}$ is a VPG representation of $\bar{G}[W]$ completely contained in R, where each $Q(v)$ starts and ends at the boundary of R and has at most $3 + 2k$ bends, where k is the number of back neighbors of v in W.

Now we have defined for each pair of color classes $V_i \cup V_j$ a VPG-representation of $\bar{G}[V_i \cup V_j]$. For every vertex $v \in V$ we have defined three Q-paths, one for each colors class that v is not in. In total the three Q-paths for the same vertex v have at most $9 + 2k \leq 19$ bends, where $k \leq 5$ is the back degree of v. It remains to place the six representations of $\bar{G}[V_i \cup V_j]$ non-overlapping and to "connect" the three Q-paths for each vertex in such a way that connections for vertices of different color do not intersect. This can easily be done with two extra bends per paths, basically because K_4 is planar (we refer to Figure 5 for one way to do this). Finally, note that the first and last segment of every path in the representation can be omitted, yielding the claimed bound. □

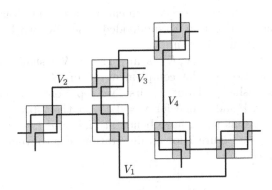

Fig. 5. Interconnecting the VPG representations of $\bar{G}[V_i \cup V_j]$ by adding at most two bends for each vertex. The set of paths corresponding to color class V_i is indicated by a single path labeled V_i, $i = 1, 2, 3, 4$.

5   Proof of Theorem 5

Proof. Let G be any graph and H arise from G by subdividing each edge at least once. Without loss of generality we may assume that every edge of G is subdivided exactly once or twice. Indeed, if an edge e of G is subdivided three times or more, then H can be seen as a full subdivision of the graph G' that arises from G by subdividing e once.

(i) Assuming that G is planar, we shall find an $\{L\}$-representation of H as follows. Without loss of generality G is maximally planar. We consider a bar visibility representation of G, i.e., vertices of G are disjoint horizontal segments in the plane and edges are disjoint vertical segments in the plane whose endpoints are contained in the two corresponding vertex segments and which are disjoint from all other vertex segments. Such a representation for a planar triangulation exists e.g. by [23]. See Figure 6 for an illustration.

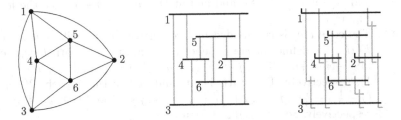

Fig. 6. A planar graph G on the left, a bar visibility representation of G in the center, and an $\{L\}$-representation of a full division of G on the right. Here, the edges $\{1, 2\}$, $\{1, 3\}$ and $\{3, 6\}$ are subdivided twice.

It is now easy to interpret every segment as an L, and replace an segment corresponding to edge that is subdivided twice by two L-shapes. Let us simply refer to Figure 6 again.

(ii) Now assume that $G = (V, E)$ is any graph. We shall construct a B_2-VPG representation of the complement $\bar{H}$ of $H = (V \cup W, E')$ with monotone increasing paths only. First, we represent the clique $\bar{H}[V]$. Let $V = \{v_1, \ldots, v_n\}$ and define for $i = 1, \ldots, n$ the 2-bend path $P(v_i)$ for vertex v_i to start at $(i, 0)$, have bends at (i, i) and $(i + n, i)$, and end at $(i + n, n + 1)$. See Figure 7 for an illustration. For convenience, let us call these paths v-paths.

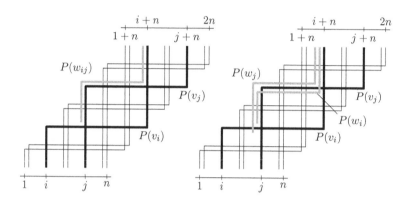

Fig. 7. Left: Inserting the path $P(w_{ij})$ for a single vertex w_{ij} subdividing the edge $v_i v_j$ in G. Right: Inserting the paths $P(w_i)$ and $P(w_j)$ for two vertices w_i, w_j subdividing the edge $v_i v_j$ in G.

Next, we define for every edge of G the 2-bend paths for the one or two corresponding subdivision vertices in $\bar{H}$. We call these paths w-paths. So let $v_i v_j$ be any edge of G with $i < j$. We distinguish two cases.

Case 1. The edge $v_i v_j$ is subdivided by only one vertex w_{ij} in H. We define the w-path $P(w_{ij})$ to start at $(j - \frac{1}{4}, i + \frac{1}{4})$, have bends at $(j - \frac{1}{4}, j + \frac{1}{4})$ and $(i + n - \frac{1}{4}, j + \frac{1}{4})$, and end at $(i + n - \frac{1}{4}, n + 1)$, see the left of Figure 7.

Case 2. The edge $v_i v_j$ is subdivided by two vertices w_i, w_j with $v_i w_i, v_j w_j \in E(H)$. We define the start, bends and end of the w-path $P(w_i)$ to be $(j - \frac{1}{4}, i + \frac{1}{4})$, $(j - \frac{1}{4}, j - \frac{1}{4})$, $(i + n - \frac{1}{4}, j - \frac{1}{4})$ and $(i + n - \frac{1}{4}, n + 1)$, respectively. The start, bends and end of the w-path $P(w_j)$ are $(j - \frac{1}{2}, i - \frac{1}{4})$, $(j - \frac{1}{2}, j + \frac{1}{4})$, $(i + n - \frac{1}{2}, j + \frac{1}{4})$ and $(i + n - \frac{1}{2}, n + 1)$, respectively. See the right of Figure 7.

It is easy to see that every w-path $P(w)$ intersects every v-path, except for the one or two v-paths corresponding to the neighbors of w in H. Moreover, the two w-paths in Case 2 are disjoint. It remains to check that the w-paths for distinct edges of G mutually intersect. To this end, note that every w-path for edge $v_i v_j$ starts near (j, i), bends near (j, j) and $(i + n, j)$ and ends near $(i+n, n)$. Consider two w-paths P and P' that start at (j, i) and (j', i'), respectively, and bend near (j, j) and (j', j'), respectively. If $j = j'$ then it is easy to check that P and P' intersect near (j, j). Otherwise, let $j' > j$. Now if $j > i'$, then P and P' intersect near (j', i), and if $j \leq i'$, then P and P' intersect near $(i + n, j')$.

Hence we have found a B_2-VPG-representation of $\bar{H}$, as desired. Let us remark, that in this representation some w-paths intersect non-trivially along some horizontal or vertical lines, i.e., share more than a finite set of points. However, this can be omitted by a slight and appropriate perturbation of endpoints and bends of w-paths. □

6   Conclusions and Open Problems

Motivated by Middendorf and Pfeiffer's theorem (Theorem 1 in [24]) that every $\{L, \Gamma\}$-representation can be stretched into a SEG-representation, we considered the question which subclasses of SEG-graphs are actually $\{L, \Gamma\}$-graphs, or even $\{L\}$-graphs. We proved that this is indeed the case for several graph classes related to planar graphs. We feel that the question whether PLANAR $\subset \{L, \Gamma\}$, as already conjectured [9], is of particular importance. After all, this, together with Theorem 1, would give a new proof for the fact that PLANAR $\subset$ SEG.

Open Problem 1. *Each of the following is open.*

(i) When can a B_1-VPG-representation be stretched into a combinatorially equivalent SEG-representation?

(ii) Is $\{L, \Gamma\} = SEG \cap B_1$-VPG?

(iii) Is every planar graph an $\{L\}$-graph, or B_1-VPG-graph?

(iv) Does every planar graph admit an even subdivision whose complement is an $\{L\}$-graph, or B_1-VPG-graph?

(v) Recognizing B_k-VPG graphs is known to be NP-complete for each $k \geq 0$ [7]. What is the complexity of recognizing $\{L\}$-graphs, or $\{L, \Gamma\}$-graphs?

References

1. Aerts, N., Felsner, S.: Straight line triangle representations. In: Wismath, S., Wolff, A. (eds.) GD 2013. LNCS, vol. 8242, pp. 119–130. Springer, Heidelberg (2013)
2. Alon, N., Scheinerman, E.: Degrees of freedom versus dimension for containment orders. Order 5, 11–16 (1988)
3. Asinowski, A., Cohen, E., Golumbic, M.C., Limouzy, V., Lipshteyn, M., Stern, M.: Vertex intersection graphs of paths on a grid. J. Graph Algorithms Appl. 16(2), 129–150 (2012)
4. Cabello, S., Cardinal, J., Langerman, S.: The clique problem in ray intersection graphs. Discrete & Computational Geometry 50(3), 771–783 (2013)
5. Chalopin, J., Gonçalves, D., Ochem, P.: Planar graphs have 1-string representations. Discrete & Computational Geometry 43(3), 626–647 (2010)

6. Chalopin, J., Gonçalves, D.: Every planar graph is the intersection graph of segments in the plane: extended abstract. In: Proceedings of the 41st Annual ACM Symposium on Theory of Computing, STOC 2009, pp. 631–638 (2009)

7. Chaplick, S., Jelínek, V., Kratochvíl, J., Vyskočil, T.: Bend-bounded path intersection graphs: Sausages, noodles, and waffles on a grill. In: Golumbic, M.C., Stern, M., Levy, A., Morgenstern, G. (eds.) WG 2012. LNCS, vol. 7551, pp. 274–285. Springer, Heidelberg (2012)

8. Chaplick, S., Kobourov, S.G., Ueckerdt, T.: Equilateral L-contact graphs. In: Brandstädt, A., Jansen, K., Reischuk, R. (eds.) WG 2013. LNCS, vol. 8165, pp. 139–151. Springer, Heidelberg (2013)

9. Chaplick, S., Ueckerdt, T.: Planar graphs as VPG-graphs. Journal of Graph Algorithms and Applications 17(4), 475–494 (2013)

10. de Fraysseix, H., Ossona de Mendez, P.O.: Representations by contact and intersection of segments. Algorithmica 47(4), 453–463 (2007)

11. de Fraysseix, H., de Mendez, P.O.: Stretching of Jordan arc contact systems. Discrete Applied Mathematics 155(9), 1079–1095 (2007)

12. De Fraysseix, H., Pach, J., Pollack, R.: How to draw a planar graph on a grid. Combinatorica 10(1), 41–51 (1990)

13. Felsner, S., Knauer, K., Mertzios, G.B., Ueckerdt, T.: Intersection graphs of L-shapes and segments in the plane. arXiv preprint arXiv:1405.1476 (2014)

14. Francis, M.C., Kratochvíl, J., Vyskočil, T.: Segment representation of a subclass of co-planar graphs. Discrete Mathematics 312(10), 1815–1818 (2012)

15. Garey, M.R., Johnson, D.S.: Computers and Intractability: A guide to the theory of NP-completeness. W.H. Freeman (1979)

16. Golumbic, M.C., Lipshteyn, M., Stern, M.: Edge intersection graphs of single bend paths on a grid. Networks 54(3), 130–138 (2009)

17. Golumbic, M.C., Rotem, D., Urrutia, J.: Comparability graphs and intersection graphs. Discrete Mathematics 43(1), 37–46 (1983)

18. Gyárfás, A.: Problems from the world surrounding perfect graphs. Zastos. Mat. 19(3-4), 413–441 (1987)

19. Heldt, D., Knauer, K., Ueckerdt, T.: Edge-intersection graphs of grid paths: The bend-number. Discrete Appl. Math. 167, 144–162 (2014)

20. Kobourov, S.G., Ueckerdt, T., Verbeek, K.: Combinatorial and geometric properties of planar Laman graphs. In: SODA, pp. 1668–1678. SIAM (2013)

21. Kratochvíl, J., Kuběna, A.: On intersection representations of co-planar graphs. Discrete Mathematics 178(1-3), 251–255 (1998)

22. Kratochvíl, J., Matousek, J.: Intersection graphs of segments. Journal of Combinatorial Theory, Series B 62(2), 289–315 (1994)

23. Luccio, F., Mazzone, S., Wong, C.K.: A note on visibility graphs. Discrete Mathematics 64(2-3), 209–219 (1987)

24. Middendorf, M., Pfeiffer, F.: The max clique problem in classes of string-graphs. Discrete Mathematics 108(1), 365–372 (1992)

25. Pawlik, A., Kozik, J., Krawczyk, T., Lasoń, M., Micek, P., Trotter, W.T., Walczak, B.: Triangle-free intersection graphs of line segments with large chromatic number. Journal of Combinatorial Theory, Series B (2013)

26. Scheinerman, E.R.: Intersection classes and multiple intersection parameters of graphs. PhD thesis, Princeton University (1984)

27. Warren, H.E.: Lower bounds for approximation by nonlinear manifolds. Trans. Amer. Math. Soc. 133, 167–178 (1968)

Autoreducibility and Mitoticity
of Logspace-Complete Sets
for NP and Other Classes*

Christian Glaßer and Maximilian Witek

Julius-Maximilians-Universität Würzburg, Germany
{glasser,witek}@informatik.uni-wuerzburg.de

Abstract. We study the autoreducibility and mitoticity of complete sets for NP and other complexity classes, where the main focus is on logspace reducibilities. In particular, we obtain:

- For NP and all other classes of the PH: Each $\leq_{m}^{\log}$-complete set is $\leq_{T}^{\log}$-autoreducible.
- For P, Δ_{k}^{p}, NEXP: Each $\leq_{m}^{\log}$-complete set is a disjoint union of two $\leq_{2\text{-tt}}^{\log}$-complete sets.
- For PSPACE: Each $\leq_{dtt}^{p}$-complete set is a disjoint union of two $\leq_{dtt}^{p}$-complete sets.

1 Introduction

Complete sets for NP and other complexity classes are one of the main objects of research in theoretical computer science. However, basic questions regarding complete sets are still open. For instance: Is it possible to split each complete set of a certain class into two complete sets? If the answer is yes, then all complete sets of this class are in some sense redundant. In this paper we study two types of redundancy of sets.

- *Autoreducibility* of A: $A(x)$ can be efficiently computed from $A(y)$ for $y \neq x$.
- *Mitoticity* of A: A is a disjoint union of two sets that are equivalent to A.

There are several notions of autoreducibility depending on the computing resources and the number of values $A(y)$ that we can ask for. For each reducibility $\leq$, a set A is $\leq$-autoreducible, if there exists a $\leq$-reduction from A to A that on input x does not query x. Similarly, there are several notions of mitoticity depending on the notion of equivalence that is used. For each reducibility $\leq$, a set A is weakly $\leq$-mitotic, if there exists a set S such that A, $A \cap S$, and $A \cap \overline{S}$ are pairwise $\leq$-equivalent. If S has low complexity, then A is called $\leq$-mitotic.

Typical complete problems for NP, PSPACE, and other classes are not only polynomial-time-complete, but even logspace-complete, which brings us to the main question of this paper:

Does logspace-completeness imply logspace-autoreducibility or even logspace-mitoticity?

---

* Proofs omitted in this version can be found in the technical report [14].

E. Csuhaj-Varjú et al. (Eds.): MFCS 2014, Part II, LNCS 8635, pp. 311–323, 2014.

We study this question for general complexity classes and conclude results for the classes P, NP, Δ_k^P, Σ_k^P, Π_k^P, and NEXP.

Related Work. The notions of autoreducibility and mitoticity were originally studied in computability theory. Trakhtenbrot [20] defined a set A to be *autoreducible* if there exists an oracle Turing machine M such that $A = L(M^A)$ and M on input x never queries x. Ladner [16] defined a set A to be *mitotic* if it is the disjoint union of two sets of the same degree. He showed that a computably enumerable set is mitotic if and only if it is autoreducible.

Motivated by the hope to gain insight in the structure of sets in NP, Ambos-Spies [1] introduced and studied the variants of autoreducibility and mitoticity that are defined by polynomial-time many-one reducibility ($\leq_m^P$) and polynomial-time Turing reducibility ($\leq_T^P$). Moreover, he introduced the distinction between mitoticity (splitting by some $S \in P$) and weak mitoticity (splitting by an arbitrary S). For the study of sets inside P one needs refined notions of autoreducibility and mitoticity, which we obtain by using logspace reducibilities [11].

It is easy to see that in general, mitoticity implies autoreducibility. Ambos-Spies [1] showed that $\leq_T^P$-autoreducibility does not imply $\leq_T^P$-mitoticity. Moreover, $\leq_m^P$-autoreducibility and $\leq_m^P$-mitoticity are equivalent [13]. The same paper showed that $\leq_T^P$-autoreducibility does not imply weak $\leq_T^P$-mitoticity.

A matter of particular interest is the question of whether complete sets are autoreducible or mitotic. Ladner [16] showed that there are Turing-complete sets for RE that are not mitotic. Over the years, researchers showed the polynomial-time mitoticity or at least the polynomial-time autoreduciblity of complete sets of prominent complexity classes: Beigel and Feigenbaum [2] proved that $\leq_T^P$-complete sets for all levels of the polynomial hierarchy and PSPACE are $\leq_T^P$-autoreducible. The same result holds for $\leq_m^P$ reducibility [12,13]. Buhrman, Hoene, and Torenvliet [6] showed that $\leq_m^P$-complete sets for EXP are weakly $\leq_m^P$-mitotic, which was later improved to $\leq_m^P$-mitotic [7]. Buhrman et al. [5] proved that all $\leq_T^P$-complete sets for EXP are $\leq_T^P$-autoreducible. Moreover, the same paper contains interesting negative results like the existence of polynomial-time bounded-truth-table complete sets in EXP that are not polynomial-time bounded-truth-table autoreducible. Nguyen and Selman [19] showed negative autoreducibility results for NEXP. These and other results for polynomial-time reducibilities are summarized in Table 2. In a recent paper [11], the authors studied autoreducibility and mitoticity also for logspace reducibilities (cf. Table 1).

Our Contribution. We prove the following general results on the autoreducibility and mitoticity of complete sets. Let $\mathcal{C} \supseteq (\text{DSPACE}(\log \cdot \log^{(c)}) \cap P)$ be a complexity class that is closed under intersection, for some $c > 0$.

(a) $A \leq_m^{\log}$-complete for $\mathcal{C}$ and $\leq_m^{\log}$-autoreducible $\implies A$ weakly $\leq_{2\text{-dtt}}^{\log}$-mitotic.
(b) $A \leq_{1\text{-tt}}^{\log}$-complete for $\mathcal{C}$ and $\leq_{1\text{-tt}}^{\log}$-autoreducible $\implies A$ weakly $\leq_{2\text{-tt}}^{\log}$-mitotic.
(c) $A \leq_T^P$-hard for P and $\leq_{tt}^{\log}$-autoreducible $\implies A \leq_T^{\log}$-autoreducible.

The results (a) and (b) are particularly interesting for P, the levels Δ_k^P of the polynomial hierarchy, and NEXP. Previously it was known that $\leq_m^{\log}$-complete sets are $\leq_m^{\log}$-autoreducible (resp., $\leq_{1\text{-tt}}^{\log}$-autoreducible). From (a) and (b) it follows:

(d) All $\leq_m^{\log}$-complete sets for P and all other Δ_k^P levels are weakly $\leq_{2\text{-tt}}^{\log}$-mitotic.

(e) All $\leq_m^{\log}$-complete sets for NEXP are weakly $\leq_{2\text{-dtt}}^{\log}$-mitotic.

So each of these sets is a disjoint union of two $\leq_{2\text{-tt}}^{\log}$-complete sets (resp., $\leq_{2\text{-dtt}}^{\log}$-complete sets). It remains an open question whether this can be improved to $\leq_m^{\log}$-mitoticity. This question is interesting, since in contrast to $\leq_m^P$ reducibility, we do not know whether $\leq_m^{\log}$-autoreducibility and $\leq_m^{\log}$-mitoticity are equivalent (they are inequivalent relative to some oracle [10]).

The result (c) is particularly interesting for NP, coNP and the other Σ_k^P and Π_k^P levels of the polynomial hierarchy, where only the polynomial-time autoreducibility and mitoticity of complete sets was known. Here we obtain logspace Turing autoreducibility:

(f) All $\leq_{\text{dtt}}^{\log}$-complete or $\leq_{1\text{-tt}}^{\log}$-complete sets for NP, coNP, Σ_k^P, Π_k^P are $\leq_T^{\log}$-autoreducible.

Finally, with our technique we also obtain a new result for the polynomial-time setting. Previously it was known that all $\leq_{\text{dtt}}^P$-complete sets for PSPACE are $\leq_{\text{dtt}}^P$-autoreducible. We obtain:

(g) All $\leq_{\text{dtt}}^P$-complete sets for PSPACE are weakly $\leq_{\text{dtt}}^P$-mitotic.

Again this means that each such set is a disjoint union of two $\leq_{\text{dtt}}^P$-complete sets.

Table 1 and Table 2 summarize known results for logspace and polynomial-time complete sets and emphasize the new results we obtained in this paper.

2 Preliminaries

We use standard notation for intervals of natural numbers, i.e., $[a, b] = \{a, a + 1, \ldots, b\}, [a, b) = \{a, a + 1, \ldots, b - 1\}, (a, b] = \{a + 1, a + 2, \ldots, b\}$, and $(a, b) = \{a + 1, a + 2, \ldots, b - 1\}$ for $a, b \in \mathbb{N}$. We call a set *trivial* if it is either finite or cofinite, and *non-trivial* otherwise. We only consider non-trivial sets. For a set A let c_A denote its characteristic function, i.e., $c_A(x) = 1 \iff x \in A$. We denote the Boolean exclusive or by $\oplus$. For functions f and g, by $(f \circ g)$ we denote the composition of the functions, i.e., $(f \circ g)(x) := f(g(x))$. Let $f^{(i)}$ denote the i-th iteration of the function f, i.e., $f^{(0)}(x) := x$, and $f^{(i)}(x) := f(f^{(i-1)}(x))$ for $i > 0$. For a function f and some x, we refer to the sequence $f^{(0)}(x), f^{(1)}(x), f^{(2)}(x), \ldots$ as f's *trace on* x. For $k \geq 1$, we say that a set S is a k-*ruling set (for f)* if for every x there exists some $i \leq k$ with $c_S(x) \neq c_S(f^{(i)}(x))$. Let $\log$ denote the logarithm to base 2. We will often use the iterated logarithm $\log^{(k)}$ for some $k > 0$. For the sake of simplicity, we define $\log(x) := 0$ for all $x < 1$, hence $\log$ and its iterations are total functions that are always greater than or equal to 0. For every x, by $|x|$ we denote the length of x's binary representation, by $\text{abs}(x)$ we denote its absolute value, and by $\text{sgn}(x)$ we denote the sign of x. We say that a function f is polynomially length-bounded if there exists a polynomial p such that $|f(x)| \leq p(|x|)$ holds for all x. When we use functions s and t as space and time bounds, we assume that s and t are monotone functions.

Table 1. Redundancy of logspace complete sets, where $l \geq 1$ and $k \geq 2$. For the cell in row $\leq_r$ and column $\mathcal{C}$, the entry A_s means that every $\leq_r$-complete set for $\mathcal{C}$ is $\leq_s$-autoreducible. Analogously, the entry W_s means that every $\leq_r$-complete set for $\mathcal{C}$ is weakly $\leq_s$-mitotic, and the entry M_s means that every $\leq_r$-complete set for $\mathcal{C}$ is $\leq_s$-mitotic. For the cells marked with X_1 and X_2, negative results are known: There is a $\leq^{\log}_{btt}$-complete set for PSPACE that is not $\leq^{\log}_{btt}$-autoreducible [11] and a $\leq^{\log}_{btt}$-complete set for EXP that is not $\leq^{p}_{btt}$-autoreducible (and hence not $\leq^{\log}_{btt}$-autoreducible) [5]. Results implied by universal relations between reductions are omitted, and results obtained in this paper are framed. For the definitions of the reductions, see section 2.

$\leq$	P	NP, coNP Δ^p_k	Σ^p_k, Π^p_k	PSPACE	EXP	NEXP
$\leq^{\log}_{m}$	$A^{\log}_{1\text{-}tt}$, $\boxed{W^{\log}_{2\text{-}tt}}$	$A^{\log}_{1\text{-}tt}$, $\boxed{W^{\log}_{2\text{-}tt}}$	$M^{\log}_{m}$	$M^{\log}_{m}$	$A^{\log}_{m}$, $\boxed{W^{\log}_{2\text{-}dtt}}$	
$\leq^{\log}_{l\text{-}ctt}$	$A^{\log}_{l\text{-}tt}$	$A^{\log}_{l\text{-}tt}$		$M^{\log}_{l\text{-}ctt}$	$M^{\log}_{l\text{-}ctt}$	$A^{\log}_{l\text{-}ctt}$
$\leq^{\log}_{l\text{-}dtt}$	$A^{\log}_{l\text{-}tt}$	$A^{\log}_{l\text{-}tt}$		$M^{\log}_{l\text{-}dtt}$	$M^{\log}_{l\text{-}dtt}$	$A^{\log}_{l\text{-}dtt}$
$\leq^{\log}_{ctt}$				$M^{\log}_{ctt}$	$M^{\log}_{ctt}$	$A^{\log}_{ctt}$
$\leq^{\log}_{dtt}$		$\boxed{A^{\log}_{T}}$	$\boxed{A^{\log}_{T}}$	$M^{\log}_{dtt}$	$M^{\log}_{dtt}$	$A^{\log}_{dtt}$
$\leq^{\log}_{1\text{-}tt}$	$A^{\log}_{2\text{-}tt}$	$\boxed{A^{\log}_{T}}$	$\boxed{A^{\log}_{T}}$	$M^{\log}_{m}$	$M^{\log}_{m}$	$A^{\log}_{m}$, $\boxed{W^{\log}_{2\text{-}dtt}}$
$\leq^{\log}_{2\text{-}tt}$				$M^{\log}_{2\text{-}tt}$	$M^{\log}_{2\text{-}tt}$	$A^{\log}_{2\text{-}tt}$
$\leq^{\log}_{btt}$	$A^{\log[1]}_{\log\text{-}T}$	$A^{\log[1]}_{\log\text{-}T}$		X_1	X_2	
$\leq^{\log}_{tt}$	$A^{\log}_{tt}$	$A^{\log}_{tt}$				

Oracle Access. There are several possibilities to define oracle access of space-bounded oracle Turing machines. We use the multi-tape oracle access model proposed by Lynch [18], where a space-bounded oracle Turing machine consists of a single read-write working tape subject to the space bounds and an arbitrary but fixed number of write-only oracle tapes not subject to the space bounds. In each step, the oracle Turing machine may ask the query that is written on some particular oracle tape, after which the oracle Turing machine enters an answer state accordingly, and erases the particular oracle tape again. Note that for logspace oracle Turing machines, there implicitly exists a polynomial space bound on the oracle tapes.

Ladner and Lynch [17] considered the above model with only one oracle tape. They showed that for every such logspace oracle Turing maching there exists an equivalent logspace oracle Turing machine that asks queries non-adaptively. Lynch [18] argued that this does not hold for the general case, where adaptive queries are more powerful than non-adaptive queries.

Reductions. For sets A and B we say that A is polynomial-time Turing reducible to B ($A \leq^p_T B$), if there exists a polynomial-time oracle Turing machine that accepts A with B as its oracle. If M on input x asks at most $O(\log|x|)$ queries, then A is polynomial-time log-Turing reducible to B ($A \leq^p_{\log\text{-}T} B$). If M's queries are non-

Table 2. Redundancy of polynomial-time complete sets, where $l \geq 1$, $l' = l(l^2 + l + 1)$, and $k \geq 2$. The entries of the table are read analogously to the entries in Table 1. Recall that there is a $\leq_{\text{btt}}^{\log}$-complete (and hence also $\leq_{\text{btt}}^{\text{P}}$-complete) set for EXP that is not $\leq_{\text{btt}}^{\text{P}}$-autoreducible [5], hence we have a negative result marked by X_3.

$\leq$	NP	Δ_k^{P}	$\Sigma_k^{\text{P}}, \Pi_k^{\text{P}}$	PSPACE	EXP	NEXP
$\leq_{\text{m}}^{\text{P}}$	M_{m}^{P}	M_{m}^{P}	M_{m}^{P}	M_{m}^{P}	M_{m}^{P}	M_{m}^{P}
$\leq_{\text{1-tt}}^{\text{P}}$	$M_{\text{1-tt}}^{\text{P}}$	$M_{\text{1-tt}}^{\text{P}}$	$M_{\text{1-tt}}^{\text{P}}$	$M_{\text{1-tt}}^{\text{P}}$	M_{m}^{P}	M_{m}^{P}
$\leq_{\text{2-tt}}^{\text{P}}$					$M_{\text{2-tt}}^{\text{P}}$	$A_{\text{2-tt}}^{\text{P}}$
$\leq_{l\text{-ctt}}^{\text{P}}$		$A_{l\text{-tt}}^{\text{P}}$	$A_{l\text{-tt}}^{\text{P}}$		$M_{l\text{-ctt}}^{\text{P}}$	$A_{l\text{-ctt}}^{\text{P}}$
$\leq_{l\text{-dtt}}^{\text{P}}$	$A_{l\text{-dtt}}^{\text{P}}$	$A_{l\text{-dtt}}^{\text{P}}$	$A_{l\text{-dtt}}^{\text{P}}$	$A_{l\text{-dtt}}^{\text{P}}$, $\boxed{W_{l'\text{-dtt}}^{\text{P}}}$	$M_{l\text{-dtt}}^{\text{P}}$	$A_{l\text{-dtt}}^{\text{P}}$
$\leq_{\text{btt}}^{\text{P}}$					X_3	
$\leq_{\text{ctt}}^{\text{P}}$				A_{tt}^{P}	$M_{\text{ctt}}^{\text{P}}$	$A_{\text{ctt}}^{\text{P}}$
$\leq_{\text{dtt}}^{\text{P}}$	$A_{\text{dtt}}^{\text{P}}$	$A_{\text{dtt}}^{\text{P}}$	$A_{\text{dtt}}^{\text{P}}$	$A_{\text{dtt}}^{\text{P}}$, $\boxed{W_{\text{dtt}}^{\text{P}}}$	$M_{\text{dtt}}^{\text{P}}$	$A_{\text{dtt}}^{\text{P}}$
$\leq_{\text{tt}}^{\text{P}}$	$A_{\text{tt}}^{\text{BPP}}$	A_{tt}^{P}			$A_{\text{tt}}^{\text{BPP}}$	
$\leq_{\text{T}}^{\text{P}}$	A_{T}^{P}	A_{T}^{P}	A_{T}^{P}	A_{T}^{P}	A_{T}^{P}	

adaptive (i.e., independent of the oracle), then A is polynomial-time truth-table reducible to B ($A \leq_{\text{tt}}^{\text{P}} B$). If M asks at most k nonadaptive queries, then A is polynomial-time k-truth-table reducible to B ($A \leq_{k\text{-tt}}^{\text{P}} B$). A is polynomial-time bounded-truth-table reducible to B ($A \leq_{\text{btt}}^{\text{P}} B$), if $A \leq_{k\text{-tt}}^{\text{P}} B$ for some k. A is polynomial-time disjunctive-truth-table reducible to B ($A \leq_{\text{dtt}}^{\text{P}} B$), if there exists a polynomial-time-computable function f such that for all x, $f(x) = \langle y_1, y_2, \ldots, y_n \rangle$ for some $n \geq 1$ and $c_A(x) = \max\{c_B(y_1), c_B(y_2), \cdots, c_B(y_n)\}$. If n is bounded by some constant k, then A is polynomial-time k-disjunctive-truth-table reducible to B ($A \leq_{k\text{-dtt}}^{\text{P}} B$). A is polynomial-time bounded-disjunctive-truth-table reducible to B ($A \leq_{\text{bdtt}}^{\text{P}} B$), if $A \leq_{k\text{-dtt}}^{\text{P}} B$ for some k. The polynomial-time conjunctive-truth-table reducibilities $\leq_{\text{ctt}}^{\text{P}}$, $\leq_{k\text{-ctt}}^{\text{P}}$, and $\leq_{\text{bctt}}^{\text{P}}$ are defined analogously. A is polynomial-time many-one reducible to B ($A \leq_{\text{m}}^{\text{P}} B$), if there exists a polynomial-time-computable function f such that $c_A(x) = c_B(f(x))$. We also use the following logspace reducibilities, which are defined analogously in terms of logspace oracle Turing machines and logspace-computable functions:

$$\leq_{\text{T}}^{\log}, \leq_{\log\text{-T}}^{\log}, \leq_{\text{tt}}^{\log}, \leq_{k\text{-tt}}^{\log}, \leq_{\text{btt}}^{\log}, \leq_{\text{dtt}}^{\log}, \leq_{k\text{-dtt}}^{\log}, \leq_{\text{bdtt}}^{\log}, \leq_{\text{ctt}}^{\log}, \leq_{k\text{-ctt}}^{\log}, \leq_{\text{bctt}}^{\log}, \leq_{\text{m}}^{\log}$$

It is easy to see that the truth-table reductions can be performed by oracle Turing machines that use only one oracle tape. So multiple oracle tapes are only significant for logspace Turing reductions with adaptive queries. We define $A \leq_{\text{T}}^{\log[k]} B$ if $A \leq_{\text{T}}^{\log} B$ via some logspace oracle Turing machine with k oracle tapes. Furthermore, if $A \leq_{\text{T}}^{\log[k]} B$ with an oracle machine that on input x asks at most $O(\log(|x|))$ queries, then we write $A \leq_{\text{log-T}}^{\log[k]} B$. By Ladner and Lynch [17] it holds that $A \leq_{\text{T}}^{\log[1]} B$ if and only if $A \leq_{\text{tt}}^{\log} B$.

Definition 1. 1. A set A is called $\leq_{\text{T}}^{\text{P}}$-autoreducible if $A \leq_{\text{T}}^{\text{P}} A$ via some polynomial-time oracle Turing machine that on input x never queries x.
2. A set A is called $\leq_{\text{dtt}}^{\text{P}}$-autoreducible if $A \leq_{\text{dtt}}^{\text{P}} A$ via some $f \in \text{FP}$ where from $f(x) = \langle y_1, y_2, \ldots, y_n \rangle$ it follows that $x \notin \{y_1, y_2, \ldots, y_n\}$ for all x.

We define autoreducibility for the remaining reductions analogously, where the reduction oracle machine or the reduction function has to be chosen accordingly.

Definition 2. A set A is called weakly $\leq_{\text{T}}^{\text{P}}$-mitotic if $A \equiv_{\text{T}}^{\text{P}} A \cap S \equiv_{\text{T}}^{\text{P}} A \cap \overline{S}$ for some set S. We refer to S as a separator. If in addition it holds that $S \in \text{P}$, we call $A \leq_{\text{T}}^{\text{P}}$-mitotic.

We define mitoticity and weak mitoticity for the remaining reductions analogously, where for logspace mitoticity, the separator must be inside L. If the reduction is not transitive, then the sets A, $A \cap S$, and $A \cap \overline{S}$ must be pairwise equivalent.

In general, mitoticity implies autoreducibility, while the converse does not always hold. We are hence interested in the general question whether autoreducibility implies mitoticity or at least weak mitoticity.

3   Ruling Sets for Autoreductions

For transforming many-one autoreducibility into mitoticity, we consider the trace of words obtained by the repeated application of the autoreduction function of some language to the input x. Clearly all words on the trace of x have the same membership to the language. The challenge is to define a set S of low complexity such that when we follow such a trace for r steps, then we visit at least one word in S and at least one word in $\overline{S}$. Cole and Vishkin [8] developed the deterministic coin tossing, which is a technique for the construction of such S. In their terminology, the set S is called an r-ruling set.

In a recent paper [10], the author shows that for every non-trivial set L with autoreduction $f \in \text{FL}$ there is an autoreduction $g \in \text{FSPACE}(\log \cdot \log^{(2)})$ with a 1-ruling set $S \in \text{L}$. We generalize this proof and obtain the following lemma.

Lemma 3. Let f be polynomially length-bounded with $f(x) \neq x$ for all x. For all $k \geq 1$ there is a set S, a constant c_0, and a polynomial q such that:

1. For all x there is some $i \leq c_0 \cdot (\log^{(k)}(|x|) + 1)$ such that $c_S(x) \neq c_S(f^{(i)}(x))$, and for all $j \leq i$ it holds that $|f^{(j)}(x)| \leq q(|x|)$.
2. $f \in \text{FSPACE}(s) \implies S \in \text{DSPACE}(s)$ $(s \geq \log)$
3. $f \in \text{FTIME}(t) \implies S \in \text{DTIME}(O(t \circ q))$ $(t \geq n)$

Given the set S from Lemma 3, for every x we know that after $O(\log^{(k)}(|x|))$ steps of f there must be a change in the membership to S. We can hence define a set S' by an algorithm that on input x finds the smallest i such that $c_S(f^{(i)}(x)) > c_S(f^{(i+1)}(x))$ and accepts if i is even. Since the algorithm for S' has to decide S and compute f at most $O(\log^{(k)}(|x|))$ times, its complexity slightly increases. However, S' is almost a 1-ruling set for f, and the inputs x with $c_{S'}(x) = c_{S'}(f(x))$ can easily be avoided: if $x \in S$ and $f(x) \notin S$, jump to $f(x)$. We hence obtain the following lemma.

Lemma 4. *Let f be polynomially length-bounded with $f(x) \neq x$ for all x. For all $k \geq 1$ there is a set S, a function g, and a polynomial q such that:*

1. *For all x it holds that $c_S(g(x)) \neq c_S(f(g(x)))$ and $g(x) \in \{x, f(x)\}$.*
2. *$f \in \mathrm{FSPACE}(s) \implies S \in \mathrm{DSPACE}(s \cdot \log^{(k)}) \wedge g \in \mathrm{FSPACE}(s)$ (s $\geq$ log)*
3. *$f \in \mathrm{FTIME}(t) \implies S \in \mathrm{DTIME}(O(t \circ q)) \wedge g \in \mathrm{FTIME}(O(t \circ q))$ (t $\geq$ n)*

4 Weak Mitoticity

We show that autoreducibility of complete sets for general classes implies weak mitoticity. This gives progress towards the general question of whether complete sets are mitotic.

General Approach. Given a many-one autoreduction f for some set A complete for some class $\mathcal{C}$, we apply the results of the previous section to generate a 2-ruling set S for f. Since the complexity of S is only slightly higher than the complexity of f, we obtain $A \cap S \in \mathcal{C}$ and $A \cap \overline{S} \in \mathcal{C}$. Considering some input x, we then find two elements y, z on f's trace on x with the same membership to A as x such that exactly one is contained in S. Hence, y, z form a 2-dtt-reduction from A to $A \cap S$ and $A \cap \overline{S}$. So A is many-one complete for $\mathcal{C}$, and $A \cap S$ and $A \cap \overline{S}$ are 2-dtt complete, which shows weak 2-dtt mitoticity of A.

This approach can be generalized to further reducibility notions, including disjunctive truth-table reductions and reductions with exactly one query.

4.1 Many-One Complete Sets

We first consider logspace many-one autoreducible, complete sets for classes that contain the intersection of P and some space class slightly higher than L. Since the classes will be closed under intersection, the intersection of the complete set with the separator and its complement remain in the same complexity class.

Theorem 5. *Let $\mathcal{C} \supseteq (\mathrm{DSPACE}(\log \cdot \log^{(c)}) \cap \mathrm{P})$ for some $c \geq 1$ be closed under intersection. If A is $\leq_m^{\log}$-complete for $\mathcal{C}$ and $\leq_m^{\log}$-autoreducible, then A is weakly $\leq_{2\text{-dtt}}^{\log}$-mitotic.*

Proof. Let $f \in \mathrm{FL}$ be a $\leq_m^{\log}$-autoreduction for A. From Lemma 4 we obtain a set $S \in (\mathrm{DSPACE}(\log \cdot \log^{(c)}) \cap \mathrm{P})$ and a function $g \in \mathrm{FL}$ such that for all x

it holds that $c_S(g(x)) \neq c_S(f(g(x)))$ and $g(x) \in \{x, f(x)\}$. We will show that $A \cap S$ and $A \cap \overline{S}$ are $\leq^{\log}_{2\text{-dtt}}$-complete for $\mathcal{C}$.

Note that $(\text{DSPACE}(\log \cdot \log^{(c)}) \cap P)$ is closed under complementation, hence we have $S \in \mathcal{C}$ and $\overline{S} \in \mathcal{C}$. Since $\mathcal{C}$ is closed under intersection, we obtain $A \cap S \in \mathcal{C}$ and $A \cap \overline{S} \in \mathcal{C}$. So it remains to show the $\leq^{\log}_{2\text{-dtt}}$-hardness of $A \cap S$ and $A \cap \overline{S}$ for $\mathcal{C}$. Since A is $\leq^{\log}_{m}$-hard for $\mathcal{C}$, it suffices to show $A \leq^{\log}_{2\text{-dtt}} A \cap S$ and $A \leq^{\log}_{2\text{-dtt}} A \cap \overline{S}$.

Observe that $c_A(x) = c_A(g(x)) = c_A(f(g(x)))$ and $\{g(x), f(g(x))\} \cap S \neq \emptyset$. Let $h(x) := \{g(x), f(g(x))\}$. If $x \in A$, then $h(x) \subseteq A$, hence $h(x) \cap (S \cap A) = (h(x) \cap A) \cap S = h(x) \cap S \neq \emptyset$. If $x \notin A$, then $h(x) \cap (A \cap S) \subseteq h(x) \cap A = \emptyset$. Hence, h shows that $A \leq^{\log}_{2\text{-dtt}} A \cap S$. Analogously, h shows that $A \leq^{\log}_{2\text{-dtt}} A \cap \overline{S}$. So, A is $\leq^{\log}_{m}$-complete, and $A \cap S$ and $A \cap \overline{S}$ are $\leq^{\log}_{2\text{-dtt}}$-complete for $\mathcal{C}$.   □

Note that every $\leq^{\log}_{1\text{-tt}}$-complete set for NEXP is $\leq^{\log}_{m}$-complete, and every $\leq^{\log}_{m}$-complete set for NEXP is $\leq^{\log}_{m}$-autoreducible [11]. Since NEXP clearly satisfies the requirements of Theorem 5, we obtain the following corollary.

Corollary 6. *1. Every $\leq^{\log}_{m}$-complete set for NEXP is weakly $\leq^{\log}_{2\text{-dtt}}$-mitotic.*
2. Every $\leq^{\log}_{1\text{-tt}}$-complete set for NEXP is weakly $\leq^{\log}_{2\text{-dtt}}$-mitotic.

4.2   Truth-Table Complete Sets with One Query

We generalize our approach for many-one autoreductions to truth-table autoreductions that ask exactly one query. We can think of such a truth-table autoreduction for some set A as two functions f, f', where f' maps to the set of all unary Boolean functions, such that for all x it holds that $f(x) \neq x$ and $c_A(x) = f'(x)(c_A(f(x)))$. For non-trivial A we can modify f' such that it never maps to a constant function. We further modify the autoreduction such that on each input it either has a long part in its trace that behaves like a many-one autoreduction, or ends up after a few steps in a small cycle. We treat cycles directly and proceed on long many-one parts of the trace similar to the many-one case.

Theorem 7. *Let $\mathcal{C} \supseteq (\text{DSPACE}(\log \cdot \log^{(c)}) \cap P)$ for some $c \geq 1$ be closed under intersection. If A is $\leq^{\log}_{1\text{-tt}}$-complete for $\mathcal{C}$ and $\leq^{\log}_{1\text{-tt}}$-autoreducible, then A is weakly $\leq^{\log}_{2\text{-tt}}$-mitotic.*

Proof (Sketch). For non-trivial A, choose $f, f' \in \text{FL}$ such that f' maps to the set of unary Boolean functions $\{\text{id}, \text{non}\}$, and such that for all x it holds that $x \neq f(x)$ and $c_A(x) = f'(x)(c_A(f(x)))$. Define g such that for each x,

$g(x) := f^{(k)}(x)$ for $k \leq 3$ minimal with $x \neq f^{(k)}(x)$ and $c_A(x) = c_A(f^{(k)}(x))$, if such a k exists, and

$g(x) := f(x)$ otherwise.

Observe that $g \in \text{FL}$ and $g(x) \neq x$. We can show [14] that for all x, if $g^{(2)}(x) \neq g^{(4)}(x)$, then $c_A(g^{(k)}(x)) = c_A(g^{(k+1)}(x)) = c_A(g^{(k+2)}(x))$ for some $k \leq 2$.

From Lemma 4 we obtain a set $S \in (\mathrm{DSPACE}(\log \cdot \log^{(c)}) \cap \mathrm{P})$ and a function $h \in \mathrm{FL}$ such that $c_S(h(x)) \neq c_S(g(h(x)))$ and $h(x) \in \{x, g(x)\}$ for all x. We define $S' := \{x \in S \mid x \neq g^{(2)}(x)\} \cup \{x \mid x = g^{(2)}(x) \text{ and } x < g(x)\}$. Then, $S' \in (\mathrm{DSPACE}(\log \cdot \log^{(c)}) \cap \mathrm{P})$, hence $A \cap S' \in \mathcal{C}$. We show that $A \cap S'$ is $\leq_{2\text{-tt}}^{\log}$-complete for $\mathcal{C}$ by proving $A \leq_{2\text{-tt}}^{\log} A \cap S'$ (completeness of $A \cap \overline{S'}$ works analogously). On input x, if $g^{(2)}(x) = g^{(4)}(x)$, then we obtain $i \in \{2, 3\}$ with $c_{S'}(g^{(i)}(x)) = 1$, hence $c_A(x) = b \oplus c_{A \cap S'}(g^{(i)}(x))$, where b can be computed by looking at f'. If $g^{(2)}(x) \neq g^{(4)}(x)$, then $c_A(g^{(k)}(x)) = c_A(g^{(k+1)}(x)) = c_A(g^{(k+2)}(x))$ for some $k \leq 2$. Determine k and let $y = h(g^{(k)}(x))$ and $z = g(y)$. Then, $c_{S'}(y) \neq c_{S'}(z)$. Compute b with $b \oplus c_A(x) = c_A(y) = c_A(z)$ by looking at f'. By case distinction one can see that $c_A(x) = b \oplus \max\{c_{A \cap S'}(y), c_{A \cap S'}(z)\}$.   □

If some class is closed under complement, then all its $\leq_{\mathrm{m}}^{\log}$-complete sets are $\leq_{1\text{-tt}}^{\log}$-autoreducible. Hence we obtain the following corollaries.

Corollary 8. *Let $\mathcal{C} \supseteq (\mathrm{DSPACE}(\log \cdot \log^{(c)}) \cap \mathrm{P})$ for some $c \geq 1$ be closed under intersection and complementation. If A is $\leq_{\mathrm{m}}^{\log}$-complete for $\mathcal{C}$, then A is weakly $\leq_{2\text{-tt}}^{\log}$-mitotic.*

Corollary 9. *Every $\leq_{\mathrm{m}}^{\log}$-complete set for P and every $\leq_{\mathrm{m}}^{\log}$-complete set for the levels Δ_k^{P} of the PH is weakly $\leq_{2\text{-tt}}^{\log}$-mitotic.*

4.3 Disjunctive Truth-Table Complete Sets for PSPACE

We further generalize our approach to disjunctive truth-table autoreductions of complete sets for some higher complexity classes. Here, we consider the reduction graph of some disjunctive truth-table autoreduction f. If we grant the separator enough resources, then for each input x, it can determine the smallest equivalent $y \in f(x)$ and hence treat f like a many-one reduction. While for most higher classes, a diagonalization method leads to (strong) mitoticity results, for PSPACE in the polynomial-time reducibility setting, only autoreducibility results are known. Our approach as described above shows weak mitoticity.

Lemma 10. *Let $A \in \mathrm{PSPACE}$ and let $f \in \mathrm{FP}$ be a $\leq_{\mathrm{dtt}}^{\mathrm{P}}$-autoreduction for A such that f never maps to the empty set. Then there exists a set $S \in \mathrm{PSPACE}$ such that for all x there exist $y \in f(x)$ and $z \in f(y)$ with the following properties:*

1. $c_A(x) = c_A(y) = c_A(z)$
2. $\emptyset \neq (\{x, y, z\} \cap S) \neq \{x, y, z\}$

Proof. We consider the function g with

$$g(x) := \begin{cases} y_i & \text{if } f(x) = \langle y_1, \ldots, y_k \rangle \wedge y_i \in A \wedge y_j \notin A \text{ for all } j < i \text{, and} \\ y_1 & \text{if } f(x) = \langle y_1, \ldots, y_k \rangle \wedge y_j \notin A \text{ for all } j \leq k, \end{cases}$$

for all x. Since $A \in \mathrm{DSPACE}(p)$ for some polynomial p, there exists a polynomial q such that $g \in \mathrm{FSPACE}(q)$. Furthermore, since g maps to values of f, we have $g(x) \neq x$ for all x, and we can modify q such that $|g(x)| \leq q(|x|)$.

We apply Lemma 4 and obtain a set $S \in \mathrm{DSPACE}(q \cdot \log^{(c)}) \subseteq \mathrm{PSPACE}$ (where $c \geq 1$ is some constant) and a function h such that $h(x) \in \{x, g(x)\}$ and $c_S(h(x)) \neq c_S(g(h(x)))$ for all x. Choose $y := g(x)$ and $z := g(y)$. Hence $y \in f(x)$ and $z \in f(y)$, and $c_A(x) = c_A(y) = c_A(z)$. Furthermore, $h(x) \in \{x, y\}$, so we either have $c_S(x) \neq c_S(y)$, or $c_S(y) \neq c_S(z)$. $\square$

Theorem 11. *Let $k \geq 2$ and $k' = (k^2 + k + 1)$.*

1. *All $\leq^{\mathrm{P}}_{k\text{-dtt}}$-complete sets for PSPACE are weakly $\leq^{\mathrm{P}}_{(k \cdot k')\text{-dtt}}$-mitotic.*
2. *All $\leq^{\mathrm{P}}_{\mathrm{bdtt}}$-complete sets for PSPACE are weakly $\leq^{\mathrm{P}}_{\mathrm{bdtt}}$-mitotic.*
3. *All $\leq^{\mathrm{P}}_{\mathrm{dtt}}$-complete sets for PSPACE are weakly $\leq^{\mathrm{P}}_{\mathrm{dtt}}$-mitotic.*

Proof. If L is $\leq^{\mathrm{P}}_{k\text{-dtt}}$-complete for PSPACE, then L is $\leq^{\mathrm{P}}_{k\text{-dtt}}$-autoreducible [12]. Let f be some $\leq^{\mathrm{P}}_{k\text{-dtt}}$-autoreduction for L. We assume that f never maps to the empty set. From Lemma 10 we obtain $S \in \mathrm{PSPACE}$ with the specified properties. We show that $L \cap S$ is $\leq^{\mathrm{P}}_{k \cdot k'\text{-dtt}}$-complete for PSPACE, the completeness of $L \cap \overline{S}$ is shown analogously.

Clearly, $L \cap S \in \mathrm{PSPACE}$, so it remains to show hardness. For arbitrary $A \in \mathrm{PSPACE}$ we already know that $A \leq^{\mathrm{P}}_{k\text{-dtt}} L$, hence it suffices to show $L \leq^{\mathrm{P}}_{k'\text{-dtt}} L \cap S$. On input x, return $Q_x := \{x\} \cup f(x) \cup \bigcup_{y \in f(x)} f(y)$, which can be computed in polynomial time. The number of the elements in the output is bounded by $(1 + k + k^2) = k'$. To show that Q_x is a reduction as claimed above, choose y, z as in the lemma. If $x \in L$, then $\{x, y, z\} \subseteq L$, and, since $\{x, y, z\} \cap S \neq \emptyset$ and $\{x, y, z\} \subseteq Q_x$ we obtain $(L \cap S) \cap Q_x \supseteq (L \cap S) \cap \{x, y, z\} = S \cap \{x, y, z\} \neq \emptyset$. If $x \notin L$, then $(L \cap S) \cap Q_x \subseteq L \cap Q_x = \emptyset$. This shows the $\leq^{\mathrm{P}}_{k \cdot k'\text{-dtt}}$-hardness.

We have shown item 1. The other items are shown analogously. $\square$

5   Logspace Autoreducibility for NP

In this section we consider logspace complete sets for NP. In this setting, neither can we apply diagonalization (here, NP is too weak to diagonalize against logspace reductions), nor can we trace entire computation paths in the nondeterministic computation tree (because logspace reductions have not enough storage). However, we know that logspace complete sets for NP are redundant in the polynomial-time setting, which gives us access to particular deterministic polynomial-time computations and the transcripts of those computations.

Theorem 12. *Let A be $\leq^{\log[k]}_{\mathrm{T}}$-hard for P. If A is $\leq^{\mathrm{P}}_{\mathrm{tt}}$-autoreducible, then A is $\leq^{\log[2k+1]}_{\mathrm{T}}$-autoreducible.*

Proof (Sketch). Choose autoreduction functions $f, g \in \mathrm{FP}$ such that for all x there is some m with $f(x) = \langle y_1, \ldots, y_m \rangle$, $x \notin \{y_1, \ldots, y_m\}$ and $c_A(x) = g(x, c_A(y_1), \ldots, c_A(y_m))$. Consider the transcripts (i.e., bit string representations of the sequence of configurations on some input, starting with the input itself, and ending on the function value computed) of polynomial-time Turing transducers that compute f and g, respectively. Given such transcripts, we can verify the consistency of each bit in space $\log(n)$ by looking at constantly many previous bits of the transcript. The transcript bits are computable in polynomial time and can hence be reduced to A in logspace with k oracle tapes. On input x we compute and verify the transcript bits with oracle $A \cup \{x\}$. If some verification fails, $A \cup \{x\}$ is the wrong oracle, hence $x \notin A$. Otherwise we obtain $c_A(x)$ by looking at the last transcript bit.

The bits of $y_1, \ldots, y_m$ can be computed with k oracle tapes. Hence each y_i can be written on oracle tape $(k+1)$, and we obtain $y = (x, c_A(y_1), \ldots, c_A(y_m))$ with $k+1$ oracle tapes. To obtain $c_A(x) = g(y)$, we have to look at bits of the transcript for g on y, which can be done in logspace by a recomputation with k additional oracle tapes. So we need k tapes for bitwise computing $\langle y_1, \ldots, y_m \rangle$, one tape for storing an y_i, and k tapes for bitwise computing $g(y)$.   □

Note that if A is $\leq^{\log}_T$-hard for P, then there exists a $\leq^{\log}_m$-complete set B for P and some k such that $B \leq^{\log[k]}_T A$, hence A is $\leq^{\log[k]}_T$-hard for P.

Corollary 13. *Let A be $\leq^{\log}_T$-hard for P. If A is $\leq^P_{tt}$-autoreducible, then A is $\leq^{\log}_T$-autoreducible.*

Theorem 14 ([12]). *Let r be one of the reductions $\leq^P_m$, $\leq^P_{1\text{-tt}}$, $\leq^P_{dtt}$, $\leq^P_{l\text{-dtt}}$ for $l \geq 2$. Then every nontrivial set that is r-complete for one of the following classes is r-autoreducible: PSPACE, Σ^P_k, Π^P_k, Δ^P_k, 1NP, the levels of the Boolean hierarchy over NP, the levels of the MODPH hierarchy.*

Note that each of the classes mentioned in Theorem 14 contains P, so here we can apply Corollary 13. While for PSPACE and the Δ-levels of the PH, autoreducibility and mitoticity results are known, we obtain new autoreducibility results for the Σ-levels and the Π-levels of the PH, including NP and coNP.

Corollary 15. *Let r be one of the reductions $\leq^{\log}_m$, $\leq^{\log}_{1\text{-tt}}$, $\leq^{\log}_{dtt}$, $\leq^{\log}_{l\text{-dtt}}$ for $l \geq 2$. Then every nontrivial set that is r-complete set for one of the following classes is $\leq^{\log}_T$-autoreducible: NP, coNP, Σ^P_k, Π^P_k, 1NP, the levels of the Boolean hierarchy over NP, the levels of the MODPH hierarchy.*

6 Summary and Conclusion

We obtained that for complete sets of most complexity classes of interest, $\leq^{\log}_m$-autoreducibility implies weak $\leq^{\log}_{2\text{-dtt}}$-mitoticity, and $\leq^{\log}_{1\text{-tt}}$-autoreducibility implies weak $\leq^{\log}_{2\text{-tt}}$-mitoticity. These results apply to classes such as P, the Δ-levels of the PH, and NEXP, where the logspace mitoticity of complete sets is not known. Our proof technique follows traces of autoreduction functions and labels their elements with a separator set of low complexity in such a way that after very few steps, there must be a change in the separator membership. With small modifications, our technique also shows that $\leq^P_{dtt}$-complete sets for PSPACE are weakly $\leq^P_{dtt}$-mitotic. It remains an open question whether this technique can be improved to show (strong) mitoticity results or many-one mitoticity results.

The latter would be particularly interesting for NEXP, where $\leq_m^{\log}$-mitoticity of complete sets is still open. Also, can we obtain analogous results for $\leq_{ctt}^p$-complete sets for PSPACE?

We further obtain that all $\leq_m^{\log}$-complete sets for NP and all other classes of the PH are $\leq_T^{\log}$-autoreducible, and this can be extended to further reducibilities. Our proof builds on known results on polynomial-time autoreducibility and mitoticity. It seems to be difficult to obtain a short self-contained proof, because on the one hand, the classes of the polynomial hierarchy are too weak to simulate arbitrary logspace reductions, and hence diagonalization techniques do not apply here, yet the classes are complex enough such that logspace reductions cannot verify their computations (for instance, in logspace, we cannot simulate an accepting NP computation path). Note that the $\leq_T^{\log}$-autoreduction uses multiple oracle tapes. It remains open whether the number of oracle tapes can be reduced such that we obtain $\leq_{tt}^{\log}$-autoreducibility.

References

1. Ambos-Spies, K.: P-mitotic sets. In: Börger, E., Rödding, D., Hasenjaeger, G. (eds.) Rekursive Kombinatorik 1983. LNCS, vol. 171, pp. 1–23. Springer, Heidelberg (1984)
2. Beigel, R., Feigenbaum, J.: On being incoherent without being very hard. Computational Complexity 2, 1–17 (1992)
3. Berman, L.: Polynomial Reducibilities and Complete Sets. PhD thesis, Cornell University, Ithaca, NY (1977)
4. Buhrman, H.: Resource Bounded Reductions. PhD thesis, University of Amsterdam (1993)
5. Buhrman, H., Fortnow, L., van Melkebeek, D., Torenvliet, L.: Separating complexity classes using autoreducibility. SIAM J. Comput. 29(5), 1497–1520 (2000)
6. Buhrman, H., Hoene, A., Torenvliet, L.: Splittings, robustness, and structure of complete sets. SIAM J. Comput. 27(3), 637–653 (1998)
7. Buhrman, H., Torenvliet, L.: A Post's program for complexity theory. Bulletin of the EATCS 85, 41–51 (2005)
8. Cole, R., Vishkin, U.: Deterministic coin tossing with applications to optimal parallel list ranking. Information and Control 70(1), 32–53 (1986)
9. Ganesan, K., Homer, S.: Complete problems and strong polynomial reducibilities. SIAM J. Comput. 21(4), 733–742 (1992)
10. Glaßer, C.: Space-efficient informational redundancy. J. Comput. System Sci. 76(8), 792–811 (2010)
11. Glaßer, C., Nguyen, D.T., Reitwießner, C., Selman, A.L., Witek, M.: Autoreducibility of complete sets for log-space and polynomial-time reductions. In: Fomin, F.V., Freivalds, R., Kwiatkowska, M., Peleg, D. (eds.) ICALP 2013, Part I. LNCS, vol. 7965, pp. 473–484. Springer, Heidelberg (2013)
12. Glaßer, C., Ogihara, M., Pavan, A., Selman, A.L., Zhang, L.: Autoreducibility, mitoticity, and immunity. J. Comput. System Sci. 73(5), 735–754 (2007)
13. Glaßer, C., Pavan, A., Selman, A.L., Zhang, L.: Splitting NP-complete sets. SIAM J. Comput. 37(5), 1517–1535 (2008)
14. Glaßer, C., Witek, M.: Autoreducibility and mitoticity of logspace-complete sets for NP and other classes. Electronic Colloquium on Computational Complexity (ECCC) 20, 188 (2013)

15. Homer, S., Kurtz, S.A., Royer, J.S.: On 1-truth-table-hard languages. Theoretical Computer Science 115(2), 383–389 (1993)
16. Ladner, R.E.: Mitotic recursively enumerable sets. Journal of Symbolic Logic 38(2), 199–211 (1973)
17. Ladner, R.E., Lynch, N.A.: Relativization of questions about log space computability. Mathematical Systems Theory 10, 19–32 (1976)
18. Lynch, N.A.: Log space machines with multiple oracle tapes. Theoretical Computer Science 6, 25–39 (1978)
19. Nguyen, D.T., Selman, A.L.: Non-autoreducible sets for NEXP. In: Proceedings of the 31st Symposium on Theoretical Aspects of Computer Science (STACS). Leibniz International Proceedings in Informatics (LIPIcs), pp. 590–601. Springer (2014)
20. Trakhtenbrot, B.: On autoreducibility. Dokl. Akad. Nauk SSSR 192(6), 1224–1227 (1970); Translation in Soviet Math. Dokl. 11(3), 814–817 (1970)

Editing to a Connected Graph of Given Degrees[*]

Petr A. Golovach[1,2]

[1] Department of Informatics, University of Bergen, Norway
[2] Steklov Institute of Mathematics at St.Petersburg,
Russian Academy of Sciences, Russia
petr.golovach@ii.uib.no

Abstract. The aim of edge editing or modification problems is to change a given graph by adding and deleting of a small number of edges in order to satisfy a certain property. We consider the EDGE EDITING TO A CONNECTED GRAPH OF GIVEN DEGREES problem that for a given graph G, non-negative integers d, k and a function $\delta \colon V(G) \to \{1, \ldots, d\}$, asks whether it is possible to obtain a connected graph G' from G such that the degree of v is $\delta(v)$ for any vertex v by at most k edge editing operations. As the problem is NP-complete even if $\delta(v) = 2$, we are interested in the parameterized complexity and show that EDGE EDITING TO A CONNECTED GRAPH OF GIVEN DEGREES admits a polynomial kernel when parameterized by $d + k$. For the special case $\delta(v) = d$, i.e., when the aim is to obtain a connected d-regular graph, the problem is shown to be fixed parameter tractable when parameterized by k only.

1 Introduction

The aim of graph editing or modification problems is to change a given graph as little as possible by applying specified operations in order to satisfy a certain property. Standard operations are vertex deletion, edge deletion, edge addition and edge contraction, but other operations are considered as well. Various problems of this type are well-known and widely investigated. For example, such problems as CLIQUE, INDEPENDENT SET, FEEDBACK (EDGE OR VERTEX) SET, CLUSTER EDITING and many others can be seen as graph editing problems. Probably the most extensively studied variants are the problems for hereditary properties. In particular, Lewis and Yannakakis [7] proved that for any non-trivial (in a certain sense) hereditary property, the corresponding vertex-deletion problem is NP-hard. The edge-deletion problems were considered by Yannakakis [13] and Alon, Shapira and Sudakov [1]. The case where edge additions and deletions are allowed and the property is the inclusion in some hereditary graph class was considered by Natanzon, Shamir and Sharan [10] and Burzyn, Bonomo and Durán [2]. The results by Cai [3] and Khot and

[*] The research leading to these results has received funding from the European Research Council under the European Union's Seventh Framework Programme (FP/2007-2013) / ERC Grant Agreement n. 267959 and and the Government of the Russian Federation (grant 14.Z50.31.0030).

E. Csuhaj-Varjú et al. (Eds.): MFCS 2014, Part II, LNCS 8635, pp. 324–335, 2014.
© Springer-Verlag Berlin Heidelberg 2014

Raman [6] give a characterization of the parameterized complexity. For non-hereditary properties, a great deal less is known.

Moser and Thilikos in [9] and Mathieson and Szeider [8] initiated a study of the parameterized complexity of graph editing problems where the aim is to obtain a graph that satisfies degree constraints. Mathieson and Szeider [8] considered different variants of the following problem:

EDITING TO A GRAPH OF GIVEN DEGREES
   Instance: A graph G, non-negative integers d, k and a function
      $\delta\colon V(G) \to \{0,\dots,d\}$.
Parameter 1: d.
Parameter 2: k.
   Question: Is it possible to obtain a graph G' from G such that
      $d_{G'}(v) = \delta(v)$ for each $v \in V(G')$ by at most k operations
      from the set S?

They classified the parameterized complexity of the problem for

$$S \subseteq \{\text{vertex deletion}, \text{edge deletion}, \text{edge addition}\}.$$

In particular, they proved that if all the three operations are allowed, then EDITING TO A GRAPH OF GIVEN DEGREES is *Fixed Parameter Tractable* (FPT) when parameterized by d and k. Moreover, the FPT result holds for a more general version of the problem where vertices and edges have costs and the degree constraints are relaxed: for each $v \in V(G')$, $d_{G'}(v)$ should be in a given set $\delta(v) \subseteq \{1,\dots,d\}$. Mathieson and Szeider also showed that EDITING TO A GRAPH OF GIVEN DEGREES is polynomial time solvable even if d and k are a part of the input when only edge deletions and edge additions are allowed.

We are interested in the following natural variant:

EDGE EDITING TO A CONNECTED GRAPH OF GIVEN DEGREES
   Instance: A graph G, non-negative integers d, k and a function
      $\delta\colon V(G) \to \{0,\dots,d\}$.
Parameter 1: d.
Parameter 2: k.
   Question: Is it possible to obtain a *connected* graph G' from G such that
      $d_{G'}(v) = \delta(v)$ for each $v \in V(G')$ by at most k edge deletion
      and edge addition operations?

We show that this problem is FPT when parameterized by d and k in Section 3 by demonstrating a polynomial kernel of size $O(kd^3(k+d)^2)$. For the special case $\delta(v) = d$ for $v \in V(G)$, we call the problem EDGE EDITING TO A CONNECTED REGULAR GRAPH. We prove that this problem is FPT even if it is parameterized by k only in Section 4. Due to space restrictions, proofs are either omitted or just sketched in this extended abstract. The full version of the paper is available at [5].

2   Basic Definitions and Preliminaries

Graphs. We consider only finite undirected graphs without loops or multiple edges. The vertex set of a graph G is denoted by $V(G)$ and the edge set is denoted by $E(G)$.

For a set of vertices $U \subseteq V(G)$, $G[U]$ denotes the subgraph of G induced by U, and by $G - U$ we denote the graph obtained from G by the removal of all the vertices of U, i.e., the subgraph of G induced by $V(G) \setminus U$. For a non-empty set U, $\binom{U}{2}$ is the set of unordered pairs of distinct elements of U. Also for $S \subseteq \binom{V(G)}{2}$, we say that $G[S]$ is induced by S, if S is the set of edges of $G[S]$ and the vertex set of $G[S]$ is the set of vertices of G incident to the pairs from S. By $G - S$ we denote the graph obtained from G by the removal of all the edges of $S \cap E(G)$. Respectively, for $S \subseteq \binom{V(G)}{2}$, $G + S$ is the graph obtained from G by the addition the edges that are elements of $S \setminus E(G)$. If $S = \{a\}$, then for simplicity, we write $G - a$ or $G + a$.

For a vertex v, we denote by $N_G(v)$ its *(open) neighborhood*, that is, the set of vertices which are adjacent to v, and for a set $U \subseteq V(G)$, $N_G(U) = (\cup_{v \in U} N_G(v)) \setminus U$. The *closed neighborhood* $N_G[v] = N_G(v) \cup \{v\}$, and for a positive integer r, $N_G^r[v]$ is the set of vertices at distance at most r from v. For a set $U \subseteq V(G)$ and a positive integer r, $N_G^r[U] = \cup_{v \in U} N_G^r[u]$, and $N_G^r(U) = N_G^r[U] \setminus N_G^{r-1}[U]$ if $r \geq 2$. The *degree* of a vertex v is denoted by $d_G(v) = |N_G(v)|$, and $\Delta(G)$ is the maximum degree of G.

A *trail* in G is a sequence $P = v_0, e_1, v_1, e_2, \ldots, e_s, v_s$ of vertices and edges of G such that $v_0, \ldots, v_s \in V(G)$, $e_1, \ldots, e_s \in E(G)$, the edges $e_1, \ldots, e_s$ are pairwise distinct, and for $i \in \{1, \ldots, s\}$, $e_i = v_{i-1} v_i$; v_0, v_s are the *end-vertices* of the trail. A trail is *closed* if $v_0 = v_s$. For $0 \leq i < j \leq s$, we say that $P' = v_i, e_{i+1}, \ldots, e_j, v_j$ is a *segment* of P. A trail is a *path* if $v_0, \ldots, v_s$ are pairwise distinct except maybe v_0, v_s. Sometimes we write $P = v_0, \ldots, v_s$ to denote a trail $P = v_0, e_1, \ldots, e_s, v_s$ omitting edges.

A set of vertices U is a *cut set* of G if $G - U$ has more components than G. A vertex v is a *cut vertex* if $S = \{v\}$ is a cut set. An edge uv is a *bridge* of a connected graph G if $G - uv$ is disconnected. A graph is said to be *unicyclic* if it has exactly one cycle.

A set M of pairwise non-adjacent edges is called a *matching*, and for a bipartite graph G with the given bipartition X, Y of $V(G)$, a matching M is *perfect* (with respect to X) if each vertex of X is incident to an edge of M.

Parameterized Complexity. Parameterized complexity is a two dimensional framework for studying the computational complexity of a problem. One dimension is the input size n and the other is a parameter k. It is said that a problem is *fixed parameter tractable* (or FPT), if it can be solved in time $f(k) \cdot n^{O(1)}$ for some function f. A *kernelization* for a parameterized problem is a polynomial algorithm that maps each instance (x, k) with the input x and the parameter k to an instance (x', k') such that i) (x, k) is a YES-instance if and only if (x', k') is a YES-instance of the problem, and ii) the size of x' is bounded by $f(k)$ for a computable function f. The output (x', k') is called a *kernel*. The function

f is said to be a *size* of a kernel. Respectively, a kernel is *polynomial* if f is polynomial. We refer to the books of Flum and Grohe [4] and Niedermeier [11] for detailed introductions to parameterized complexity.

Solutions of Edge Editing to a Connected Graph of Given Degrees. Let (G, δ, d, k) be an instance of EDGE EDITING TO A CONNECTED GRAPH OF GIVEN DEGREES. Suppose that a connected graph G' is obtained from G by at most k edge deletions and edge additions such that $d_{G'}(v) = \delta(v)$ for $v \in V(G')$. Denote by D the set of deleted edges and by A the set of added edges. We say that (D, A) is a *solution* of EDGE EDITING TO A CONNECTED GRAPH OF GIVEN DEGREES. We also say that the graph $G' = G - D + A$ is obtained by editing with respect to (D, A).

We need the following structural observation about solutions of EDGE EDITING TO A CONNECTED GRAPH OF GIVEN DEGREES. Let (D, A) be a solution for (G, δ, d, k) and let $G' = G - D + A$. We say that a trail $P = v_0, e_1, v_1, e_2, \ldots, e_s, v_s$ in G' is (D, A)-*alternating* if $e_1, \ldots, e_s \subseteq D \cup A$ and for any $i \in \{2, \ldots, s\}$, either $e_{i-1} \in D, e_i \in A$ or $e_{i-1} \in A, e_i \in D$. We say that a (D, A)-alternating trail $P = v_0, e_1, v_1, e_2, \ldots, e_s, v_s$ is *closed*, if $v_0 = v_s$ and s is even. Notice that if $v_0 = v_s$ but s is odd, then such a trail is not closed. Let $H(D, A)$ be the graph with the edge set $D \cup A$, and the vertex set of H consists of vertices of G incident to the edges of $D \cup A$. Let also $Z = \{v \in V(G) | d_G(v) \neq \delta(v)\}$.

Lemma 1. *For any solution* (D, A), *the following holds.*

i) $Z \subseteq V(H(D, A))$.

ii) For any $v \in V(H(D, A)) \setminus Z$, $|\{e \in D | e$ *is incident to* $v\}| = |\{e \in A | e$ *is incident to* $v\}|$.

iii) For any $z \in Z$, $d_G(z) - \delta(z) = |\{e \in D | e$ *is incident to* $z\}| - |\{e \in A | e$ *is incident to* $z\}|$.

iv) The graph $H(D, A)$ *can be covered by a family of edge-disjoint* (D, A)-*alternating trails* $\mathcal{T}$ *(i.e., each edge of* $D \cup A$ *is in the unique trail of* $\mathcal{T}$*) and each non-closed trail in* $\mathcal{T}$ *has its end-vertices in* Z. *Also* $\mathcal{T}$ *can be constructed in polynomial time.*

Hardness of Edge Editing to a Connected Graph of Given Degrees. As we are interested in FPT results, we conclude this section by the observations about the classical complexity of the considered problems. Recall that Mathieson and Szeider proved in [8] that EDITING TO A GRAPH OF GIVEN DEGREES is polynomial time solvable even if d and k are a part of the input when only edge deletions and edge additions are allowed. But if the obtained graph should be connected then the problem becomes NP-complete by an easy reduction from the HAMILTONICITY problem.

Proposition 1. *For any fixed* $d \geq 2$, EDGE EDITING TO A CONNECTED REGULAR GRAPH *is* NP-*complete.*

3 Polynomial Kernel for Edge Editing to a Connected Graph of Given Degrees

In this section we prove the following theorem

Theorem 1. EDGE EDITING TO A CONNECTED GRAPH OF GIVEN DEGREES *has a kernel of size* $O(kd^3(k+d)^2)$.

Proof. Due to space restrictions, we only sketch the construction of the kernel.

Let (G, δ, d, k) be an instance of EDGE EDITING TO A CONNECTED GRAPH OF GIVEN DEGREES. We assume without loss of generality that $d \geq 3$ (otherwise, we simply set $d = 3$). Let $Z = \{v \in V(G) | d_G(v) \neq \delta(v)\}$ and $s = \sum_{v \in V(G)} |d_G(v) - \delta(v)|$.

First, we apply the following rule.

Rule 1. If $|Z| > 2k$ or s is odd or $s > 2k$ or G has at least $k + 2$ components, then stop and return a NO-answer.

It is straightforward to see that the rule is safe, because each edge deletion (addition respectively) decreases (increases respectively) the degrees of two its end-vertices by one. Also it is clear that if G has at least $k+2$ components, then at least $k + 1$ edges should be added to obtain a connected graph.

From now without loss of generality we assume that $|Z| \leq 2k$, $s \leq 2k$ and G has at most $k + 1$ components. Denote by $G_1, \ldots, G_p$ the components of $G - Z$. We show that if a component G_i has a matching M of size at least $k/3$ with end-vertices of the edges at distance at least 3 from Z such that the deletion of M does not destroy the connectivity of G_i, then G_i can be replaced by a small gadget. Notice that we can always find such a matching if $|E(G_i)| - |V(G_i)|$ is sufficiently large, i.e., G_i has many cycles with edges outside $N_G^2[Z]$. The proof is based on Lemma 1 and uses the fact that for a solution (D, A), $H(D, A)$ can be covered by edge-disjoint (D, A)-alternating trails.

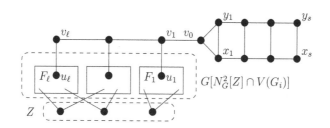

Fig. 1. Modification of G_i by Rule 2

Rule 2. Consider the component G_i of $G - Z$. Let $F_1, \ldots, F_\ell$ be the components of $G[N_G^2[Z] \cap V(G_i)]$. Notice that it can happen that $N_G^2[Z] \cap V(G_i) = \emptyset$, and it is assumed that $\ell = 0$ in this case. If $|V(G_i)| - |N_G^2[Z] \cap V(G_i)| > \ell + 2k + 1$, then do the following.

i) Construct spanning trees of $F_1, \ldots, F_\ell$ and then construct a spanning tree T of G_i that contains the constructed spanning trees of $F_1, \ldots, F_\ell$ as subgraphs.

ii) Let R be the set of edges of $E(G_i) \backslash E(T)$ that are not incident to the vertices of $N_G^2[Z]$. Find a maximum matching M in $G[R]$.

iii) If $|M| \geq k/3$, then modify G and the function δ as follows (see Fig. 1):

  * delete the vertices of $V(G_i) \setminus N_G^2[Z]$;
  * construct vertices $v_0, \ldots, v_\ell, x_1, \ldots, x_s$ and $y_1, \ldots, y_s$ for $s = \lfloor k/3 \rfloor$;
  * for $j \in \{1, \ldots \ell\}$, choose a vertex u_j in F_j adjacent to some vertex in $G_i - V(F_i)$;
  * construct edges $u_1 v_1, \ldots, u_\ell v_\ell, v_0 v_1, \ldots, v_{\ell-1} v_\ell, v_0 x_1, v_0 y_1, x_1 x_2, \ldots,$ $x_{s-1} x_s, y_1 y_2, \ldots, y_{s-1} y_s$ and $x_1 y_1, \ldots, x_s y_s,$
  * set $\delta(v_\ell) = \delta(x_s) = \delta(y_s) = 2$, $\delta(v_0) = \ldots = \delta(v_{\ell-1}) = \delta(x_1) = \ldots = \delta(x_{s-1}) = \delta(y_1) = \ldots = \delta(y_{s-1}) = 3$, $\delta(v) = d_G(v)$ (in the modified graph G) for $V(G_i) \cap N_G^2[Z]$, and δ has the same values as before for all other vertices of G.

We apply Rule 2 for all $i \in \{1, \ldots, p\}$. To simplify notations, assume that (G, δ, d, k) is the obtained instance of EDGE EDITING TO A CONNECTED GRAPH OF GIVEN DEGREES and $G_1, \ldots, G_p$ are the components of $G - Z$. The next rule is applied to components of G that are trees without vertices adjacent to Z. Let $i \in \{1, \ldots, p\}$. We show that if such a component is sufficiently large, then it can be replaced by a path of bounded length.

Rule 3. If G_i is a tree with at least $kd/2 + 1$ vertices and $N_G(Z) \cap V(G_i) = \emptyset$, then replace G_i by a path $P = u_1, \ldots, u_k$ on k vertices and set $\delta(u_1) = \delta(u_k) = 1$ and $\delta(u_2) = \ldots = \delta(u_{k-1}) = 2$.

The next rule is applied to components of G that are unicyclic graphs without vertices adjacent to Z. Let $i \in \{1, \ldots, p\}$. We prove that if such a component is sufficiently large, then it can be replaced by a cycle of bounded length.

Rule 4. If G_i is a unicyclic graph with at least $kd/2$ vertices and $N_G(Z) \cap V(G_i) = \emptyset$, then replace G_i by a cycle $C = u_0, \ldots, u_k$ on k vertices, $u_0 = u_k$, and set $\delta(u_1) = \ldots = \delta(u_k) = 2$.

The Rules 3 and 4 are applied for all $i \in \{1, \ldots, p\}$. We again assume that (G, δ, d, k) is the obtained instance of EDGE EDITING TO A CONNECTED GRAPH OF GIVEN DEGREES and $G_1, \ldots, G_p$ are the components of $G - Z$.

We construct the set of *branch* vertices $B = B_1 \cup B_2$. Let $\hat{G}$ be the graph obtained from G by the recursive deletion of vertices $V(G) \setminus N_G[Z]$ of degree one or zero. A vertex $v \in V(\hat{G})$ is included in B_1 if $d_{\hat{G}}(v) \geq 3$ or $v \in N_{\hat{G}}[Z]$, and v is included in B_2 if $v \notin B_1$, $d_{\hat{G}}(v) = 2$ and there are $x, y \in B_1$ (possibly $x = y$) such that v is in a (x, y)-path of length at most 6.

We apply the following Rules 5 and 6 for each $v \in B$. To construct Rule 5, we show that if a sufficiently large forest without vertices of Z is attached to v as is shown in Fig. 2 (a), then this tree can be replaced by a path of bounded length with one end-vertex in v. For Rule 6, we show that if a sufficiently large tree

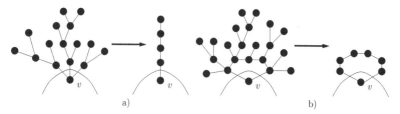

Fig. 2. Modification of G by Rule 5 and Rule 6

without vertices of Z is attached to v by two vertices as it shown in Fig. 2 (b), then this tree can be replaced by a cycle of bounded length that goes through v.

Rule 5. If $v \in B$ is a cut vertex of G, then find all components of $T_1, \ldots, T_\ell$ of $G - v$ such that for $i \in \{1, \ldots, \ell\}$, i) T_i is a tree, ii) $V(T_i) \subseteq V(G) \setminus B$, and iii) T_i has the unique vertex v_i adjacent to v. Let $T = G[V(T_1) \cup \ldots \cup V(T_\ell) \cup \{v\}]$. If the tree T has at least $kd/2 + d^2$ vertices, then replace $T_1, \ldots, T_\ell$ by a path $P = u_0, u_1, \ldots, u_k$, join v and u_0 by an edge, and set $\delta(u_0) = \delta(u_{k-1}) = 2$, $\delta(u_k) = 1$, and $\delta(v) = d_G(v) - \ell + 1$.

Rule 6. If $v \in B$ is a cut vertex of G and there is a component T of $G - v$ such that i) T is a tree, ii) $V(T) \subseteq V(G) \setminus B$, iii) T has exactly two vertices adjacent to v, and iv) $|V(T)| \geq (k/2+2)d+1$, then replace T by a path $P = u_0, \ldots, u_{k+1}$, join v and u_0, u_{k+1} by edges, and set $\delta(u_0) = \ldots = \delta(u_{k+1}) = 2$.

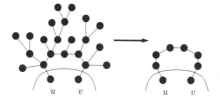

Fig. 3. Modification of G by Rule 7

The next rule is applied to pairs of distinct vertices $u, v \in B$. To construct it, we show that if a sufficiently large tree without vertices of Z is attached to u and v as is shown in Fig. 3, then this tree can be replaced by a path of bounded length with its end-vertices in u and v.

Rule 7. If $\{u, v\} \in B$ is a cut set of G and there is a component T of $G - \{u, v\}$ such that i) T is a tree, ii) $V(T) \subseteq V(G) \setminus B$, iii) T has a unique vertex adjacent to u and a unique vertex adjacent to v and has no vertices adjacent to both u and v, and iv) $|V(T)| \geq (k/2 + 2)d + 1$, then replace T by a path $P = u_0, \ldots, u_{k+1}$, join u with u_0 and v with u_{k+1} by edges, and set $\delta(u_0) = \ldots = \delta(u_{k+1}) = 2$.

It is straightforward to see that Rules 1–7 can be applied in polynomial time. We show that we obtain an equivalent instance of EDGE EDITING TO A CONNECTED GRAPH OF GIVEN DEGREES. To show that we have a polynomial kernel, it can be proved that if (G', δ', d, k) is an instance of EDGE EDITING TO A CONNECTED GRAPH OF GIVEN DEGREES obtained from (G, δ, d, k) by the application of Rules 1–7, then $|V(G')| = O(kd^3(k + d)^2)$. $\qquad \qquad \square$

4   FPT Algorithm for Edge Editing to a Connected Regular Graph

In this section we construct an FPT-algorithm for EDGE EDITING TO A CONNECTED REGULAR GRAPH with the parameter k (i.e., d is not considered to be a parameter). Because of space restrictions, we only sketch the algorithm.

We need the result obtained by Mathieson and Szeider in [8]. Let G be a graph, and let $\rho \colon \binom{V(G)}{2} \to \mathbb{N}$ be a *cost* function that for any two distinct vertices u, v defines the cost $\rho(uv)$ of the addition or deletion of the edge uv. For a set of unordered pairs $X \subseteq \binom{V(G)}{2}$, $\rho(X) = \sum_{uv \in X} \rho(uv)$. Suppose that a graph G' is obtained from G by some edge deletions and additions. Then the *editing cost* is $\rho((E(G) \setminus E(G')) \cup (E(G') \setminus E(G)))$. Mathieson and Szeider considered the following problem:

> EDGE EDITING TO A GRAPH OF GIVEN DEGREES WITH COSTS
> *Instance:* A graph G, a non-negative integer k, a degree function
> $\qquad \delta \colon V(G) \to \mathbb{N}$ and a cost function $\rho \colon \binom{V(G)}{2} \to \mathbb{N}$.
> *Question:* Is it possible to obtain a graph G' from G such that
> $\qquad d_{G'}(v) = \delta(v)$ for each $v \in V(G')$ by edge deletions and
> additions with editing cost at most k?

They proved that EDGE EDITING TO A GRAPH OF GIVEN DEGREES WITH COSTS can be solved in polynomial time.

We also need some results about graphic sequences for bipartite graphs. Let $\alpha = (\alpha_1, \ldots, \alpha_p)$ and $\beta = (\beta_1, \ldots, \beta_q)$ be non-increasing sequences of positive integers. We say that the pair (α, β) is a *bipartite graphic pair* if there is a bipartite graph G with the bipartition of the vertex set $X = \{x_1, \ldots, x_p\}$, $Y = \{y_1, \ldots, y_q\}$ such that $d_G(x_i) = \alpha_i$ for $i \in \{1, \ldots, p\}$ and $d_G(y_j) = \beta_j$ for $j \in \{1, \ldots, q\}$. It is said that G *realizes* (α, β).

Gale and Ryser [12] gave necessary and sufficient conditions for (α, β) to be a bipartite graphic pair. It is more convenient to give them in the terms of partitions of integers. Recall that a non-increasing sequence of positive integers $\alpha = (\alpha_1, \ldots, \alpha_p)$ is a *partition* of n if $\alpha_1 + \ldots + \alpha_p = n$. A sequence $\alpha = (\alpha_1, \ldots, \alpha_p)$ *dominates* $\beta = (\beta_1, \ldots, \beta_q)$ if $\alpha_1 + \ldots + \alpha_i \geq \beta_1 + \ldots + \beta_i$ for all $i \geq 1$; to simplify notations, we assume that $\alpha_i = 0$ ($\beta_i = 0$ respectively) if $i > p$ ($i > q$ respectively). We write $\alpha \trianglerighteq \beta$ to denote that α dominates β. Clearly, if $\alpha \trianglerighteq \beta$ and $\beta \trianglerighteq \gamma$, then $\alpha \trianglerighteq \gamma$. For a partition $\alpha = (\alpha_1, \ldots, \alpha_p)$ of n, the partition $\alpha^* = (\alpha_1^*, \ldots, \alpha_{\alpha_1}^*)$ of n, where $\alpha_j^* = |\{h | 1 \leq h \leq p, \alpha_h \geq j\}|$ for $j \in \{1, \ldots, \alpha_1\}$, is called the *conjugate* partition for α. Notice that $\alpha^{**} = \alpha$. Gale and Ryser [12] proved that a pair of non-increasing sequences of positive integers (α, β) is a bipartite graphic pair if and only if α and β are partitions of some positive integer n and $\alpha^* \trianglerighteq \beta$.

Now we sketch the proof of the following theorem.

Theorem 2. EDGE EDITING TO A CONNECTED REGULAR GRAPH *can be solved in time* $O^*(k^{O(k^3)})$.

Proof. Let (G, d, k) be an instance of EDGE EDITING TO A CONNECTED REGULAR GRAPH. We assume that $k \geq 1$, as otherwise the problem is trivial. If $d \leq 3k + 1$, then we solve the problem in time $O^*(k^{O(k)})$ by Theorem 1. From now it is assumed that $d > 3k + 1$. Let $Z = \{v \in V(G) | d_G(v) \neq d\}$.

First, we check whether $|Z| \leq 2k$ and stop and return a NO-answer otherwise using the observation that each edge deletion (addition respectively) decreases (increases respectively) the degrees of two its end-vertices by one. From now we assume that $|Z| \leq 2k$. Denote by $G_1, \ldots, G_p$ the components of $G - Z$.

We say that two components G_i, G_j have the *same type*, if for any $z \in Z$, either $|N_G(z) \cap V(G_i)| = |N_G(z) \cap V(G_j)| \leq k$ or $|N_G(z) \cap V(G_i)| > k$ and $|N_G(z) \cap V(G_j)| > k$. Denote by $\Theta_1, \ldots, \Theta_t$ the partition of $\{G_1, \ldots, G_p\}$ into classes according to this equivalence relation. Observe that the number of distinct types is at most $(k + 2)^{2k}$. Notice also that for any solution (D, A), the graph $H(D, A)$ contains vertices of at most $2k$ components $G_1, \ldots, G_p$.

The general idea of the algorithm is to guess the structure of a possible solution (D, A) (if it exists). We guess the edges of D and A that join the vertices of Z. Then we guess the number and the types of components of $G - Z$ that contain vertices of $H(D, A)$. For them, we guess the number of edges that join these components with each other and with each vertex of Z. Notice that the edges of A between distinct components of $G - Z$ should form a bipartite graph. Hence, we guess some additional conditions that ensure that such a graph can be constructed. Then for each guess, we check in polynomial time whether we have a solution that corresponds to it. The main ingredient here is the fact that we can modify the components of $G - Z$ without destroying their connectivity. We construct partial solutions for some components of $G - Z$ and then "glue" them together.

Let $Z = \{z_1, \ldots, z_r\}$. We define records $L = (s, \Theta, C, R, D_Z, A_Z)$, where

- $0 \leq s \leq \min\{2k, p\}$ is an integer,
- Θ is an s-tuple $(\tau_1, \ldots, \tau_s)$ of integers and $1 \leq \tau_1 \leq \ldots \leq \tau_s \leq t$;
- C is a $s \times s$ table of bipartite graphic pairs $(\alpha_{j,h}, \beta_{j,h})$ with the sum of elements of $\alpha_{j,h}$ denoted $c_{j,h}$ such that $\alpha_{j,h} = \beta_{h,j}$, $0 \leq c_{j,h} \leq k$ and $c_{j,j} = 0$ for $j, h \in \{1, \ldots, s\}$, notice that it can happen that $c_{j,h} = 0$ and it is assumed that $(\alpha_{j,h}, \beta_{j,h}) = (\emptyset, \emptyset)$ in this case;
- R is $r \times s$ integer matrix with the elements $r_{j,h}$ such that $-k \leq r_{j,h} \leq k$ for $j \in \{1, \ldots, r\}$ and $h \in \{1, \ldots, s\}$;
- $D_Z \subseteq E(G[Z])$; and
- $A_Z \subseteq \binom{Z}{2} \setminus E(G[Z])$.

Let (D, A) be a solution for (G, d, k). We say that (D, A) *corresponds* to L if

i) the graph $H(D, A)$ contains vertices from exactly s components $G_{i_1}, \ldots, G_{i_s}$ of $G - Z$;

ii) $G_{i_j} \in \Theta_{\tau_j}$ for $j \in \{1, \ldots, s\}$;

iii) for $j, h \in \{1, \ldots, s\}$, A has exactly $c_{j,h}$ edges between G_{i_j} and G_{i_h} if $j \neq h$;

iv) for any $j \in \{1, \ldots, r\}$ and $h \in \{1, \ldots, s\}$, $|\{z_j x \in A | x \in V(G_{i_h})\}| - |\{z_j x \in D | x \in V(G_{i_h})\}| = r_{j,h}$;

v) $D \cap E(G[Z]) = D_Z$;

vi) $A \cap \binom{Z}{2} = A_Z$.

It is straightforward to verify that the number of all possible records L is at most $k^{O(k^3)}$. We consider all such records, and for each L, we check whether (G, d, k) has a solution that corresponds to L. If we find a solution for some L, then we stop and return it. Otherwise, if we fail to find any solution, we return a NO-answer. From now we assume that L is given.

For $i \in \{1, \ldots, p\}$, a given r-tuple $Q = (q_1, \ldots, q_r)$ and ℓ-tuple $Q' = (q'_1, \ldots, q'_\ell)$, where $-k \leq q_1, \ldots, q_r \leq k$, $\ell \leq k$ and $1 \leq w_1, \ldots, w_\ell \leq k$, we consider an auxiliary instance $\Pi(i, Q, Q')$ of EDGE EDITING TO A GRAPH OF GIVEN DEGREES WITH COSTS defined as follows. We consider the graph $G[Z \cup V(G_i)]$, delete the edges between the vertices of Z, and add a set of ℓ isolated vertices $W = \{w_1, \ldots, w_\ell\}$. Each vertex w_j, we say that it *corresponds* to q'_j for $j \in \{1, \ldots, \ell\}$. Denote the obtained graph by F_i. We set $\delta(v) = d$ if $v \in V(G_i)$, $\delta(z_j) = d_{F_i}(z_j) + q_j$ for $j \in \{1, \ldots, r\}$, and $\delta(w_j) = q'_j$ for $j \in \{1, \ldots, \ell\}$. We set $\rho(uv) = k + 1$ if $u, v \in Z \cup W$, and $\rho(uv) = 1$ for all other pairs of vertices of $\binom{V(F_i)}{2}$. Observe that it can happen that $\delta(z_j) < 0$ for some $j \in \{1, \ldots, r\}$. In this case we assume that $\Pi(i, Q, Q')$ is a NO-instance. In all other cases we solve $\Pi(i, Q, Q')$ and find a solution of minimum editing cost $c(i, Q, Q')$ using the result of Mathieson and Szeider in [8]. If we have a NO-instance or $c(i, Q, Q') > k$, then we set $c(i, Q, Q') = +\infty$. We need the following property of the solutions.

Claim 1. *If $c(i, Q, Q') \leq k$, then any solution for $\Pi(i, Q, Q')$ of cost at most k has no edges between vertices of $Z \cup W$ and there is a solution (A, D) for $\Pi(i, Q, Q')$ of cost $c(i, Q, Q') \leq k$ such that if $F' = F_i - D + A$, then any $u, v \in V(G_i)$ are in the same component of F'. Moreover, such a solution can be found in polynomial time.*

Now we are ready to describe the algorithm that for a record $L = (s, \Theta, C, R, D_Z, A_Z)$, checks whether (G, d, k) has a solution that corresponds to L.

First, we check whether the modification of G with respect to L would satisfy the degree restrictions for Z, as otherwise we have no solution. Also the number of edges between $G_1, \ldots, G_p$ should be at most k.

Step 1. Let $\hat{G}$ be the graph obtained from G by the deletion of the edges of D_Z and the addition the edges of A_Z. If for any $j \in \{1, \ldots, r\}$, $d_{\hat{G}}(z_j) + \sum_{h=1}^{s} r_{j,h} \neq d$, then stop and return a NO-answer.

Step 2. If $\sum_{1 \leq j < h \leq s} c_{j,h} > k$, then stop and return a NO-answer.

From now we assume that the degree restrictions for Z are fulfilled and the number of added edges between the components of $G - Z$ should be at most k.

Step 3. Construct an auxiliary weighted bipartite graph F, where $X = \{x_1, \ldots, x_s\}$ and $Y = \{y_1, \ldots, y_p\}$ is the bipartition of the vertex set. For $i \in \{1, \ldots, s\}$ and $j \in \{1, \ldots, p\}$, we construct an edge $x_i y_j$ if $G_j \in \Theta_{\tau_i}$. To define the weight $w(x_i y_j)$, we consider $\Pi(j, Q_j, Q'_j)$ where $Q_j = (r_{1,i}, \ldots, r_{r,i})$ and Q'_j is the sequence obtained by the concatenation of non-empty sequences $\alpha^*_{j,1}, \ldots, \alpha^*_{j,s}$. Denote by $W_{j,h}$ the set of vertices of the graph in $\Pi(j, Q_j, Q'_j)$ corresponding to the elements of $\alpha^*_{j,h}$. Notice that by Step 2, Q'_j has at most k elements. We set $w(x_i y_j) = c(j, Q_j, Q'_j)$. Observe that some edges can have infinite weights.

Step 4. Find a perfect matching M with respect to X in F of minimum weight. If F has no perfect matching of finite weight, then the algorithm stops and returns a NO-answer. Assume that $M = \{x_1 y_{j_1}, \ldots x_s y_{j_s}\}$ is a perfect matching of minimum weight $\mu < +\infty$. If $\mu - \sum_{1 \leq j < h \leq s} c_{j,h} + |D_Z| + |A_Z| > k$, then we stop and return a NO-answer.

Now we assume that M has weight at most k.

Step 5. Consider the solutions (D_i, A_i) of cost $c(j, Q_{j_i}, Q'_{j_i})$ for $\Pi(j_i, Q_{j_i}, Q'_{j_i})$ for $i \in \{1, \ldots, s\}$.
    Set $D = D_Z \cup (\cup_{i=1}^s D_i)$.
    Construct a set A as follows. For $i \in \{1, \ldots, s\}$, denote by A'_i the set of edges of A_i with the both end-vertices in $V(G_{j_i}) \cup Z$, and let $A_{i,h}$ be the subset of edges that join G_{j_i} with W_{j_i, j_h} for $h \in \{1, \ldots, s\}$, $h \neq j$. Initially we include in A the set $\cup_{i=1}^s A'_i$. For each pair of indices $i, h \in \{1, \ldots, s\}$, such that $i < h$ and $c_{j_i, j_h} > 0$, consider graphs induced by $A_{i,h}$ and $A_{h,i}$ respectively, and denote by $\alpha'_{i,h}$ and $\alpha'_{h,i}$ respectively the degree sequences of these graphs for the vertices in G_{j_i} and G_{j_h} respectively. By the construction of the problems $\Pi(j, Q_j, Q'_j)$, $(\alpha'_{i,h}, \alpha^*_{i,h})$ and $(\alpha'_{h,i}, \alpha^*_{h,i})$ are bipartite graphic pairs. Recall that $\beta_{i,h} = \alpha_{h,i}$ and $(\alpha_{i,h}, \beta_{i,h})$ is a bipartite graphic pair. We show that $(\alpha'_{i,h}, \alpha'_{h,i})$ is also a bipartite graphic pair. Construct a bipartite graph that realizes $(\alpha'_{i,h}, \alpha'_{h,i})$ and denote its set of edges by $A'_{i,h}$. We use the vertices of G_{j_i} and G_{j_h} incident with the vertices of $A_{i,h}$ and $A_{h,i}$ as the sets of bipartition and construct our bipartite graph in such a way that for each vertex, the number of edges of $A'_{i,h}$ incident to it is the same as the number of edges of $A_{i,h}$ or $A_{h,i}$ respectively incident to this vertex. Then we include the edges of $A'_{i,h}$ in A.

Step 6. For each $i \in \{1, \ldots, s\}$, do the following. Consider the set of vertices $W_i = \{w_1, \ldots, w_\ell\}$ of $\cup_{h=1}^s V(G_{j_h}) \setminus V(G_{j_i})$ incident to the edges of A that join G_{j_i} with these vertices and let q'_h be the number of edges of A that join G_{j_i} with w_h for $h \in \{1, \ldots, \ell\}$. Consider $\Pi(j_i, Q_i, Q'_i)$ where $Q_i = (r_{1,j_i}, \ldots, r_{r,j_i})$ and $Q'_i = (q'_1, \ldots, q'_\ell)$. Using Claim 1, find a solution (D_i, A_i) for $\Pi(j_i, Q_i, Q'_i)$ of minimum cost. Modify (D, A) by replacing the edges of D and A incident to the vertices of G_{j_i} by the edges of D_i and A_i respectively identifying the set W_i and the set of vertices W in $\Pi(j_i, Q_i, Q'_i)$.

Step 7. Let $G' = G - D + A$. If G is connected, then return (D, A). Otherwise return a NO-answer.   □

5 Conclusions

We proved that EDITING TO A CONNECTED GRAPH OF GIVEN DEGREES has a polynomial kernel of size $O(kd^3(k + d)^2)$. It is natural to ask whether the size can be improved. Also, is the problem FPT when parameterized by k only? We proved that it holds for the special case $\delta(v) = d$, i.e., for EDGE EDITING TO A CONNECTED REGULAR GRAPH. Another open question is whether EDITING TO A GRAPH OF GIVEN DEGREES (or EDGE EDITING TO A CONNECTED REGULAR GRAPH) has a polynomial kernel with the size that depends on k only.

References

1. Alon, N., Shapira, A., Sudakov, B.: Additive approximation for edge-deletion problems. In: FOCS, pp. 419–428. IEEE Computer Society (2005)
2. Burzyn, P., Bonomo, F., Durán, G.: NP-completeness results for edge modification problems. Discrete Applied Mathematics 154(13), 1824–1844 (2006)
3. Cai, L.: Fixed-parameter tractability of graph modification problems for hereditary properties. Inf. Process. Lett. 58(4), 171–176 (1996)
4. Flum, J., Grohe, M.: Parameterized complexity theory. Texts in Theoretical Computer Science. An EATCS Series. Springer, Berlin (2006)
5. Golovach, P.A.: Editing to a connected graph of given degrees. CoRR abs/1308.1802 (2013)
6. Khot, S., Raman, V.: Parameterized complexity of finding subgraphs with hereditary properties. Theor. Comput. Sci. 289(2), 997–1008 (2002)
7. Lewis, J.M., Yannakakis, M.: The node-deletion problem for hereditary properties is np-complete. J. Comput. Syst. Sci. 20(2), 219–230 (1980)
8. Mathieson, L., Szeider, S.: Editing graphs to satisfy degree constraints: A parameterized approach. J. Comput. Syst. Sci. 78(1), 179–191 (2012)
9. Moser, H., Thilikos, D.M.: Parameterized complexity of finding regular induced subgraphs. J. Discrete Algorithms 7(2), 181–190 (2009)
10. Natanzon, A., Shamir, R., Sharan, R.: Complexity classification of some edge modification problems. Discrete Applied Mathematics 113(1), 109–128 (2001)
11. Niedermeier, R.: Invitation to fixed-parameter algorithms. Oxford Lecture Series in Mathematics and its Applications, vol. 31. Oxford University Press, Oxford (2006)
12. Ryser, H.J.: Combinatorial mathematics. The Carus Mathematical Monographs, vol. 14. Published by The Mathematical Association of America (1963)
13. Yannakakis, M.: Node- and edge-deletion NP-complete problems. In: Lipton, R.J., Burkhard, W.A., Savitch, W.J., Friedman, E.P., Aho, A.V. (eds.) STOC, pp. 253–264. ACM (1978)

Circuit Complexity of Properties of Graphs with Constant Planar Cutwidth

Kristoffer Arnsfelt Hansen[1,*], Balagopal Komarath[2,**], Jayalal Sarma[2],
Sven Skyum[1], and Navid Talebanfard[1,*]

[1] Aarhus University, Denmark
[2] IIT Madras, Chennai, India

Abstract. We study the complexity of several of the classical graph decision problems in the setting of bounded cutwidth and how imposing planarity affects the complexity. We show that for 2-coloring, for bipartite perfect matching, and for several variants of disjoint paths, the straightforward NC^1 upper bound may be improved to $\mathsf{AC}^0[2]$, ACC^0, and AC^0 respectively for bounded planar cutwidth graphs. We obtain our upper bounds using the characterization of these circuit classes in tems of finite monoids due to Barrington and Thérien. On the other hand we show that 3-coloring and Hamilton cycle remain hard for NC^1 under projection reductions, analogous to the NP-completeness for general planar graphs. We also show that 2-coloring and (non-bipartite) perfect matching are hard under projection reductions for certain subclasses of $\mathsf{AC}^0[2]$. In particular this shows that our bounds for 2-coloring are quite close.

1 Introduction

We consider several of the classical graph decision problems, namely those of deciding existence of 2- and 3-colorings, perfect matchings, Hamiltonian cycles, and disjoint paths. For these problems we are interested in their complexity in the setting of bounded *planar cutwidth*. The *cutwidth* of a graph $G = (V, E)$ with $n = |V|$ vertices is defined in terms of linear arrangements of the vertices. A linear arrangement is simply a 1-1 map $f : V \to \{1, \ldots, n\}$, and its cutwidth is the maximum over i of the number of edges between $V_i = \{v \in V \mid f(v) \leq i\}$ and $V \setminus V_i$. The cutwidth of G is the minimum cutwidth of a linear arrangement. Similarly, if the graph G is planar we can define a notion of *planar cutwidth*. Given a linear arrangement f we consider a planar embedding where vertex v is placed at coordinate $(f(v), 0)$. The planar cutwidth of this embedding is then the maximum number of edge-crossings at a vertical line in the plane. We define the planar cutwidth as the minimum planar cutwidth of such a linear arrangement

---

[*] Hansen and Talebanfard acknowledge support from the Danish National Research Foundation and The National Science Foundation of China (under the grant 61061130540) for the Sino-Danish Center for the Theory of Interactive Computation, within which this work was performed.

[**] Supported by the TCS PhD Fellowship.

E. Csuhaj-Varjú et al. (Eds.): MFCS 2014, Part II, LNCS 8635, pp. 336–347, 2014.

and an embedding. We wish to stress that the planar cutwidth of a planar graph is in general not the same as its cutwidth. In particular there are simple examples of n-vertex graphs of constant cutwidth having planar cutwidth $\Omega(n)$.

All the problems we consider can be decided in NC^1 for graphs of bounded cutwidth, and they are in fact NC^1-complete under projection reductions. Imposing planarity, or more precisely considering graphs of bounded planar cutwidth, we are able to place several of the problems in smaller classes such as AC^0, $AC^0[2]$, and ACC^0, while for some problems they remain NC^1-complete.

Before stating our results we review known complexity results about the graph problems without restriction on cutwidth and the consequences of imposing planarity, for comparison with our results in the bounded cutwidth setting. The 2-coloring problem is in L, as an easy consequence of Reingold's algorithm for undirected connectivity [15], whereas 3-coloring is NP-complete and remains so for planar graphs by the existence of a cross-over gadget [10]. The complexity of deciding if a graph has a perfect matching is still not known. It belongs to P, but it is an open problem whether it belongs to NC. For planar graphs the problem is known to be in NC as shown by Vazirani based on work of Kasteleyn [13,18]; for planar bipartite graphs the problem was shown to be in UL by Datta et al. [6]. The Hamiltonian cycle problem is NP-complete and as shown by Garey et al. it remains so for planar graphs [11], and Itai et al. showed it is NP-hard even for grid graphs [12].

The disjoint paths problem has numerous variations. In the general setting we are given pairs of vertices $(s_1, t_1), \ldots, (s_k, t_k)$ in a graph G, and are to decide whether disjoint paths between s_i and t_i for each i exists. Here disjoint may mean either vertex-disjoint or edge-disjoint, but either variant is reducible to the other. We shall consider only the case of constant k. When G is an undirected graph a polynomial time algorithm was given by Robertson and Seymour [16], as a result arising from their seminal work on graph minors. When G is a directed graph the problem is NP-complete already for $k = 2$ as shown by Fortune et al. [9]. On the other hand, when G is planar Schrijver [17] gave a polynomial time algorithm for the vertex-disjoint paths problem. The reduction between the vertex-disjoint and the edge-disjoint versions of the problem does not preserve planarity, and it is an open problem whether the edge-disjoint paths problem in planar directed graphs is NP-complete or solvable in polynomial time [5]. It can however be solved in polynomial time for (not necessarily planar) directed acyclic graphs [9].

1.1   Results and Techniques

A convenient way to obtain an NC^1 upper bound is through monadic second order (MSO) logic. Elberfeld et al. [7] showed that MSO-definable problems can be decided in NC^1 when restricted to input structures of bounded treewidth, and when a tree decomposition of bounded width is supplied in the so-called term representation. We shall not formally define the tree-width of a graph, but we will note that the treewidth of a graph is bounded from above by the cutwidth of the graph [4]. Furthermore, given as input a linear arrangement of bounded cutwidth k, a tree decomposition of tree width k can be constructed by an AC^0 circuit. Thus we have the following meta-theorem as an easy consequence.

Theorem 1. *Any graph property definable in monadic second order logic with quantification over sets of vertices and edges can be decided by NC^1 circuits on graphs of bounded cutwidth if a linear arrangement of bounded cutwidth is supplied as auxiliary input.*

All the graph properties we consider can easily be expressed in monadic second order logic, thereby establishing NC^1 upper bounds. We can show that all these problems are in fact also hard for NC^1 under projection reductions. This is based on Barrington's characterization of NC^1 in terms of bounded width permutation branching programs [2].

We first discuss the general technique behind our upper bounds that improve upon the generic NC^1 bound. Namely our upper bounds are based on reducing to *word problems* on appropriately defined finite monoids. By results of Barrington and Thérien, we then directly get circuit upper bounds depending on the group structure of the given monoid. For general graphs the improved complexity bounds obtained when imposing planarity are obtained by very different algorithms. In our setting of constant cutwidth, when imposing planarity we instead obtain the improvements in a uniform way by obtaining an *algebraic* understanding of the respective problems. The general idea is as follows. We consider grid-layered planar graphs (defined later) of a fixed width w for which we want to decide a certain graph property, and we may view these as a free semigroup under concatenation. We then define an appropriate finite monoid $\mathcal{M}$. For each grid-layered planar graph G we associate a monoid element $G^{\mathcal{M}}$. In the simplest setting we will be able to determine if the graph property under consideration holds for the graph G directly from the monoid element $G^{\mathcal{M}}$. We will also have defined the elements of $\mathcal{M}$ and the monoid operation in such a way that the map $G \mapsto G^{\mathcal{M}}$ is a homomorphism. What then remains is to analyze the groups inside $\mathcal{M}$. For the disjoint paths problem we show that all groups are trivial, and this gives AC^0 circuits. For 2-coloring we characterize the groups as being isomorphic to groups of the form $\mathbb{Z}_2^l$, and this gives $\mathsf{AC}^0[2]$ circuits. For perfect matching in bipartite graphs we are not able to fully analyze the groups of the corresponding monoid. We are however able to rule out groups of order 2, and thus by the celebrated Feit-Thompson theorem all remaining groups must be solvable, and this gives ACC^0 circuits.

When considering the graph properties for graphs of bounded planar cutwidth we supply as additional input the corresponding embedding of bounded cutwidth of the graph. But before dealing with this issue, we consider special classes of such graphs where such an embedding is implicit. We consider a grid $\Lambda = \{1, \ldots, \ell\} \times \{1, \ldots, w\}$ of *width* w and *length* l. A *grid graph* $G = (V, E)$ of width w and length ℓ is a graph where $V \subseteq \Lambda$ and all edges are of Euclidean length 1. We think of the vertices with the same first coordinate to be in the same *layer*. A grid graph with (planar) diagonals allows edges of Euclidean length < 2, but no crossing edges. We relax these requirements further, defining the class of constant width *grid-layered planar* graphs. A grid-layered planar graph

$G = (V, E)$ of width w and length ℓ is a graph embedded in the plane with no edge-crossings, with $V \subseteq \Lambda$ and if two vertices (a, b) and (c, d) are connected by an edge, then $|a - c| \leq 1$ and the edge is fully contained in the region $[a - 1, a] \times [1, w]$ or the region $[a, a + 1] \times [1, w]$. If we consider bipartite grid-layered planar graphs we assume that the bipartition is defined by the parities of the sums of coordinates of each vertex. All our lower bounds hold for grid graphs or grid graphs with diagonals, and all our circuit upper bounds hold for grid-layered planar graphs.

Just as 3-coloring and Hamiltonian cycle remain NP-complete for planar graphs, 3-coloring remains hard for NC^1 on constant width grid graphs with diagonals and Hamiltonian cycle remains hard for NC^1 on constant width grid graphs. We show that 2-coloring on constant width grid-layered planar graphs is in $AC^0[2]$. This is complemented by an AND $\circ$ XOR $\circ$ AC^0 lower bound for grid graphs with diagonals. This lower bound is in some sense not far from the $AC^0[2]$ upper bound. Namely by the approach of Razborov [14] we have that quasipolynomial size randomized XOR$\circ$AND is equal to quasipolynomial $AC^0[2]$. Furthermore Allender and Hertrampf [1] show that in fact quasipolynomial size AND $\circ$ OR $\circ$ XOR $\circ$ AND is equal to quasipolynomial size $AC^0[2]$.

We show that perfect matching is in ACC^0 for bipartite grid-layered planar graphs, and we have an AC^0 lower bound. For non-bipartite grid-layered planar graphs we have a AND $\circ$ OR $\circ$ XOR $\circ$ AND lower bound. For the disjoint paths problem in constant width grid-layered planar graphs we give AC^0 upper bounds for the following 3 settings: (1) node-disjoint paths in directed graphs. (2) edge-disjoint paths in upward planar graphs. (3) edge-disjoint paths in undirected graphs. We leave open the case of edge-disjoint paths in directed graphs. For all the settings we have an AC^0 lower bound. All these results are summarized in the following table.

Problem	Upper bound	Lower bound (projections)
2-coloring	$AC^0[2]$	AND $\circ$ XOR $\circ$ AC^0
3-coloring	NC^1	NC^1
Bipartite perfect matching	ACC^0	AC^0
Perfect matching	NC^1	AND $\circ$ OR $\circ$ XOR $\circ$ AC^0
Hamiltonian cycle	NC^1	NC^1
Disjoint paths variants	AC^0	AC^0

We now briefly discuss extending the upper bounds above from constant width grid-layered planar graphs to the larger classes of graphs of bounded planar cutwidth. Whereas the embedding was implicitly given for grid-layered planar graphs, for graphs of bounded planar cutwidth we will supply a representation of the embedding in addition to the linear arrangement of the vertices. A simple way to represent both the linear arrangement and the planar embedding of a graph $G = (V, E)$ of bounded planar cutwidth is to provide instead a grid-layered planar graph $G' = (V', E')$ with $V \subseteq V'$, where the vertices V are placed on a horizontal line and the vertices $V' \setminus V$ are dummy vertices describing the embedding of the edges. When given this representation as input, our upper bounds are easily adapted. Namely, for the disjoint paths problems an edge can be replaced by a path, and we may simply promote the dummy vertices to regular vertices. For 2-coloring and perfect matching an edge can be replaced by a path of odd length, but this can be done by an AC^0 circuit making locally use of the coordinates of vertices. Namely, we can just ensure that the path alternates between vertices of the implicit bipartition, except possibly at the end.

In this extended abstract we shall cover only the upper bounds for 2-coloring and bipartite perfect matching as well as outline some ideas for the disjoint paths problems. We refer to the full version of the paper for the remaining upper bounds as well as the lower bounds.

2   Preliminaries

Boolean circuits. We give here standard definitions of the Boolean functions and circuit classes we consider. As is usual, when considering a Boolean function $f : \{0,1\}^n \to \{0,1\}$, unless otherwise specified we always have a family of such functions in mind, one for each input length. AC^0 is the class of polynomial size constant depth circuits built from unbounded fanin AND and OR gates. $AC^0[m]$ allows in addition the function MOD_m given by $MOD_m(x_1, \ldots, x_k) = 1$ if and only if $\sum_{i=1}^{k} x_i \not\equiv 0 \pmod{m}$. We shall also denote the function MOD_2 by XOR. The union of $AC^0[m]$ for all m is the class ACC^0. NC^1 is the class of polynomial size circuits of depth $O(\log n)$ built from fanin 2 AND and OR gates.

A class of Boolean functions immediately defines a class of Boolean circuits as families of single gate circuits. Given two classes of circuits C_1 and C_2 we denote by $C_1 \circ C_2$ the class of circuits consisting of circuits from C_1 that is fed as inputs the output of circuits from C_2. For instance, $AND \circ XOR \circ AC^0$ is the class of polynomial size constant depth circuits that has an AND gate at the output, followed by XOR gates that in turn take as inputs the output of AC^0 circuits.

Semigroups, monoids and programs. A semigroup is a set S with an associative binary operation. A monoid $\mathcal{M}$ is a semigroup with a two-sided identity. A subset $\mathcal{G}$ of $\mathcal{M}$ is a group in $\mathcal{M}$ if it is a group with respect to the operation of $\mathcal{M}$. We also say that $\mathcal{M}$ contains $\mathcal{G}$. A monoid is aperiodic if every group it contains is trivial; it is solvable if every group it contains is solvable. A monoid which is not solvable is called unsolvable.

Barrington and Thérien [3] showed that several circuit classes are exactly captured by so-called programs over finite monoids of polynomial length. We shall use only one direction of this characterization, and for this reason it is convenient to reformulate as follows. Let $\mathcal{M}$ be a finite monoid. The *word problem* over $\mathcal{M}$ is to compute the product $x_1 \cdots x_m$ when given as input $x_1, \ldots, x_m \in \mathcal{M}$. Then the results of Barrington and Thérien have the following consequences: When $\mathcal{M}$ is aperiodic the word problem is in AC^0, when $\mathcal{M}$ is solvable and all groups in $\mathcal{M}$ have orders dividing an m-power[1] the word problem is in $AC^0[m]$, when $\mathcal{M}$ is solvable the word problem is in ACC^0, and we always have the word problem is in NC^1.

---

[1] This consequence is not stated explicitly in [3], but follows from the given proof.

3   Upper Bounds

We first state a geometric lemma that we shall make use of in our results about bipartite matching and disjoint paths. Consider a piecewise smooth infinite simple curve C such that C is contained entirely in the strip $\{(x, y) \mid 1 \le y \le w\}$. We say that C is periodic with period p if the horizontally shifted curve $C+(p, 0)$ coincides with C.

Lemma 2. *Let C be a curve that is periodic with period p and let $C' = C+(q, 0)$ be a horizontal shift of the curve C. Then C and C' intersect.*

3.1   2-Coloring

We prove here our upper bound for 2-coloring.

Theorem 3. *Testing whether a given grid-layered planar graph is 2-colorable can be done in $\mathsf{AC}^0[2]$.*

We prove this result by reducing 2-coloring to the word problem over a finite monoid $\mathcal{M}$. We then show that all groups in $\mathcal{M}$ are solvable and of order a power of 2. By the results of Barrington and Thérien this gives the $\mathsf{AC}^0[2]$ upper bound.

Reduction to a Monoid Word Problem. A grid-layered planar graph G gives rise to a binary relation $\mathcal{R}(G) \subseteq 2^{\{1,\dots,w\}} \times 2^{\{1,\dots,w\}}$. Here $\{1,\dots,w\}$ are the numbering of the vertices in the layers of the graph. We have that $(S, T) \in \mathcal{R}(G)$ if and only if there is a two-coloring of G such that the vertices in the first layer colored 1 is the set S and the vertices in the last layer colored 1 is the set T. Let $\mathcal{M}$ be the monoid of all such relations under normal composition of relations. Let $G \circ H$ denote the concatenation of the graphs G and H. It is not hard to see that $\mathcal{R}(G \circ H) = \mathcal{R}(G)\mathcal{R}(H)$ which shows the soundness of the reduction to the word problem over $\mathcal{M}$. The proof of the upper bound is now completed by the following result.

Proposition 4. *Every group $\mathcal{G} \subseteq \mathcal{M}$ is isomorphic to $\mathbb{Z}_2^\ell$ for some ℓ.*

Proof. For a graph G, let us identify the nodes in the first layer with the set $\{1,\dots,w\}$ and the nodes in the last layer with the set $\{1',\dots,w'\}$.

The observation that makes the proof possible is the following: Suppose G and H are both 2-colorable graphs and $\mathcal{R}(G) = \mathcal{R}(H)$. Then for any $u, v \in \{1,\dots,w\} \cup \{1',\dots,w'\}$ we have that u and v are connected in G if and only if u and v are connected in H. Furthermore if u and v are connected in G by an odd (even) length path then u and v are connected in H by an odd (even) length path.

Assume that G is 2-colorable. Then every connected component of G can be 2-colored in exactly two different ways. This means that $\mathcal{R}(G)$ can be reconstructed from only the following information about G: Which nodes from $\{1,\dots,w\} \cup \{1',\dots,w'\}$ are connected in addition to a single 2-coloring of those vertices.

Let $\mathcal{G} \subseteq \mathcal{M}$ be a nontrivial group with identity E. For any element A in $\mathcal{G}$ fix a grid-layered planar graph $G(A)$ such that $\mathcal{R}(G(A)) = A$. Note that each such $G(A)$ is 2-colorable. Otherwise A is the empty relation, and since that behaves as a zero element in $\mathcal{M}$, the group $\mathcal{G}$ would be the trivial group consisting just of the empty relation (since each element of $\mathcal{G}$ must have an inverse).

Let $(i \sim j) \in G$ denote that i and j are connected via a path in G.

Claim 1. Let $A, B \in \mathcal{G}$. Then $(i \sim j) \in G(A)$ if and only if $(i \sim j) \in G(B)$, and $(i' \sim j') \in G(A)$ if and only if $(i' \sim j') \in G(B)$.

Proof. It is enough to prove the claim for the case when B is just the identity element E. Assume $(i \sim j) \notin G(A)$. Since $EA = A$ and $\mathcal{R}(G(E) \circ G(A)) = EA = A$, there cannot be a path between i and j in $G(E) \circ G(A)$ since otherwise the colors of i and j will depend on each other and we know that this is not the case since $(i \sim j) \notin G(A)$. Therefore $(i \sim j) \notin G(E) \circ G(A)$ which in particular implies $(i \sim j) \notin G(E)$. For the other direction assume that $(i \sim j) \notin G(E)$. We have $AA^{-1} = E$. Note that $\mathcal{R}(G(A) \circ G(A^{-1})) = AA^{-1} = E$. This implies that there is no path between i and j in $G(A) \circ G(A^{-1})$ since otherwise the colors of i and j will depend on each other in $G(E)$ which we know is not the case. Therefore $(i \sim j) \notin G(A) \circ G(A^{-1})$ and hence $(i \sim j) \notin G(A)$. This shows that $(i \sim j) \in G(E)$ if and only if $(i \sim j) \in G(A)$. To show that $(i' \sim j') \in G(A)$ if and only if $(i' \sim j') \in G(E)$ we consider equations $AE = A$ and $A^{-1}A = E$ and use a similar argument as above. $\square$

Let $A \in \mathcal{G}$ and consider the graph $G(A)$. For any set S of vertices of $G(A)$ let $\overleftarrow{V}(S) = S \cap \{1, \ldots, w\}$ and $\overrightarrow{V}(S) = S \cap \{1', \ldots, w'\}$. We define $L(A)$ to be the set of all connected components C in $G(A)$ such that $\overleftarrow{V}(C) \neq \emptyset$ and $\overrightarrow{V}(C) = \emptyset$. Similarly let $R(A)$ denote the set of all connected components C such that $\overleftarrow{V}(C) = \emptyset$ and $\overrightarrow{V}(C) \neq \emptyset$, and finally define $M(A)$ as the set of connected components C such that $\overleftarrow{V}(C) \neq \emptyset$ and $\overrightarrow{V}(C) \neq \emptyset$. We now let $V_L(A) = \{\overleftarrow{V}(C) : C \in L(A)\}$ and $V_R(A) = \{\overrightarrow{V}(C) : C \in R(A)\}$. Likewise we define $V_L^M(A) = \{\overleftarrow{V}(C) : C \in M(A)\}$ and $V_R^M(A) = \{\overrightarrow{V}(C) : C \in M(A)\}$.

Claim 2. The following properties hold.

(i) $V_L(A) = V_L(E)$ and $V_R(A) = V_R(E)$. Furthermore, for any pair of i and j that are in the same component in $L(A)$, the lengths of all paths between i and j in $G(A)$ have the same parity and that is the same as in $G(E)$. Similarly for every pair i' and j' that are in the same connected component in $R(A)$, the length of all paths between i' and j' are of the same parity and that is the same as in $G(E)$.

(ii) $V_L^M(A) = V_L^M(E)$ and $V_R^M(A) = V_R^M(E)$.

All these follow again from $A^{-1}A = AA^{-1} = E$ and $EA = AE = A$ as in the proof of Claim 1, and we omit further details.

For each component in $M(A)$ we pick two representatives, one from each side. We pick the left representatives $i_1 < \ldots < i_m$ arbitrarily. But for the right representatives if i_k is connected to i'_k then we pick i'_k as the representative of the k'th component, otherwise we pick an arbitrary node in the component. Let the right representatives be $j'_1 < \ldots < j'_m$. We map the left representative of a component to its right representative. By the planar embedding of $G(A)$, for every k we have that i_k is mapped to j'_k. This means that we can rename the components in $M(A)$ by $C_1, \ldots, C_m$ such that all vertices in C_i appear after all vertices in C_{i-1}.

Furthermore we know by above claim that in a group, these components are the same on the boundaries of the graph of each group element. For any $A \in \mathcal{G}$ and any $1 \le k \le m$ let π_k^A be the parity of the length of all paths between i_k and j'_k in $G(A)$. We show that there exists a sequence $\epsilon_1, \ldots, \epsilon_m \in \{0,1\}^m$ such that for any $A, B \in \mathcal{G}$ and all $1 \le k \le m$, the parity of the paths between i_k and j'_k in $G(A) \circ G(B)$ is given by $\pi_k^A \oplus \pi_k^B \oplus \epsilon_k$. To see this consider the graph $G(A) \circ G(B)$ and rename the i_k and j'_k on the side where $G(A)$ and $G(B)$ meet as $i^{(1)}$ and $i^{(2)}$ ($i^{(1)} = i_k$ and $i^{(2)} = j'_k$). If $j'_k = i'_k$ we set $\epsilon_k = 0$. This clearly satisfies the desired property, since to get from i_k on the left layer of $G(A)$ to i'_k on the right layer of $G(B)$ we can first go to i'_k on the right layer of $G(A)$ and then to i'_k on the right layer of $G(B)$. Any such path clearly has parity $\pi_k^A + \pi_k^B$. If $j'_k \ne i'_k$ we note that the parity between $i^{(1)}$ and $i^{(2)}$ is exactly the same as in $G(E) \circ G(E)$ by Claim 2. We denote this by ϵ_j. Now to color $G(A) \circ G(B)$ if we use color 0 on i_j then we are forced to use color π_j^A on $i^{(2)}$, and hence $\pi_j^A \oplus \epsilon_j$ on $i^{(1)}$ and finally we should use $\pi_j^A \oplus \epsilon_j \oplus \pi_j^B$ on i'_j. This means that the parity between i_j and i'_j is $\pi_j^A \oplus \pi_j^B \oplus \epsilon_j$ as claimed.

We define a group $\mathbb{Z}_2^{(\epsilon_1, \ldots, \epsilon_m)}$ as follows. The elements are just the same as $\mathbb{Z}_2^m$, and the group operation is defined as $\mathbb{Z}_2^m$ but then adding the vector $(\epsilon_1, \ldots, \epsilon_m)$ to the result. It is clear that $\mathbb{Z}_2^{(\epsilon_1, \ldots, \epsilon_m)}$ and $\mathbb{Z}_2^m$ are isomorphic. The above argument shows that $\mathcal{G}$ is isomorphic to $\mathbb{Z}_2^{(\epsilon_1, \ldots, \epsilon_m)}$ and hence to $\mathbb{Z}_2^m$.   □

3.2 Bipartite Matching

We prove here our upper bound for bipartite matching.

Theorem 5. *Given a bipartite grid-layered planar graph G, we can decide whether G has a perfect matching in* ACC^0.

Reduction to a Monoid Word Problem. For each grid-layered planar graph G of odd length ℓ that has no vertical edges in the rightmost layer, we define the corresponding monoid element $G^{\mathcal{M}}$ as the triple (X, Y, R) where $X \subseteq [w]$ is the set of vertices in the leftmost layer of G, $Y \subseteq [w]$ is the set of vertices in the rightmost layer of G and $R \subseteq 2^X \times 2^Y$ is a binary relation such that for any $X_1 \subseteq X$, $X_2 \subseteq Y$ we have $(X_1, X_2) \in R$ if and only if G has a matching that matches all vertices in G except $\overline{X_1}$ in the leftmost layer and X_2 in the rightmost layer. The monoid product is defined as $(X_1, X_2, R)(X_3, X_4, S) = (X_1, X_4, R \circ S)$ when $X_2 = X_3$ and $\circ$ is the usual composition of binary relations. When $X_2 \ne X_3$, we define the product to be an element 0 for which $0x = x0 = 0$ for any x in the monoid. Now define the monoid $\mathcal{M} = \{G^{\mathcal{M}} : G$ is an odd length bipartite grid-layered planar graph$\} \cup \{0\} \cup \{1\}$, where 1 is an added identity. It is easy to see that the monoid operation described corresponds to concatenation of graphs (by merging the vertices in the rightmost layer of first graph with the vertices in the leftmost layer of the second graph). So a perfect matching exists in G if and only if $G^{\mathcal{M}} = (X_1, X_2, R)$ and R contains the element $(X_1, \overline{X_2})$.

To show that $\mathcal{M}$ is solvable we will prove that it does not contain any group of order two. We then use Proposition 6 and Theorem 7 below to conclude that $\mathcal{M}$ is solvable.

Proposition 6. *If $\mathcal{G}$ is a finite group of order $2k$ for some $k \geq 1$, then there exists $a \in \mathcal{G}$ such that $a \neq e$ and $a^2 = e$, where e is the identity of $\mathcal{G}$.*

Theorem 7 (Feit-Thompson [8]). *Every group of odd order is solvable.*

Proposition 8. *The monoid $\mathcal{M}$ is solvable.*

Proof. We begin by considering an arbitrary group $\mathcal{G} \subset \mathcal{M}$, such that $\mathcal{G} \neq \{0\}$ and $\mathcal{G} \neq \{1\}$. First, observe that for any two elements (X_1, X_2, R) and (X_3, X_4, S) in the group $X_1 = X_2 = X_3 = X_4$ as $0 \notin \mathcal{G}$. So we can identify any element (X_1, X_2, R) of the group by simply using R. Suppose now that $\mathcal{G} = \{E, R\}$ is of order 2, where E is the identity element in $\mathcal{G}$. We will show that $E \subseteq R$. This then means that $R = ER \subseteq R^2 = E$, contradicting the existence of $\mathcal{G}$. Let $k = 2^w$. Suppose $(Y_0, Y_{k+1}) \in E$. Since $E^{k+1} = E$ there exists $Y_1, \ldots, Y_k$ such that $(Y_i, Y_{i+1}) \in E$ for all i. Thus there must exist some $X_1 = Y_i$ such that $(Y_0, X_1) \in E$, $(X_1, X_1) \in E$, and $(X_1, Y_{k+1}) \in E$. We shall show that $(X_1, X_1) \in R$, and since $R = ERE$ this shows also $(Y_0, Y_{k+1}) \in R$.

Since $(X_1, X_1) \in E$ and $R^2 = E$ there exists X_2 such that $(X_1, X_2) \in R$ and $(X_2, X_1) \in R$. Consider a graph G defining R, and let M_1 be a matching in G corresponding to $(X_1, X_2) \in R$ and let M_2 be a matching in G corresponding to $(X_2, X_1) \in R$. Consider now the graph $S = M_1 \cup M_2$. The graph S^n is obtained by concatenating n copies of S. We note that for any odd (even) n, the graph S^n is a union of two matchings. The matching M obtained by the concatenation of matchings $M_1 M_2 \ldots$ and the matching N obtained by the concatenation of matchings $M_2 M_1 \ldots$.

We label the vertices on the left side on the ith copy of S as $1_{(i)}, \ldots, k_{(i)}$. The rightmost vertices in S^n are labelled $1_{(n+1)}, \ldots, k_{(n+1)}$. A path in S^n is called a *blocking path* if it connects some vertex in the leftmost layer to some vertex in the rightmost layer.

We will show below that for $n > w$ the graph S^n does not have a blocking path, but first we show that this will complete the proof. Suppose that S^n in fact does not have a blocking path, and assume without loss of generality that n is even. Consider the set V_L of all vertices in S^n that are reachable from some vertex in the left end and the set V_R of all vertices in S^n that are reachable from some vertex in the right end. Put any remaining vertices in the set V_L. Since there is no blocking path V_L and V_R are disjoint. Now we can obtain a matching corresponding to (X_1, X_2) in $R^n = E$ by using the matching M_1 on the vertices in V_L and using the matching M_2 on V_R. Since $(X_2, X_1) \in R$ this means $(X_1, X_1) \in R$ as should be shown.

We say that a path P crosses a boundary in S^n if it has two consecutive edges e_1 and e_2 such that they belong to different copies of S in S^n. Note that e_1 and e_2 must belong to the same matching M_1 or M_2. If they do not, the vertex common to those edges must be in $X_1 \cap \overline{X_1}$ or $X_2 \cap \overline{X_2}$.

Claim 3. For any n, the graph S^n cannot have a path from $v_{(i)}$ to $v_{(i+1)}$ for any i and v.

Proof. To simplify the proof, for a graph corresponding to a given monoid element we attach length 2 horizontal paths to the vertices in the left and right side through two new layers. Notice that this does not change the monoid element since the vertices in the graph corresponding to the monoid element which were originally matched inside the graph remains matched inside the graph itself and vice versa.

Suppose such a path P from $v_{(i)}$ to $v_{(i+1)}$ exists. Suppose also that P connects to both these vertices from the same side, left or right. Consider the shifted version P' of P in S^{n+1} from $v_{(i+1)}$ to $v_{(i+2)}$. These path thus share an edge, but they must diverge at some vertex. This means there exist a vertex of degree at least 3 in S^{n+1} which is impossible since S^{n+1} is a union of two matchings. Thus P must connect to the two vertices $v_{(i)}$ to $v_{(i+1)}$ from opposite sides. This means that it crosses boundaries an even number of times. By bipartiteness the path is of even length, and together this means that the first and last edge can not be from the same matching. This implies that $v \in X_1 \cap \overline{X_1}$ or $v \in X_2 \cap \overline{X_2}$, contradicting the existence of P.   □

The following last claim completes the proof that $\mathcal{M}$ is solvable.

Claim 4. S^n does not have a blocking path for $n \geq w$.

Proof. Assume that S^n has a blocking path for $n \geq w$. This blocking path must pass each of the $n + 1$ boundaries (including left and right ends) at least once. Therefore we can find integers i and j such that this blocking path has a segment P connecting $v_{(i)}$ to $v_{(j)}$ for some $1 \leq v \leq w$. By Claim 3, we have $j > i + 1$. Now consider the graph S^{n+1}. This graph also has this path P from $v_{(i)}$ to $v_{(j)}$ and also a path P' from $v_{(i+1)}$ to $v_{(j+1)}$ that is simply a "shifted" version of P. By Claim 3, these paths are vertex disjoint. Because if they intersect then we can construct a path from $v_{(i)}$ to $v_{(i+1)}$ in S^{n+1}. By using Lemma 2 we conclude that the paths P and P' must intersect. This concludes the proof.   □

3.3 Disjoint Paths

We consider several different variants of the disjoint paths problem, but there is significant overlap in the different approaches. In each case we define a monoid

$\mathcal{M}$ and show it is aperiodic. We can thus compute the word problem over $\mathcal{M}$ by AC^0 circuits and we can use these to solve the disjoint paths problem. We describe the definition of the monoids, how to reduce the disjoint paths problem to the word problem over to monoids, and outline some ideas of the proof that the monoids are aperiodic. We are able to do this in the following 3 settings: (1) node-disjoint paths in directed graphs. (2) edge-disjoint paths in upward planar graphs. (3) edge-disjoint paths in undirected graphs.

The monoids. We describe here the monoid in general terms. Elements of $\mathcal{M}$ consist of a (downward closed) family of sets of edges between the set of vertices $W = \{1, \ldots, w\} \cup \{1', \ldots, w'\}$. Consider a grid-layered planar graph G. This may be either undirected or directed. We construct a monoid-element $G^{\mathcal{M}}$ from G as follows, by letting every set of disjoint paths in G between vertices from W give rise to a set of corresponding edges in $G^{\mathcal{M}}$. Depending on the setting these paths may be vertex-disjoint or edge-disjoint, and if the graph is directed the edges are directed accordingly. The operation of the monoid will be the natural operation that makes the map $G \mapsto G^{\mathcal{M}}$ a homomorphism. Note that if $A \subseteq A'$ and $B \subseteq B'$ then $AB \subseteq A'B'$.

Reduction to monoid product. Let G be a grid-layered planar directed graph with pairs of terminals $(s_1, t_1), \ldots, (s_k, t_k)$. Consider the partition of G into at most $2k + 1$ segments obtained by dividing at every layer containing a terminal. For each segment we divide the graph into segments of length 1, translate these to monoid elements and compute the product of these. This results in at most $2k+1$ monoid elements describing all possible disjoint paths connecting endpoints of every segment. Since k is fixed this is a fixed amount of information from which it can then be directly decided whether disjoint paths exist between all pairs of terminals.

Showing aperiodicity of the monoid. The approach we will use in all cases is as follows. Let $\mathcal{G}$ be a group in $\mathcal{M}$ with identity E, and let A be any element of $\mathcal{G}$. We shall then prove that $E \subseteq A$. Note then that this means $A^{-1} = EA^{-1} \subseteq AA^{-1} = E$, and hence $A = E$. Showing this for all A implies that $\mathcal{G}$ is trivial.

Let $\mathcal{G}$ be a group in $\mathcal{M}$ with identity E. Let A be an element of $\mathcal{G}$ of order $p \geq 2$. A central step is that we have a number of paths in the graph $G(A)^p$ that we can think of as being induced by infinite paths $P_1, \ldots, P_c$ with period p in an infinite concatenation of the graph $G(A)$. We then consider paths $P'_1, \ldots, P'_c$ obtained by shifting by the length of one graph $G(A)$ to the right. By Lemma 2 P_i will intersect P'_i. The idea is then to construct new disjoint paths in $R_1, \ldots, R_c$ in $G(A)^{2pc+1}$ for some c. These paths start out following along the paths $P_1, \ldots, P_c$ but end up following along $P'_1, \ldots, P'_c$. These paths will then be present also in $G(A)$ since $A^{pc+1} = A$, and this will mean that $E \subseteq A$.

4   Conclusion

We have obtained new upper and lower bounds for several classical graph decision problems in the setting of bounded planar cutwidth graphs, providing insight into the computational power of the circuit classes AC^0, $AC^0[2]$, ACC^0, and NC^1. Several open problems remain, most notably for perfect matching where we conjecture that the upper bounds can be significantly improved: In the general case from NC^1 to $AC^0[2]$ and in the bipartite case from ACC^0 to AC^0.

References

1. Allender, E., Hertrampf, U.: Depth reduction for circuits of unbounded fan-in. Information and Computation 112(2), 217–238 (1994)
2. Barrington, D.A.M.: Bounded-width polynomial-size branching programs recognize exactly those languages in NC^1. J. Comput. Syst. Sci. 38(1), 150–164 (1989)
3. Barrington, D.A.M., Thérien, D.: Finite monoids and the fine structure of NC^1. J. ACM 35(4), 941–952 (1988)
4. Bodlaender, H.L.: Some classes of planar graphs with bounded treewidth. Bulletin of the EATCS 36, 116–126 (1988)
5. Cygan, M., Marx, D., Pilipczuk, M., Pilipczuk, M.: The planar directed k-vertex-disjoint paths problem is fixed-parameter tractable. In: FOCS, pp. 197–206. IEEE Computer Society (2013)
6. Datta, S., Gopalan, A., Kulkarni, R., Tewari, R.: Improved bounds for bipartite matching on surfaces. In: STACS. LIPIcs, vol. 14, pp. 254–265. Schloss Dagstuhl - Leibniz-Zentrum fuer Informatik (2012)
7. Elberfeld, M., Jakoby, A., Tantau, T.: Algorithmic meta theorems for circuit classes of constant and logarithmic depth. In: STACS. LIPIcs, vol. 14, pp. 66–77. Schloss Dagstuhl - Leibniz-Zentrum fuer Informatik (2012)
8. Feit, W., Thompson, J.G.: Solvability of groups of odd order. Pacific J. Math. 13(3), 775–1029 (1963)
9. Fortune, S., Hopcroft, J.E., Wyllie, J.: The directed subgraph homeomorphism problem. Theor. Comput. Sci. 10, 111–121 (1980)
10. Garey, M.R., Johnson, D.S., Stockmeyer, L.J.: Some simplified NP-complete graph problems. Theor. Comput. Sci. 1(3), 237–267 (1976)
11. Garey, M.R., Johnson, D.S., Tarjan, R.E.: The planar hamiltonian circuit problem is NP-complete. SIAM J. Comput. 5(4), 704–714 (1976)
12. Itai, A., Papadimitriou, C.H., Szwarcfiter, J.L.: Hamilton paths in grid graphs. SIAM J. Comput. 11(4), 676–686 (1982)
13. Kasteleyn, P.W.: Graph theory and crystal physics. In: Harary, F. (ed.) Graph Theory and Theoretical Physics, pp. 43–110. Academic Press (1967)
14. Razborov, A.A.: Lower bounds for the size of circuits of bounded depth with basis $(\wedge, \oplus)$. Mathematical Notes of the Academy of Science of the USSR 41(4), 333–338 (1987)
15. Reingold, O.: Undirected connectivity in log-space. J. ACM 55(4) (2008)
16. Robertson, N., Seymour, P.D.: Graph minors. XIII. the disjoint paths problem. J. Comb. Theory, Ser. B 63(1), 65–110 (1995)
17. Schrijver, A.: Finding k disjoint paths in a directed planar graph. SIAM J. Comput. 23(4), 780–788 (1994)
18. Vazirani, V.V.: NC algorithms for computing the number of perfect matchings in $K_{3,3}$-free graphs and related problems. Inf. Comput. 80(2), 152–164 (1989)

On Characterizations of Randomized Computation Using Plain Kolmogorov Complexity

Shuichi Hirahara and Akitoshi Kawamura

The University of Tokyo, Japan

Abstract. Allender, Friedman, and Gasarch recently proved an upper bound of PSPACE for the class DTTR_K of decidable languages that are polynomial-time truth-table reducible to the set of prefix-free Kolmogorov-random strings regardless of the universal machine used in the definition of Kolmogorov complexity. It is conjectured that DTTR_K in fact lies closer to its lower bound BPP established earlier by Buhrman, Fortnow, Koucký, and Loff. It is also conjectured that we have similar bounds for the analogous class DTTR_C defined by plain Kolmogorov randomness. In this paper, we provide further evidence for these conjectures. First, we show that the time-bounded analogue of DTTR_C sits between BPP and PSPACE $\cap$ P/poly. Next, we show that the class $\mathrm{DTTR}_{C,\alpha}$ obtained from DTTR_C by imposing a restriction on the reduction lies between BPP and PSPACE. Finally, we show that the class $\mathrm{P}/R_C^{=\log}$ obtained by further restricting the reduction to ask queries of logarithmic length lies between BPP and $\Sigma_2^p \cap \mathrm{P/poly}$.

1 Introduction

The word "randomness" is used in at least two different contexts in computation theory: One is in the theory of Kolmogorov complexity, which measures randomness of a finite string in terms of incompressibility. The other randomness refers to coin flips used for efficient computation, and gives rise to (among others) the complexity class BPP. Despite the remote origins of these two notions of randomness, some interesting links between them have been found in recent work by Allender et al. [3,2]: they conjectured, and gave some evidence, that BPP is characterized as the class of languages truth-table reducible to the set of Kolmogorov random strings, that is, those decidable in deterministic polynomial time by asking the oracle non-adaptively whether a string is random. Our aim is to strengthen this connection between Kolmogorov randomness and BPP.

Let $C_U(x)$ denote the Kolmogorov complexity of a string $x \in \{0,1\}^*$, i.e., the length of the shortest description of x when a fixed universal machine U is used as a decoder (see Section 2.1). Depending on whether or not the machine is required to be prefix-free, this complexity is called the *prefix-free* or *plain* complexity, and for the former it is customary to write K_U instead of C_U. We consider the set R_{C_U} (or R_{K_U}) of *random* strings, i.e., those that have no description shorter than themselves: $R_f = \{\, x \in \{0,1\}^* \mid f(x) \geq |x| \,\}$.

E. Csuhaj-Varjú et al. (Eds.): MFCS 2014, Part II, LNCS 8635, pp. 348–359, 2014.

Buhrman, Fortnow, Koucký, and Loff [5] showed that every language in BPP reduces to the set of random strings via polynomial-time truth-table reducibility (denoted $\leq_{tt}^p$ in the following theorem, see Definition 7), regardless of whether the machine defining the random strings is plain universal or prefix-free universal:

Theorem 1 ([5]). *For every language* $L \in$ BPP*, we have* $L \leq_{tt}^p R_{C_U}$ *for any universal machine* U*, and* $L \leq_{tt}^p R_{K_U}$ *for any prefix-free universal machine* U.

Theorem 1 gives a lower bound of BPP about the class of languages that reduce to R_{C_U} or R_{K_U}, but this bound is obviously not "tight," as the sets R_{C_U} and R_{K_U} are undecidable [7]. However, we obtain an interesting upper bound if we take the intersection over all universal machines U. That is, let DTTR$_K$ be the class of decidable languages L such that $L \leq_{tt}^p R_{K_U}$ for all universal prefix-free machines U. Allender, Friedman, and Gasarch [3] recently showed:

Theorem 2 ([3]). DTTR$_K \subseteq$ PSPACE.

Combining Theorems 1 and 2, we have BPP $\subseteq$ DTTR$_K \subseteq$ PSPACE. The question then arises: to which does DTTR$_K$ sit closer, BPP or PSPACE? Allender [1] has conjectured the following:

Conjecture 3 ([1]). BPP $=$ DTTR$_K$.

The conjecture is more plausible when we consider time-bounded prefix-free Kolmogorov complexity. Allender, Buhrman, Friedman, and Loff [2] defined the classes called TTRT$_K$ and TTRT$_K'$, which are time-bounded versions of DTTR$_K$, and showed the following ("$/\alpha$" means advice strings of length α [4, Definition 6.16]):

Theorem 4 ([2])

1. BPP $\subseteq$ TTRT$_K \subseteq$ PSPACE$/\alpha \cap$ P$/$poly *for any nondecreasing unbounded computable function* $\alpha: \mathbb{N} \to \mathbb{N}$.
2. BPP $\subseteq$ TTRT$_K' \subseteq$ PSPACE $\cap$ P$/$poly.

Less is known about the analogous classes DTTR$_C$, TTRT$_C$, TTRT$_C'$ for plain Kolmogorov complexity (see Definitions 16 and 10). We know that these classes all contain BPP by Theorem 1 or by a similar argument. For the upper bound, Allender et al. [3] conjecture that DTTR$_C \subseteq$ PSPACE (similarly to Theorem 2), but in reality we do not even know whether DTTR$_C$ equals all decidable languages. The techniques used by Allender et al. [3,2] to prove Theorems 2 and 4 make use of the coding theorem and thus cannot be directly applied to C_U (see the discussion in [3, Section 5]).

In Section 3, we prove the same inclusions for TTRT$_C$ and TTRT$_C'$ as Theorem 4 states for prefix-free complexity.

In Section 4, we consider a weaker reduction $\leq_{tt_\alpha}^p$ (see Definition 13) in which we cannot query a string of length less than $\alpha(n)$, where n is the input length. We prove that for the corresponding classes TTRT$_{C,\alpha}$ and DTTR$_{C,\alpha}$, the upper bound can be improved to PSPACE (while the BPP lower bound remains true).

In Section 5, we restrict the reduction further, allowing only queries of exactly logarithmic length. This defines the subclass $P/R_C^{=\log} \subseteq \text{TTRT}_{C,\log}$, which still contains BPP. We prove that $P/R_C^{=\log} \subseteq \Sigma_2^p \cap P/\text{poly}$. The upper bound of Σ_2^p comes from a 1-round game played between two players. Together with the BPP lower bound, this gives an alternative explanation of the well-known inclusion $\text{BPP} \subseteq \Sigma_2^p$.

In view of this upper bound of $\Sigma_2^p \cap P/\text{poly}$, we believe that the following characterization of BPP is quite likely, and propose this as a step towards Conjecture 3.

Conjecture 5. $\text{BPP} = P/R_C^{=\log}$.

Another Characterization of BPP

We mention in passing that there is a simple characterization of randomized computation using Kolmogorov complexity (see the full version for a proof).

Proposition 6 (forklore). *Let* BPP'_U *denote the class of languages L for which there exist a polynomial-time Turing machine M and a polynomial p such that for every $n \in \mathbb{N}$, $x \in \{0,1\}^n$ and $r \in R_{C_U+2}^{=p(n)}$, we have $M(x,r) = L(x)$. Then* $\text{BPP} = \text{BPP}'_U$ *for any universal Turing machine U.*

The rest of this paper is also about bounding BPP using R_{C_U}, although we will consider reductions to (not a single random string, but) the oracle that tells us whether a given string is random or not. We hope that this connection to Kolmogorov randomness helps us better understand BPP.

2   Preliminaries and Notations

We regard the set of strings $\{0,1\}^*$ as equipped with the length-increasing lexicographical order. For a set of strings $A \subseteq \{0,1\}^*$ and a natural number $n \in \mathbb{N}$, $A^{=n}$, $A^{\leq n}$, and $A^{<n}$ are defined as $A \cap \{0,1\}^n$, $\bigcup_{m \leq n} A^{=m}$, and $\bigcup_{m < n} A^{=m}$, respectively. We abbreviate $(\{0,1\}^*)^{\leq n}$ as $\{0,1\}^{\leq n}$. For a set A, $\overline{A}$ denotes the complement of A.

For functions $s,t \colon \mathbb{N} \to \mathbb{N}$, we write $s \leq t$ if $s(n) \leq t(n)$ for all $n \in \mathbb{N}$. A nondecreasing and time-constructible function $t \colon \mathbb{N} \to \mathbb{N}$ is called a *time bound*. For a nondecreasing and unbounded function $t \colon \mathbb{N} \to \mathbb{N}$, define t^{-1} as $t^{-1}(n) = \min\{\, m \in \mathbb{N} \mid t(m) > n \,\}$. Then $(t^{-1})^{-1} \leq t$ holds because $(t^{-1})^{-1}(n) = \min\{m \mid t^{-1}(m) > n\}$ and $t^{-1}(t(n)) = \min\{m \mid t(m) > t(n)\} > n$.

We assume that the reader is familiar with basics of complexity theory [4].

In this paper, we mainly discuss truth-table (i.e., non-adaptive) reductions to the set of random strings.

Definition 7. *For languages $A, B \subseteq \{0,1\}^*$, we write $A \leq_{tt}^p B$ if there exists a polynomial time Turing machine M that, on input x, outputs an encoding of a circuit λ and a list of queries $q_1, \ldots, q_m$ so that $A(x) = \lambda\big(B(q_1), \ldots, B(q_m)\big)$.*

2.1 Kolmogorov Complexity

We review some basic definitions and facts about Kolmogorov complexity. For details, see Li and Vitányi [7].

The *Kolmogorov complexity* of a string x on a Turing machine U is defined as the length $C_U(x) = \min\{\,|d| \mid U(d) = x\,\} \in \mathbb{N} \cup \{\infty\}$ of the shortest description d of x when using U as a decoder. We also consider the time-bounded version: for a time bound t, we write $C_U^t(x)$ for the length of the shortest input d that causes U to outputs x in at most $t(|x|)$ time steps (note that time is measured in terms of the output length).

As mentioned at the beginning, it is sometimes required that the domain of U is *prefix-free*, i.e., if $U(d)$ is defined, then $U(d')$ is undefined for any proper prefix d' of d. In this case, it is customary to write K_U instead of C_U (and call it *prefix-free complexity*).

We will be interested in the set $R_f = \{\, x \in \{0,1\}^* \mid f(x) \geq |x|\,\}$ of *random* (or *incompressible*) strings where $f = C_U, C_U^t$, etc.

Of course, Kolmogorov complexity, and hence the set of random strings, depend on U. It is therefore important to use the "best" machine U in the following sense:

Definition 8. *A Turing machine U is said to be* universal *if for each Turing machine M, there exists c_M such that for any $x \in \{0,1\}^*$, $C_U(x) \leq C_M(x) + c_M$.*

It is known that there exists a Turing machine U which simulates every Turing machine M (see [4]), and it is not hard to see that U is universal. Thus, we usually fix one such machine U and discuss complexity with respect to it.

For prefix-free Kolmogorov complexity, the Turing machines U and M in Definition 8 are required to be a prefix-free machine.

For time-bounded Kolmogorov complexity, some slowdown is needed when a universal Turing machine U simulates each machine M. Therefore, we introduce a parameter $f : \mathbb{N} \to \mathbb{N}$, which means that U must simulate M within $f(s)$ steps if M halts within s steps. This is the notion of *f-efficient universality*:

Definition 9 ([2]). *For a time bound $f : \mathbb{N} \to \mathbb{N}$, a Turing machine U is said to be f-efficient universal if U is universal, and there exists a constant c_M for each Turing machine M, such that for any time bounds t and t', for all but finitely many x, $t(|x|) \geq f(t'(|x|))$ implies $C_U^t(x) \leq C_M^{t'}(x) + c_M$. We say that U is* time-efficient universal *if it is p-efficient universal for some polynomial p.*

Time-efficient universality means that the machine U is allowed to simulate each machine with a polynomial slowdown.

3 Bounds for Time-Bounded Kolmogorov Complexity

Our goal for this section is to prove an analogue of Theorem 4 for plain Kolmogorov complexity. First, let us define the classes TTRT_C and TTRT_C' by using time-efficient universality and f-efficient universality, respectively.

Definition 10 ([2])

1. TTRT_C *contains all the languages* L *such that for all sufficiently large* t *and for any time-efficient universal Turing machine* U, $L \leq^p_{tt} R_{C^t_U}$. *In short,* $\mathrm{TTRT}_C = \{L \subseteq \{0,1\}^* \mid \exists t_0, \forall t \geq t_0, \forall U, L \leq^p_{tt} R_{C^t_U}\}$.
2. TTRT'_C *contains all the languages* L *such that for each computable function* f, *there exists* t_0 *such that for all* $t \geq t_0$ *and any* f*-efficient universal Turing machine* U, $L \leq^p_{tt} R_{C^t_U}$.

It is easy to see that $\mathrm{TTRT}'_C \subseteq \mathrm{TTRT}_C$ by letting $f(n) = 2^n$. We have the following:

Theorem 11. 1. $\mathrm{BPP} \subseteq \mathrm{TTRT}_C \subseteq \mathrm{PSPACE}/\alpha \cap \mathrm{P}/\mathrm{poly}$ *for any nondecreasing unbounded computable function* $\alpha \colon \mathbb{N} \to \mathbb{N}$.
2. $\mathrm{BPP} \subseteq \mathrm{TTRT}'_C \subseteq \mathrm{PSPACE} \cap \mathrm{P}/\mathrm{poly}$.

Proof (of Theorem 11.1). We have $\mathrm{BPP} \subseteq \mathrm{TTRT}'_C \subseteq \mathrm{TTRT}_C \subseteq \mathrm{P}/\mathrm{poly}$ because this was proved in [5,2] regardless of whether the Kolmogorov complexity defining the classes is plain or prefix-free. All we have to show is that $\mathrm{TTRT}_C \subseteq \mathrm{PSPACE}/\alpha$.

For simplicity, we show that $\mathrm{TTRT}_C \subseteq \mathrm{PSPACE}/2\alpha$ for any function α that satisfies the stated property. Assume, by way of contradiction, that $L \in \mathrm{TTRT}_C \setminus \mathrm{PSPACE}/2\alpha$. Then $L \leq^p_{tt} R_{C^{t_0}_U}$ holds for some time bound t_0 and some Turing machine U. Since $R_{C^{t_0}_U}$ is decidable, L is also decidable. Let t_L denote a time bound such that it is decidable whether $x \in L$ or not in time $t_L(|x|)$.

Given any time bound t_0, we will show that there exist a time-efficient universal Turing machine U and a time bound $t^* \geq t_0$ such that $L \not\leq^p_{tt} R_{C^{t^*}_U}$. This means that $L \notin \mathrm{TTRT}_C$, which contradicts the assumption. In order to show it, a time-efficient universal Turing machine U is constructed by combining two Turing machines, U_0 and M. The following lemma states that if U_0 is time-efficient universal, then U is.

Lemma 12. *Let* U_0 *be a time-efficient universal Turing machine. Suppose that for a time bound* t, M *is a Turing machine such that, for all but finitely many* x, M *either halts in* $t(|x|)^2$ *steps or does not halt. Then there exists a Turing machine* U *such that for some* $t^* = \Theta(t^2)$, $C^{t^*}_U(x) = \min\{C^t_{U_0}(x)+2, C^{t^2}_M(x)+1\}$ *for all large* $|x|$. *Moreover,* U *is again a time-efficient universal Turing machine.*

Proof. Define U as follows: On input $00d$, U runs U_0 on input d. Suppose that U_0 outputs x after s steps. Then U outputs x after it waits s^2 steps. On input $1d$, U outputs $M(d)$. Otherwise U does not halt. Note that the steps to calculate $U_0(d)$ can be counted by adding some tapes to U.

Let $p(s)$ be the steps it takes for U to halt on input $00d$, where s denotes the number of steps to calculate $U_0(d)$. Note that $p(s)$ includes the steps it takes to run U_0, to compute s^2, and so on. We can assume that $p(s) \geq s^2$ and $p(s)$ does not depend on d, and that $t^*(n) := p(t(n)) = \Theta(t(n)^2)$ is again a time bound.

Then $U(00d) = x$ in at most $t^*(|x|)$ steps if and only if $U_0(d) = x$ in at most $t(|x|)$ steps for any time bound t. One can verify that the lemma follows by this property.   □

Fix a standard time-efficient universal Turing machine U_0 for reference. A Turing machine M will be defined later. Let U be the time-efficient universal machine defined in Lemma 12 for some time bound t fixed later in Claim 2. In order to derive a contradiction, M is defined so that it fools each Turing machine that computes a truth-table reduction $\leq_{tt}^p$. In the course of computation of M, M holds an approximation of $R_{C_U^{t*}}$ denoted by R. More precisely, R is first set to be $\{0,1\}^*$, always $R \supseteq R_{C_U^{t*}}$ holds, and $\max\{k \in \mathbb{N} \mid R^{<k} = R_{C_U^{t*}}^{<k}\}$ increases as calculation proceeds. Since $\overline{R}$ is finite, we can store it. Note that, by the property of Lemma 12, $R_{C_U^{t*}} = R_{C_{U_0}^t + 2} \cap R_{C_M^{t^2} + 1}$ except finitely many strings.

Let us efficiently enumerate all polynomial-time Turing machines that compute polynomial-time truth-table reductions, namely $\gamma_1, \gamma_2, \cdots$. This can be done by putting a clock of $n^e + e$ on γ_e and by regarding the output of $\gamma_e(x)$ as a circuit $\lambda_{e,x}$ and queries $q_1, \cdots, q_k$ (see Definition 7). Since U_0 is out of our control, we will play a game against U_0 in order to make the output of the circuit $\lambda_{e,x}$ differ from $L(x)$. We can always do so for some x; otherwise L is equal to a winning side, and thus $L \in \mathrm{PSPACE}$, which contradicts the assumption. This idea is originally due to Allender et al. [3]. Now let us precisely define the game $\mathcal{G}_{e,x}$ for some γ_e and some x.

Description of the Game. The game $\mathcal{G}_{e,x}$ is played between two players, the YES player and the NO player. Before the construction of the game, let $R := R \cap R_{C_{U_0}^t + 2}^{<\alpha(|x|)}$ to fix randomness of the strings of length less than $\alpha(|x|)$. Each player does not disturb $R^{<\alpha(|x|)}$ anymore. During the game, R is going to be altered by moves of the players. The current value of the game, denoted by $\mathrm{val}(R)$, is defined as $\lambda_{e,x}(R(q_1), \cdots, R(q_k))$. Note that $\mathrm{val}(R)$ is equal to the value of $\gamma_e(x)$ regarded as a truth-table reduction to R.

Let $l := \max\{l, |q_1|, \cdots, |q_k|\}$, where l denotes a variable of the algorithm of M. The game has rounds from $\alpha(|x|)$ to l. In the rth round, where $\alpha(|x|) \leq r \leq l$, the YES player first decides $Z_1 \subseteq \{0,1\}^r$ such that $|Z_1| \leq 2^{r-2}$. This means that the YES player declares that each element of Z_1 is not random. Next, the NO player decides $Z_0 \subseteq \{0,1\}^r$ such that $|Z_0| \leq 2^{r-2}$. Then let $R := R \setminus (Z_0 \cup Z_1)$ at the end of the round. After all the rounds end, the YES player wins if $\mathrm{val}(R) = 1$ and the NO player wins otherwise. Since the game is finite and deterministic, exactly one of the players has a winning strategy. Let $\mathrm{WIN}_{e,x} \in \{0,1\}$ denote which player is going to win the game $\mathcal{G}_{e,x}$.

The $\mathrm{WIN}_{e,x}$ side plays the game optimally according to a winning strategy against U_0. Suppose that the player of the $\mathrm{WIN}_{e,x}$ side chooses $Z \subseteq \{0,1\}^r$ in the rth round and $|d| = r - 2$, where d is the input of M. Let $Z = \{z_1 < z_2 < \cdots < z_k\}$, where $k \leq 2^{r-2}$. If, for $i \leq k$, the input d is the ith string in $\{0,1\}^{r-2}$ in the lexicographical order, then M outputs z_i and halts. This makes

z_i not random since $M(d) = z_i$ and $|d| < |z_i| - 1$. If not, M does not halt, and continues computation.

On the other hand, U_0 plays the role of the opponent of the $\mathrm{WIN}_{e,x}$ side. The opponent always moves $Z = \left(\overline{R_{C^t_{U_0}+2}}\right)^{=r}$ in the rth round. Note that $|Z| = \left|\left(\overline{R_{C^t_{U_0}+2}}\right)^{=r}\right| \leq 2^{r-2}$ holds (see [7, Definition 2.2.1 and Theorem 2.2.1]).

The overall algorithm of M is shown in Algorithm 1. Note that what M outputs is determined by a move of the player who has a winning strategy.

Algorithm 1. Algorithm of M

Input: $d \in \{0,1\}^*$
Output: $z_* \in \{0,1\}^*$ such that $|z_*| = |d| + 2$
  $R := \{0,1\}^*$
  $l := -1$
  for $e = 1$ to ∞ **do**
    Find the lexicographically first string x such that $\alpha(|x|) > l$ and $\mathrm{WIN}_{e,x} \neq L(x)$.
    $l := \max(\{|q_i|\}_i \cup \{l\})$, where $\{q_i\}_i$ denotes the queries of $\gamma_e(x)$
    for $r = \alpha(|x|)$ to l **do**
      Let the $\mathrm{WIN}_{e,x}$ side play the game $\mathcal{G}_{e,x}$ for the rth round according to a winning strategy, and let the opponent choose $\left(\overline{R_{C^t_{U_0}+2}}\right)^{=r}$. If the player of the $\mathrm{WIN}_{e,x}$ side chooses $Z = \{z_1 < \cdots < z_k\}$ and the input $d \in \{0,1\}^{r-2}$ is the ith string in $\{0,1\}^{r-2}$, then output z_i and halt.
      Update R according to the moves in this round.
    end for
  end for

Analysis. Let us move on to analysis of the algorithm. We first claim that there always exists a "witness" $x \in \{0,1\}^*$ such that $\mathrm{WIN}_{e,x} \neq L(x)$.

Claim 1. M can always find a string x such that $\alpha(|x|) > l$ and $\mathrm{WIN}_{e,x} \neq L(x)$.

Proof (of Claim 1). Suppose that M fails to find such a string x at some point, which implies that $\mathrm{WIN}_{e,x} = L(x)$ for all x with $|x| \geq \alpha^{-1}(l)$. We present an algorithm to decide the language $\mathrm{WIN}_{e,*} = \{x \in \{0,1\}^* \mid \mathrm{WIN}_{e,x} = 1\}$ in $\mathrm{PSPACE}/2^\alpha$. On input x, let the advice give the truth-table of $R_{C^t_{U_0}+2}^{<\alpha(|x|)}$. This can be encoded with $2^{\alpha(|x|)} - 1$ bits. We know the value $\alpha(|x|)$ by the length of the advice. Thus, this advice enables us to construct the game $\mathcal{G}_{e,x}$. Then the winning side, $\mathrm{WIN}_{e,x}$, can be computed by the standard minimax algorithm. Note that there exist at most exponentially many moves since a move Z is meaningful only if $Z \subseteq \{q_1, \cdots, q_k\}$, where k is bounded by a polynomial in $|x|$.

Since $\mathrm{WIN}_{e,x} = L(x)$ for all sufficiently large x, we can conclude that $L \in \mathrm{PSPACE}/2^\alpha$. This, however, contradicts the assumption. □

Next, we fix a fast-growing time bound t so that M halts within t^2 steps.

Claim 2. There exists a time bound t such that M halts within $t(|z|)^2$ steps for all large $|z|$ if M outputs z.

Proof (of Claim 2). Suppose that M halts at the rth round of the game $\mathcal{G}_{e,x}$. Note that $|d| + 2 = |z| = r \geq \alpha(|x|)$. M has computed $R_{C_{U_0}^t+2}^{\leq r}$ so far. It takes $O\left(\sum_{r' \leq r} t(r')2^{r'}\right)$ time steps. M has also calculated $\text{WIN}_{e,y}$ and $L(y)$ for each $y \leq x$. It takes $2^{|x|^{O(1)}} + O(2^{|x|+1}t_L(|x|))$ time steps since $\text{WIN}_{e,*} \in \text{EXP}$. M needs to compute $\alpha(m)$ for each $m \leq |x|$. Let $t_\alpha(m)$ denote the number of steps to compute $\alpha(m)$.

The length of x is bounded by $\alpha^{-1}(r)$ since $\alpha(|x|) \leq r$. Then the number of overall time steps is roughly less than

$$2^{\alpha^{-1}(r)^{O(1)}} t_L(\alpha^{-1}(r)) \quad + \sum_{0 \leq r' \leq r} t(r')2^{r'} \quad + \sum_{0 \leq m \leq \alpha^{-1}(r)} t_\alpha(m). \quad (1)$$

The second term is linear in $t(r)$, and the other terms do not depend on t. Thus t can be defined as a sufficiently fast growing function $\geq t_0$ so that (1) is less than $t(r)^2$ for all large r.   □

Claim 2 ensures that R converges to $R_{C_U^{t*}}$ as time passes. Finally, we claim that each Turing machine is indeed fooled.

Claim 3. $L \not\leq_{tt}^p R_{C_U^{t*}}$.

Proof (of Claim 3). After the game $\mathcal{G}_{e,x}$ has been played out, since the $\text{WIN}_{e,x}$ side wins, $\text{val}(R) = \text{WIN}_{e,x} \neq L(x)$ holds at this point. Let l_e denote the value of l in the eth outer loop before the inner loop in Algorithm 1. Then, in the later game $\mathcal{G}_{e',x'}$ for $e' > e$, we have $|q_i| \leq l_e \leq l_{e'} < \alpha(|x'|)$, which implies that the later computation does not disturb the value $R(q_i)$. Therefore, $R^{\leq l}$ is actually equal to $R_{C_U^{t*}}^{\leq l}$ after $\mathcal{G}_{e,x}$ ends. It follows that γ_e does not compute a truth-table reduction to $R_{C_U^{t*}}$ since the answer of the truth-table reduction on input x is equal to $\text{val}(R)$, which is not equal to $L(x)$.   □

Claim 3 contradicts the assumption that $L \in \text{TTRT}_C$.   □

Let us review why the slight nonuniformity α is needed: To determine a time bound to calculate M, $|x|$ is bounded by the inequation $\alpha(|x|) \leq r$ in Claim 2. The inequation is derived from fixing randomness of strings whose length is less than $\alpha(|x|)$. If α is slow enough, then the initial segment $R_{C_{U_0}^t+2}^{<\alpha(|x|)}$ is decidable in PSPACE. However, since the definition of TTRT_C uses time-efficient universality, M must run in at most polynomial time in t. This time limit prevents α from growing slowly because $|x|$ becomes huge as α becomes slow.

In the case of TTRT_C', since a f-efficient universal Turing machine U is allowed to run longer than time-efficient one, the Turing machine M can also run long enough to calculate the initial segment $R_{C_{U_0}^t+2}^{<\alpha(|x|)}$. Therefore, we can eliminate the slight nonuniformity as stated in Theorem 11.2 (see the full version for a proof).

4   A Restricted Reduction

Our crucial observation is that we do not use the initial segment $R_{C_U}{}^{<\alpha(n)}$ to prove Theorem 1: Its proof relies on the hardness versus randomness framework of Impagliazzo and Wigderson [6]. That is, the set of random strings of length $\Theta(\log n)$ is highly complex, and therefore hard enough to construct the pseudorandom generator that can derandomize BPP completely. Nevertheless, as discussed in Section 3, the initial segment $R_{C_U}{}^{<\alpha(n)}$ is the very obstacle to proving Theorem 11.1 without the slight nonuniformity α.

For the purpose of characterizing BPP using Kolmogorov complexity, we should not allow reductions to query any short string of length less than $\alpha(n)$. This motivates the notion of α-restricted truth-table reductions:

Definition 13. *For languages $A, B \subseteq \{0,1\}^*$, we write $A \leq^p_{tt_\alpha} B$ if there exists a polynomial time Turing machine M that, on input x, outputs an encoding of a circuit λ and a list of queries $q_1, \ldots, q_m$, where $|q_i| \geq \alpha(|x|)$, so that $A(x) = \lambda(B(q_1), \ldots, B(q_m))$.*

Then let us restrict TTRT_C to $\mathrm{TTRT}_{C,\alpha}$ by imposing the α-restriction:

Definition 14. $\mathrm{TTRT}_{C,\alpha}$ *is the class of all languages L such that for all large t and for any time-efficient universal Turing machine U, $L \leq^p_{tt_\alpha} R_{C_U^t}$.*

At a glance, our requirement, which bounds the query lengths from below by a unbounded function (that could be very slow-growing), may seem to be an atypical, and perhaps rather weak, restriction. Nevertheless, this restriction allows us to eliminate the slight nonuniformity in the upper bound:

Theorem 15. BPP $\subseteq \mathrm{TTRT}_{C,\alpha} \subseteq$ PSPACE $\cap$ P/poly *for any nondecreasing unbounded computable function $\alpha(n) = O(\log n)$.*

Proof. As discussed, we still have BPP $\subseteq \mathrm{TTRT}_{C,\alpha}$ even if the reduction is restricted (see the full version for a proof). Thus we present only a proof of the inclusion $\mathrm{TTRT}_{C,\alpha} \subseteq$ PSPACE. Note that the proof is almost the same with that of Theorem 11.

We change a polynomial-time truth-table reduction into an α-restricted one. Thus, enumerate all polynomial-time Turing machines $\gamma_1, \gamma_2, \cdots$ and regard them as α-restricted polynomial-time truth-table reductions. That is, whenever $\gamma_e(x)$ queries a string of length less than $\alpha(|x|)$, we ignore the Turing machine γ_e because it turns out to be *not* an α-restricted reduction. More precisely, when M seeks a "witness" x in Algorithm 1, if γ_e has turned out to be not an α-restricted reduction, then M goes to the next e.

Claim 4 (Revised Claim 1). M can find a string x such that $\alpha(|x|) > l$ and $\mathrm{WIN}_{e,x} \neq L(x)$, or γ_e turns out to be not an α-restricted reduction.

Proof. We can assume that on input x, γ_e does not query a string of length less than $\alpha(|x|)$ because otherwise γ_e turns out not an α-restricted reduction. Then, in order to compute $\mathrm{WIN}_{e,*}$, the initial segment $R_{C_U^{t*}}{}^{<\alpha(|x|)}$ is not needed. Thus $\mathrm{WIN}_{e,*} \in$ PSPACE.   □

Claim 5 (Revised Claim 3). $L \not\leq^p_{\mathrm{tt}_\alpha} R_{C^{t*}_U}$.

Proof. All α-restricted reductions are enumerated. Thus, in the same way with Claim 3, it follows that any Turing machine γ_e does not compute $L \leq^p_{\mathrm{tt}_\alpha} R_{C^{t*}_U}$.
$\square$

$\square$

In fact, we have another application of α-restriction to ordinary Kolmogorov complexity. Let us define a limited class of DTTR$_C$:

Definition 16. *For a function* $\alpha : \mathbb{N} \to \mathbb{N}$, DTTR$_{C,\alpha}$ *contains all the decidable languages* L *such that for any universal Turing machine,* $L \leq^p_{\mathrm{tt}_\alpha} R_{C_U}$. *Let* DTTR$_C$ *denote* DTTR$_{C,0}$.

Although we failed to prove any upper bound for DTTR$_C$, we show that DTTR$_{C,\alpha}$ sits within PSPACE for any slow-growing function α, whose proof we defer to the full version.

Theorem 17. BPP $\subseteq$ DTTR$_{C,\alpha}$ $\subseteq$ PSPACE *for any nondecreasing unbounded computable function* $\alpha(n) = O(\log n)$.

5   Using the Set of Short Random Strings as Advice

As discussed in Section 4, all we need to show the lower bound of BPP is the hardness of $R_{C^t_U}{}^{=c \log n}$, where the function log is regarded as log : $\mathbb{N} \to \mathbb{N}$. Thus, even if we further restrict the α-restricted reduction into one that can query only strings of length $c \log n$, we still have the same lower bound. It appears natural to present a class defined by the restricted reduction as advice:

Definition 18. *A language* L *is in* P$/R_C^{=\log}$ *if for some constant* $c \in \mathbb{N}$, *for all large time bounds* t *and each time-efficient universal Turing machine* U, *there exists a polynomial-time Turing machine* M *such that for any* $n \in \mathbb{N}$, *for all* $x \in \{0,1\}^n$, $M(x, \llcorner R_{C^t_U}{}^{=c \log n} \lrcorner) = L(x)$, *where* $\llcorner A^{=m} \lrcorner \in \{0,1\}^{2^m}$ *denotes a standard encoding of* $A^{=m} \subseteq \{0,1\}^m$.

The notation P$/R_C^{=\log}$ gives some intuition, but does not express all. In particular, random strings in Definition 18 are defined by time-bounded Kolmogorov complexity.

The class P$/R_C^{=\log}$ can be regarded as a limited class of TTRT$_{C,\log}$. Therefore, by Theorem 15, we have P$/R_C^{=\log} \subseteq$ PSPACE $\cap$ P$/$poly. In fact, we can get a better uniform upper bound:

Theorem 19. BPP $\subseteq$ P$/R_C^{=\log} \subseteq \Sigma_2^p \cap$ P$/$poly.

Proof. It is obvious that P$/R_C^{=\log} \subseteq$ P$/$poly since the length of advice is at most polynomial in n. To prove BPP $\subseteq$ P$/R_C^{=\log}$, one can show that, in order to construct a pseudorandom generator, the constant c in Definition 18 can be chosen so that it depends on neither a time-efficient universal machine U nor sufficiently large time bounds t (see the full version for a proof).

The proof of the inclusion $P/R_C^{=\log} \subseteq \Sigma_2^p$ is based on that of Theorem 15. Assume, by way of contradiction, that $L \in P/R_C^{=\log} \setminus \Sigma_2^p$. Then there exists a constant c given in Definition 18. Define α as $\alpha(n) = c \log n$. The advice of random strings can be regarded as an α-restricted truth-table reduction to the set of random strings. Therefore, we can prove the upper bound in the same way with Theorem 15.

There is one change in the rule of the games. Recall that R denotes the current knowledge of $R_{C_U^{t*}}$. The current value of the game $\mathcal{G}_{e,x}$ is equal to $\gamma_e(x, \llcorner R^{=\alpha(n)} \lrcorner)$, where $n = |x|$. In the proof of Theorem 15, for $r \in \{\alpha(|x|), \cdots, l\}$, there is the rth round, where r corresponds to the length of queries. Now that there are only queries of length $\alpha(|x|)$, there is only the $\alpha(|x|)$th round. Thus the player who has a winning strategy, $WIN_{e,x}$, can be computed in Σ_2^p as shown in the proof of the next claim.

Claim 6 (Revised Claim 4). M can find a string x such that $\alpha(|x|) > l$ and $WIN_{e,x} \neq L(x)$.

Proof. First, the YES player decides a subset $Z_1 \subseteq \{0,1\}^{\alpha(n)}$, where $|Z_1| \leq 2^{\alpha(n)-2}$. Next, the NO player decides a subset $Z_0 \subseteq \{0,1\}^{\alpha(n)}$, where $|Z_0| \leq 2^{\alpha(n)-2}$. Then the winning player corresponds to $\gamma_e\left(x, \llcorner R^{=\alpha(n)} \setminus (Z_1 \cup Z_0) \lrcorner\right) =: v(Z_1, Z_0)$. Therefore, $WIN_{e,x} = 1$ if and only if $\exists Z_1, \forall Z_0, v(Z_1, Z_0) = 1$, which is decidable in Σ_2^p. If the claim is false, then $L(x) = WIN_{e,x}$, and it is decidable in Σ_2^p, which contradicts the assumption. $\square$

Claim 7 (Revised Claim 5). For any polynomial-time Turing machine γ_e, there exists a string x such that $\gamma_e(x, \llcorner R_{C_U^{t*}}^{=\alpha(n)} \lrcorner) \neq L(x)$.

Proof. By Claim 6, M can find a string x such that $WIN_{e,x} \neq L(x)$. Since the $WIN_{e,x}$ side plays the games according to a winning strategy, after the game, $\gamma_e(x, \llcorner R_{C_U^{t*}}^{=\alpha(n)} \lrcorner) = WIN_{e,x} \neq L(x)$. $\square$

The claim above contradicts. $\square$

At the end, we should point out that there is similar work related to the class $P/R_C^{=\log}$. Buhrman et al. [5] also tried to show that using random strings as a source of randomness is the only way to make use of it. They modeled the initial segment of random strings as advice with (unrealistic) hardness. While their argument is incomplete in that it relies on the unrealistic hardness, it may provide some improvement on our result. Let us illustrate their work in terms of our notation. They used the unrealistic hardness in order to show the following hypothesis that "good advice" has polynomial density:

Hypothesis 1. If a language $L \in P/R_C^{=\log}$, then there exists a Turing machine M such that for all large $n \in \mathbb{N}$ and for some $m = n^{O(1)}$,

$$\Pr_{r \in \{0,1\}^m} [\forall x \in \{0,1\}^n, M(x,r) = L(x)] \geq \frac{1}{n^{O(1)}}.$$

Under this hypothesis, by using their result [5, Theorem 15], we can show that the equality between $P/R_C^{=\log}$ and a certain class is exactly equal to the equality between BPP and it (see the full version for a proof), and hence Conjecture 5 is more plausible under Hypothesis 1.

Proposition 20. *If Hypothesis 1 is true, then for any class* $C \in \{\mathrm{NP}, \mathrm{P}^{\#\mathrm{P}},$ $\mathrm{PSPACE}, \mathrm{EXP}\}$, $C = P/R_C^{=\log}$ *if and only if* $C = \mathrm{BPP}$.

Acknowledgement. The authors thank Eric Allender for introducing the second author to the relevant line of research. They are also grateful for the advice and suggestions provided by Hiroshi Imai and the members of his group at the University of Tokyo, where part of this research was conducted as the first author's bachelor thesis. This work was supported in part by KAKENHI.

References

1. Allender, E.: Curiouser and curiouser: The link between incompressibility and complexity. In: Cooper, S.B., Dawar, A., Löwe, B. (eds.) CiE 2012. LNCS, vol. 7318, pp. 11–16. Springer, Heidelberg (2012)
2. Allender, E., Buhrman, H., Friedman, L., Loff, B.: Reductions to the set of random strings: The resource-bounded case. In: Rovan, B., Sassone, V., Widmayer, P. (eds.) MFCS 2012. LNCS, vol. 7464, pp. 88–99. Springer, Heidelberg (2012)
3. Allender, E., Friedman, L., Gasarch, W.: Limits on the computational power of random strings. Information and Computation 222, 80–92 (2013)
4. Arora, S., Barak, B.: Computational Complexity: A Modern Approach, 1st edn. Cambridge University Press (2009)
5. Buhrman, H., Fortnow, L., Koucký, M., Loff, B.: Derandomizing from random strings. In: Proceedings of the 25th Annual Conference on Computational Complexity, CCC 2010, pp. 58–63 (2010)
6. Impagliazzo, R., Wigderson, A.: P = BPP if E requires exponential circuits: Derandomizing the xor lemma. In: Proceedings of the 29th Annual ACM Symposium on Theory of Computing, STOC 1997, pp. 220–229 (1997)
7. Li, M., Vitányi, P.: An Introduction to Kolmogorov Complexity and Its Applications, 3rd edn. Springer Publishing Company (2008)

New Results for Non-Preemptive Speed Scaling

Chien-Chung Huang[1] and Sebastian Ott[2]

[1] Chalmers University, Göteborg, Sweden
villars@gmail.com
[2] Max-Planck-Institut für Informatik, Saarbrücken, Germany
ott@mpi-inf.mpg.de

Abstract. We consider the speed scaling problem introduced in the seminal paper of Yao et al. [23]. In this problem, a number of jobs, each with its own processing volume, release time, and deadline, needs to be executed on a speed-scalable processor. The power consumption of this processor is $P(s) = s^\alpha$, where s is the processing speed, and $\alpha > 1$ is a constant. The total energy consumption is power integrated over time, and the objective is to process all jobs while minimizing the energy consumption.

The preemptive version of the problem, along with its many variants, has been extensively studied over the years. However, little is known about the non-preemptive version of the problem, except that it is strongly NP-hard and allows a (large) constant factor approximation [5,7,15]. Up until now, the (general) complexity of this problem is unknown. In the present paper, we study an important special case of the problem, where the job intervals form a laminar family, and present a quasipolynomial-time approximation scheme for it, thereby showing that (at least) this special case is not APX-hard, unless NP $\subseteq$ DTIME($2^{poly(\log n)}$).

The second contribution of this work is a polynomial-time algorithm for the special case of equal-volume jobs. In addition, we show that two other special cases of this problem allow fully polynomial-time approximation schemes.

1 Introduction

Speed scaling is a widely applied technique for energy saving in modern microprocessors. Its general idea is to strategically adjust the processing speed, with the dual goals of finishing the tasks at hand in a timely manner while minimizing the energy consumption. The following theoretical model was introduced by Yao et al. in their seminal paper of 1995 [23]. We are given a set of jobs, each with its own *volume* v_j (number of CPU cycles needed for completion of this job), *release time* r_j (when the job becomes available), and *deadline* d_j (when the job needs to be finished), and a processor with power function $P(s) = s^\alpha$, where s is the processing speed, and $\alpha > 1$ is a constant (typically between two and three for modern microprocessors [12,22]). The energy consumption is power integrated over time, and the objective is to process all given jobs within their *time windows* $[r_j, d_j)$, while minimizing the total energy consumption.

Most work in the literature focuses on the *preemptive* version of the problem, where the execution of a job may be interrupted and resumed at a later point of

E. Csuhaj-Varjú et al. (Eds.): MFCS 2014, Part II, LNCS 8635, pp. 360–371, 2014.
© Springer-Verlag Berlin Heidelberg 2014

time. For this setting, Yao et al. [23] gave a polynomial-time exact algorithm to compute the optimal schedule. The *non-preemptive* model, where a job must be processed uninterruptedly until its completion, has so far received surprisingly little attention, even though it is often preferred in practice and widely used in current real-life applications. For example, most current real-time operating systems for automotive applications use non-preemptive scheduling as defined by the OSEK/VDX standard [21]. The advantage of this strategy lies in the significant lower overhead (preemption requires to memorize and restore the state of the system and the job) [5], and the avoidance of synchronization efforts for shared resources [21]. From a theoretical point of view, the non-preemptive model is of interest, since it is a natural variation of Yao et al.'s original model. So far, little is known about the complexity of the non-preemptive speed scaling problem. On the negative side, no lower bound is known, except that the problem is strongly NP-hard [5]. On the positive side, Antoniadis and Huang [5] showed that the problem has a constant factor approximation algorithm, although the obtained factor $2^{5\alpha-4}$ is rather large. Recently, Bampis et al. [7] and Cohen-Addad et al. [15] have significantly improved on the constant.

1.1 Our Results and Techniques

In this paper, we work towards better understanding the complexity of the non-preemptive speed scaling problem, by considering several special cases and presenting (near-)optimal algorithms. In the following, we give a summary of our results.

Laminar Instances: An instance is said to be *laminar* if for any two different jobs j_1 and j_2, either $[r_{j_1}, d_{j_1}) \subseteq [r_{j_2}, d_{j_2})$, or $[r_{j_2}, d_{j_2}) \subseteq [r_{j_1}, d_{j_1})$, or $[r_{j_1}, d_{j_1}) \cap [r_{j_2}, d_{j_2}) = \emptyset$. The problem remains strongly NP-hard for this case [5]. We present the first $(1 + \epsilon)$-approximation for this problem, with a quasipolynomial running time (i.e. a running time bounded by $2^{poly(\log n)}$ for any fixed $\epsilon > 0$); a so-called quasipolynomial-time approximation scheme (QPTAS). Our result implies that laminar instances are not APX-hard, unless NP $\subseteq$ DTIME($2^{poly(\log n)}$). We remark that laminar instances form an important subclass of instances that not only arise commonly in practice (e.g. when jobs are created by recursive function calls [18]), but are also of theoretical interest, as they highlight the difficulty of the non-preemptive speed scaling problem: Taking instances with an "opposite" structure, namely *agreeable instances* (here for any two jobs j_1 and j_2 with $r_{j_1} < r_{j_2}$, it holds that $d_{j_1} < d_{j_2}$), the problem becomes polynomial-time solvable [5]. On the other hand, further restricting the instances from laminar to *purely-laminar* (see next case) results in a problem that is only weakly NP-hard and admits an FPTAS.

Purely-Laminar Instances: An instance is said to be *purely-laminar* if for any two different jobs j_1 and j_2, either $[r_{j_1}, d_{j_1}) \subseteq [r_{j_2}, d_{j_2})$, or $[r_{j_2}, d_{j_2}) \subseteq [r_{j_1}, d_{j_1})$. We present a fully polynomial-time approximation scheme (FP-TAS) for this class of instances. This is the best possible result (unless P = NP), as the problem is still (weakly) NP-hard [5].

Equal-Volume Jobs: If all jobs have the same volume $v_1 = v_2 = \ldots = v_n = v$, we present a polynomial-time algorithm for computing an (exact) optimal schedule. We thereby improve upon a recent result of Bampis et al. [6], who proposed a 2^α-approximation algorithm, and answer their question for the complexity status of this problem.

Bounded Number of Time Windows: If the total number of different time windows is bounded by a constant, we present an FPTAS for the problem. This result is again optimal (unless P = NP), as the problem remains (weakly) NP-hard even if there are only two different time windows [5].

The basis of all our results is a discretization of the problem, in which we allow the processing of any job to start and end only at a carefully chosen set of *grid points* on the time axis. We then use dynamic programming to solve the discretized problem. For laminar instances, however, even computing the optimal discretized solution is hard. The main technical contribution of our QPTAS is a relaxation that decreases the exponential size of the DP-tableau without adding too much energy cost. For this, we use a kind of overly compressed representation of job sets in the bookkeeping. Roughly speaking, we "lose" a number of jobs in each step of the recursion, but we ensure that these jobs can later be scheduled with only a small increment of energy cost.

1.2   Related Work

The study of dynamic speed scaling problems for reduced energy consumption was initiated by Yao, Demers, and Shenker in 1995. In their seminal paper [23], they presented a polynomial-time algorithm for finding an optimal schedule when preemption of jobs is allowed. Furthermore, they also studied the online version of the problem (again with preemption of jobs allowed), where jobs become known only at their release times, and developed two constant-competitive algorithms called *Average Rate* and *Optimal Available*.

Over the years, a rich spectrum of variations and generalizations of the original model have been investigated, mostly with a focus on the preemptive version. Irani et al. [17], for instance, considered a setting where the processor additionally has a sleep state available. Another extension of the original model is to restrict the set of possible speeds that we may choose from, for example by allowing only a number of discrete speed levels [14,19], or bounding the maximum possible speed [8,13,16]. Variations with respect to the objective function have also been studied, for instance by Albers and Fujiwara [2] and Bansal et al. [10], who tried to minimize a combination of energy consumption and total flow time of the jobs. Finally, the problem has also been studied for arbitrary power functions [9], as well as for multiprocessor settings [1,3,11].

In contrast to this diversity of results, the non-preemptive version of the speed scaling problem has been addressed rarely in the literature. Only in 2012, Antoniadis and Huang [5] proved that the problem is strongly NP-hard, and gave a $2^{5\alpha-4}$-approximation algorithm for the general case. Recently, the approximation ratio has been improved to $2^{\alpha-1}(1+\epsilon)\tilde{B}_\alpha$, where $\tilde{B}_\alpha$ is the α-th generalized

Bell number, by Bampis et al. [7], and to $(12(1 + \epsilon))^{\alpha-1}$ by Cohen-Addad et al. [15]. For the special case where all jobs have the same volume, Bampis et al. [6] proposed a 2^{α}-approximation algorithm. Independently of our result for this setting, Angel et al. [4] also gave a polynomial-time exact algorithm for such instances, with the same complexity of $\mathcal{O}(n^{21})$.

Very recently, multi-processor non-preemptive speed scaling also started to draw the attention of researchers. See [6,15] for details.

1.3 Organization of the Paper

Our paper is organized as follows. In section 2 we give a formal definition of the problem and establish a couple of preliminaries. In section 3 we present a QPTAS for laminar instances, and in section 4 we present a polynomial-time algorithm for instances with equal-volume jobs. Our FPTASs for purely-laminar instances and instances with a bounded number of different time windows can be found in the full version of this paper. Due to space constraints, most proofs are also deferred to the full version.

2 Preliminaries and Notations

The input is given by a set $\mathcal{J}$ of n jobs, each having its own release time r_j, deadline d_j, and volume $v_j > 0$. The power function of the speed-scalable processor is $P(s) = s^{\alpha}$, with $\alpha > 1$, and the energy consumption is power integrated over time. A *schedule* specifies for any point of time (i) which job to process, and (ii) which speed to use. A schedule is called *feasible* if every job is executed entirely within its time window $[r_j, d_j)$, which we will also call the *allowed interval* of job j. Preemption is not allowed, meaning that once a job is started, it must be executed entirely until its completion. Our goal is to find a feasible schedule of minimum total energy consumption.

We use $E(S)$ to denote the total energy consumed by a given schedule S, and $E(S, j)$ to denote the energy used for the processing of job j in schedule S. Furthermore, we use OPT to denote the energy consumption of an optimal schedule. A crucial observation is that, due to the convexity of the power function $P(s) = s^{\alpha}$, it is never beneficial to vary the speed during the execution of a job. This follows from Jensen's Inequality. We can therefore assume that in an optimal schedule, every job is processed using a uniform speed.

In the following, we restate a proposition from [5], which allows us to speed up certain jobs without paying too much additional energy cost.

Proposition 1. *Let S and S' be two feasible schedules that process j using uniform speeds s and $s' > s$, respectively. Then $E(S', j) = (s'/s)^{\alpha-1} \cdot E(S, j)$.*

As mentioned earlier, all our results rely on a discretization of the time axis, in which we focus only on a carefully chosen set of time points. We call these points *grid points* and define *grid point schedules* as follows.

Definition 1 (Grid Point Schedule). *A schedule is called* grid point schedule *if the processing of every job starts and ends at a grid point.*

We use two different sets of grid points, $\mathcal{P}_{\text{approx}}$ and $\mathcal{P}_{\text{exact}}$. The first set, $\mathcal{P}_{\text{approx}}$, is more universal, as it guarantees the existence of a near-optimal grid point schedule for any kind of instances. On the contrary, the set $\mathcal{P}_{\text{exact}}$ is specialized for the case of equal-volume jobs, and on such instances guarantees the existence of a grid point schedule with energy consumption exactly OPT. We now give a detailed description of both sets. For this, let us call a time point t an *event* if $t = r_j$ or $t = d_j$ for some job j, and let $t_1 < t_2 < \ldots < t_p$ be the set of ordered events. We call the interval between two consecutive events t_i and t_{i+1} a *zone*. Furthermore, let $\gamma := 1 + \lceil 1/\epsilon \rceil$, where $\epsilon > 0$ is the error parameter of our approximation schemes.

Definition 2 (Grid Point Set $\mathcal{P}_{\text{approx}}$). *The set $\mathcal{P}_{\text{approx}}$ is obtained in the following way. First, create a grid point at every event. Secondly, for every zone (t_i, t_{i+1}), create $n^2\gamma - 1$ equally spaced grid points that partition the zone into $n^2\gamma$ many subintervals of equal length $L_i = \frac{t_{i+1}-t_i}{n^2\gamma}$. Now $\mathcal{P}_{\text{approx}}$ is simply the union of all created grid points.*

Note that the total number of grid points in $\mathcal{P}_{\text{approx}}$ is at most $\mathcal{O}\left(n^3(1+\frac{1}{\epsilon})\right)$, as there are $\mathcal{O}(n)$ zones, for each of which we create $\mathcal{O}(n^2\gamma)$ grid points.

Lemma 1. *There exists a grid point schedule $\mathcal{G}$ with respect to $\mathcal{P}_{\text{approx}}$, such that $E(\mathcal{G}) \leq (1+\epsilon)^{\alpha-1}\text{OPT}$.*

Definition 3 (Grid Point Set $\mathcal{P}_{\text{exact}}$). *For every pair of events $t_i \leq t_j$, and for every $k \in \{1,\ldots,n\}$, create $k-1$ equally spaced grid points that partition the interval $[t_i, t_j]$ into k subintervals of equal length. Furthermore, create a grid point at every event. The union of all these grid points defines the set $\mathcal{P}_{\text{exact}}$.*

Clearly, the total number of grid points in $\mathcal{P}_{\text{exact}}$ is $\mathcal{O}(n^4)$.

Lemma 2. *If all jobs have the same volume $v_1 = v_2 = \ldots = v_n = v$, there exists a grid point schedule $\mathcal{G}$ with respect to $\mathcal{P}_{\text{exact}}$, such that $E(\mathcal{G}) = \text{OPT}$.*

3   Laminar Instances

In this section, we present a QPTAS for laminar problem instances. We start with a small example to motivate our approach, in which we reuse some ideas of Muratore et al. [20] for a different scheduling problem. Consider Figure 1, where we have drawn a number of (laminar) time intervals, purposely arranged in a tree structure. Imagine that for each of those intervals I_k, we are given a set of jobs J_k whose allowed interval is equal to I_k. Furthermore, let us make the simplifying assumption that no job can "cross" the boundary of any interval I_k during its execution. Then, in any feasible schedule, the set of jobs J_1 at the root of the tree decomposes into two subsets; the set of jobs processed in the left child I_2, and the set of jobs processed in the right child I_3. Having a recursive

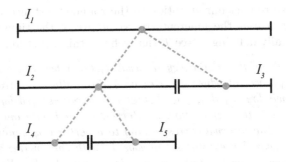

Fig. 1. Time intervals of a laminar instance, arranged in a tree structure

procedure in mind, we can think of the jobs in the root as being split up and handed down to the respective children. Each child then has a set of "inherited" jobs, plus its own original jobs to process, and both are available throughout its whole interval. Now, the children also split up their jobs, and hand them down to the next level of the tree. This process continues until we finally reach the leaves of the tree, where we can simply execute the given jobs at a uniform speed over the whole interval.

Aiming for a reduced running time, we reverse the described process and instead compute the schedules in a bottom-up manner via dynamic programming, enumerating all possible sets of jobs that a particular node could "inherit" from its ancestors. This dynamic programming approach is the core part of our QP-TAS, though it bears two major technical difficulties. The first one is that a job from a father node could also be scheduled "between" its children, starting in the interval of child one, stretching over its boundary, and entering the interval of child two. We overcome this issue by taking care of such jobs separately, and additionally listing the truncated child-intervals in the dynamic programming tableau. The second (and main) difficulty is the huge number of possible job sets that a child node could receive from its parent. Reducing this number requires a controlled "omitting" of small jobs during the recursion, and a condensed representation of job sets in the DP tableau. At any point of time, we ensure that "omitted" jobs only cause a small increment of energy cost when being added to the final schedule. We now elaborate the details, beginning with a rounding of the job volumes. Let $\mathcal{I}$ be the original problem instance.

Definition 4 (Rounded Instance). *The* rounded instance $\mathcal{I}'$ *is obtained by rounding down every job volume v_j to the next smaller number of the form $v_{\min}(1 + \epsilon)^i$, where $i \in \mathbb{N}_{\geq 0}$ and $v_{\min}$ is the smallest volume of any job in the original instance. The numbers $v_{\min}(1 + \epsilon)^i$ are called* size classes, *and a job belongs to* size class $\mathcal{C}_i$ *if its rounded volume is $v_{\min}(1 + \epsilon)^i$.*

Lemma 3. *Every feasible schedule S' for $\mathcal{I}'$ can be transformed into a feasible schedule S for $\mathcal{I}$ with $E(S) \leq (1 + \epsilon)^\alpha E(S')$.*

From now on, we restrict our attention to the rounded instance $\mathcal{I}'$. Remember that our approach uses the inherent tree structure of the time windows. We proceed by formally defining a tree T that reflects this structure.

Definition 5 (Tree T). *For every interval $[t_i, t_{i+1})$ between two consecutive events t_i and t_{i+1}, we introduce a vertex v. Additionally, we introduce a vertex for every time window $[r_j, d_j)$, $j \in \mathcal{J}$ that is not represented by a vertex yet. If several jobs share the same allowed interval, we add only one single vertex for this interval. The interval corresponding to a vertex v is denoted by I_v. We also associate a (possibly empty) set of jobs J_v with each vertex v, namely the set of jobs j whose allowed interval $[r_j, d_j)$ is equal to I_v. Finally, we specify a distinguished root node r as follows. If there exists a vertex v with $I_v = [r^*, d^*)$, where r^* is the earliest release time and d^* the latest deadline of any job in $\mathcal{J}$, we set $r := v$. Otherwise, we introduce a new vertex r with $I_r := [r^*, d^*)$ and $J_r := \emptyset$. The edges of the tree are defined in the following way. A node u is the son of a node v if and only if $I_u \subset I_v$ and there is no other node w with $I_u \subset I_w \subset I_v$. As a last step, we convert T into a binary tree by repeating the following procedure as long as there exists a vertex v with more than two children: Let v_1 and v_2 be two "neighboring" sons of v, such that $I_{v_1} \cup I_{v_2}$ forms a contiguous interval. Now create a new vertex u with $I_u := I_{v_1} \cup I_{v_2}$ and $J_u := \emptyset$, and make u a new child of v, and the new parent of v_1 and v_2. This procedure eventually results in a binary tree T with $\mathcal{O}(n)$ vertices.*

The main idea of our dynamic program is to stepwise compute schedules for subtrees of T, that is for the jobs associated with the vertices in the subtree (including its root), plus a given set of "inherited" jobs from its ancestors. Enumerating all possible sets of "inherited" jobs, however, would burst the limits of our DP tableau. Instead, we use a condensed representation of those sets via so-called *job vectors*, focusing only on a logarithmic number of size classes and ignoring jobs that are too small to be covered by any of these. To this end, let δ be the smallest integer such that $n/\epsilon \le (1+\epsilon)^\delta$, and note that δ is $\mathcal{O}(\log n)$ for any fixed $\epsilon > 0$.

Definition 6 (Job Vector). *A job vector $\overrightarrow{\lambda}$ is a vector of $\delta + 1$ integers $\lambda_0, \ldots, \lambda_\delta$. The first component λ_0 specifies a size class, namely the largest out of δ size classes from which we want to represent jobs (therefore $\lambda_0 \ge \delta - 1$). The remaining δ components take values between 0 and n each, and define a number of jobs for each of the size classes $\mathcal{C}_{\lambda_0}$, $\mathcal{C}_{\lambda_0 - 1}$, $\ldots$, $\mathcal{C}_{\lambda_0 - \delta + 1}$ in this order. For example, if $\delta = 2$, the job vector $(4, 2, 7)$ defines a set containing 2 jobs with volume $v_{\min}(1+\epsilon)^4$ and 7 jobs with volume $v_{\min}(1+\epsilon)^3$.*
We refer to the set of jobs defined by a job vector $\overrightarrow{\lambda}$ as $J(\overrightarrow{\lambda})$.

Remark: We do not associate a strict mapping from the jobs defined by a job vector $\overrightarrow{\lambda}$ to the real jobs (given as input) they represent. The jobs $J(\overrightarrow{\lambda})$ should rather be seen as dummies that are used to reserve space and can be replaced by any real job of the same volume.

Definition 7 (Heritable Job Vector). *A job vector* $\overrightarrow{\lambda} = (\lambda_0, \ldots, \lambda_\delta)$ *is heritable to a vertex v of T if:*

1. *At least λ_i jobs in* $\displaystyle\bigcup_{u \text{ ancestor of } v} J_u$ *belong to size class $C_{\lambda_0 - i + 1}$, for $1 \leq i \leq \delta$.*

2. *$\lambda_1 > 0$ or $\lambda_0 = \delta - 1$.*

The conditions on a heritable job vector ensure that for a fixed vertex v, λ_0 can take only $\mathcal{O}(n)$ different values, as it must specify a size class that really occurs in the rounded instance, or be equal to $\delta - 1$. Therefore, in total, we can have at most $\mathcal{O}(n^{\delta+1})$ different job vectors that are heritable to a fixed vertex of the tree. In order to control the error caused by the laxity of our job set representation, we introduce the concept of δ-omitted schedules.

Definition 8 (δ-omitted Schedule). *Let J be a given set of jobs. A δ-omitted schedule for J is a feasible schedule for a subset $R \subseteq J$, s.t. for every job $j \in J \setminus R$, there exists a job $\mathrm{big}(j) \in R$ with volume at least $v_j(1 + \epsilon)^\delta$ that is scheduled entirely inside the allowed interval of j. The jobs in $J \setminus R$ are called omitted jobs, the ones in R non-omitted jobs.*

Lemma 4. *Every δ-omitted schedule S' for a set of jobs J can be transformed into a feasible schedule S for all jobs in J, such that $E(S) \leq (1 + \epsilon)^\alpha E(S')$.*

The preceding lemma essentially ensures that representing the δ largest size classes of an "inherited" job set suffices if we allow a small increment of energy cost. The smaller jobs can then be added safely (i.e. without increasing the energy cost by too much) to the final schedule. We now turn to the central definition of the dynamic program. All schedules in this definition are with respect to the rounded instance $\mathcal{I}'$, and all grid points relate to the set $\mathcal{P}_{\mathrm{approx}}$.

Definition 9. *For any vertex v in the tree T, any job vector $\overrightarrow{\lambda}$ that is heritable to v, and any pair of grid points $g_1 \leq g_2$ with $[g_1, g_2] \subseteq I_v$, let $G(v, \overrightarrow{\lambda}, g_1, g_2)$ denote a minimum cost grid point schedule for the jobs in the subtree of v (including v itself) plus the jobs $J(\overrightarrow{\lambda})$ (these are allowed to be scheduled anywhere inside $[g_1, g_2)$) that uses only the interval $[g_1, g_2)$. Furthermore, let $S(v, \overrightarrow{\lambda}, g_1, g_2)$ be a δ-omitted schedule for the same set of jobs in the same interval $[g_1, g_2)$, satisfying $E\bigl(S(v, \overrightarrow{\lambda}, g_1, g_2)\bigr) \leq E\bigl(G(v, \overrightarrow{\lambda}, g_1, g_2)\bigr)$.*

Dynamic Program. Our dynamic program computes the schedules $S(v, \overrightarrow{\lambda}, g_1, g_2)$. For ease of exposition, we focus only on computing the energy consumption values $E(v, \overrightarrow{\lambda}, g_1, g_2) := E\bigl(S(v, \overrightarrow{\lambda}, g_1, g_2)\bigr)$, and omit the straightforward bookkeeping of the corresponding schedules. The base cases are the leaves of T. For a particular leaf node ℓ, we set

$$E(\ell, \overrightarrow{\lambda}, g_1, g_2) := \begin{cases} 0 & \text{if } J_\ell \cup J(\overrightarrow{\lambda}) = \emptyset \\ \dfrac{V^\alpha}{(g_2 - g_1)^{\alpha-1}} & \text{otherwise,} \end{cases}$$

where V is the total volume of all jobs in $J_\ell \cup J(\vec{\lambda})$. This corresponds to executing $J_\ell \cup J(\vec{\lambda})$ at uniform speed using the whole interval $[g_1, g_2)$. The resulting schedule is feasible, as no release times or deadlines occur in the interior of I_ℓ. Furthermore, it is also optimal by the convexity of the power function. Thus $E(\ell, \vec{\lambda}, g_1, g_2) \le E\big(G(\ell, \vec{\lambda}, g_1, g_2)\big)$.

When all leaves have been handled, we move on to the next level, the parents of the leaves. For this and also the following levels up to the root r, we compute the values $E(v, \vec{\lambda}, g_1, g_2)$ recursively, using the procedure COMPUTE in Figure 2. An intuitive description of the procedure is given below.

COMPUTE $(v, \vec{\lambda}, g_1, g_2)$:

Let v_1 and v_2 be the children of v, such that I_{v_1} is the earlier of the intervals I_{v_1}, I_{v_2}. Furthermore, let g be the grid point at which I_{v_1} ends and I_{v_2} starts.

Initialize MIN $:= \infty$.

For all gridpoints $\tilde{g}_1, \tilde{g}_2$, s.t. $g_1 \le \tilde{g}_1 < g < \tilde{g}_2 \le g_2$, and all jobs $j \in J_v \cup J(\vec{\lambda})$, **do:**

$$E := \frac{v_j{}^\alpha}{(\tilde{g}_2 - \tilde{g}_1)^{\alpha-1}}; \quad \tilde{J} := \big(J_v \cup J(\vec{\lambda})\big) \setminus \{j\}; \quad \vec{\gamma} := \text{VECTOR}(\tilde{J}).$$

$$\text{MIN} := \min\Big\{\text{MIN},$$
$$\min\{E + E(v_1, \vec{\gamma_1}, g_1, \tilde{g}_1) + E(v_2, \vec{\gamma_2}, \tilde{g}_2, g_2) : J(\vec{\gamma_1}) \cup J(\vec{\gamma_2}) = J(\vec{\gamma})\}\Big\}.$$

$\tilde{J} := J_v \cup J(\vec{\lambda}); \quad \vec{\gamma} := \text{VECTOR}(\tilde{J}).$

$a_1 := \min\{g_1, g\}; \quad a_2 := \min\{g_2, g\}; \quad b_1 := \max\{g_1, g\}; \quad b_2 := \max\{g_2, g\}.$

$$E(v, \vec{\lambda}, g_1, g_2) := \min\Big\{\text{MIN},$$
$$\min\{E(v_1, \vec{\gamma_1}, a_1, a_2) + E(v_2, \vec{\gamma_2}, b_1, b_2) : J(\vec{\gamma_1}) \cup J(\vec{\gamma_2}) = J(\vec{\gamma})\}\Big\}.$$

VECTOR $(\tilde{J})$:

Let $\mathcal{C}_\ell$ be the largest size class of any job in $\tilde{J}$.

$i := \max\{\ell, \delta - 1\}$.

For $k := i - \delta + 1, \ldots, i$ **do:** $x_k := |\{p \in \tilde{J} : p \text{ belongs to size class } \mathcal{C}_k\}|$.

Return $(i, x_i, x_{i-1}, \ldots, x_{i-\delta+1})$.

Fig. 2. Procedure for computing the remaining entries of the DP

Our first step is to iterate through all possible options for a potential "crossing" job j, whose execution interval $[\tilde{g}_1, \tilde{g}_2)$ stretches from child v_1 into the interval of child v_2. For every possible choice, we combine the optimal energy cost E for this job (obtained by using a uniform execution speed) with the best possible way to split up the remaining jobs between the truncated intervals of v_1 and v_2. Here we consider only the δ largest size classes of the remaining jobs $\tilde{J}$, and omit the smaller jobs. This omitting happens during the construction of a vector representation for $\tilde{J}$ using the procedure VECTOR. Finally, we also try the option that no "crossing" job exists and all jobs are split up between v_1 and

v_2. In this case we need to take special care of the subproblem boundaries, as $g_1 > g$ or $g_2 < g$ are also valid arguments for COMPUTE.

Lemma 5. *The schedules $S(v, \overrightarrow{\lambda}, g_1, g_2)$ constructed by the above dynamic program are δ-omitted schedules for the jobs in the subtree of v plus the jobs $J(\overrightarrow{\lambda})$. Furthermore, they satisfy $E\big(S(v, \overrightarrow{\lambda}, g_1, g_2)\big) \leq E\big(G(v, \overrightarrow{\lambda}, g_1, g_2)\big)$.*

Combining Lemmas 1, 3, 4, and 5 we can now state our main theorem. A detailed proof is provided in the full version of the paper.

Theorem 1. *The non-preemptive speed scaling problem admits a QPTAS if the instance is laminar.*

4   Equal-Volume Jobs

In this section, we consider the case that all jobs have the same volume $v_1 = v_2 = \ldots = v_n = v$. We present a dynamic program that computes an (exact) optimal schedule for this setting in polynomial time. All grid points used for this purpose relate to the set $\mathcal{P}_{\text{exact}}$.

As a first step, let us order the jobs such that $r_1 \leq r_2 \leq \ldots \leq r_n$. Furthermore, let us define an ordering on schedules as follows.

Definition 10 (Completion Time Vector). *Let $C_1, \ldots, C_n$ be the completion times of the jobs $j_1, \ldots, j_n$ in a given schedule S. The vector $\overrightarrow{S} := (C_1, \ldots, C_n)$ is called the completion time vector of S.*

Definition 11 (Lexicographic Ordering). *A schedule S is said to be lexicographically smaller than a schedule S' if the first component in which their completion time vectors differ is smaller in $\overrightarrow{S}$ than in $\overrightarrow{S'}$.*

We now elaborate the details of the DP, focusing on energy consumption values only.

Definition 12. *Let $i \in \{1, \ldots, n\}$ be a job index, and let g_1, g_2, and g_3 be grid points satisfying $g_1 \leq g_2 \leq g_3$. We define $E(i, g_1, g_2, g_3)$ to be the minimum energy consumption of a grid point schedule for the jobs $\{j_k \in \mathcal{J} : k \geq i \land g_1 < d_k \leq g_3\}$ that uses only the interval $[g_1, g_2)$.*

Dynamic Program. Our goal is to compute the values $E(i, g_1, g_2, g_3)$. To this end, we let

$$E(i, g_1, g_2, g_3) := \begin{cases} 0 & \text{if } \{j_k \in \mathcal{J} : k \geq i \land g_1 < d_k \leq g_3\} = \emptyset \\ \infty & \text{if } \exists k \geq i : g_1 < d_k \leq g_3 \land [r_k, d_k) \cap [g_1, g_2) = \emptyset. \end{cases}$$

Note that if $g_1 = g_2$, one of the above cases must apply. We now recursively compute the remaining values, starting with the case that g_1 and g_2 are consecutive grid points, and stepwise moving towards cases with more and more grid points in between g_1 and g_2. The recursion works as follows. Let $E(i, g_1, g_2, g_3)$

be the value we want to compute, and let j_q be the smallest index job in $\{j_k \in \mathcal{J} : k \geq i \wedge g_1 < d_k \leq g_3\}$. Furthermore, let $\mathcal{G}$ denote a lexicographically smallest optimal grid point schedule for the jobs $\{j_k \in \mathcal{J} : k \geq i \wedge g_1 < d_k \leq g_3\}$, using only the interval $[g_1, g_2)$. Our first step is to "guess" the grid points b_q and e_q that mark the beginning and end of j_q's execution interval in $\mathcal{G}$, by minimizing over all possible options. We then use the crucial observation that in $\mathcal{G}$, all jobs $J^- := \{j_k \in \mathcal{J} : k \geq q+1 \wedge g_1 < d_k \leq e_q\}$ are processed completely before j_q, and all jobs $J^+ := \{j_k \in \mathcal{J} : k \geq q+1 \wedge e_q < d_k \leq g_3\}$ are processed completely after j_q. For J^- this is obviously the case because of the deadline constraint. For J^+ this holds as all these jobs have release time at least r_q by the ordering of the jobs, and deadline greater than e_q by definition of J^+. Therefore any job in J^+ that is processed before j_q could be swapped with j_q, resulting in a lexicographic smaller schedule; a contradiction. Hence, we can use the following recursion to compute $E(i, g_1, g_2, g_3)$.

$$E(i, g_1, g_2, g_3) := \min \left\{ \frac{v_q{}^\alpha}{(e_q - b_q)^{\alpha-1}} + E(q+1, g_1, b_q, e_q) + E(q+1, e_q, g_2, g_3) : \right.$$

$$\left. (g_1 \leq b_q < e_q \leq g_2) \wedge (b_q \geq r_q) \wedge (e_q \leq d_q) \right\}.$$

Once we have computed all values, we output the schedule S corresponding to $E(1, r^*, d^*, d^*)$, where r^* is the earliest release time and d^* the latest deadline of any job in $\mathcal{J}$. Lemma 2 implies that $E(S) = \text{OPT}$. The running time complexity of this algorithm is $\mathcal{O}(n^{21})$: There are $\mathcal{O}(n^4)$ grid points in $\mathcal{P}_{\text{exact}}$, and thus $\mathcal{O}(n^{13})$ entries to compute. To calculate one entry, we need to minimize over $\mathcal{O}(n^8)$ different options.

Theorem 2. *The non-preemptive speed scaling problem admits a polynomial time algorithm if all jobs have the same volume.*

5   Conclusion

In this paper, we made a step towards narrowing down the complexity of the non-preemptive speed scaling problem. The most interesting open question is whether a (Q)PTAS is also possible for general instances. Some of our techniques, such as the grid point discretization or δ-omitted schedules, can also be applied to this setting. The problematic part is that our QPTAS relies on the tree structure of the time windows, which is only given in laminar instances. It is unclear whether and how this approach can be refined to deal with the general case.

References

1. Albers, S., Antoniadis, A., Greiner, G.: On multi-processor speed scaling with migration: extended abstract. In: SPAA, pp. 279–288. ACM (2011)
2. Albers, S., Fujiwara, H.: Energy-efficient algorithms for flow time minimization. In: Durand, B., Thomas, W. (eds.) STACS 2006. LNCS, vol. 3884, pp. 621–633. Springer, Heidelberg (2006)
3. Albers, S., Müller, F., Schmelzer, S.: Speed scaling on parallel processors. In: SPAA, pp. 289–298. ACM (2007)

4. Angel, E., Bampis, E., Chau, V.: Throughput maximization in the speed-scaling setting, arXiv:1309.1732
5. Antoniadis, A., Huang, C.-C.: Non-preemptive speed scaling. In: Fomin, F.V., Kaski, P. (eds.) SWAT 2012. LNCS, vol. 7357, pp. 249–260. Springer, Heidelberg (2012)
6. Bampis, E., Kononov, A., Letsios, D., Lucarelli, G., Nemparis, I.: From preemptive to non-preemptive speed-scaling scheduling. In: Du, D.-Z., Zhang, G. (eds.) COCOON 2013. LNCS, vol. 7936, pp. 134–146. Springer, Heidelberg (2013)
7. Bampis, E., Kononov, A., Letsios, D., Lucarelli, G., Sviridenko, M.: Energy efficient scheduling and routing via randomized rounding. In: FSTTCS. LIPIcs, vol. 24, pp. 449–460. Schloss Dagstuhl - Leibniz-Zentrum fuer Informatik (2013)
8. Bansal, N., Chan, H.-L., Lam, T.-W., Lee, L.-K.: Scheduling for speed bounded processors. In: Aceto, L., Damgård, I., Goldberg, L.A., Halldórsson, M.M., Ingólfsdóttir, A., Walukiewicz, I. (eds.) ICALP 2008, Part I. LNCS, vol. 5125, pp. 409–420. Springer, Heidelberg (2008)
9. Bansal, N., Chan, H.L., Pruhs, K.: Speed scaling with an arbitrary power function. In: SODA, pp. 693–701. SIAM (2009)
10. Bansal, N., Pruhs, K., Stein, C.: Speed scaling for weighted flow time. In: SODA, pp. 805–813. SIAM (2007)
11. Bingham, B.D., Greenstreet, M.R.: Energy optimal scheduling on multiprocessors with migration. In: ISPA, pp. 153–161. IEEE (2008)
12. Brooks, D., Bose, P., Schuster, S., Jacobson, H.M., Kudva, P., Buyuktosunoglu, A., Wellman, J.D., Zyuban, V.V., Gupta, M., Cook, P.W.: Power-aware microarchitecture: Design and modeling challenges for next-generation microprocessors. IEEE Micro 20(6), 26–44 (2000)
13. Chan, H.L., Chan, J.W.T., Lam, T.W., Lee, L.K., Mak, K.S., Wong, P.W.H.: Optimizing throughput and energy in online deadline scheduling. ACM Transactions on Algorithms 6(1) (2009)
14. Chen, J.-J., Kuo, T.-W., Lu, H.-I.: Power-saving scheduling for weakly dynamic voltage scaling devices. In: Dehne, F., López-Ortiz, A., Sack, J.-R. (eds.) WADS 2005. LNCS, vol. 3608, pp. 338–349. Springer, Heidelberg (2005)
15. Cohen-Addad, V., Li, Z., Mathieu, C., Mills, I.: Energy-efficient algorithms for non-preemptive speed-scaling, arXiv:1402.4111v2
16. Han, X., Lam, T.W., Lee, L.K., To, I.K.K., Wong, P.W.H.: Deadline scheduling and power management for speed bounded processors. Theor. Comput. Sci. 411(40-42), 3587–3600 (2010)
17. Irani, S., Shukla, S.K., Gupta, R.K.: Algorithms for power savings. In: SODA, pp. 37–46. ACM/SIAM (2003)
18. Li, M., Liu, B.J., Yao, F.F.: Min-energy voltage allocation for tree-structured tasks. In: Wang, L. (ed.) COCOON 2005. LNCS, vol. 3595, pp. 283–296. Springer, Heidelberg (2005)
19. Li, M., Yao, F.F.: An efficient algorithm for computing optimal discrete voltage schedules. In: Jedrzejowicz, J., Szepietowski, A. (eds.) MFCS 2005. LNCS, vol. 3618, pp. 652–663. Springer, Heidelberg (2005)
20. Muratore, G., Schwarz, U.M., Woeginger, G.J.: Parallel machine scheduling with nested job assignment restrictions. Oper. Res. Lett. 38(1), 47–50 (2010)
21. Negrean, M., Ernst, R.: Response-time analysis for non-preemptive scheduling in multi-core systems with shared resources. In: SIES, pp. 191–200. IEEE (2012)
22. Wierman, A., Andrew, L.L.H., Tang, A.: Power-aware speed scaling in processor sharing systems: Optimality and robustness. Perform. Eval. 69(12), 601–622 (2012)
23. Yao, F.F., Demers, A.J., Shenker, S.: A scheduling model for reduced cpu energy. In: FOCS, pp. 374–382. IEEE Computer Society (1995)

Lower Bounds for Splittings
by Linear Combinations⋆

Dmitry Itsykson and Dmitry Sokolov

Steklov Institute of Mathematics at St.Petersburg
27, Fontanka, St.Petersburg, Russia, 191023
`dmitrits@pdmi.ras.ru, sokolov.dmt@gmail.com`

Abstract. A typical DPLL algorithm for the Boolean satisfiability
problem splits the input problem into two by assigning the two possible values to a variable; then it simplifies the two resulting formulas. In
this paper we consider an extension of the DPLL paradigm. Our algorithms can split by an arbitrary linear combination of variables modulo
two. These algorithms quickly solve formulas that explicitly encode linear systems modulo two, which were used for proving exponential lower
bounds for conventional DPLL algorithms.

We prove exponential lower bounds on the running time of DPLL
with splitting by linear combinations on 2-fold Tseitin formulas and on
formulas that encode the pigeonhole principle.

Raz and Tzameret introduced a system $R(lin)$ which operates with
disjunctions of linear equalities with *integer coefficients*. We consider an
extension of the resolution proof system that operates with disjunctions
of linear equalities over $\mathbb{F}_2$; we call this system Res-Lin. Res-Lin can be
p-simulated in $R(lin)$ but currently we do not know any superpolynomial
lower bounds in $R(lin)$. Tree-like proofs in Res-Lin are equivalent to the
behavior of our algorithms on unsatisfiable instances. We prove that Res-
Lin is implication complete and also prove that Res-Lin is polynomially
equivalent to its semantic version.

1 Introduction

Splitting is the one of the most frequent methods for exact algorithms for NP-
hard problems. It considers several cases and recursively executes on each of that
cases. For the CNF satisfiability problem the classical splitting algorithms are so
called DPLL algorithms (by authors Davis, Putnam, Logemann and Loveland)
[6], [5] in which splitting cases are values of a variable. A very natural extension
of such algorithms is a splitting by a value of some formula. In this paper we
consider an extension of DPLL that allows splitting by linear combinations of
variables over $\mathbb{F}_2$. There is a polynomial time algorithm that check whether a
system of linear equations has a solution and whether a system of linear equations

---

⋆ The research is partially supported by the RFBR grant 14-01-00545, by the
  President's grant MK-2813.2014.1 and by the Government of the Russia (grant
  14.Z50.31.0030).

E. Csuhaj-Varjú et al. (Eds.): MFCS 2014, Part II, LNCS 8635, pp. 372–383, 2014.

contradicts a clause. Thus the running time of an algorithm that solves CNF-SAT using splitting by linear combinations (in the contrast to splitting by arbitrary functions) is at most the size of its splitting tree up to a polynomial factor.

Formulas that encode unsatisfiable systems of linear equations are hard for resolution and hence for DPLL [15], [3]. Systems of linear equations are also hard satisfiable examples for myopic and drunken DPLL algorithms [1], [9]. Hard examples for myopic algorithms with a cut heuristic are also based on linear systems [8]. We show that a splitting by linear combinations helps to solve explicitly encoded linear systems over $\mathbb{F}_2$ in polynomial time.

For every CNF formula ϕ we denote by $\phi^\oplus$ a CNF formula obtained from ϕ by substituting $x_1 \oplus x_2$ for each variable x. Urquhart shows that for unsatisfiable ϕ the running time of any DPLL algorithm on $\phi^\oplus$ is at least $2^{d(\phi)}$, where $d(\phi)$ is the minimal depth of the recursion tree of DPLL algorithms running on the input ϕ [16]. Urquhart also gives an example of Pebbling contradictions $Peb(G_n)$ such that $d(Peb(G_n)) = \Omega(n/\log n)$ and there is a DPLL algorithm that solves $Peb(G_n)$ in $O(n)$ steps. Thus $Peb^\oplus(G_n)$ is one more example that is hard for DPLL algorithms but easy for DPLL with splitting by linear combinations.

The recent algorithm by Seto and Tamaki [14] solves satisfiability of formulas over full binary basis using a splitting by linear combination of variables. The similar idea was used by Demenkov and Kulikov in the simplified lower bound $3n - o(n)$ for circuit complexity over full binary basis [7]. The common idea of [14] and [7] is that a restricting a circuit with a linear equation may significantly reduce the size of the circuit.

Our results. We prove an exponential lower bound on the size of a splitting tree by linear combinations for 2-fold Tseitin formulas that can be obtained from ordinary Tseitin formulas by substituting every variable by the conjunction of two new variables. The plan of the proof is following: let for every unsatisfiable formula ϕ a search problem $Search_\phi$ be the problem of finding falsified clause given a variable assignment. We prove that it is possible to transform a splitting tree T into a randomized communication protocol for the problem $Search_\phi$ of depth $O(\log |T| \log \log |T|)$ if some variables are known by Alice and other variables are known by Bob. And finally, we note that a lower bound on the randomized communication complexity of the problem $Search_{TS^2_{G,c}}$ for a 2-fold Tseitin formula $TS^2_{G,c}$ follows from [10] and [2].

We also give an elementary proof of the lower bound $2^{\frac{n-1}{2}}$ on the size of linear splitting trees of formulas PHP^m_n that encode the pigeonhole principle.

It is well known that the behavior of DPLL algorithms on unsatisfiable formulas corresponds to tree-like resolution proofs. We consider the extension of the resolution proof system that operates with disjunctions of linear equalities. A system Res-Lin contains the weakening rule and the resolution rule. We also consider a system Sem-Lin that is a semantic version of Res-Lin; Sem-Lin contains semantic implication rule with two premises instead of the resolution rule. We prove that this two systems are polynomially equivalent and they are implication complete. We also show that tree-like versions of Res-Lin and Sem-Lin are equivalent to linear splitting trees; the latter implies that our lower bounds hold for tree-like Res-Lin and Sem-Lin. Raz and Tzameret studied a system $R(lin)$ which operates with disjunctions of linear equalities with *integer coefficients* [12]. It is possible to p-simulate Res-Lin in $R(lin)$ but the existence of the simulation in the other direction is an open problem.

Futher research. The main open problem is to prove a superpolynomial lower bound in the DAG-like Res-Lin. One of the ways to prove a lower bound is to simulate the Res-Lin system with another system for which a superpolynomial lower bound is known. It is impossible to simulate Res-Lin with Res(k) (that extends Resolution and operate with k-DNF instead of clauses) and in PCR (Polynomial Calculus + Resolution) over field with *char* $\neq 2$ because there are known exponential lower bounds in Res(k) and PCR for formulas based on systems of linear equations [13]. It is interesting whether it is possible to simulate Res-Lin with Polynomial Calculus (or PCR) over $\mathbb{F}_2$ or with the system $R^0(lin)$ which is a subsystem of $R(lin)$ with known exponential lower bounds based on the interpolation. Another open problem is to prove lower bounds for splitting by linear combinations on satisfiable formulas, for example, for algorithms that arbitrary choose a linear combination for splitting and randomly choose a value to investigate first.

2   Preliminaries

We will use the following notation: $[n] = \{1, 2, \ldots, n\}$. Let $X = \{x_1, \ldots, x_n\}$ be a set of variables that take values from $\mathbb{F}_2$. A linear form is a polynomial $\sum_{i=1}^{n} \alpha_i x_i$ over $\mathbb{F}_2$.

Consider a binary tree T with edges labeled with linear equalities. For every vertex v of T we denote by Φ_v^T a system of all equalities that are written along the path from the root of T to v. A *linear splitting tree* for a CNF formula ϕ is a binary tree T with the following properties. Every internal node is labeled by a linear form that depends on variables from ϕ. For every internal node that is labeled by a linear form f one of the edges going to its children is labeled by $f = 0$ and the other edge is labeled by $f = 1$. For every leaf v of the tree exactly one of the following conditions hold: 1) The system Φ_v^T does not have solutions. We call such leaf degenerate. 2) The system Φ_v^T is satisfiable but contradicts a clause C of formula ϕ. We say that such leaf refutes C. 3) The system Φ_v^T has exactly one solution in the variables of ϕ and this solution satisfies the formula ϕ. We call such leaf satisfying.

A linear splitting tree may also be viewed as a recursion tree of an algorithm that searches for satisfying assignments of a CNF formula using the following recursive procedure. It gets on the input a CNF formula ϕ and a system of linear equations Φ, the goal of the algorithm is to find a satisfying assignment of $\phi \wedge \Phi$. Initially $\Phi = True$ and on every step it somehow chooses a linear form f and a value $\alpha \in \mathbb{F}_2$ and makes two recursive calls: on the input $(\phi, \Phi \wedge (f = \alpha))$ and on the input $(\phi, \Phi \wedge (f = 1 + \alpha))$. The algorithm backtracks in one of the three cases: 1) The system Φ does not have solutions (it can be verified in polynomial time); 2) The system Φ contradicts to a clause C of the formula ϕ (A system Ψ contradicts a clause $(\ell_1 \vee \ell_2 \vee \cdots \vee \ell_k)$ iff for all $i \in [k]$ the system $\Psi \wedge (\ell_i = 1)$ is unsatisfiable. Hence this condition may be verified in polynomial time); 3) The system Φ has the unique solution that satisfies ϕ (it can also be verified in polynomial time). Note that if it is enough to find just one satisfying assignment, then the algorithm may stop in the first satisfying leaf. But in the case of unsatisfiable formulas it must traverse the whole splitting tree.

Proposition 1. *For every linear splitting tree T for a formula ϕ it is possible to construct a splitting tree that has no degenerate leaves. The number of vertices in the new tree is at most the number of vertices in T.*

Proposition 2. *Let formula ϕ in CNF encode an unsatisfiable system of linear equations $\bigwedge_{i=1}^{m}(f_i = \beta_i)$ over $\mathbb{F}_2$. The i-th equation $f_i = \beta_i$ is represented by a CNF formula ϕ_i and $\phi = \bigwedge_{i=1}^{m} \phi_i$. It is possible that encodings of different formulas ϕ_i have the same clause; we assume that such clause is repeated in ϕ. Then there exists a splitting tree for ϕ of size $O(|\phi|)$.*

Proof. We will describe a binary tree T that has a path from the root to a leaf labeled by equalities $f_1 = \beta_1, f_2 = \beta_2, \ldots, f_m = \beta_m$; the leaf is degenerate since the system is unsatisfiable. For all $i \in [m]$ the i-th vertex on the path has the edge to the child u_i labeled by $f_i = \beta_i + 1$. Now we describe a subtree T_{u_i} with the root u_i; it is just a splitting tree over all variables of formula ϕ_i. Let $x_1, x_2, \ldots, x_k$ be variables that appear in f with nonzero coefficients. We sequentially make splittings on $x_1, x_2, \ldots x_k$ starting in u_i. We know that $\Phi_{u_i}^T$ contradicts ϕ_i, therefore every leaf of T_{u_i} either refutes clause of ϕ_i or is degenerate (the system contradicts $f_i = 1 + \beta_i$). T_{u_i} has 2^k leaves but it is well known that every CNF representation of $x_1 + x_2 + \cdots + x_k = \beta_i$ has at least 2^{k-1} clauses. Therefore the size of T is at most $O(|\phi|)$.

3   Lower Bound for 2-Fold Tseitin Formulas

In this section we prove a lower bound on the size of a linear splitting tree. The proof consists of two parts. At first we transform a splitting tree to a communication protocol and then we prove a lower bound on the communication complexity.

Communication protocol from linear splitting tree. Let ϕ be an unsatisfiable CNF formula. For every assignment of its variables there exists a clause of ϕ that is falsified by the assignment. By $Search_\phi$ we denote a search problem where the instances are variables assignments of and the solutions are clauses of ϕ that are falsified by the assignment.

Let's consider some function or search a problem f with inputs $\{0, 1\}^n$; the set $[n]$ is split into two disjoint sets X and Y. Alice knows bits of input corresponding to X and Bob knows bits of inputs corresponding to Y. A randomized communication protocol with public random bits and error ϵ is a binary tree such that every internal node v is labeled with a function of one of the two types: $a_v : \{0,1\}^X \times \{0,1\}^R \to \{0,1\}$ or $b_v : \{0,1\}^Y \times \{0,1\}^R \to \{0,1\}$, where R is an integer that denotes the number of random bits used by a protocol. For every internal node one of the edges to children is labeled with 0 and the other with 1, and leaves are labeled with strings (answers of a protocol). Assume that Alice knows $x \in \{0,1\}^X$ and Bob knows $y \in \{0,1\}^Y$; both of them know a random string $r \in \{0,1\}^R$. Alice and Bob communicate according to the protocol in the following way: initially they put a token in the root of the tree. Every time if the node with the token is labeled by a function of type a_v, then Alice computes the value of $a_v(x,r)$ and sends the result to Bob; and if the node is labeled by a function of type b_v, then Bob computes the value of $b_v(x,r)$ and sends the result to Alice. After this, both players move the token to the child that corresponds to the sent bit. The communication stops whenever the token moves to a leaf. The label in the leaf is the result of the communication with a given string of random bits r. It is required that with probability at least $1 - \epsilon$ over random choice of the string $r \leftarrow \{0,1\}^R$ the result of the protocol is a correct answer to the problem f. The complexity of a communication protocol is a depth of the tree or, equivalently, the number of bits that Alice and Bob must send in the worst case. By a randomized communication complexity with error ϵ of the problem f we call a number $R_\epsilon^{pub}(f)$ that equals the minimal complexity of a protocol that solves f. See [11] for more details.

Let $EQ : \{0,1\}^{2n} \to \{0,1\}$, and for all $x, y \in \{0,1\}^n$, $EQ(x,y) = 1$ iff $x = y$. When we study the communication complexity of EQ we assume that Alice knows x and Bob knows y.

Lemma 1 ([11]). $R_\delta^{pub}(EQ) \leq \lceil \log \frac{1}{\delta} \rceil + 1$.

Theorem 1. *Let ϕ be an unsatisfiable CNF formula and T be a linear splitting tree for ϕ. Then for every distribution of variables of ϕ between Alice and Bob, $R_{1/3}^{pub}(Search_\phi) = O(\log |T| \log \log |T|)$.*

Proof. We construct a communication protocol from the tree T without degenerate leaves. Alice and Bob together know an assignment π of variables of ϕ (Alice knows some bits of π and Bob knows the other bits of π). The assignment π determines a path ℓ_π in T that corresponds to edges with labels that are satisfied by π. This path contains a leaf that refutes some clause C_π of ϕ. The protocol that we are describing with high probability returns the clause C_π.

The protocol has $O(\log |T|)$ randomized rounds. In the analisys of the next round we will assume that all previous rounds do not contain errors. Thus the total error may be estimated as a sum of errors of the individual rounds. Both Alice and Bob at the beginning of the i-th round know a tree T_i that is a connected subgraph of T; $T_1 = T$. Since T_i is a connected subgraph of T, we may assume that the root of T_i is its highest vertex in T. Under the assumption that all previous rounds were correct we will ensure that T_i contains the part of the path ℓ_π that goes from the root of T_i to the leaf that refutes C_π. We also maintain inequality $|T_{i+1}| \leq \frac{2}{3}|T_i|$. Thus if T_i has only one vertex it would be the leaf of T that refutes C_π, therefore Alice and Bob will know C_π.

Let $|T_i| > 1$, then there exists such a vertex v of T_i that the size of the subtree of T_i with the root v (we denote it by $T_i^{(v)}$) is at least $\frac{1}{3}|T_i|$ and at most $\frac{2}{3}|T_i|$. The tree T_{i+1} equals $T_i^{(v)}$ if v belongs to the path ℓ_π and equals $T_i \setminus T_i^{(v)}$ otherwise. Alice and Bob, using a fixed algorithm, find the vertex v; now they have to verify whether v belongs to ℓ_π. The vertex v belongs to the path ℓ_π iff π satisfies all equalities that are written along the path from the root of T_i to v. Assume that we have to verify t equalities. When we verify the j-th equality, Alice have to compute the sum of her variables and Bob computes the sum of his variable. And we should verify that the sum of the results of Alice and Bob equals the right hand side of the equality α_j. Let the sum of Alice variables of the j-th equality plus α_i equals z_j and the sum of variables of Bob of the j-th equality equals y_j. All equalities are satisfied by π iff $EQ(z_1 z_2 \ldots z_t, y_1 y_2 \ldots y_t) = 1$. In order to compute EQ we use a protocol for EQ from the Lemma 1 with $\delta = \frac{1}{3\lceil \log_{3/2}|T|\rceil}$. Since the number of rounds is at most $\lceil \log_{3/2}|T|\rceil$, the total error of the protocol is at most $\frac{1}{3}$. The total depth of the protocol is at most the number of rounds $\lceil \log_{3/2}|T|\rceil$ times the depth of the EQ protocol $O(\log\log|T|)$. □

Lower bound on communication complexity. A Tseitin formula $TS_{G,c}$ can be constructed from an arbitrary graph $G(V,E)$ and a function $c : V \to \mathbb{F}_2$; variables of $TS_{G,c}$ correspond to edges of G. The formula $TS_{G,c}$ is a conjunction of the following conditions encoded in CNF for every vertex v: the parity of the number of edges incident to v that have value 1 is the same as the parity of $c(v)$. It is well known that $TS_{G,c}$ is unsatisfiable if and only if $\sum_{v \in V} c(v) = 1$.

A k-fold Tseitin formula $TS_{(G,c)}^k$ [2] can be obtained from Tseitin formula $TS_{G,c}$ if we substitute every variable x_i by a conjunction of k new variables $(z_{i1} \wedge z_{i2} \wedge \cdots \wedge z_{ik})$ and translate the resulting formula into CNF. Note that if the maximal degree of G is bounded by a constant, then for every constant k the formula $TS_{(G,c)}^k$ has CNF representation of size polynomial in $|V|$.

Theorem 2. *In time polynomial in n one may construct a graph $G(V,E)$ on n vertices with maximal degree bounded by a constant and a function $c : V \to \mathbb{F}_2$ such that $TS_{(G,c)}^2$ is unsatisfiable and $R_{1/3}^{pub}(Search_{TS_{(G,c)}^2}) = \Omega\left(\frac{n^{1/3}}{(\log(n)\log\log(n))^2}\right)$.*

Corollary 1. *In the condition of the Theorem 2 the size of any linear splitting tree of $TS_{(G,c)}^2$ is at least $\Omega\left(2^{n^{1/3}/\log^3(n)}\right)$.*

Proof (Proof of Corollary 1). Follows from Theorem 2 and Theorem 1. □

We define a function $DISJ_{n,2} : \{0,1\}^n \times \{0,1\}^n \to \{0,1\}$ that for all $x, y \in \{0,1\}^n$ $DISJ_{n,2}(x,y) = 1$ iff $x_i \wedge y_i = 0$ for all $i \in [n]$.

Theorem 3. *([2], Section 5) Let $m = \frac{n^{1/3}}{\log(n)}$, then in time polynomial in n one may construct a graph $G(V, E)$ on n vertices with maximal degree bounded by a constant and a function $c : V \to \mathbb{F}_2$ such that $TS^2_{(G,c)}$ is unsatisfiable and*

$$R^{pub}_\epsilon(DISJ_{m,2}) = O\left(R^{pub}_\epsilon(Search_{TS^2_{(G,c)}})\log(n)(\log\log(n))^2\right).$$

Lemma 2. *([10]) $R^{pub}_{1/3}(DISJ_{n,2}) = \Omega(n)$.*

Proof (Proof of Theorem 2). Let $m = \frac{n^{1/3}}{\log(n)}$. By Lemma 2, $R^{pub}_\epsilon(DISJ_{m,2}) = \Omega(m)$, then by theorem 3 it is possible to construct G and c such that $R^{pub}_{1/3}\left(Search_{TS^2_{(G,c)}}\right) = \Omega\left(\frac{n^{1/3}}{(\log(n)\log\log(n))^2}\right)$. ☐

4   Lower Bound for Pigeonhole Principle

In this section we prove a lower bound on the size of linear splitting trees for formulas PHP^m_n that encode the pigeonhole principle. Formula PHP^m_n has variables $p_{i,j}$, where $i \in [m]$, $j \in [n]$; $p_{i,j}$ states that i-th pigeon is in the j-th hole. A formula has the two types of clauses: 1) Long clauses that encode that every pigeon is in some hole: $p_{i,1} \vee p_{i,2} \cdots \vee p_{i,n}$ for all $i \in [m]$; 2) Short clauses that encode that every hole contains at most one pigeon: $\neg p_{i,k} \vee \neg p_{j,k}$ for all $i \neq j \in [m]$ and all $k \in [n]$. If $m > n$ then PHP^m_n is unsatisfiable.

We call an assignment of values of variables $p_{i,j}$ *acceptable* if it satisfies all short clauses. In other words in every acceptable assignment there are no holes with two or more pigeons.

Lemma 3. *Let a linear system $Ap = b$ from variables $p = (p_{i,j})_{i\in[m],j\in[n]}$ have at most $\frac{n-1}{2}$ equations and let it have an acceptable solution. Then for every $i \in [m]$ this system has an acceptable solution that satisfies the long clause $p_{i,1} \vee p_{i,2} \vee \cdots \vee p_{i,n}$.*

Proof. Note that if we change 1 to 0 in an acceptable assignment, then it remains acceptable. Let the system have k equations; we know that $k \leq \frac{n-1}{2}$. We consider an acceptable solution π of the system $Ap = b$ with the minimum number of ones. We prove that the number of ones in π is at most k. Let the number of ones is greater than k. Consider $k+1$ variables that take value 1 in π: $p_{j_1}, p_{j_2}, \ldots, p_{j_{k+1}}$. Since the matrix A has k rows, the columns that correspond to variables $p_{j_1}, p_{j_2}, \ldots, p_{j_{k+1}}$ are linearly depended. Therefore there exists a nontrivial solution π' of the homogeneous system $Ap = 0$ such that every variable with value one in π' is from the set $\{p_{j_1}, p_{j_2}, \ldots, p_{j_{k+1}}\}$. The assignment $\pi' + \pi$ is also a solution of $Ap = b$ and is acceptable because $\pi' + \pi$ can be obtained from π by changing ones to zeroes. Since π' is nontrivial, the number of ones in $\pi' + \pi$ is less than the number of ones in π and this contradicts the minimality of π.

The fact that π has at most k ones implies that π has at least $n-k$ empty holes. From the statement of the lemma we know that $n - k \geq k + 1$; we choose $k + 1$

empty holes with numbers $\ell_1, \ell_2, \ldots, \ell_{k+1}$. We fix $i \in [m]$; the columns of A that correspond to variables $p_{i,\ell_1}, \ldots, p_{i,\ell_{k+1}}$ are linearly depended, therefore there exists a nontrivial solution τ of the system $Ap = 0$ such that every variable with value 1 in τ is from the set $\{p_{i,\ell_1}, \ldots, p_{i,\ell_{k+1}}\}$. The assignment $\pi + \tau$ is a solution of $Ap = b$; $\pi + \tau$ is acceptable since holes with numbers $\ell_1, \ell_2, \ldots, \ell_{k+1}$ are empty in π, and τ puts at most one pigeon to them (if τ puts a pigeon in a hole, then this is the i-th pigeon). The assignment $\pi + \tau$ satisfies $p_{i,1} \vee p_{i,2} \vee \cdots \vee p_{i,n}$ because τ is nontrivial. $\qquad\Box$

Theorem 4. *For all $m > n$ every linear splitting tree for PHP_n^m has size at least $2^{\frac{n-1}{2}}$.*

Proof. We say that the equality $f = \alpha$ is acceptably implied from a linear system Φ if every acceptable solution of Φ satisfies $f = \alpha$.

We consider a linear splitting tree T for PHP_n^m. Remove from T all the vertices v for which Φ_v^T has no acceptable solutions. The resulting graph is a tree since if we remove a vertex, then we should remove its subtree, and the root of T is not removed. We denote this tree by T'. Note that it is impossible that a leaf of T' is not a leaf in T. Indeed, assume that v is labeled in T by a linear form f, then every acceptable assignment that satisfies Φ_v^T also satisfies one of the systems $\Phi_v^T \wedge (f = 1)$ or $\Phi_v^T \wedge (f = 0)$, so one of the children is not removed. Hence in every leaf ℓ of T' the system Φ_ℓ^T refutes a clause of PHP_n^m. Since there exists an acceptable assignment that satisfies Φ_ℓ^T, then Φ_ℓ^T can't refute short clause, therefore it refutes a long clause.

Consider a vertex v of T' with the only child u, let the edge (u, v) be labeled by $f = \alpha$. We know that the system $\Phi_v^T \wedge (f = 1 + \alpha)$ has no acceptable solutions. Hence the equality $f = \alpha$ is acceptably implied from Φ_v^T; and the sets of acceptable solutions of $\Phi_u^{T'}$ and $\Phi_v^{T'}$ are equal.

Let T' contain a vertex v with the only child u; we merge u and v in one vertex and remove the edge (u, v) with its label. We repeat this operation while the current tree has vertices with the only child. We denote the resulting tree by T''. Let V' be the set of vertices of T', and V'' be the set of vertices of T''. We define a surjective mapping $\mu : V' \to V''$ that maps a vertex from T' to a vertex of T'' into which it was merged. We know that for all $u \in T'$ the sets of acceptable solutions of $\Phi_u^{T'}$ and $\Phi_{\mu(u)}^{T''}$ are equal.

For every leaf ℓ'' of T'' there exists a leaf ℓ' of T' such that $\mu(\ell') = \ell''$, the system $\Phi_{\ell'}^{T'}$ refutes some long clause $p_{i,1} \vee \cdots \vee p_{i,n}$, therefore the system $\Phi_{\ell''}^{T''}$ has no acceptable solutions that satisfy $p_{i,1} \vee \cdots \vee p_{i,n}$. By construction all internal nodes of T'' have two children. Lemma 3 implies that the depth of all leaves in T'' is at least $\frac{n-1}{2}$, hence the size of T'' is at least $2^{(n-1)/2}$. $\qquad\Box$

5   Proof Systems Res-Lin and Sem-Lin

A linear clause is a disjunction of linear equalities $\bigvee_{i=1}^k (f_i = \alpha_i)$, where f_i is a linear form and $\alpha_i \in \mathbb{F}_2$. Equivalently we may rewrite a linear clause as a negation of a system of linear equalities $\neg \bigwedge_{i=1}^{n}(f_i = 1 + \alpha_i)$. A trivial linear clause is a linear clause that is identically true. A clause $\neg \bigwedge_{i=1}^{n}(f_i = \alpha_i)$ is trivial iff the system $\bigwedge_{i=1}^{n}(f_i = \alpha_i)$ has no solutions.

A linear CNF formula is a conjunction of linear clauses. We say that propositional formula ϕ is semantically implied form the set of formulas $\psi_1, \psi_2, \ldots, \psi_k$ if every assignment that satisfies ψ_i for all $i \in [k]$ also satisfies ϕ.

Definition 1. *We define a proof system Res-Lin that can be used to prove that a linear CNF formula is unsatisfiable. This system has two rules: 1)The weakening rule allows to derive from a linear clause C any linear clause D such that C semantically implies D. 2)The resolution rule allows to derive from linear clauses $(f = 0) \vee D$ and $(f = 1) \vee D'$ the linear clause $D \vee D'$.*

A derivation of a linear clause C from a linear CNF ϕ in the Res-Lin system is a sequence of linear clauses that ends with C and every clause is either a clause of ϕ or it may be obtained from previous clauses by a derivation rule. The proof of the unsatisfiability of a linear CNF is a derivation of the empty clause (contradiction). The Sem-Lin system differs from Res-Lin by the second rule. It is replaced by a semantic rule that allows to derive from linear clauses C_1, C_2 any linear clause C_0 such that C_1 and C_2 semantically imply C_0.

In order to verify that systems Sem-Lin and Res-Lin are proof systems in the sence of [4] we have to ensure that it is possible to verify a correctness of a proof in polynomial time. It is enough to verify a correctness of applications of rules. The correctness of the resolution rule is easy to verify, and for the verification of the other rules we use the following proposition.

Proposition 3. *It is possible to verify in polynomial time: 1) whether a linear clause $C_0 = \neg \bigwedge_{i \in I}(f_i = \alpha_i)$ is a result of the weakening rule of $C_1 := \neg \bigwedge_{i \in J}(g_i = \beta_i)$; 2) whether a linear clause $C_0 := \neg \bigwedge_{i \in J}(g_i = \beta_i)$ is semantically implied from $C_1 := \neg \bigwedge_{i \in J}(g_i = \beta_i)$ and $C_2 = \neg \bigwedge_{i \in K}(h_i = \gamma_i)$.*

Remark 1. We note that the weakening may be simulated by polynomial applications of the following pure syntactic rules: 1) Simplification rule that allows to derive D from $D \vee (0 = 1)$; 2) Syntactic weakening rule that allows to derive $D \vee (f = \alpha)$ from D; 3) Addition rule that allows to derive $D \vee (f_1 = \alpha_1) \vee (f_1 + f_2 = \alpha_1 + \alpha_2 + 1)$ from $D \vee (f_1 = \alpha_1) \vee (f_2 = \alpha_2)$.

We show that systems Sem-Lin and Res-Lin are polynomially equivalent. It means that any proof in one system may be translated to the proof in other system in polynomial time. Every proof in Res-Lin is also a proof in Sem-Lin; the next proposition is about the opposite translation.

Proposition 4. *Let nontrivial linear clause $C_0 := \neg\{f_i = \alpha_i\}_{i \in I}$ be a semantic implication of $C_1 := \neg\{g_i = \beta_i\}_{i \in J}$ and $C_2 := \neg\{h_i = \gamma_i\}_{i \in L}$. Then C_0 can be obtained from C_1 and C_2 by applications of at most one resolution rule and several weakening rules.*

Before we start a proof we consider an example that shows how the linear clause $(x + y = 0)$ can be derived from $(x = 0)$ and $(y = 0)$ in Res-Lin: 1) Apply weakening rule to $(x = 0)$ and get $(x + y = 0) \vee (y = 1)$; 2) Apply resolution rule to $(x + y = 0) \vee (y = 1)$ and $(y = 0)$ and get $(x + y = 0)$.

We will use the following well known lemma:

Lemma 4. *If for a matrix $A \in \mathbb{F}_2^{m \times n}$ and a vector $b \in \mathbb{F}_2^m$ the linear system $Ax = b$ has no solutions, then there exists a vector $y \in \mathbb{F}_2^m$ such that $y^T A = 0$ and $y^T b = 1$. In other words if a linear system over $\mathbb{F}_2$ is unsatisfiable then it is possible to sum several equations and get a contradiction $0 = 1$.*

Proof (Proof of Proposition 4). Both C_1 and C_2 can't be trivial since in this case C_0 must be trivial. If C_i for $i \in \{1, 2\}$ is trivial, then C_0 is a weakening of C_{2-i}. So we assume that C_1 and C_2 are not trivial.

For all $j \in J$ and $l \in L$ the system $\bigwedge_{i \in I}(f_i = \alpha_i) \wedge (g_j = 1 + \beta_j) \wedge (h_l = 1 + \gamma_l)$ is unsatisfiable. Since the system $\bigwedge_{i \in I}(f_i = \alpha_i)$ is satisfiable, one of the following holds: 1)$\bigwedge_{i \in I}(f_i = \alpha_i)$ becomes unsatisfiable if we add just one equality (for example $g_j = 1 + \beta_j$). Then by Lemma 4 the negation of this equality can be obtained as a linear combination of equalities from $\bigwedge_{i \in I}(f_i = \alpha_i)$. 2) The system $\bigwedge_{i \in I}(f_i = \alpha_i)$ becomes unsatisfiable only if we add both equalities ($g_j = 1 + \beta_j) \wedge (h_l = 1 + \gamma_l)$. By Lemma 4 the equality $g_j + h_l = \beta_j + \gamma_l + 1$ may be obtained as a linear combination of equalities from the system $\bigwedge_{i \in I}(f_i = \alpha_i)$. Note that if equalities $g_j = 1 + \beta_j$ and $h_l = 1 + \gamma_l$ contradict each other (i.e. $g_j = h_l$ and $\beta_j = 1 + \gamma_l$), then the equality $g_j + h_l = \beta_j + \gamma_l + 1$ is just $0 = 0$.

We split J into two disjoint sets J' and J'', where $j \in J''$ iff the system $\bigwedge_{i \in I}(f_i = \alpha_i) \wedge (g_j = \beta_j + 1)$ is unsatisfiable. Similarly we define a splitting $L = L' \cup L''$. Note that if $J = J''$, then $\neg \bigwedge_{i \in I}(f_i = \alpha_i)$ is a weakening of $\neg \bigwedge_{j \in J}(g_j = \beta_j)$, similarly if $L = L''$, then $\neg \bigwedge_{i \in I}(f_i = \alpha_i)$ is a weakening of $\neg \bigwedge_{i \in L}(h_i = \gamma_i)$. Thus in what follows we assume that $J' \neq \emptyset$ and $L' \neq \emptyset$.

We get that C_0 is a weakening of $D := \neg(\bigwedge_{i \in J''}(g_i = \beta_i) \wedge \bigwedge_{i \in L''}(h_i = \gamma_i) \wedge \bigwedge_{i \in J', j \in L'}(g_i + h_j = \beta_i + \beta_j + 1)$. It remains to show that D can be obtained from C_1 and C_2 by application of one resolution rule and several weakening rules.

Let $j_0 \in J'$ and $l_0 \in L'$.

1) Apply the weakening rule to C_1 and get $D_1 :=$ $\neg((g_{j_0} = \beta_{j_0}) \wedge \bigwedge_{i \in J'}(g_i + h_{l_0} = \beta_i + \gamma_{l_0} + 1) \wedge \bigwedge_{i \in J''}(g_i = \beta_i))$;

2) Apply the weakening rule to C_2 and get $D_2 :=$ $\neg((g_{j_0} = \beta_{j_0} + 1) \wedge \bigwedge_{i \in L'}(h_i + g_{j_0} = \beta_{j_0} + \gamma_{l_0} + 1) \wedge \bigwedge_{i \in L''}(h_i = \gamma_i\}))$;

3) Apply the resolution rule to D_1 and D_2, and get

$$D_3 := \neg \left(\bigwedge_{i \in J'}(g_i + h_{l_0} = \beta_i + \gamma_{l_0} + 1) \wedge \bigwedge_{i \in L'}(h_i + g_{j_0} = \beta_{j_0} + \gamma_{l_0} + 1) \wedge \right.$$

$$\left. \bigwedge_{i \in J''}(g_i = \beta_i) \wedge \bigwedge_{i \in L''}(h_i = \gamma_i) \right)$$

4) Apply the weakening rule to D_3 and get D.

$\square$

Tree-like Res-Lin and linear splitting trees. A proof in Res-Lin (or Sem-Lin) is tree-like if all clauses can be put in the nodes of a rooted tree in such a way that 1) the empty clause is in the root; 2) the clauses of an initial formula are in the leaves; 3) a clause in every internal node is a result of a rule of its children.

Linear splitting trees are naturally generalized to linear CNFs.

Lemma 5. *1) Every linear splitting tree for an unsatisfiable linear CNF may be translated into a tree-like Res-Lin proof and the size of the resulting proof is at most twice the size of the splitting tree. 2) Every tree-like Res-Lin proof of an unsatisfiable formula ϕ may be translated to a linear splitting tree for ϕ without increasing the size of the tree.*

Corollary 2. *1) For all $m > n$ every tree-like proof in Res-Lin and Sem-Lin of PHP_n^m has size $2^{\Omega(n)}$. 2) In the conditions of Theorem 2 the size of any tree-like resolution proof in Res-Lin and Sem-Lin of $TS_{(G,c)}^2$ is at least $\Omega(2^{n^{\frac{1}{3}}/\log^3(n)})$.*

Proof. Follows from Lemma 5, Proposition 4, Theorem 4 and Theorem 2.   □

Now we prove that Res-Lin is implication complete.

Lemma 6. *1) If a linear clause D is a weakening of a linear clause C, then for every linear clause E the clause $D \vee E$ is a weakening of $C \vee E$. 2) If a linear clause D is a semantic implication of (or a result of the resolution rule applied for) C and F, then for every linear clause E the clause $D \vee E$ is a semantic implication (or a result of the resolution rule applied for) $C \vee E$ and $F \vee E$.*

Theorem 5. *If a linear clause C_0 is a semantic implication of $C_1, C_2, \ldots, C_k$, then C_0 may be derived from $C_1, C_2, \ldots, C_k$ in Res-Lin.*

Proof. The plan of the proof is following: we construct a list of linear clauses $\mathcal{D}$ such that the conjunction of clauses from $\mathcal{D}$ is unsatisfiable. Since Res-Lin is complete (Res-Lin is complete because every linear CNF has a splitting tree with splitting over all variables), then there exists a derivation of the empty clause from $\mathcal{D}$. By Lemma 6 from the list $\mathcal{D}' := \{D \vee C_0 \mid D \in \mathcal{D}\}$ it is possible to derive C_0. After this we show that every clause in $\mathcal{D}'$ is a weakening of some clause among $C_1, C_2, \ldots, C_k$.

We construct the list $\mathcal{D}$ step by step; initially $\mathcal{D}$ consists of clauses $C_1, C_2, \ldots, C_k$. Note that if an assignment π refutes C_0, then by the statement of the theorem it also refutes one of the clauses $C_1, C_2, \ldots, C_k$, hence it refutes their conjunction. Let $C_1 := \bigvee_{i=1}^n (f_i = \alpha_i)$ and $C_0 := \bigvee_{i=1}^m (g_i = \beta_i)$ While there exists such an assignment π that satisfies C_0 and satisfies all clauses from $\mathcal{D}$, we add to the list $\mathcal{D}$ a new clause C^π. Since π satisfies C_0, then there exists i such that π satisfies $g_i = \beta_i$. Let's denote $I := \{i \mid \pi \text{ satisfies} f_i = \alpha_i\}$ and let the clause C^π equal $\bigvee_{i \in I}(f_i + g_i = \alpha_i + \beta_i + 1) \vee \bigvee_{i \notin I}(f_i = \alpha_i)$. By construction π refutes C^π. Finally, for every assignment of variables there exists such a clause in the list $\mathcal{D}$ that is not satisfied by the assignment. Hence the conjunction of clauses from $\mathcal{D}$ is unsatisfiable. We have to show that for all $D \in \mathcal{D}$ the clause $D \vee C_0$ is a weakening of some clause among $C_1, C_2, \ldots, C_k$. If D equals one clause from $C_1, C_2, \ldots, C_k$, we are done. Let $D = C^\pi$, then $D \vee C_0$ is a weakening of $C_1 \vee C_0$ and therefore is a weakening of C_1.   □

Acknowledgements. The authors are grateful to Jan Krajíček, Edward A. Hirsch and Alexander Knop for fruitful discussions, to Alexander Shen for the suggestion to simplify the presentation of the first lower bound and to anonymous reviewers for multiple helpful comments.

References

1. Alekhnovich, M., Hirsch, E.A., Itsykson, D.: Exponential lower bounds for the running time of DPLL algorithms on satisfiable formulas. J. Autom. Reason. 35(1-3), 51–72 (2005)
2. Beame, P., Pitassi, T., Segerlind, N.: Lower bounds for lovász-schrijver systems and beyond follow from multiparty communication complexity. SIAM Journal on Computing 37(3), 845–869 (2007)
3. Ben-Sasson, E., Wigderson, A.: Short proofs are narrow — resolution made simple. Journal of ACM 48(2), 149–169 (2001)
4. Cook, S.A., Reckhow, R.A.: The relative efficiency of propositional proof systems. The Journal of Symbolic Logic 44(1), 36–50 (1979)
5. Davis, M., Logemann, G., Loveland, D.: A machine program for theorem-proving. Communications of the ACM 5, 394–397 (1962)
6. Davis, M., Putnam, H.: A computing procedure for quantification theory. Journal of the ACM 7, 201–215 (1960)
7. Demenkov, E., Kulikov, A.S.: An elementary proof of a $3n - o(n)$ lower bound on the circuit complexity of affine dispersers. In: Murlak, F., Sankowski, P. (eds.) MFCS 2011. LNCS, vol. 6907, pp. 256–265. Springer, Heidelberg (2011)
8. Itsykson, D., Sokolov, D.: The complexity of inversion of explicit goldreich's function by DPLL algorithms. In: Kulikov, A., Vereshchagin, N. (eds.) CSR 2011. LNCS, vol. 6651, pp. 134–147. Springer, Heidelberg (2011)
9. Itsykson, D.: Lower bound on average-case complexity of inversion of goldreich's function by drunken backtracking algorithms. Theory Comput. Syst. 54(2), 261–276 (2014)
10. Kalyanasundaram, B., Schintger, G.: The probabilistic communication complexity of set intersection. SIAM J. Discret. Math. 5(4), 545–557 (1992)
11. Kushilevitz, E., Nisan, N.: Communication Complexity. Cambridge University Press, New York (1997)
12. Raz, R., Tzameret, I.: Resolution over linear equations and multilinear proofs. Ann. Pure Appl. Logic 155(3), 194–224 (2008)
13. Razborov, A.A.: Pseudorandom generators hard for k-dnf resolution and polynomial calculus resolution. Technical report (2003)
14. Seto, K., Tamaki, S.: A satisfiability algorithm and average-case hardness for formulas over the full binary basis. Computational Complexity 22(2), 245–274 (2013)
15. Tseitin, G.S.: On the complexity of derivation in the propositional calculus. Zapiski Nauchnykh Seminarov LOMI 8, 234–259 (1968); English translation of this volume: Consultants Bureau, N.Y., pp. 115–125 (1970)
16. Urquhart, A.: The depth of resolution proofs. Studia Logica 99(1-3), 249–364 (2011)

On the Complexity of List Ranking
in the Parallel External Memory Model

Riko Jacob[1], Tobias Lieber[1], and Nodari Sitchinava[2]

[1] Institute for Theoretical Computer Science, ETH Zürich, Switzerland
{rjacob,lieberto}@inf.ethz.ch
[2] Department of Information and Computer Sciences, University of Hawaii, USA
nodari@hawaii.edu

Abstract. We study the problem of *list ranking* in the parallel external memory (PEM) model. We observe an interesting dual nature for the hardness of the problem due to limited information exchange among the processors about the structure of the list, on the one hand, and its close relationship to the problem of permuting data, which is known to be hard for the external memory models, on the other hand.

By carefully defining the power of the computational model, we prove a permuting lower bound in the PEM model. Furthermore, we present a stronger $\Omega(\log^2 N)$ lower bound for a special variant of the problem and for a specific range of the model parameters, which takes us a step closer toward proving a non-trivial lower bound for the list ranking problem in the bulk-synchronous parallel (BSP) and MapReduce models. Finally, we also present an algorithm that is tight for a larger range of parameters of the model than in prior work.

1 Introduction

Analysis of massive graphs representing social networks using distributed programming models, such as MapReduce and Hadoop, has renewed interests in distributed graph algorithms. In the classical RAM model, depth-first search traversal of the graph is the building block for many graph analysis solutions. However, no efficient depth-first search traversal is known in the parallel/ distributed setting. Instead, list ranking serves as such a building block for parallel solutions to many problems on graphs.

The *list ranking* problem is defined as follows: given a linked list compute for each node the length of the path to the end of the list. In the classic RAM model, the list can be ranked in linear time by traversing the list. However, in the PRAM model (the parallel analog of the RAM model) it took almost a decade from the first solution by Wyllie [1] till it was solved optimally [2].

The problem is even more intriguing in the models that study block-wise access to memory. For example, in the *external memory* (EM) model of Aggarwal and Vitter [3] list ranking is closely related to the problem of permuting data in an array. The EM model studies the *input/output (I/O) complexity* – the number of transfers an algorithm has to perform between a disk that contains

E. Csuhaj-Varjú et al. (Eds.): MFCS 2014, Part II, LNCS 8635, pp. 384–395, 2014.
© Springer-Verlag Berlin Heidelberg 2014

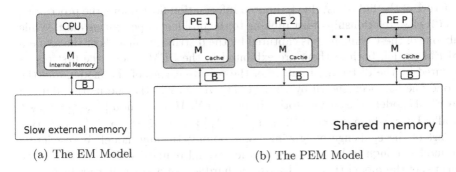

(a) The EM Model   (b) The PEM Model

Fig. 1. The sequential and parallel external memory models

the input and a fast internal memory of size M. Each transfer is performed in blocks of B contiguous elements. In this model, permuting and, consequently, list ranking require I/O complexity which is closely related to sorting [4], rather than the linear complexity required in the RAM model.

In the distributed models, such as bulk-synchronous parallel (BSP) [5] or MapReduce [6] models, the hardness of the list ranking problem is more poorly understood. These models consist of P processors, each with a private memory of size M. With no other data storage, typically $M = \Theta(N/P)$. The data is exchanged among the processors during the *communication rounds* and the number of such rounds defines the complexity metric of these models.

One established modeling of today's commercial data centers running MapReduce is to assume $P = \Theta(N^\epsilon)$ and $M = \Theta(N^{1-\epsilon})$ for a constant $0 < \epsilon < 1$. Since network bandwidth is usually the limiting factor of these models, $O(\log_M P) = O(1)$ communication rounds is the ultimate goal of computation on such models [7]. Indeed, if each processor is allowed to send up to $M = N/P$ items to any subset of processors, permuting of the input can be implemented in a single round, while sorting takes $O(\log_M P) = O(1)$ rounds [8,9]. On the other hand, the best known solution for list ranking is via the simulation results of Karloff et al. [7] by simulating the $O(\log N)$ time PRAM algorithm [2], yielding $O(\log P)$ rounds, which is strictly worse than both sorting and permuting. Up to now, no non-trivial lower bounds (i.e. stronger than $\Omega(\log_M P) = \Omega(1)$) are known in the BSP and MapReduce models.

In this paper we study lower bounds for the list ranking problem in the *parallel external memory (PEM)* model. The PEM model was introduced by Arge et al. [10] as a parallel extension of the EM model to capture the hierarchical memory organization of modern multicore processors. The model consists of P processing units, each containing a private cache of size M, and a shared (external) memory of conceptually unlimited size (see Figure 1b). The data is concurrently transferred between the processors' caches and shared memory in blocks of size B. The model measures the *parallel I/O complexity* – the number of parallel block transfers. From the discussion above, it appears that the hardness of list ranking stems from two factors: (1) limited speed of discovery of the structure of the linked list due to limited information flow among the processors, and (2) close relationship of list ranking to the problem of permuting data. While only one of these challenges is captured by the distributed models or the sequential EM model, both of them are exhibited in the PEM model: the first one is captured by the distributed nature of the private caches of the model, and the second one has been shown by Greiner [11] who proves that permuting data in the PEM model takes asymptotically $\mathrm{perm}_P(N, M, B) = \min\left\{\frac{N}{P}, \frac{N}{PB}\overline{\log}_d \frac{N}{B}\right\}$ parallel I/Os, where $d = \max\left\{2, \min\left\{\frac{M}{B}, \frac{N}{PB}\right\}\right\}$ and $\overline{\log}(x) = \max\{1, \log(x)\}$.

Part of the challenge of proving lower bounds (in any model) is restricting the model enough to be able to prove non-trivial bounds, while identifying the features of the model that emphasize the hardness of a particular problem.

An example of such restriction in the external memory models (both sequential and parallel) is the so-called *indivisibility assumption* [3]. The assumption states that each item is processed as a whole, and no information can be obtained from a part of the input, for example, by combining several items into one. To our knowledge, without the indivisibility assumption, it is not clear how to prove lower bounds in the PEM model exceeding the information-theoretic lower bounds of $\Omega\left(\log P\right)$ parallel I/Os [12,10].

1.1   Our Contributions

In this paper, we address the precise formulation of the power of the PEM model for the list ranking problem. In Section 2 we present the atomic PEM model which formalizes the indivisibility assumption in the PEM model. It can be viewed as the parallel analog of the model for proving permuting lower bounds in the sequential EM model [3]. We extend this basic model by allowing an algorithm to perform operations on the atoms that create new atoms. While we always keep the indivisibility of the atoms, the precise operations and the information the algorithm has about the content of the atom varies.

In the sequential EM model, Chiang et al. [4] sketch a lower bound for list ranking via a reduction to the *proximate neighbors problem*. However, since there is no equivalent to Brent's scheduling principle [13] in the PEM model [11], the lower bound does not generalize to the PEM model.

Therefore, in Section 3 we derive a lower bound of $\Omega(\mathrm{perm}_P(N, M, B))$ parallel I/Os for the proximate neighbor problem, and two problems which are related to the list ranking problem. Our lower bounds hold for both deterministic and randomized algorithms. In the process we provide an alternative proof for the proximate neighbor lower bound in the sequential external memory model, matching the result of Chiang et al. [4]. Those lower bounds essentially exploit the fact that the same problem can be represented as input in many different layouts.

The discussion in Section 1 about the dual nature of hardness of the list ranking problem might hint at the fact that the result above is only part of the picture and a stronger lower bound might be achievable. Part of the challenge in proving a stronger lower bound lies in the difficulty of combining the indivisibility assumption with the restrictions on how the structure of the linked list is shared among the processors without giving the model too much power, thus, making the solutions trivial. We address this challenge by defining the *interval PEM* model and defining the *guided interval fusion (GIF)* problem (Section 4). We prove that GIF requires $\Omega(\log^2 N)$ parallel I/Os in the interval PEM. Our lower bound for GIF in the PEM model implies a $\Omega(\log N)$ lower bound for the number of rounds for GIF in the distributed models when $P = \Theta(M) = \Theta(\sqrt{N})$.

GIF captures the way how all currently known algorithms use information to solve list ranking in all parallel/distributed models. Therefore, if this lower bound could be broken for the list ranking problem, it will require completely new algorithmic techniques. Thus, our result brings us a step closer to proving the unconditional $\Omega(\log P)$ lower bound in the BSP and MapReduce models.

Finally, in Section 5 we improve the PEM list ranking algorithm of Arge et al. [14] to work efficiently for a larger range of parameters in the PEM model.

2   Modeling

We extend the description of the PEM model given in Section 1 to define the PEM model more precisely. Initially, the data resides in the shared memory. To process any element, it must be present in the corresponding processor's cache. The shared memory is partitioned into blocks of B contiguous elements and the transfer of data between the shared memory and caches is performed by transferring these blocks as units. Each transfer, an *input-output operation*, or simply *I/O*, can transfer one block of B items between the main memory and each processors' cache. Thus, up to P blocks can be transferred in each *parallel I/O* operation. The complexity measure of a PEM algorithm is the number of parallel I/Os that the algorithm performs.

Similar to the PRAM model, there are several policies in the PEM model, for handling simultaneous accesses by multiple processors to a block in shared memory. In this paper we consider the CREW PEM model, using a block wise concurrent read, exclusive write policy.

In order to prove lower bounds we make the definition of the model more precise by stating what an algorithm is able to do in each step. In particular, we assume that each element of the input is an indivisible unit of data, an *atom*, which consumes one memory cell in the cache or shared memory. Such atoms come into existence either as input atoms, or by an operation, as defined later, on two atoms. A program or an algorithm has limited knowledge about the content of an atom. In this paper an atom does not provide any information. Furthermore, the *atomic PEM* is limited to the following operations: an I/O operation reads or writes one block of up to B atoms in the shared memory, and atoms can be copied or deleted. Formal definitions of similar PEM machines can be found in [10,11].

For providing lower bounds for different problems, the concept of the atomic PEM is extended in later sections.

In the following, we distinguish between algorithms and programs in the following way: In an *algorithm* the control flow might depend on the input, i.e., there are conditional statements (and therefore loops). In contrast, a *program* has no conditional statements and is a sequence of valid instructions for a PEM model, independent of the input (atoms). For a given instance of a computational task, a program can be seen as an instantiation of an algorithm to which all results of conditional statements and index computations are presented beforehand. Note, that in the problems considered in Section 3 the copying and the deletion operation of the atomic PEM do not help at all, since a program can be stripped down to operations which operate on atoms which influence the final result.

3   Counting Lower Bounds to the List Ranking Problem

In this section we prove the lower bound for the list ranking problem by showing the lower bound to the proximate neighbors problem [4] and reducing it to the problems of semigroup evaluation, edge contraction and, finally, list ranking.

3.1   Proximate Neighbors Problem in PEM

Definition 1 ([4]). *A* block permutation *describes the content of the shared memory of a PEM configuration as a set of at most B atoms for each block.*

Definition 2. *An* instance of the proximate neighbors problem *of size N consists of atoms x_i for $i \in [N]$. All atoms are labeled by a labeling function $\lambda \colon [N] \mapsto [\frac{N}{2}]$ with $|\lambda^{-1}(i)| = 2$. An output block permutation solves the problem if for every $i \in [\frac{N}{2}]$ the two neighboring atoms $\lambda^{-1}(i)$ are stored in the same block. The blocks in such an output may contain less than B atoms.*

Lemma 3. *Let A be a computational problem of size N for which an algorithm has to be capable of generating at least $\left(\frac{N}{eB}\right)^{cN}$ block permutations, for a constant $c > 0$. Then in the CREW atomic PEM model with $P \leq \frac{N}{B}$ processors, at least half of the input instances of A require $\Omega\left(\mathrm{perm}_P(N, M, B)\right)$ parallel I/Os.*

Proof. Straightforward generalization of the proof to Theorem 2.7 in [11].   □

Theorem 4. *At least half of the instances of the proximate neighbors problem of size N require $\Omega\left(\mathrm{perm}_P(N, M, B)\right)$ parallel I/Os in the CREW atomic PEM model with $P < N/B$ processors.*

Proof. Any algorithm solving an instance of the proximate neighbors problem must be capable of generating at least $\frac{(N/2)!}{(B/2)^{\frac{N}{2}}}$ block permutations (see the full version of this paper [15] for a complete proof). The theorem follows from Lemma 3 and the observation that $\frac{(N/2)!}{(B/2)^{\frac{N}{2}}} \geq \left(\frac{N}{eB}\right)^{\frac{N}{2}}$.   □

Note, that the bound holds even if a program has full access to the labeling function λ and thus is fully optimized for an input. The origin of the complexity of the problem rather is permuting the atoms to the output block permutation that solves the problem.

3.2   Semigroup Evaluation in the PEM Model

Consider the problem of evaluating a very simple type of expressions, namely that of a semigroup, in the PEM model.

Definition 5 (Semigroup Evaluation). *Let S be a semigroup with its associative binary operation $\cdot : S \times S \to S$. The semigroup evaluation problem is defined as evaluating the expression $\prod_{i=1}^{N} a_i$, with $a_i = x_{\pi(i)}$, for the array of input atoms $x_i \in S$ for $1 \le i \le N$ and where π is a permutation over $[N]$.*

To be able to solve the semigroup evaluation problem, algorithms must be able to apply the semigroup operation to atoms. Thus, we extend the atomic PEM model to the *semigroup PEM* model by the following additional operation: if two atoms x and y are in the cache of a processor, a new atom $z = x \cdot y$ can be created.

We say that a program is correct in the semigroup PEM if it computes the correct result for any input and any semigroup.

Theorem 6. *At least one instance of the semigroup evaluation problem of size N requires $\Omega\left(\text{perm}_P(N, M, B)\right)$ parallel I/Os in the CREW semigroup PEM model with $P \le \frac{N}{B}$ processors.*

Proof (Sketch). Let I_λ^P be an instance of the proximate neighbors problem over the input atoms $X = \{x_i | i \in [N]\}$ with its labeling function λ. We consider an instance I_π^S of the semigroup evaluation problem over the semigroup on the set X^2 with the semigroup operation $(a, b) \cdot (c, d) = (a, d)$, where $a, b, c, d \in X$. Furthermore, the instance I_π^S is defined over the input atoms $a_i = (x_i, x_i)$, with $1 \le i \le N$. The permutation π of I_π^S is one of the permutations such that for all $i \in \left[\frac{N}{2}\right]$, $\{\pi(2i - 1), \pi(2i)\} = \lambda^{-1}(i)$ holds.

Then, the key idea is to write for each application of the semigroup operation $(a, b) \cdot (c, d)$ in a program solving I_π^S, the pair $\{b, c\}$ as a result for I_λ^P to the output. This would yield an efficient program for I_λ^P, and therefore yields the lower bound by Theorem 4. The full argument can be found in [15]. $\square$

3.3   Atomic Edge Contraction in the PEM Model

Definition 7. *The input of the atomic edge contraction problem of size N consists of atoms x_i, $1 \le i \le N$, which represent directed edges e_i on a $(N+1)$-vertex path between vertices s and t. Initially, the edges are located in arbitrary locations of the shared memory. The instance is solved if an atom representing the edge (s, t) is created and written to shared memory.*

To prove the lower bound for the atomic edge contraction problem, we extend the atomic PEM with an additional operation: two atoms representing a pair of edges (a, b) and (b, c) can be removed and replaced by a new atom representing a new edge (a, c). We call the resulting model *edge-contracting PEM*.

Theorem 8. *There is at least one instance of the atomic edge contraction problem of size N which requires $\Omega\left(\text{perm}_P(N, M, B)\right)$ parallel I/Os in the CREW edge-contracting PEM model with $P \le \frac{N}{B}$ processors.*

Proof (Sketch). An instance I_π^S of the semigroup evaluation problem can be reduced to an instance $I^\mathcal{E}$ of the atomic edge contraction problem, by defining the atom $x_{\pi(i)}$ of $I^\mathcal{E}$, initially stored at location $\pi(i)$, as $e_{\pi(i)} = (\pi(i), \pi(i+1))$, where π is the permutation of I_π^S. The full argument can be found in [15].   □

3.4   Randomization and Relation to the List Ranking Problem

Observe that the expected number of parallel I/Os of a randomized algorithm for an instance is a convex combination of the number of parallel I/Os of programs. Combining this observation with the $\Omega(\log P)$ lower bound of [12,10] mentioned in Section 1 we obtain:

Theorem 9. *For the proximate neighbors, semigroup evaluation, and the atomic edge contraction problems, there exists at least one instance that requires at least $\Omega(\text{perm}_P(N, M, B) + \log P)$ expected parallel I/Os by any randomized algorithm in the corresponding PEM model with $P \leq \frac{N}{B}$ processors.*

Although our semigroup PEM and edge-contracting PEM models might seem too restrictive at a first glance. To the best of our knowledge all current parallel solutions to list ranking utilize pointer hopping, which can be reduced to atomic edge contraction and thus the lower bound applies.

4   The Guided Interval Fusion Problem (GIF)

In this section we prove for the GIF problem, which is very similar to the atomic edge contraction problem, a lower bound of $\Omega(\log^2 N)$ in the PEM model with parameters $P = M$ and $B = M/2$ for inputs of size $N = PM = 2^x$ for some $x \in \mathbb{N}$. In contrast to the atomic edge contraction problem, in the GIF problem an algorithm is not granted unlimited access to the permutation π.

The chosen parameters of the PEM model complement the upper bounds of Section 5 at one specific point in the parameter range. Note that with careful modifications the $\Omega(\log^2 N)$ bound can even be proven for $N = M^{\frac{3}{2}+\varepsilon}$.

Definition 10. *The* interval PEM *is an extension of the atomic PEM: Two atoms x and y representing closed intervals I_x and I_y, located in one cache, can be fused if $I_x \cap I_y \neq \emptyset$. Fusing creates a new atom z representing the interval $I_z = I_x \cup I_y$. We say z is derived from x if z is the result of zero or more fusing operations starting from atom x.*

Definition 11. *The* guided interval fusion problem *(GIF) is a game between an algorithm and an adversary, played on an interval PEM. The algorithm obtains a GIF instance $\mathcal{G}$ in the first N cells of the shared memory, containing N uniquely named atoms x_i, $1 \leq i \leq N$. Each initial atom x_i represents the (invisible to the algorithm) closed interval $I_{x_i} = [k-1, k]$ for $k = \pi(i)$ with $1 \leq k \leq N$.*

The permutation π is gradually revealed by the adversary in form of boundaries $p = (i, j)$ meaning that the (initial) atoms x_i and x_j represent neighboring intervals $(\pi(j) = \pi(i) + 1)$. We say that the boundary point p for x_i and x_j is revealed. The adversary must guarantee that at any time, for all existing atoms, at least one boundary is revealed. The game ends as soon as an atom representing $[0, N]$ exists.

Note the following: The algorithm may try to fuse two atoms, even though by the revealed boundaries this is not guaranteed to succeed. If this attempt is successful because their intervals share a point, a new atom is created and the algorithm solved a boundary. We call this phenomenon a *chance encounter*. Since the interval PEM extends the atomic PEM, copying of atoms is allowed.

For the lower bound, we assume that the algorithm is omniscient. More precisely, we assume there exists a central processing unit, with unlimited computational power, that has all presently available information on the location of atoms and what is known about boundaries. This unit can then decide on how atoms are moved and fused.

Thus, as soon as all boundary information is known to the algorithm, the instance is solvable with $\mathcal{O}(\log N)$ parallel I/Os: The central unit can virtually list rank the atoms, group the atoms by rank into P groups, and then by permuting move to every processor $\mathcal{O}(M)$ atoms which then can be fused with $\mathcal{O}(1)$ I/Os to the solving atom.

Hence, the careful revealing of the boundary information is crucial. To define the revealing process for GIF instances, the atoms and boundaries of a GIF instance $\mathcal{G}$ of size N are related to a perfect binary tree $\mathcal{T}_{\mathcal{G}}$. The tree $\mathcal{T}_{\mathcal{G}}$ has N leaves, $N-1$ internal nodes and every leaf is at distance $h = \log N$ from the root. More precisely, each leaf $i \in [N]$ corresponds to the atom representing the interval $[i-1, i]$. And each internal vertex v_p corresponds to the boundary $p = (i, j)$ where i corresponds to the rightmost leaf of its left subtree, and j to the leftmost leaf of the right subtree. The levels of $\mathcal{T}_{\mathcal{G}}$ are numbered bottom up: the leaves have level 1 and the root vertex level $\log N$ (corresponding to the revealing order of boundaries).

The protocol for boundary announcement, shows for a random GIF instance $\mathcal{G}$, that a deterministic algorithm takes $\Omega(\log^2 N)$ parallel I/Os.

Definition 12. *The tree $\mathcal{T}_{\mathcal{G}}$ is the* guide *of $\mathcal{G}$ if boundaries are revealed in the following way. Let x be an atom of $\mathcal{G}$ representing the interval $I = [a, b]$. If neither of the boundaries a and b are revealed, the boundary whose node in $\mathcal{T}_{\mathcal{G}}$ has smaller level, is revealed. If both have the same level, a is revealed.*

Note that for the analysis it is irrelevant how to break ties (in situations when the two invisible boundaries have the same level). By the assumption that the algorithm is omniscient, when p is revealed, immediately all intervals having p as boundary know that they share this boundary. Thus, the guide ensures that at any time each atom knows at least one initial atom with which it can be fused.

A node $v \in \mathcal{T}_{\mathcal{G}}$ is called *solved*, if there is an atom of $\mathcal{G}$ representing an interval that contains the intervals of the leaves of the subtree of v. The boundary p is only revealed by the guide if at least one child of v_p is solved.

An easy (omniscient) algorithm solving a GIF instance can be implemented: in each of $O(\log N)$ rounds, permute the atoms such that for every atom there is at least one atom, known to be fuseable, which resides in the same cache. Fuse all pairs of neighboring atoms, reducing the number of atoms by a factor of at least 2, revealing new boundaries. Repeat permuting and fusing until the instance is solved. Because the permuting step can be achieved with $\mathcal{O}(\log N)$ parallel I/Os, this algorithm finishes in $\mathcal{O}(\log^2 N)$ parallel I/Os. Note that solving a boundary resembles bridging out one element of an independent set in the classical list ranking scheme. Thus, most list ranking algorithms use information as presented to an algorithm solving a GIF instance $\mathcal{G}$ guided by $\mathcal{T}_\mathcal{G}$. Hence, this natural way of solving a GIF instance can be understood as solving in every of the $\log N$ rounds a proximate neighbors instance, making the $\Omega(\log^2 N)$ lower bound reasonable.

In the following we prove the lower bound for GIF by choosing k and showing that $s = k - 1 = \mathcal{O}(\log N) = \mathcal{O}(\log M)$ *progress stages* are necessary to solve all nodes W of level k of $\mathcal{T}_\mathcal{G}$ (thus, $|W| = 2^{h+1-k}$). For each stage we show in Lemma 15 that it takes $\Omega(\log N)$ parallel I/Os to compute.

To measure the progress of a stage, configurations of interval PEM machines are used. The *configuration* C^t after the interval PEM machine performed t I/Os followed by fusing operations consists of sets of atoms. For each cache and each block of the shared memory, there is one set of atoms.

For $e \in W$, let T_e be the subtree of e in $\mathcal{T}_\mathcal{G}$, and $\mathcal{B}_e$ be all boundaries in T_e. The progress measure towards solving e is the highest level of a solved boundary in $\mathcal{B}_e$. More precisely, T_e is *unsolved* on level i, if all boundaries of level i of $\mathcal{B}_e$ are unsolved. Initially every T_e is unsolved on level 2. The solved level increases by one at a time if only revealed boundaries are solved, but chance encounters may increase it faster.

The execution of a deterministic algorithm A defines the following s progress stages: Let $s = \frac{2\log M}{16}$ and $X = \frac{|W|}{s}$. In each stage $1 < i \leq s$, at least X elements increase their level to i. Over time, the number of elements that are unsolved on level i decreases, and we define t_i to be the last time where in C^{t_i} the number of elements of W that are unsolved on level $i + 1$ is at least $|W| - iX$. Further, let W_i be the elements (at least X of them) that in stage i get solved on level i or higher (in the time-frame from t_{i-1} to $t_i + 1$). We choose $k - 1 = \frac{h}{16}$ such that $X = \frac{2^{h+1-k}}{s} \geq 2^{\frac{15h}{16}}/s = M^{\frac{15}{8}}/s > M^{\frac{7}{4}}$ because $s = \frac{2\log M}{16} < M^{\frac{1}{8}}$ for $M \in \mathbb{N}$.

In the beginning of stage i, for each $v \in W_i$ the level of v is at most $i - 1$, and hence all level i nodes are not announced to the algorithm. Let P_i be the set of boundaries for which progress is traced: For every $e \in W_i$, there is a node v_{p_e} of level i with boundary p_e that is solved first (brake ties arbitrarily). Then P_i consists of those boundaries. We define a_e and b_e to be the two level 1 atoms (original intervals) defining the boundary p_e. Then all intervals having boundary p_e are derived of a_e or b_e. Solving the boundary p_e means fusing any interval derived of a_e with any interval derived of b_e. Furthermore a traced boundary is considered solved if in its interval (the one corresponding to an element of W) a chance encounter solves a boundary of level greater than i.

To trace the progress of the algorithm towards fusing the atoms of one stage, we define the graph $G_t^i = (V, E_t^i)$ from the configuration C^{t_i+t}. There is one vertex for each cache and each block of the shared memory (independent of t). There is an edge (self-loops allowed) $\{u, v\} \in E_t^i$ if for some $e \in W_i$ some atom derived of a_e is at u and some atom derived of b_e is at v or vice versa. The *multiplicity* of an edge counts the number of such e. The multiplicity of the graph is the maximal multiplicity of an edge.

Note that solving a node v_e requires that it counts as a self-loop somewhere. Hence the sum of the multiplicities of self-loops are an upper bound on the number of solved nodes, and for the stage to end, i.e., at time $t_{i+1} + 1$, the sum of the multiplicities of loops must be at least X. After each parallel I/O chance encounters may happen. Thus, the number of chance encounters is given by P times the multiplicity of self-loops at the beginning of the stage.

We say that two nodes of $\mathcal{T}_{\mathcal{G}}$ are indistinguishable if they are on the same level and exchanging them could still be consistent with the information given so far. Let l_e (derived of a_e) and r_e (derived of b_e) be the two children of $v_e \in P_i$. By definition, at time t_i both l_e and r_e are unsolved and hence v_e is not revealed.

Boundaries corresponding to nodes of level higher than k may be announced or solved (not only due to chance encounters). To account for that, we assume that all such boundaries between the intervals corresponding to W are solved. Hence the algorithm is aware of the leftmost and rightmost solved interval belonging to these boundaries, and this may extend to other intervals by revealed boundaries. Only the nodes of level i that correspond to this leftmost or rightmost interval might be identifiable to the algorithm, all other nodes of level i are indistinguishable. Because $i < k$, for all traced pairs a_e, b_e at least one of the elements belongs to this big set of indistinguishable nodes. We mark identifiable nodes. Hence, at stage i the algorithm has to solve a random matching of the traced pairs where all marked nodes are matched with unmarked ones (and unmarked ones might be matched with marked or unmarked ones).

The next lemma derives a high-probability upper bound on the multiplicity of a graph. *All* remaining proofs of this section are deferred to [15]. The proof of the following lemma uses the Hoeffding inequality.

Lemma 13. *Consider a deterministic GIF algorithm operating on a uniformly chosen permutation π defining $\mathcal{T}_{\mathcal{G}}$ for the GIF instance $\mathcal{G}$. Let $p(i, M)$ be the probability that G_0^i has multiplicity at most $M^{\frac{5}{8}}$ (where P, N, t_j, and k depend on parameter M). Then there is a M' such that for all $M \geq M'$ and for each $i \leq k$ it holds $p(i, M) \geq 1 - \frac{1}{M^2}$.*

Fundamental insights on identifying two pairs of a K_4 show that the progress achieved with one I/O can not be too large.

Lemma 14. *If the graph G_t^i has multiplicity at most m, then G_{t+1}^i has multiplicity at most $4m$.*

By the two previous lemmas we obtain the following result.

Lemma 15. *Let A be an algorithm solving an GIF instance $\mathcal{G}$ guided by $\mathcal{T}_{\mathcal{G}}$ of height h traced at level $k - 1 = h/16$. Each progress stage $j < s = k - 1$ of A, assuming $t_j < \log^2 M$, takes time $t_j - t_{j-1} = \Omega(\log M)$.*

There are $\mathcal{O}(\log N)$ stages, each taking at least $\Omega(\log N)$ I/Os, yielding with a union bound over all G_0^i for all progress stages $i < s$:

Lemma 16. *Consider a deterministic GIF algorithm operating on a uniformly chosen permutation π defining $\mathcal{T}_{\mathcal{G}}$ for the GIF instance $\mathcal{G}$. Then there is a M' such that for all $M \geq M'$, solving $\mathcal{G}$ in the interval PEM takes with high probability $(p > 1 - 1/M)$ at least $\Omega(\log^2 N)$ parallel I/Os.*

By Yao's principle [16] this can be transferred to randomized algorithms:

Theorem 17. *The expected number of parallel I/Os to solve a GIF instance of size $N = PM$ on an interval PEM with $M = P$ is $\Omega(\log^2 N)$.*

A simple reduction yields:

Theorem 18. *Solving the GIF problem in the BSP or in the MapReduce model with $N = PM$ and $P = \Theta(M) = \Theta(\sqrt{N})$ takes $\Omega(\log N)$ communication rounds.*

GIF is an attempt to formulate how the known algorithms for list ranking distribute information by attaching it to atoms of the PEM model. Most known algorithms for list ranking use fusing of edges on an independent set of edge-pairs (bridging out edges). This means that every edge is used (if at all) either as first or second edge in the fusing. This choice of the algorithm is taken without complete information, and hence we take it as reasonable to replace it by an adversarial choice, leading to the definition of the guide of a GIF instance.

Additionally, the PEM lower bound shows that there is no efficient possibility to perform different stages (matchings) in parallel, showing that (unlike in the efficient PRAM sorting algorithms) no pipelining seems possible. At this stage, our lower bound is hence more a bound on a class of algorithms, and it remains a challenge to formulate precisely what this class is. Additionally, it would be nice to show lower bounds in a less restrictive setting.

5   Upper Bounds

Improvements in the analysis [11] of the PEM merge sort algorithm [10] yield:

Lemma 19 ([11]). *The I/O complexity of sorting N records with the PEM merge sort algorithm using $P \leq \frac{N}{B}$ processors is $\mathrm{sort}_P(N, M, B) = \mathcal{O}\left(\frac{N}{PB} \overline{\log}_d \frac{N}{B}\right)$ for $d = \max\{2, \min\{\frac{N}{PB}, \frac{M}{B}\}\}$.*

We use it to extend the parameter range for the randomized list ranking algorithm [14] from $P \leq \frac{N}{B^2}$, and $M = B^{\mathcal{O}(1)}$ to:

Theorem 20. *The expected number of parallel I/Os, needed to solve the list ranking problem of size N in the CREW PEM model with $P \leq \frac{N}{B}$ is*

$$\mathcal{O}\left(\text{sort}_P(N, M, B) + (\log P) \overline{\log} \frac{B}{\log P}\right)$$

which is for $B < \log P$ just $\text{sort}_P(N, M, B)$.

In order to make the standard recursive scheme work, two algorithms are used. By Lemma 19, the randomized algorithm of [14], which is based on [13], can be used whenever $N \geq P \min\{\log P, B\}$. This yields the $(\log P) \overline{\log} \frac{B}{\log P}$ term, if $B > \log P$. Otherwise ($N \leq P \min\{\log P, B\}$) a simulation of a work-optimal PRAM algorithm [2] is used. A careful implementation and analysis [15] yield Theorem 20.

References

1. Wyllie, J.: The Complexity of Parallel Computation. PhD thesis, Cornell University (1979)
2. Anderson, R.J., Miller, G.L.: Deterministic parallel list ranking. In: Reif, J.H. (ed.) AWOC 1988. LNCS, vol. 319, pp. 81–90. Springer, Heidelberg (1988)
3. Aggarwal, A., Vitter, J.S.: The input/output complexity of sorting and related problems. Commun. ACM 31(9), 1116–1127 (1988)
4. Chiang, Y.J., Goodrich, M., Grove, E., Tamassia, R., Vengroff, D.E., Vitter, J.S.: External-memory graph algorithms. In: Proceedings of SODA 1995, pp. 139–149 (1995)
5. Valiant, L.G.: A bridging model for parallel computation. Commun. ACM 33(8), 103–111 (1990)
6. Dean, J., Ghemawat, S.: Mapreduce: Simplified data processing on large clusters. Commun. ACM 51(1), 107–113 (2008)
7. Karloff, H.J., Suri, S., Vassilvitskii, S.: A model of computation for mapreduce. In: Charikar, M. (ed.) SODA, pp. 938–948. SIAM (2010)
8. Goodrich, M.: Communication-efficient parallel sorting. SIAM J. Comput. 29(2), 416–432 (1999)
9. Goodrich, M., Sitchinava, N., Zhang, Q.: Sorting, searching, and simulation in the mapreduce framework. In: Asano, T., Nakano, S.-I., Okamoto, Y., Watanabe, O. (eds.) ISAAC 2011. LNCS, vol. 7074, pp. 374–383. Springer, Heidelberg (2011)
10. Arge, L., Goodrich, M., Nelson, M., Sitchinava, N.: Fundamental parallel algorithms for private-cache chip multiprocessors. In: SPAA 2008, pp. 197–206 (2008)
11. Greiner, G.: Sparse Matrix Computations and their I/O Complexity. Dissertation, Technische Universität München, München (2012)
12. Karp, R.M., Ramachandran, V.: Handbook of theoretical computer science, pp. 869–941 (1990)
13. Vishkin, U.: Randomized speed-ups in parallel computation. In: STOC, pp. 230–239 (1984)
14. Arge, L., Goodrich, M., Sitchinava, N.: Parallel external memory graph algorithms. In: IPDPS, pp. 1–11. IEEE (2010)
15. Jacob, R., Lieber, T., Sitchinava, N.: On the complexity of list ranking in the parallel external memory model. CoRR abs/1406.3279 (2014)
16. Yao, A.C.C.: Probabilistic computations: Toward a unified measure of complexity (extended abstract). In: FOCS, pp. 222–227. IEEE Computer Society (1977)

Knocking Out P_k-free Graphs

Matthew Johnson, Daniël Paulusma⋆, and Anthony Stewart

School of Engineering and Computing Sciences, Durham University,
South Road, Durham, DH1 3LE, UK
{matthew.johnson2,daniel.paulusma,a.g.stewart}@durham.ac.uk

Abstract. A parallel knock-out scheme for a graph proceeds in rounds in each of which each surviving vertex eliminates one of its surviving neighbours. A graph is KO-reducible if there exists such a scheme that eliminates every vertex in the graph. The PARALLEL KNOCK-OUT problem is to decide whether a graph G is KO-reducible. This problem is known to be NP-complete and has been studied for several graph classes since MFCS 2004. We show that the problem is NP-complete even for split graphs, a subclass of P_5-free graphs. In contrast, our main result is that it is linear-time solvable for P_4-free graphs (cographs).

1 Introduction

We consider *parallel knock-out schemes* for finite undirected graphs with no self-loops and no multiple edges. These schemes, which were introduced by Lampert and Slater [14], proceed in rounds. In the first round each vertex in the graph selects exactly one of its neighbours, and then all the selected vertices are eliminated simultaneously. In subsequent rounds this procedure is repeated in the subgraph induced by those vertices not yet eliminated. The scheme continues until there are no vertices left, or until an isolated vertex is obtained (since an isolated vertex will never be eliminated). A graph is called *KO-reducible* if there exists a parallel knock-out scheme that eliminates the whole graph. The *parallel knock-out number* of a graph G, denoted by pko(G), is the minimum number of rounds in a parallel knock-out scheme that eliminates every vertex of G. If G is not KO-reducible, then pko(G) = ∞.

Examples. Every graph G with a hamiltonian cycle has pko(G) = 1, as each vertex can select its successor on a hamiltonian cycle C of G after fixing some orientation of C. Also every graph G with a perfect matching has pko(G) = 1, as each vertex can select its matching neighbour in the perfect matching. In fact it is not difficult to see [2] that a graph G has pko(G) = 1 if and only if G contains a *[1,2]-factor*, that is, a spanning subgraph in which every component is either a cycle or an edge.

We study the computational complexity of the PARALLEL KNOCK-OUT problem, which is the problem of deciding whether a given graph is KO-reducible.

⋆ Supported by EPSRC grant EP/K025090/1.

E. Csuhaj-Varjú et al. (Eds.): MFCS 2014, Part II, LNCS 8635, pp. 396–407, 2014.
© Springer-Verlag Berlin Heidelberg 2014

The main motivation for doing so stems from the close relation to cycles and matchings as illustrated by the above examples. We also consider the variant in which the number of rounds permitted is fixed. This problem is known as the k-PARALLEL KNOCK-OUT problem, which has as input a graph G and ask whether $pko(G) \leq k$ for some fixed integer k (i.e. that is not part of the input).

Known Results. The 1-PARALLEL KNOCK-OUT problem is polynomial-time solvable, because it is equivalent [2] to testing whether a graph has a $[1,2]$-factor, which is well-known to be polynomial-time solvable (see e.g. [3] for a proof). However, both the problems PARALLEL KNOCK-OUT and k-PARALLEL KNOCK-OUT with $k \geq 2$ are NP-complete even for bipartite graphs [3]. On the other hand, it is known that PARALLEL KNOCK-OUT and k-PARALLEL KNOCK-OUT (for all $k \geq 1$) can be solved in $O(n^{3.5} \log^2 n)$ time on trees [2]. These results were later extended to graph classes of bounded treewidth [3]. It remains *open* whether a further generalization is possible to graph classes of bounded clique-width. Broersma et al. in [4] gave an $O(n^{5.376})$ time algorithm for solving PARALLEL KNOCK-OUT on n-vertex claw-free graphs. Afterward this was improved to an $O(n^2)$ time algorithm for almost claw-free graphs (which generalize the class of claw-free graphs) [13]. The latter paper also gives a full characterization of connected almost claw-free graphs that are KO-reducible. In particular it shows that every KO-reducible almost claw-free graph has parallel knock-out number at most 2. In general, KO-reducible graphs (even KO-reducible trees [2]) may have an arbitrarily large parallel knock-out number. Broersma et al. [4] showed that a KO-reducible n-vertex graph G has $pko(G) \leq \min\{-\frac{1}{2} + (2n - \frac{7}{4})^{\frac{1}{2}}, \frac{1}{2} + (2\alpha - \frac{7}{4})^{\frac{1}{2}}\}$ (where α denotes the size of a largest independent set in G). This bound is asymptotically tight for complete bipartite graphs [2]. Broersma et al. [4] also showed that every KO-reducible graph with no induced $(p + 1)$-vertex star $K_{1,p}$ has parallel knock-out number at most $p - 1$.

Our Results. We address the open problem of whether PARALLEL KNOCK-OUT is polynomial-time solvable on graph classes whose clique-width is bounded by a constant. This seems a very challenging problem, and in this paper we focus on graphs of clique-width at most 2. It is known that a graph has clique-width at most 2 if and only if it is a cograph [7]. Cographs are also known as P_4-free graphs (a graph is called P_k-free if it has no induced k-vertex path).

In Section 3 we give a linear-time algorithm for solving the PARALLEL KNOCK-OUT problem on cographs. The first step of the algorithm is to compute the cotree of a cograph. It then traverses the cotree twice. The first time to compute to what extent "large" subgraphs can be reduced by themselves and how many free "firings" from outside are available. The second time to check whether the number of free external firings is sufficient to knock them out. In this way it will be verified whether the whole graph is KO-reducible. In Section 4 we prove that both the PARALLEL KNOCK-OUT problem and the k-PARALLEL KNOCK-OUT problem ($k \geq 2$) are NP-complete even for split graphs. Because split graphs are P_5-free, our results imply a dichotomy result for the computational complexity of the PARALLEL KNOCK-OUT problem restricted to P_k-free graphs, as shown in Section 5, where we also give some (other) open problems.

2   Preliminaries

We denote a graph by $G = (V(G), E(G))$ and write $|G| = |V(G)|$ to denote the order of G. An edge joining vertices u and v is denoted by uv. If not stated otherwise a graph is assumed to be finite, undirected and simple.

Let $G = (V, E)$ be a graph. The *neighbourhood* of $u \in V$, that is, the set of vertices adjacent to u is denoted by $N_G(u) = \{v \mid uv \in E\}$. For a subset $S \subseteq V$, we let $G[S]$ denote the *induced* subgraph of G, which has vertex set S and edge set $\{uv \in E \mid u, v \in S\}$. A set $I \subseteq V$ is called an *independent set* of G if no two vertices in I are adjacent to each other. A subset $C \subseteq V$ is called a *clique* of G if any two vertices in C are adjacent to each other. A subset $D \subseteq V$ is a *dominating set* of a graph $G = (V, E)$ if every vertex of G is in D or adjacent to a vertex in D.

The *union* of two graphs G and H is the graph with vertex set $V(G) \cup V(H)$ and edge set $E(G) \cup E(H)$. If $V(G) \cap V(H) = \emptyset$, then we say that the union of G and H is *disjoint* and write $G + H$. We denote the disjoint union of r copies of G by rG.

For $n \geq 1$, the graph P_n denotes the *path* on n vertices, that is, $V(P_n) = \{u_1, \ldots, u_n\}$ and $E(P_n) = \{u_i u_{i+1} \mid 1 \leq i \leq n-1\}$. For $n \geq 3$, the graph C_n denotes the *cycle* on n vertices, that is, $V(C_n) = \{u_1, \ldots, u_n\}$ and $E(C_n) = \{u_i u_{i+1} \mid 1 \leq i \leq n-1\} \cup \{u_n u_1\}$. The graph K_n denotes the *complete graph* on n vertices, that is, the n-vertex graph whose vertex set is a clique. A graph is *complete bipartite* if its vertex set can be partitioned into two classes such that two vertices u and v are adjacent if and only if u and v belong to different classes. The graph $K_{p,q}$ is the *complete bipartite graph* with partition classes of sizes p and q, respectively.

Let G be a graph and let $\{H_1, \ldots, H_p\}$ be a set of graphs. We say that G is $(H_1, \ldots, H_p)$-*free* if G has no induced subgraph isomorphic to a graph in $\{H_1, \ldots, H_p\}$. If $p = 1$ we may write H_1-free instead of (H_1)-free. A P_4-free graph is also called a *cograph*. A graph G is a *split graph* if its vertex set can be partitioned into a clique and an independent set. Split graphs coincide with $(2K_2, C_4, C_5)$-free graphs [9].

We also need some formal terminology for parallel knock-out schemes. For a graph $G = (V, E)$, a *KO-selection* is a function $f : V \to V$ with $f(v) \in N(v)$ for all $v \in V$. If $f(v) = u$, we say that vertex v *fires* at vertex u, or that u *is knocked out* by a firing of v. If $u \in U$ for some $U \subseteq V$ then the firing is said to be *internal* with respect to U if $v \in U$; otherwise it is said to be *external* (with respect to U).

For a KO-selection f, we define the corresponding *KO-successor* of G as the subgraph of G that is induced by the vertices in $V \setminus f(V)$; if G' is the KO-successor of G we write $G \rightsquigarrow G'$. Note that every graph without isolated vertices has at least one KO-successor. A sequence

$$G \rightsquigarrow G^1 \rightsquigarrow G^2 \rightsquigarrow \cdots \rightsquigarrow G^s,$$

is called a *parallel knock-out scheme* or *KO-scheme*. A KO-scheme in which G^s is the null graph $(\emptyset, \emptyset)$ is called a *KO-reduction scheme*; in that case G is also called *KO-reducible*. A single step in a KO-scheme is called a *(firing) round*. Recall that the parallel knock-out number of G, pko(G), is the smallest number of rounds of any KO-reduction scheme, and that if G is not KO-reducible then pko(G) = ∞.

We will use the following result of Broersma et al. [2].

Lemma 1 ([2]). *Let p and q be two integers with $0 < p \leq q$. Then $K_{p,q}$ is KO-reducible if and only if $pko(K_{p,q}) \leq p$ if and only if $q \leq \frac{1}{2}p\,(p+1)$.*

3   Cographs

In this section we show that PARALLEL KNOCK-OUT can be solved in linear time for cographs. For doing so we need to introduce some extra notation and terminology.

Let G_1 and G_2 be two disjoint graphs. The *join* operation $\otimes$ adds an edge between every vertex of G_1 and every vertex of G_2. The *union* operation $\oplus$ creates the disjoint union of G_1 and G_2 (note that we may also write $G_1 + G_2$ instead of $G_1 \oplus G_2$).

It is well known (see, for example, [1]) that a graph G is a cograph if and only if G can be generated from K_1 by a sequence of operations, where each operation is either a join or a union. Such a sequence corresponds to a decomposition tree, which has the following properties:

1. its root r corresponds to the graph $G_r = G$;
2. every leaf x of it corresponds to exactly one vertex of G, and vice versa, implying that x corresponds to a unique single-vertex graph G_x;
3. every internal node x has at least two children, is either labeled $\oplus$ or $\otimes$, and corresponds to an induced subgraph G_x of G defined as follows:
   - if x is a $\oplus$-node, then G_x is the disjoint union of all graphs G_y where y is a child of x;
   - if x is a $\otimes$-node, then G_x is the join of all graphs G_y where y is a child of x.

A cograph G may have more than one such tree but has exactly one unique tree [5], called a *cotree*, if the following additional property is required:

4. Labels of internal nodes on the (unique) path from any leaf to r alternate between $\oplus$ and $\otimes$.

We denote the cotree of a cograph G by T_G and use the following result of Corneil, Perl and Stewart [6] as a lemma.

Lemma 2 ([6]). *Let G be a graph with n vertices and m edges. Deciding if G is a cograph and constructing T_G (if it exists) can be done in time $O(n + m)$.*

We now present our algorithm, which we call Cograph-PKO, for solving PARALLEL KNOCK-OUT on cographs.

Sketch. We start by giving some intuition. Let G be a cograph. We may assume without loss of generality that G is connected, as otherwise we could consider each connected component of G separately. We first construct the cotree T_G. Because G is connected, the root r of T_G is a $\otimes$-node. Recall that $G_r = G$ by definition. Consider a partition (X, Y) of the set of children of r such that

$$p = \sum_{x \in X} |G_x| \leq \sum_{y \in Y} |G_y| = q.$$

Note that G has a spanning complete bipartite graph with partition classes $\bigcup_{x \in X} V(G_x)$ and $\bigcup_{y \in Y} V(G_y)$. Hence, if $q \leq \frac{1}{2}p(p+1)$ then G is KO-reducible by Lemma 1. However, such a partition (X, Y) need not exist, but G might still be KO-reducible. In order to find out, we must analyze the cotree of G at lower levels.

The main idea behind our algorithm is as follows. As mentioned above, the graph G_x corresponding to a join node x has at least one spanning complete bipartite subgraph. We will show that it is sufficient to consider only bipartitions, in which one bipartition class corresponds to a single child z of x. We chose z in such a way that if the corresponding complete bipartite subgraph is unbalanced (with respect to the ratio prescribed in Lemma 1) then the vertices of G_z correspond to a "large" bipartition class. We will then try to reduce G_z as much as possible by internal firings only. If G_z cannot be reduced to the empty graph, then external firings are needed. In particular, some of these external firings will be internal firings for supergraphs of G_z. Hence, we first traverse T_G from top to bottom, starting with the root r, to determine the number of external firings for each graph G_z. Afterward we can then use a bottom-up approach, starting with the leaves of T_G, to determine the number of vertices a graph G_z can be reduced to by internal firings only. If this number is zero for r then G is KO-reducible; otherwise it is not.

Full Description. Let G be a connected cograph, and let $x \in V(T_G)$. We say that $|G_x|$ is the *size* of x. We fix a *largest* child of x, that is, a child of x with largest size over all children of x. We denote this child by $z(x)$ (if there is more than one largest child we pick an arbitrary largest one). Let $C(x)$ consist of all other children of x in T_G (so excluding z). We write $F(x) = \sum_{y \in C(x)} G_y$.

In our algorithm we recursively define two functions f and l that assign a positive integer to the nodes of $V(T_G)$. We write $f(x) = \bot$ or $l(x) = \bot$ if we have not yet assigned an integer $f(x)$ or $l(x)$ to node x; for some nodes x our algorithm might never do this (as we shall see, l will define an integer to a node x if and only if f has previously done so). The meaning of these two functions will be made more clear later. In particular, we will show that $f(x)$ (if defined) is the the number of vertices in $V(G) \setminus V_x$ adjacent to each vertex of V_x. This function will help us in determining how many additional internal firing rounds we have when we expand G_x to a larger subgraph of G by moving up the tree. The integer $l(x)$ (if defined) is, as we will prove, equal to the smallest number of vertices in G_x that cannot be knocked out internally (that is, within G_x) by any KO-scheme of G. We will show that $l(r)$ is defined, that is, $l(r) \neq \bot$.

Hence, there exists a KO-scheme that knocks out all vertices of $V(G_r) = V(G)$ if and only if $l(r) = 0$.

Cograph-PKO

input : a connected cograph G
output : yes if G is KO-reducible; no otherwise

Step 1. Compute the size $|G_x|$ for all $x \in V(T_G)$.

Step 2. Recursively define a function f. Initially set $f(x) := \bot$ for all $x \in V(T_G)$. Set $f(r) := 0$. Now let x be a vertex in T_G with $f(x) \neq \bot$.

2a. If x is a $\oplus$-node: $f(y) := f(x)$ for all $y \in C(x) \cup \{z(x)\}$.

2b. If x is a $\otimes$-node: $f(z(x)) := f(x) + |F(x)|$.

Step 3. Let $B = \{\ell \mid \ell \text{ is a leaf of } T_G \text{ with } f(\ell) \neq \bot\}$.

Step 4. Recursively define a function l. Initially set $l(x) := \bot$ for all $x \in V(T_G)$. Set $l(\ell) := 1$ for all $\ell \in B$. Now let x be a vertex in T that is either a $\oplus$-node with $l(y) \neq \bot$ for all $y \in C(x) \cup \{z(x)\}$ or a $\otimes$-node with $l(z(x)) \neq \bot$.

4a. If x is a $\oplus$-node: $l(x) := l(z(x)) + \sum_{y \in C(x)} l(y)$.

4b. If x is a $\otimes$-node: $l(x) := \max\{0, \ l(z(x)) - f(x) \cdot |F(x)| - \frac{1}{2}|F(x)|(|F(x)| + 1)\}$.

Step 5. If $l(r) = 0$ then return yes; otherwise return no.

Note that for some $x \in V(T_G)$, it may happen indeed that $f(x) = \bot$ or $l(x) = \bot$ holds (for example, if x is a leaf node not in B then $l(x) = \bot$).

We need some new terminology and a number of lemmas. Let x be a node in T_G. From now on we write $V_x = V(G_x)$. We say that a vertex $v \in V(G)$ is *complete* to a set $U \subseteq V(G)$ with $v \notin U$ if v is adjacent to all vertices of U.

Lemma 3. *Let* $x \in V(T_G)$ *with* $f(x) \neq \bot$. *The following two statements hold:*

(i) any vertex in $V(G) \setminus V_x$ *adjacent to a vertex of* V_x *is complete to* V_x;
(ii) the number of vertices in $V(G) \setminus V_x$ *complete to* V_x *is equal to* $f(x)$.

Proof. Let $x \in V(T_G)$ with $f(x) \neq \bot$. Statement (i) follows from the definition of T_G. We prove (ii) as follows. Let $\text{dist}(x, r)$ denote the distance between x and r in T_G. We use induction on $\text{dist}(x, r)$. The claim is true for $\text{dist}(x, r) = 0$ because in that case $x = r$ and $V(G) \setminus V_x = \emptyset$.

Let $\text{dist}(x, r) \geq 1$. Then x has a parent in T_G. Denote this parent by x'. By the induction hypothesis, $f(x')$ is equal to the number of vertices not in $G_{v'}$ that are complete to $V_{x'}$. Because V_x is contained in $V_{x'}$, these vertices are complete to V_x as well. Suppose that x is a $\oplus$-node. Then x' is a $\otimes$-node. This means that all vertices in $F(x')$ are complete to V_x. Hence, the total number of vertices in $V(G) \setminus V_x$ that are complete to V_x is equal to $f(x') + |F(x')| = f(x)$. Suppose that x is a $\otimes$-node. Then x' is a $\oplus$-node. This means that no vertex in $F(x')$ is adjacent to a vertex in V_x. Hence, the total number of vertices in $V(G) \setminus V_x$ that are complete to V_x is equal to $F(x') = f(x)$. ☐

The following lemma follows directly from the construction of our algorithm.

Lemma 4. *Let $x \in V(T_G)$. Then $l(x) \neq \perp$ if and only if $V(G_x) \cap B \neq \emptyset$.*

Let x be a node in T_G. An x-*pseudo-KO-selection* of G is a function $V_x \to V(G)$ with $f(v) \in N(v)$ for all $v \in V$. We copy some terminology. If $f(v) = u$, we say that v fires at u, or that u is knocked-out by a firing of v. Note that every KO-selection of G_x is an x-pseudo-KO-selection of G (but the reverse implication is not necessarily true). For an x-pseudo-KO-selection, we define the x-*pseudo-KO-successor* of G as the subgraph of G induced by $V(G) \setminus f(V)$. We write $G \rightsquigarrow^x G'$ to denote that G' is an x-pseudo-KO-successor of G. We call a sequence $G \rightsquigarrow^x G^1 \rightsquigarrow^x \cdots \rightsquigarrow^x G^s$ an x-*pseudo-KO-scheme* (where each single step is called a *round*) if in addition there is no vertex of V_x that fires at a vertex of V_x in some round and at a vertex $V(G) \setminus V_x$ in some later round. We say that G is x-*pseudo-reducible* to G^s. Define pseudo(x) as the number of vertices in a smallest graph to which G is x-pseudo-reducible and say that a corresponding x-pseudo-KO-scheme is *optimal*.

Lemma 5. *The cograph G is KO-reducible if and only if* pseudo$(r) = 0$.

Proof. Recall that $V_r = V(G)$. Then the statement of the lemma holds because every KO-reduction scheme of G (if there exists one) is an r-pseudo-KO-scheme with pseudo$(r) = 0$, and vice versa. ☐

The following lemma is crucial for the correctness of our algorithm.

Lemma 6. *Let $x \in V(T_G)$ be a $\otimes$-node with $l(x) \neq \perp$. Then $l(x) =$ pseudo(x).*

Proof. Let $x \in V(T_G)$ be a $\otimes$-node with $l(x) \neq \perp$. By Lemma 4, $V(G_x) \cap B \neq \emptyset$. We write $z = z(x)$. Let $|V_z| = q$ and $|F(x)| = p$. This enables us to write:

$$l(x) = \max\{0, l(z) - f(x) \cdot |F(x)| - \tfrac{1}{2}|F(x)|(|F(x)| + 1)\}$$
$$= \max\{0, l(z) - f(x) \cdot p - \tfrac{1}{2}p(p + 1)\}.$$

Note that $q \geq 1$ and $p \geq 1$ by the definition of a $\otimes$-node. Let d denote the number of $\otimes$-nodes on the longest path from x to a leaf in the subtree of T_G rooted at x. We prove the lemma by induction on d.

Let $d = 0$. Then every child of x is a leaf or otherwise all children of that child are leaves.

First suppose z is a leaf. Because $V(G_x) \cap B \neq \emptyset$, we find that $z \in B$. Hence, $l(z) = 1$. Then, as $p \geq 1$, we find that $l(z) - f(x) \cdot p - \tfrac{1}{2}p(p + 1) \leq 0$. Hence, $l(x) = 0$. Note that $q = 1$. Because z is a largest child of x, all children of x are leaves. Hence, G_x is a complete graph on $p+1$ vertices. This means that G_x is KO-reducible. We conclude that pseudo$(x) = 0 = l(x)$.

Now suppose z is not a leaf. Then z has at least two children (which are all leaves). Hence, $q \geq 2$. Because $V(G_x) \cap B \neq \emptyset$, every child of z is in B, that is, $V_z = B$ is an independent set, in particular, $q = |B|$. Because $l(\ell) = 1$ for every $\ell \in B$, this means that $l(z) = |B| = q$. We distinguish three cases.

Case 1. $q < p$.
Then $l(z) - f(x) \cdot p - \frac{1}{2}p(p+1) = q - f(x) \cdot p - \frac{1}{2}p(p+1) \leq 0$. Hence, $l(x) = 0$.

Let $y_1, \ldots, y_r, z$ be the children of x for some $r \geq 0$. In fact, because $2 \leq q < p$ and z is the largest child of x, we find that $r \geq 2$. Assume that $|V_{y_1}| \geq \cdots \geq |V_{y_r}|$. By definition, $q = |V_z| \geq |V_{y_1}|$. Because $q < p$, we can pick a set D of $q - |V_{y_1}| \geq 0$ vertices of $V(F(x)) \setminus V_{y_1}$. We define $T_1 = V_{y_1} \cup D$ and $T_i = V_{y_i} \setminus D$ for $i = 2, \ldots, r$. Note that $|T_1| = q$. Let $\{|T_1|, \ldots, |T_r|\} = \{j_1, \ldots, j_s\}$ for some $s \leq r$, where $j_1 \geq \cdots \geq j_s$. Because $|T_1| = q$, we find that $j_1 = q$. We partition $V(G_x)$ into s subsets. For the first subset we pick j_s vertices from V_z and also j_s vertices from each non-empty T_i. The graph induced by the union of all these vertices has a hamilton cycle, as $q < p$, so besides T_1 at least one other set T_i is nonempty. We remove all chosen vertices. Then, for the second subset of our partition, we pick $j_{s-1} - j_s$ vertices from V_z and also $j_{s-1} - j_s$ vertices from each T_i that is not yet empty. The graph induced by the union of all chosen vertices has a hamilton cycle if there are two non-empty sets T_i and a perfect matching otherwise. We repeat this procedure until all sets T_i are empty. In this way we have found a $[1,2]$-factor of G_x. Consequently, pko$(G_x) = 0$. Hence, pseudo$(x) = 0 = l(x)$.

Case 2. $q \geq p$ and $l(x) = 0$.
As $l(z) = q$, the assumption that $l(x) = 0$ implies that $q - f(x) \cdot p \leq \frac{1}{2}p(p+1)$. By Lemma 3, all vertices in $V(G) \setminus V_x$ that are adjacent to V_x are complete to V_x and moreover, the number of such vertices is equal to $f(x)$. This enables us to define the following x-pseudo-KO-scheme. Let all vertices of $F(x)$ fire at different vertices in V_z for the first $f(x)$ rounds. Let all vertices in V_z fire at the same vertex of $V(G) \setminus V_x$ for the first $f(x)$ rounds. Note that q decreases in this way. However, we may not need to perform all these rounds: after each round we check whether $p \leq q \leq \frac{1}{2}p(p+1)$. Because $q - f(x) \cdot p \leq \frac{1}{2}p(p+1)$, it will eventually happen that $q \leq \frac{1}{2}p(p+1)$. If it turns out that $q < p$, we slightly adjust the previous round by letting a sufficient number of vertices of $F(x)$ fire at the same vertex in V_z instead of at different vertices, in order to get $p \leq q \leq \frac{1}{2}p(p+1)$. We then apply Lemma 1 to knock out the remaining vertices of V_x in at most p additional rounds. Hence pseudo$(x) = 0 = l(x)$.

Case 3. $q \geq p$ and $l(x) > 0$.
As $l(z) = q$, the assumption that $l(x) > 0$ implies that $q > f(x) \cdot p + \frac{1}{2}p(p+1)$. Recall that, by Lemma 3, all vertices in $V(G) \setminus V_x$ that are adjacent to V_x are complete to V_x, and moreover, the number of such vertices is equal to $f(x)$. This enables us to define the following x-pseudo-KO-scheme. Let all vertices of $F(x)$ fire at different vertices in V_z for the first $f(x)$ rounds. Let all vertices in V_z fire at the same vertex of $V(G) \setminus V_x$ for the first $f(x)$ rounds. Afterward we can

reduce the number of vertices of V_x by at most $\frac{1}{2}p(p+1)$ by letting all vertices of $F(x)$ fire at different vertices in V_z, whereas all vertices in V_z fire at the same vertex of $F(x)$ until $F(x) = \emptyset$. Because $F(x) = \emptyset$ in the end, and in each round we have reduced the maximum number of vertices of the independent set V_z, we find that pseudo$(x) = q - f(x) \cdot p - \frac{1}{2}p(p+1) = l(z) - f(x) \cdot p - \frac{1}{2}p(p+1) = l(x)$.

Let $d \geq 1$. Then z is not a leaf as otherwise all children of x are leaves, which contradicts $d \geq 1$. Consequently, z is a $\oplus$-node. We distinguish two cases.

Case 1. $q < p$.
Observe that $l(z) \leq q$. Then $l(z) - f(x) \cdot p - \frac{1}{2}p(p+1) \leq q - f(x) \cdot p - \frac{1}{2}p(p+1) \leq 0$. Hence, $l(x) = 0$. We repeat the same arguments as for the corresponding case for $d = 0$ to obtain that pseudo$(x) = 0 = l(x)$. So Case 1 is proven.

Before we consider Case 2, we first analyze the subtree of T_G rooted at x. Let $s_1, \ldots, s_p$ be the children of z with $l(s_i) > 0$ for $i = 1, \ldots, p$ (if such children exist) and let $t_1, \ldots, t_q$ be the children of z with $l(t_i) = 0$ for $i = 1, \ldots, q$ (if such children exist). Note that all children of z are either leaves or $\otimes$-nodes. Let z' be a child of z. If z' is a leaf, then pseudo$(z') = 1 = l(z')$. If z' is a $\otimes$-node, we may apply the induction hypothesis to find that pseudo$(z') = l(z')$. In other words, pseudo$(s_i) = l(s_i)$ for $i = 1, \ldots, p$ and pseudo$(t_i) = l(t_i)$ for $i = 1, \ldots, q$. Then, because $G_z = G_{s_1} + \cdots + G_{s_p} + G_{t_1} + \cdots + G_{t_q}$, we find that an optimal z-pseudo-KO-scheme mimics the optimal s_i-pseudo-KO-schemes and optimal t_j-pseudo-KO-schemes (we may assume without loss of generality that all external firings outside G_z in a round are always at a single vertex). Hence, pseudo$(z) = l(s_1) + \cdots + l(s_p) + l(t_1) + \cdots + l(t_q) = l(z)$.

Case 2. $q \geq p$.
We define the following x-pseudo-KO-selection scheme. The firing rounds for the vertices in G_z are according to an optimal z-pseudo-KO-scheme under the following conditions. For the first $f(x)$ rounds any external firings outside G_z are at a single vertex, which is not in G_x. Note that this is possible by Lemma 3. Afterward any external firing outside G_z must be in $F(x)$ and also for such firings we require that they are at a single vertex in every round. The vertices in $F(x)$ fire in each round at different vertices of G_x that are in $G_{s_1} + \cdots + G_{s_p}$ and that are not being fired at by vertices in G_x. They stop firing in a graph G_{s_i} as soon as they have knocked out $l(s_i)$ of its vertices. Note that we are guaranteed a budget of exactly $f(x) \cdot p + \frac{1}{2}p(p+1)$ firings from vertices outside G_z into G_z.

First suppose that $l(z) - f(x) \cdot p \leq \frac{1}{2}p(p+1)$, so $l(x) = 0$. Then we can knock out all $l(z)$ vertices of G_z that cannot be knocked out by internal firings inside G_z. As we still need to knock out the vertices of $F(x)$, we check after each round whether q has decreased such that $p \leq q \leq \frac{1}{2}p(p+1)$ holds. Because $q - f(x) \cdot p \leq \frac{1}{2}p(p+1)$, it will eventually happen that $q \leq \frac{1}{2}p(p+1)$. If it turns out that $q < p$, we slightly adjust the previous round as we did in Case 2 for $d = 0$, in order to get $p \leq q \leq \frac{1}{2}p(p+1)$. We then apply Lemma 1 to knock out the remaining vertices of V_x in at most p additional rounds. We conclude that pseudo$(x) = 0 = l(x)$.

Now suppose that $l(z) - f(x) \cdot p > \frac{1}{2}p(p+1)$, so $l(x) > 0$. Then, by the definition of our x-pseudo-KO-reduction scheme, all vertices in $F(x)$ have fired at different vertices in every round for $f(x) \cdot p + \frac{1}{2}p(p+1)$ rounds. Moreover, all vertices in $F(x)$ are knocked out afterward. Because pseudo$(z) = l(z)$ and we mimicked an optimal z-pseudo-KO-scheme as regards the firings of the vertices of G_z in each round, we cannot improve. We conclude that pseudo$(x) = l(z) - f(x) \cdot p - \frac{1}{2}p(p+1) = l(x)$. This completes the proof of Lemma 6.   □

Theorem 1. *The* PARALLEL KNOCK-OUT *problem can be solved in* $O(n + m)$ *time on cographs with n vertices and m edges.*

Proof. Let G be a cograph with n vertices and m edges. If G is disconnected we consider each connected component of G separately. Hence, assume that G is connected. We construct T_G. Run `Cograph-PKO` with input G. By Lemma 4, we find that $l(r) \neq \bot$. Hence, we may apply Lemma 6 to find that $l(r) = $ pseudo(r). By Lemma 5, we find that G is KO-reducible if and only if pseudo$(r) = 0$. As `Cograph-PKO` outputs a yes-answer if and only if $l(r) = 0$, we find it is correct. It remains to show that it runs in linear time. We can perform Step 1 in a bottom-up approach starting from the leaves of T_G. So, Steps 1-3 each visit each node at most once. This means that every node of x is visited at most three times in total. Because every co-tree has at most $n + n - 1 = 2n - 1$ vertices, we find that the running time of `Cograph-PKO` is $O(n)$. Because constructing T_G costs time $O(n + m)$ by Lemma 2, the total running time is $O(n + m)$.   □

4   Split Graphs

We show the following result, the proof of which is (partially) based on the NP-hardness proof of 2-PARALLEL KNOCK-OUT for bipartite graphs from [3].

Theorem 2. *The* PARALLEL KNOCK-OUT *problem and, for any $k \geq 2$, the* k-PARALLEL KNOCK-OUT *problem are* NP-*complete for split graphs.*

Proof. First consider the PARALLEL KNOCK-OUT problem. We reduce from the DOMINATING SET problem, which is well known to be NP-complete (see [10]). This problem takes as input a graph $G = (V, E)$ and a positive integer p. We may assume without loss of generality that $p \leq |V|$. The question is whether G has a dominating set of cardinality at most p.

From an instance (G, p) of DOMINATING SET we construct a split graph G' as follows. Let $V(G) = \{v_1, \ldots, v_n\}$. We let $V(G')$ consist of three mutually disjoint sets: the set $V = \{v_1, \ldots, v_n\}$, a set $V' = \{v'_1, \ldots, v'_n\}$ and a set $W = \{w_1, \ldots, w_r\}$ where $r = \frac{1}{2}(n - p)(n - p + 1)$. We define $E(G')$ as follows. First we add the edges $v_i v'_i$ for $i = 1, \ldots, n$. For all $i \neq j$, we add the edges $v_i v'_j$ and $v_j v'_i$ if and only if $v_i v_j$ is an edge in $E(G)$. We also add an edge between every v_i and every w_j. Finally, we add an edge between any two vertices in V. Observe that G' is indeed a split graph in which V is a clique of size n and $V' \cup W$ is an independent set of size $n + r$. We claim that G has a dominating set of size at most p if and only if G' is KO-reducible.

First suppose G has a dominating set D of size at most p. Because $p \leq |V|$, we may assume without loss of generality that $D = \{v_1, \ldots, v_p\}$. We construct a KO-reduction scheme of G' as follows. In the first round let every vertex $v_i \in V$ fire at $v_i' \in V'$. For $i = 1, \ldots, p$, let v_i' fire at v_i. For $i = p + 1, \ldots, n$ let v_i' fire at an arbitrary vertex in D, which is possible because D is a dominating set of G. Finally, let every vertex in W fire at an arbitrary vertex in D as well; this is possible by the construction of G'. The resulting (split) graph G'' consists of a clique $V \setminus D$ of size $n - p$ and the independent set W of size $\frac{1}{2}(n - p)(n - p)$. Because there is an edge between every vertex in V and every vertex in W, we find that G'' is KO-reducible by Lemma 1.

Now suppose G' is KO-reducible. Consider a KO-reduction scheme of G'. Let D be the subset of vertices that are knocked out in the first round. Because each vertex must fire at a neighbour, D is a dominating set of G. We claim that $|D| \leq p$. For contradiction, suppose that $|D| \geq p + 1$. Let $V_1 = V \setminus D$ be the subset of V consisting of vertices not knocked out in the first round. Because $|D| \geq p + 1$, we obtain $|V_1| = |V| - |D| \leq n - p - 1$. Let V^* and W^* be the subsets of V' and W, respectively, that consist of vertices not knocked out in the first round. Vertices in $V' \cup W$ can only be knocked out by vertices of V. Moreover, the total number of vertices that V can knock out in the first round is at most $|V| = n$. This means that $V^* \cup W^*$ is an independent set of size

$$|V^* \cup W^*| = |V^*| + |W^*| \geq |V'| + |W| - n = \tfrac{1}{2}(n - p)(n - p + 1).$$

However, as in every round the size of V_1 is reduced by at least 1, the maximum number of vertices in $V^* \cup W^*$ that V_1 can knock out is at most $(n - p - 1) + (n - p - 2) + \cdots + 1 < \frac{1}{2}(n - p)(n - p + 1)$. Hence, the scheme is not a KO-reduction scheme of G'. This is a contradiction, and we have completed the proof for PARALLEL KNOCK-OUT.

Now let $k \geq 2$ and consider the k-PARALLEL KNOCK-OUT problem. We use the same reduction and the same arguments as for PARALLEL KNOCK-OUT after changing the size of W into $r := (n - p) + (n - p - 1) + \cdots + (n - p - k + 2)$.   □

5   Conclusions

We have shown in Theorem 1 that PARALLEL KNOCK-OUT is linear-time solvable for P_4-free graphs (whether it is possible to compute pko(G) in polynomial time for cographs is still open). We have also shown in Theorem 2 that PARALLEL KNOCK-OUT and, for any $k \geq 2$, k-PARALLEL KNOCK-OUT are NP-complete for split graphs. Because split graphs are $(2K_2, C_4, C_5)$-free [9], they are P_5-free. Hence, Theorems 1 and 2 have the following consequence.

Corollary 1. *The* PARALLEL KNOCK-OUT *problem restricted to P_r-free graphs is linear-time solvable if $r \leq 4$ and* NP-complete *if $r \geq 5$.*

We recall that our long standing goal is to determine the complexity of PARALLEL KNOCK-OUT on graph classes of bounded clique-width and that cographs are exactly those graphs that have clique-width at most 2 [7]. Can we solve PARALLEL KNOCK-OUT in polynomial time for graphs of clique-width at most 3?

For this we could start by considering the class of distance-hereditary graphs, which have clique-width at most 3 [11]. We also do not know whether there is a constant c such that PARALLEL KNOCK-OUT is NP-complete for graphs of clique-width at most c. However, it is known that the related NP-complete problem HAMILTON CYCLE, which tests whether a graph has a hamiltonian cycle, is polynomial-time solvable on any graph class whose clique-width is bounded by a constant (this follows from combining results of [12,15], also see [8]).

A different direction from above for extending our results would be to classify the complexity of PARALLEL KNOCK-OUT restricted to H-free graphs. The complexity status is open even for small graphs $H \in \{4P_1, 2P_1 + 2P_2, P_1 + P_3, K_{1,4}\}$.

References

1. Brandstädt, A., Le, V.B., Spinrad, J.: Graph Classes: A Survey. SIAM Monographs on Discrete Mathematics and Applications (1999)
2. Broersma, H.J., Fomin, F.V., Královič, R., Woeginger, G.J.: Eliminating graphs by means of parallel knock-out schemes. Discrete Applied Mathematics 155, 92–102 (2007); See also Broersma, H., Fomin, F.V., Woeginger, G.J.: Parallel knock-out schemes in networks. In: Fiala, J., Koubek, V., Kratochvíl, J. (eds.) MFCS 2004. LNCS, vol. 3153, pp. 204–214. Springer, Heidelberg (2004)
3. Broersma, H.J., Johnson, M., Paulusma, D., Stewart, I.A.: The computational complexity of the parallel knock-out problem. Theoretical Computer Science 393, 182–195 (2008)
4. Broersma, H.J., Johnson, M., Paulusma, D.: Upper bounds and algorithms for parallel knock-out numbers. Theoretical Computer Science 410, 1319–1327 (2008)
5. Corneil, D.G., Lerchs, H., Stewart Burlingham, L.: Complement reducible graphs. Discrete Applied Mathematics 3, 163–174 (1981)
6. Corneil, D.G., Perl, Y., Stewart, L.K.: A linear recognition algorithm for cographs. SIAM Journal on Computing 14, 926–934 (1985)
7. Courcelle, B., Olariu, S.: Upper bounds to the clique width of graphs. Discrete Applied Mathematics 101, 77–144 (2000)
8. Espelage, W., Gurski, F., Wanke, E.: How to solve NP-hard graph problems on clique-width bounded Graphs in polynomial time. In: Brandstädt, A., Le, V.B. (eds.) WG 2001. LNCS, vol. 2204, pp. 117–128. Springer, Heidelberg (2001)
9. Földes, S., Hammer, P.L.: Split graphs, 8th South–Eastern Conf. on Combinatorics. Graph Theory and Computing, Congressus Numerantium 19, 311–315 (1977)
10. Garey, M.R., Johnson, D.S.: Computers and Intractability: A Guide to the Theory of NP-Completeness. Freeman (1979)
11. Golumbic, M.C., Rotics, U.: On the clique-width of some perfect graph classes. International Journal of Foundations of Computer Science 11, 423–443 (2000)
12. Johansson, O.: Clique-decomposition, NLC-decomposition, and modular decomposition – relationships and results for random graphs. Congressus Numerantium 132, 39–60 (1998)
13. Johnson, M., Paulusma, D., Wood, C.: Path factors and parallel knock-out schemes of almost claw-free graphs. Discrete Mathematics 310, 1413–1423 (2010)
14. Lampert, D.E., Slater, P.J.: Parallel knockouts in the complete graph. American Mathematical Monthly 105, 556–558 (1998)
15. Wanke, E.: k-NLC graphs and polynomial algorithms. Discrete Applied Mathematics 54, 251–266 (1994)

Relating the Time Complexity of Optimization Problems in Light of the Exponential-Time Hypothesis

Peter Jonsson, Victor Lagerkvist, Johannes Schmidt, and Hannes Uppman

Department of Computer and Information Science, Linköping University, Sweden
{peter.jonsson,victor.lagerkvist,johannes.schmidt,hannes.uppman}@liu.se

Abstract. Obtaining lower bounds for NP-hard problems has for a long time been an active area of research. Recent algebraic techniques introduced by Jonsson et al. (SODA 2013) show that the time complexity of the parameterized $\mathrm{SAT}(\cdot)$ problem correlates to the lattice of strong partial clones. With this ordering they isolated a relation R such that $\mathrm{SAT}(R)$ can be solved at least as fast as any other NP-hard $\mathrm{SAT}(\cdot)$ problem. In this paper we extend this method and show that such languages also exist for the *max ones problem* ($\mathrm{MAX\text{-}ONES}(\Gamma)$) and the *Boolean valued constraint satisfaction problem* over finite-valued constraint languages ($\mathrm{VCSP}(\Delta)$). With the help of these languages we relate $\mathrm{MAX\text{-}ONES}$ and VCSP to the exponential time hypothesis in several different ways.

1   Introduction

A superficial analysis of the NP-complete problems may lead one to think that they are a highly uniform class of problems: in fact, under polynomial-time reductions, the NP-complete problems may be viewed as a *single* problem. However, there are many indications (both from practical and theoretical viewpoints) that the NP-complete problems are a diverse set of problems with highly varying properties, and this becomes visible as soon as one starts using more refined methods. This has inspired a strong line of research on the "inner structure" of the set of NP-complete problem. Examples include the intensive search for faster algorithms for NP-complete problems [20] and the highly influential work on the *exponential time hypothesis* (ETH) and its variants [14]. Such research might not directly resolve whether P is equal to NP or not, but rather attempts to explain the seemingly large difference in complexity between NP-hard problems and what makes one problem harder than another. Unfortunately there is still a lack of general methods for studying and comparing the complexity of NP-complete problems with more restricted notions of reducibility. Jonsson et al. [9] presented a framework based on *clone theory*, applicable to problems that can be viewed as "assigning values to variables", such as constraint satisfaction problems, the vertex cover problem, and integer programming problems. To analyze and relate the complexity of these problems in greater detail we utilize polynomial-time reductions which increase the number of variables by a constant factor (*linear variable reductions* or *LV-reductions*) and reductions which increases the amount of variables by a constant (*constant variable reductions* or *CV-reductions*). Note the following: (1) if a problem A is solvable in $O(c^n)$ time (where n denotes the number of variables) for all $c > 1$ and if problem B is LV-reducible to A then B is also solvable in $O(c^n)$ time for all $c > 1$ and (2) if A is solvable

E. Csuhaj-Varjú et al. (Eds.): MFCS 2014, Part II, LNCS 8635, pp. 408–419, 2014.
© Springer-Verlag Berlin Heidelberg 2014

in time $O(c^n)$ and if B is CV-reducible to A then B is also solvable in time $O(c^n)$. Thus LV-reductions preserve subexponential complexity while CV-reductions preserve exact complexity. Jonsson et al. [9] exclusively studied the Boolean satisfiability SAT$(\cdot)$ problem and identified an NP-hard SAT$(\{R\})$ problem CV-reducible to all other NP-hard SAT$(\cdot)$ problems. Hence SAT$(\{R\})$ is, in a sense, the *easiest* NP-complete SAT$(\cdot)$ problem since if SAT(Γ) can be solved in $O(c^n)$ time, then this holds for SAT$(\{R\})$, too. With the aid of this result, they analyzed the consequences of subexponentially solvable SAT$(\cdot)$ problems by utilizing the interplay between CV- and LV-reductions. As a by-product, Santhanam and Srinivasan's [16] negative result on sparsification of infinite constraint languages was shown not to hold for finite languages.

We believe that the existence and construction of such easiest languages forms an important puzzle piece in the quest of relating the complexity of NP-hard problems with each other, since it effectively gives a lower bound on the time complexity of a given problem with respect to constraint language restrictions. As a logical continuation on the work on SAT$(\cdot)$ we pursue the study of CV- and LV-reducibility in the context of Boolean optimization problems. In particular we investigate the complexity of MAX-ONES$(\cdot)$ and VCSP$(\cdot)$ and introduce and extend several non-trivial methods for this purpose. The results confirms that methods based on universal algebra are indeed useful when studying broader classes of NP-complete problems. The MAX-ONES$(\cdot)$ problem [11] is a variant of SAT$(\cdot)$ where the goal is to find a satisfying assignment which maximizes the number of variables assigned the value 1. This problem is closely related to the 0/1 LINEAR PROGRAMMING problem. The VCSP$(\cdot)$ problem is a function minimization problem that generalizes the MAX-CSP and MIN-CSP problems [11]. We treat both the unweighted and weighted versions of these problems and use the prefix U to denote the unweighted problem and W to denote the weighted version. These problems are well-studied with respect to separating tractable cases from NP-hard cases [11] but much less is known when considering the weaker schemes of LV-reductions and CV-reductions. We begin (in Section 3.1) by identifying the easiest language for W-MAX-ONES$(\cdot)$. The proofs make heavy use of the *algebraic method* for constraint satisfaction problems [7,8] and the *weak base method* [18]. The algebraic method was introduced for studying the computational complexity of constraint satsifaction problems up to polynomial-time reductions while the weak base method was shown by Jonsson et al. [9] to be useful for studying CV-reductions. To prove the main result we however need even more powerful reduction techniques based on *weighted primitive positive implementations* [19]. For VCSP$(\cdot)$ the situation differs even more since the algebraic techniques developed for CSP$(\cdot)$ are not applicable — instead we use *multimorphisms* [3] when considering the complexity of VCSP$(\cdot)$. We prove (in Section 3.2) that the binary function $f_{\neq}$ which returns 0 if its two arguments are different and 1 otherwise, results in the easiest NP-hard VCSP$(\cdot)$ problem. This problem is very familiar since it is the MAX CUT problem slightly disguised. The complexity landscape surrounding these problems is outlined in Section 3.3.

With the aid of the languages identified in Section 3, we continue (in Section 4) by relating MAX-ONES and VCSP with LV-reductions and connect them with the ETH. Our results imply that (1) if the ETH is true then no NP-complete U-MAX-ONES(Γ), W-MAX-ONES(Γ), or VCSP(Δ) is solvable in subexponential time and (2) that if the ETH is false then U-MAX-ONES(Γ) and U-VCSP$_d(\Delta)$ are solvable in subexponential time for every choice of Γ and Δ and $d \geq 0$. Here $\text{U-VCSP}_d(\Delta)$ is the $\text{U-VCSP}(\Delta)$ problem restricted to instances where the sum to minimize contains at most dn terms. Thus, to disprove the ETH, our result implies that it is sufficient to find a single language Γ or a set of cost functions Δ such that $\text{U-MAX-ONES}(\Gamma)$, $\text{W-MAX-ONES}(\Gamma)$ or $\text{VCSP}(\Delta)$ is NP-hard and solvable in subexponential time.

2 Preliminaries

Let Γ denote a finite set of finitary relations over $\mathbb{B} = \{0,1\}$. We call Γ a *constraint language*. Given $R \subseteq \mathbb{B}^k$ we let $\text{ar}(R) = k$ denote its arity, and similarly for functions. When $\Gamma = \{R\}$ we typically omit the set notation and treat R as a constraint language.

2.1 Problem Definitions

The *constraint satisfaction problem* over Γ ($\text{CSP}(\Gamma)$) is defined as follows.

INSTANCE: A set V of variables and a set C of constraint applications $R(v_1, \ldots, v_k)$ where $R \in \Gamma$, $k = \text{ar}(R)$, and $v_1, \ldots, v_k \in V$.
QUESTION: Is there a function $f : V \to \mathbb{B}$ such that $(f(v_1), \ldots, f(v_k)) \in R$ for each $R(v_1, \ldots, v_k)$ in C?

For the Boolean domain this problem is typically denoted as $\text{SAT}(\Gamma)$. By $\text{SAT}(\Gamma)$-B we mean the $\text{SAT}(\Gamma)$ problem restricted to instances where each variable can occur in at most B constraints. This restricted problem is occasionally useful since each instance contains at most Bn constraints. The *weigthed maximum ones problem* over Γ ($\text{W-MAX-ONES}(\Gamma)$) is an optimization version of $\text{SAT}(\Gamma)$ where we for an instance on variables $\{x_1, \ldots, x_n\}$ and weights $w_i \in \mathbb{Q}_{\geq 0}$ want to find a solution h for which $\sum_{i=1}^{n} w_i h(x_i)$ is maximal. The *unweigthed maximum ones problem* ($\text{U-MAX-ONES}(\Gamma)$) is the $\text{W-MAX-ONES}(\Gamma)$ problem where all weights have the value 1. A *finite-valued cost function* on $\mathbb{B}$ is a function $f : \mathbb{B}^k \to \mathbb{Q}_{\geq 0}$. The *valued constraint satisfaction problem* over a finite set of finite-valued cost functions Δ ($\text{VCSP}(\Delta)$) is defined as follows.

INSTANCE: A set $V = \{x_1, \ldots, x_n\}$ of variables and the objective function $f_I(x_1, \ldots, x_n)$ $= \sum_{i=1}^{q} w_i f_i(\mathbf{x}^i)$ where, for every $1 \leq i \leq q$, $f_i \in \Delta$, $\mathbf{x}^i \in V^{\text{ar}(f_i)}$, and $w_i \in \mathbb{Q}_{\geq 0}$ is a weight.
GOAL: Find a function $h : V \to \mathbb{B}$ such that $f_I(h(x_1), \ldots, h(x_n))$ is minimal.

When the set of cost functions is singleton $\text{VCSP}(\{f\})$ is written as $\text{VCSP}(f)$. We let U-VCSP be the VCSP problem without weights and U-VCSP_d (for $d \geq 0$) denote the U-VCSP problem restricted to instances containing at most $d\,|\text{Var}(I)|$ constraints. Many optimization problems can be viewed as $\text{VCSP}(\Delta)$ problems for suitable Δ: well-known examples are the $\text{MAX-CSP}(\Gamma)$ and $\text{MIN-CSP}(\Gamma)$ problems where the number of satisfied constraints in a CSP instance are maximized or minimized. For each Γ, there obviously exists sets of cost functions $\Delta_{\min}, \Delta_{\max}$ such that $\text{MIN-CSP}(\Gamma)$ is polynomial-time equivalent to $\text{VCSP}(\Delta_{\min})$ and $\text{MAX-CSP}(\Gamma)$ is polynomial-time equivalent to $\text{VCSP}(\Delta_{\max})$. We have defined U-VCSP, VCSP, U-MAX-ONES and W-MAX-ONES as optimization problems, but to obtain a more uniform treatment we often view them as decision problems, i.e. given k we ask if there is a solution with objective value k or better.

2.2   Size-Preserving Reductions and Subexponential Time

If A is a computational problem we let $I(A)$ be the set of problem instances and $\|I\|$ be the size of any $I \in I(A)$, i.e. the number of bits required to represent I. Many problems can in a natural way be viewed as problems of assigning values from a fixed finite set to a collection of variables. This is certainly the case for $\text{SAT}(\cdot)$, $\text{MAX-ONES}(\cdot)$ and $\text{VCSP}(\cdot)$ but it is also the case for various graph problems such as MAX-CUT and $\text{MAX INDEPENDENT SET}$. We call problems of this kind *variable problems* and let $\text{Var}(I)$ denote the set of variables of an instance I.

Definition 1. *Let A_1 and A_2 be variable problems in NP. The function f from $I(A_1)$ to $I(A_2)$ is a* many-one linear variable reduction *(LV-reduction) with parameter $C \geq 0$ if: (1) I is a yes-instance of A_1 if and only if $f(I)$ is a yes-instance of A_2, (2) $|\text{Var}(f(I))| = C \cdot |\text{Var}(I)| + O(1)$, and (3) $f(I)$ can be computed in time $O(\text{poly}(\|I\|))$.*

LV-reductions can be seen as a restricted form of SERF-reductions [6]. The term CV-reduction is used to denote LV-reductions with parameter 1, and we write $A_1 \leq^{CV} A_2$ to denote that the problem A_1 has an CV-reduction to A_2. If A_1 and A_2 are two NP-hard problems we say that A_1 is *at least as easy* as (or *not harder than*) A_2 if A_1 is solvable in $O(c^{|\text{Var}(I)|})$ time whenever A_1 is solvable in $O(c^{|\text{Var}(I)|})$ time. By definition if $A_1 \leq^{CV} A_2$ then A_1 is not harder than A_2 but the converse is not true in general. A problem solvable in time $O(2^{c|\text{Var}(I)|})$ for all $c > 0$ is a *subexponential problem*, and SE denotes the class of all variable problems solvable in subexponential time. It is straightforward to prove that LV-reductions preserve subexponential complexity in the sense that if A is LV-reducible to B then $A \in \text{SE}$ if $B \in \text{SE}$. Naturally, SE can be defined using other complexity parameters than $|\text{Var}(I)|$ [6].

2.3   Clone Theory

An operation $f : \mathbb{B}^k \to \mathbb{B}$ is a *polymorphism* of a relation R if for every $\mathbf{t}^1, \ldots, \mathbf{t}^k \in R$ it holds that $f(\mathbf{t}^1, \ldots, \mathbf{t}^k) \in R$, where f is applied element-wise. In this case R is *closed*, or *invariant*, under f. For a set of functions F we define $\text{Inv}(\mathsf{F})$ (often abbreviated as IF) to be the set of all relations invariant under all functions in F. Dually $\text{Pol}(\Gamma)$ for a set of relations Γ is defined to be the set of polymorphisms of Γ. Sets of the form $\text{Pol}(\Gamma)$ are known as *clones* and sets of the form $\text{Inv}(\mathsf{F})$ are known as *co-clones*. The reader unfamiliar with these concepts is referred to the textbook by Lau [13]. The relationship between these structures is made explicit in the following *Galois connection* [13].

Theorem 2. *Let Γ, Γ' be sets of relations. Then $\text{Inv}(\text{Pol}(\Gamma')) \subseteq \text{Inv}(\text{Pol}(\Gamma))$ if and only if $\text{Pol}(\Gamma) \subseteq \text{Pol}(\Gamma')$.*

Co-clones can equivalently be described as sets containing all relations R definable through *primitive positive* (p.p.) implementations over a constraint language Γ, i.e. definitions of the form $R(x_1, \ldots, x_n) \equiv \exists y_1, \ldots, y_m . R_1(\mathbf{x}^1) \wedge \ldots \wedge R_k(\mathbf{x}^k)$, where each $R_i \in \Gamma \cup \{\text{eq}\}$ and each $\mathbf{x}^i$ is a tuple over $x_1, \ldots, x_n, y_1, \ldots, y_m$ and where $\text{eq} = \{(0,0),(1,1)\}$. As a shorthand we let $\langle \Gamma \rangle = \text{Inv}(\text{Pol}(\Gamma))$ for a constraint language Γ, and as can be verified this is the smallest set of relations closed under p.p. definitions over Γ. In this case Γ is said to be a *base* of $\langle \Gamma \rangle$. It is known that if Γ' is finite and $\text{Pol}(\Gamma) \subseteq \text{Pol}(\Gamma')$ then

$CSP(\Gamma')$ is polynomial-time reducible to $CSP(\Gamma)$ [7]. With this fact and Post's classification of all Boolean clones [15] Schaefer's dichotomy theorem [17] for $SAT(\cdot)$ follows almost immediately. The reader is referred to Böhler et al. [2] for a visualization of the Boolean clone lattice and a complete list of bases. The complexity of $MAX\text{-}ONES(\Gamma)$ is also preserved under finite expansions with relations p.p. definable in Γ, and hence follow the standard Galois connection [11]. Note however that $Pol(\Gamma') \subseteq Pol(\Gamma)$ does not imply that $CSP(\Gamma') \leq^{CV} CSP(\Gamma)$ or that $CSP(\Gamma')$ LV-reduces to $CSP(\Gamma)$ since the number of constraints is not necessarily linearly bounded by the number of variables.

To study these restricted classes of reductions we are therefore in need of Galois connections with increased granularity. In Jonsson et al. [9] the $SAT(\cdot)$ problem is studied with the Galois connection between closure under p.p. definitions without existential quantification and *strong partial clones*. Here we concentrate on the relational description and instead refer the reader to Schnoor [18] for definitions of partial polymorphisms and the aforementioned Galois connection. If R is an n-ary Boolean relation and Γ a constraint language then R has a *quantifier-free primitive positive* (q.p.p.) implementation in Γ if $R(x_1,\ldots,x_n) \equiv R_1(\mathbf{x}^1) \wedge \ldots \wedge R_k(\mathbf{x}^k)$, where each $R_i \in \Gamma \cup \{eq\}$ and each $\mathbf{x}^i$ is a tuple over $x_1,\ldots,x_n$. We use $\langle \Gamma \rangle_{\not\exists}$ to denote the smallest set of relations closed under q.p.p. definability over Γ. If $IC = \langle IC \rangle_{\not\exists}$ then IC is a *weak partial co-clone*. In Jonsson et al. [9] it is proven that if $\Gamma' \subseteq \langle \Gamma \rangle_{\not\exists}$ and if Γ and Γ' are both finite constraint languages then $CSP(\Gamma') \leq^{CV} CSP(\Gamma)$. It is not hard to extend this result to the $MAX\text{-}ONES(\cdot)$ problem since it follows the standard Galois connection, and therefore we use this fact without explicit proof. A *weak base* R_w of a co-clone IC is then a base of IC with the property that for any finite base Γ of IC it holds that $R_w \in \langle \Gamma \rangle_{\not\exists}$ [18]. In particular this means that $SAT(R_w)$ and $MAX\text{-}ONES(R_w)$ CV-reduce to $SAT(\Gamma)$ and $MAX\text{-}ONES(\Gamma)$ for any base Γ of IC, and R_w can therefore be seen as the easiest language in the co-clone. See Table 1 for a list of weak bases for the co-clones where $MAX\text{-}ONES(\cdot)$ is NP-hard. A full list of weak bases for all Boolean co-clones can be found in Lagerkvist [12]. In addition these weak bases can be implemented without the equality relation [12].

Table 1. Weak bases for all Boolean co-clones where $MAX\text{-}ONES(\cdot)$ is NP-hard

Co-clone	Weak base	Co-clone	Weak base
$IS_1^n, n \geq 2$	$NAND^n(x_1,\ldots,x_n) \wedge F(c_0)$	IL_2	$EVEN_{3\neq}^3(x_1,\ldots,x_6) \wedge F(c_0) \wedge T(c_1)$
$IS_{12}^n, n \geq 2$	$NAND^n(x_1,\ldots,x_n) \wedge F(c_0) \wedge T(c_1)$	IL_3	$EVEN_{4\neq}^4(x_1,\ldots,x_8)$
$IS_{11}^n, n \geq 2$	$NAND^n(x_1,\ldots,x_n) \wedge (x \to x_1 \cdots x_n) \wedge F(c_0)$	IE_0	$(x_1 \leftrightarrow x_2 x_3) \wedge (x_2 \vee x_3 \to x_4) \wedge F(c_0)$
$IS_{10}^n, n \geq 2$	$NAND^n(x_1,\ldots,x_n) \wedge (x \to x_1 \cdots x_n) \wedge F(c_0) \wedge T(c_1)$	IE_2	$(x_1 \leftrightarrow x_2 x_3) \wedge F(c_0) \wedge T(c_1)$
ID_2	$OR_{2\neq}^2(x_1,x_2,x_3,x_4) \wedge F(c_0) \wedge T(c_1)$	II_0	$(\bar{x_1} \vee \bar{x_2}) \wedge (\bar{x_1}\bar{x_2} \leftrightarrow \bar{x_3}) \wedge F(c_0)$
IN_2	$EVEN_{4\neq}^4(x_1,\ldots,x_8) \wedge x_1 x_4 \leftrightarrow x_2 x_3$	II_2	$R_{3\neq}^{1/3}(x_1,\ldots,x_6) \wedge F(c_0) \wedge T(c_1)$
IL_0	$EVEN^3(x_1,x_2,x_3) \wedge F(c_0)$		

2.4   Operations and Relations

An operation f is called *arithmetical* if $f(y,x,x) = f(y,x,y) = f(x,x,y) = y$ for every $x,y \in \mathbb{B}$. The max function is defined as $\max(x,y) = 0$ if $x = y = 0$ and 1 otherwise. We often express a Boolean relation R as a logical formula whose satisfying assignment corresponds to the tuples of R. F and T are the two constant relations $\{(0)\}$ and $\{(1)\}$ while neq denotes inequality, i.e. the relation $\{(0,1),(1,0)\}$. The relation EVEN^n is defined as $\{(x_1,\ldots,x_n) \in \mathbb{B}^n \mid \sum_{i=1}^n x_i \text{ is even}\}$. The relation ODD^n is defined dually. The relations OR^n and NAND^n are the relations corresponding to the clauses $(x_1 \vee \ldots \vee x_n)$ and $(\bar{x_1} \vee \ldots \vee \bar{x_n})$. For any n-ary relation and R we let $R_{m\neq}$, $1 \leq m \leq n$, denote the $(n+m)$-ary relation defined as $R_{m\neq}(x_1,\ldots,x_{n+m}) \equiv R(x_1,\ldots,x_n) \wedge \text{neq}(x_1,x_{n+1}) \wedge \ldots \wedge \text{neq}(x_n,x_{n+m})$. We let $R^{1/3} = \{(0,0,1),(0,1,0),(1,0,0)\}$. Variables are typically named $x_1,\ldots,x_n$ or x except when they occur in positions where they are forced to take a particular value, in which case they are named c_0 and c_1 respectively to explicate that they are in essence constants. As convention c_0 and c_1 always occur in the last positions in the arguments to a predicate. We now see that $R_{\text{II}_2}(x_1,\ldots,x_6,c_0,c_1) \equiv R_{3\neq}^{1/3}(x_1,\ldots,x_6) \wedge F(c_0) \wedge T(c_1)$ and $R_{\text{IN}_2}(x_1,\ldots,x_8) \equiv \text{EVEN}_{4\neq}^4(x_1,\ldots,x_8) \wedge (x_1 x_4 \leftrightarrow x_2 x_3)$ from Table 1 are the two relations (where the tuples in the relations are listed as rows)

$$R_{\text{II}_2} = \left\{ \begin{smallmatrix} 0 & 0 & 1 & 1 & 1 & 0 & 0 & 1 \\ 0 & 1 & 0 & 1 & 0 & 1 & 0 & 1 \\ 1 & 0 & 0 & 0 & 1 & 1 & 0 & 1 \end{smallmatrix} \right\} \quad \text{and} \quad R_{\text{IN}_2} = \left\{ \begin{smallmatrix} 0 & 0 & 0 & 0 & 1 & 1 & 1 & 1 \\ 0 & 0 & 1 & 1 & 1 & 1 & 0 & 0 \\ 0 & 1 & 0 & 1 & 1 & 0 & 1 & 0 \\ 1 & 1 & 1 & 0 & 0 & 0 & 0 & 0 \\ 1 & 1 & 0 & 0 & 0 & 0 & 1 & 1 \\ 1 & 0 & 1 & 0 & 0 & 1 & 0 & 1 \end{smallmatrix} \right\}.$$

3 The Easiest NP-Hard MAX-ONES and VCSP Problems

We will now study the complexity of W-MAX-ONES and VCSP with respect to CV-reductions. We remind the reader that constraint languages Γ and sets of cost functions Δ are always finite. We prove that for both these problems there is a single language which is CV-reducible to every other NP-hard language. Out of the infinite number of candidate languages generating different co-clones, the language $\{R_{\text{II}_2}\}$ defines the easiest W-MAX-ONES$(\cdot)$ problem, which is the same language as for SAT$(\cdot)$ [9]. This might be contrary to intuition since one could be led to believe that the co-clones in the lower parts of the co-clone lattice, generated by very simple languages where the corresponding SAT$(\cdot)$ problem is in P, would result in even easier problems.

3.1 The MAX-ONES Problem

Here we use a slight reformulation of Khanna et al.'s [11] complexity classification of the MAX-ONES problem expressed in terms of polymorphisms.

Theorem 3 ([11]). *Let Γ be a finite Boolean constraint language. MAX-ONES(Γ) is in P if and only if Γ is 1-closed, max-closed, or closed under an arithmetical operation.*

The theorem holds for both the weighted and the unweighted version of the problem and showcases the strength of the algebraic method since it not only eliminates all constraint languages resulting in polynomial-time solvable problems, but also tells us exactly which cases remain, and which properties they satisfy.

Theorem 4. U-MAX-ONES$(R) \leq^{\text{CV}}$ U-MAX-ONES(Γ) *for some* $R \in \{R_{\text{IS}_1^2}, R_{\text{II}_2}, R_{\text{IN}_2},$ $R_{\text{IL}_0}, R_{\text{IL}_2}, R_{\text{IL}_3}, R_{\text{ID}_2}\}$ *whenever* U-MAX-ONES(Γ) *is NP-hard.*

Proof. By Theorem 3 in combination with the bases of Boolean clones in Böhler et al. [2] it follows that U-MAX-ONES(Γ) is NP-hard if and only if $\langle\Gamma\rangle \supseteq \mathsf{IS}_1^2$ or if $\langle\Gamma\rangle \in \{\mathsf{IL}_0, \mathsf{IL}_3, \mathsf{IL}_2, \mathsf{IN}_2\}$. In principle we then for every co-clone have to decide which language is CV-reducible to every other base of the co-clone, but since a weak base always have this property, we can eliminate a lot of tedious work and directly consult the precomputed relations in Table 1. From this we first see that $\langle R_{\mathsf{IS}_1^2}\rangle_{\not=} \subset \langle R_{\mathsf{IS}_1^n}\rangle_{\not=}$, $\langle R_{\mathsf{IS}_{12}^2}\rangle_{\not=} \subset \langle R_{\mathsf{IS}_{12}^n}\rangle_{\not=}$, $\langle R_{\mathsf{IS}_{11}^2}\rangle_{\not=} \subset \langle R_{\mathsf{IS}_{11}^n}\rangle_{\not=}$ and $\langle R_{\mathsf{IS}_{10}^2}\rangle_{\not=} \subset \langle R_{\mathsf{IS}_{10}^n}\rangle_{\not=}$ for every $n \geq 3$. Hence in the four infinite chains IS_1^n, IS_{12}^n, IS_{11}^n, IS_{10}^n we only have to consider the bottom-most co-clones IS_1^2, IS_{12}^2, IS_{11}^2, IS_{10}^2. Observe that if R and R' satisfies $R(x_1,\ldots,x_k) \Rightarrow \exists y_0, y_1.R'(x_1,\ldots,x_k,y_0,y_1) \wedge F(y_0) \wedge T(y_1)$ and $R'(x_1,\ldots,x_k,y_0,y_1) \Rightarrow R(x_1,\ldots,x_k) \wedge F(y_0)$, and $R'(x_1,\ldots,x_k,y_0,y_1) \in \langle\Gamma\rangle_{\not=}$, then U-MAX-ONES($R$) $\leq^{CV}$ U-MAX-ONES(Γ), since we can use y_0 and y_1 as global variables and because an optimal solution to the instance we construct will always map y_1 to 1 if the original instance is satisfiable. For $R_{\mathsf{IS}_1^2}(x_1,x_2,c_0)$ we can q.p.p. define predicates $R'_{\mathsf{IS}_1^2}(x_1,x_2,c_0,y_0,y_1)$ with $R_{\mathsf{IS}_{12}^2}, R_{\mathsf{IS}_{11}^2}, R_{\mathsf{IS}_{10}^2}, R_{\mathsf{IE}_2}, R_{\mathsf{IE}_0}$ satisfying these properties as follows:

- $R'_{\mathsf{IS}_1^2}(x_1,x_2,c_0,y_0,y_1) \equiv R_{\mathsf{IS}_{12}^2}(x_1,x_2,c_0,y_1) \wedge R_{\mathsf{IS}_{12}^2}(x_1,x_2,y_0,y_1)$,
- $R'_{\mathsf{IS}_1^2}(x_1,x_2,c_0,y_0,y_1) \equiv R_{\mathsf{IS}_{11}^2}(x_1,x_2,c_0,c_0) \wedge R_{\mathsf{IS}_{11}^2}(x_1,x_2,y_0,y_0)$,
- $R'_{\mathsf{IS}_1^2}(x_1,x_2,c_0,y_0,y_1) \equiv R_{\mathsf{IS}_{10}^2}(x_1,x_2,c_0,c_0,y_1) \wedge R_{\mathsf{IS}_{10}^2}(x_1,x_2,c_0,y_0,y_1)$,
- $R'_{\mathsf{IS}_1^2}(x_1,x_2,c_0,y_0,y_1) \equiv R_{\mathsf{IE}_2}(c_0,x_1,x_2,c_0,y_1) \wedge R_{\mathsf{IE}_2}(c_0,x_1,x_2,y_0,y_1)$,
- $R'_{\mathsf{IS}_1^2}(x_1,x_2,c_0,y_0,y_1) \equiv R_{\mathsf{IE}_0}(c_0,x_1,x_2,y_1,c_0) \wedge R_{\mathsf{IE}_0}(y_0,x_1,x_2,y_1,y_0)$,

and similarly a relation R'_{II_2} using R_{II_0} as follows $R'_{\mathsf{II}_2}(x_1,x_2,x_3,x_4,x_5,x_6,c_0,c_1,y_0,y_1) \equiv R_{\mathsf{II}_0}(x_1,x_2,x_3,c_0) \wedge R_{\mathsf{II}_0}(c_0,c_1,y_1,y_0) \wedge R_{\mathsf{II}_0}(x_1,x_4,y_1,y_0) \wedge R_{\mathsf{II}_0}(x_2,x_5,y_1,y_0) \wedge R_{\mathsf{II}_0}(x_3,x_6, y_1,y_0)$. We then see that the only remaining cases for Γ when $\langle\Gamma\rangle \supset \mathsf{IS}_1^2$ is when $\langle\Gamma\rangle = \mathsf{II}_2$ or when $\langle\Gamma\rangle = \mathsf{ID}_2$. This concludes the proof. $\square$

Using q.p.p. implementations to further decrease the set of relations in Theorem 4 appears difficult and we therefore make use of more powerful implementations. Let Optsol(I) be the set of all optimal solutions of a W-MAX-ONES(Γ) instance I. A relation R has a *weighted p.p. definition* (w.p.p. definition) [19] in Γ if there exists an instance I of W-MAX-ONES(Γ) on variables V such that $R = \{(\phi(v_1),\ldots,\phi(v_m)) \mid \phi \in$ Optsol(I)$\}$ for some $v_1,\ldots,v_m \in V$. The set of all relations w.p.p. definable in Γ is denoted $\langle\Gamma\rangle_w$ and we furthermore have that if $\Gamma' \subseteq \langle\Gamma\rangle_w$ is finite then W-MAX-ONES(Γ') is polynomial-time reducible to W-MAX-ONES(Γ) [19]. If there is a W-MAX-ONES(Γ) instance I on V such that $R = \{(\phi(v_1),\ldots,\phi(v_m)) \mid \phi \in$ Optsol(I)$\}$ for $v_1,\ldots,v_m \in V$ satisfying $\{v_1,\ldots,v_m\} = V$, then we say that R is q.w.p.p. definable in Γ. We use $\langle\Gamma\rangle_{\not=,w}$ for set of all relations q.w.p.p. definable in Γ. It is not hard to check that if $\Gamma' \subseteq \langle\Gamma\rangle_{\not=,w}$, then every instance is mapped to an instance of equally many variables — hence W-MAX-ONES(Γ') is CV-reducible to W-MAX-ONES(Γ) whenever Γ' is finite.

Theorem 5. *Let Γ be a constraint language such that W-MAX-ONES(Γ) is NP-hard. Then it holds that W-MAX-ONES(R_{II_2}) $\leq^{CV}$ W-MAX-ONES(Γ).*

Proof. We utilize q.w.p.p. definitions and note that the following holds.

$R_{II_2} = \arg\max_{\mathbf{x}\in\mathbb{B}^8:(\mathbf{x}_7,\mathbf{x}_1,\mathbf{x}_2,\mathbf{x}_6,\mathbf{x}_8,\mathbf{x}_4,\mathbf{x}_5,\mathbf{x}_3)\in R_{IN_2}}\mathbf{x}_8,$

$R_{II_2} = \arg\max_{\mathbf{x}\in\mathbb{B}^8:(\mathbf{x}_5,\mathbf{x}_4,\mathbf{x}_2,\mathbf{x}_1,\mathbf{x}_7,\mathbf{x}_8),(\mathbf{x}_6,\mathbf{x}_4,\mathbf{x}_3,\mathbf{x}_1,\mathbf{x}_7,\mathbf{x}_8),(\mathbf{x}_6,\mathbf{x}_5,\mathbf{x}_3,\mathbf{x}_4,\mathbf{x}_7,\mathbf{x}_8)\in R_{ID_2}}(\mathbf{x}_1+\mathbf{x}_2+\mathbf{x}_3),$

$R_{II_2} = \arg\max_{\mathbf{x}\in\mathbb{B}^8:(\mathbf{x}_4,\mathbf{x}_5,\mathbf{x}_6,\mathbf{x}_1,\mathbf{x}_2,\mathbf{x}_3,\mathbf{x}_7,\mathbf{x}_8)\in R_{IL_2}}(\mathbf{x}_4+\mathbf{x}_5+\mathbf{x}_6),$

$R_{IL_2} = \arg\max_{\mathbf{x}\in\mathbb{B}^8:(\mathbf{x}_7,\mathbf{x}_1,\mathbf{x}_2,\mathbf{x}_3,\mathbf{x}_8,\mathbf{x}_4,\mathbf{x}_5,\mathbf{x}_6)\in R_{IL_3}}\mathbf{x}_8,$

$R_{IL_2} = \arg\max_{\mathbf{x}\in\mathbb{B}^8:(\mathbf{x}_4,\mathbf{x}_5,\mathbf{x}_6,\mathbf{x}_7),(\mathbf{x}_8,\mathbf{x}_1,\mathbf{x}_4,\mathbf{x}_7),(\mathbf{x}_8,\mathbf{x}_2,\mathbf{x}_5,\mathbf{x}_7),(\mathbf{x}_8,\mathbf{x}_3,\mathbf{x}_6,\mathbf{x}_7)\in R_{IL_0}}\mathbf{x}_8,$

$R_{II_2} = \arg\max_{\mathbf{x}\in\mathbb{B}^8:(\mathbf{x}_1,\mathbf{x}_2,\mathbf{x}_7),(\mathbf{x}_1,\mathbf{x}_3,\mathbf{x}_7),(\mathbf{x}_2,\mathbf{x}_3,\mathbf{x}_7),(\mathbf{x}_1,\mathbf{x}_4,\mathbf{x}_7),(\mathbf{x}_2,\mathbf{x}_5,\mathbf{x}_7),(\mathbf{x}_3,\mathbf{x}_6,\mathbf{x}_7)\in R_{IS_1^2}}(\mathbf{x}_1+\ldots+\mathbf{x}_8).$

Hence, $R_{II_2} \in \langle R\rangle_{\not\exists,w}$ for every $R \in \{R_{IS_1^2}, R_{IN_2}, R_{IL_0}, R_{IL_2}, R_{IL_3}, R_{ID_2}\}$ which by Theorem 4 completes the proof. □

3.2 The VCSP Problem

Since VCSP does not adhere to the standard Galois connection in Theorem 2, the weak base method is not applicable and alternative methods are required. For this purpose we use *multimorphisms* from Cohen et al. [3]. Let Δ be a set of cost functions on $\mathbb{B}$, let p be a unary operation on $\mathbb{B}$, and let f,g be binary operations on $\mathbb{B}$. We say that Δ admits the binary *multimorphism* (f,g) if it holds that $v(f(x,y)) + v(g(x,y)) \leq v(x) + v(y)$ for every $v \in \Delta$ and $x,y \in \mathbb{B}^{\mathrm{ar}(v)}$. Similarly Δ admits the unary *multimorphism* (p) if it holds that $v(p(x)) \leq v(x)$ for every $v \in \Delta$ and $x \in \mathbb{B}^{\mathrm{ar}(v)}$. Recall that the function $f_{\neq}$ equals $\{(0,0) \mapsto 1, (0,1) \mapsto 0, (1,0) \mapsto 0, (1,1) \mapsto 1\}$ and that the minimisation problem VCSP$(f_{\neq})$ and the maximisation problem MAX CUT are trivially CV-reducible to each other. We will make use of (a variant of) the concept of *expressibility* [3]. We say that a cost function g is $\not\exists$-*expressible* in Δ if $g(x_1,\ldots,x_n) = \sum_i w_i f_i(\mathbf{s}^i) + w$ for some tuples $\mathbf{s}^i$ over $\{x_1,\ldots,x_n\}$, weights $w_i \in \mathbb{Q}_{\geq 0}$, $w \in \mathbb{Q}$ and $f_i \in \Delta$. It is not hard to see that if every function in a finite set Δ' is $\not\exists$-expressible in Δ, then VCSP$(\Delta') \leq^{CV}$ VCSP(Δ). Note that if the constants 0 and 1 are expressible in Δ then we may allow tuples $\mathbf{s}^i$ over $\{x_1,\ldots,x_n,0,1\}$, and still obtain a CV-reduction.

Theorem 6. *Let Δ be a set of finite-valued cost functions on $\mathbb{B}$. If the problem* VCSP(Δ) *is NP-hard, then* VCSP$(f_{\neq}) \leq^{CV}$ VCSP(Δ).

Proof. Since VCSP(Δ) is NP-hard (and since we assume P $\neq$ NP) we know that Δ does not admit the unary (0)-multimorphism or the unary (1)-multimorphism [3]. Therefore there are $g,h \in \Delta$ and $\mathbf{u} \in \mathbb{B}^{\mathrm{ar}(g)}$, $\mathbf{v} \in \mathbb{B}^{\mathrm{ar}(h)}$ such that $g(\mathbf{0}) > g(\mathbf{u})$ and $h(\mathbf{1}) > h(\mathbf{v})$. Let $\mathbf{w} \in \arg\min_{\mathbf{x}\in\mathbb{B}^b}(g(\mathbf{x}_1,\ldots,\mathbf{x}_a) + h(\mathbf{x}_{a+1},\ldots,\mathbf{x}_b))$ and then define $o(x,y) = g(z_1,\ldots,z_a) + h(z_{a+1},\ldots,z_b)$ where $z_i = x$ if $\mathbf{w}_i = 0$ and $z_i = y$ otherwise. Clearly $(0,1) \in \arg\min_{\mathbf{x}\in\mathbb{B}^2} o(\mathbf{x})$, $o(0,1) < o(0,0)$, and $o(0,1) < o(1,1)$. We will show that we always can force two fresh variables v_0 and v_1 to 0 and 1, respectively. If $o(0,0) \neq o(1,1)$, then assume without loss of generality that $o(0,0) < o(1,1)$. In this case we force v_0 to 0 with the (sufficiently weighted) term $o(v_0,v_0)$. Set $g'(x) = g(z_1,\ldots,z_{\mathrm{ar}(g)})$ where $z_i = x$ if $u_i = 1$ and $z_i = v_0$ otherwise. Note that $g'(1) < g'(0)$ which means that we can force v_1 to 1. Otherwise $o(0,0) = o(1,1)$. If $o(0,1) = o(1,0)$, then $f_{\neq} = \alpha_1 o + \alpha_2$, otherwise assume without loss of generality that $o(0,1) < o(1,0)$. In this case v_0, v_1 can be forced to 0, 1 with the help of the (sufficiently weighted) term $o(v_0, v_1)$.

By [3], since VCSP(Δ) is NP-hard by assumption, we know that Δ does not admit the (min, max)-multimorphism. Hence, there exists a k-ary function $f \in \Delta$ and $\mathbf{s}, \mathbf{t} \in \mathbb{B}^k$ such that $f(\min(\mathbf{s},\mathbf{t})) + f(\max(\mathbf{s},\mathbf{t})) > f(\mathbf{s}) + f(\mathbf{t})$. Let $f_1(x) = \alpha_1 o(v_0, x) + \alpha_2$ for some $\alpha_1 \in \mathbb{Q}_{\geq 0}$ and $\alpha_2 \in \mathbb{Q}$ such that $f_1(1) = 0$ and $f_1(0) = 1$. Let also $g(x,y) = f(z_1, \ldots, z_k)$ where $z_i = v_1$ if $\min(\mathbf{s}_i, \mathbf{t}_i) = 1$, $z_i = v_0$ if $\max(\mathbf{s}_i, \mathbf{t}_i) = 0$, $z_i = x$ if $s_i > t_i$ and $z_i = y$ otherwise. Note that $g(0,0) = f(\min(\mathbf{s},\mathbf{t}))$, $g(1,1) = f(\max(\mathbf{s},\mathbf{t}))$, $g(1,0) = f(\mathbf{s})$ and $g(0,1) = f(\mathbf{t})$. Set $h(x,y) = g(x,y) + g(y,x)$. Now $h(0,1) = h(1,0) < \frac{1}{2}(h(0,0) + h(1,1))$. If $h(0,0) = h(1,1)$, then $f_{\neq} = \alpha_1 h + \alpha_2$ for some $\alpha_1 \in \mathbb{Q}_{\geq 0}$ and $\alpha_2 \in \mathbb{Q}$. Hence, we can without loss of generality assume that $h(1,1) - h(0,0) = 2$. Note now that $h'(x,y) = f_1(x) + f_1(y) + h(x,y)$ satisfies $h'(0,0) = h'(1,1) = \frac{1}{2}(h(0,0) + h(1,1) + 2)$ and $h'(0,1) = h'(1,0) = \frac{1}{2}(2 + h(0,1) + h(1,0))$. Hence, $h'(0,0) = h'(1,1) > h'(0,1) = h'(1,0)$. So $f_{\neq} = \alpha_1 h' + \alpha_2$ for some $\alpha_1 \in \mathbb{Q}_{\geq 0}$ and $\alpha_2 \in \mathbb{Q}$. $\qquad\square$

3.3 The Broader Picture

Theorems 5 and 6 does not describe the relative complexity between the SAT($\cdot$), MAX-ONES($\cdot$) and VCSP($\cdot$) problems. However we readily see (1) that SAT(R_{II_2}) $\leq^{\text{CV}}$ W-MAX-ONES(R_{II_2}), and (2) that W-MAX-ONES(R_{II_2}) $\leq^{\text{CV}}$ W-MAX INDEPENDENT SET since W-MAX INDEPENDENT SET can be expressed by W-MAX-ONES(NAND^2). The problem W-MAX-ONES(NAND^2) is in turn expressible by MAX-CSP({NAND^2, T,F}). To show that W-MAX INDEPENDENT SET $\leq^{\text{CV}}$ VCSP($f_{\neq}$) it is in fact, since MAX-CSP(neq) and VCSP($f_{\neq}$) is the same problem, sufficient to show that MAX-CSP({NAND^2,T,F}) $\leq^{\text{CV}}$ MAX-CSP(neq). We do this as follows. Let v_0 and v_1 be two global variables. We force v_0 and v_1 to be mapped to different values by assigning a sufficiently high weight to the constraint neq(v_0, v_1). It then follows that $T(x) =$ neq(x, v_0), $F(x) =$ neq(x, v_1) and NAND$^2(x,y) = \frac{1}{2}(\text{neq}(x,y) + F(x) + F(y))$ and we are done. It follows from this proof that MAX-CSP({NAND^2,T,F}) and VCSP($f_{\neq}$) are mutually CV-interreducible. Since MAX-CSP({NAND^2,T,F}) can also be formulated as a VCSP it follows that VCSP($\cdot$) does not have a unique easiest set of cost functions. The complexity results are summarized in Figure 1. Some trivial inclusions are omitted in the figure: for example it holds that SAT(Γ) $\leq^{\text{CV}}$ W-MAX-ONES(Γ) for all Γ.

4   Subexponential Time and the Exponential-Time Hypothesis

The exponential-time hypothesis states that 3-SAT $\notin$ SE [5]. We remind the reader that the ETH can be based on different size parameters (such as the number of variables or the number of clauses) and that these different definitions often coincide [6]. In this section we investigate the consequences of the ETH for the U-MAX-ONES and U-VCSP problems. A direct consequence of Section 3 is that if there exists any finite constraint language Γ or set of cost functions Δ such that W-MAX-ONES(Γ) or VCSP(Δ) is NP-hard and in SE, then SAT(R_{II_2}) is in SE which implies that the ETH is false [9]. The other direction is interesting too since it highlights the likelihood of subexponential time algorithms for the problems, relative to the ETH.

Lemma 7. *If* U-MAX-ONES(Γ) *is in* SE *for some finite constraint languages* Γ *such that* U-MAX-ONES(Γ) *is NP-hard, then the ETH is false.*

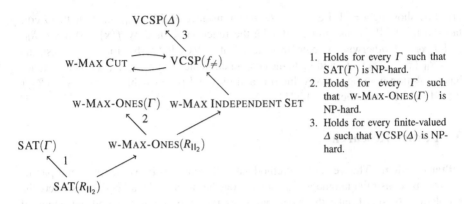

Fig. 1. The complexity landscape of some Boolean optimization and satisfiability problems. A directed arrow from one node A to B means that $A \leq^{CV} B$.

Proof. From Jonsson et al. [9] it follows that 3-SAT $\in$ SE if and only if SAT(R_{II_2})-2 $\in$ SE. Combining this with Theorem 4 we only have to prove that SAT(R_{II_2})-2 LV-reduces to U-MAX-ONES(R) for $R \in \{R_{IS_1^2}, R_{IN_2}, R_{IL_0}, R_{IL_2}, R_{IL_3}, R_{ID_2}\}$. We provide an illustrative reduction from SAT(R_{II_2})-2 to U-MAX-ONES$(R_{IS_1^2})$; the remaining reductions are presented in Lemmas 11–15 in the extended preprint of this paper [10]. Since $R_{IS_1^2}$ is the NAND relation with one additional constant column, the U-MAX-ONES$(R_{IS_1^2})$ problem is basically the maximum independent set problem or, equivalently, the maximum clique problem in the complement graph. Given an instance I of CSP(R_{II_2})-2 we create for every constraint 3 vertices, one corresponding to each feasible assignment of values to the variables occurring in the constraint. We add edges between all pairs of vertices that are not inconsistent and that do not correspond to the same constraint. The instance I is satisfied if and only if there is a clique of size m where m is the number of constraints in I. Since $m \leq 2n$ this implies that the number of vertices is $\leq 2n$.   □

The proofs of the following two lemmas are omitted due to space constraints and can be found in the extended electronic preprint of this paper [10].

Lemma 8. *If the ETH is false, then* U-MAX-ONES$(\Gamma) \in$ SE *for every finite Boolean constraint language* Γ.

Lemma 9. *If* U-MAX-ONES$(\Gamma) \in$ SE *for every finite Boolean constraint language* Γ *then* U-VCSP$_d(\Delta) \in$ SE *for every finite set of Boolean cost functions* Δ *and* $d \geq 0$.

Theorem 10. *The following statements are equivalent.*

1. *The exponential-time hypothesis is false.*
2. U-MAX-ONES$(\Gamma) \in$ SE *for every finite* Γ.
3. U-MAX-ONES$(\Gamma) \in$ SE *for some finite* Γ *such that* U-MAX-ONES(Γ) *is NP-hard.*
4. U-VCSP$(\Delta)_d \in$ SE *for every finite set of finite-valued cost functions* Δ *and* $d \geq 0$.

Proof. The implication 1 $\Rightarrow$ 2 follows from Lemma 8, 2 $\Rightarrow$ 3 is trivial, and 3 $\Rightarrow$ 1 follows by Lemma 7. The implication 2 $\Rightarrow$ 4 follows from Lemma 9. We finish the proof by showing $4 \Rightarrow 1$. Let $I = (V, C)$ be an instance of $\mathrm{SAT}(R_{\mathrm{II}_2})$-2. Note that I contains at most $2|V|$ constraints. Let f be the function defined by $f(\mathbf{x}) = 0$ if $\mathbf{x} \in R_{\mathrm{II}_2}$ and $f(\mathbf{x}) = 1$ otherwise. Create an instance of $\mathrm{U\text{-}VCSP}_2(f)$ by, for every constraint $C_i = R_{\mathrm{II}_2}(x_1, \ldots, x_8) \in C$, adding to the cost function the term $f(x_1, \ldots, x_8)$. This instance has a solution with objective value 0 if and only if I is satisfiable. Hence, $\mathrm{SAT}(R_{\mathrm{II}_2})$-$2 \in \mathrm{SE}$ which contradicts the ETH [9]. $\qquad \square$

5   Future Research

Other problems. The weak base method naturally lends itself to other problems parameterized by constraint languages. In general, one has to consider all co-clones where the problem is NP-hard, take the weak bases for these co-clones and find out which of these are CV-reducible to the other cases. The last step is typically the most challenging — this was demonstrated by the U-MAX-ONES problems where we had to introduce q.w.p.p. implementations. An example of an interesting problem where this strategy works is the *non-trivial* SAT problem $(\mathrm{SAT}^*(\Gamma))$, i.e. the problem of deciding whether a given instance has a solution in which not all variables are mapped to the same value. This problem is NP-hard in exactly six cases [4] and by following the aforementioned procedure one can prove that the relation R_{II_2} results in the easiest NP-hard $\mathrm{SAT}^*(\Gamma)$ problem. Since $\mathrm{SAT}^*(R_{\mathrm{II}_2})$ is in fact the same problem as $\mathrm{SAT}(R_{\mathrm{II}_2})$ this shows that restricting solutions to non-trivial solutions does not make the satisfiability problem easier. This result can also be extended to the co-NP-hard *implication problem* [4] and we believe that similar methods can also be applied to give new insights into the complexity of e.g. *enumeration*, which also follows the same complexity classification [4].

Weighted versus Unweighted Problems. Theorem 10 only applies to unweighted problems and lifting these results to the weighted case does not appear straightforward. We believe that some of these obstacles could be overcome with generalized sparsification techniques and provide an example proving that if any NP-hard $\mathrm{W\text{-}MAX\text{-}ONES}(\Gamma)$ problem is in SE, then MAX-CUT can be approximated within a multiplicative error of $(1 \pm \varepsilon)$ (for any $\varepsilon > 0$) in subexponential time. Assume that $\mathrm{W\text{-}MAX\text{-}ONES}(\Gamma) \in \mathrm{SE}$ is NP-hard, and arbitrarily choose $\varepsilon > 0$. Let $\mathrm{MAX\text{-}CUT}_c$ be the MAX-CUT problem restricted to graphs $G = (V, E)$ where $|E| \le c \cdot |V|$. We first prove that $\mathrm{MAX\text{-}CUT}_c$ is in SE for arbitrary $c \ge 0$. By Theorem 5, we infer that $\mathrm{W\text{-}MAX\text{-}ONES}(R_{\mathrm{II}_2}) \in \mathrm{SE}$. Given an instance (V, E) of $\mathrm{MAX\text{-}CUT}_c$, one can introduce one fresh variable x_v for each $v \in V$ and one fresh variable x_e for each edge $e \in E$. For each edge $e = (v, w)$, we then constrain the variables x_v, x_w and x_e as $R(x_v, x_w, x_e)$ where $R = \{(0, 0, 0), (0, 1, 1), (1, 0, 1), (1, 1, 0)\} \in \langle R_{\mathrm{II}_2} \rangle$. It can then be verified that, for an optimal solution h, that the maximum value of $\sum_{e \in E} w_e h(x_e)$ (where w_e is the weight associated with the edge e) equals the weight of a maximum cut in (V, E). This is an LV-reduction since $|E| = c \cdot |V|$. Now consider an instance (V, E) of the unrestricted MAX-CUT problem. By Batson et al. [1], we can (in polynomial time) compute a *cut sparsifier* (V', E') with only $D_\varepsilon \cdot n / \varepsilon^2$ edges (where D_ε is a constant depending only on ε), which approximately preserves the value of the maximum cut of (V, E) to within a multiplicative error of $(1 \pm \varepsilon)$. By using the LV-reduction above from $\mathrm{MAX\text{-}CUT}_{D_\varepsilon / \varepsilon^2}$ to $\mathrm{W\text{-}MAX\text{-}ONES}(\Gamma)$, it follows that we can approximate the maximum cut of (V, E) within $(1 \pm \varepsilon)$ in subexponential time.

References

1. Batson, J., Spielman, D.A., Srivastava, N.: Twice-ramanujan sparsifiers. SIAM Journal on Computing 41(6), 1704–1721 (2012)
2. Böhler, E., Creignou, N., Reith, S., Vollmer, H.: Playing with Boolean blocks, part I: Post's lattice with applications to complexity theory. ACM SIGACT-Newsletter 34(4), 38–52 (2003)
3. Cohen, D.A., Cooper, M.C., Jeavons, P.G., Krokhin, A.A.: The complexity of soft constraint satisfaction. Artificial Intelligence 170(11), 983–1016 (2006)
4. Creignou, N., Hébrard, J.J.: On generating all solutions of generalized satisfiability problems. Informatique Théorique et Applications 31(6), 499–511 (1997)
5. Impagliazzo, R., Paturi, R.: On the complexity of k-SAT. Journal of Computer and System Sciences 62(2), 367–375 (2001)
6. Impagliazzo, R., Paturi, R., Zane, F.: Which problems have strongly exponential complexity? Journal of Computer and System Sciences 63(4), 512–530 (2001)
7. Jeavons, P.: On the algebraic structure of combinatorial problems. Theoretical Computer Science 200, 185–204 (1998)
8. Jeavons, P., Cohen, D., Gyssens, M.: Closure properties of constraints. Journal of the ACM 44(4), 527–548 (1997)
9. Jonsson, P., Lagerkvist, V., Nordh, G., Zanuttini, B.: Complexity of SAT problems, clone theory and the exponential time hypothesis. In: Khanna, S. (ed.) SODA 2013, pp. 1264–1277. SIAM (2013)
10. Jonsson, P., Lagerkvist, V., Schmidt, J., Uppman, H.: Relating the time complexity of optimization problems in light of the exponential-time hypothesis. CoRR, abs/1406.3247 (2014)
11. Khanna, S., Sudan, M., Trevisan, L., Williamson, D.: The approximability of constraint satisfaction problems. SIAM Journal on Computing 30(6), 1863–1920 (2000)
12. Lagerkvist, V.: Weak bases of Boolean co-clones. Information Processing Letters 114(9), 462–468 (2014)
13. Lau, D.: Function Algebras on Finite Sets: Basic Course on Many-Valued Logic and Clone Theory. Springer Monographs in Mathematics. Springer-Verlag New York, Inc., Secaucus (2006)
14. Lokshtanov, D., Marx, D., Saurabh, S.: Lower bounds based on the exponential time hypothesis. Bulletin of the EATCS 105, 41–72 (2011)
15. Post, E.: The two-valued iterative systems of mathematical logic. Annals of Mathematical Studies 5, 1–122 (1941)
16. Santhanam, R., Srinivasan, S.: On the limits of sparsification. In: Czumaj, A., Mehlhorn, K., Pitts, A., Wattenhofer, R. (eds.) ICALP 2012, Part I. LNCS, vol. 7391, pp. 774–785. Springer, Heidelberg (2012)
17. Schaefer, T.J.: The complexity of satisfiability problems. In: Proceedings 10th Symposium on Theory of Computing, pp. 216–226. ACM Press (1978)
18. Schnoor, I.: The weak base method for constraint satisfaction. PhD thesis, Gottfried Wilhelm Leibniz Universität, Hannover, Germany (2008)
19. Thapper, J.: Aspects of a Constraint Optimisation Problem. PhD thesis, Linköping University, The Institute of Technology (2010)
20. Woeginger, G.: Exact algorithms for NP-hard problems: A survey. In: Jünger, M., Reinelt, G., Rinaldi, G. (eds.) Combinatorial Optimization - Eureka, You Shrink! LNCS, vol. 2570, pp. 185–207. Springer, Heidelberg (2003)

Affine Consistency and the Complexity of Semilinear Constraints*

Peter Jonsson[1] and Johan Thapper[2]

[1] Department of Computer and Information Science, Linköping University, Sweden
peter.jonsson@liu.se
[2] LIGM, Université Paris-Est Marne-la-Vallée, France
thapper@u-pem.fr

Abstract. A semilinear relation is a finite union of finite intersections of open and closed half-spaces over, for instance, the reals, the rationals or the integers. Semilinear relations have been studied in connection with algebraic geometry, automata theory, and spatiotemporal reasoning, just to mention a few examples. We concentrate on relations over the reals and rational numbers. Under this assumption, the computational complexity of the constraint satisfaction problem (CSP) is known for all finite sets Γ of semilinear relations containing the relations $R_+ = \{(x, y, z) \mid x + y = z\}$, $\leq$, and $\{1\}$. These problems correspond to extensions of LP feasibility. We generalise this result as follows. We introduce an algorithm, based on computing affine hulls, which solves a new class of semilinear CSPs in polynomial time. This allows us to fully determine the complexity of CSP(Γ) for semilinear Γ containing R_+ and satisfying two auxiliary conditions. Our result covers all semilinear Γ such that $\{R_+, \{1\}\} \subseteq \Gamma$. We continue by studying the more general case when Γ contains R_+ but violates either of the two auxiliary conditions. We show that each such problem is equivalent to a problem in which the relations are finite unions of homogeneous linear sets and we present evidence that determining the complexity of these problems may be highly non-trivial.

1 Introduction

Let $X = \mathbb{Q}$ or $X = \mathbb{R}$. We say that a relation $R \subseteq X^k$ is *semilinear* if it can be represented as a finite union of finite intersections of open and closed half-spaces. Alternatively, R is semilinear if it is first-order definable in $\{R_+, \leq, \{1\}\}$ where $R_+ = \{(x, y, z) \in X^3 \mid x + y = z\}$. Semilinear relations appear in many different contexts within mathematics and computer science: they are, for instance, frequently encountered in algebraic geometry, automata theory, spatiotemporal reasoning, and computer algebra. Semilinear relations have also attained a fair amount of attention in connection with *constraint satisfaction problems* (CSPs). Here, we are given a set of variables that take their values from a (finite or infinite) domain and a set of constraints (e.g. relations) that constrain the values different

---

* Partially supported by the Swedish Research Council (VR) under grant 621-2012-3239.

E. Csuhaj-Varjú et al. (Eds.): MFCS 2014, Part II, LNCS 8635, pp. 420–431, 2014.
© Springer-Verlag Berlin Heidelberg 2014

variables can take, and the question is whether the variables can be assigned values such that all constraints are satisfied or not. CSPs are often parameterized by a finite set Γ of allowed relations, known as a *constraint language*. All constraints in the input of CSP(Γ) must be members of Γ. This way of parameterizing constraint satisfaction problems has proved to be very fruitful for CSPs over both finite and infinite domains. Since Γ is finite, the computational complexity of such a problem does not depend on the actual representation of constraints. The complexity of finite-domain CSPs has been studied for a long time and a powerful algebraic toolkit has gradually formed [6]. Much of this work has been devoted to the *Feder-Vardi conjecture* [7], i.e., that every finite-domain CSP is either polynomial-time solvable or NP-complete. Infinite-domain CSPs, on the other hand, constitute a much more diverse set of problems: every computational problem is polynomial-time equivalent to an infinite-domain CSP [1]. Obtaining a full understanding of their computational complexity is thus extremely difficult and we have to contain ourselves to studying restricted cases. The main motivation behind this paper is the following result by Bodirsky et al. [2].

Theorem 1. *If Γ is a finite set of semilinear relations over $\mathbb{R}$ or $\mathbb{Q}$ that contains R_+, $\leq$, and $\{1\}$, then CSP(Γ) is either polynomial-time solvable or NP-complete.*

Characterizing the polynomial-time solvable cases is fairly simple: we say that a relation $R \subseteq \mathbb{R}^k$ is *essentially convex* if for all $p, q \in R$ there are only finitely many points on the line segment between p and q that are not in R. If Γ contains essentially convex relations only, then CSP(Γ) is in P by exploiting linear programming, and the problem is NP-complete otherwise. One may suspect that there are semilinear constraint languages Γ such that CSP(Γ) $\in P$ but Γ is not essentially convex. This is indeed true and we identify two such cases.

In the first case, we encounter polynomial-time solvable classes that, informally speaking, contain relations with large "cavities". It is not surprising that the algorithm for essentially convex relations (and the ideas behind it) cannot be applied in the presence of this kind of highly non-convex relations. Thus, we introduce (in Section 3.1) a new algorithm based on computing affine hulls.

In the second case, we have polynomial-time solvable classes where the relations look essentially convex when viewed from the origin. That is, any points p and q that witnesses that the relations are not essentially convex lie on a line that does not pass through the origin. We show (in Section 3.2) that we can remove the holes witnessed by p and q and find an equivalent language that is essentially convex, and thereby solve the problem.

Combining these algorithmic results with certain hardness results (that are collected in Section 4) yields a dichotomy: if a semilinear constraint language Γ contains R_+ and satisfies two additional properties (P$_0$) and (P$_\infty$), then CSP(Γ) is either polynomial-time solvable or NP-hard. Actually, CSP(Γ) is always in NP for a semilinear constraint language Γ, cf. Theorem 5.2 in Bodirsky et al. [2].

This result immediately generalizes Theorem 1 since it implies that semilinear constraint languages that contain R_+ and $\{1\}$ (but not necessarily $\leq$) exhibit a dichotomy. These results and their proofs together with formal definitions of the properties (P$_0$) and (P$_\infty$) can be found in Section 5.

A natural goal at this point would be to determine the complexity when one or more of the side conditions are not met. In Section 6, we prove that if $\{R_+\} \subseteq \Gamma$ does not satisfy (P$_0$) and/or (P$_\infty$), then CSP(Γ) is equivalent to a problem CSP(Γ') where Γ' contains *homogeneous* semilinear relations only. By a homogeneous semilinear relation, we mean that it can be defined in terms of homogeneous inequalities. How hard it is to determine the complexity of CSP(Γ') is difficult to say and, consequently, we discuss this issue in some detail.

2 Preliminaries

2.1 Constraint Satisfaction Problems

Let $\Gamma = \{R_1, \ldots, R_n\}$ be a finite set of finitary relations over some domain D (which will usually be infinite). We refer to Γ as a *constraint language*. A first-order formula is called *primitive positive* if it is of the form $\exists x_1, \ldots, x_n.\psi_1 \wedge \cdots \wedge \psi_m$, where ψ_i are formulas of the form $x = y$ or $R(x_{i_1}, \ldots, x_{i_k})$ with R the relation symbol for a k-ary relation from Γ. We call such a formula a *pp-formula*. The conjuncts in a pp-formula Φ are also called the *constraints* of Φ. The *constraint satisfaction problem for* Γ (CSP(Γ) for short) is the computational problem to decide whether a given primitive positive sentence Φ is true in Γ.

Definition 1. *The problem CSP(Γ) is* tractable *(or polynomial-time solvable) if for every finite $\Gamma' \subseteq \Gamma$, CSP(Γ') is solvable in polynomial time. We say that CSP(Γ) is NP-hard if CSP(Γ') is NP-hard for some finite $\Gamma' \subseteq \Gamma$.*

A relation $R(x_1, \ldots, x_k)$ is *pp-definable from* Γ if there exists a quantifier-free formula φ over Γ such that $R(x_1, \ldots, x_k) \equiv \exists y_1, \ldots, y_n.\varphi(x_1, \ldots, x_k, y_1, \ldots, y_n)$. The set of all relations that are pp-definable over Γ is denoted by $\langle \Gamma \rangle$. The following simple but important result explains the importance of primitive positive definability for the constraint satisfaction problem. We will use it extensively in the sequel without making explicit references to it.

Lemma 1. *Let Γ be a constraint language and $\Gamma' = \Gamma \cup \{R\}$ where $R \in \langle \Gamma \rangle$. Then CSP($\Gamma$) is polynomial-time equivalent to CSP(Γ').*

2.2 Semilinear Relations

The domain, X, of every relation in this paper will either be the set of reals, $\mathbb{R}$, or the set of rationals, $\mathbb{Q}$. In all cases, the set of coefficients, Y, will be the set of rationals, but in order to avoid confusion, we will still make this explicit in our notation. Let $LE_X[Y]$ denote the set of linear equalities over X with coefficients in Y and $LI_X[Y]$ denote the set of (strict and non-strict) linear inequalities over X with coefficients in Y. Sets defined by finite conjunctions of inequalities from $LI_X[Y]$ are called *linear sets*. The set of *semilinear sets*, $SL_X[Y]$, is defined to be the set of finite unions of linear sets. We will refer to $SL_\mathbb{Q}[\mathbb{Q}]$ and $SL_\mathbb{R}[\mathbb{Q}]$ as semilinear relations over $\mathbb{R}$ and $\mathbb{Q}$, respectively.

The following lemma is a direct consequence of our definitions: this particular property is often referred to as *o-minimality* in the literature.

Lemma 2. *Let $R \in SL_X[Y]$ be a unary semilinear relation. Then, R can be written as a finite union of open, half-open, and closed intervals with endpoints in $Y \cup \{-\infty, \infty\}$ together with a finite set of points in Y.*

Due to the alternative definition of a semilinear relation as a relation that is first-order definable in $\{R_+, \leq, \{1\}\}$, the set $SL_X[Y]$ is closed under pp-definitions. Consequently, Lemma 2 is applicable to all relations discussed in this paper.

Given a relation R of arity k, let $R|_X = R \cap X^k$ and $\Gamma|_X = \{R|_X \mid R \in \Gamma\}$. We demonstrate that $\mathrm{CSP}(\Gamma)$ and $\mathrm{CSP}(\Gamma|_\mathbb{Q})$ are equivalent as constraint satisfaction problems whenever $\Gamma \subseteq SL_\mathbb{R}[\mathbb{Q}]$. Thus, we can exclusively concentrate on relations from $SL_\mathbb{Q}[\mathbb{Q}]$ in the sequel. Let $\Gamma \subseteq SL_\mathbb{R}[\mathbb{Q}]$ and let I be an instance of $\mathrm{CSP}(\Gamma)$. Construct an instance I' of $\mathrm{CSP}(\Gamma|_\mathbb{Q})$ by replacing each occurrence of R in I by $R|_\mathbb{Q}$. If I' has a solution, then I has a solution since $R|_\mathbb{Q} \subseteq R$ for each $R \in \Gamma$. If I has a solution, then it has a rational solution by Lemma 3.7 in Bodirsky et al. [2] so I' has a solution, too.

Lemma 3 (Lemma 4.3 in Bodirsky et al. [3]). *Let $r_1, \ldots, r_k, r \in \mathbb{Q}$. The relation $\{(x_1, \ldots, x_k) \in \mathbb{Q}^k \mid r_1 x_1 + \ldots + r_k x_k = r\}$ is pp-definable in $\{R_+, \{1\}\}$ and it is pp-definable in $\{R_+\}$ if $r = 0$.*

It follows that $LE_\mathbb{Q}[\mathbb{Q}] \subseteq \langle\{R_+, \{1\}\}\rangle$ and $LI_\mathbb{Q}[\mathbb{Q}] \subseteq \langle\{R_+, \leq, \{1\}\}\rangle$.

2.3 Unary Semilinear Relations

Given a relation $R \subseteq \mathbb{Q}^k$ and two distinct points $a, b \in \mathbb{Q}^k$, we define

$$\mathcal{L}_{R,a,b}(y) \equiv \exists x_1, \ldots, x_k. R(x_1, \ldots, x_k) \wedge \bigwedge_{i=1}^{k} x_i = y a_i + (1 - y) b_i.$$

The relation $\mathcal{L}_{R,a,b}$ is a parameterisation of the intersection between the relation R and a line through the points a and b. Note that $\mathcal{L}_{R,a,b}$ is a member of $\langle LE_\mathbb{Q}[\mathbb{Q}] \cup \{R\}\rangle$ and that $\mathcal{L}_{R,a,b}$ is pp-definable in $\{R_+, \{1\}, R\}$ by Lemma 3.

A k-ary relation R is *bounded* if there exists an $a \in \mathbb{Q}$ such that $R \subseteq [-a, a]^k$. If $k = 1$, then we say that R is *unbounded in one direction* if there exists $a \in \mathbb{Q}$ such that exactly one of the following hold: for all $b \leq a$, there exists a $c \leq b$ such that $c \in R$; or for all $b \geq a$, there exists a $c \geq b$ such that $c \in R$.

A unary relation is called a *bnu* (for *bounded, non-constant, and unary*) if it is bounded and contains more than one point.

Lemma 4. *Let U be a unary relation in $SL_\mathbb{Q}[\mathbb{Q}]$ that is unbounded in one direction. Then, $\langle\{R_+, \{1\}, U\}\rangle$ contains a bnu.*

For a unary semilinear relation $T \subseteq \mathbb{Q}$, and a rational $\delta > 0$, let $T + \mathcal{I}(\delta)$ denote the set of unary semilinear relations U such that $T \subseteq U$ and for all $x \in U$, there exists a $y \in T$ with $|x - y| < \delta$.

Lemma 5. *Let $U \neq \varnothing$ be a bounded unary semilinear relation such that $U \cap (-\infty, \varepsilon) = \varnothing$ for some $\varepsilon > 0$. Then, $\langle R_+, U\rangle$ contains a relation $U_\delta \in \{1\} + \mathcal{I}(\delta)$, for every $\delta > 0$.*

Lemma 6. *Let U be a bounded unary semilinear relation such that $U \cap (-\varepsilon, \varepsilon) = \varnothing$ for some $\varepsilon > 0$ and $U \cap -U \neq \varnothing$. Then, $\langle R_+, U\rangle$ contains a relation $U_\delta \in \{-1, 1\} + \mathcal{I}(\delta)$, for every $\delta > 0$.*

2.4   Essential Convexity

Let R be a k-ary relation over $\mathbb{Q}^k$. The relation R is *convex* if for all $p, q \in R$, R contains all points on the line segment between p and q. We say that R *excludes an interval* if there are $p, q \in R$ and real numbers $0 < \delta_1 < \delta_2 < 1$ such that $p + (q - p)y \notin R$ whenever $\delta_1 \leq y \leq \delta_2$. Note that we can assume that δ_1, δ_2 are rational numbers, since we can choose any two distinct rational numbers $\gamma_1 < \gamma_2$ between δ_1 and δ_2 instead of δ_1 and δ_2. We say that R is *essentially convex* if for all $p, q \in R$ there are only finitely many points on the line segment between p and q that are not in R. If R is *not* essentially convex, and if p and q are such that there are infinitely many points on the line segment between p and q that are not in R, then p and q *witness* that R is not essentially convex. A semilinear relation is essentially convex if and only if it does not exclude an interval.

Theorem 2 (Theorem 5.1 and 5.4 in Bodirsky et al. [2]). *If Γ is an essentially convex semilinear constraint language, then $CSP(\Gamma)$ is tractable.*

3   Tractability

In this section, we present our two main sources of tractability. In Section 3.1, we introduce a new algorithm for semilinear constraint languages Γ containing $\{R_+, \{1\}\}$ and such that $\langle \Gamma \rangle$ does not contain a bnu. In Section 3.2, we show that the algorithm in Theorem 2 has a wider applicability.

3.1   Affine Consistency

For a subset $X \subseteq \mathbb{Q}^n$, let $\mathrm{aff}(X)$ denote the *affine hull of X in $\mathbb{Q}^n$*: $\mathrm{aff}(X) = \{\sum_{i=1}^{k} c_i x_i \mid k \geq 1, c_i \in \mathbb{Q}, x_i \in X, \sum_{i=1}^{k} c_i = 1\}$. An *affine subspace* is a subset $X \subseteq \mathbb{Q}^n$ for which $\mathrm{aff}(X) = X$. The points $p_1, \ldots, p_k \in \mathbb{Q}^n$ are said to be *affinely independent* if $a_1 p_1 + \cdots + a_k p_k = 0$ with $a_1 + \cdots + a_k = 0$ implies $a_1 = \cdots = a_k = 0$. The dimension, $\dim(X)$, of a set $X \subseteq \mathbb{Q}^n$ is defined to be one less than the maximum number of affinely independent points in X.

We define a notion of consistency for sets of semilinear constraints which we call affine consistency. Let V be a finite set of variables. A set of constraints $R_i(x_{i_1}, \ldots, x_{i_k})$ with $\{x_{i_1}, \ldots, x_{i_k}\} \subseteq V$ is *affinely consistent* with respect to an affine subspace $A \subseteq \mathbb{Q}^V$ if $\mathrm{aff}(\hat{R}_i \cap A) = A$ for all i, where $\hat{R}_i := \{(x_1, \ldots, x_n) \in \mathbb{Q}^V \mid (x_{i_1}, \ldots, x_{i_k}) \in R_i\}$.

Algorithm 1 establishes affine consistency for a set of constraints and answers "yes" if the resulting affine subspace is non-empty and "no" otherwise. In the rest of this section, we show that this algorithm correctly solves $CSP(\Gamma)$ when $\{R_+, \{1\}\} \subseteq \Gamma$ is a semilinear constraint language such that $\langle \Gamma \rangle$ does not contain a bnu. Furthermore, we show how for such constraint languages, the algorithm can be implemented to run in polynomial time.

Lemma 7. *Let $P = P_1 \cup \cdots \cup P_k, Q = Q_1 \cup \cdots \cup Q_l \in SL_{\mathbb{Q}}[\mathbb{Q}]$ be two n-ary relations such that neither $\langle LE_{\mathbb{Q}}[\mathbb{Q}] \cup \{P\} \rangle$ nor $\langle LE_{\mathbb{Q}}[\mathbb{Q}] \cup \{Q\} \rangle$ contains a bnu. If $\mathrm{aff}(P) = \mathrm{aff}(Q) =: A$, then $\mathrm{aff}(P_i \cap Q_j) = A$ for some i and j.*

Algorithm 1. Affine consistency

Input: A set of constraints $\{R_i(x_{i_1}, \ldots, x_{i_k})\}$ over variables V
Output: "yes" if the resulting affine subspace is non-empty, "no" otherwise

1 $U := \mathbb{Q}^V$
2 **repeat**
3 | **foreach** *constraint* $R_i(x_{i_1}, \ldots, x_{i_k})$ **do**
4 | | $U := \text{aff}(\hat{R}_i \cap U)$
5 | **end**
6 **until** U *does not change*
7 **if** $U \neq \varnothing$ **then return** "yes" **else return** "no"

Proof. The proof is by induction on the dimension $d = \dim(A)$. For $d = 0$, both P and Q consist of a single point p. Clearly, $P_i = \{p\}$ for some i and $Q_j = \{p\}$ for some j. Now assume that $d > 0$ and that the lemma holds for all P', Q' with $\text{aff}(P') = \text{aff}(Q') = A'$ and $\dim(A') < d$. Let $p_0, p_1, \ldots, p_d$ be $d+1$ affinely independent points in P and let $q_0, q_1, \ldots, q_d$ be $d+1$ affinely independent points in Q. For $1 \leq i \leq d$, consider the lines L_i^p through p_0 and p_i, and the lines L_i^q through q_0 and q_i. Let $H = \{y \in \mathbb{Q}^n \mid \alpha \cdot y = 0\}$ ($\alpha \in \mathbb{Q}^n$) be a hyperplane in $\mathbb{Q}^n$ through the origin that is not parallel to any of the lines L_i^p or L_i^q. Then, H intersects each of the $2d$ lines. Let $H(c) = \{y \in \mathbb{Q}^n \mid \alpha \cdot y = c\}$ and let $B(c) = \{y \in \mathbb{Q}^n \mid \alpha \cdot y \notin [-c, c]\}$.

Express the line L_i^p as $\{y \in \mathbb{Q}^n \mid y = a \cdot x + b, x \in \mathbb{Q}\}$, for some $a, b \in \mathbb{Q}^n$. Define a unary relation T by the formula $\varphi(x) \equiv \exists y \in \mathbb{Q}^n.P(y) \wedge y = a \cdot x + b$. Note that $T \in \langle LE_{\mathbb{Q}}[\mathbb{Q}] \cup \{P\}\rangle$. Since T contains p_0 and p_i, it follows that T is not a constant and hence unbounded. By Lemma 4, T is unbounded in both directions. By Lemma 2, $B(c_i^p) \cap L_i^p \subseteq T \subseteq P$, for some positive constant c_i^p. An analogous argument shows that that $B(c_j^q) \cap L_j^q \subseteq Q$, for some positive constant c_j^q. Let c' be a positive constant such that $p_0, q_0 \notin B(c')$ and let $c = \max\{c'\} \cup \{c_i^p, c_j^q \mid 1 \leq i, j \leq d\}$. This ensures that for any $x > c$, $H(x) \cap P$ intersects the lines L_i^p in d affinely independent points, and that $H(x) \cap Q$ intersects the lines L_j^q in d affinely independent points.

We now have $\text{aff}(H(x) \cap P) = \text{aff}(H(x) \cap Q) = A'(x)$ with $\dim(A'(x)) = \dim(A) - 1$, for every $x > c$. By induction on $H(x) \cap P = (H(x) \cap P_1) \cup \cdots \cup (H(x) \cap P_k)$ and $H(x) \cap Q = (H(x) \cap Q_1) \cup \cdots \cup (H(x) \cap Q_l)$, it follows that $\text{aff}(H(x) \cap P_{i(x)} \cap Q_{j(x)}) = A'(x)$ for some $i(x)$ and $j(x)$. This holds for all $x > c$, hence there exist distinct $x_1, x_2 > c$ with $i(x_1) = i(x_2) = i'$ and $j(x_1) = j(x_2) = j'$. Since $A'(x_1), A'(x_2) \subseteq \text{aff}(P_{i'} \cap Q_{j'})$, $A'(x_1) \cap A'(x_2) = \varnothing$, and $\dim(A'(x_2)) = d - 1 \geq 0$, it follows that $\text{aff}(P_{i'} \cap Q_{j'})$ strictly contains $A'(x_1)$, so we have $A' \subset \text{aff}(P_{i'} \cap Q_{j'}) \subseteq A$, and $\dim(A'(x_1)) = \dim(A) - 1$. Therefore we have the equality $\text{aff}(P_{i'} \cap Q_{j'}) = A$. The lemma follows. $\square$

For a semilinear relation R, we let $\text{size}(R)$ denote the *representation size* of R, i.e., the number of bits needed to describe the arities and coefficients of each inequality in some fixed definition of R.

Lemma 8. *Let $R \in SL_{\mathbb{Q}}[\mathbb{Q}]$ be a relation such that $\langle LE_{\mathbb{Q}}[\mathbb{Q}] \cup \{R\}\rangle$ does not contain a bnu and let $U \subseteq \mathbb{Q}^n$ be an affine subspace. Algorithm 2 computes a set*

Algorithm 2. Calculate aff$(R \cap U)$

Input: A semilinear relation $R = R_1 \cup \cdots \cup R_k$ and an affine subspace U.
Output: A set of inequalities defining aff$(R \cap U)$.

1 Find i that maximises $d_i := \dim(\text{aff}(R_i \cap U))$.
2 **if** aff$(R_i \cap U) = \varnothing$ **then return** $\bot$
3 Let I be the set of inequalities for R_i and J be the set of inequalities for U.
4 $S := I \cup J$
5 **foreach** $ineq \in I \cup J$ **do**
6   **if** $\dim(\text{aff}(S \setminus \{ineq\})) = d_i$ **then**
7         $S := S \setminus \{ineq\}$
8   **end**
9 **end**
10 **return** S

---

of linear inequalities S defining aff$(R \cap U)$ *in time polynomial in* size(R)+size(U) *and with* size$(S) \leq$ size$(R) +$ size(U).

Proof. Let $R = R_1 \cup \cdots \cup R_k$ be the representation of R as the union of (convex) linear sets R_i. By Lemma 7, there exists an i such that aff$(R \cap U) =$ aff$(R_i \cap U)$ and since aff$(R_j \cap U) \subseteq$ aff$(R \cap U)$ for all j, the algorithm will find such an i on line 1 by simply comparing the dimensions of these sets. If aff$(R \cap U) =$ aff$(R_i \cap U) = \varnothing$, then the algorithm returns $\bot$, signalling that the affine hull is empty.

Otherwise, the affine hull of a non-empty polyhedron can always be obtained as a subset of its defining inequalities (see for example [11, Section 8.2]). Here, some of the inequalities may be strict, but it is not hard to see that removing them does not change the affine hull. If $ineq \in I \cup J$ is an inequality that cannot be removed without increasing the dimension of the affine hull, then it is clear that $ineq$ still cannot be removed after the loop. Hence, after the loop, no inequality in S can be removed without increasing the dimension of the affine hull. It follows that S itself defines an affine subspace, U_S, and $U_S =$ aff$(U_S) =$ aff$(R_i \cap U) =$ aff$(R \cap U)$.

Using the ellipsoid method, we can determine the dimension of the affine hull of a polyhedron defined by a system of linear inequalities in time polynomial in the representation size of the inequalities [11, Corollary 14.1f]. To handle strict inequalities on line 1, we can perturb these by a small amount, while keeping the representation sizes polynomial, to obtain a system of non-strict inequalities with the same affine hull. The algorithm does at most $|I \cup J| + k$ affine hull calculations. The total time is thus polynomial in size$(R) +$ size(U). Finally, the set S is a subset of $I \cup J$, so size$(U_S) \leq$ size$(R) +$ size(U). $\qquad\square$

Theorem 3. *Let $\{R_+, \{1\}\} \subseteq \Gamma \subseteq SL_{\mathbb{Q}}[\mathbb{Q}]$. If there is no bnu in $\langle \Gamma \rangle$, then Algorithm 1 solves $CSP(\Gamma)$ in polynomial time.*

Proof. Assume that each relation $R \in \Gamma$ is given as $R = R_1 \cup \cdots \cup R_k$, where R_i is a (convex) linear set for each i. First, we show that the algorithm terminates with U equal to the affine hull of the solution space of the constraints.

Assume that the input consists of the constraints $R_i(x_{i_1}, \ldots, x_{i_k})$ over variables V, $i = 1, \ldots, m$. Let $Z = \bigcap_{i=1}^m \hat{R}_i$ denote the solution space of the instance. It is clear that Z is contained in U throughout the execution of the algorithm. Therefore, $\text{aff}(Z) = \text{aff}(Z \cap U)$ so it suffices to show that $\text{aff}(Z \cap U) = U$. We will show that $\text{aff}(\bigcap_{i=1}^j \hat{R}_i \cap U) = U$ for all $j = 1, \ldots, m$. When the algorithm terminates, we have $\text{aff}(\hat{R}_i \cap U) = U$ for every $i = 1, \ldots, m$. In particular, the claim holds for $j = 1$. Now assume that the claim holds for $j - 1$. Then, $P = \bigcap_{i=1}^{j-1} \hat{R}_i \cap U$ and $Q = \hat{R}_j \cap U$ satisfy the requirements of Lemma 7 with $\text{aff}(P) = \text{aff}(Q) = U$. Therefore, we can use this lemma to conclude that $\text{aff}(\bigcap_{i=1}^j \hat{R}_i \cap U) = \text{aff}(P \cap Q) = U$.

Finally, we show that the algorithm can be implemented to run in polynomial time. The call to Algorithm 2 in the inner loop is carried out at most mn times, where $n = |V|$. The size of $\hat{R}$ is at most $\text{size}(R) + \log n$, so the size of U never exceeds $\mathcal{O}(mn(\text{size}(R) + \log n))$, where R is a relation with maximal representation size. Therefore, each call to Algorithm 2 takes polynomial time and consequently, the entire algorithm runs in polynomial time. □

3.2 Essential Convexity

The dimension of a set is defined with respect to its affine hull, as in Section 3.1. We give a result on the structure of certain not necessarily essentially convex relations. It is based on the intuition that even if we do not have the constant relation $\{1\}$ to help us identify excluded intervals, we are still able to see excluded full-dimensional holes. We follow this up by showing that we can remove certain lower-dimensional holes and thus recover an equivalent essentially convex CSP.

Lemma 9. Let $U \in \{1\} + \mathcal{I}(c)$ for some $0 < c < 1$ and assume that $R \in SL_{\mathbb{Q}}[\mathbb{Q}]$ is a semilinear relation such that every unary relation in $\langle \{R_+, U, R\} \rangle$ is essentially convex. Then, R can be defined by a formula $\varphi_0 \wedge \neg\varphi_1 \wedge \cdots \wedge \neg\varphi_k$, where $\varphi_0, \ldots, \varphi_k$ are conjunctions over $LI_{\mathbb{Q}}[\mathbb{Q}]$, and $\varphi_1, \ldots, \varphi_k$ define convex sets of dimensions strictly lower than the set defined by φ_0.

Theorem 4. Let $U \in \{1\} + \mathcal{I}(c)$ for some $0 < c < 1$ and assume that $\{R_+, U\} \subseteq \Gamma \subseteq SL_{\mathbb{Q}}[\mathbb{Q}]$ is a constraint language such that every unary relation in $\langle \Gamma \rangle$ is essentially convex. Then, $CSP(\Gamma)$ is equivalent to $CSP(\Gamma')$ for an essentially convex constraint language $\Gamma' \subseteq SL_{\mathbb{Q}}[\mathbb{Q}]$.

Proof. If Γ is essentially convex, then there is nothing to prove. Assume therefore that Γ is not essentially convex. By Lemma 9, each $R \in \Gamma$ can be defined by a formula $\varphi_0 \wedge \neg\varphi_1 \wedge \cdots \wedge \neg\varphi_k$, where $\varphi_0, \varphi_1, \ldots, \varphi_k$ are conjunction over $LI_{\mathbb{Q}}[\mathbb{Q}]$, and $\varphi_1, \ldots, \varphi_k$ define sets whose affine hulls are of dimensions strictly lower than that of the set defined by φ_0. Assume additionally that the formulas are numbered so that the affine hulls of the sets defined by $\varphi_1, \ldots \varphi_m$ do not contain $(0, \ldots, 0)$ and that the affine hulls of the sets defined by $\varphi_{m+1}, \ldots, \varphi_k$ do contain $(0, \ldots, 0)$. Define R' by $\varphi \wedge \neg\varphi'_1 \wedge \cdots \wedge \neg\varphi'_m \wedge \neg\varphi_{m+1} \wedge \cdots \wedge \neg\varphi_k$, where φ'_i defines the affine hull of the set defined by φ_i. Then, the constraint language $\Gamma' = \{R' \mid R \in \Gamma\}$ is essentially convex since witnesses of an excluded interval only occur inside an affine subspace not containing $(0, \dots, 0)$; otherwise we could use such a witness to pp-define a unary relation excluding an interval.

Let I be an arbitrary instance of $CSP(\Gamma)$ over the variables V and construct an instance I' of $CSP(\Gamma')$ by replacing each occurrence of a relation R in I by R'. Clearly, if I' is satisfiable, then so is I. Conversely, let $s \in \mathbb{Q}^V$ be a solution to I and assume that I' is not satisfiable. Let L be the line in $\mathbb{Q}^V$ through $(0, \dots, 0)$ and s and let U be the unary relation $\mathcal{L}_{I,s,(0,\dots,0)} \in \langle \Gamma \rangle$. All tuples in U correspond to solutions of I that are not solutions to I'.

Fix a constraint $R(x_1, \dots, x_l)$ in I and consider the points in U that satisfy this constraint but not $R'(x_1, \dots, x_l)$. These are the points $p \in \mathbb{Q}^V$ on L for which $(p(x_1), \dots, p(x_l))$ satisfies $(\varphi'_1 \vee \cdots \vee \varphi'_m) \wedge \neg(\varphi_1 \vee \cdots \vee \varphi_m)$. For each $1 \leq i \leq m$, φ'_i satisfies at most one point on L since otherwise the affine hull of the relation defined by φ_i would contain $(0, \dots, 0)$. Hence, each constraint in I can account for at most a finite number of points in U, so U is finite.

There are two cases to consider: (1) U contains more than one point and therefore excludes an interval; or (2) U is the constant $\{1\}$. Since Γ is assumed not to be essentially convex, the relation U can then be used to pp-define a unary relation that is not essentially convex. In either case, there is a contradiction to the assumption that every unary relation in $\langle \Gamma \rangle$ is essentially convex. It follows that I' must be satisfiable. $\qquad\square$

4   NP-Hardness

We now prove the necessary hardness result. It is based on the following simple reduction from the NP-hard problem Not-All-Equal 3SAT [10]. We then show that having a bnu T that excludes an interval and that is bounded away from 0 is a sufficient condition for $CSP(\{R_+, T\})$ to be NP-hard. This is a unified hardness condition for all CSPs classified in this paper.

Lemma 10. Let $T \in \{-1, 1\} + \mathcal{I}(\frac{1}{2})$. Then, $CSP(\{R_+, T\})$ is NP-hard.

For a rational c, and a unary relation U, let $c \cdot U = \{c \cdot x \mid x \in U\} \in \langle\{R_+, U\}\rangle$.

Lemma 11. Let $T \neq \varnothing$ be a bounded unary relation such that $T \cap (-\varepsilon, \varepsilon) = \varnothing$, for some $\varepsilon > 0$. Then, either $\langle R_+, T \rangle$ contains a unary relation $U_\delta \in \{1\} + \mathcal{I}(\delta)$ for every $\delta > 0$; or $\langle R_+, T \rangle$ contains a unary relation $U_\delta \in \{-1, 1\} + \mathcal{I}(\delta)$, for every $\delta > 0$.

Proof. If $T \cap -T \neq \varnothing$, then the result follows from Lemma 6. Otherwise, by Lemma 2, there exists a constant $c^+ > 0$ such that the set $T^+ = \{x \in T \mid |x| \geq c^+\}$ is non-empty and contains points that are either all positive or all negative. Similarly, there exists a constant $c^- > 0$ such that $T^- = \{x \in T \mid |x| \leq c^-\}$ is non-empty and contains points that are either all positive or all negative. Let $a \in T^+$ and $b \in T^-$. Assume that both sets contain positive points only or that both sets contain negative points only. Then, the result follows using Lemma 5 with the relation $U = a^{-1} \cdot T \cap b^{-1} \cdot T$ (or $-U$ if the points of U are negative). The case when the one set contains positive points and the other contains negative points is handled similarly using the relation $U' = a^{-1} \cdot T \cap b^{-1} \cdot (-T)$. $\qquad\square$

Lemma 12. *Let T be a bnu such that $T \cap (-\varepsilon, \varepsilon) = \varnothing$, for some $\varepsilon > 0$, and U be a unary relation that excludes an interval. Then, $CSP(\{R_+, T, U\})$ is NP-hard.*

Proof. We show that $\langle R_+, T, U \rangle$ contains a unary relation $\{-1, 1\} + \mathcal{I}(\frac{1}{2})$. The result then follows from Lemma 10. If already $\langle R_+, T \rangle$ contains such a relation, then we are done. Otherwise, by Lemma 11, $\langle R_+, T \rangle$ contains a unary relation $U_\delta \in \{1\} + \mathcal{I}(\delta)$, for every $\delta > 0$. Since U excludes an interval, there are points $p, q \in U$ and $0 < \delta_1 < \delta_2 < 1$ such that $p + (q - p)y \notin U$ whenever $\delta_1 \le y \le \delta_2$. Furthermore, p and q can be chosen so that $\delta_1 < 1/2 < \delta_2$, and by scaling U, we may assume that $|q - p| = 2$. Let $m = (p+q)/2$. Note that $T \cap (m - \varepsilon', m + \varepsilon') = \varnothing$, for some $\varepsilon' > 0$. Similarly, possibly by first scaling T, let $p', q' \in T$ be distinct points with $|q' - p'| = 2$ and let $m' = (p' + q')/2$.

Now, define the unary relation $T_0(x) \equiv \exists y \exists z. U_\delta(y) \wedge z = x - y \cdot m \wedge U(z)$, and the unary relation $T_\infty(x) \equiv \exists y' \exists z'. U_\delta(y') \wedge z' = x - y' \cdot m' \wedge T(z')$. The relations T_0 and T_∞ are roughly translations of U and T, where the constant relation $\{1\}$ has been approximated by the relation U_δ. Since $1 \in U_\delta$, we have $\{-1, 1\} \subseteq T_0, T_\infty$. Hence, if δ is chosen small enough, then the relation $T_0 \cap T_\infty \in \langle R_+, T, U \rangle$ will satisfy the conditions of Lemma 6. This finishes the proof. $\qquad\square$

5   Expansions of $\{R_+\}$

We now study the class of semilinear constraint languages containing R_+ and having the properties (P_0) and (P_∞). These properties are defined as follows.

- (P_0) There is a unary relation U in $\langle \Gamma \rangle$ that contains a positive point and satisfies $U \cap (0, \varepsilon) = \varnothing$ for some $\varepsilon > 0$.
- (P_∞) There is a unary relation U in $\langle \Gamma \rangle$ that contains a positive point and satisfies $U \cap (M, \infty) = \varnothing$ for some $M < \infty$.

A relation is *0-valid* if it contains the tuple $(0, \ldots, 0)$ and a constraint language is *0-valid* if every relation in it is 0-valid. We now state our main result and get a complete classification for semilinear constraint languages containing R_+ and $\{1\}$ as an immediate corollary.

Theorem 5. *Let $\{R_+\} \subseteq \Gamma \subseteq SL_{\mathbb{Q}}[\mathbb{Q}]$ be a finite constraint language that satisfies (P_0) and (P_∞).*

- *If Γ is 0-valid, then $CSP(\Gamma)$ is trivially tractable;*
- *otherwise, if Γ does not contain a bnu, then $CSP(\Gamma)$ is tractable by establishing affine consistency;*
- *otherwise, if all unary relations in $\langle \Gamma \rangle$ are essentially convex, then $CSP(\Gamma)$ is tractable via a reduction to an essentially convex constraint language;*
- *otherwise, $CSP(\Gamma)$ is NP-hard.*

Proof. Let $\mathcal{U}$ be the set of all bounded, non-empty unary relations U in $\langle \Gamma \rangle$ such that $U \cap (-\varepsilon, \varepsilon) = \varnothing$ for some $\varepsilon > 0$. Assume that Γ is not 0-valid and let R be a relation in Γ such that $(0, \ldots, 0) \notin R$ and $a \in R$. Then, the relation

$\mathcal{L}_{R,a,(0,\ldots,0)} \in \langle\{R_+, R\}\rangle$ is unary, does not contain 0 but does contain 1. Let $U_0 \in \langle\Gamma\rangle$ be a unary relation witnessing (P$_0$) and let $U_\infty \in \langle\Gamma\rangle$ be a unary relation witnessing (P$_\infty$). Scale U_0 and U_∞ so that some positive point from each coincides with 1 and let $T = \mathcal{L}_{R,a,(0,\ldots,0)} \cap U_0 \cap U_\infty$. If T does not contain a negative point, then $T \in \mathcal{U}$. Otherwise, T contains a negative point b. It follows that $T \cap b \cdot T \in \mathcal{U}$. Hence, the set $\mathcal{U}$ is non-empty.

Assume that $\langle\Gamma\rangle$ does not contain a bnu. Then, neither does $\mathcal{U}$ and hence $\mathcal{U}$ contains only constants. It follows by Theorem 3 that establishing affine consistency solves CSP(Γ).

Otherwise, $\mathcal{U}$ contains a bnu. If all unary relations of $\langle\Gamma\rangle$ are essentially convex, then by Lemma 11 and Theorem 4, CSP(Γ) is equivalent to CSP(Γ') for an essentially convex constraint language Γ'. Tractability follows from Theorem 2.

Finally, if $\mathcal{U}$ contains a bnu and $\langle\Gamma\rangle$ contains a unary relation that excludes an interval, then NP-hardness follows from Lemma 12. $\square$

Theorem 6. *Let $\{R_+, \{1\}\} \subseteq \Gamma \subseteq SL_\mathbb{Q}[\mathbb{Q}]$ be a finite constraint language. If $\langle\Gamma\rangle$ contains a bnu and $\langle\Gamma\rangle$ contains a relation that is not essentially convex, then CSP(Γ) is NP-hard. Otherwise, CSP(Γ) is tractable.*

6   Discussion and Future Work

We have determined the complexity of CSP(Γ) for all finite semilinear Γ containing R_+ and satisfying (P$_0$) and (P$_\infty$). Clearly, one would like to obtain a full classification for semilinear constraint languages without side conditions but this appears to be an extremely hard task. One may instead study cases when (P$_0$) and (P$_\infty$) hold (and $\{R_+\} \not\subseteq \Gamma$) or when only $\{R_+\} \subseteq \Gamma$ is required (and (P$_0$) or (P$_\infty$) do not hold). We will discuss these two possibilities below.

For simplicity, let SL^1 denote the set of semilinear constraint languages such that $\{\{1\}\} \subseteq \Gamma$ and $\{R_+\} \not\subseteq \Gamma$. The languages in SL^1 satisfy both (P$_0$) and (P$_\infty$). A straightforward modification of the construction in Sec. 6.3 of [9] gives the following: for every finite constraint language Γ' over a finite domain, there exists a $\Gamma \in SL^1$ such that CSP(Γ') and CSP(Γ) are polynomial-time equivalent problems. Hence, a complete classification would give us a complete classification of finite-domain CSPs, and such a classification is a major open question within the CSP community [7,8]. We also observe that for every finite *temporal constraint language* (i.e., languages that are first-order definable in $(\mathbb{Q}; <)$), there exists a $\Gamma \in SL^1$ such that CSP(Γ') and CSP(Γ) are polynomial-time equivalent problems. This follows from the fact that every temporal constraint language Γ' admits a polynomial-time reduction from CSP($\Gamma' \cup \{\{1\}\}$) to CSP(Γ'): simply equate all variables appearing in $\{1\}$-constraints and note that any solution can be translated into a solution such that this variable is assigned the value 1. The complexity of temporal constraint languages is fully determined [4] and the polynomial-time solvable cases fall into nine different categories. The proof is complex and it is based on the universal-algebraic approach for studying CSPs.

Let us now consider the set SL^+ of all semilinear constraint languages containing R_+ and not satisfying (P$_0$) or (P$_\infty$). If either (P$_0$) or (P$_\infty$) is violated, then we can show that the languages in SL^+ are of a particular restricted type. Let $HSL_\mathbb{Q}[\mathbb{Q}]$ denote the set of *homogeneous* semilinear relations, i.e., relations $R \subseteq \mathbb{Q}^n$ that are finite unions of homogeneous linear sets.

Theorem 7. *Arbitrarily choose $\Gamma \in SL^+$. $CSP(\Gamma)$ is equivalent to $CSP(\Gamma')$ for a finite constraint language $\{R_+\} \subseteq \Gamma' \subseteq HSL_\mathbb{Q}[\mathbb{Q}]$.*

It seems like a difficult task to classify the complexity of subsets of $HSL_\mathbb{Q}[\mathbb{Q}]$ since this would imply the previously mentioned classification of temporal constraint problems and also the constraint problems studied by Bodirsky et al. [3]. It is also closely connected with CSPs over domains of size 3 as demonstrated by the following proposition.

Proposition 1. *Let A be a finite constraint language over the domain $\{-1, 0, 1\}$. There is a $\Gamma \subseteq HSL_\mathbb{Q}[\mathbb{Q}]$ such that $CSP(A)$ and $CSP(\Gamma)$ are polynomial-time equivalent.*

The complexity of CSPs on three-element domains is fully determined [5] and the lengthy proof is based on machinery from universal algebra. One has to note, though, that we may not need to completely classify the complexity of $HSL_\mathbb{Q}[\mathbb{Q}]$ in order to classify the complexity of SL^+: we have the additional condition that R_+ is a member of the languages under consideration. It is plausible that this would simplify the task.

References

1. Bodirsky, M., Grohe, M.: Non-dichotomies in constraint satisfaction complexity. In: Aceto, L., Damgård, I., Goldberg, L.A., Halldórsson, M.M., Ingólfsdóttir, A., Walukiewicz, I. (eds.) ICALP 2008, Part II. LNCS, vol. 5126, pp. 184–196. Springer, Heidelberg (2008)
2. Bodirsky, M., Jonsson, P., von Oertzen, T.: Essential convexity and complexity of semi-algebraic constraints. Logical Methods in Computer Science 8(4) (2012)
3. Bodirsky, M., Jonsson, P., von Oertzen, T.: Horn versus full first-order: Complexity dichotomies in algebraic constraint satisfaction. J. Log. Comput. 22(3), 643–660 (2012)
4. Bodirsky, M., Kára, J.: The complexity of temporal constraint satisfaction problems. J. ACM 57(2) (2010)
5. Bulatov, A.: A dichotomy theorem for constraint satisfaction problems on a 3-element set. J. ACM 53(1), 66–120 (2006)
6. Bulatov, A., Jeavons, P., Krokhin, A.: Classifying the computational complexity of constraints using finte algebras. SIAM J. Comput. 34(3), 720–742 (2005)
7. Feder, T., Vardi, M.Y.: Monotone monadic SNP and constraint satisfaction. In: Proceedings of the 25th ACM Symposium on Theory of Computing (STOC 1993), pp. 612–622 (1993)
8. Hell, P., Nešetřil, J.: Colouring, constraint satisfaction, and complexity. Computer Science Review 2(3), 143–163 (2008)
9. Jonsson, P., Lööw, T.: Computational complexity of linear constraints over the integers. Artif. Intell. 195, 44–62 (2013)
10. Schaefer, T.J.: The complexity of satisfiability problems. In: Proceedings of the 10th ACM Symposium on Theory of Computing (STOC 1978), pp. 216–226 (1978)
11. Schrijver, A.: Theory of linear and integer programming. John Wiley & Sons (1986)

Small Complexity Classes
for Computable Analysis

Akitoshi Kawamura and Hiroyuki Ota

University of Tokyo, Japan

Abstract. Type-two Theory of Effectivity (TTE) provides a general framework for Computable Analysis. To refine it to polynomial-time computability while keeping as much generality as possible, Kawamura and Cook recently proposed a modification to TTE using machines that have random access to an oracle and run in time depending on the "size" of the oracle. They defined type-two analogues of P, NP, PSPACE and applied them to real functions and operators. We further refine their model and study computation below P: type-two analogues of the classes L, NC, and P-completeness under log-space reductions. The basic idea is to use second-order polynomials as resource bounds, as Kawamura and Cook did, but we need to make some nontrivial (yet natural, as we will argue) choices when formulating small classes in order to make them well-behaved. Most notably: we use a modification of the constant stack model of Aehlig, Cook and Nguyen for query tapes in order to allow sufficient oracle accesses without interfering with space bounds; representations need to be chosen carefully, as computational equivalence between them is now finer; uniformity of circuits must be defined with varying sizes of oracles taken into account. As prototypical applications, we recast several facts (some in a stronger form than was known) about the complexity of numerical problems into our framework.

1 Introduction

Computable Analysis [4, 11, 16] studies problems involving real numbers, real functions and other objects in analysis from the viewpoint of computability on digital machines. Elements of uncountable sets (such as real numbers) are represented through approximations (such as sequences of rational numbers) and processed by Turing machines. Such approximation can be presented to the machine as infinite strings on the tape or as oracles (i.e., functions taking strings to strings), and this choice does not make much difference as long as we only discuss computability. But when we want to pay attention to bounds on time and space, Kawamura and Cook [10] recently pointed out that it is more convenient to use oracles with random access, and moreover, to allow the running time to depend on the "size" of the oracle. Employing *type-two complexity theory* and using *second-order polynomials* to bound time and space, they formulated

E. Csuhaj-Varjú et al. (Eds.): MFCS 2014, Part II, LNCS 8635, pp. 432–444, 2014.
© Springer-Verlag Berlin Heidelberg 2014

analogues of complexity classes P, NP, PSPACE and applied them to some typical operators in analysis[1]. The basic definitions will be reviewed in Section 2.

One benefit of this was a greater variety of objects for which we can define complexity. For example, with the second-order formulation we obtain a canonical representation of the space $C[0, 1]$ of continuous real functions, so that we can discuss the complexity of an operator $F\colon C[0, 1] \to C[0, 1]$. This extends the previously accepted notion of complexity of $f \in C[0, 1]$ (which could also be formulated in the infinite string model) in a natural way, so that many known non-uniform results of the form

if f is in the complexity class X, then $F(f)$ is in the class Y, and
there is f in X such that $F(f)$ is hard for the complexity class Z,

can now be transformed into a stronger, uniform statement of the form

F is in the complexity class $\mathcal{Y}$, and
F is hard for $\mathcal{Z}$ under the $\mathcal{X}$ reduction,

where $\mathcal{X}, \mathcal{Y}, \mathcal{Z}$ are type-two classes analogous to X, Y, Z. See [2, 7, 14, 15, 18] for further discussion (including some criticism), applications and extensions of this approach, as well as connection to other approaches.

We continue their research and proceed down into P. In Section 3, we formulate and study analogues of L (logarithmic space), NC (poly-logarithmic depth circuits; "efficiently parallelizable"), and P-completeness ("inherently sequential"). While the fundamental idea, i.e., that of using second-order polynomials as resource bounds, is common to [10], there are several choices that we need to make carefully in implementing it, due to the subtleties pertaining to the interaction of small complexity classes with oracles, as we will explain. In particular, we use a modified version of the *constant stack machine* [1], since it is consistent with relativized circuit complexity classes and still makes some elemental operation log-space computable. Formulation of uniform circuit family also requires careful consideration to accommodate function oracles.

In Section 4, we apply this framework to a few problems in analysis. We present several examples of real functions and operators that are already essentially known (but non-trivially) to be, in our terminology, in L and NC. We then take up Hoover's theorem about fixed points of contractions [8] and Ko's theorem about inverting a function [11], which both state hardness for P in a sense, and reformulate them into our framework as stronger uniform statements (the original versions come as corollaries). Our proofs are not technically hard or involve new algorithmic ideas (they are either relatively easy using known versions or obtained by minor modifications); rather, the benefit of having the results stated in the TTE framework is that it clarifies (through the choices of

---

[1] These type-two analogues are denoted in boldface letters, such as **P**, **NP**, **PSPACE**. Also, in view of the importance of being precise about the representation in this context, they attach prefix "(γ, δ)-" when talking about the classes of functions on spaces represented by the encodings γ and δ. We will stick to this convention, also for our smaller classes.

representations) which information exactly is needed for the computation and how the amount of required computational resource depends on it.

Notation. Let $\mathbb{N}$ and $\mathbb{R}$ denote the sets of nonnegative integers and real numbers, respectively. When we talk about polynomials as bounds on time or space, we always assume that they are increasing functions.

We consider computational problems as *multi-valued functions* (or *multi-functions*). A multi-function F from X to Y (written $F :\subseteq X \rightrightarrows Y$) is formally a subset of $X \times Y$. The set of $x \in X$ such that there is $y \in Y$ with $(x, y) \in F$ is called the *domain of definition* or the *promise* of F and denoted $\mathrm{dom}\, F$. For $x \in \mathrm{dom}\, F$, we write $F[x]$ for the (nonempty) set of all such y. If $F[x]$ is a singleton, we write $F(x)$ for its unique element. If this is the case for all $x \in \mathrm{dom}\, F$, we say that F is *single-valued*, or is a *partial function*. When $\mathrm{dom}\, F = X$, we say that F is *total*. A single-valued total multi-function is called a *function*.

The intuitive meaning is that F specifies a problem where, given any $x \in \mathrm{dom}\, F$, one is required to output *some* element of $F[x]$. Thus, the specification becomes stricter as $\mathrm{dom}\, F$ gets bigger or as $F[x]$ gets smaller. We choose to regard problems as multi-functions, despite the fact that usually each computing device (a Turing machine or a circuit) yields a single-valued (partial or total) function on strings. This is because, when describing a problem (rather than a specific implementation), it is often convenient and natural to avoid specifying values unnecessarily strictly. In particular, when one object has several different representations, it is natural to ask for *any one* of them (see Definition 2.3.2 below), and this underspecification is sometimes essential for making the computation feasible, especially in our context involving non-discrete objects [4].

2   TTE with Second-Order Polynomials

We review the Type-two Theory of Effectivity (TTE), a powerful framework for computable analysis [16] (extended by [10] for complexity considerations). We use string functions to encode various objects, and use *oracle Turing machines* (henceforth just *machines*) to work on them. Section 2.1 defines polynomial-time computability on these string functions, and Section 2.2 explains how to apply it to real functions (and other objects) through representations.

2.1   Type-Two Machines

As mentioned above, we will use (a class of) string functions to encode objects. A (total) function $\varphi \colon \Sigma^* \to \Sigma^*$ is *length-monotone* if $|\varphi(u)| \le |\varphi(v)|$ whenever $|u| \le |v|$. We denote the set of length-monotone functions by Σ^{**}. We restrict attention to Σ^{**} (rather than using all functions from Σ^* to Σ^*) to keep the notion of their *size* (to be defined shortly) simple. We write $M^\varphi(u)$ for the output string when a machine M is given $\varphi \in \Sigma^{**}$ as oracle and $u \in \Sigma^*$ as input. We say that M *computes* a multi-function $A :\subseteq \Sigma^{**} \rightrightarrows \Sigma^{**}$ if for any $\varphi \in \mathrm{dom}\, A$, there is $\psi \in A[\varphi]$ such that $M^\varphi(u) = \psi(u)$ for all $u \in \Sigma^*$ (Figure 1).

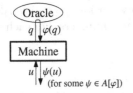

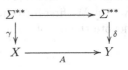

Fig. 1. A machine computing a multi-function $A :\subseteq \Sigma^{**} \rightrightarrows \Sigma^{**}$

Fig. 2. Computing a multi-function A under representations γ and δ

The *size* of $\varphi \in \Sigma^{**}$, denoted $|\varphi|$, is a (non-decreasing) function from $\mathbb{N}$ to $\mathbb{N}$ defined by $|\varphi|(|u|) = |\varphi(u)|$. This is well-defined since a length-monotone function maps strings of the same length to strings of the same length. To define the class **FP** of multi-functions from Σ^{**} to Σ^{**} computable in polynomial time, we bound the running time by *second-order polynomials* $P(|\varphi|)(|x|)$ in the sizes of the oracle φ and string x given to the machine. A second-order polynomial $P(L)(n)$ is an expression built from the number n using $+$, $\times$, *and application of the function* $L \colon \mathbb{N} \to \mathbb{N}$; for example, $5L(L(n)^3 + n^2) + 2n + 4$.

Definition 2.1. *We write* **FP** *for the class of multi-functions from* Σ^{**} *to* Σ^{**} *computed by a machine that runs in second-order polynomial time.*

We use bold letters (such as **FP**) for classes of multi-functions from Σ^{**} to Σ^{**}, as opposed to the usual complexity classes, such as FP, which we regard as consisting of multi-functions from Σ^* to Σ^*. The following is immediate.

Lemma 2.2. *Functions in* **FP** *map elements of* FP $\cap \Sigma^{**}$ *into* FP.

For $\varphi, \psi \in \Sigma^{**}$, we define $\langle \varphi, \psi \rangle \in \Sigma^{**}$ by $\langle \varphi, \psi \rangle(0u) = \varphi(u)10^{|\psi(u)|}$ and $\langle \varphi, \psi \rangle(1u) = \psi(u)10^{|\varphi(u)|}$ (we pad 0s to make $\langle \varphi, \psi \rangle$ length-monotone). We write $\langle \varphi, \psi, \theta \rangle$ for $\langle \varphi, \langle \psi, \theta \rangle \rangle$, and so on. We also write $\langle \varphi, u \rangle$, etc., for $\varphi \in \Sigma^{**}$ and a string $u \in \Sigma^*$, by identifying u with the constant function in Σ^{**} with value u.

2.2   Representations

A *representation* of a set X is a partial function γ from Σ^{**} to X such that for every $x \in X$ there is φ with $\gamma(\varphi) = x$. We call φ a γ-*name* of x if $x = \gamma(\varphi)$.

Definition 2.3. *1. Let C be a class of functions (or multi-functions) from Σ^* to Σ^*, and let γ be a representation of a set X. We write γ-C for the set of $x \in X$ that have a γ-name in C.*

2. Let C be a class of multi-functions from Σ^{**} to Σ^{**}, and γ and δ be representations of sets X and Y, respectively. A multi-function $A :\subseteq X \rightrightarrows Y$ is in (γ, δ)-C if the multi-function $\delta^{-1} \circ A \circ \gamma :\subseteq \Sigma^{**} \rightrightarrows \Sigma^{**}$ defined by $\mathrm{dom}(\delta^{-1} \circ A \circ \gamma) = \{ \varphi \in \mathrm{dom}\,\gamma \mid \gamma(\varphi) \in \mathrm{dom}\,A \}$ and $(\delta^{-1} \circ A \circ \gamma)[\varphi] = \{ \psi \in \mathrm{dom}\,\delta \mid \delta(\psi) \in A[\gamma(\varphi)] \}$ (Figure 2) is in C.

For real numbers and real functions, we use representations $\rho_{\mathbb{R}}$ and $\delta_\square$, defined as follows [10]. We first introduce an encoding of dyadic numbers. For each $n \in \mathbb{N}$, let $\mathbb{D}_n$ denote the set of strings of the form

$$sx/1\underbrace{00\ldots0}_{n}, \tag{1}$$

where s is the sign and $x \in \{0,1\}^*$. We write $\mathbb{D}$ for the union $\bigcup_n \mathbb{D}_n$. We regard $u \in \mathbb{D}$ as a fraction of binary integers, and write u also for the number it encodes.

We define the representation $\rho_{\mathbb{R}}$ of $\mathbb{R}$ by saying that $\varphi \in \Sigma^{**}$ is a $\rho_{\mathbb{R}}$-name of $x \in \mathbb{R}$ if $\varphi(0^i) \in \mathbb{D}$ and $|\varphi(0^i) - x| \le 2^{-i}$ for all $i \in \mathbb{N}$. We write $\rho_{\mathbb{R}}|^{[0,1]}$ for the representation of the interval $[0,1]$ obtained by restricting $\rho_{\mathbb{R}}$. The class $(\rho_{\mathbb{R}}|^{[0,1]}, \rho_{\mathbb{R}})$-**FP** equals the polynomial-time computable functions by Ko [11].

We call $\mu \colon \mathbb{N} \to \mathbb{N}$ a *modulus of continuity* of $f \in C[0,1]$ if for all $n \in \mathbb{N}$ and $x, y \in [0,1]$ with $|x - y| \le 2^{-\mu(n)}$, we have $|f(x) - f(y)| \le 2^{-n}$. It is not hard to verify the following.

Lemma 2.4. *A real function $f \colon [0,1] \to \mathbb{R}$ is in $(\rho_{\mathbb{R}}|^{[0,1]}, \rho_{\mathbb{R}})$-**FP** if and only if it has a polynomial modulus of continuity and there is a function $\varphi \in$ FP such that*

$$|\varphi(d, 0^n) - f(d)| \le 2^{-n} \tag{2}$$

for all $d \in \mathbb{D} \cap [0,1]$ and $n \in \mathbb{N}$.

The following representation $\delta_\square$ of $C[0,1]$ is inspired by this. For a non-decreasing function $\mu \colon \mathbb{N} \to \mathbb{N}$, we write $\overline{\mu} \in \Sigma^{**}$ for the function that maps each string u to $0^{\mu(|u|)}$. A $\delta_\square$-name of $f \in C[0,1]$ is a pair $\langle \overline{\mu}, \varphi \rangle$ (see the end of Section 2.1 for the pairing function) of $\varphi \in \Sigma^{**}$ and $\mu \colon \mathbb{N} \to \mathbb{N}$ such that μ is a modulus of continuity of f and φ satisfies (2). Lemma 2.4 implies that $(\rho_{\mathbb{R}}|^{[0,1]}, \rho_{\mathbb{R}})$-**FP** equals $\delta_\square$-**FP**.

3   Small Type-Two Classes

In this section, we introduce type-two complexity classes corresponding to log-space L and circuit complexity NC based on the framework we reviewed in the previous section. We also define P-completeness under log-space reductions.

3.1   Logarithmic Space

As reviewed in the previous section, we use oracle Turing machines in the definition of type-two classes such as **FP**. The machine has an input tape, output tape, work tape, query tape and answer tape. When the machine enters the query state, and the string on the query tape at that time is u, then in the next transition, the machine will be in the after-query state, and the string on the answer tape will be $\varphi(u)$, where φ is the oracle (although this exact mechanism for oracle access did not matter very much for **FP** and larger classes). For the definition of the logarithmic-space class **FL**, we will basically continue using this model, with space bound $\log(P(|\varphi|)(|u|))$ for the computation on oracle φ and string input u. To make this sub-linear space restriction meaningful, the standard convention (already for the type-one class FL) is that the input and output tapes are not counted towards the space limit, and that (to avoid exploiting these tapes as memory) the input tape is read-only and the output tape is write-only. For our purpose of type-two computation, we need to have similar consideration for query and answer tapes, and this raises some issues we should be careful about. (See also the discussion in the slightly different context of relativized classes [1, 5, 12, 17].)

Note that even in this logarithmic-space setting, we want the machine to be able to ask queries of polynomial length. For example, when computing a real function f under the $\rho_{\mathbb{R}}$ representation, it is reasonable to give the machine access to the input real number with polynomially many digits of precision (in the number of digits we want of the output)—with logarithmically many digits, we would only compute constant functions.

Thus, the right formulation of logarithmic-space computation with oracle access requires careful resource control, by which the mechanism for oracle access is exempted from the space limit, while "other parts" of computation are limited to logarithmic space. For some special cases, this is achieved by simply adopting the convention that the query and answer tapes are not counted towards the space. In fact, Ko's logarithmic-space computable real functions on $[0,1]$ [11, Chapter 4] is defined roughly in this way (see the discussion in [11, pp. 121–122]), using a representation similar to our $\rho_{\mathbb{R}}$. We also obtain the equivalent definition by the infinite string model, because it does not count the input and output tapes, and it forbids reading from the output tape.

The reason this definition successfully led to a reasonable class of real functions seems to be because, when we were dealing only with $\rho_{\mathbb{R}}$-names, the queries were always of the simple form 0^m. But in general, the queries are polynomially long binary string, which the machine cannot even store by itself on the work tape. This subtlety calls for the following modifications on the model:

> Our model is an oracle Turing machine with a stack of query tapes (write-only) and an answer tape (read-only). The machine can write a symbol on the top query tape, or *push* a new query tape on the top of the stack (and start writing on it). When the machine issues a query, the stack is *popped* automatically; that is, if the string on the query tape at the top of the stack was u, the oracle φ writes the string $\varphi(u)$ on the answer tape, and at the same time the top query tape is removed from the stack. We put the restriction that the height of the stack of a machine is bounded by some constant for all inputs and oracles. We also assume that the answer tape is erased after each push operation.

This model resolves two issues arising from the aforementioned tension between allowing the machine to issue long queries while disallowing it to write them down.

The first issue is that the machine may need to make nested queries (queries depending on answers to previous queries), and it may need to do so by writing a query halfway and then issuing other queries in order to complete the remaining part of the query. Observe that the stack model above makes such queries possible. The ability to ask nested queries (with constant depth of nesting) is needed naturally for our applications in real-number computation (see e.g. the comments on Theorem 4.4 below), and is also essential if we want the class to contain FAC^0 (see Theorem 3.4 below). The idea of using a stack of query tapes in order to obtain a reasonable relativization of logspace computation is due to Wilson [17] and Aehlig, Cook and Nguyen [1] (but note that these are about predicate oracles).

The second issue (which arises because we are dealing with function oracles) is that we do not want the machine to cheat by using the query and answer tapes as extra memory. This is why we required that the answer tape is erased after each push operation. This makes our model equivalent to accessing the oracle φ by questions of the form $(u, 0^i)$ asking for the ith bit of $\varphi(u)$ (or an explicit error message when i exceeds the length of $\varphi(u)$). The erasure of the answer tape also ensures that the nested depth of queries is restricted to a constant. Without this restriction, a log-space constant-stack machine with a function oracle φ with $|\varphi|(n) = n$ could compute the result of iterating the function φ polynomially many times. On the other hand, there is a function $\varphi \in \Sigma^{**}$ such that $|\varphi|(n) = n$ and iteration of φ is not in NC relative to φ [1], and hence the inclusion $\mathsf{L} \subseteq \mathsf{NC}$ would not relativize.

Definition 3.1. *A machine (with the oracle access convention described above) runs in (second-order) logarithmic space if there is a second-order polynomial P such that, given oracle $\varphi \in \Sigma^{**}$ and string $u \in \Sigma^*$, it visits at most $\log(P(|\varphi|)(|u|))$ cells in the work tape. We write* **FL** *for the set of multi-functions[2] from Σ^{**} to Σ^{**} computed by such a machine.*

Lemma 3.2. *Functions in* **FL** *map elements of* $\mathsf{FL} \cap \Sigma^{**}$ *into* FL.

3.2   Circuits of Bounded Depth

In this section, we discuss the type-two analogue of AC^d and NC by considering circuits with oracle gates. Since the oracle in our setting has variable (second-order) sizes, and the resource (circuit size and depth) depends (second-order) polynomially on their size, the circuit family will be indexed not only by the size of the input string but also by the size of the oracle. We will also see that the circuit classes are in the right containment with the Turing machine-based classes **L** and **P** from the previous sections.

Let $n, m \in \mathbb{N}$ and let $L \colon \mathbb{N} \to \mathbb{N}$ be a non-decreasing function. A *circuit with size-L oracle gates* is a circuit C (with several inputs and several outputs) built from the standard logical gates NOT, OR, AND (the latter two with unbounded fan-in) and oracle gates, where each oracle gate with k inputs has $L(k)$ outputs.

---

[2] We can of course define a class **L**, analogous to **P** in [10], of multi-functions in **FL** whose values are functions in Σ^{**} that are $\{0, 1\}$-valued, but we will not use such classes in this paper.

The *size* of C is the number of gates. The *depth* of C is the length of the longest path in C. For an input $x \in \{0, 1\}^*$, an oracle $\varphi \in \Sigma^{**}$, and a $|x|$-input m-output circuit C with size-$|\varphi|$ oracle gates, we write $C^{\varphi}(x)$ for the m-bit output of C.

As mentioned above, we consider *oracle circuit families* $C = (C_{L,n})_{L,n}$, indexed by $n \in \mathbb{N}$ and non-decreasing functions $L \colon \mathbb{N} \to \mathbb{N}$, such that each $C_{L,n}$ is an n-input circuit with size-L oracle gates. Such a family C is said to *compute* a multi-function[3] $A \colon\subseteq \Sigma^{**} \rightrightarrows \Sigma^{**}$ if for all $\varphi \in \mathrm{dom}\, A$, there is $\psi \in A[\varphi]$ satisfying $\psi(x) = C^{\varphi}_{|\varphi|,|x|}(x)$ for all $x \in \Sigma^*$.

This family $(C_{L,n})_{L,n}$ is said to have *polynomial size* if there is a second-order polynomial P such that the size of $C_{L,n}$ is bounded by $P(L)(n)$. Likewise, it has *k-logarithmic depth* if there is a second-order polynomial P such that each $C_{L,n}$ has depth bounded by $(\log(P(L)(n)))^k$.

The definition of uniformity also takes into account the oracle size: we say that the family $(C_{L,n})_{L,n}$ is **L**-*uniform* if there is a function $A \in$ **FL** such that for all $n \in \mathbb{N}$ and non-decreasing $L \colon \mathbb{N} \to \mathbb{N}$, the string $A(\overline{L})(0^n)$ is (a description of) the circuit $C_{L,n}$ (recall from the end of Section 2.2 that $\overline{L}$ is the function taking strings u to $0^{L(|u|)}$). Hereafter, we assume log-space uniformity on all (type-one and type-two) circuit classes, so we write just "uniform" or omit "**L**-uniform".

Definition 3.3. *For each $k \in \mathbb{N}$, we write* **FAC**^{k} *for the class of multi-functions from Σ^{**} to Σ^{**} computed by a (uniform) circuit family $(C_{L,n})_{L,n}$ of polynomial size and k-logarithmic depth[4]. We write* **FNC** $= \bigcup_{k \in \mathbb{N}}$ **FAC**^{k}.

Similarly to the type-one FP, we can show that a multi-function from Σ^{**} to Σ^{**} is in **FP** if and only if it is computed by a polynomial-size uniform circuit family (for one direction, we modify the standard argument for the type-one class, observing that we may assume any machine for **FP** to be "oblivious," i.e., its head motions depend only on the sizes of the input string and the oracle).

The relativization using the stack model by Aehlig, Cook and Nguyen preserves the inclusion of non-relativized classes $\mathrm{AC}^0 \subseteq \mathrm{L} \subseteq \mathrm{AC}^1 \subseteq \mathrm{NC}$. Since we define the type-two log-space class by extending the stack model, the analogous inclusions can be shown for our type-two classes by a similar argument. The (straightforward) proof is omitted in the present version.

Theorem 3.4. **FAC**$^0 \subseteq$ **FL** $\subseteq$ **FAC**$^1 \subseteq$ **FNC** $\subseteq$ **FP**.

Lemma 3.5. *1. Functions in* **FAC**^{i} *map elements of* $\mathrm{FAC}^j \cap \Sigma^{**}$ *into* FAC^{i+j}.
2. Functions in **FNC** *map elements of* $\mathrm{FNC} \cap \Sigma^{**}$ *into* FNC.

---

[3] As mentioned at the end of Section 1, the family C actually specifies a (single-valued) function, and thus the computed multi-functions are exactly those that underspecify this function. Nevertheless, we find it convenient to allow multi-functions in this definition, when we use it in combination with e.g. Definition 2.3.2.

[4] We are thus defining the analogue of NC as the union of AC^i rather than of NC^i. This is just for simplicity of presentation.

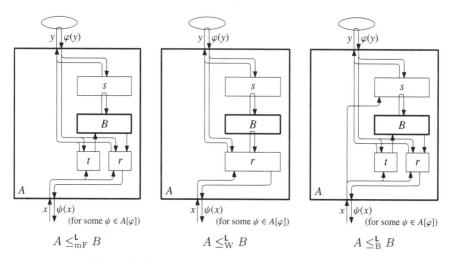

$A \leq^{\mathsf{L}}_{\mathrm{mF}} B$ $A \leq^{\mathsf{L}}_{\mathrm{W}} B$ $A \leq^{\mathsf{L}}_{\mathrm{B}} B$

Fig. 3. Reductions between multi-functions on Σ^{**}

3.3 Reductions and Completeness

The formulation of reduction and hardness is mostly analogous to what was done for larger classes [10], simply using weaker (logspace) reduction this time. The following logspace reductions $\leq^{\mathsf{L}}_{\mathrm{mF}}$, $\leq^{\mathsf{L}}_{\mathrm{W}}$, $\leq^{\mathsf{L}}_{\mathrm{B}}$ (Figure 3) are analogues of the polynomial-time reductions $\leq^2_{\mathrm{mF}}$, $\leq^2_{\mathrm{W}}$ in Kawamura and Cook [10] and the "many-one reduction" in Beame et al. [3].

Definition 3.6. *Let* A, $B :\subseteq \Sigma^{**} \rightrightarrows \Sigma^{**}$ *be multi-functions. We write*

- $A \leq^{\mathsf{L}}_{\mathrm{mF}} B$ *(*A *is* many-one log-space reducible *to* B*) if there are functions* $r, s, t \in \mathbf{FL}$ *such that for all* $\varphi \in \mathrm{dom}\, A$*, we have* $s(\varphi) \in \mathrm{dom}\, B$ *and for each* $\theta \in B[s(\varphi)]$*, the function that maps* $x \in \Sigma^*$ *to* $r(\varphi)(x, \theta(t(\varphi)(x)))$ *is in* $A[\varphi]$*.*
- $A \leq^{\mathsf{L}}_{\mathrm{W}} B$ *(*A *is* Weihrauch log-space reducible *to* B*) if there are functions* $r, s \in \mathbf{FL}$ *such that for all* $\varphi \in \mathrm{dom}\, A$*, we have* $s(\varphi) \in \mathrm{dom}\, B$ *and for each* $\theta \in B[s(\varphi)]$*, we have* $r(\langle \varphi, \theta \rangle) \in A[\varphi]$*.*
- $A \leq^{\mathsf{L}}_{\mathrm{B}} B$ *if there are functions* $r, s, t \in \mathbf{FL}$ *such that for all* $\varphi \in \mathrm{dom}\, A$*, the function* σ *that maps* $y \in \Sigma^*$ *to* $s(\varphi)(x, y)$ *belongs to* $\mathrm{dom}\, B$ *and for each* $\theta \in B[\sigma]$*, the function that maps* $x \in \Sigma^*$ *to* $r(\varphi)(x, \theta(t(\varphi)(x)))$ *is in* $A[\varphi]$*.*

For a class $\mathcal{C}$ of multi-functions from Σ^{**} to Σ^{**} and a reduction $\leq$, we say that $B :\subseteq \Sigma^{**} \rightrightarrows \Sigma^{**}$ is $\mathcal{C}$-$\leq$-*hard* if $A \leq B$ for all $A \in \mathcal{C}$, and B is $\mathcal{C}$-$\leq$-*complete* if B is $\mathcal{C}$-$\leq$-hard and in $\mathcal{C}$.

Note that $A \leq^{\mathsf{L}}_{\mathrm{mF}} B$ implies $A \leq^{\mathsf{L}}_{\mathrm{W}} B$ and $A \leq^{\mathsf{L}}_{\mathrm{B}} B$. The reduction $\leq^{\mathsf{L}}_{\mathrm{B}}$ is stronger than $\leq^{\mathsf{L}}_{\mathrm{mF}}$ in that s can read the input string x when answering queries from B. In many applications to analysis, the (thus easier) proof of $\leq^{\mathsf{L}}_{\mathrm{B}}$-hardness already seems to capture the essential complexity of the operator (e.g. Theorem 4.8 in our case), but we need the stronger hardness (with $\leq^{\mathsf{L}}_{\mathrm{mF}}$ or $\leq^{\mathsf{L}}_{\mathrm{W}}$) if we want to derive the non-uniform versions just from the statement.

See [10] for more discussion on these reductions, most of which applies to our logarithmic-space setting. In particular, we have the analogue of [10, Lemma 3.6] for completeness with respect to $\leq^{\mathsf{L}}_{\mathrm{mF}}$, and thus, e.g., a **FP**-$\leq^{\mathsf{L}}_{\mathrm{mF}}$-hard function maps some input in FL to a FP-$\leq^{\mathsf{L}}_{\mathrm{mF}}$-hard output, where $\leq^{\mathsf{L}}_{\mathrm{mF}}$ is the usual many-one log-space reduction between type-one multi-functions. We omit the details here due to limited space.

3.4 Representations

Carrying over the above argument to represented spaces is analogous to what has been done [10, Section 3.4] for larger classes, and is omitted in the present version due to lack of space.

4 Applications

In Section 4.1, we show some examples of real functions and operators that are in **FL** and **FNC** (under suitable representations). In Section 4.2, we show the **P**-completeness of the inverse operation and the fix-point operation. Due to lack of space, many of the results will be stated without proofs. They will be included in the full version of this paper.

4.1 Within FL and FNC

We start with real numbers and real functions, using the representation $\rho_{\mathbb{R}}$ (see Section 2.2). It can be shown that Ko's class of *log-space real functions* in C[0, 1] [11] equals our $(\rho_{\mathbb{R}}|^{[0,1]}, \rho_{\mathbb{R}})$-**FL**, despite our choice of constant-stack log-space machine (Ko uses the obvious log-space oracle Turing machine with no stack).

For representations γ_0, γ_1 of spaces X_0, X_1, respectively, we define the representation $[\gamma_0, \gamma_1]$ of the Cartesian product $X_0 \times X_1$ by setting $[\gamma_0, \gamma_1](\langle \varphi_0, \varphi_1 \rangle) = (\gamma_0(\varphi_0), \gamma_1(\varphi_1))$ for each $\varphi_0, \varphi_1 \in \Sigma^{**}$ (see the end of Section 2.1 for $\langle \cdot, \cdot \rangle$).

Example 4.1. Binary addition and binary multiplication are in $([\rho_{\mathbb{R}}, \rho_{\mathbb{R}}], \rho_{\mathbb{R}})$-**FL**.

Ko also defines the class of NC real functions in C[0, 1] as $\delta_{\square}$-**FNC** in our terminology [11], which easily equals $(\rho_{\mathbb{R}}|^{[0,1]}, \rho_{\mathbb{R}})$-**FNC**. The NC-real functions from $\mathbb{R}$ to $\mathbb{R}$ defined by Hoover [8] are exactly those in $(\rho_{\mathbb{R}}, \rho_{\mathbb{R}})$-**FNC**.

An important update to our knowledge after the publication of Ko's book [11] was the discovery by Chiu, Davida and Litow [6] of a logarithmic space algorithm for division and iterated multiplication. This is especially important in our context of computable analysis, because it enables us to evaluate a fast-converging Taylor series, by which many real numbers and functions are defined.

Example 4.2. The circle ratio π is in $\rho_{\mathbb{R}}$-**FL**.

Example 4.3. The sine function $\sin \colon \mathbb{R} \to \mathbb{R}$ is in $(\rho_{\mathbb{R}}, \rho_{\mathbb{R}})$-**FL**.

Now we consider computation under the representation $\delta_\square$ (see Section 2.2) of real functions. The following states that $\delta_\square$ is the weakest representation that makes function evaluation logspace computable, and gives evidence that $\delta_\square$ is the natural choice (relative to $\rho_\mathbb{R}$) as a basic representation of $C[0,1]$. It can be proved similarly to the analogous claim for **FP** [9].

Theorem 4.4. *Define Apply*: $C[0,1] \times [0,1] \to \mathbb{R}$ *by* $Apply(f)(x) = f(x)$. *For a representation* δ *of* $C[0,1]$, *we have* $Apply \in ([\delta, \rho_\mathbb{R}|^{[0,1]}], \rho_\mathbb{R})$-**FL** *if and only if* δ *is logspace translable to* $\delta_\square$, *i.e., there is a function* $F \in$ **FL** *such that* $F(\varphi) \in \operatorname{dom}\delta_\square$ *and* $\delta(\varphi) = \delta_\square(F(\varphi))$ *for all* $\varphi \in \operatorname{dom}\delta$.

Using $\rho_\mathbb{R}$, $\delta_\square$ and suitable representations for other spaces, we can formulate in our terminology some known results about log-space and NC computability of functions and operators involving real numbers (which have been often formulated for, say, rational inputs). For example, Neff's result [13] that "the roots of a complex polynomial can be found in NC" can be seen as a claim that the root-finding operator is in **FNC** under certain natural representations. Details will be included in a full version of this paper.

4.2   P-complete Operations

Here we state uniform versions of Hoover's and Ko's P-hardness results about operators on real functions.

Inverting a Function. Fix $a, b \in \mathbb{R}$ with $a < b$. Let M be the set of one-to-one functions $f \in C[0,1]$ whose range is $[a,b]$. We define the function $Inv: M \to C[a,b]$ by saying that $Inv(f) = f^{-1}$ is the inverse function of f.

Ko proved the following non-uniform theorem about the complexity of inversion. Recall that Ko's polynomial-time and log-space computability of a real function is equivalent to our $(\rho_\mathbb{R}|^{[0,1]}, \rho_\mathbb{R})$-**FP** and $(\rho_\mathbb{R}|^{[0,1]}, \rho_\mathbb{R})$-**FL**.

Theorem 4.5 ([11, Corollary 4.7 and Theorem 4.18]).

1. *Assume that* $f \in M$ *is polynomial-time computable. If* f^{-1} *has a polynomial modulus of continuity, then* f^{-1} *is polynomial-time computable.*
2. *There is a log-space computable* $f \in M$ *such that* f^{-1} *has a polynomial modulus of continuity but is not log-space computable unless* P = L.

We define a representation $\delta_{\square INV}$ of M as follows: $\delta_{\square INV}(\langle \varphi, \overline{\mu} \rangle) = f$ if and only if φ is a $\delta_\square$-name of f and μ is a modulus of continuity of f^{-1}. We add a modulus of continuity of f^{-1} to $\delta_\square$ since without this information, there is no upper bound on the complexity of the inverse operation, as the following fact [11, Theorem 4.4] shows: for any recursive $x \in [0,1]$, there exists a strictly increasing function $f \in C[0,1]$ such that f is polynomial-time computable and $x = f^{-1}(0)$. The following is a uniform version of Theorem 4.5, and it implies Theorem 4.5.

Theorem 4.6. *Inv is* $(\delta_{\square INV}, \delta_\square)$-**FP**-$\leq_{mF}^{L}$-*complete.*

We omit the proof that this follows from Theorem 4.8 below.

Fixed Points of Contractions. A function $g\colon K \to \mathbb{R}$ on a set $K \subseteq \mathbb{R}$ is called q-*Lipschitz*, for $q > 0$, if $|g(x) - g(y)| \leq q \cdot |x - y|$ for all $x, y \in K$. A *contraction* on K is a function $g\colon K \to \mathbb{R}$ which is q-Lipschitz for some $q < 1$ and whose values are in K. The Banach fixpoint theorem states that every contraction has a unique fixed point. Hoover's theorem about the complexity of finding this fixed point can be written in our terminology as follows:

Theorem 4.7 ([8, Theorem 4.5]). *There is a function $f\colon \mathbb{R} \to \mathbb{R}$ in $(\rho_\mathbb{R}, \rho_\mathbb{R})$-* **FNC** *whose restriction $f|_{[2k,2k+1]}$ is a contraction for each $k \in \mathbb{N}$ and which has the following property:* $\mathsf{NC} = \mathsf{P}$ *if and only if there is a function (from strings to strings) in* **FNC** *mapping a pair of a number $k \in \mathbb{N}$ (written in binary) and the string 0^n to a 2^{-n}-approximation of the fixed point of $f|_{[2k,2k+1]}$.*

For our formulation in TTE, let C be the set of contractions on $[0,1]$. We define its representation $\delta_{\square\mathrm{CM}}$ by saying that a $\delta_{\square\mathrm{CM}}$-*name* of $g \in C$ is a pair $\langle \varphi, q \rangle$ of a $\delta_\square$-name $\varphi \in \Sigma^{**}$ of g and a dyadic number $q \in \mathbb{D}$ such that g is q-Lipschitz. Let $\mathit{Fix}\colon C \to \mathbb{R}$ be the function taking each contraction on $[0,1]$ to its fixed point.

Theorem 4.8. *The operator Fix is $(\delta_{\square\mathrm{CM}}, \rho_\mathbb{R})$-***FP**-$\leq_\mathsf{B}^\mathsf{L}$-*complete. It remains so even if* $\mathrm{dom}\, \mathit{Fix}$ *is restricted to contractions that are $1/2$-Lipschitz.*

Due to lack of space, we omit the proof that this implies Theorem 4.7.

Acknowledgements. Some of the work presented here was done as part of the master thesis of the second author at the University of Tokyo. He thanks his advisor Hiroshi Imai for his generous support and guidance. This work was supported in part by KAKENHI 24106002 and 26700001.

References

[1] Aehlig, K., Cook, S., Nguyen, P.: Relativizing small complexity classes and their theories. In: Duparc, J., Henzinger, T.A. (eds.) CSL 2007. LNCS, vol. 4646, pp. 374–388. Springer, Heidelberg (2007)

[2] Ambos-Spies, K., Brandt, U., Ziegler, M.: Real benefit of promises and advice. In: Bonizzoni, P., Brattka, V., Löwe, B. (eds.) CiE 2013. LNCS, vol. 7921, pp. 1–11. Springer, Heidelberg (2013)

[3] Beame, P., Cook, S., Edmonds, J., Impagliazzo, R., Pitassi, T.: The relative complexity of NP search problems. Journal of Computer and System Sciences 57(1), 3–19 (1998)

[4] Brattka, V., Hertling, P., Weihrauch, K.: A tutorial on computable analysis. In: New Computational Paradigms. Springer (2008)

[5] Buss, J.F.: Relativized alternation and space-bounded computation. Journal of Computer and System Sciences 36(3), 351–378 (1988)

[6] Chiu, A., Davida, G., Litow, B.: Division in logspace-uniform NC^1. RAIRO-Theoretical Informatics and Applications 35(03), 259–275 (2001)

[7] Férée, H., Hoyrup, M.: Higher-order complexity in analysis. In: Proc. CCA 2013, pp. 22–35 (2013)

[8] Hoover, H.J.: Real functions, contraction mappings, and P-completeness. Information and Computation 93(2), 333–349 (1991)

[9] Kawamura, A.: On function spaces and polynomial-time computability. Dagstuhl Seminar 11411: Computing with Infinite Data (2011)

[10] Kawamura, A., Cook, S.: Complexity theory for operators in analysis. ACM Transactions on Computation Theory 4(2), Article 5 (2012)

[11] Ko, K.I.: Complexity theory of real functions. Birkhauser Boston Inc. (1991)

[12] Ladner, R.E., Lynch, N.A.: Relativization of questions about log space computability. Theory of Computing Systems 10(1), 19–32 (1976)

[13] Neff, C.A.: Specified precision polynomial root isolation is in NC. Journal of Computer and System Sciences 48(3), 429–463 (1994)

[14] Rettinger, R.: Computational complexity in analysis (extended abstract). In: CCA 2013, pp. 100–109 (2013)

[15] Rösnick, C.: Closed sets and operators thereon: representations, computability and complexity. In: Proc. CCA 2013, pp. 110–121 (2013)

[16] Weihrauch, K.: Computable Analysis: An Introduction. Springer (2000)

[17] Wilson, C.B.: A measure of relativized space which is faithful with respect to depth. Journal of Computer and System Sciences 36(3), 303–312 (1988)

[18] Ziegler, M.: Real computation with least discrete advice: A complexity theory of nonuniform computability with applications to effective linear algebra. Annals of Pure and Applied Logic 163(8), 1108–1139 (2012)

Two Results about Quantum Messages[*,**]

Hartmut Klauck[1] and Supartha Podder[2]

[1] Centre for Quantum Technologies and Nanyang Technological University, Singapore
hklauck@gmail.com
[2] Centre for Quantum Technologies and National University of Singapore, Singapore
supartha@gmail.com

Abstract. We prove two results about the relationship between quantum and classical messages. Our first contribution is to show how to replace a quantum message in a one-way communication protocol by a deterministic message, establishing that for all partial Boolean functions $f : \{0,1\}^n \times \{0,1\}^m \to \{0,1\}$ we have $D^{A \to B}(f) \leq O(Q^{A \to B,*}(f) \cdot m)$. This bound was previously known for total functions, while for partial functions this improves on results by Aaronson [1,2], in which either a log-factor on the right hand is present, or the left hand side is $R^{A \to B}(f)$, and in which also no entanglement is allowed.

In our second contribution we investigate the power of quantum proofs over classical proofs. We give the first example of a scenario in which quantum proofs lead to exponential savings in computing a Boolean function, for quantum verifiers. The previously only known separation between the power of quantum and classical proofs is in a setting where the input is also quantum [3].

We exhibit a partial Boolean function f, such that there is a one-way quantum communication protocol receiving a quantum proof (i.e., a protocol of type QMA) that has cost $O(\log n)$ for f, whereas every one-way quantum protocol for f receiving a classical proof (protocol of type QCMA) requires communication $\Omega(\sqrt{n}/\log n)$.

1 Introduction

The power of using quantum messages over classical messages is a central topic in information and communication theory. It is always good to understand such questions well in the simplest settings where they arise. An example is the setting of one-way communication complexity, which is rich enough to lead to many interesting results, yet accessible enough for us to show results about deep questions like the relationship between different computational modes, e.g. quantum versus classical or nondeterministic versus deterministic.

---

[*] This work is funded by the Singapore Ministry of Education (partly through the Academic Research Fund Tier 3 MOE2012-T3-1-009) and by the Singapore National Research Foundation.

[**] Supported by a CQT Graduate Scholarship.

E. Csuhaj-Varjú et al. (Eds.): MFCS 2014, Part II, LNCS 8635, pp. 445–456, 2014.
© Springer-Verlag Berlin Heidelberg 2014

1.1   One-Way Communication Complexity

Perhaps the simplest question one can ask about the power of quantum messages is the relationship between quantum and classical one-way protocols. Alice sends a message to Bob in order to compute the value of a function $f : \{0,1\}^n \times \{0,1\}^m \to \{0,1\}$. Essentially, Alice communicates a quantum state and Bob performs a measurement, both depending on their respective inputs. Though deceptively simple, this scenario is not at all fully understood. Let us just mention the following open problem: what is the largest complexity gap between quantum and classical protocols of this kind for computing a total Boolean function? The largest gap known is a factor of 2, as shown by Winter [23], but for all we know there could be examples where the gap is exponential, as it indeed is for certain partial functions (i.e., functions that are only defined on a subset of $\{0,1\}^n \times \{0,1\}^m$) [7].

An interesting bound on such speedups can be found by investigating the effect of replacing quantum by classical messages. Let us sketch the proof of such a result. Suppose a total Boolean function f has a quantum one-way protocol with communication c, namely Alice sends c qubits to Bob, who can decide f with error $1/3$ by measuring Alice's message. We allow Alice and Bob to share an arbitrary input-independent entangled state. Extending Nayak's random access code bound [18] Klauck [13] showed that $Q^{A \to B,*}(f) \geq \Omega(VC(f))$, where $Q^{A \to B,*}(f)$ denotes the entanglement-assisted quantum one-way complexity of f, and $VC(f)$ the Vapnik-Chervonenkis dimension of the communication matrix of f. Together with Sauer's Lemma [21] this implies that $D^{A \to B}(f) \leq O(Q^{A \to B,*}(f) \cdot m)$, where m is the length of Bob's input. See also [12] for a related result.

A result such as the above is much more interesting in the case of partial functions. The reason is that for total functions a slightly weaker statement follows from a weak version of the random access code bound, which can be (and indeed has been [5]) established by the following argument: boost the quantum protocol for f until the error is below 2^{-2m}, where m is Bob's input length. Measure the message sent by Alice with *all* the measurements corresponding to Bob's inputs (this can be done with small total error) in order to determine Alice's row of the communication matrix and hence her input. This is a hard task by standard information theory facts (Holevo's bound). When considering partial functions the proof breaks down: Bob does not know for which of his inputs y the value $f(x, y)$ is defined. If Bob measures the message for x with the observable for y and $f(x, y)$ is undefined any acceptance probability is possible and the message state can be destroyed.

Aaronson [1] circumvented this problem in the following way: Bob now tries to learn Alice's message. He starts with a guess (the totally mixed state) and keeps a classical description of his guess. Alice also always knows what Bob's guess is. Bob can simulate quantum measurements by brute-force calculation: for any measurement operator Bob can simply calculate the result from his classical description. Alice can do the same. Since Bob has some 2^m measurements he is possibly interested in, Alice can just tell him on which of these he will be wrong. Bob can then adjust his quantum state accordingly, and Aaronson's main argument is that he does not have to do this too often before he reaches an approximation of the message state. Note that Bob might never learn the message state if it so happens that all measurements are approximately correct on his guess. But if he makes a certain number of adjustments he will learn the message state and no further adjustments are needed.

Let us state Aaronson's result from [1].

Fact 1. $D^{A \to B}(f) \leq O(Q^{A \to B}(f) \cdot \log(Q^{A \to B}(f)) \cdot m)$ for all partial Boolean $f : \{0,1\}^n \times \{0,1\}^m \to \{0,1\}$.

Aaronson later proved the following result, that removes the log-factor at the expense of having randomized complexity on the left hand side.

Fact 2. $R^{A \to B}(f) \leq O(Q^{A \to B}(f) \cdot m)$ for all partial Boolean $f : \{0,1\}^n \times \{0,1\}^m \to \{0,1\}$.

Our first result is the following improvement.

Result 1. $D^{A \to B}(f) \leq O(Q^{A \to B,*}(f) \cdot m)$ for all partial Boolean $f : \{0,1\}^n \times \{0,1\}^m \to \{0,1\}$.

Hence we remove the log-factor, and we allow the quantum communication complexity on the right hand side to feature prior entanglement between Alice and Bob. Arguably, looking into the entanglement-assisted case (which is interesting for our second main result) led us to consider a more systematic progress measure than in Aaronson's proof, which in turn allowed us to analyze a different update rule for Bob that also works for protocols with error $1/3$, instead of extremely small error as used in [1], which is the cause of the lost log-factor.

We note that this result can be used to slightly improve on the "quantum-classical" simultaneous message passing lower bound for the Equality function by Gavinsky et al. [9], establishing a tight $\Omega(\sqrt{n})$ lower bound on the complexity of Equality in a model where quantum Alice and classical Bob (who do not share a public coin or entanglement) each send messages to the referee. The tight lower bound has also recently been established via a completely different and simpler method [8] (as well as generalized to a nondeterministic setting). Our result (as well as the one in [8]) allows a generalization to a slightly stronger model: Alice and the referee may share entanglement.

1.2   The Power of Quantum Proofs

We now turn to the second result of our paper, which is philosophically the more interesting. Interactive proof systems are a fundamental concept in computer science. Quantum proofs have a number of disadvantages: reading them may destroy them, errors may occur during verification, verification needs some sort of quantum machine, and it may be much harder to provide them than classical proofs. The main hope is that quantum proofs can in some situations be verified using fewer resources than classical proofs. Until now such a hope has not been verified formally. In the fully interactive setting Jain et al. have shown that the set of languages recognizable in polynomial time with the help of a quantum prover is equal to the set where the prover and verifier are classical (i.e., IP=QIP [11]).

The question remains open in the noninteractive setting. A question first asked by Aharonov and Naveh [4] and meriting much attention, is whether proofs that are quantum states can ever be easier to verify than classical proofs (by quantum machines) in the absence of interaction, i.e., whether the class QMA is larger than its analogue with classical proofs but quantum verifiers, known as QCMA. An indication that quantum proofs may be powerful was given by Watrous [22], who described an efficient QMA black-box algorithm for deciding nonmembership in a subgroup. However, Aaronson and Kuperberg [3] later showed how to solve the same problem efficiently using a classical witness, giving a QCMA black-box algorithm for the problem. They also introduced a quantum problem, for which they show that QMA black-box algorithms are more efficient than QCMA black-box algorithms. Using a quantum problem to show hardness for algorithms using classical proofs seems unfair though, and a similar separation has remained open for Boolean problems.

In our second main result we compare the two modes of noninteractive proofs and quantum verification for a Boolean function in the setting of one-way communication complexity. More precisely we exhibit a partial Boolean function f, such that the following holds. f can be computed in a protocol where a prover who knows x, y can provide a quantum proof to Alice, and Alice sends quantum message to Bob, such that the total message length (proof plus message Alice to Bob) is $O(\log n)$. In the setting where a prover Merlin (still knowing all inputs) sends a classical proof to Alice, who sends a quantum message to Bob, the total communication is $\Omega(\sqrt{n}/\log n)$.

Result 2. *There is a partial Boolean function f such that $QMA^{A \to B}(f) = O(\log n)$, while $QCMA^{A \to B,*}(f) = \Omega(\sqrt{n}/\log n)$.*

We note that this is the first known exponential gap between computing Boolean functions in a QCMA and a QMA mode in any model of computation. Also, the lower bound is not too far from being tight, since there is an obvious upper bound of $O(\sqrt{n}\log n)$ for the problem.

So where does the power of quantum proofs come from in our result? Raz and Shpilka [20] show that QMA one-way protocols are as powerful as QMA two-way protocols. Their proof uses a quantum witness that is a superposition over the messages of different rounds. We show that for a certain problem with an efficient QMA protocol there is no efficient one-way QCMA protocol. So in this case the weakness of classical proofs is due to the inability of Merlin to supply a randomized message about (only) Bob's input to Alice in a verifiable way. If the message is sent in superposition Bob can later test that the superposition is close to the correct one, while this is not possible for classical messages (for which Merlin can always cheat by choosing the best deterministic message).

2   Preliminaries

2.1   Quantum

For basic quantum background we refer to [19].

2.2   Communication Complexity Models

We assume familiarity with communication complexity, referring to [17] for more details about classical communication complexity and [24] for quantum communication complexity.

For a partial Boolean function $f : \{0,1\}^n \times \{0,1\}^m \times \{0,1,\bot\}$, where $\bot$ stands for "undefined" the communication matrix A_f has rows labeled by $x \in \{0,1\}^n$ and columns labeled by $y \in \{0,1\}^m$, and entries $f(x,y)$. A protocol for f is correct, if it gives the correct output for all x,y with $f(x,y) \neq \bot$ (with certainty for deterministic protocols, and with probability $2/3$ for quantum protocols). A protocol is one-way, if Alice sends a message to Bob, who computes the function value, or vice versa. We denote by $D^{A \rightarrow B}(f)$ the deterministic one-way communication complexity of a function f, when Alice sends the message to Bob.

Two rows x, x' of A_f are *distinct*, if there is a column y, such that $f(x,y) = 1$ and $f(x',y) = 0$ or vice versa, i.e., the function values differ on some defined input. Note that being not distinct is not an equivalence relation: x, x' can be not distinct, as well as x, x'', while x, x' are distinct. Nevertheless a one-way protocol for f needs to group inputs x into messages such that no two distinct x, x' share the same message.[1]

Similar to the above, we denote by $Q^{A \rightarrow B}(f)$ the quantum one-way communication complexity of f with error $1/3$. This notion is of course asymptotically robust when it comes to changing the error to any other constant. $Q^{A \rightarrow B,*}(f)$ denotes the complexity if Alice and Bob share entanglement.

We now define some more esoteric modes of communication that extend the standard nondeterministic mode to the quantum case. We restrict our attention to one-way protocols.

Definition 1. *In a one-way MA-protocol there are 3 players Merlin, Alice, Bob. Merlin sends a classical message to Alice, who sends a classical message to Bob, who gives the output. Alice and Bob share a public coin, which is not seen by Merlin. For a Boolean function $f : \{0,1\}^n \times \{0,1\}^m \rightarrow \{0,1\}$ the protocol is correct, if for all 1-inputs there is a message from Merlin, such that with probability $2/3$ Bob will accept, whereas for all 0-inputs, and all messages from Merlin, Bob*

---

[1] It is instructive to consider the function $f(x,i;y,j) = x_{i \oplus j}$ under the promise that $x = y$. This function has only n *distinct* rows and columns, and $D^{A \rightarrow B}(f) = D^{B \rightarrow A}(f) \leq O(\log n)$. Nevertheless A_f has many more actual rows and columns. Trying to reduce the number of actual columns to a set of distinct columns increases the number of distinct rows. Hence one has to be careful when considering partial functions.

will reject with probability 2/3. The communication cost of the protocol on an input is the sum of the lengths of the two messages from Merlin and Alice. The communication complexity is defined as usual and denoted by $MA^{A \to B}(f)$.

A one-way QCMA-protocol is defined similarly, but whereas Merlin's message is still classical, Alice can send a quantum message to Bob, and Alice and Bob may share entanglement. The complexity with shared entanglement is denoted $QCMA^{A \to B,*}(f)$.

In a one-way QMA-protocol also Merlin's message may be quantum. The complexity is denoted by $QMA^{A \to B}(f)$ in the case where no entanglement is allowed.

2.3   Quantum Information Measures

In this paper we need only a few well established notions of information and distinguishability. A density matrix is a positive semidefinite matrix of trace 1. Density matrices will also be referred to as quantum states in this paper.

Definition 2. *The von Neumann entropy of a quantum state ρ is defined as $S(\rho) = -\operatorname{Tr}\rho \log \rho$.*

The relative von Neumann entropy of quantum states ρ, σ is defined by $S(\rho||\sigma) = \operatorname{Tr}\rho \log \rho - \operatorname{Tr}\rho \log \sigma$ if supp $\rho \subseteq$ supp σ, otherwise $S(\rho||\sigma) = \infty$.

The relative min-entropy of ρ, σ is defined as $S_\infty(\rho||\sigma) = \inf\{c : \sigma - \rho/2^c \text{ is positive semidefinite}\}$.

It is easy to see that $S(\rho||\sigma) \leq S_\infty(\rho||\sigma)$, see [6] for a proof. An important measure of how far apart quantum states are is the trace distance.

Definition 3. *The trace norm of a Hermitian operator ρ is defined as $||\rho||_t = \operatorname{Tr}\sqrt{\rho\rho^\dagger}$.*

The trace distance between ρ and σ is $||\rho - \sigma||_t$.

We list two well known facts. First, Uhlmann monotonicity.

Fact 3. *If $\tilde{\rho}, \tilde{\sigma}$ result from measuring ρ, σ then $S(\tilde{\rho}||\tilde{\sigma}) \leq S(\rho||\sigma)$.*

Secondly, the quantum Pinsker inequality [10], see also [15].

Fact 4. $||\rho - \sigma||_t \leq \sqrt{2 \ln 2\, S(\rho||\sigma)}$.

We note that any two states that are close in trace distance are hard to distinguish by any measurement, namely the classical total variation distance between the measurement results is at most the trace distance of the measured states.

3   Making Quantum Messages Deterministic

Theorem 5. *For every partial Boolean function $f : \{0,1\}^n \times \{0,1\}^m \to \{0,1\}$ we have $D^{A \to B}(f) \leq O(Q^{A \to B,*}(f) \cdot m)$.*

We note that in the case of total functions the theorem follows from a result in [13] combined with Sauer's lemma [21], and that two weaker versions of the theorem have been proved by Aaronson: in [1] he shows the result with an additional log-factor on the right hand side, and without allowing entanglement, in [2] with $R^{A \to B}(f)$ on the left hand side (and no additional log-factor), but again without entanglement.

Our proof follows Aaronson's main approach in [1], in which Bob maintains a classical description of a quantum state as his guess for the message he should have received, and Alice informs him about inputs on which this state will perform badly, so that he can adjust his guess. His goal is to either get all measurement results approximately right, or to learn the message state. We will refer to these states as the current guess state, and the target state.

We deviate from Aaronson's proof in two ways. First, we work with a different progress measure that is more transparent than Aaronson's, namely the relative entropy between the target state and the current guess. This already allows us to work in the entanglement-assisted case.

Secondly, we modify the rule by which Bob updates his guess. In Aaronson's proof Bob projects his guess state onto the subspace on which the target state has a large projection (because the message is accepted by the corresponding measurement with high probability). This has the drawback that one cannot use the actual message state of the protocol as the target state, because that state usually has considerable projection onto the orthogonal complement of the subspace, making the relative entropy infinitely large! Hence Aaronson uses a boosted and projected message state as the target state. This state is close to the actual message state thanks to the boosting, and projection of the guess state now properly decreases the relative entropy, since the target state is fully inside the subspaces. The boosting step costs exactly the log-factor we aim to remove.

So in the situation where Bob wants to update his guess state σ, knowing that the target state ρ will be accepted with probability $1 - \epsilon$ when measuring the observable consisting of subspace V_y and its complement, we let Bob replace σ with the mixture of $1 - \epsilon$ times the projection onto V_y and ϵ times the projection onto $V_y^{\perp}$. The main part of the proof is then to show that this decreases the relative entropy $S(\rho||\sigma)$ given that $\text{Tr}(V_y \sigma) < 1 - 10\sqrt{\epsilon}$, i.e., in case σ was not good enough already. Eventually either all measurements can be done by Bob giving the correct result, or the current guess state σ satisfies $S(\rho||\sigma) \leq 5\sqrt{\epsilon}$, in which case ρ and σ are also close in the trace distance meaning that any future measurement will give almost the same results on both states.

Proof. Fix any entanglement-assisted one-way protocol with quantum communication $q = Q^{A \to B, *}(f)$. Using standard boosting we may assume that the error of the protocol is at most $\epsilon = 10^{-6}$ for any input x, y. This increases the communication by a small constant factor at most.

Using teleportation we can replace the quantum communication by $2q$ classical bits of communication at the expense of adding q EPR-pairs to the shared entangled state. Let $|\phi\rangle$ denote the entangled state shared by the new protocol. We can assume this is a pure state, because if this is not the case we may consider any purification, and Alice and Bob can ignore the purification part. Note that we do not restrict the number of qubits used in $|\phi\rangle$.

In the protocol, for a given input x Alice has to perform a unitary transformation on her part of $|\phi\rangle$ (we assume that any extra space used is also included in $|\phi\rangle$ and that measurements are replaced by unitaries) and then sends a classical message. Bob first applies the unitary from the teleportation protocol (which only depends on the classical message). Let's denote the state shared by Alice and Bob at this point by $|\phi_x\rangle$. Following this Bob performs a measurement (depending on his input y) on his part of $|\phi_x\rangle$. This measurement determines the output of the protocol on x, y. We may assume by standard techniques that Bob's measurements are projection measurements, and that the subspaces used in the projection measurements have dimension $d/2$, where d is the dimension of the underlying Hilbert space.

Recall that $|\phi\rangle$ and $|\phi_x\rangle$ are bipartite states shared by Alice and Bob. Let $\rho = \text{Tr}_A |\phi_x\rangle\langle\phi_x|$ and $\sigma_1 = \text{Tr}_A |\phi\rangle\langle\phi|$, i.e., the states when Alice's part is traced out. Bob wants to learn ρ in order to be able to determine the results of all of his measurements (for inputs y) on ρ. We show how to do this with $O(m \cdot q)$ bits of deterministic communication from Alice: Bob either gets to know an approximate classical description of ρ, or he will know the result of the measurements for all inputs y (such that $f(x, y)$ is defined). Note that the state σ_1 is known to Bob in the sense that he knows its classical description.

Since Alice's local operations do not change Bob's part of $|\phi\rangle\langle\phi|$, the difference between ρ and σ_1 is introduced via the correction operations in the teleportation protocol that Bob applies after he receives Alice's message. But with probability 2^{-2q} Bob does not have to do anything, i.e., when Alice's message is the all 0-s string. This implies that

$$\sigma_1 = \frac{1}{2^{2q}}\rho + \theta,$$

for some positive semidefinite θ with trace $1 - 1/2^{2q}$. Hence we get that

$$S(\rho\|\sigma_1) \leq S_\infty(\rho\|\sigma_1) \leq 2q.$$

In other words, Bob's target ρ and initial guess σ_1 have small relative entropy.

We can now describe the protocol. Bob starts with the classical description of σ_1. This state is also known by Alice, since it does not depend on the input. Throughout the protocol Bob will hold states σ_i, which will be updated when needed, using information provided by Alice. Bob also has a set of measurement operators $P_y, I - P_y$ for all his inputs y. Bob and Alice each loop over his inputs y, and compute $p_y = \text{Tr}(P_y \sigma_i)$. This is the acceptance probability, if σ_i is measured with the measurement for his input y. Alice also computes $p'_y = \text{Tr}(P_y \rho)$ which is the acceptance probability of the quantum protocol. If p_y and p'_y are too far apart, Alice will notify Bob of the correct acceptance probability on y (with precision ϵ^2), which takes $m + O(1)$ bits of communication.

Alice does not send a message if $f(x, y)$ is undefined, because the acceptance probability on such inputs is irrelevant. Suppose $p'_y = 1 - \epsilon_y$, where $\epsilon_y \leq \epsilon$ is the error on x, y (and $f(x, y) = 1$), but $p_y = 1 - a$ for some $1 > a \geq 10\sqrt{\epsilon}$. If this is not the case the measurement for y applied to σ_i already yields the correct result and no information from Alice is needed. So if p_y, p'_y are far apart Alice will send y (using m bits) and ϵ_y as a floating point number with precision ϵ^2 (using $O(1)$ bits).

Bob then adjusts σ_i to obtain a state σ_{i+1}. Suppose he knows the correct ϵ_y (the difference between ϵ_y and its approximation sent by Alice will be irrelevant). This means that $\mathsf{Tr}(P_y\sigma_i) = 1 - a$ but $\mathsf{Tr}(P_y\rho) = 1 - \epsilon_y$. P_y is the projector onto a subspace V_y. We have assumed that each V_y has dimension $d/2$ if d is the dimension of the underlying Hilbert space. Let B_i denote an orthonormal basis, in which the first $d/2$ elements span V_y, and the remaining $d/2$ span $V_y^{\perp}$, and in which the upper left and lower right quadrants of σ_i are diagonal. Hence

$$\sigma_i = \begin{pmatrix} A & B \\ B^* & D \end{pmatrix},$$

where A, D are diagonal and $\mathsf{Tr}(A) = 1 - a$ and $\mathsf{Tr}(D) = a$. We define

$$\sigma_{i+1} = \begin{pmatrix} \frac{1-\epsilon_y}{1-a}A & 0 \\ 0 & \frac{\epsilon_y}{a}D \end{pmatrix}.$$

σ_{i+1} is diagonal in our basis B_i. Clearly σ_{i+1} would perform exactly as desired on measurement $P_y, I - P_y$. But Bob already knows the function value on y and can carry on with the next y.

Before we continue we have to argue that the case $a = 1$ can never happen. Since $\mathsf{Tr}(P_y\rho) = 1 - \epsilon_y > 0$ the state ρ has a nonzero projection onto V_y. If $a = 1$ then σ_i sits entirely in $V_y^{\perp}$, and hence $S(\rho||\sigma_i) = \infty$. But since we start with a finite $S(\rho||\sigma_1)$ and only decrease that value the situation $a = 1$ is impossible.

Coming back to the protocol, it is obvious that Bob will learn the correct value of $f(x, y)$ for all y such that $f(x, y)$ is defined. Hence the protocol is deterministic and correct. The remaining question is how many times Alice has to send a message to Bob. We will show that this happens at most $O(Q^{A \to B, *}(f)/\sqrt{\epsilon})$ times, which establishes our theorem.

The main claim that remains to be shown is the following. For the proof we refer to the full version of this paper [16].

Claim. $S(\rho||\sigma_i) \geq S(\rho||\sigma_{i+1}) + a/2$ if $a \geq 10\sqrt{\epsilon}$.

This establishes the upper bound on the number of messages, because the relative entropy, which starts at $2q$ is decreased by $a/2 \geq 5\sqrt{\epsilon}$ for each message. After at most $2q/(5\sqrt{\epsilon})$ iterations the protocol has either ended (in which case Bob might never learn ρ, but will still know all measurement results), or we have $S(\rho||\sigma_T) \leq 5\sqrt{\epsilon}$.

To see this assume we are still in the situation of the claim. The claim states that the relative entropy can be reduced by $a/2$ as long as $a \geq 10\sqrt{\epsilon}$. So the process stops (assuming we don't run out of suitable y's) no earlier than when $S(\rho||\sigma_i) < a/2 \leq 5\sqrt{\epsilon}$.

But then by the quantum Pinsker inequality we have that at the final time T: $||\rho - \sigma_T||_t \leq \sqrt{10 \ln 2 \sqrt{\epsilon}} < 0.1$ in the end, and hence for *all* measurements their results are close. Hence no more than $O(q/\sqrt{\epsilon}) = O(q)$ messages have to be sent. □

4   Quantum versus Classical Proofs

Let us first define the problem for which we prove our separation result.

Definition 4. *The function* $\mathsf{Majlx}(x, I)$, *where* $I = \{i_1, \ldots, i_{\sqrt{n}}\}$, *each* $i_j \in \{1, \ldots, n\}$, *and* $x \in \{0, 1\}^n$ *is defined as follows:*

1. *if* $|\{j : x_{i_j} = 1\}| = \sqrt{n}$ *then* $\mathsf{Majlx}(x, I) = 1$,
2. *if* $|\{j : x_{i_j} = 1\}| \leq 0.9\sqrt{n}$ *then* $\mathsf{Majlx}(x, I) = 0$,
3. *otherwise* $\mathsf{Majlx}(x, I)$ *is undefined.*

The function has been studied in [14], where it is shown that one-way MA protocols for the problem need communication $\Omega(\sqrt{n})$. Our main technical result here is to extend this to one-way QCMA protocols.

It is obvious on the other hand, that there is a cheap protocol when Bob can send a message to Alice.

Lemma 1. $R^{B \to A}(\mathsf{Majlx}) = O(\log n)$.

Raz and Shpilka [20] show that any problem with $QMA(f) = c$ (i.e., QMA protocol where Alice and Bob can interact over many rounds) has a QMA protocol of cost $poly(c)$ in which Merlin sends a message to Alice, who sends a message to Bob. By inspection of their proof the polynomial overhead can be removed in the case of constant rounds of interaction between Alice and Bob.

Lemma 2. *If* $QMA(f) = c$ *and this cost can be achieved by a protocol with* $O(1)$ *rounds, then* $QMA^{A \to B}(f) = O(c)$.

We give more details in the full version [16]. The lemma immediately implies the following.

Theorem 6. $QMA^{A \to B}(\mathsf{Majlx}) = O(\log n)$.

We also give a self-contained proof of this fact in the full version [16]. Our protocol has completeness 1, hence even the one-sided error version of $QMA^{A \to B}$ is separated from $QCMA^{A \to B}$ by the following lower bound.

Theorem 7. $QCMA^{A \to B,*}(\mathsf{Majlx}) \geq \Omega(\sqrt{n}/\log n)$.

Hence we can conclude the following.

Corollary 1. *There is a partial Boolean function* f *such that* $QMA^{A \to B}(f) = O(\log n)$, *while* $QCMA^{A \to B,*}(f) = \Omega(\sqrt{n}/\log n)$.

Proof of Theorem 7. Fix any QCMA protocol $\mathcal{P}$ for Majlx. Furthermore define a distribution on inputs as follows. Fix any error correcting code $C \subseteq \{0,1\}^n$ with distance $n/4$ (i.e., every two codewords have Hamming distance at least $n/4$). Such codes of size $2^{\Omega(n)}$ exist by the Gilbert-Varshamov bound. We do not care about the complexity of decoding and encoding for our code. Furthermore we require the code to be balanced, i.e., that any codeword has exactly $n/2$ ones. This can also be achieved within the stated size bound. For our distribution on inputs first choose $x \in C$ uniformly, and then uniformly choose I among all subsets of $\{1, \ldots, n\}$ of size $\sqrt{n}$. Note that the probability of 1-inputs under the distribution μ just defined is between $2^{-\sqrt{n}}$ and $2^{-\sqrt{n}-1}$ due to the balance condition on the code.

If the cost (i.e., communication from Merlin plus communication from Alice) of $\mathcal{P}$ is c, then there are at most 2^c different classical proofs sent by Merlin. We identify such proofs p with the set of 1-inputs that are accepted by the protocol with probability at least $2/3$ when using the proof p. Hence there must be a proof P containing 1-inputs of measure at least $2^{-\sqrt{n}-c-1}$, because for every 1-input there is a proof with which it is accepted with probability $2/3$ or more. Furthermore, given P, no 0-input is accepted with probability larger than $1/3$. Note that inputs outside of the promise, or 1-inputs outside of P can be accepted with any probability between 0 and 1. Denote by f_P the partial function $\{0,1\}^n \times \{0,1\}^m \to \{0,1,\bot\}$, in which all inputs in P are accepted, and all 0-inputs of f are rejected, and the remaining inputs have undefined function value ($\bot$). $m = \Theta(\sqrt{n} \log n)$ is the length of Bob's input.

Obviously f_P can be computed by a one-way quantum protocol (without prover) using communication c (and possibly using shared entanglement between Alice and Bob). Now due to Theorem 5 this implies that $D^{A \to B}(f_P) \leq O(c \cdot m)$. We prove in the full version [16] that on the other hand $D^{A \to B}(f_P) \geq \Omega(n)$, and hence $c \geq \Omega(n/m) = \Omega(\sqrt{n}/\log n)$, which is our theorem.

$\square$

Acknowledgements. The authors thank Dmitry Gavinsky for discussions and the anonymous referees for helpful suggestions.

References

1. Aaronson, S.: Limitations of quantum advice and one-way communication. Theory of Computing 1, 1–28 (2005); Earlier version in Complexity 2004 (2004), quant-ph/0402095
2. Aaronson, S.: The learnability of quantum states. Proceedings of the Royal Society of London A463(2088) (2007), quant-ph/0608142
3. Aaronson, S., Kuperberg, G.: Quantum versus classical proofs and advice. Theory of Computing 3(1), 129–157 (2007)
4. Aharonov, D., Naveh, T.: Quantum NP - a survey (2002), quant-ph/0210077
5. Ambainis, A., Nayak, A., Ta-Shma, A., Vazirani, U.: Dense quantum coding and a lower bound for 1-way quantum automata. In: Proceedings of 31st ACM STOC, pp. 697–704 (1999)

6. Datta, N.: Min- and max- relative entropies and a new entanglement monotone. IEEE Transactions on Information Theory 55, 2816–2826 (2009)
7. Gavinsky, D., Kempe, J., Kerenidis, I., Raz, R., de Wolf, R.: Exponential separation for one-way quantum communication complexity, with applications to cryptography. SIAM J. Comput. 38(5), 1695–1708 (2008)
8. Gavinsky, D., Klauck, H.: Equality, Revisited. Manuscript (2014)
9. Gavinsky, D., Regev, O., de Wolf, R.: Simultaneous Communication Protocols with Quantum and Classical Messages. Chicago Journal of Theoretical Computer Science 7 (2008)
10. Hiai, F., Ohya, M., Tsukada, M.: Sufficiency, KMS condition and relative entropy in von Neumann algebras. Pacific J. Math. 96, 99–109 (1981)
11. Jain, R., Ji, Z., Upadhyay, S., Watrous, J.: QIP = PSPACE. J. ACM 58(6) (2011)
12. Jain, R., Zhang, S.: New bounds on classical and quantum one-way communication complexity. Theoretical Computer Science 410(26), 2463–2477 (2009)
13. Klauck, H.: On quantum and probabilistic communication: Las Vegas and one-way protocols. In: Proceedings of 32nd ACM STOC, pp. 644–651 (2000)
14. Klauck, H.: On Arthur Merlin Games in Communication Complexity. In: IEEE Conference on Computational Complexity, pp. 189–199 (2011)
15. Klauck, H., Nayak, A., Ta-Shma, A., Zuckerman, D.: Interaction in Quantum Communication. IEEE Transactions on Information Theory 53(6), 1970–1982 (2007)
16. Klauck, H., Podder, S.: Two Results about Quantum Messages (Full Version). arXiv:1402.4312 (2014)
17. Kushilevitz, E., Nisan, N.: Communication Complexity. Cambridge University Press (1997)
18. Nayak, A.: Optimal lower bounds for quantum automata and random access codes. In: Proceedings of 40th IEEE FOCS, pp. 369–376 (1999), quant-ph/9904093
19. Nielsen, M.A., Chuang, I.L.: Quantum Computation and Quantum Information. Cambridge University Press (2000)
20. Raz, R., Shpilka, A.: On the power of quantum proofs. In: Proceedings of Computational Complexity, pp. 260–274 (2004)
21. Sauer, N.: On the density of families of sets. J. Combin. Theory Ser. A 13, 145–147 (1972)
22. Watrous, J.: Succinct quantum proofs for properties of finite groups. In: Proceedings of 41st IEEE FOCS, pp. 537–546 (2000), quant-ph/0011023
23. Winter, A.: Quantum and classical message identification via quantum channels. In: Festschrift A.S. Holevo 60, pp. 171–188 (2004)
24. de Wolf, R.: Quantum communication and complexity. Theoretical Computer Science 287(1), 337–353 (2002)

Parameterized Approximations
via d-SKEW-SYMMETRIC MULTICUT

Sudeshna Kolay[1], Pranabendu Misra[1], M. S. Ramanujan[2],
and Saket Saurabh[1,2]

[1] The Institute of Mathematical Sciences, Chennai, India
{skolay,pranabendu,saket}@imsc.res.in
[2] University of Bergen, Bergen, Norway
Ramanujan.Sridharan@ii.uib.no

Abstract. In this paper we design polynomial time approximation
algorithms for several parameterized problems such as ODD CYCLE
TRANSVERSAL, ALMOST 2-SAT, ABOVE GUARANTEE VERTEX COVER
and DELETION q-Horn BACKDOOR SET DETECTION. Our algorithm proceeds by first reducing the given instance to an instance of the d-SKEW-
SYMMETRIC MULTICUT problem, and then computing an approximate
solution to this instance. Our algorithm runs in polynomial time and
returns a solution whose size is bounded quadratically in the parameter,
which in this case is the solution size, thus making it useful as a first
step in the design of kernelization algorithms. Our algorithm relies on
the properties of a combinatorial object called (L, k)-set, which builds
on the notion of (L, k)-components, defined by a subset of the authors
to design a linear time FPT algorithm for ODD CYCLE TRANSVERSAL.
The main motivation behind the introduction of this object in their work
was to replicate in skew-symmetric graphs, the properties of important
separators introduced by Marx [2006] which has played a very significant role in several recent parameterized tractability results. Combined
with the algorithm of Reed, Smith and Vetta, our algorithm also gives
an alternate linear time algorithm for ODD CYCLE TRANSVERSAL. Furthermore, our algorithm significantly improves upon the running time of
the earlier parameterized approximation algorithm for DELETION q-Horn
BACKDOOR SET DETECTION which had an exponential dependence on
the parameter; albeit at a small cost in the approximation ratio.

1   Introduction

Let $\rho : \mathbb{N} \to \mathbb{R}_{\geq 1}$ be a computable function[1]. We say that a parameterized
problem Π admits a $\rho(k)$-parameterized approximation if there exists an algorithm $\mathbb{A}$ that, given an input (x, k) to Π, either outputs y such that
$|y| \geq k/\rho(k)$ (if Π is a maximization problem) or $|y| \leq k\rho(k)$ (if Π is a minimization problem) or returns that "$(x, k) \notin \Pi$'. The algorithm $\mathbb{A}$ is allowed to

---

[1] Where $\mathbb{N}$ and $\mathbb{R}_{\geq 1}$ denote the set of natural numbers and the set of real numbers at
least 1, respectively.

E. Csuhaj-Varjú et al. (Eds.): MFCS 2014, Part II, LNCS 8635, pp. 457–468, 2014.
© Springer-Verlag Berlin Heidelberg 2014

take $f(k)|x|^{\mathcal{O}(1)}$ time. However, in this paper, we will be interested in obtaining parameterized approximation algorithms where the function f is actually polynomial.

There are several problems parameterized by the solution size, for which the best known approximation algorithm is in fact also a parameterized approximation algorithm. For example, the polynomial time approximation algorithm for DIRECTED FEEDBACK VERTEX SET [4] as well as the constant factor approximation algorithms. Furthermore, there are problems whose parameterized complexity is unknown or it is known to be W[1]-hard, but admit a parameterized approximation. We cite the examples of DIRECTED DISJOINT CYCLE PACKING, which is both hard to approximate and W[1]-hard but admits a parameterized approximation [9], and CLIQUEWIDTH, whose parameterized complexity is unknown but has a $(2^{k+1} - 1)$-parameterized approximation [10]. It is also important to point out some recent approximations for crossing number, and euler-genus of an input graph and for the EDGE PLANARIZATION and VERTEX PLANARIZATION problems [2,3].

Recently, *polynomial time* parameterized approximation algorithms have also been used for the purpose of kernelization. For instance, the polynomial kernelization algorithm for ODD CYCLE TRANSVERSAL (OCT) by Kratsch and Wahlstrom [12] requires an approximate solution for the problem whose size is bounded polynomially in the size of the parameter. In order to obtain such a solution, they combined an existing $2^{\mathcal{O}(k)}n^{\mathcal{O}(1)}$ fixed parameter tractable (FPT) algorithm [15] with the $\mathcal{O}(\sqrt{\log n})$ approximation algorithm of Agarwal et al. [1]. In general, this approach requires a pair of an FPT algorithm and an approximation algorithm such that, the complexity of the FPT algorithm "matches" the approximation factor. For instance, if for a parameterized problem Π, the best known FPT algorithm runs in time $f(k)n^{\mathcal{O}(1)}$ where $f(.)$ is a tower of ℓ 2's, then the approximation factor required would by $\log \ldots \log n$, where the depth is ℓ. It is possible that such a pair of algorithms may not even exist for the problem Π. This motivates the question of designing polynomial time parameterized approximation algorithms. Furthermore, many of the existing approximation algorithms (e.g. for OCT) require the use of fairly sophisticated techniques and tools. However it might be possible to exploit the combinatorial structure of the parameterized problem to design a much simpler algorithm.

With these motivations in mind, we design an approximation algorithm for the d-SKEW-SYMMETRIC MULTICUT problem which runs in polynomial time such that any solution returned by the algorithm has size bounded quadratically in the parameter. This algorithm, combined with known parameter preserving reductions from OCT, ALMOST 2-SAT, ABOVE GUARANTEE VERTEX COVER and DELETION q-Horn BACKDOOR SET DETECTION ([14]) gives parameterized approximation algorithms for all these problems. In particular, since OCT and ALMOST 2-SAT are special cases of 1-SKEW-SYMMETRIC MULTICUT, we get a factor - $2(k + 1)$ approximation and since DELETION q-Horn BACKDOOR SET DETECTION is known to be a special case of 3-SKEW-SYMMETRIC MULTICUT, we get a factor - $6(k + 1)$ approximation. Gaspers et al. [6] showed that a special case of d-SKEW-SYMMETRIC MULTICUT, DELETION q-Horn BACK-DOOR SET DETECTION, has a factor-$(k + 1)$approximation algorithm running in time $\mathcal{O}(6^k n^2)$. Observe that our corollary for DELETION q-Horn BACKDOOR SET DETECTION improves on the exponential dependence on the parameter in the running time of the algorithm of Gaspers et al. [6] significantly with only a constant factor increase in the approximation factor. Furthermore, the algorithm for OCT we get from our theorem can be patched with the algorithm of Reed, Smith and Vetta to give an alternate linear time FPT algorithm for OCT albeit with a worse dependence on the parameter compared with those in [14,11].

A skew-symmetric graph $(D = (V, A), \sigma)$ is a directed graph D with an involution σ on the set of vertices and arcs, that is for all $x \in V \cup A$, $\sigma(\sigma(x)) = x$. Flows on skew-symmetric graphs have been used to generalize maximum flow and maximum matching problems on graphs, initially by Tutte [16], and later by Goldberg and Karzanov [7]. The d-SKEW-SYMMETRIC MULTICUT problem was introduced recently in [14] where it was shown to generalize several well studied classical problems including OCT, ALMOST 2-SAT, DELETION q-Horn BACKDOOR SET DETECTION for fixed values of d. The same work included a linear time FPT algorithm for this problem. The d-SKEW-SYMMETRIC MULTICUT problem is a variant of the MULTICUT problem on skew-symmetric graphs, defined as follows.

---

d-SKEW-SYMMETRIC MULTICUT   **Parameter:** k
Input: A skew-symmetric graph $D = ((V, A), \sigma)$, a family $\mathcal{T}$ of d-sets of vertices, integer k.
Question: Is there a set $S \subseteq A$ such that $S = \sigma(S)$, $|S| \leq 2k$, and for any d-set $\{v_1, \ldots, v_d\}$ in $\mathcal{T}$, there is a vertex v_i such that v_i and $\sigma(v_i)$ lie in distinct strongly connected components of $D \setminus S$?

The set S in the above definition is called a *skew-symmetric multicut* for the given instance. Our main result in this paper is the following.

Theorem 1. *There is an algorithm that, given an instance $(D = (V, A, \sigma), \mathcal{T}, k)$ of d-SKEW-SYMMETRIC MULTICUT, runs in time $\mathcal{O}(k^5(\ell + m + n))$ and either returns a skew-symmetric multicut of size at most $2d(k^2 + k)$ or correctly concludes that no such set of size at most $2k$ exists. Here $m = |A|$, $n = |V|$, and ℓ, the length of the family $\mathcal{T}$, is defined as $d \cdot |\mathcal{T}|$.*

Overview of the Algorithm. At a high level, we follow the approach of Gaspers et al. [6]. However, here we need to introduce a combinatorial object that is specifically tailored to skew-symmetric graphs, called an (L, k)-set which has two structural properties that point us to an approximate solution.

- Every (L, k)-set has a small "boundary" and intersects every solution for the given instance.
- Removing the "boundary" of any (L, k)-set reduces the size of an optimal solution for the residual instance.

The notion of (L, k)-sets builds upon that of (L, k)-components defined in [14]. However, the notion of (L, k)-components is not strong enough to guarantee either of the above properties. We will show that given these two properties, if we can always find such an (L, k)-set (if it exists) in polynomial time, then we can get an algorithm which either concludes that there is no solution of the required size or returns an approximate solution whose size is bounded by $\mathcal{O}(k^2)$. Therefore, once we prove both the structural properties, we give an algorithm to compute *one* such (L, k)-set at every step. However, the running time of the algorithm to compute this set must have only a polynomial dependence on the input as well as the parameter. For this, we build upon a lemma from [14] which can only check for the existence of a special kind of (L, k)-sets and generalize this algorithm to check for the existence of an (L, k)-set and compute it if it exists.

2 Preliminaries

In this section we give some basic definitions and set up the notations for the paper.

Digraphs. Let $D = (V, A)$ be a directed graph. For an arc $(u, v) \in A$, we refer to u as the *tail* of this arc and we refer to v as the *head* of this arc. For a set of vertices V', we let $A[V']$ denote the set of arcs with both end points in the set V'. For a set of vertices V', we let $\delta^+(V')$ denote the set of arcs which have their tail in V' and their head in $V \setminus V'$. Similarly, we let $\delta^-(V')$ denote the set of arcs which have their head in V' and their tail in $V \setminus V'$. Given two disjoint vertex sets X and Y, we define an X-Y path as a directed path from a vertex $x \in X$ to a vertex $y \in Y$ whose internal vertices are disjoint from $X \cup Y$.

Skew-Symmetric Graphs. This notation is from [8]. A *skew-symmetric graph* is a digraph $D = (V, A)$ and an involution $\sigma : V \cup A \to V \cup A$ such that:

1. for each $x \in V \cup A$, $\sigma(x) \neq x$
2. for each $v \in V$, $\sigma(v) \in V$
3. for each $a = (v, w) \in A$, $\sigma(a) = (\sigma(w), \sigma(v))$

We call $\sigma(x)$ *symmetric* to x and also refer to x and $\sigma(x)$ as *conjugates*. For ease of description, we let x' denote the conjugate of an element x and we let S' denote the set of conjugates of the elements in the set S. We say that a set S is *regular* if $S \cap S' = \emptyset$ and *irregular* otherwise. A set S is called *self-conjugate* if $S = S'$.

3 Skew-Symmetric Graphs, Separators and Components

We require the following observations and definitions from [14] regarding skew-symmetric graphs.

Observation 1. *Let $D = ((V, A), \sigma)$ be a skew-symmetric graph and let $u, v \in V$. There is a path from v to u in D if and only if there is a path from u' to v'.*

Definition 1. *Let $D = (V, A)$ be a directed graph and let X, Y be disjoint subsets of V. A set $S \subseteq A$ is an X-Y separator if there is no directed path from X to Y in the graph $D \setminus S$. We say that S is a minimal X-Y separator if no proper subset of S is an X-Y separator.*

Definition 2. *Let $D = ((V, A), \sigma)$ be a skew-symmetric graph and let L be a regular set of vertices. Let $X \subseteq A$ be a self-conjugate set of arcs of D. We call X an L-L' **self-conjugate separator** if X is a (not necessarily minimal) L-L' separator. We call X a minimal L-L' self-conjugate separator if there is no self-conjugate strict subset of X which is also an L-L' separator.*

Definition 3. *Let $D = ((V, A), \sigma)$ be a skew-symmetric graph and let L be a regular set of vertices. Let X be an L-L' self-conjugate separator. We denote by $R(L, X)$ the set of vertices of D that can be reached from L via directed paths in $D \setminus X$, and we denote by $\bar{R}(L, X)$ the set of vertices of D which have a directed path to L in $D \setminus X$.*

Observation 2. *Let $D = ((V, A), \sigma)$ be a skew-symmetric graph and let L be a regular set of vertices. Let X be an L-L' self-conjugate separator. Then, the sets $R(L, X)$ and $\bar{R}(L', X)$ are also regular and $\sigma(R(L, X)) = \bar{R}(L', X)$.*

Proof. Since deleting a self-conjugate set of arcs from a skew-symmetric graph results in a skew-symmetric graph, we know that there is a path from u to v in $D \setminus X$ if and only if there is a path from v' to u' in $D \setminus X$. Therefore, if $R(L, X)$ is irregular, then there is a path from L to y and y' for some vertex y, which is disjoint from X, which implies a path from L to L' in $D \setminus X$, which is a contradiction. Therefore, $R(L, X)$ and $\bar{R}(L', X)$ are regular and since $D \setminus X$ is a skew-symmetric graph, they are conjugates. $\qquad\square$

3.1 The Notion of (L, k)-Sets

The following definition defines an object called (L, k)-set, whose properties and the computation of which will form the main part of our paper.

Definition 4. *Let $D = ((V, A), \sigma)$ be a skew-symmetric graph and $k \in \mathbb{N}$. Let $L \subseteq V$ be a regular set of vertices. A set of vertices $Z \subseteq V$ is called an (L, k)-set if it satisfies the following properties.*

1. $L \subseteq Z$
2. Z is regular
3. Z is reachable from L in $D[Z]$
4. *The size of a minimum Z-Z' separator is at most $2k$.*
5. Z is inclusion-wise maximal among the sets satisfying the above properties.

The above definition generalizes the notion of (L, k)-components because it follows from the definition of (L, k) components (see Definition 4, [14]) that a set $Z \subseteq V$ is an (L, k)-component if it is an (L, k)-set and the size of a minimum Z-Z' separator is equal to the size of a minimum L-L' separator. A reader familiar with the notion of important separators [13] can interpret an (L, k)-component as a "skew-symmetric" analogue of a smallest important separator between L and L'.

Observation 3. *Given a skew-symmetric graph $D = ((V, A), \sigma)$ and a regular set L of vertices, L satisfies properties 1-3 of Definition 4.*

The following lemma shows that the *neighborhood* of an (L, k)-component is also a minimum L-L' separator.

Lemma 1. *Given a skew-symmetric graph $D = ((V, A), \sigma)$ and a regular set L of vertices, let Z be a (L, k) component. Then $\delta^+(Z)$ is a minimum Z-Z' separator.*

Proof. From the definition of (L, k) components we know that a minimum Z-Z' separator is also a minimum L-L' separator. Let X be such a minimum separator. Also, $\delta^+(Z)$ is a Z-Z' separator.

If there is a conjugate pair $(u, v), (v', u') \in \delta^+(Z)$ then $u, v' \in Z$ and $u', v \in Z'$ and any Z-Z' separator must contain both arcs of the conjugate pair. Now suppose that $(u, v) \in \delta^+(Z) \setminus X$ but $(v', u') \notin \delta^+(Z)$. By definition of skew symmetric graphs, $(v', u') \in \delta^-(Z')$. Therefore, X must hit all v-v' paths in D. In other words, X is a $Z \cup \{v\}$-$Z' \cup \{v'\}$ separator. It is easy to verify that $Z \cup \{v\}$ satisfies properties 1-4 of Definition 4. Since X is a $Z \cup \{v\}$-$Z' \cup \{v'\}$ separator, this contradicts the fact that Z was a (L, k) component. Hence, (v', u') must be X. But this implies that $|\delta^+(Z)| \leq |X|$. As X is a minimum Z-Z' separator, $\delta^+(Z)$ is in fact a minimum Z-Z' separator.   □

4   Structural Properties and Computation of (L, k)-Sets

In this section, we first give a formal proof of the utility of (L, k)-sets. Following this, we give an algorithm for the computation of an (L, k)-set.

Lemma 2. *Let $(D = (V, A), \sigma, \mathcal{T}, k)$ be a YES instance of d-SKEW-SYMMETRIC MULTICUT and let L be a regular set of vertices such that there is an L-L' path in D. If there is a solution S for the given instance which is an L-L' self-conjugate separator in D, then the following hold.*

1. *If Z is an (L, k)-set, then $S \cap (A[Z] \cup \delta^+(Z)) \neq \emptyset$ and the instance $(D \setminus (\delta^+(Z) \cup \delta^-(Z')), \sigma, \mathcal{T}, k - 1)$ is a YES instance.*
2. *If X is an irregular L-L' separator, then $S \cap (A[Z] \cup X) \neq \emptyset$ where $Z = R(L, X \cup X')$ and the instance $(D \setminus (X \cup X'), \sigma, \mathcal{T}, k - 1)$ is a YES instance.*

Proof. For the first statement, suppose that there are no arcs of the solution S in the set $A[Z] \cup \delta^+(Z)$ and let $K = R(L, S)$. Clearly, $K \supseteq Z \cup N^+(Z)$, and satisfies the first 4 properties of an (L, k)-set, which implies that K contradicts the maximality of Z. Therefore, we conclude that $S \cap (A[Z] \cup \delta^+(Z))$ is non-empty and denote this set by F.

Suppose that $S \cap A[Z]$ is empty. Then, $F \cap \delta^+(Z)$ is non-empty and therefore, the set $\hat{S} = S \setminus (F \cup F')$ has size at most $2k - 2$ and is clearly a skew-symmetric multicut for the instance $(D \setminus (\delta^+(Z) \cup \delta^-(Z')), \sigma, \mathcal{T}, k - 1)$.

Now, suppose that $S \cap A[Z]$ is non-empty. Then, the set $\hat{S} = S \setminus (F \cup F')$ has size at most $2k - 2$. Furthermore, if $\hat{S}$ were not a skew-symmetric multicut for the instance $(\hat{D} = D \setminus (\delta^+(Z) \cup \delta^-(Z')), \sigma, \mathcal{T}, k - 1)$, then there is a violating set $T \in \mathcal{T}$ such that for every $t \in T$, t and t' are in the same strongly connected component of $\hat{D} \setminus \hat{S}$. Fix any such $t \in T$. Since S is a solution for the given instance, it must be the case that there is a closed walk in the graph $\hat{D} \setminus \hat{S}$ containing t and t' which also intersects the set $F \cup F'$. Let $(a, b) \in F$ which appears in this closed walk. Observe that $(a, b) \in F$ implies that $(a, b) \in A[Z]$, which in turn implies that a is in the set $R(L, \delta^+(Z) \cup \delta^-(Z'))$. Since a lies on a closed walk containing t and t', we conclude that there is a path from L to t and t' in the graph $D \setminus (\delta^+(Z) \cup \delta^-(Z'))$. But, by Observation 1, we get a path from t' to L' in the same graph, which is a contradiction since we now have a path from L to L' in D which is also disjoint from the set $\delta^+(Z)$. This completes the proof of the first statement.

The proof of the second statement is similar. Suppose that $S \cap A[Z] = \emptyset$. By Lemma 4, $|\delta^+(Z)| = |X|$. Consider those arcs in $\delta^+(Z)$ such that the conjugate is not in $\delta^+(Z)$, but the pair in contained in X. Let this set of arcs be Y. It is easy to verify that $|\delta^+(Z) \setminus A| = |X \setminus (Y \cup Y')|$. If the set Y is nonempty then $|\delta^+(Z)| < |X|$, which is not possible. So, if an arc and its conjugate is contained in X then the pair is also contained in $\delta^+(Z)$. Since X is irregular, there is an arc $(u, v) \in X$ such that $(u, v), (v', u') \in X$, which implies that $(u, v), (v', u') \in \delta^+(Z)$. In fact, by definition of a skew symmetric graph, $(u, v), (v', u') \in \delta^+(Z) \cap \delta^-(Z')$. Therefore, we have that u is in Z and v is in Z'. Since $S \cap A[Z] = \emptyset$ and S is an L-L' separator, it must be the case that $(u, v), (v', u') \in S$, that is $X \cap S \neq \emptyset$. Observe that in this case, the set $\hat{S} = S \setminus (X \cup X')$ has size at most $k - 2$ and is a skew symmetric multicut for the instance $\hat{I} = I \setminus (X \cup X')$.

Observe that if $S \cap A[Z] = F \neq \emptyset$, then the set $\hat{S} = S \setminus (F \cup F')$ has size at most $k - 2$. Furthermore, we claim that $\hat{S}$ is a skew symmetric multicut for the instance $(\hat{D} = D \setminus (X \cup X'), \sigma, \mathcal{T}, k - 1)$. If this were not the case, then there is a violating set $T \in \mathcal{T}$ such that for every $t \in T$, t and t' are in the same strongly connected component of $\hat{D} \setminus \hat{S}$. Fix a $t \in T$. Since S is a solution for the given instance, it must be the case that there is a closed walk containing t and t' which also intersects the set $F \subseteq A[Z]$. That is, there is an arc $(a, b) \in F$ such that it is contained in a closed walk along with t and t'. Since $a \in Z = R(L, X \cup X')$, we have that there are paths from L to t and t' in the graph $\hat{D}$, implying a path from L to L' in D disjoint from X, a contradiction. This completes the proof of the lemma. □

4.1   Computation of (L, k)-Sets

In this subsection, we give an algorithm (Lemma 3) that in linear time either computes an (L, k)-set or finds an irregular L-L' separator.

Proposition 1. ([14]) *Let* $D = ((V, A), \sigma)$ *be a skew-symmetric graph and* $k \in \mathbb{N}$. *Let* $L \subseteq V$ *be a regular set of vertices such that there is an* $L\text{-}L'$ *path in* D. *There is an algorithm which runs in time* $\mathcal{O}(k^3(m+n))$ *and*

- *correctly concludes that no* (L, k)-*component exists or*
- *returns an* (L, k)-*component or*
- *returns an irregular minimum* $L\text{-}L'$ *separator (that is, the separator is irregular)*

where $m = |A|$ *and* $n = |V|$.

The following is the main lemma of this section.

Lemma 3. *Let* $D = ((V, A), \sigma)$ *be a skew-symmetric graph and* $k \in \mathbb{N}$. *Let* $L \subseteq V$ *be a regular set of vertices such that there is an* $L\text{-}L'$ *path in* D. *There is an algorithm which runs in time* $\mathcal{O}(k^4(m+n))$ *and*

- *correctly concludes that no* (L, k)-*set exists or*
- *returns an* (L, k)-*set or*
- *returns an irregular* $L\text{-}L'$ *separator of size at most* $2k$

where $m = |A|$ *and* $n = |V|$.

Proof. **Description of the Algorithm.** We begin by running the algorithm of Lemma 1 to find an (L, k)-component. If the algorithm concluded that there is no (L, k)-component, then from Onservation 3 we know that the size of a minimum $L\text{-}L'$ separator exceeds $2k$. So we can also conclude that there is no (L, k)-set in D. Similarly, if this algorithm returned an irregular minimum $L\text{-}L'$ separator (which must be of size at most $2k$ since an (L, k)-component *does* exist), then we are done. Therefore, assume that the algorithm returned an (L, k)-component Z. We then check if there is an $a \in N^+(Z)$ such that the size of a minimum $Z \cup \{a\}\text{-}Z' \cup \{a'\}$ separator is at most $2k$. If there is such an a, then we recursively compute and return either a $(Z \cup \{a\}, k)$-set H or an irregular $Z \cup \{a\}\text{-}Z' \cup \{a'\}$ separator X of size at most $2k$ (one of which must exist). If there is no such a, then we return Z.

Proof of Correctness. Suppose that there is no $a \in N^+(Z)$ such that the size of a minimum $Z \cup \{a\}\text{-}Z' \cup \{a'\}$ separator is at most $2k$. This implies that there is no (L, k)-set which strictly contains Z, since for any $K \supset Z$, a $K\text{-}K'$ separator is also a $Z\text{-}Z'$ separator. Therefore, the algorithm is correct in concluding that Z itself is an (L, k)-set. Now, suppose that for some $a \in N^+(Z)$ the size of a minimum $Z \cup \{a\}\text{-}Z' \cup \{a'\}$ separator is at most $2k$. Suppose that we recursively obtained an irregular $Z \cup \{a\}\text{-}Z' \cup \{a'\}$ separator of size at most $2k$. Then this is also clearly an irregular $L\text{-}L'$ separator of size at most $2k$. Finally, suppose that we recursively obtained a $(Z \cup \{a\}, k)$-set K. We claim that K is also an (L, k)-set. If this were not the case, then there is a set $H \supset K$ which is an (L, k)-set. However, this implies that H is also a $(Z \cup \{a\}, k)$-set, which contradicts the maximality of K as $(Z \cup \{a\}, k)$-set. This completes the proof of correctness.

Running Time Analysis.   Observe that the time taken at each step of the recursion is bounded by the time required to apply Lemma 1, which is $\mathcal{O}(k^3(m+n))$ and the time to check for each $a \in N^+(Z)$ if there is a minimum $Z \cup \{a\}$-$Z' \cup \{a'\}$ separator of size at most $2k$. The latter requires $|N^+(Z)|$ applications, of $2k + 1$ steps of the Ford-Fulkerson Algorithm [5], each of which takes time $\mathcal{O}(k(m + n))$. Therefore checking for the presence of an $a \in N^+(Z)$ such that there is a minimum $Z \cup \{a\}$-$Z' \cup \{a'\}$ separator of size at most $2k$ can be done in time $\mathcal{O}(k^2(m + n))$, since $|N^+(Z)| \leq 2k$ (follows from Lemma 1). Now, it only remains for us to bound the recursion depth. Observe that due to the maximality of Z as an (L, k)-component, for any $a \in N^+(Z)$, the size of a minimum $Z \cup \{a\}$-$Z' \cup \{a'\}$ separator is strictly greater than the size of a minimum L-L' separator. Therefore, the depth of the recursion is bounded by $2k$ and hence the running time of the algorithm is $\mathcal{O}(k^4(m + n))$. This completes the proof of the lemma.
□

The following property of the algorithm of the above lemma will be required to prove the correctness of our approximation algorithm.

Lemma 4. *Suppose that the algorithm of Lemma 3 returned an irregular L-L' separator X of size at most $2k$ and let $Z = R(L, X \cup X')$. Then, $\delta^+(Z)$ is a minimum Z-Z' separator and $|\delta^+(Z)| = |X|$.*

Proof. Since $X \cup X'$ is a self-conjugate L-L' separator, from Observation 2, Z is regular and $Z' = \bar{R}(L', X)$. From definition of Z, we know that $\delta^+(Z) \subseteq X \cup X'$. Suppose there is a conjugate pair of arcs $(u, v), (v', u') \in \delta^+(Z)$. Then $u, v' \in Z$ and $v, u' \in Z'$. From the definition of Z, there is a path P_1 from L to u and a P_2 from L to v'. From Observation 1, there is a path P_1' from u' to L' and P_2' from v to L'. To hit the paths $P_1 u v P_2'$ and $P_2 v' u' P_1'$ both (u, v) and (v', u') have to belong to X. Therefore, for any arc a, where $a \in X, a' \notin X$, we can have at most one of a or a' in $\delta^+(Z)$. This shows that $|\delta^+(Z)| \leq |X|$.

From the algorithm there are two ways in which we can get an irregular L-L' separator as an output:

- The algorithm of Lemma 1 returns X in the very first step of execution. Then X is a minimum irregular L-L' separator. Moreover, we know that an Z-Z' separator is also an L-L' separator. In particular, $\delta^+(Z)$ is an L-L' separator. From the above argument that $|\delta^+(Z)| \leq |X|$, which is a Z-Z' separator, and the optimality of X as an L-L' separator, we know that $\delta^+(Z)$ is a minimum Z-Z' separator and that $|\delta^+(Z)| = |X|$, otherwise we contradict the optimality of X as a L-L' separator.
- The algorithm of Lemma 1 returns a (L, k)-component Y_0 in the first step of execution and $\exists a_0 \in N^+(Y_0)$ such that a minimum $(Y_0 \cup \{a_0\})$-$(Y_0' \cup \{a_0'\})$ separator is of size at most $2k$. Recursively, Lemma 3 returns an irregular $(Y_0 \cup \{a_0\})$-$(Y_0' \cup \{a_0'\})$ separator X after ℓ steps of recursion. In each step i of recursion, there is a vertex $a_{i-1} \in N^+(Y_{i-1})$ such that a minimum $(Y_{i-1} \cup \{a_{i-1}\})$-$(Y_{i-1}' \cup \{a_{i-1}'\})$ separator is of size at most $2k$. Also, for every $i < \ell$, Y_i is a $(Y_{i-1} \cup \{a_{i-1}\}, k)$ component. Finally, in the ℓth step of

recursion an irregular minimum $(Y_{\ell-1} \cup \{a_{\ell-1}\})$-$(Y'_{\ell-1} \cup \{a'_{\ell-1}\})$ separator X is returned. $(Y_{\ell-1} \cup \{a_{\ell-1}\}) \subseteq Z$; so a Z-Z' separator is also a $(Y_{\ell-1} \cup \{a_{\ell-1}\})$-$(Y'_{\ell-1} \cup \{a'_{\ell-1}\})$ separator. Again, by optimality of $(Y_{\ell-1} \cup \{a_{\ell-1}\})$-$(Y'_{\ell-1} \cup \{a'_{\ell-1}\})$ separator X and from $|\delta^+(Z)| \leq |X|$, we obtain that $\delta^+(Z)$ is a minimum Z-Z' separator and that $|\delta^+(Z)| = |X|$.

This completes the proof of the lemma.   □

5 Approximation Algorithm for d-SKEW-SYMMETRIC MULTICUT

In this section we design our approximation algorithm for d-SKEW-SYMMETRIC MULTICUT.

From this point on, we assume that an instance of d-SKEW-SYMMETRIC MULTICUT is of the form $(D = (V, A), \sigma, T, k, L)$ where L is a regular set of vertices and the question is to check *if there is a solution for the given instance which is an L-L' self-conjugate separator*. To solve the problem on the given input instance, we simply solve it on the instance $(D = (V, A), \sigma, \mathcal{T}, k, \emptyset)$.

We are now ready to prove Theorem 1 by giving an approximation algorithm for d-SKEW-SYMMETRIC MULTICUT.

Proof (of Theorem 1). **Description of Algorithm.** The input to our algorithm for d-SKEW-SYMMETRIC MULTICUT is an instance $(D = (V, A), \sigma, \mathcal{T} = \{J_1, \ldots, J_r\}, k, L)$ where $J_i = \{v_{i_1}, \ldots, v_{i_d}\}$ and the algorithm either returns a skew-symmetric multicut of size at most $2d(k^2 + k)$ which is an L-L' self-conjugate separator in D or concludes correctly that no skew-symmetric multicut of size at most k exists. In order to solve the problem on the given instance of d-SKEW-SYMMETRIC MULTICUT, the algorithm is invoked on the input $(D = (V, A), \sigma, \mathcal{T}, k, \emptyset)$. At any step, if the current instance is already solved (which can be tested by simply computing the strongly connected components of D), then the algorithm returns the empty set. Otherwise, the algorithm selects a violating set T and for each $t \in T$, it computes the set C_t and $C_{t'}$ returned by the invocations of the algorithm of Lemma 3 on input $(D, \sigma, k, \{t\})$ and $(D, \sigma, k, \{t'\})$ respectively. If for every $t \in T$, the algorithm outputs that no $(\{t\}, k)$ set or $(\{t'\}, k)$ set is present, then we stop and say NO. Finally, the algorithm returns the set $C = \cup_{t \in T}(C_t \cup C_{t'})$ along with the output of the recursion of the same algorithm on the instance $\hat{I} = (D \setminus C, \sigma, T, k - 1)$.

Correctness. Observe that in each recursive step, if the budget is r, we add a set of size at most $4dr$. Since the initial budget is k and it drops by 1 in each recursion the set returned finally is bounded by $2d(k^2 + k)$. That the set returned is in fact a skew symmetric multicut follows from Lemma 2 and the fact that for any set $T \in \mathcal{T}$, for any $t \in T$, any skew-symmetric multicut for the given instance is either a $\{t\}$-$\{t'\}$ separator or a $\{t'\}$-$\{t\}$ separator.

Suppose for some violating set T, for every $t \in T$ the algorithm of Lemma 3 takes $(D, \sigma, k, \{t\})$ and $(D, \sigma, k, \{t'\})$ as inputs, and outputs that there is neither a ($\{t\}, k$)-set nor a ($\{t'\}, k$)-set in the current graph. From Observation 3 it follows that the minimum $\{t\}$-$\{t'\}$ cut exceeds $2k$. We know that for any solution to d-Skew-Symmetric Multicut, there is a $t \in T$ such that the solution contains either a $\{t\}$-$\{t'\}$ separator or a $\{t'\}$-$\{t\}$ separator. Therefore, we can safely conclude that there is no solution of size k to this instance of d-Skew-Symmetric Multicut.

Running Time. Since the time taken at each recursive step is bounded by the time required to compute a violating set and apply the algorithm of Lemma 3 at most $2d$ times and there are at most k recursions, the algorithm in total takes time $\mathcal{O}(dk^5(m + n + \ell))$.   □

Corollaries. Next we give a few corollaries of Theorem 1. It is known that there are polynomial time parameter preserving reductions from a number of problems to d-Skew-Symmetric Multicut (see for example [14]). In particular, the Almost 2-SAT problem has such a reduction to the 1-Skew-Symmetric Multicut problem and Deletion q-Horn Backdoor Set Detection problem has such a reduction to the 3-Skew-Symmetric Multicut problem.

Lemma 5 ([14]). *Let (F, k) be an instance of* Almost *2-SAT and $D(F)$ be the implication graph built on F. There is a polynomial time reduction from* Almost *2-SAT to* 1-Skew-Symmetric Multicut *such that (F, k) is a* Yes *instance of* Almost *2-SAT if and only if $(D(F), \mathcal{T} = \{\{x_1\}, \ldots, \{x_n\}\}, k)$ is a* Yes *instance of* 1-Skew-Symmetric Multicut.

Lemma 6 ([14]). *Let (F, k) be an instance of* Deletion q-Horn Backdoor Set Detection. *There is a parameter-preserving reduction from* Deletion q-Horn Backdoor Set Detection *to* 3-Skew-Symmetric Multicut *such that (F, k) is a* Yes *instance of* Deletion q-Horn Backdoor Set Detection *if and only if the reduced instance is a* Yes *instance of* 3-Skew-Symmetric Multicut. *If the length of the F was ℓ then the reduction runs in $\mathcal{O}(k\ell)$ time and returns a skew-symmetric graph with $\mathcal{O}(k\ell)$ arcs.*

Therefore, we immediately get the following corollaries.

Corollary 1. *There is an algorithm that, given an instance (F, k) of* Almost 2-SAT, *runs in time $\mathcal{O}(k^5(m+n))$ and either returns an* Almost 2-SAT *solution of size at most $2(k^2 + k)$ or correctly concludes that no such set of size at most k exists. Here, m is the number of clauses and n is the number of variables in the formula F.*

Corollary 2. *There is an algorithm that, given an instance (F, k) of* Deletion q-Horn Backdoor Set Detection, *runs in time $\mathcal{O}(k^6(n + \ell))$ and either returns a* Deletion q-Horn Backdoor Set Detection *solution of size at most $6(k^2 + k)$ or correctly concludes that no such set of size at most k exists. Here, m is the number of clauses, n is the number of variables in the formula F and ℓ is the length of the formula F.*

Results for OCT, EDGE BIPARTIZATION, ABOVE GUARANTEE VERTEX
COVER and other problems mentioned in the introduction follow from the known
polynomial time reductions to ALMOST 2-SAT. We define all the problems considered in this paper and reducibility among them in the full version.

References

1. Agarwal, A., Charikar, M., Makarychev, K., Makarychev, Y.: $O(\sqrt{\log n})$ approximation algorithms for min uncut, min 2cnf deletion, and directed cut problems. In: STOC, pp. 573–581 (2005)
2. Chekuri, C., Sidiropoulos, A.: Approximation algorithms for euler genus and related problems. In: FOCS, pp. 167–176 (2013)
3. Chuzhoy, J., Makarychev, Y., Sidiropoulos, A.: On graph crossing number and edge planarization. In: Randall, D. (ed.) SODA, pp. 1050–1069. SIAM (2011)
4. Even, G., Naor, J., Schieber, B., Sudan, M.: Approximating minimum feedback sets and multicuts in directed graphs. Algorithmica 20(2), 151–174 (1998)
5. Ford Jr., L.R., Fulkerson, D.R.: Maximal flow through a network. Canadian J. Math. 8, 399–404 (1956)
6. Gaspers, S., Ordyniak, S., Ramanujan, M.S., Saurabh, S., Szeider, S.: Backdoors to q-horn. In: STACS, pp. 67–79 (2013)
7. Goldberg, A.V., Karzanov, A.V.: Path problems in skew-symmetric graphs. Combinatorica 16(3), 353–382 (1996)
8. Goldberg, A.V., Karzanov, A.V.: Maximum skew-symmetric flows and matchings. Math. Program. 100(3), 537–568 (2004)
9. Grohe, M., Grüber, M.: Parameterized approximability of the disjoint cycle problem. In: Arge, L., Cachin, C., Jurdziński, T., Tarlecki, A. (eds.) ICALP 2007. LNCS, vol. 4596, pp. 363–374. Springer, Heidelberg (2007)
10. Hlinený, P., Oum, S.I.: Finding branch-decompositions and rank-decompositions. SIAM J. Comput. 38(3), 1012–1032 (2008)
11. Iwata, Y., Oka, K., Yoshida, Y.: Linear-time fpt algorithms via network flow. In: SODA, pp. 1749–1761 (2014)
12. Kratsch, S., Wahlström, M.: Representative sets and irrelevant vertices: New tools for kernelization. In: FOCS, pp. 450–459 (2012)
13. Marx, D.: Parameterized graph separation problems. Theor. Comput. Sci. 351(3), 394–406 (2006)
14. Ramanujan, M.S., Saurabh, S.: Linear time parameterized algorithms via skew-symmetric multicuts. In: SODA, pp. 1739–1748 (2014)
15. Reed, B., Smith, K., Vetta, A.: Finding odd cycle transversals. Oper. Res. Lett. 32(4), 299–301 (2004)
16. Tutte, W.T.: Antisymmetrical digraphs. Canadian J. Math. 19, 1101–1117 (1967)

On the Clique Editing Problem*

Ivan Kováč[1], Ivana Selečéniová[1,2], and Monika Steinová[2]

[1] Department of Computer Science, Comenius University, Bratislava, Slovakia
{ikovac,seleceniova}@dcs.fmph.uniba.sk
[2] Department of Computer Science, ETH Zürich, Switzerland
{ivana.seleceniova,monika.steinova}@inf.ethz.ch

Abstract. We study the hardness and approximability of the problem CLIQUEEDITING, where the goal is to edit a given graph G into a graph consisting of a clique and a set of isolated vertices while using a minimum number of editing operations. The problem is interesting from both practical and theoretical points of view, and it belongs to the well-studied family of graph modification problems. We prove that the problem is NP-complete and construct a 3.524-approximation algorithm. Furthermore, we prove an existence of a PTAS for the still NP-complete version of the problem restricted to bipartite graphs, and the existence of a polynomial-time algorithm for the problem restricted to planar graphs.

1 Introduction

In graph modification problems one has to modify an input graph into a target graph, using as little changes as possible. The target graph is usually a graph with some prescribed property, such as a graph consisting of a set of disjoint cliques, a 2-connected graph, or, as in our case, a graph consisting of exactly one clique and a set of isolated vertices. Allowed operations are usually combinations of vertex deletions, edge deletions, edge insertions, and edge contractions. In our case only edge insertions and deletions are allowed. The cost of a solution is measured by the number of altered edges.

Graph modification problems have applications in several areas, such as molecular biology (see e.g. [9]), numerical algebra [12], circuit design [6], and machine learning [2]. These problems have been extensively studied in the past 30 years. Most of the practically motivated classes of target graphs have hereditary properties, that is, they are closed under removal of vertices. This is also the case in our problem. The study of CLIQUEEDITING was originally motivated as a clustering problem in a noisy data set [4].

Graph modification problems are an important and well-studied part of graph theory. Already in 1979, Garey and Johnson mentioned 18 versions of these problems (both edge and vertex modification) in their collection of NP-hard problems [8]. More recent results on the hardness of these problems can be found

---

* This work was partially supported by grants VEGA 1/0979/12, VEGA 1/0671/11 and by the SNF grant 200021-146372.

E. Csuhaj-Varjú et al. (Eds.): MFCS 2014, Part II, LNCS 8635, pp. 469–480, 2014.

in [1], [3], or [11]. The problem addressed in this paper, CLIQUEEDITING, was introduced in [4]. In their paper, the authors conjectured the NP-completeness of the decision version of CLIQUEEDITING and provided a proof that the problem is in the class of subexponential fixed-parameter tractable problems (SUBEPT, see [7]). The NP-completeness has already been stated as an open question at IWOCA 2013 by Peter Damaschke. In this paper, we prove the proposed conjecture, and investigate the approximability of the problem.

This paper is organized as follows. Section 2 contains basic notations of graph theory used in the paper, a formal definition of CLIQUEEDITING, and an elementary lemma which is of substantial importance for the rest of the paper. In Section 3, we investigate the hardness of the problem and prove NP-completeness even for the decision version of the problem restricted to bipartite graphs. In Section 4, we present a polynomial-time algorithm for the problem restricted to planar graphs. The approximability results, namely a PTAS on bipartite graphs and a 3.524-approximation in the general case, are addressed in Section 5.

2 Preliminaries

Throughout this paper, we consider only simple undirected graphs. We use the standard notation of graph theory: a *clique* is a complete subgraph of a graph, and a *bi-clique* is a complete bipartite subgraph of a graph. A complete graph on n vertices is denoted by K_n, a complete bipartite graph with i vertices in one shore and j vertices in the other shore is denoted by $K_{i,j}$. The set of all vertices of a graph G is denoted by $V(G)$, the set of all edges of G is denoted by $E(G)$. The subgraph of G induced by a vertex set S is denoted by $G[S]$, and the set of edges of $G[S]$ is denoted by $E(G[S])$. A *non-edge* of a graph G is any pair of vertices $\{u, v\} \notin E(G)$. The *complement* $\overline{G}$ of a graph G is a graph on the same set of vertices whose edge set consists of all non-edges of G. Furthermore, a set of edges in G incident with a vertex v is denoted by $E_v(G)$.

We now formally define the CLIQUEEDITING problem and its decision version, DECCLIQUEEDITING.

Definition 1 (CLIQUEEDITING). *The input of* CLIQUEEDITING *is a graph G. The output is a partition of $V(G)$ into two sets C and I. For the sake of brevity, we shall call C a solution. The I-part of the solution can be easily determined from C, since $I = V(G) - C$. The goal is to minimize the number of edges one needs to add or remove from G in order to create a clique on C and to isolate every vertex from I. Formally, let us define the cost $\mathrm{cost}_G(C)$ of a solution C by*

$$\mathrm{cost}_G(C) = |E(\overline{G}[C])| + |\{\{u, v\} \in E(G) \mid u \in I \vee v \in I\}| \,.$$

We abbreviate $\mathrm{cost}_G(C)$ by $\mathrm{cost}(C)$ whenever G is clear from the context.

Definition 2 (DECCLIQUEEDITING). *The input for the decision version of* CLIQUEEDITING*, denoted by* DECCLIQUEEDITING*, is a pair (G, k), where G is a graph and $k \in \mathbb{N}$. The goal is to determine whether a solution C with $\mathrm{cost}_G(C) \leq k$ exists.*

We start with a simple but essential lemma, which states that every vertex v in an *optimal* solution C_{OPT} has at least $(|C_{OPT}| - 1)/2$ neighbours in $G[C_{OPT}]$.

Lemma 1. *Let C_{OPT} be an optimal solution on the input graph G. For every vertex $v \in C_{OPT}$ it holds that $|E_v(G[C_{OPT}])| \geq |E_v(\overline{G}[C_{OPT}])|$. In other words, $|E_v(G[C_{OPT}])| \geq (|C_{OPT}| - 1)/2$.*

Proof. Since C_{OPT} is an optimal solution, any other solution has the cost at least as large as C_{OPT}, which implies $\text{cost}(C_{OPT} \setminus \{v\}) - \text{cost}(C_{OPT}) \geq 0$. If we look at the definition of the cost function, the difference between the first summands of $\text{cost}(C_{OPT} \setminus \{v\})$ and $\text{cost}(C_{OPT})$ is exactly $-|E_v(\overline{G}[C_{OPT}])|$. The difference between the second summands is exactly $|E_v(G[C_{OPT}])|$. Putting it all together concludes the proof. □

3   CliqueEditing on Bipartite Graphs and NP-completeness

In this section we investigate some properties of CliqueEditing on bipartite graphs and prove the NP-hardness of this subproblem. Note that DecCliqueEditing is in NP, even on general graphs – it is sufficient to non-deterministically guess a solution, and verify that its cost is at most the constant given as a part of an input. Therefore, the NP-completeness of the restricted problem implies the NP-completeness of the unrestricted problem as well.

We attack the question of NP-hardness in the following manner. Informally speaking, we start by proving that any optimal solution is a bi-clique. Subsequently, we prove the existence of an optimal solution that is balanced, i.e., has the same number of vertices in both shores. Finally, the NP-hardness is proven by a polynomial-time reduction from the problem of finding a balanced bi-clique in a bipartite graph, which is known to be NP-hard [10].

Lemma 2. *Let G be a bipartite graph with shores P and Q. Let C_{OPT} be an optimal solution containing p vertices from the shore P and q vertices from the shore Q. Without loss of generality, let $p \leq q$. Then $p \geq q - 1$.*

Proof. Let $v \in C_{OPT}$ be a vertex from Q. In the induced subgraph $G[C_{OPT}]$, the vertex v can be connected only to vertices from P (at most p vertices), which implies $|E_v(G[C_{OPT}])| \leq p$. Furthermore, it cannot be connected to vertices from Q, which contains at least $q - 1$ vertices (not counting v), that is, $|E_v(\overline{G}[C_{OPT}])| \geq q - 1$. Using Lemma 1, we obtain

$$p \geq |E_v(G[C_{OPT}])| \geq |E_v(\overline{G}[C_{OPT}])| \geq q - 1,$$

which proves our claim. □

Corollary 1. *Let G be a bipartite graph. An optimal solution for G contains either p vertices from each of the two shores, or p vertices from one and $p + 1$ vertices from the other one, for some $p \in \mathbb{N}$.*

Lemma 3. *Let G be a bipartite graph. For every optimal solution C_{OPT}, the graph $G[C_{OPT}]$ is either $K_{p,p}$ or $K_{p,p+1}$, for some $p \in \mathbb{N}$.*

Proof. Assume the contrary, that is, $G[C_{OPT}]$ is not a complete bipartite graph. Let us denote the vertices of C_{OPT} from the shores P and Q by C_{OPT}^P and C_{OPT}^Q, respectively. Furthermore, let $|C_{OPT}^P| = p$ and $|C_{OPT}^Q| = q$, where $p \leq q$. Due to Lemma 2, it either holds $p = q$ or $p + 1 = q$.

Since $G[C_{OPT}]$ is not $K_{p,q}$ (i.e., neither $K_{p,p}$ nor $K_{p,p+1}$), there must exist a non-edge $\{u, v\}$, such that $u \in C_{OPT}^P$ and $v \in C_{OPT}^Q$. Because of the missing edge, we can deduce that $|E_v(G[C_{OPT}])|$ is smaller than or equal to $p - 1$ and $|E_v(\overline{G}[C_{OPT}])|$ is greater than or equal to $q - 1 + 1$. By Lemma 1, we get

$$p - 1 \geq |E_v(G[C_{OPT}])| \geq |E_v(\overline{G}[C_{OPT}])| \geq q \geq p \,,$$

which is a contradiction.   □

Theorem 1. *For every bipartite graph G, there exists an optimal solution C_{OPT} such that $G[C_{OPT}]$ is $K_{p,p}$, where $p \in \mathbb{N}$.*

Proof. Lemma 3 implies that the subgraph induced by an optimal solution is $K_{p,p}$ or $K_{p,p+1}$. In the first case the claim is trivial. In the second case, one can remove any vertex v from the larger shore (of $G[C_{OPT}]$), obtaining a solution with equal cost, because the change in the cost of the solution is equal to $|E_v(G[C_{OPT}])| - |E_v(\overline{G}[C_{OPT}])|$, where $|E_v(G[C_{OPT}])| = |E_v(\overline{G}[C_{OPT}])| = p$. □

Lemma 4. *Let G be a bipartite graph. The cost of the solution of type $K_{p,p}$ is $|E(G)| - p$.*

Proof. By inspecting the summands of the cost in the definition, we get

$$\left| E\left(\overline{G}[C]\right) \right| = 2 \cdot \frac{p(p-1)}{2} = p^2 - p \,,$$

$$|\{\{u, v\} \in E(G) \mid u \in I_{OPT} \vee v \in I_{OPT}\}| = |E(G)| - |E(G[C])| = |E(G)| - p^2 \,.$$

Adding both parts together, we obtain the cost

$$\text{cost}(C_{OPT}) = |E(G)| - p^2 + p^2 - p = |E(G)| - p \,,$$

which completes the proof.   □

In order to prove the NP-hardness of DECCLIQUEEDITING, we describe a reduction from the NP-complete problem BALANCED COMPLETE BIPARTITE SUBGRAPH (BCBS for short), which we now define formally.

Definition 3 (BCBS). *The input of the BCBS is a bipartite graph G and $k \in \mathbb{N}$. The goal is to determine whether there is a complete bipartite subgraph $K_{k,k}$ of the graph G.*

Theorem 2 (Johnson [10]). BCBS *is NP-complete.*

Theorem 3. *There exists a polynomial-time reduction from* BCBS *to* DECCLI-QUEEDITING *on bipartite graphs.*

Proof. For an instance $\mathcal{I}_1 = (G, k_1)$ of BCBS we create an instance of CLIQUE-EDITING $\mathcal{I}_2 = (G, |E(G)| - k_1)$. Clearly, this reduction can be done in polynomial time, as we only change the number k_1 to the number $|E(G)| - k_1$. Let us divide the proof of the fact that this is indeed a valid reduction into two parts.

If the answer for $\mathcal{I}_1$ is yes, then G contains a subgraph K_{k_1,k_1}, which, based on Lemma 4, has cost $|E(G)| - k_1$. Thus an optimal solution for $\mathcal{I}_2$ is less than or equal to $|E(G)| - k_1$, hence the answer for $\mathcal{I}_2$ is yes.

If the answer for $\mathcal{I}_2$ is yes, then there exists a solution with cost at most $|E(G)| - k_1$. From Theorem 1, there is an optimal solution K_{k_2,k_2}. From Lemma 4, its cost is $|E(G)| - k_2$. Hence, $|E(G)| - k_1 \geq |E(G)| - k_2$, which means $k_2 \geq k_1$. Therefore, there exists a subgraph K_{k_2,k_2} of G, such that $k_2 \geq k_1$, which trivially implies the existence of K_{k_1,k_1}. The answer for $\mathcal{I}_1$ is yes. □

Corollary 2. *The problem* DECCLIQUEEDITING *is an* NP-*complete problem on both bipartite graphs and general graphs..*

Proof. As we have already mentioned, DECCLIQUEEDITING is in NP. The corollary then follows from the existence of reduction from BCBS as described in Theorem 3 and from the NP-completeness of BCBS due to Theorem 2. □

4   CLIQUEEDITING on Planar Graphs

As proved above, CLIQUEEDITING is in its general case an NP-hard problem. In this section we investigate this problem on planar graphs, and show that it is polynomially tractable in this case. The proof of the correctness of the algorithm is based on the well-known fact that the average degree of a planar graph is less than six.[1]

Theorem 4. *Let* C_{OPT} *be an optimal solution for* CLIQUEEDITING *on a planar graph* G. *Then* $|C_{OPT}| \leq 11$.

Proof. The graph $G[C_{OPT}]$ is planar, because G is planar. Hence, the average degree of $G[C_{OPT}]$ is less than 6 and there must exist a vertex $v \in C_{OPT}$ with degree less than 6 (in $G[C_{OPT}]$). From this, $|E_v(G[C_{OPT}])| = \deg_{G[C_{OPT}]}(v) \leq 5$ follows and from Lemma 1 we have $|E_v(G[C_{OPT}])| \geq (|C_{OPT}| - 1)/2$. Putting these two inequalities together, we obtain $|C_{OPT}| \leq 11$. □

Theorem 5. *There exists a polynomial-time algorithm for* CLIQUEEDITING *on planar graphs.*

Proof. It suffices to check every set $C \subseteq V(G)$ with size at most eleven. The total number of such sets is $\sum_{i=1}^{11} \binom{n}{i} \in \mathcal{O}(n^{11})$, where $n = |V(G)|$. □

---

[1] This follows from Euler's formula for planar graphs, $|V| - |E| + |F| = 2$. Based on the fact that every face is incident with at least 3 edges, and every edge is in at most 2 faces, we get the inequality $|E| \leq 3|V| - 6$. Since the sum of degrees is equal to $2|E|$, the average degree is $6 - 12/|V| < 6$. For more details, see [5].

5   Approximation Algorithms

If not stated otherwise, we assume that G contains at least one edge, since in any graph without edges it is easy to find an optimal solution (containing only one vertex).

5.1   PTAS on Bipartite Graphs

Consider a trivial approximation algorithm, which always chooses a solution C', such that $|C'| = 1$. We will show that this algorithm is a $(1 + \mathcal{O}(1/\sqrt{|E(G)|}))$-approximation. It is clear that $\text{cost}(C') = |E(G)|$, because every edge has to be removed in order to isolate all vertices in $V(G) \setminus C'$. From Lemma 4 we know that an optimal solution C_{OPT} has cost $|E(G)| - \ell$, where $G[C_{\text{OPT}}] = K_{\ell,\ell}$. Since the number of edges in $K_{\ell,\ell}$ is ℓ^2, the number ℓ is bounded from above by $\sqrt{|E(G)|}$. By a simple calculation we get that the approximation ratio of the solution C' is

$$\frac{|E(G)|}{|E(G)| - \ell} = 1 + \frac{\ell}{|E(G)| - \ell} \leq 1 + \frac{\sqrt{|E(G)|}}{|E(G)| - \sqrt{|E(G)|}} = 1 + \frac{1}{\sqrt{|E(G)|} - 1}.$$

We use this algorithm in our effort to construct a polynomial-time approximation scheme for CLIQUEEDITING on bipartite graphs. Suppose we want to obtain an approximation ratio of $1 + \varepsilon$, for some $\varepsilon > 0$. If

$$1 + \varepsilon \geq 1 + \frac{1}{\sqrt{|E(G)|} - 1},$$

we can use the algorithm described above. For graphs where $\varepsilon < 1/(\sqrt{|E(G)|} - 1)$, we have $|E(G)| < (1 + 1/\varepsilon)^2$. Hence, these graphs have a constant number of edges (for a fixed ε), and a constant number of non-isolated vertices. We can solve these cases using a brute-force approach, that is, trying every feasible solution that contains only non-isolated vertices.

Theorem 6. CLIQUEEDITING *on bipartite graphs admits a PTAS.*

5.2   Constant Approximation Algorithm on General Graphs

In this part we analyze an algorithm GREEDYREMOVAL, which starts with a solution $V(G)$ and iteratively removes from the solution a vertex with minimum degree, if it is beneficial. A formal definition of the algorithm is stated below. Note that $\delta(G)$ denotes the minimum degree of the graph G.

Algorithm 1. GREEDYREMOVAL

1: $C \leftarrow V(G)$
2: **while** $\delta(G[C]) < (|C| - 1)/2$ **do**
3:     $C \leftarrow C \setminus \{v\}$, where v is a vertex with minimum degree in $G[C]$.
4: **end while**
5: **return** C

From the algorithm description it is obvious that, if the minimum degree of an n-vertex graph G is at least $(n-1)/2$, the solution produced by this algorithm is $C = V(G)$. First, we show some upper bounds on the cost of solutions produced by this algorithm. Next, we prove an approximation ratio of 2 if the minimum degree of a graph G is at least $(n-1)/2$.

Lemma 5. *Let G be a graph with n vertices and C_{ALG} be a solution produced by the algorithm* GREEDYREMOVAL. *Then* $\mathrm{cost}(C_{\mathrm{ALG}}) \leq |E(G)|$.

Proof. In the case that the minimum degree is at least $(n-1)/2$, the claim is obvious – the cost of the algorithm is the number of non-edges, and there are at most $n(n-1)/4$ non-edges, due to the fact that every vertex has degree at least $(n-1)/2$. Therefore, for every vertex, there are at most $(n-1)/2$ edges missing. For the same reason, the number of edges in the graph G is at least $n(n-1)/4$ and hence $\mathrm{cost}(V(G)) \leq |E(G)|$.

In the case that the minimum degree is less than $(n-1)/2$, we use induction on the number of vertices. As the base case, consider a graph with only one vertex. Both $E(G)$ and the cost of the solution of the algorithm are equal to zero, therefore the base case holds. For the inductive step, assume the claim holds for every graph with at most i vertices. If the graph G has $i+1$ vertices, and its minimum degree is less than $(n-1)/2$, the algorithm removes a vertex v with minimum degree from the solution. If we denote the cost of the solution of the algorithm on the graph $G \setminus \{v\}$ by c_i, and the one on the graph G by c_{i+1}, we have

$$c_{i+1} = c_i + \deg(v) \leq |E(G \setminus \{v\})| + \deg(v) = |E(G)|\,,$$

which concludes the proof. □

Lemma 6. *Let G be a graph with n vertices and C_{ALG} be a solution produced by the algorithm* GREEDYREMOVAL. *Then* $\mathrm{cost}(C_{\mathrm{ALG}}) \leq |E(\overline{G})|$.

Proof. The proof is analogous to the proof of Lemma 5. In the case that the minimum degree is at least $(n-1)/2$, the cost of the solution of the algorithm is exactly $|E(\overline{G})|$. In the other case, we can again use induction and the fact that $E(\overline{G} \setminus \{v\}) + \deg(v) \leq |E(\overline{G})|$, as $\deg(v)$ is less than $(n-1)/2$. □

Lemma 7. *Let G be a graph with n vertices and with minimum degree at least $(n-1)/2$. Then the solution $C_{\mathrm{ALG}} = V(G)$ produced by* GREEDYREMOVAL *is a 2-approximation.*

Proof. Let $(C_{\mathrm{OPT}}, I_{\mathrm{OPT}})$ be an optimal solution. The proof is divided into two cases. First assume that $|C_{\mathrm{OPT}}| \leq n/2$, which implies $|I_{\mathrm{OPT}}| \geq n/2$. Furthermore, the cost of the optimal solution is bounded from below by the number of edges that have to be removed, which is at least $(|I_{\mathrm{OPT}}| \cdot (n-1)/2)/2$. (This number corresponds to the minimum number of edges incident with vertices from I_{OPT}, when the minimum degree is at least $(n-1)/2$.) All in all, the cost of the optimal solution is at least $n(n-1)/8$. Based on Lemmas 5 and 6, we can bound the cost of the solution produced by the algorithm by $\min(E(G), E(\overline{G})) \leq n(n-1)/4$, hence the approximation ratio is at most 2.

It remains to resolve the case of $|C_{\text{OPT}}| > n/2$. In this case $|I_{\text{OPT}}| < n/2$. Let us denote the number of edges that have to be inserted into G in order to make $G[C_{\text{OPT}}]$ a clique by m and the number of edges existing in $G[I_{\text{OPT}}]$ by ℓ. The number ℓ can be bounded from above by

$$\ell \leq \frac{|I_{\text{OPT}}|(|I_{\text{OPT}}| - 1)}{2}.\tag{1}$$

Let us estimate the cost of the optimal solution. Clearly, m edges need to be added into the graph. Furthermore, the edges incident with vertices from I_{OPT} have to be removed. Since the minimum degree of G is at least $(n-1)/2$, the number of these edges is bounded from below by $(|I_{\text{OPT}}|(n-1)/2) - \ell$. (In the first part of the expression, the edges within I_{OPT} are counted twice.) Thus, the cost of the optimal solution is bounded from below by

$$\text{cost}(C_{\text{OPT}}) \geq m + \frac{|I_{\text{OPT}}|(n-1)}{2} - \ell.\tag{2}$$

The algorithm GREEDYREMOVAL produces the solution $C_{\text{ALG}} = V(G)$. In order to make the whole graph a clique, the following edges have to be inserted: m edges that make $G[C_{\text{OPT}}]$ a clique and all edges incident with I_{OPT}, which do not yet exist in G. Let us count the non-edges in G, (edges in $\overline{G}$) which are incident with I_{OPT}. The sum of all *degrees* of I_{OPT} in $\overline{G}$ is at most $|I_{\text{OPT}}|(n-1)/2$, as the maximal degree in $\overline{G}$ is at most $(n-1)/2$. Every edge incident with I_{OPT} is counted in this sum, however, if an edge has both endpoints in I_{OPT}, the edge is counted twice. Therefore, we have to subtract the number of non-edges within $G[I_{\text{OPT}}]$, which is $|I_{\text{OPT}}|(|I_{\text{OPT}}| - 1)/2 - \ell$. Hence, the cost of C_{ALG} satisfies

$$\text{cost}(C_{\text{ALG}}) \leq m + \frac{|I_{\text{OPT}}|(n-1)}{2} - \frac{|I_{\text{OPT}}|(|I_{\text{OPT}}| - 1)}{2} + \ell$$
$$= m + \frac{|I_{\text{OPT}}|(n - |I_{\text{OPT}}|)}{2} + \ell.\tag{3}$$

Applying the bound of ℓ from (1) into (2), and using the bound (3), we can easily compute the approximation ratio, leading to

$$\frac{\text{cost}(C_{\text{ALG}})}{\text{cost}(C_{\text{OPT}})} \leq \frac{m + \frac{|I_{\text{OPT}}|(n-|I_{\text{OPT}}|)}{2} + \ell}{m + \frac{|I_{\text{OPT}}|(n-|I_{\text{OPT}}|)}{2}} = 1 + \frac{\ell}{m + \frac{|I_{\text{OPT}}|(n-|I_{\text{OPT}}|)}{2}}$$
$$\leq 1 + \frac{\frac{|I_{\text{OPT}}|(|I_{\text{OPT}}|-1)}{2}}{\frac{|I_{\text{OPT}}|(n-|I_{\text{OPT}}|)}{2}} = \frac{n-1}{|C_{\text{OPT}}|} \leq 2 \cdot \frac{n-1}{n} \leq 2,$$

which concludes the proof. □

Before we proceed to the proof of the main theorem of this section, we need another auxiliary result, which claims the non-existence of any instance with particular properties.

Lemma 8. *Let c be a constant such that $c^3 - 3c^2 - c - 1 > 0$.[2] Let G be a graph with n vertices such that the following properties hold:*

1. *the minimum degree of G is less than $(n-1)/2$,*
2. *there exists a vertex v of minimum degree (in G) such that v belongs to every optimal solution,*
3. *the cost of an optimal solution is less than $|E(G)|/c$, and*
4. *the number of vertices n is greater than $1 + 2(c^2 - c)/(c^3 - 3c^2 - c - 1)$.*

Then $|C_{\mathrm{OPT}}| > 1 + \deg(v)(c+1)/c$, where C_{OPT} is an optimal solution.

Proof. By contradiction, assume that $|C_{\mathrm{OPT}}| \leq 1 + \deg(v)(c+1)/c$. Using the first property, we can bound the size of the optimal solution from above by $|C_{\mathrm{OPT}}| < 1 + (n-1)(c+1)/(2c)$. Hence, the number of vertices that are not contained in the optimal solution $|I_{\mathrm{OPT}}| = |V(G)| - |C_{\mathrm{OPT}}|$ is bounded by

$$|I_{\mathrm{OPT}}| = n - |C_{\mathrm{OPT}}| > n - \frac{(c+1)(n-1)}{2c} - 1 = \frac{(c-1)(n-1)}{2c}.$$

Let us denote the number of edges incident with I_{OPT} by $\ell_{I_{\mathrm{OPT}}}$. Clearly, since v is a vertex of minimum degree, $\ell_{I_{\mathrm{OPT}}} \geq |I_{\mathrm{OPT}}| \deg(v)/2$. Furthermore, the cost of the optimal solution is bounded from below by $\ell_{I_{\mathrm{OPT}}}$, because every edge incident with a vertex from I_{OPT} has to be removed in order to isolate all vertices from I_{OPT}.

Let us count the number of all edges in the graph. There are edges incident with v and edges incident with I_{OPT} – the number of these edges is at most $\ell_{I_{\mathrm{OPT}}} + \deg(v)$. Moreover, there are at most $(|C_{\mathrm{OPT}}| - 1)(|C_{\mathrm{OPT}}| - 2)/2$ edges not incident with any vertex from $I_{\mathrm{OPT}} \cup \{v\}$. Therefore,

$$|E(G)| \leq \ell_{I_{\mathrm{OPT}}} + \deg(v) + \frac{(|C_{\mathrm{OPT}} - 1|)(|C_{\mathrm{OPT}}| - 2)}{2}.$$

Applying the previous results, we get

$$\ell_{I_{\mathrm{OPT}}} \leq \mathrm{cost}(C_{\mathrm{OPT}}) \leq \frac{|E(G)|}{c} \leq \frac{\ell_{I_{\mathrm{OPT}}} + \deg(v) + \frac{(|C_{\mathrm{OPT}}|-1)(|C_{\mathrm{OPT}}|-2)}{2}}{c}.$$

The first inequality holds due to the second paragraph of the proof, the second one is due to the third property from the statement of this lemma, and the last one holds due to the above bound of the number of edges in the graph. Using simple arithmetics, we obtain

$$\ell_{I_{\mathrm{OPT}}} \leq \frac{\deg(v)}{c-1} + \frac{|C_{\mathrm{OPT}}|^2 - 3|C_{\mathrm{OPT}}| + 2}{2(c-1)}.$$

Using the assumption that $|C_{\mathrm{OPT}}| \leq 1 + \deg(v)(c+1)/c$ and the fact that $\ell_{I_{\mathrm{OPT}}} \geq |I_{\mathrm{OPT}}| \deg(v)/2$, we get

$$\frac{|I_{\mathrm{OPT}}| \deg(v)}{2} \leq \frac{2 \deg(v) - \frac{c+1}{c} \deg(v)}{2(c-1)} + \frac{(c+1)^2 \deg(v)^2}{2c^2(c-1)},$$

---

[2] That is, c is greater than approximately 3.383.

which can be further simplified to

$$|I_{\mathrm{OPT}}| \le \frac{1}{c} + \frac{(c+1)^2 \deg(v)}{c^2(c-1)}.$$

Since the degree of the vertex v is smaller than $(n-1)/2$, and we have a lower bound on the number of vertices in I_{OPT}, it holds that

$$\frac{(c-1)(n-1)}{2c} \le \frac{1}{c} + \frac{(c+1)^2(n-1)}{2c^2(c-1)}.$$

We continue by simplifying the inequality, and we finally obtain

$$(n-1)(c^3 - 3c^2 - c - 1) \le 2c(c-1),$$

which contradicts the last property, i.e., $n > 1 + (2c^2 - 2c)/(c^3 - 3c^2 - c - 1)$.   □

Theorem 7. GREEDYREMOVAL *is a 3.524-approximation.*

Proof. Let $c := 3.524$. We do an induction on the number of vertices of the graph G.

Base Case: Checking all possible inputs with a brute force program we have found out that the claim holds for all graphs on at most 9 vertices.

Inductive hypothesis: Assume that GREEDYREMOVAL is a c-approximation for all graphs on i vertices, where $i \ge 9$.

Inductive step: Let G be a graph on $i+1$ vertices. We have to prove that GREEDYREMOVAL is a c-approximation on G. If the minimum degree of G is at least $i/2$, the claim follows from Lemma 7. Otherwise, we consider two cases.

Case 1: The minimum degree is less than $i/2$ and, for all $v \in V(G)$ which have the minimum degree, there exists an optimal solution C_{OPT} such that $v \notin C_{\mathrm{OPT}}$.

Let v be a vertex with minimum degree removed by the algorithm in the first step, let $C_{\mathrm{OPT}}^{\neg v}$ be an optimal solution which does not contain v. Furthermore, let $C_{\mathrm{ALG}(G')}$ be a solution produced by the algorithm on a graph G'. By the definition of the algorithm,

$$\mathrm{cost}_G(C_{\mathrm{ALG}(G)}) = \mathrm{cost}_{G \setminus \{v\}}(C_{\mathrm{ALG}(G \setminus \{v\})}) + \deg(v).$$

Since v is not contained in $C_{\mathrm{OPT}}^{\neg v}$, the edges incident with v are counted in the cost of $C_{\mathrm{OPT}}^{\neg v}$ and

$$\mathrm{cost}_G(C_{\mathrm{OPT}}^{\neg v}) = \mathrm{cost}_{G \setminus \{v\}}(C_{\mathrm{OPT}}^{\neg v} \setminus \{v\}) + \deg(v).$$

Using the inductive hypothesis in the form

$$\mathrm{cost}_{G \setminus \{v\}}(C_{\mathrm{ALG}(G \setminus \{v\})}) \le c \cdot \mathrm{cost}_{G \setminus \{v\}}(C_{\mathrm{OPT}}^{\neg v} \setminus \{v\}),$$

we can compute the approximation ratio as

$$\frac{\mathrm{cost}_G(C_{\mathrm{ALG}(G)})}{\mathrm{cost}_G(C_{\mathrm{OPT}}^{\neg v})} \le \frac{c \cdot \mathrm{cost}_{G \setminus \{v\}}(C_{\mathrm{OPT}}^{\neg v} \setminus \{v\}) + \deg(v)}{\mathrm{cost}_{G \setminus \{v\}}(C_{\mathrm{OPT}}^{\neg v} \setminus \{v\}) + \deg(v)} \le c.$$

Case 2: The minimum degree is less than $i/2$ and there is a vertex $v \in V(G)$ with minimum degree that is contained in any optimal solution C_{OPT}.

Fix any optimal solution (C_{OPT}, I_{OPT}). If $\text{cost}(C_{OPT}) \geq |E(G)|/c$, the approximation ratio is at most c due to Lemma 5. Hence, for the rest of the proof, we can assume $\text{cost}(C_{OPT}) < |E(G)|/c$. As the graph G has at least ten vertices, all the requirements from Lemma 8 are met, and thus $|C_{OPT}| > 1 + \deg(v)(c+1)/c$. Analogously to the first case, we can bound the cost of the algorithm from above as $\text{cost}_G(C_{ALG(G)}) \leq \text{cost}_{G \setminus \{v\}}(C_{ALG(G \setminus \{v\})}) + \deg(v)$.

To estimate the cost of the optimal solution with respect to the cost of an optimal solution $C_{OPT(G \setminus \{v\})}$ on the graph $G \setminus \{v\}$, let us take a look at $\text{cost}_{G \setminus \{v\}}(C_{OPT} \setminus \{v\})$. Recall that $|E_v(G[C_{OPT}])|$ is the number of edges within C_{OPT} incident with v. Clearly $|E_v(G[C_{OPT}])| \leq \deg(v)$. Furthermore, there are $|C_{OPT}| - 1 - |E_v(G[C_{OPT}])|$ edges that have to be inserted into G in order to connect v to the other vertices in C_{OPT}, and $\deg(v) - |E_v(G[C_{OPT}])|$ edges that have to be removed in order to isolate v from vertices not contained in C_{OPT}. From the above, it follows that

$$\text{cost}_G(C_{OPT}) = \text{cost}_{G \setminus \{v\}}(C_{OPT} \setminus \{v\}) + |C_{OPT}| - 1 - 2|E_v(G[C_{OPT}])| + \deg(v)$$
$$\geq \text{cost}_{G \setminus \{v\}}(C_{OPT} \setminus \{v\}) + |C_{OPT}| - 1 - \deg(v).$$

Using the optimality of $C_{OPT(G \setminus \{v\})}$ on $G \setminus \{v\}$ we have

$$\text{cost}_{G \setminus \{v\}}(C_{OPT} \setminus \{v\}) \geq \text{cost}_{G \setminus \{v\}}(C_{OPT(G \setminus \{v\})}),$$

and by the previously proven fact that $|C_{OPT}| > 1 + \deg(v)(c+1)/c$, we obtain

$$\text{cost}_G(C_{OPT}) > \text{cost}_{G \setminus \{v\}}(C_{OPT(G \setminus \{v\})}) + \frac{\deg(v)}{c}.$$

Hence, from the inductive hypothesis, for the approximation ratio it holds that

$$\frac{\text{cost}(C_{ALG(G)})}{\text{cost}(C_{OPT})} \leq \frac{\text{cost}_{G \setminus \{v\}}(C_{ALG(G \setminus \{v\})}) + \deg(v)}{\text{cost}_{G \setminus \{v\}}(C_{OPT(G \setminus \{v\})}) + \frac{\deg(v)}{c}} \leq c,$$

which concludes the proof.   □

6   Conclusion

In the presented paper, we proved the NP-completeness of the decision version of the problem CLIQUEEDITING on both bipartite and general graphs. We also stated that the problem is solvable in polynomial time on planar graphs and admits a PTAS on bipartite graphs. Moreover, we constructed a constant approximation algorithm for the general case, and proved that the approximation ratio is at most 3.524. This particular constant emerged solely from our ability to verify the base case in the proof of Theorem 7 for 9 vertices. Using a manual verification for the base case, one can prove that the algorithm is a 3.708-approximation, considering graphs on at most 4 vertices.

We conjecture that this algorithm is a 3.383-approximation. This is supported by Lemma 8 – using an analogous proof as in Theorem 7 with an extended base case (in the case of an approximation ratio close to 3.385, it would be 612 vertices) we expect that one could prove an approximation ratio close to 3.383.

Acknowledgement. The authors would like to express their thanks to Hans-Joachim Böckenhauer and Dennis Komm for valuable discussions and help.

References

1. Alon, N., Stav, U.: Hardness of edge-modification problems. Theor. Comput. Sci. 410(47-49), 4920–4927 (2009)
2. Bansal, N., Blum, A., Chawla, S.: Correlation clustering. Mach. Learn. 56(1-3), 89–113 (2004)
3. Burzyn, P., Bonomo, F., Durán, G.: NP-completeness results for edge modification problems. Discrete Applied Mathematics 154(13), 1824–1844 (2006)
4. Damaschke, P., Mogren, O.: Editing the simplest graphs. In: Pal, S.P., Sadakane, K. (eds.) WALCOM 2014. LNCS, vol. 8344, pp. 249–260. Springer, Heidelberg (2014)
5. Diestel, R.: Graph Theory, 4th edn. Graduate texts in mathematics, vol. 173. Springer (2012)
6. El-Mallah, E.S., Colbourn, C.J.: The complexity of some edge deletion problems. IEEE Transactions on Circuits and Systems 35(3), 354–362 (1988)
7. Flum, J., Grohe, M.: Parameterized Complexity Theory, 1st edn. Texts in Theoretical Computer Science. An EATCS Series1 edition. Springer (March 2006)
8. Garey, M.R., Johnson, D.S.: Computers and Intractability: A Guide to the Theory of NP-Completeness. W. H. Freeman & Co., New York (1979)
9. Goldberg, P.W., Golumbic, M.C., Kaplan, H., Shamir, R.: Four strikes against physical mapping of dna. J. Comput. Biol. 2(1), 139–152 (1995)
10. Johnson, D.S.: The NP-Completeness Column: An Ongoing Guide. J. Algorithms 8(3), 438–448 (1987)
11. Natanzon, A., Shamir, R., Sharan, R.: Complexity classification of some edge modification problems. Discrete Applied Mathematics 113(1), 109–128 (2001)
12. Rose, D.J.: A graph-theoretic study of the numerical solution of sparse positive definite systems of linear equations. In: Graph Theory and Computing, pp. 183–217. Academic Press, New York (1973)

On the Complexity of Symbolic Verification and Decision Problems in Bit-Vector Logic*

Gergely Kovásznai[1], Helmut Veith[1], Andreas Fröhlich[2], and Armin Biere[2]

[1] Formal Methods in Systems Engineering Group,
Vienna University of Technology, Wien, Austria
[2] Institute for Formal Models and Verification,
Johannes Kepler University, Linz, Austria

Abstract. We study the complexity of decision problems encoded in bit-vector logic. This class of problems includes word-level model checking, i.e., the reachability problem for transition systems encoded by bit-vector formulas. Our main result is a generic theorem which determines the complexity of a bit-vector encoded problem from the complexity of the problem in explicit encoding. In particular, NL-completeness of graph reachability directly implies PSPACE-completeness and ExpSpace-completeness for word-level model checking with unary and binary arity encoding, respectively. In general, problems complete for a complexity class C are shown to be complete for an exponentially harder complexity class than C when represented by bit-vector formulas with unary encoded scalars, and further complete for a double exponentially harder complexity class than C with binary encoded scalars. We also show that multi-logarithmic succinct encodings of the scalars result in completeness for multi-exponentially harder complexity classes. Technically, our results are based on concepts from descriptive complexity theory and related techniques for OBDDs and Boolean encodings.

1 Introduction

Symbolic encodings of decision problems by Boolean formalisms are well-known to increase the problem complexity [1,2,3,4,5,6,7,8,9,10,11,12]. In particular, the literature has studied graph problems and other relational problems whose adjacency relation is given by a Boolean formula, circuit or BDD. As Tab. 1 shows, the complexity of these problems typically rises by an exponential, e.g., from NL to PSPACE, from NP to NExpTime, etc. In this paper, we show that symbolic encodings by quantifier-free bit-vector logic (QF_BV) will in general also lead to a complexity increase which ranges from exponential to multi-exponential. Interestingly, the increase depends on a *single* factor, namely how the bit-width of bit-vectors is encoded. For unary encoding, bit-vector logic shows the same complexity behavior as Boolean logic, and for binary encoding, the complexity

---

* Supported by the NFN grant S11403-N23 (RiSE) of the Austrian Science Fund (FWF) and by the grant ICT10-050 (PROSEED) of the Vienna Science and Technology Fund (WWTF).

E. Csuhaj-Varjú et al. (Eds.): MFCS 2014, Part II, LNCS 8635, pp. 481–492, 2014.

increase is double exponential. We can generalize the latter encoding, and call it "ν-logarithmic": encode the bit-width $2^{2^{\cdots 2^c}}$ as c in binary form, where the degree of exponentiation is $\nu - 2$. We achieve a ν-exponential increase in this case. Importantly, hardness already holds for bit-vector logics with the simple operators $\wedge, \vee, \sim, =$, and the increment operator $+_1$. Membership holds for *all* bit-vector operators which allow log-space computable bit-blasting. Note that $\wedge, \vee, \sim, =, +_1$ defines a very weak logic: $\wedge, \vee, \sim, =$ are contained in all reasonable logics, and the increment operator $+_1$ can be defined from other operators easily [13]. Therefore, our results determine the complexity of decision problems for a large class of bit-vector logics.

Table 1. Examples of complexity increase by symbolic encoding. New results are indicated in boldface. All membership results hold for logics whose operators allow log-space computable bit-blasting. Hardness requires the operators $\wedge, \vee, \sim, =, +_1$. The column with ν holds for all $\nu > 1$.

Encoding → ↓ *Problem*	explicit	Boolean circ./formula, BDD	unary QF_BV	binary QF_BV	ν-logarithmic QF_BV
Word-Level MC, Reachability	NL	PSPACE	**PSPACE**	**EXPSPACE**	$(\nu-1)$-**EXPSPACE**
Circuit Value, Alternating Reachability	P	EXPTIME	**EXPTIME**	**2-EXPTIME**	ν-**EXPTIME**
Clique, 3-SAT, SAT, Knapsack	NP	NEXPTIME	**NEXPTIME**	**2-NEXPTIME**	ν-**NEXPTIME**
k-QBF	Σ_k^P	NE$^{\Sigma_k^P}$	**NE$^{\Sigma_k^P}$**	**2-NE$^{\Sigma_k^P}$**	ν-**NE$^{\Sigma_k^P}$**

Bit-Vector Logic. The theory of *fixed-width bit-vector logics* (i.e., logics where each bit-vector has a given fixed bit-width) is investigated in several scientific works [14,15,16,17,18], and even concrete formats for specifying such bit-vector problems exist, e.g., the SMT-LIB format [19] or the BTOR format [20]. In this paper, we restrict ourselves to *quantifier-free bit-vector* (QF_BV [19]) logics.

As discussed below, bit-vector logics have attracted significant interest in computer-aided verification and SMT solvers. From a theory perspective, bit-vector logics are very succinct logics to express Boolean functions. In contrast to Boolean logic, BDDs, and QBF, they are based on variables for *bit-vectors* rather than variables for individual bits. Thus, for instance $x^{[32]} = y^{[32]}$ expresses that two bit-vectors x and y of bit-width 32 are equal. Bit-vector operators are therefore defined for arbitrary bit-width n, for instance bitwise and/or/xor, shift operators, etc. This has important consequences: (1) a bit-vector logic is given by a list of operators, (2) there is an infinite number of bit-vector logics, and (3) there is no finite functionally complete set of operators from which all other operators can be defined. Moreover, it is evident that the encoding of scalars such as the number 32 in the above simple example is related to the complexity of bit-vector logic.

In previous work by some of the authors [21,13], we investigated the complexity of satisfiability checking of bit-vector formulas. For instance, we showed in [21] that satisfiability checking of QF_BV is NP-complete resp. NExpTime-complete if unary resp. binary encoding of scalars is used and any standard operator of the SMT-LIB [19] is allowed. (All these operators allow log-space computable bit-blasting.) In the binary case, we further analyzed what happened if we restricted the operator set; e.g., if only *bitwise operators, equality*, and *left shift by one* are allowed, then the complexity turns out to be PSPACE-complete [13]. In fact, it is easy to see that also the logic of the operators $\wedge, \vee, \sim, =, +_1$ has a satisfiability problem in PSPACE.

Word-Level Model Checking and Decision Problems. In hardware and software verification, bit-vector logics are a natural framework for word-level system descriptions; e.g., registers in digital circuits and variables in software can be represented by bit-vectors, and word-level operators, such as bitwise ones and arithmetic ones, can be applied to them. The main practical motivation for our work is *word-level model checking*, a bit-vector encoded problem that is of importance in practice. With word-level model checking, we refer to the problem of reachability in a transition system where a state is given by a valuation of one or more bit-vectors, and the transition relation over the states is expressed as a bit-vector formula. Such a representation provides a natural encoding for design information captured at a higher level than that of individual wires and primitive gates. In the past, there has been lots of research on bit-level model checking [22] as well as bit-vector formula decision procedures [23,24]. Comparatively few work has yet been published on word-level model checking. However, with increasing performance of state-of-the-art model checkers [25] and SMT solvers [26,27], also the interest in word-level model checking is growing [28,20,29]. While there are some practical approaches to attack word-level model checking [28,20,29], we are not aware of any work that is dealing with the complexity of the underlying decision problem. Row 1 of Tab. 1 shows that we determine the complexity of word-level model checking for a large class of operators and scalar encodings.

Beyond word-level model checking, we also address the complexity of other decision problems. Rows 2-4 of Tab. 1 give examples of the complexity results for well-known decision problems in bit-vector encoding.

Technical Contribution. Instead of individual complexity results, the paper presents a generic technique to *lift* known complexity results for explicit encodings to the case of bit-vector encodings. Similar techniques were previously developed for symbolic encodings by circuits [7,8,9], Boolean formulas [10], and OBDDs [30]. Lifting membership for a complexity class is the easier part, for which we give a general result in Thm. 1. Lifting hardness requires more effort. Similarly as in [10,30], our method assumes that the problems in explicit encoding are hard under *quantifier-free reductions*, a notion of reduction introduced in descriptive complexity theory [31]. Note that the problems in Tab. 1 fulfill this requirement. The key theorem is Thm. 2, from which a general hardness result is implied in Corr. 2.

Discussion. The results of this paper show that the complexity of bit-vector encoded problems depends crucially on the formalism to represent the bit-width of the bit-vectors. At first sight, these results may seem unexpected, e.g., a small part of the formalism clearly dominates the complexity. From an algorithmic perspective, however, this is not surprising: executing a for-loop from 0 to INT_MAX on architectures with bit-width 16, 2^{16} or $2^{2^{16}}$ will result in drastically different runtimes!

It may also be surprising that QF_BV fragments with PSPACE satisfiability and fragments with NEXPTIME satisfiability have the same complexity, e.g., for word-level model checking. This is however a common phenomenon: Boolean logic has an NP satisfiability problem, while satisfiability of BDDs is constant time. Nevertheless, the model checking problem for both of them is PSPACE-complete [10,30].

Using unary and binary encodings for scalars draws a connection to previous work [21]. Intuitively, results for the unary case measure complexity in terms of bit-widths, and those for the binary case measure complexity in the classical sense, i.e., in terms of formula size. The ν-logarithmic encoding also manifests itself in practice, such as the one in the SMB-LIB to declare arrays by writing (Array idx elem), where *idx* is the sort for array indexes, and *elem* is the sort for array elements. If *idx* is a bit-vector sort (_ BitVec n), where n is encoded w.l.o.g. in binary form, the size of the array is double exponential in the length of the binary encoding of n.

We finally note that hardness for the unary case can also be concluded from an analysis of the proofs in [10] using the definitions of symbolic encodings in [30]. The current paper gives a direct proof for the unary case which is independent of the predecessor papers.

2   Preliminaries

Let $\mathbb{N}$ be the set of natural numbers $\{0, 1, 2, \dots\}$, while $\mathbb{N}^+$ denotes $\mathbb{N}\setminus\{0\}$. $\mathbb{B} = \{0, 1\}$ is the Boolean domain. Given $i \in \mathbb{N}$, let us define the repeated exponentiation function $exp_i : \mathbb{N} \mapsto \mathbb{N}$ as follows: $exp_0(n) = n$ and $exp_{i+1}(n) = 2^{exp_i(n)}$. Given a logical formula ϕ (in either bit-vector, first-order, or Boolean logic), if $x_1, \dots, x_k$ are all the free variables that occur in ϕ, we indicate this by writing $\phi(x_1, \dots, x_k)$.

Complexity Classes. We assume that the reader is familiar with standard complexity classes such as NL, P, EXPTIME, etc., as listed in Tab 1. For simplicity, we will refer to these complexity classes as "standard complexity classes". For a standard complexity class, it is natural to define the exponentially harder complexity class: $EXP_1(L) = EXP_1(NL) = PSPACE$, $EXP_2(NL) = EXP_1(PSPACE) = EXPSPACE$, etc. Similarly, $EXP_1(P) = EXPTIME$, $EXP_2(P) = EXP_1(EXPTIME) = 2\text{-}EXPTIME$, etc., and analogously for other standard complexity classes. For a formal definition of this concept (which is beyond the scope and goal of this paper) one can use the concept of leaf languages [9,2].

Computational Problems in Descriptive Complexity Theory. A *relational sig-nature* is a tuple $\tau = (P_1^{a_1}, \ldots, P_k^{a_k})$ of relation symbols of arity $a_1, \ldots, a_k$, respectively. A finite *structure* over τ is a tuple $\mathcal{A} = (U, \widehat{P}_1^{a_1}, \ldots, \widehat{P}_k^{a_k})$ where U is a nonempty finite set (called the universe of $\mathcal{A}$) and each $\widehat{P}_i^{a_i} \subseteq U^{a_i}$ is a relation over U. The class of all finite structures over τ is denoted by $Struct\,(\tau)$. A *computational problem* over τ is a class $A \subseteq Struct\,(\tau)$, such that A is closed under isomorphism. In this paper, we assume *convex* problems, as introduced in [30], and similarly in [32]. A problem is convex if adding isolated elements to the universe of a structure does not change membership in the problem. In Sec. 4 we will show that the model checking problem is naturally presented in this framework. For background on descriptive complexity see [33].

3   Bit-Vector Logic

A *bit-vector*, or word, is a sequence of bits (i.e. 0 or 1). In this paper, we consider bit-vectors of a fixed size $n \in \mathbb{N}^+$, where n is called the *bit-width* of the bit-vector. We assume the usual syntax and semantics for *quantifier-free bit-vector logic* (QF_BV), cf. the SMT-LIB format [19] and the literature [14,15,16,17,18]. Basically, a bit-vector formula contains bit-vector variables and bit-vector con-stants, each of which is of a certain bit-width specified next to the variable resp. constant, and uses certain bit-vector operators whose semantics is a priori de-fined. For example, $x^{[16]} \neq y^{[16]} \wedge \left(u^{[32]} + v^{[32]} = (x^{[16]} \circ y^{[16]}) \ll 1^{[32]}\right)$ is a bit-vector formula with variables x and y of bit-width 16, u and v of bit-width 32, and operators for addition, shifting, concatenation, and comparison.

Note that, in bit-vector formulas, there exist such components which them-selves do *not* represent bit-vectors, but rather carry additional *numerical* infor-mation to the bit-vectors. We call them *scalars*. Bit-width is a scalar, and there might be also other types of scalars in a formula[1]. This paper demonstrates the effect of encoding the scalars in different ways. For instance, scalars could be encoded as unary numbers or w.l.o.g. binary numbers, or we could choose even more succinct encodings, such as the binary encoding of the logarithm of the scalar. Formally, we represent those encodings by an integer $\nu \in \mathbb{N}^+$, i.e., ν denotes how $n \in \mathbb{N}$ is obtained from a scalar s: (1) if $\nu = 1$, then s is a *unary* number encoding of n; (2) if $\nu > 1$, then s is a *binary* number encoding of a number $d \in \mathbb{N}$ such that $n = \exp_{\nu-2}(d)$. Let $encode_\nu\,(n)$ denote the scalar that ν-encodes the number n, and let $decode_\nu\,(s)$ denote the number that is ν-encoded by the appropriate scalar s.

Now we give a formal definition of bit-vector formulas with the operators we use throughout in the rest of the paper. Let us suppose that an encoding ν is fixed. A *bit-vector term* t of bit-width n is denoted by $t^{[s]}$ where $s = encode_\nu\,(n)$, and defined inductively as follows:

---

[1] For example, the common operators *extraction* and *zero/sign extensions* use scalar arguments as well, cf. [19,14,15,16,17,18].

	term	condition	bit-width
constant:	$c^{[s]}$	$c \in \mathbb{N}, 0 \le c < 2^n$	n
variable:	$x^{[s]}$	x is an identifier	n
bitwise negation:	$\sim t^{[s]}$	$t^{[s]}$ is a term	n
bitwise and/or/xor, addition: $\bullet \in \{\&, |, \oplus, +\}$	$(t_1^{[s]} \bullet t_2^{[s]})$	$t_1^{[s]}, t_2^{[s]}$ are terms	n
equality, unsigned less than: $\bullet \in \{=, <_\mathsf{u}\}$	$(t_1^{[s]} \bullet t_2^{[s]})$	$t_1^{[s]}, t_2^{[s]}$ are terms	1

Note that the value c of a bit-vector constant is *not* a scalar, therefore it is always encoded as a binary number, regardless of ν. By a *bit-vector formula* we mean a term of bit-width 1, since this case can be considered as the Boolean case. For better readability, we write $\neg, \wedge, \vee$ instead of $\sim, \&, |$ for bit-width 1, respectively. Given a bit-vector operator set Ω, let $\mathcal{BV}_\nu^\Omega$ denote the fragment of QF_BV that applies the encoding ν to scalars and only uses operators from Ω. *Bit-blasting*, or flattening [34], interprets bit-vector variables as strings of Boolean variables and translates bit-vector operations into Boolean formulas. By denoting the Boolean logic as $\mathcal{BO}$, we give a formal definition.

Definition 1 (Bit-blasting). *Given an operator set Ω, a bit-blasting function $bblast_\nu^\Omega : \mathcal{BV}_\nu^\Omega \mapsto \mathcal{BO}$ is defined as follows:*

$$bblast_\nu^\Omega \left(\psi(x_1^{[s_1]}, \dots, x_k^{[s_k]}) \right) = \phi(y_1^1, \dots, y_1^{n_1}, \dots, y_k^1, \dots, y_k^{n_k}, z_1, \dots, z_l)$$

where $n_i = decode_\nu(s_i)$, such that $\forall d_1 \in \mathbb{B}^{n_1}, \dots, d_k \in \mathbb{B}^{n_k}$

$$\psi(d_1, \dots, d_k) = true \text{ iff}$$
$$\exists! e_1, \dots, e_l \in \mathbb{B} . \phi(d_1^1, \dots, d_1^{n_1}, \dots, d_k^1, \dots, d_k^{n_k}, e_1, \dots, e_l) = true$$

where d_i^j denotes the jth bit of d_i.

Note that the additional Boolean values $e_1, \dots, e_l$ are uniquely existentially quantified. Therefore, in fact, each Boolean variable z_i can rather be considered as a bit-vector function $f_i(x_1^{[s_1]}, \dots, x_k^{[s_k]}) : \mathbb{B}^{n_1} \times \cdots \times \mathbb{B}^{n_k} \mapsto \mathbb{B}$. Thus, ψ and ϕ encode the same $\left(\sum_{i=1}^k n_i \right)$-ary relation over $\mathbb{B}$.

We say that $bblast_\nu^\Omega$ is *log-space computable in bit-width* if it is log-space computable in $\sum_{i=1}^k n_i$. Let Π denote the set of all the bit-vector operators such that $bblast_\nu^\Pi$ is log-space computable in bit-width, for all $\nu \in \mathbb{N}^+$. Note that all the common bit-vector operators [19] fall into Π.

4   Motivating Example: Word-Level Model Checking

We now demonstrate that our generic main results can be applied to the important example of reachability analysis in model checking, as to establish the complexity of reachability in word-level model checking.

The model checking problem has a natural representation with a relational signature $\tau = (I^1, T^2, P^1)$. In model checking terminology, I represents the set of initial states, T the transition relation, and P the condition to check, i.e., the set of states whose reachability we want to verify. Thus, a structure $\mathcal{A} = (U, \widehat{I}^1, \widehat{T}^2, \widehat{P}^1)$ is essentially a Kripke structure. *Reachability analysis* in $\mathcal{A}$ means to check if there exists a reachable $\widehat{P}$-state in the defined transition system, i.e., if $\exists s_0, s_1, \ldots, s_k \in U$ such that   (1) $s_0 \in \widehat{I}$, (2) $\forall i \in [1,k]$. $(s_{i-1}, s_i) \in \widehat{T}$, and (3) $s_k \in \widehat{P}$. We call $MC = \{\mathcal{A} \in Struct(\tau) \mid \exists$ a reachable $\widehat{P}$-state in $\mathcal{A}\}$ the (explicit) *model checking* problem. Since MC is a simple variant of graph reachability, we know from [31] that MC is NL-complete under quantifier-free reductions.

The *word-level* encoding of MC means to encode the states by tuples of bit-vectors, and to define the relations $\widehat{I}, \widehat{T}, \widehat{P}$ by bit-vector formulas. The corresponding decision problem is called $bv_\nu^\Omega(MC)$, where ν specifies the scalar encoding and Ω is a set of bit-vector operators that are allowed in the formulas. We will formally define this problem in Sec. 5.

Our results require the following assumptions on Ω: (1) Ω contains only such operators for which bit-blasting is log-space computable in bit-width and (2) Ω contains all the simple operators $\wedge, \vee, \sim, =, +_1$. In particular, Ω may contain all common bit-vector operators [19] that are used in practice.

Then we obtain the following results as a direct consequence of Thm. 1, Cor. 2, and the NL-completeness of MC:

Corollary 1. *Let $\Omega \subseteq \Pi$. The decision problem $bv_\nu^\Omega(MC)$ is*

1. *PSPACE-complete, if $\nu = 1$ and $\Omega \supseteq \{\wedge, \vee, \neg\}$,*
2. *$(\nu - 1)$-ExpSpace-complete, if $\nu > 1$ and $\Omega \supseteq \{\wedge, \vee, \sim, =, +_1\}$,*

under log-space reductions.

In practice, the term *word-level model checking* usually refers to the problem $bv_2^\Omega(MC)$, i.e., all scalars in the formulas are encoded as w.l.o.g. binary numbers. Thus, our results show that word-level model checking is ExpSpace-complete.

5   Bit-Vector Representation of Problems

Our intention is to represent instances of computational problems as bit-vector formulas. More precisely, given a relational signature $\tau = (P_1^{a_1}, \ldots, P_k^{a_k})$, we define what the bit-vector definition of a corresponding relation $\widehat{P}_i^{a_i}$ looks like and what structure these definitions generate.

In order to simplify the presentation, we introduce the concept of *term vectors*. A term vector is a sequence $t_1^{[s_1]}, \ldots, t_l^{[s_l]}$ of bit-vector terms. We write term vectors in boldface, i.e., $\mathbf{t} = t_1^{[s_1]}, \ldots, t_l^{[s_l]}$, and say that $\mathbf{t}$ has the bit-width signature $s_1, \ldots, s_l$. We distinguish the special case when terms are variables, by denoting *variable vectors* as $\mathbf{x}, \mathbf{y}, \mathbf{z}$.

Word-level model checking can again serve as motivation here, since it represents the states of a transition system by the same set of bit-vector variables

$x_1^{[s_1]}, \ldots, x_l^{[s_l]}$. I.e., a state is in fact can be represented as the valuation of terms $t_1^{[s_1]}, \ldots, t_l^{[s_l]}$ assigned to those variables. Therefore, it is important that each state must have the same bit-width signature $s_1, \ldots, s_l$.

Definition 2. *Let $\mathbf{x}_1, \ldots, \mathbf{x}_a$ be variable vectors each of which has the bit-width signature $s_1, \ldots, s_l$. Let ν be a scalar encoding, and let $n_i = decode_\nu(s_i)$ denote the actual bit-widths. A bit-vector formula $\psi(\mathbf{x}_1, \ldots, \mathbf{x}_a)$ defines the a-ary relation*

$$gen_\nu^a(\psi) = \{(d_1, \ldots, d_a) \in (\mathbb{B}^{n_1} \times \cdots \times \mathbb{B}^{n_l})^a \mid \psi(d_1, \ldots, d_a) = true\}.$$

Let $\tau = (P_1^{a_1}, \ldots, P_k^{a_k})$ be a relational signature. The tuple of definitions

$$\Psi = \Big(P_1(\mathbf{x}_1^1, \ldots, \mathbf{x}_{a_1}^1) := \psi_1(\mathbf{x}_1^1, \ldots, \mathbf{x}_{a_1}^1),$$
$$\ldots,$$
$$P_k(\mathbf{x}_1^k, \ldots, \mathbf{x}_{a_k}^k) := \psi_k(\mathbf{x}_1^k, \ldots, \mathbf{x}_{a_k}^k)\Big)$$

where each ψ_i is a bit-vector formula and each $\mathbf{x}_j^i$ is a variable vector that has the bit-width signature $s_1, \ldots, s_l$, defines the τ-structure

$$gen_\nu^\tau(\Psi) = \big(\mathbb{B}^{n_1} \times \cdots \times \mathbb{B}^{n_l}, gen_\nu^{a_1}(\psi_1), \ldots, gen_\nu^{a_k}(\psi_k)\big).$$

The bit-vector representation of a computational problem consists of all the bit-vector representations of all the structures in the problem. Besides the definitions Ψ of relations, it is also necessary to include the scalar encoding ν to use, as follows.

Definition 3. *Let $A \subseteq Struct(\tau)$ be a problem, ν a scalar encoding, and Ω a set of bit-vector operators. Then we define*

$$bv_\nu^\Omega(A) = \{(\Psi, \nu) \mid gen_\nu^\tau(\Psi) \in A, \text{ and } \Psi \text{ contains only } \mathcal{BV}_\nu^\Omega \text{ formulas}\}.$$

In order to show how *membership* for a standard complexity class C can be automatically lifted when bit-vector representation is used, we give a necessary, although not very strong, criterion on the operator set. This criterion is based on bit-blasting, and requires to use operators from Π, i.e., those which allow log-space computable bit-blasting in bit-width.

Theorem 1. *Given a problem A, a standard complexity class C, and an operator set $\Omega \subseteq \Pi$, if $A \in C$, then $bv_\nu^\Omega(A) \in \text{Exp}_\nu(C)$.*

6   Lifting Hardness

The main contribution of this paper is to show how hardness for a standard complexity class C can also be automatically lifted. Our most important theorem, Thm. 2 gives a rather general hardness result, from which we derive Cor. 2 to show hardness of bv_ν^Ω for $\text{Exp}_\nu(C)$, where $\Omega \supseteq \{\wedge, \vee, \sim, =, +_1\}$.

Our proofs employ the framework of *descriptive complexity theory* [31]. In particular, we use the standard assumption that all structures are equipped with a binary successor relation. Thus, the universe of a structure can be naturally seen as an initial segment of the natural numbers. Our complexity results for bit-vector encoded problems assume that the problems in explicit encoding are hard under *quantifier-free reductions*, i.e., quantifier-free interpretations with equality and the successor relation. Examples of such problems including those in Tab. 1 can be found in [35,36,37,38,39]. For natural problems, it is usually not difficult to rephrase an existing reduction as a quantifier-free reduction. Let $A \leqslant_{\mathrm{qf}} B$ resp. $A \leqslant_{\mathrm{L}} B$ denote that the problem A has a *quantifier-free* resp. *log-space* reduction to the problem B. Note that quantifier-free reductions are weaker than log-space reductions, i.e., $A \leqslant_{\mathrm{qf}} B$ implies $A \leqslant_{\mathrm{L}} B$. For exact background material and definitions, see [31].

The key steps for Thm. 2 are two lemmas. Lemma 1 ("Conversion Lemma") shows that a quantifier-free reduction between A and B can be lifted to a log-space reduction between $bv_\nu(A)$ and $bv_\nu(B)$. Lemma 2 shows that A is log-space reducible to $bv_\nu(long_\nu(A))$ where $long_\nu(\cdot)$ is an operator which *decreases* the complexity ν-exponentially. From these two lemmas, Thm. 2 follows easily. The methodology of this paper is closest to [30], which contains a more thorough discussion of related work, descriptive complexity, and complexity theoretic background.

Lemma 1 (Conversion Lemma). *Let $\Omega \supseteq \{\wedge, \vee, \sim, =, +_1\}$. Given two problems $A \subseteq Struct\,(\sigma)$ and $B \subseteq Struct\,(\tau)$, if $A \leqslant_{\mathrm{qf}} B$, then $bv_\nu^\Omega(A) \leqslant_{\mathrm{L}} bv_\nu^\Omega(B)$, for any ν.*

The role of the following definition is to obtain from a problem A another problem $long_\nu(A)$ of ν-exponentially lower complexity. In order to construct this latter problem, we are going to "blow up" the size of a structure in a potentially ν-exponential way. To this end, we view a structure $\mathcal{A}$ as a bit string, and interpret the bit string as a binary number $char(\mathcal{A})$. The bit string is obtained from the characteristic sequences of the relations in $\mathcal{A}$, i.e., for each tuple in lexicographic order, a single bit indicates whether the tuple is in the relation. Due to the presence of the successor relation, this notion is well defined.

Definition 4. *Given a structure $\mathcal{A} = (U, \widehat{P}_1, \ldots, \widehat{P}_k)$, let $char(\widehat{P}_i)$ denote the characteristic sequence of the tuples in $\widehat{P}_i$ in lexicographical order. Let $char(\mathcal{A})$ denote the binary number obtained by concatenating a leading 1 with the concatenation of $char(\widehat{P}_1), \ldots, char(\widehat{P}_k)$.*

We define $long_\nu(\mathcal{A}) = \{(V, \widehat{R}^1) \mid |V| = \exp_{\nu-1}(char(\mathcal{A}))$ and $|\widehat{R}| = |V|\}$. For a problem A, let $long_\nu(A) = \bigcup_{\mathcal{A} \in A} long_\nu(\mathcal{A})$. For a complexity class C, let $long_\nu(C) = \bigcup_{A \in C} long_\nu(A)$.

The next lemma shows that encoding the problem $long_\nu(A)$ as bit-vector formulas applying ν-encoding to scalars gives a ν-exponentially more succinct representation, to which, consequently, the original problem A can be reduced.

Lemma 2. *Given a problem A, $A \leqslant_L bv_\nu^\Omega(long_\nu(A))$ if one of the following conditions holds:*

1. *$\nu = 1$ and $\Omega \supseteq \{<_u\}$*
2. *$\nu > 1$ and $\Omega \supseteq \{=\}$*

Theorem 2 (Upgrading Theorem). *Let C_1 and C_2 be complexity classes such that $long_\nu(C_1) \subseteq C_2$. If a problem A is C_2-hard under quantifier-free reductions, then $bv_\nu^\Omega(A)$ is C_1-hard under log-space reductions if one of the following conditions holds:*

1. *$\nu = 1$ and $\Omega \supseteq \{\wedge, \vee, \sim, =, +_1, <_u\}$*
2. *$\nu > 1$ and $\Omega \supseteq \{\wedge, \vee, \sim, =, +_1\}$*

Proof. For any $B \in C_1$, by assumption $long_\nu(B) \in C_2$, and hence $long_\nu(B) \leqslant_{qf} A$. By Lemma 1, it follows that $bv_\nu^\Omega(long_\nu(B)) \leqslant_L bv_\nu^\Omega(A)$, regardless of the additional operator $<_u$ in the unary case. Furthermore, by Lemma 2, it holds that $B \leqslant_L bv_\nu^\Omega(long_\nu(B))$. To put them together, $B \leqslant_L bv_\nu^\Omega(long_\nu(B)) \leqslant_L bv_\nu^\Omega(A)$ and, therefore, $bv_\nu^\Omega(A)$ is C_1-hard. □

As we discussed before, the case of $\nu = 1$ shows the same complexity behavior as Boolean logic. Of course, this is no wonder, since all the operators in $\Omega = \{\wedge, \vee, \sim, =, +_1, <_u\}$, or more precisely, $\mathcal{BV}_1^\Omega$ allows *log-space computable* bit-blasting in bit-width, and also *in formula size*, since bit-widths are now encoded in unary form. Thus, $\mathcal{BV}_1^\Omega$ is log-space reducible to $\mathcal{BV}_1^{\{\wedge, \vee, \neg\}}$, since $\{\wedge, \vee, \neg\}$ is a functionally complete set of Boolean operators. As a consequence, one can strengthen the first statement of Thm. 2 further as follows: $bv_1^{\Omega'}(A)$ is C_1-hard for any $\Omega' \supseteq \{\wedge, \vee, \neg\}$. Note that this is consistent with corresponding results in [10,30]. As a direct consequence, we can give the following corollary.

Corollary 2. *Given a standard complexity class C and a problem A, if A is C-hard under quantifier-free reductions, then $bv_\nu^\Omega(A)$ is $\text{Exp}_\nu(C)$-hard under log-space reductions if one of the following conditions holds:*

1. *$\nu = 1$ and $\Omega \supseteq \{\wedge, \vee, \neg\}$*
2. *$\nu > 1$ and $\Omega \supseteq \{\wedge, \vee, \sim, =, +_1\}$*

7 Conclusion

This paper gives a generic method for asserting the complexity of bit-vector logic encoded problems. As corollary we obtain a new complexity result for word-level model checking, an important practical problem. Since all complexity classes with complete problems have problems complete under quantifier-free reductions [11], we obtain a comprehensive picture of the worst case complexity of problems in bit-vector encoding. Note that our results do not apply to *satisfiability* of bit-vector logic, because "existence of a solution" is not hard for a complexity class, and thus the assumption of the Conversion Lemma is not satisfied. Nevertheless, we expect that the complexity of satisfiability for multi-logarithmic encodings shows a similar behavior as the problems studied here. We leave an analysis of this question to future work.

References

1. Balcázar, J.L., Lozano, A., Torán, J.: The complexity of algorithmic problems on succinct instances. Computer Science, 351–377 (1992)
2. Borchert, B., Lozano, A.: Succinct circuit representations and leaf language classes are basically the same concept. Inf. Process. Lett. 59(4), 211–215 (1996)
3. Das, B., Scharpfenecker, P., Torán, J.: Succinct encodings of graph isomorphism. In: Dediu, A.-H., Martín-Vide, C., Sierra-Rodríguez, J.-L., Truthe, B. (eds.) LATA 2014. LNCS, vol. 8370, pp. 285–296. Springer, Heidelberg (2014)
4. Feigenbaum, J., Kannan, S., Vardi, M.Y., Viswanathan, M.: Complexity of problems on graphs represented as OBDDs. Chicago Journal of Theoretical Computer Science 5(5) (1999)
5. Galperin, H., Wigderson, A.: Succinct representations of graphs. Information and Control 56(3), 183–198 (1983)
6. Gottlob, G., Leone, N., Veith, H.: Succinctness as a source of complexity in logical formalisms. Annals of Pure and Applied Logic 97(1), 231–260 (1999)
7. Lozano, A., Balcázar, J.L.: The complexity of graph problems for succinctly represented graphs. In: Nagl, M. (ed.) WG 1989. LNCS, vol. 411, pp. 277–286. Springer, Heidelberg (1990)
8. Papadimitriou, C.H., Yannakakis, M.: A note on succinct representations of graphs. Information and Control 71(3), 181–185 (1986)
9. Veith, H.: Succinct representation, leaf languages, and projection reductions. In: IEEE Conference on Computational Complexity, pp. 118–126 (1996)
10. Veith, H.: Languages represented by boolean formulas. Inf. Process. Lett. 63(5), 251–256 (1997)
11. Veith, H.: Succinct representation, leaf languages, and projection reductions. Information and Computation 142(2), 207–236 (1998)
12. Wagner, K.W.: The complexity of combinatorial problems with succinct input representation. Acta Informatica 23(3), 325–356 (1986)
13. Fröhlich, A., Kovásznai, G., Biere, A.: More on the complexity of quantifier-free fixed-size bit-vector logics with binary encoding. In: Bulatov, A.A., Shur, A.M. (eds.) CSR 2013. LNCS, vol. 7913, pp. 378–390. Springer, Heidelberg (2013)
14. Barrett, C.W., Dill, D.L., Levitt, J.R.: A decision procedure for bit-vector arithmetic. In: Proc. DAC 1998, pp. 522–527 (1998)
15. Bjørner, N.S., Pichora, M.C.: Deciding fixed and non-fixed size bit-vectors. In: Steffen, B. (ed.) TACAS 1998. LNCS, vol. 1384, pp. 376–392. Springer, Heidelberg (1998)
16. Bruttomesso, R., Sharygina, N.: A scalable decision procedure for fixed-width bit-vectors. In: ICCAD, pp. 13–20. IEEE (2009)
17. Cyrluk, D., Möller, O., Rueß, H.: An efficient decision procedure for a theory of fixed-sized bitvectors with composition and extraction. In: Grumberg, O. (ed.) CAV 1997. LNCS, vol. 1254, pp. 60–71. Springer, Heidelberg (1997)
18. Franzén, A.: Efficient Solving of the Satisfiability Modulo Bit-Vectors Problem and Some Extensions to SMT. PhD thesis, University of Trento (2010)
19. Barrett, C., Stump, A., Tinelli, C.: The SMT-LIB standard: Version 2.0. In: Proc. SMT 2010 (2010)
20. Brummayer, R., Biere, A., Lonsing, F.: BTOR: bit-precise modelling of word-level problems for model checking. In: Proc. 1st International Workshop on Bit-Precise Reasoning, pp. 33–38. ACM, New York (2008)

21. Kovásznai, G., Fröhlich, A., Biere, A.: On the complexity of fixed-size bit-vector logics with binary encoded bit-width. In: Proc. SMT 2012, pp. 44–55 (2012)
22. Clarke Jr., E.M., Grumberg, O., Peled, D.A.: Model Checking. MIT Press, Cambridge (1999)
23. Bryant, R.E., Lahiri, S.K., Seshia, S.A.: Modeling and verifying systems using a logic of counter arithmetic with lambda expressions and uninterpreted functions. In: Brinksma, E., Larsen, K.G. (eds.) CAV 2002. LNCS, vol. 2404, pp. 78–92. Springer, Heidelberg (2002)
24. Manolios, P., Srinivasan, S.K., Vroon, D.: BAT: The bit-level analysis tool. In: Damm, W., Hermanns, H. (eds.) CAV 2007. LNCS, vol. 4590, pp. 303–306. Springer, Heidelberg (2007)
25. Bradley, A.R.: Understanding IC3. In: Cimatti, A., Sebastiani, R. (eds.) SAT 2012. LNCS, vol. 7317, pp. 1–14. Springer, Heidelberg (2012)
26. Brummayer, R., Biere, A.: Boolector: An efficient SMT solver for bit-vectors and arrays. In: Kowalewski, S., Philippou, A. (eds.) TACAS 2009. LNCS, vol. 5505, pp. 174–177. Springer, Heidelberg (2009)
27. de Moura, L., Bjørner, N.S.: Z3: An Efficient SMT Solver. In: Ramakrishnan, C.R., Rehof, J. (eds.) TACAS 2008. LNCS, vol. 4963, pp. 337–340. Springer, Heidelberg (2008)
28. Bjesse, P.: A practical approach to word level model checking of industrial netlists. In: Gupta, A., Malik, S. (eds.) CAV 2008. LNCS, vol. 5123, pp. 446–458. Springer, Heidelberg (2008)
29. Bjorner, N., McMillan, K., Rybalchenko, A.: Program verification as satisfiability modulo theories. In: Proc. SMT 2012, pp. 3–11 (2013)
30. Veith, H.: How to encode a logical structure by an OBDD. In: Proc. 13th Annual IEEE Conference on Computational Complexity, pp. 122–131. IEEE (1998)
31. Immerman, N.: Languages that capture complexity classes. SIAM Journal on Computing 16(4), 760–778 (1987)
32. Schwentick, T.: Padding and the expressive power of existential second-order logics. In: Nielsen, M. (ed.) CSL 1997. LNCS, vol. 1414, pp. 461–477. Springer, Heidelberg (1998)
33. Immerman, N.: Descriptive complexity. Springer (1999)
34. Kroening, D., Strichman, O.: Decision Procedures: An Algorithmic Point of View. Texts in Theoretical Computer Science. Springer (2008)
35. Stewart, I.A.: Complete problems involving boolean labelled structures and projection transactions. Journal of Logic and Computation 1(6), 861–882 (1991)
36. Stewart, I.A.: On completeness for NP via projection translations. In: Martini, S., Börger, E., Kleine Büning, H., Jäger, G., Richter, M.M. (eds.) CSL 1992. LNCS, vol. 702, pp. 353–366. Springer, Heidelberg (1993)
37. Stewart, I.A.: Using the Hamiltonian path operator to capture NP. Journal of Computer and System Sciences 45(1), 127–151 (1992)
38. Stewart, I.A.: On completeness for NP via projection translations. Mathematical Systems Theory 27(2), 125–157 (1994)
39. Stewart, I.A.: Complete problems for monotone NP. Theoretical Computer Science 145(1), 147–157 (1995)

Computational Complexity of Covering Three-Vertex Multigraphs

Jan Kratochvíl[1,*], Jan Arne Telle[2,**], and Marek Tesař[1,***]

[1] Department of Applied Mathematics, Faculty of Mathematics and Physics,
Charles University, Prague, Czech Republic
{honza,tesar}@kam.mff.cuni.cz
[2] Department of Informatics, University of Bergen,
Bergen, Norway
telle@ii.uib.no

Abstract. A covering projection from a graph G to a graph H is a mapping of the vertices of G to the vertices of H such that, for every vertex v of G, the neighborhood of v is mapped bijectively to the neighborhood of its image. Moreover, if G and H are multigraphs, then this local bijection has to preserve multiplicities of the neighbors as well. The notion of covering projection stems from topology, but has found applications in areas such as the theory of local computation and construction of highly symmetric graphs. It provides a restrictive variant of the constraint satisfaction problem with additional symmetry constraints on the behavior of the homomorphisms of the structures involved.

We investigate the computational complexity of the problem of deciding the existence of a covering projection from an input graph G to a fixed target graph H. Among other partial results this problem has been shown to be NP-hard for simple regular graphs H of valency greater than 2, and a full characterization of computational complexity has been shown for target multigraphs with 2 vertices. We extend the previously known results to the ternary case, i.e., we give a full characterization of the computational complexity in the case of multigraphs with 3 vertices. We show that even in this case a P/NP-completeness dichotomy holds.

Keywords: Computational Complexity, Graph Homomorphism, Covering Projection.

1 Introduction

The concept of covering spaces or covering projections stems from topology, but has attracted a lot of attention in algebra, combinatorics, and also the theory of computation. For instance, it is used in algebraic graph theory as a very useful tool for the construction of highly symmetric graphs. The applications in

---

* Supported by Czech research grant P202/12/G061 - CE-ITI.
** Supported by the Norwegian Research Council, project PARALGO.
*** Supported by Charles University by the grant SVV-2014-260103.

E. Csuhaj-Varjú et al. (Eds.): MFCS 2014, Part II, LNCS 8635, pp. 493–504, 2014.
© Springer-Verlag Berlin Heidelberg 2014

computability include the theory of local computations (cf. [2] and [7]). A lot of interest has been paid to graphs that allow finite planar covers. This class of graphs is closed in the minor order and hence recognizable in polynomial time, yet despite a lot of effort no concrete recognition algorithm is known, since the obstruction set has not been determined yet. The class has been conjectured to be equal to the class of projective planar graphs by Negami [19] (for the most recent results cf. [11,12]).

In [1], Abello et al. raised another complexity question, asking about the computational complexity of deciding the existence of a covering projection from an input graph G to a fixed graph H (hoping for a characterization giving a P/NP-completeness dichotomy depending on H). A similar question when both G and H are part of the input was shown NP-complete by Bodlaender already in 1989 [4]. The dichotomy asked for by Abello et al. seems to be hard to obtain and only very partial results are known. The most general NP-completeness result states that for every simple regular graph H of valency at least 3, the problem is NP-complete [17]. No plausible conjecture on the borderline between polynomially solvable and NP-complete instances has been published so far, yet it is believed that a P/NP-completeness dichotomy will hold, as in the case of the constraint satisfaction problem (CSP).

The relation to CSP is worth mentioning in more detail. As shown in [9], for every fixed graph H, the H-COVER problem can be reduced to CSP, but mostly to NP-complete cases of CSP, so this reduction does not help. In a sense a covering projection is itself a variant of CSP, but with further constraints of local symmetry. Thus the dichotomy conjecture for H-COVER does not follow from the well-known Feder-Vardi dichotomy conjecture for CSP (cf. [8]).

In [16] it is shown that in order to fully understand the H-COVER problem for simple graphs, one has to understand its generalization for colored mixed multigraphs. For this reason we are dealing with multigraphs (undirected) in this paper. Kratochvil et al. [16] completely characterized the computational complexity of the H-COVER problem for colored mixed multigraphs on two vertices. The aim of this paper is to extend this characterization to 3-vertex multigraphs (in the undirected and monochromatic case). The characterization is described in the next section. It is more involved than the case of 2-vertex multigraphs, but this should not be surprising as ternary structures tend to be substantially more difficult than their binary counterparts. An analogue in CSP is the dichotomy of binary CSP proved by Schaefer in the 70's [20] followed by the characterization of CSP into ternary structures by Bulatov almost 30 years later [5].

2   Preliminaries and Statement of Our Results

For the sake of brevity we reserve the term "graph" for a multigraph. We denote the set of vertices of a graph G by $V(G)$ and the set of edges by $E(G)$. For two vertices u, v of G we denote the number of distinct edges between u and v by $m_G(u, v)$ and we say that uv is an $m_G(u, v)$-edge. The degree of vertex v of G

is denoted by $\deg_G(v)$ (recall that in multigraphs, the degree of a vertex v is defined as the number of edges going to other vertices plus twice the number of loops at v, i.e. $deg_G(v) = 2m_G(v,v) + \sum_{u \neq v} m_G(u,v)$). By $N_G(v)$ we denote the multiset of neighbors of vertex v in G where the multiplicity of v in $N_G(v)$ is $2m_G(v,v)$ and for every $u \neq v$ the multiplicity is $m_G(u,v)$. We omit G in the subscript if G is clear from the context.

Suppose A and B are two multisets. Let A', resp. B' be the set of different elements from A, resp. B. We say that a mapping $g: A' \to B'$ is a bijection from A to B if for every $b' \in B'$ the sum of multiplicities of all elements from $g^{-1}(b')$ in A equals the multiplicity of b' in B (note that g is not necessarily a bijection between sets A' and B'). If C' is a set then by $A \cap C'$ we mean a multiset that contains only elements from $A' \cap C'$ with the multiplicities corresponding to A. We denote the sum of multiplicities of all elements in A by $|A|$.

Let G and H be graphs. A *homomorphism* $f : V(G) \to V(H)$ is an edge preserving mapping from $V(G)$ to $V(H)$. A homomorphism f is a *covering projection* if $N_G(v)$ is mapped to $N_H(f(v))$ bijectively for every $v \in V(G)$ (here we consider the multiset bijection). Note that by the definition a covering projection is not necessarily surjective. The notion of a covering projection is also known as a *locally bijective homomorphism* or simply a *cover*. In this paper we denote a covering projection f from G to H by $f: G \to H$.

Strictly speaking, a covering projection (as the notion follows from topology) should be defined by a pair of mappings – one on the vertices and one on the edges of the graphs involved. But it was shown in [16] (using König's theorem and 2-factorization of $2k$-regular graphs) that every cover (defined as above) can be extended to a topological covering projection $f : V(G) \cup E(G) \to V(H) \cup E(H)$.

In this paper we consider the following decision problem.

Problem: H-COVER
Parameter: Fixed graph H.
Input: Graph G.
Task: Does there exist a covering projection $f : G \to H$?

Note that the problem H-COVER belongs to NP as we can guess a mapping $f: V(G) \to V(H)$ and verify if f is a covering projection in polynomial time. This means that in our NP-completeness results we only prove the NP-hardness part.

An equitable partition of a graph G is a partition of its vertex set into blocks $B_1, \ldots, B_d$ such that for every $i, j = 1, \ldots, d$ and every vertex v in B_i it holds that $|N_G(v) \cap B_j| = r_{i,j}$ (recall that $N_G(v)$ is generally a multiset). We call the matrix $M = (r_{i,j})$ corresponding to the coarsest equitable partition $B_1, \ldots, B_d$ of G (ordered in some canonical way; see Corneil and Gotlieb [6]) the degree refinement matrix of G, denoted by $drm(G)$, and we say that G is a d-block graph. Note that 1-block graphs are exactly regular graphs (despite the fact that vertices can contain a different number of loops).

It is also known that if G covers H via a covering f, then $drm(G) = drm(H)$. In particular, f preserves the coarsest equitable partition of G, i.e., if $B'_1, \ldots, B'_d$, resp. $B_1, \ldots, B_d$ are the blocks in the partition of G, resp. H then $f(B'_i) = B_i$ for every $i = 1, \ldots, d$. Since the matrix $drm(G)$ can be computed in time polynomial in the size of G, in this paper, we assume that $drm(G) = drm(H)$.

For every quadruplet of non-negative integers k, l, x, y we define a graph $S(k, l, x, y)$ on the vertex set $\{a, b, c\}$ with the following edge multiplicities (see Figure 1):

- $m(a, c) = m(b, c) = k$ • $m(c, c) = l$
- $m(a, a) = m(b, b) = x$ • $m(a, b) = y$

In this paper we focus on graphs H having exactly three vertices. For such graphs we give the full computational complexity characterization of H-COVER. More precisely, we show the following P/NP-completeness dichotomy.

Observation 1. *Let H be a 3-block graph on three vertices. Then H-COVER is polynomially solvable.*

Theorem 1. *Let H be a 2-block graph on three vertices. If H is isomorphic to $S(k', l, x, 0), S(k', l, 0, y)$ or $S(2, l, 0, 0)$, where $k' \in \{0, 1\}$ and $l, x, y \geq 0$, then H-COVER is polynomially solvable. Otherwise H-COVER is NP-complete.*

Theorem 2. *Let H be a t-regular graph on three vertices. If H is disconnected or $t \leq 2$, then H-COVER is polynomially solvable. Otherwise, H-COVER is NP-complete.*

Note that whenever H-COVER is polynomially solvable then we are able to find a corresponding covering projection in polynomial time, as well. That follows directly from the proofs of Observation 1, Theorem 1, and Theorem 2.

Observation 1 follows from the fact that if $drm(G) = drm(H)$ then the only mapping $f: V(G) \to V(H)$ that preserves the blocks is a covering projection.

In Section 3 we state the necessary lemmata for the proof of Theorem 1. Section 4 is devoted to the proof of Theorem 2. All polynomial cases are covered by Lemma 5. We then introduce a new decision problem - H-COVER*. We prove that this problem is NP-complete for all connected t-regular graphs H with $t \geq 4$. The proof is based on mathematical induction where we are able to use a stronger induction hypothesis than with simple H-COVER. NP-hardness of H-COVER then follows from the fact that H-COVER* is reducible to H-COVER in polynomial time. Note that due to space limitation only the full version of the paper will contain all necessary lemmata and proofs.

Let us give a few more technical definitions and notations. Throughout the rest of the paper we reserve the letter H for a graph on 3 vertices a, b, and c.

Let m, n, z be integers such that $m \geq n > 0$ and $z \geq 0$. We define a graph $H(m, n, z)$ to be the graph on the vertex set $\{a, b, c\}$ such that (see Figure 1):

- $m(a, a) = m$ • $m(b, b) = n$
- $m(a, b) = z$ • $m(b, c) = z + 2m$
- $m(a, c) = z + 2n$ • $m(c, c) = 0$

Let G, F and H be graphs. From the definition of a covering projection it is easy to show that if $f: G \to F$ and $g: F \to H$ are two covering projections then the composition $g \circ f: G \to H$ is also a covering projection. Since every graph isomorphism is a covering projection, every time we investigate the complexity

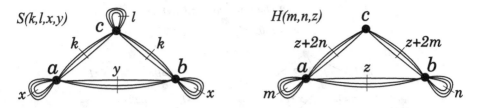

Fig. 1. The graphs $S(k, l, x, y)$ and $H(m, n, z)$

of H-COVER where H is isomorphic to $S(k, l, x, y)$ or $H(m, n, z)$, we can and we will assume that $H = S(k, l, x, y)$ or $H = H(m, n, z)$.

By a boundary $\delta_G(F)$ of an induced subgraph F of a graph G we mean the subset of vertices of F that are adjacent to at least one vertex outside F.

Let A, B be sets and let $f: A \to B$ be a mapping. Then we define $f(A) = \bigcup_{a \in A}\{f(a)\}$. If $f(A)$ contains only one element, say x, then we simply write $f(A) = x$ instead of $f(A) = \{x\}$.

3   Complexity for 2-Block Graphs on Three Vertices

In this section we provide the proof of Theorem 1. We will assume that H is a 2-block graph with the blocks $\{a, b\}$ and $\{c\}$. From the definition of an equitable partition we have $deg_H(a) = deg_H(b) \neq deg_H(c)$. The next proposition shows the connection between graphs $S(k, l, x, y)$ and 2-block graphs.

Proposition 1. *Every 2-block graph H on three vertices is isomorphic to some $S(k, l, x, y)$, where $2x + y \neq 2l + k$.*

Proof. Since we cannot distinguish vertices a and b in the block $\{a, b\}$ we have $m(a, a) = m(b, b) = x$ and $m(a, c) = m(b, c) = k$. This means that H is isomorphic to $S(k, l, x, y)$, where $l = m(c, c)$ and $y = m(a, b)$. The inequality $2x + y \neq 2l + k$ then follows directly from the fact that $deg_H(a) \neq deg_H(c)$. □

Before we proceed to the proof of Theorem 1 we split all 2-block graphs into three subsets and show the complexity separately for each subset. Figure 2 shows how we split these graphs, and shows also the computational complexity of H-COVER for the graphs H in the corresponding subset.

Lemma 1. *Let H be a 2-block graph on three vertices. If H is isomorphic to $S(k', l, x, 0)$, $S(k', l, 0, y)$ or $S(2, l, 0, 0)$ for some $k' \in \{0, 1\}$ and $l, x, y \geq 0$ then H-COVER is polynomially solvable.*

Proof. Let G be the input to H-COVER and let AB, resp. C be the block of G that corresponds to the block $\{a, b\}$, resp. $\{c\}$ of H.

First suppose that H is isomorphic to $S(k', l, x, 0)$ or $S(k', l, 0, y)$. We will construct a conjunctive normal form boolean formula φ_G with clauses of size 2, such that φ_G is satisfiable if and only if G covers H.

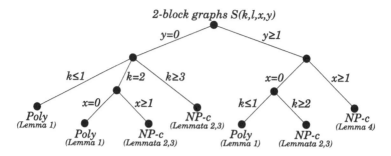

Fig. 2. The partition of 2-block graphs. Leaf vertices denote the computational complexity of H-COVER for the corresponding graph H.

Let the variables of φ_G be $\{x_u | u \in AB\}$ and for each $u, v \in AB$ we add to φ_G the following clauses:

- $(x_u \vee x_v)$ and $(\neg x_u \vee \neg x_v)$, if $u \neq v$ and u, v share a neighbor in C
- $(x_u \vee \neg x_v)$ and $(\neg x_u \vee x_v)$, if $uv \in E(G)$ and $H = S(k', l, x, 0)$
- $(x_u \vee x_v)$ and $(\neg x_u \vee \neg x_v)$, if $uv \in E(G)$ and $H = S(k', l, 0, y)$

Suppose that φ_G is satisfiable and fix one satisfying evaluation of variables. Define a mapping $f \colon V(G) \to V(H)$ by:

- $f(u) = a$, if $u \in AB$ and x_u is positive
- $f(u) = b$, if $u \in AB$ and x_u is negative
- $f(u) = c$, if $u \in C$

It is a routine check to show that f is a covering projection from G to H. On the other hand, if $f \colon G \to H$ is a covering projection then we can define an evaluation of φ_G such that x_u is positive if and only if $f(u) = a$. Such an evaluation satisfies the formula φ_G since there is exactly one positive literal in every clause. The fact that the size of φ_G is polynomial in the size of G and 2-SAT is polynomially solvable implies that H-COVER is polynomially solvable.

In the rest of the proof we suppose that $H = S(2, l, 0, 0)$. In this case the graph G covers H if and only if we can color the vertices of AB by two colors, say black and white, in such a way that for each $u \in C$ exactly two out of four vertices from $N_G(u) \cap AB$ are black.

We construct an auxiliary 4-regular graph G'. Let $V(G') = C$ and let edges of G' correspond to the vertices of AB, and connect its two neighbors in C. Note that G' can generally contain loops and multi-edges.

Then the coloring of vertices of AB in G corresponds to the coloring of edges of G' such that the black edges induce a 2-factor of G'. The problem of deciding the existence of a 2-factor in a 4-regular graph can be solved in polynomial time. In fact, such a 2-factor always exists and can be find in polynomial time.   □

In Lemma 2 we deduce NP-hardness of H-COVER from the following problem.

Problem: m-IN-$2m$-SAT$_q$
Input: A formula φ in CNF where every clause contains exactly $2m$ variables without negation and every variable occurs in φ exactly q times.

Task: Does there exist an evaluation of the variables of φ such that every clause contains exactly m positively valued variables?

Kratochvíl [14, Corollary 1] shows that this problem is NP-complete for every $q \geq 3$ and $m \geq 2$. If formula φ is a positive instance of m-in-$2m$-SAT$_q$ we simply say that φ is m-in-$2m$ satisfiable.

For the purposes of our NP-hardness deductions in Lemma 2 we will build a specific gadget according to the following needs:

Definition 1 (Variable gadget). Let $H = S(k, l, x, y)$ and let F be a graph with $2q$ specified vertices $S = \{s_1, \ldots, s_q\}$ and $S' = \{s'_1, \ldots, s'_q\}$ of degree one. Let V, resp. V' be the set of neighbors of vertices in S, resp. S' in F. Suppose that whenever F is an induced subgraph of G with $\delta_G(F) \subseteq S \cup S'$ and $f : G \to H$ is a covering projection then $f(S \cup S') = c$ and one of the following occurs:

$$i)\ f(V) = a \text{ and } f(V') = b \qquad iii)\ f(V \cup V') = a$$
$$ii)\ f(V) = b \text{ and } f(V') = a \qquad iv)\ f(V \cup V') = b$$

Furthermore, suppose that any mapping $f : S \cup S' \cup V \cup V' \to V(H)$ *such that* $f(S \cup S') = c$ *and satisfying* i) *or* ii) *can be extended to* $V(F)$ *in such a way that for each* $u \in V(F) \setminus (S \cup S')$ *the restriction of* f *to* $N_F(u)$ *is a bijection to* $N_H(f(u))$.
We denote such F *by* $VG_H(q)$ *and we call it a* variable gadget *of size* q.

The next lemma shows how we use variable gadgets while Lemma 3 proves that $VG_H(q)$ exists for some graphs $S(k, l, x, 0)$, $S(2, l, x, 0)$, and $S(k, l, 0, y)$. Note that in Definition 1 and Lemma 2 we do not use the fact that H is a 2-block graph. Hence, we can use this lemma also in Section 4.

Lemma 2. *Let* $k \geq 2$ *and let* $H = S(k, l, x, y)$. *If for some* $q \geq 3$ *there exists a variable gadget* $VG_H(q)$ *then* H-COVER *is NP-complete.*

Proof. We deduce NP-hardness of H-COVER from k-IN-$2k$-SAT$_q$. Let φ be an instance of k-IN-$2k$-SAT$_q$. Let $x_1, x_2, \ldots, x_n$, resp. $C_1, C_2, \ldots, C_m$ be the variables, resp. clauses of φ. For every clause C_i denote the variables in C_i by $l_i^1, \ldots, l_i^{2k}$ (recall that all variables have only positive appearances in φ). We construct a graph G_φ such that G_φ covers H if and only if φ is k-in-$2k$ satisfiable.

We start the construction of G_φ by taking vertices $c_1, \ldots, c_m, c'_1, \ldots, c'_m$ (corresponding to the clauses of φ) and we add l loops to each of them. For every variable x_i we take a copy $VG^i(q)$ of variable gadget $VG_H(q)$. Denote the copy of S, S', V, resp. V' in $VG^i(q)$ simply by S^i, S'^i, V^i, reps. V'^i. For every occurrence of x_i in C_j we identify one vertex from S^i, resp. S'^i with c_j, resp. c'_j. We do it in such a way that every vertex from $S^i \cup S'^i$ is identified exactly once, see Figure 3.

We claim that G_φ covers H if and only if φ is k-in-$2k$ satisfiable.

Suppose that there exists a covering projection $f : G_\varphi \to H$. We define an evaluation of the variables of φ such that x_i is true if and only if $f(V^i) = a$.

From the properties of a variable gadget we know that $f(c_j) = c$ for every $j = 1, \ldots, m$. Then $|N_{G_\varphi}(c_j) \cap f^{-1}(a)| = |N_{G_\varphi}(c_j) \cap f^{-1}(b)| = k$. This means that in every clause of φ there is exactly k positive as well as negative variables.

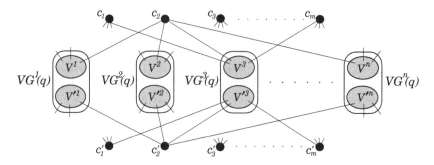

Fig. 3. The construction of the graph G_φ for $k = 2$ and $q = 3$. In this example φ contains a clause $C_2 = (x_1 \wedge x_2 \wedge x_3 \wedge x_n)$ and a variable x_3 appears in clauses C_1, C_2 and C_m.

For the opposite implication we fix one satisfying evaluation of φ. We define a mapping $f \colon V(G_\varphi) \to V(H)$ in the following way:

- $f(c_j) = f(c'_j) = c$, for all $j = 1, \ldots, m$
- $f(V^i) = a$ and $f(V'^i) = b$, if x_i is a positive variable
- $f(V^i) = b$ and $f(V'^i) = a$, if x_i is a negative variable

Then for each $i = 1, \ldots, n : f(S^i) = c$ and $f(V^i) \neq f(V'^i)$. By the definition of a variable gadget we know that f can be extended to every $VG^i(q)$ in such a way that for each $u \in V(VG^i(q)) \setminus (S^i \cup S'^i)$: the restriction of f to $N_{G_\varphi}(u)$ is a bijection to $N_H(f(u))$. It is a routine check to show that such a mapping f is a covering projection from G_φ to H. $\square$

Lemma 3. *If a 2-block graph H is one of the following:*

a) $S(k, l, x, 0)$, where $k \geq 3$, $l \geq 0$ and $x \geq 0$
b) $S(2, l, x, 0)$, where $l \geq 0$ and $x \geq 1$
c) $S(k, l, 0, y)$, where $k \geq 2$, $l \geq 0$ and $y \geq 1$

then there exists a variable gadget $VG_H(q)$ for some $q \geq 3$.

Proof. Depending on which of $a), b)$ and $c)$ holds for the graph H, we define $VG_H(q)$ and the corresponding sets S and S' as depicted in Figure 4. Note that in the case $a), b)$, resp. $c)$ we have that q is equal to $k, 4$, resp. $2k$.

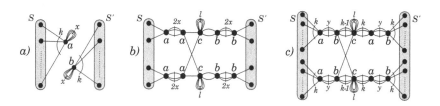

Fig. 4. Examples of the variable gadgets for the cases $a), b)$ and $c)$

The fact that the depicted graphs are really variable gadgets follows from a case analysis. Figure 4 also shows how one particular mapping $f \colon S \cup S' \cup V \cup V' \to H$ (where V, resp. V' are the neighbors of S, resp. S') can be extended to all vertices of $VG_H(q)$. Other conditions from the definition of $VG_H(q)$ follow from the fact that H has two blocks. □

Lemma 4. *Let $H = S(k, l, x, y)$ be a 2-block graph where $k, l \geq 0$ and $x, y \geq 1$. Then H-COVER is NP-complete.*

Proof. Kratochvíl et al. [16, Theorem 11] proved that if H' is a graph on two vertices L and R such that $x = m_{H'}(L, L) = m_{H'}(R, R) \geq 1$ and $y = m_{H'}(L, R) \geq 1$, then H'-COVER is NP-complete.

We deduce NP-hardness of H-COVER from H'-COVER. Let G' be an instance of H'-COVER. We construct a graph G such that G covers H if and only if G' covers H'.

We start the construction of G by taking two copies G^1 and G^2 of G'. Denote the copy of vertex $v \in V(G')$ in G^1, resp. G^2 by v^1, resp. v^2. For every $v \in V(G')$ we add to G a new vertex u_v with l loops and k-edges $v^1 u_v$ and $v^2 u_v$.

Suppose that $f \colon G \to H$ is a covering projection. Then $f(u_v) = c$ for every $v \in V(G')$ and f restricted to G^1 is a covering projection to H'. This means that G' covers H'.

For the opposite implication suppose that $f' \colon G' \to H'$ is a covering projection. We define a mapping $f \colon V(G) \to V(H)$ in the following way:

- $f(u_v) = c$
- $f(v^1) = f'(v)$
- $f(v^2) = a$ if $f(v^1) = b$, and $f(v^2) = b$ otherwise

for every $v \in V(G')$. It is a routine check to show that f is a covering projection from G to H. □

Next we proceed to the proof of Theorem 1.

Proof (of Theorem 1). The polynomial cases are settled by Lemma 1. The cases where $x, y \geq 1$ follow from Lemma 4. All other cases follow from Lemmata 2 and 3 (see Figure 2). □

4   Complexity for 1-Block Graphs on Three Vertices

In this section we focus on 1-block graphs H, i.e. regular graphs. We provide several definitions and lemmata that help us prove Theorem 2. The next lemma settles the polynomial cases.

Lemma 5. *Let H be a t-regular graph on three vertices. If H is disconnected or $t \leq 2$, then H-COVER is polynomially solvable.*

Proof. Let G be a t-regular graph. Let us first suppose that H is disconnected. Without loss of generality suppose that $m_H(a, c) = m_H(b, c) = 0$. We define a mapping $f\colon V(G) \to V(H)$ by $f(u) = c$ for every $u \in V(G)$. Then mapping f is a covering projection from G to H by the definition.

If H is connected and $t \leq 2$, then $t = 2$ and H is a triangle. A 2-regular graph G covers the triangle if and only if G consists of disjoint cycles of lengths divisible by 3. This condition can be easily verified in linear time.   □

For the NP-hardness part of Theorem 2 we use a reduction from a problem we call H-Cover*. To define H-Cover* we need the following definitions.

Definition 2. *Let G be a graph on $3n$ vertices and let $\mathcal{A} = \{A_1, A_2, \ldots, A_n\}$ be a partition of its vertices into n sets of size 3. Then we say that $\mathcal{A}$, resp. pair $(G, \mathcal{A})$ is a 3-partition, resp. graph 3-partition. Moreover, if $f\colon V(G) \to \{a, b, c\}$ is a mapping such that $f(A_i) = \{a, b, c\}$ for every $A_i \in \mathcal{A}$ then we say that f respects the 3-partition $\mathcal{A}$.*

Definition 3. *We say that a graph 3-partition $(G, \mathcal{A})$ covers* graph H if there exists a covering projection $f^*\colon G \to H$ that respects the 3-partition $\mathcal{A}$. We denote such a mapping by "$\to^*$" and call it a* covering projection* *or simply a* cover*.*

Definition 4. *Let $(G, \mathcal{A})$ be a graph 3-partition and let H be a graph. If the existence of a covering projection $f\colon G \to H$ implies the existence of a covering projection* $f^*\colon (G, \mathcal{A}) \to^* H$, then we say that $(G, \mathcal{A})$ is* nice *for H.*

Note it follows from these definitions that if G does not cover H then any graph 3-partition $(G, \mathcal{A})$ is nice for H.

> **Problem:** H-Cover*
> **Parameter:** Fixed graph H.
> **Input:** Nice graph 3-partition $(G, \mathcal{A})$ for H.
> **Task:** Does there exist a covering projection* $f\colon (G, \mathcal{A}) \to^* H$?

Similarly as H-Cover also the H-Cover* problem belongs to the class NP. This means that to show NP-completeness of H-Cover* we only need to prove NP-hardness.

Observation 2. *Let H be a graph. Then H-Cover* is polynomially reducible to H-Cover.*

Proof. Suppose that $(G, \mathcal{A})$ is an instance of H-Cover*. Since $(G, \mathcal{A})$ is nice for H we know that $(G, \mathcal{A})$ covers* H if and only if G covers H, which concludes the proof.   □

This observation allows us to prove NP-hardness of H-Cover* instead of H-Cover. We do this by mathematical induction. The key advantage of H-Cover* is that we can use a stronger induction hypothesis.

Theorem 3. *Let H be a connected t-regular graph on three vertices and $t \geq 4$. Then H-Cover* is NP-complete.*

In the rest of the paper we prove Theorem 3. We assume that H is a connected t-regular graph and $t \geq 4$.

The following lemma deduces NP-hardness of H-COVER* for a very special graph H, and will serve as an illustration of such deductions. NP-hardness of H-COVER* is deduced from a 3-edge coloring problem. Holyer [13] proved that this problem is NP-complete even for simple cubic graphs. Denote the 3-edge coloring problem for cubic graphs by 3-ECOL.

Lemma 6. *Let* $H = S(1, 1, 1, 1)$. *Then* H-COVER* *is NP-complete.*

Proof. We reduce the NP-hard problem 3-ECOL to H-COVER*. For every simple cubic graph F we construct a graph 3-partition $(G_F, \mathcal{A})$ such that $(G_F, \mathcal{A})$ covers* H if and only if F is 3-edge colorable.

For every vertex $u \in V(F)$ we insert to G_F vertices u_1, u_2, u_3 and we add 1-edges u_1u_2, u_2u_3 and u_3u_1. For every edge $uv \in E(F)$ we choose vertices u_i and v_j and we add 2-edge u_iv_j to G_F. We choose indices the i and j in such a way that the final graph G_F is 4-regular. We define the 3-partition $\mathcal{A}$ as $\bigcup_{u \in V(F)} \{\{u_1, u_2, u_3\}\}$.

We prove that $(G_F, \mathcal{A})$ is nice for H. Let $f : G_F \to H$ be a covering projection. Clearly all 2-edges of G_F must be mapped by f to loops of H. This implies that for every $u \in V(F)$ is $f(u_1, u_2, u_3) = \{a, b, c\}$ and so f respects $\mathcal{A}$.

Suppose that $f^* : (G_F, \mathcal{A}) \to^* H$ is a covering projection*. We know that every 2-edge u_iv_j corresponds to an edge uv of F and $f^*(u_i) = f^*(v_j)$. We define a coloring $col : E(F) \to V(H)$ by $c(uv) = f^*(u_i)$. The fact that f respects the 3-partition $\mathcal{A}$ implies that col is a proper 3-edge coloring of F.

In the rest of the proof suppose that $col : E(F) \to V(H)$ is a proper 3-edge coloring of F. We show that there exists a covering projection* $f^* : (G_F, \mathcal{A}) \to^* H$. For every 2-edge u_iv_j of G_F we define $f^*(u_i) = f^*(v_j) = col(uv)$. Since col is a proper 3-edge coloring of F we have $\{f^*(u_1), f^*(u_2), f^*(u_3)\} = \{a, b, c\}$ for every $u \in V(F)$. This means that f^* respects the 3-partition $\mathcal{A}$. It is a routine check to show that f^* is a covering and consequently a covering projection*. $\square$

As already mentioned, due to space limitation we have in this extended abstract removed the remainder of the lemmata needed for the proof of Theorem 3. These can be found in the full version of the paper. We proceed to the proof of Theorem 2 that handles the complexity of H-COVER for all 1-block graphs H on three vertices.

Proof (of Theorem 2). Lemma 5 covers all polynomial cases while Theorem 3 with Observation 2 covers the NP-complete cases. $\square$

5   Conclusion

We have settled the computational complexity of H-COVER for all multigraphs on three vertices. Not surprisingly, the characterization is substantially more involved than the characterization of the 2-vertex case. These results constitute an important step towards the goal of a full dichotomy for complexity of H-COVER of simple graphs, a goal that requires a full dichotomy also for colored mixed multigraphs, as shown in [16], and in particular a dichotomy for the multigraphs handled in this paper.

References

1. Abello, J., Fellows, M.R., Stillwell, J.C.: On the complexity and combinatorics of covering finite complexes. Australian Journal of Combinatorics 4, 103–112 (1991)
2. Angluin, D.: Local and global properties in networks of processors. In: Proceedings of the 12th ACM Symposium on Theory of Computing, pp. 82–93 (1980)
3. Angluin, D., Gardiner, A.: Finite common coverings of pairs of regular graphs. Journal of Combinatorial Theory, Series B 30(2), 184–187 (1981)
4. Bodlaender, H.L.: The Classification of Coverings of Processor Networks. Journal of Parallel and Distributed Computing 6, 166–182 (1989)
5. Bulatov, A.A.: A dichotomy theorem for constraint satisfaction problems on a 3-element set. J. ACM 53(1), 66–120 (2006)
6. Corneil, D.G., Gotlieb, C.C.: An Efficient Algorithm for Graph Isomorphism. J. ACM 17(1), 51–64 (1970)
7. Courcelle, B., Métivier, Y.: Coverings and Minors: Application to Local Computations in Graphs. European Journal of Combinatorics 15(2), 127–138 (1994)
8. Feder, T., Vardi, M.Y.: The computational structure of monotone monadic SNP and constraint satisfaction: a study through Datalog and group theory. SIAM Journal of Computing 1, 57–104 (1998)
9. Fiala, J., Kratochvíl, J.: Locally constrained graph homomorphisms-structure, complexity, and applications. Computer Science Review 2, 97–111 (2008)
10. Hell, P., Nešetřil, J.: Graphs and Homomorphisms. Oxford University Press (2004)
11. Hliněný, P.: $K_{4,4} - e$ Has No Finite Planar Cover. Journal of Graph Theory 21(1), 51–60 (1998)
12. Hliněný, P., Thomas, R.: On possible counterexamples to Negami's planar cover conjecture. Journal of Graph Theory 46(3), 183–206 (2004)
13. Holyer, I.: The NP-Completeness of Edge-Coloring. SIAM J. Comput. 10(4), 718–720 (1981)
14. Kratochvíl, J.: Complexity of Hypergraph Coloring and Seidel's Switching. In: Bodlaender, H.L. (ed.) WG 2003. LNCS, vol. 2880, pp. 297–308. Springer, Heidelberg (2003)
15. Kratochvíl, J., Proskurowski, A., Telle, J.A.: Covering Regular Graphs. Journal of Combinatorial Theory, Series B 71(1), 1–16 (1997)
16. Kratochvíl, J., Proskurowski, A., Telle, J.A.: Complexity of colored graph covers I. Colored directed multigraphs. In: Möhring, R.H. (ed.) WG 1997. LNCS, vol. 1335, pp. 242–257. Springer, Heidelberg (1997)
17. Kratochvíl, J., Proskurowski, A., Telle, J.A.: Complexity of graph covering problems. Nordic Journal of Computing 5, 173–195 (1998)
18. Litovsky, I., Métivier, Y., Zielonka, W.: The power and the limitations of local computations on graphs. In: Mayr, E.W. (ed.) WG 1992. LNCS, vol. 657, pp. 333–345. Springer, Heidelberg (1993)
19. Negami, S.: Graphs which have no planar covering. Bulletin of the Institute of Mathematics, Academia Sinica 4, 377–384 (1988)
20. Schaefer, T.J.: The complexity of satisfiability problems. In: Proceedings of the Tenth Annual ACM Symposium on Theory of Computing, STOC 1978, pp. 216–226 (1978)

Finding Maximum Common Biconnected Subgraphs in Series-Parallel Graphs*

Nils Kriege and Petra Mutzel

Dept. of Computer Science, Technische Universität Dortmund, Germany
{nils.kriege,petra.mutzel}@tu-dortmund.de

Abstract. The complexity of the maximum common subgraph problem in partial k-trees is still largely unknown. We consider the restricted case, where the input graphs are k-connected partial k-trees and the common subgraph is required to be k-connected. For biconnected outerplanar graphs this problem is solved and the general problem was reported to be tractable by means of tree decomposition techniques. We discuss key obstacles of tree decompositions arising for common subgraph problems that were ignored by previous algorithms and do not occur in outerplanar graphs. We introduce the concept of *potential separators*, i.e., separators of a subgraph to be searched that not necessarily are separators of the input graph. We characterize these separators and propose a polynomial time solution for series-parallel graphs based on SP-trees.

Keywords: Tree decomposition, maximum common subgraph, series-parallel graph.

1 Introduction

The complexity of the subgraph isomorphism problem in partial k-trees is exceptionally well studied [9,7,5,8], while this does not apply to the same extent to the related maximum common subgraph problem (MCS). Subgraph isomorphism is known to be tractable in partial k-trees that are k-connected or have bounded degree [9,6]. However, it is $\mathcal{NP}$-complete in partial k-trees if the smaller graph is not k-connected or has more than k vertices of unbounded degree [7]. Subgraph isomorphism even remains $\mathcal{NP}$-complete in connected outerplanar graphs [12]. Recently, polynomial time solutions were presented for connected MCS in outerplanar graphs under the additional restriction that blocks, i.e., maximal biconnected subgraphs, and bridges of the input graphs must be preserved (BBP-MCS) [11,3]. Based on this result it was shown that the problem can still be solved in polynomial time even when this restriction is dropped, but graphs have bounded degree [3]. The problem is also tractable for almost trees of bounded degree [1] and when one of the input graphs is a bounded-degree partial k-tree and the other is a connected graph with a polynomial number of possible spanning trees [13]. These results show that subgraph isomorphism and MCS both can be solved for certain restricted graph classes in polynomial time. However, only recently

---

* Research supported by the German Research Foundation (DFG), priority programme "Algorithms for Big Data" (SPP 1736).

E. Csuhaj-Varjú et al. (Eds.): MFCS 2014, Part II, LNCS 8635, pp. 505–516, 2014.

it was proven that MCS is $\mathcal{NP}$-complete in vertex-labeled partial 11-trees of bounded degree [2] — a class of graphs which allows for polynomial time subgraph isomorphism algorithms [9,6]. Essentially the complexity of MCS is unknown for graphs of bounded-degree that are not outerplanar and have tree width less than 11.

Furthermore, the problem with the requirement that the connectivity and tree width of the input graphs and the common subgraph coincide, is open.

Definition 1 (k-MCS). *Given two k-connected partial k-trees G and H, determine the maximum number of vertices in a common k-connected induced subgraph of G and H.*

Note that 1-MCS is equivalent to the maximum common subtree problem that is well known to be solvable in polynomial time [10]. A slightly different variant of MCS, where the common subgraph has a maximum number of edges and is not required to be induced, is referred to as *maximum common edge subgraph* (MCES). For outerplanar graphs, a subclass of the series-parallel graphs or partial 2-trees, 2-MCES was solved as a subproblem of BBP-MCES [11,3]. For arbitrary k the positive result for subgraph isomorphism in k-connected partial k-trees was reported to be transferred to MCS by means of normalized tree decompositions [4,8].[1] However, we show that approaches directly based on (normalized) tree decompositions fail and identify a key obstacle inherent to common subgraph problems that does not occur for subgraph isomorphism. We analyze these difficulties in the context of tree decompositions and present a solution for 2-MCS based on the SP-tree data structure. Notably these obstacles do not arise for trees and outerplanar graphs. Our solution can easily be extended to solve BBP-MCS in series-parallel graphs and may form the basis for solving k-MCS with $k > 2$.

2   Preliminaries

Given two undirected graphs G and H, a bijection $\phi : V(G) \rightarrow V(H)$ is a *(graph) isomorphism* if $\forall u, v \in V(G) : (u, v) \in E(G) \Leftrightarrow (\phi(u), \phi(v)) \in E(H)$. We write $G \simeq H$ if there is an isomorphism between G and H. A *subgraph isomorphism* from G to H is an isomorphism between G and $H[W]$, $W \subseteq V(H)$, where $H[W]$ is the subgraph induced by W in H. A *common subgraph isomorphism* between two graphs G and H is an isomorphism ϕ between $G[U]$, $U \subseteq V(G)$, and $H[W]$, $W \subseteq V(H)$. Then $C \simeq G[V] \simeq H[W]$ is said to be a *common subgraph*. The isomorphism ϕ induces the subgraph isomorphisms $\phi_G : V(C) \rightarrow V(G)$ and $\phi_H : V(C) \rightarrow V(H)$ from the common subgraph to the two input graphs G and H, respectively. A common subgraph C is *maximum* if there is no common subgraph C' with $|V(C')| > |V(C)|$.

Separators and Tree Decompositions. A set $S \subseteq V(G)$ is called $|S|$-*separator* or *separator* of a connected graph G if $G \setminus S := G[V(G) \setminus S]$ consists of at least two connected components. A separator S is called (a, b)-*separator* if $G \setminus S$ contains two disjoint connected components C and D, such that $a \in V(C)$ and $b \in V(D)$. A separator S is called *minimal* if there are vertices $a, b \in V(G)$, such that S is an (a, b)-separator, but there is no (a, b)-separator S' with $S' \subset S$. A separator S is said to *cross* another separator T if $G \setminus T$ contains components C, D such that $S \cap V(C) \neq \emptyset$ and

---

[1] In both references the result is attributed to an unpublished manuscript by F.J. Brandenburg that was kindly provided to the authors of this article.

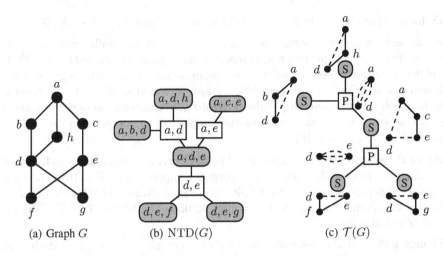

(a) Graph G (b) $\mathrm{NTD}(G)$ (c) $\mathcal{T}(G)$

Fig. 1. A biconnected partial 2-tree G (a), a normalized tree decomposition of G (b) and a SP-tree of G (c) with the associated skeleton graphs (dashed lines represent virtual edges)

$S \cap V(D) \neq \emptyset$. Let S, T be minimal separators, then S crosses T iff T crosses S; two non-crossing separators are said to be *parallel*. For example, in Fig. 1(a) the separator $\{a, e\}$ crosses $\{c, d\}$ and vice versa. A graph is called k-connected if it does not contain a j-separator with $j < k$, and it has connectivity $\kappa(G) = k$ if it is k-connected, but not $(k + 1)$-connected.

A *tree decomposition* of a graph G is a pair $(T, \mathcal{X})$, where T is a tree and $\mathcal{X} = (X_i)_{i \in V(T)}$ a family of vertex subsets $X_i \subseteq V(G)$ called *bags* satisfying:

T1 $V(G) = \bigcup_{i \in V(T)} X_i$,

T2 for every edge $(u, v) \in E(G)$ there is a node $i \in V(T)$ with $u, v \in X_i$,

T3 for every $i, j, k \in V(T)$, if j lies on the unique path with endpoints i and k then $X_i \cap X_k \subseteq X_j$.

The *width* of a tree decomposition $(T, \mathcal{X})$ is defined as $\max\{|X_i| - 1 : i \in V(T)\}$ and the *tree width* $\mathrm{tw}(G)$ of a graph G is the least width of any tree decomposition of G. The graphs with tree width $\leq k$ are also known as *partial k-trees*. Partial 2-trees correspond to *series-parallel* graphs and include the class of outerplanar graphs, i.e., graphs that can be drawn in the plane without edge crossings such that all vertices touch the outer face. It is well known that for any two adjacent nodes $i, j \in V(T)$ the intersection of the bags $X_i \cap X_j$ is a separator of G. Therefore, $\kappa(G) \leq \mathrm{tw}(G)$ and the equation $\kappa(G) = \mathrm{tw}(G) = k$ is satisfied for k-connected partial k-trees only.

To solve the subgraph isomorphism problem in k-connected partial k-trees a normalized tree decomposition was proposed in [6] making k-separators explicit by introducing separator nodes which are distinguished from clique nodes. Here, we call a tree decomposition $(T, \mathcal{X})$ *normalized* if $V(T)$ can be divided into the two disjoint sets S and C of separator and clique nodes, respectively, such that

N1 S and C is a bipartition of T and all leaves of T are clique nodes,

N2 all separator and clique nodes have bags of size k and $k + 1$, respectively,

N3 for each path (i, j, k) in T: $X_j = X_i \cup X_k$ if $j \in C$ and $X_j = X_i \cap X_k$ if $j \in S$.

The bags associated with separator nodes form a set of pairwise parallel separators. Figure 1(b) shows an example of a normalized tree decomposition, denoted by $\mathrm{NTD}(G)$. Note that in general a (normalized) tree decomposition is not unique for a given graph. Therefore, a so-called *tree decomposition graph* was used in [6] that is a directed acyclic graph incorporating all possible normalized tree decompositions. Normalized tree decompositions and tree decomposition graphs can be computed in time $\mathcal{O}(n^2)$ and $\mathcal{O}(n^{k+2})$, respectively [6].

SP-tree Data Structure. For biconnected partial 2-trees, SP-trees are a well-known data structure, which reflects their series parallel composition, cf. Fig. 1(c). We use a notation and definition common for SPQR-trees, a generalization of SP-trees. Let G be a biconnected partial 2-tree with at least three vertices. The *SP-tree* $\mathcal{T} = \mathcal{T}(G)$ is the smallest tree such that

S1 each node μ of $\mathcal{T}$ is associated with a *skeleton* graph $S_\mu = (V_\mu, E_\mu)$. Each edge $e = (u, v)$ of E_μ is either a *real* edge, i.e., $e \in E(G)$, or a *virtual* edge where $\{u, v\}$ forms a 2-separator of G.
S2 $\mathcal{T}$ has two different node types with the following skeleton structures:
    S: The skeleton S_μ is a simple cycle, i.e., μ represents a series composition.
    P: The skeleton S_μ consists of two nodes and multiple parallel edges between them, i.e., μ represents a parallel composition.
S3 For two adjacent nodes μ and ν the skeleton S_μ contains a virtual edge e_ν that represents S_ν and vice versa. The node ν is called *pertinent* to e_ν.
S4 The original graph G can be obtained by merging the skeletons of adjacent nodes, where the virtual edges e_μ and e_ν are not part of the resulting graph and their common endpoints are merged for all $(\mu, \nu) \in E(\mathcal{T})$.

The S-nodes $V_S(\mathcal{T})$ and the P-nodes $V_P(\mathcal{T})$ form a bipartition of $\mathcal{T}$. Let $r = (u, v)$ be an arbitrary edge in G. The SP-tree *rooted at* r is obtained by rooting $\mathcal{T}$ at the node τ with $r \in E(S_\tau)$. Let μ be a child of ν with respect to the parent-child relationship induced by the root τ. The virtual edge e_ν in S_μ is called *reference edge* of μ, denoted by $\mathrm{ref}(\mu)$. The edge r is considered the reference edge of τ. For a node μ of the SP-tree $\mathcal{T}(G)$, there is a direct correspondence between the virtual edges of the skeleton graph S_μ, the connected components of $\mathcal{T} \setminus \mu$ and the connected components of $G \setminus V(S_\mu)$.

A normalized tree decomposition can be obtained from the SP-tree by successively separating S-nodes that have skeletons with more than three vertices. For an arbitrary separator a new P-node is created whose skeleton has exactly two virtual edges associated with the two parts of the original skeleton graph, one of which becomes the skeleton of a newly created S-node. When there is no S-node left with a skeleton with more than three vertices, the tree corresponds to a normalized tree decomposition, whereas P-nodes map to separator nodes and S-nodes to clique nodes. Conversely, the SP-tree can be obtained from a normalized tree decomposition of a partial 2-tree by merging each separator node s with its neighbors if it is (i) adjacent to exactly two clique nodes and (ii) associated with a bag X_s of vertices that are not adjacent in G.

Note that for all P-nodes ν of $\mathcal{T}(G)$ the set $V(S_\nu)$ is a bag of some node of every normalized tree decomposition of G. SP-trees are unique in contrast to normalized tree decompositions.

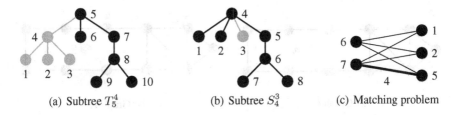

(a) Subtree T_5^4 (b) Subtree S_4^3 (c) Matching problem

Fig. 2. Two rooted subtrees (a) and (b) and the associated matching problem (c). Gray vertices and edges are not part of the rooted subtrees, edges without label in (c) have weight 1.

3 Tree Decompositions and Common Subgraph Problems

A key property of tree decompositions is their close relation to separators, which allows to systematically divide graphs into subgraphs along the tree structure. Based on solutions for these parts, typically a solution for the input graphs is computed.

We briefly review the Edmonds/Matula[2] algorithm [10] for 1-MCS, which is based on bipartite matching. A tree T is decomposed into rooted subtrees. For an edge $(u, v) \in E(T)$ we denote by T_v^u the subtree rooted at v, where the child u and its descendants are deleted, see Fig. 2. For every pair of rooted subtrees of the two input graphs G and H the MCS under the restriction that the roots are mapped to each other is computed and stored in a table D by dynamic programming. Let G_s^i and H_t^j be two rooted subtrees and $M_s = \{s_1, \ldots, s_n\}$ and $M_t = \{t_1, \ldots, t_m\}$ the children of s in G_s^i and t in H_t^j, respectively. Then $D(G_s^i, H_t^j) = 1 + \text{MWBMATCHING}(M_s, M_t, w)$, where MWBMATCHING is the size of a maximum weight bipartite matching in the complete bipartite graph with vertex set $M_s \cup M_t$ and edge weights w. The edge weights correspond to the results for pairs of smaller rooted subtrees and are determined according to $w(s_k, t_l) = D(G_{s_k}^s, H_{t_l}^t)$, $k \in \{1, \ldots, n\}$, $l \in \{1, \ldots, m\}$. The matching defines the mapping of the children of the two roots, cf. Fig. 2(c). Filling the table ordered by increasing size of subtrees eventually allows to determine the result by combining all pairs of corresponding rooted subtrees, i.e., G_s^i and G_i^s with H_t^j and H_j^t. A key observation for correctness of the approach is the following.

Observation 1. *Let C be a common subtree of G and H with isomorphism ϕ. If v is a 1-separator of C, then $\phi_G(v)$ and $\phi_H(v)$ are 1-separators of G and H, respectively.*

This fact allows to determine a solution based on results for pairs of subtrees of the input graphs determined by the separating vertices. Figure 3 illustrates that this observation does not hold in series-parallel graphs: The maximum common subgraph in the given example consists of the parts of the graphs depicted in black. Note that in every normalized tree decomposition of G (H) there is a separator node with bag $\{u, v\}$ ($\{s', t'\}$). This directly follows from the fact that separator nodes are closely related to P-nodes of the SP-tree that are unique[3], cf. Sect. 2. For a maximum common subgraph

---

[2] Note that a similar approach for subtree isomorphism was proposed by Matula and the variation for MCS in trees is attributed to Edmonds [10], but was not published.

[3] This also follows from [5], Theorem 3.3.

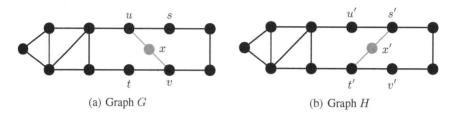

(a) Graph G (b) Graph H

Fig. 3. Example where straight-forward algorithms based on tree decomposition fail

C with isomorphism ϕ we must have $\phi(\mathbf{y}) = \mathbf{y}'$ for $\mathbf{y} \in \{u, v, s, t\}$. Let $a = \phi_G^{-1}(u)$, $b = \phi_G^{-1}(v)$, then $\{a, b\}$ is a separator of C, but $\{\phi_H(a) = u', \phi_H(b) = v'\}$ is not a separator of H. When computing common subgraphs based on normalized tree decompositions, typically all vertices of a bag in $\mathrm{NTD}(G)$ are matched to the vertices of a bag in $\mathrm{NTD}(H)$. One obstacle here is that normalized tree decompositions are not unique. This can be overcome by means of tree decomposition graphs. However, in the example, there are no bags with corresponding vertices in *any* normalized tree decomposition. Therefore, it is difficult to apply normalized tree decompositions to solve common subgraph problems. In contrast to the subgraph isomorphism problem, where normalized tree decompositions have been successfully applied, a key obstacle here is that the subgraph to be searched may be a subgraph of both input graphs and parts of the input graph that are not contained in the common subgraph constrain the possible normalized tree decompositions. This is the reason why the approaches mentioned in [4,8] do not work.

To overcome this issue, in Sect. 4 we introduce the concept of a *potential separator*, i.e., a separator of a subgraph to be searched that not necessarily is a separator of the input graph. For a given potential separator P we characterize the maximal subgraph that is separated by P and propose a decomposition of series-parallel graphs into split graphs based on all possible potential separators analogously to the approach by Edmonds/Matula for trees. In Sect. 4.1 we provide an algorithm based on SP-trees that solves 2-MCS by implicitly matching the split graphs of the two input graphs.

4   A Polynomial-Time Algorithm for 2-MCS

We say $P = \{u, v\}$ is a *potential separator* of a biconnected partial 2-tree G if there is a biconnected induced subgraph $G' \subseteq G$ with separator P.

Observation 2. *Let C be a biconnected common subgraph of the biconnected partial 2-trees G and H with isomorphism ϕ. If $\{u, v\}$ is a 2-separator of C, then $\{\phi_G(u), \phi_G(v)\}$ and $\{\phi_H(u), \phi_H(v)\}$ are potential separators of G and H, respectively.*

In the following we characterize the maximal biconnected induced subgraph G_P^* that is separated by a potential separator P. Since a potential separator not necessarily separates G, there may be parts of G keeping $G \setminus P$ connected, which can consequently not be contained in G_P^*. We show that these parts are associated with specific separators of G that we refer to as critical. We call a 2-separator S of a biconnected partial 2-tree G

compulsive if every normalized tree decomposition $NTD(G)$ contains a separator bag i with $X_i = S$. Note that S may be a compulsive separator of G, but not necessarily is for an induced biconnected subgraph $G' \subseteq G$. In this case we call S *critical*.

Lemma 1. *Let G be a biconnected partial 2-tree and $S = \{u, v\} \subset V(G)$. The following statements are equivalent:*
  (i) *The set S is a critical separator of G,*
  (ii) *S is compulsive for G and $(u, v) \notin E(G)$,*
  (iii) *there is a P-node ν in the SP-tree $\mathcal{T}(G)$ with $V(S_\nu) = S$ and $(u, v) \notin E(G)$,*
  (iv) *the graph $G \setminus S$ has at least three connected components and $(u, v) \notin E(G)$.*

The graph in Fig. 1(a), for example, contains the compulsive separators $S_1 = \{a, d\}$ and $S_2 = \{d, e\}$ that are both critical. Extending the symmetric relation of crossing separators, we say a 2-separator S *crosses* a set of two vertices $T = \{u, v\}$ iff S is an (u, v)-separator. Lemma 2 provides the key for our characterization of the maximal biconnected subgraph of G that is separated by a potential separator P (see Corollar 1).

Lemma 2. *Let G be a biconnected partial 2-tree and S a critical separator that crosses a potential separator $P = \{u, v\}$. Let C_u and C_v denote the components of $G \setminus S$ containing the vertex u and v, respectively. Every biconnected induced subgraph $G' \subseteq G$ separated by P is a subgraph of the biconnected graph $G_P^S = G[V(C_u) \cup V(C_v) \cup S]$.*

Corollar 1. *Let P be a potential separator of G and $\mathcal{S} = \{S_1, \ldots, S_l\}$ the set of critical separators that cross P. The graph $G_P^* := G[\bigcap_{S \in \mathcal{S}} V(G_P^S)]$ is the maximal biconnected subgraph of G with separator P.*

In other words: The graph G_P^* is the subgraph in which for all critical separators S_i that cross P the vertices in connected components of $G \setminus S_i$ neither containing u nor v are removed. This fact allows us to decompose biconnected partial 2-trees at the potential separators and can be used algorithmically to compute partial solutions for well-defined subgraphs. For a potential separator $P = \{u, v\}$ and a distinguished edge $r \in E(G_P^*)$, $r \neq (u, v)$, we say P *splits* G into the two subgraphs G_{uv}^r and $\overline{G_{uv}^r}$, referred to as *split graphs*. Let $C_1, \ldots, C_l$ be the connected components of $G_P^* \setminus P$ and w.l.o.g. let C_1 contain at least one endpoint of the distinguished edge r. Then $\overline{G_{uv}^r}$ denotes the subgraph $G[V(C_1) \cup \{u, v\}]$ and $G_{uv}^r := G[\bigcup_{i=2}^l V(C_i) \cup \{u, v\}]$, where u and v are referred to as *base vertex*. In case of a potential separator P that is not a separator of G, i.e., $G_P^* \neq G$, the operation is referred to as *shear split*. Figure 4 illustrates the different split operations. When the vertices u, v are not a potential separator, but adjacent, G_{uv}^r is defined as the edge $e = (u, v)$, if $e \neq r$, and the graph $(V, E \setminus e)$ otherwise. Note that we may assume split graphs to be biconnected which can be achieved by inserting a virtual edge between the two base vertices as illustrated by dashed lines in Fig. 4. The split graphs shown in Fig. 4(b) are determined by the potential separator $P = \{b, g\}$ that is not a separator of G. Note that the critical separators $S_1 = \{a, d\}$ and $S_2 = \{d, e\}$ both cross P and hence the vertices h and f are not contained in G_P^* and the split graphs associated with P.

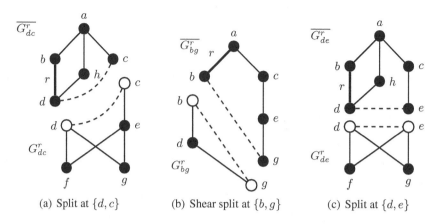

(a) Split at $\{d, c\}$ (b) Shear split at $\{b, g\}$ (c) Split at $\{d, e\}$

Fig. 4. Splitting the graph shown in Fig. 1(a)

4.1 Solving 2-MCS with SP-trees

We present a polynomial-time algorithm for 2-MCS that is based on the SP-tree data structure and considers the split graphs of the input graphs defined by potential separators. Since a vertex v of a graph may occur in multiple skeleton graphs of the SP-tree $\mathcal{T}$, we denote by $\mu(v)$ the representative of v in the skeleton S_μ. Let $\Upsilon(v) = \{\mu \in V(\mathcal{T}) \mid u \in V(S_\mu), \mu(v) = u\}$ be the set of *allocation nodes* of a vertex v, i.e., the nodes of $\mathcal{T}$ whose skeleton contains a representative of v. We define the *shear path* $P(u, v)$ as the shortest path in the SP-tree between an S-node $\mu_1 \in \Upsilon(u)$ and an S-node $\mu_l \in \Upsilon(v)$. Lemma 3 characterizes the potential separators of a biconnected partial 2-tree by means of the SP-tree.

Lemma 3. *Let $P(u, v) = (\mu_1, \nu_1, \ldots, \nu_{l-1}, \mu_l)$ be a shear path, then $T = \{u, v\}$ is*
  (i) *a potential separator of G iff there is no P-node ν_i, $i \in \{1, \ldots, l-1\}$, with S_{ν_i} containing a real edge,*
  (ii) *a separator of G iff $l = 1$.*
In case (i), T is crossed by the critical separators $V(S_{\nu_i})$, $i \in \{1, \ldots, l-1\}$.

The maximal biconnected subgraph G_P^* of G that is separated by a potential separator $P = \{u, v\}$ of G can be obtained based on the shear path by keeping the components determined by μ_i and μ_{i+1} at each critical separator $V(S_{\nu_i})$. Note that a representative contained in a skeleton graph is associated with a unique node of the SP-tree. We extend the definition of split path and split graphs to representatives from skeleton graphs of S-nodes. Given two representatives $P = \{u, v\}$, the end vertices of the shear path $P(u, v)$ are the unique allocation nodes of u and v. We can still use Lemma 3 to determine if P is a potential separator and define the graph G_P^* and the split graphs accordingly based on the critical separators. Further split graphs emerge for representatives u, v in a skeleton S_μ, where μ is an S-node with $\mathrm{ref}(\mu) = (u, v)$, e.g., in Fig. 5(b) the split graph $G_{d_7 e_4}^r$ would consist of S_μ. Note that all split graphs defined by potential separators can as well be obtained using certain representatives. Figure 5 illustrates a shear split defined by representatives as well as the corresponding split graphs and provides an example

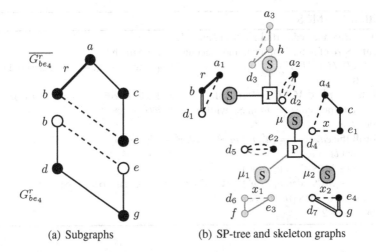

(a) Subgraphs   (b) SP-tree and skeleton graphs

Fig. 5. The graph $\overline{G^r_{be_4}}$ shown with filled black vertices and edges in (b) represents the vertices already considered and $G^r_{be_4}$, non-filled in (b), the unmapped vertices of the graph G, cf. Fig. 1(a). Parts no longer considered for solutions are shown in gray. In the example we have $cS(x) = \{\mu_1, \mu_2\}$, $pS(d_7) = pS(d_6) = d_4$ and $\mu(a) = \mu(a_i) = a_4, i \in \{1, \ldots, 4\}$.

for the additional notation: For a virtual edge e in the skeleton of an S-node we denote the children of the P-node pertinent to e by $cS(e)$. For a vertex v in a skeleton S_μ we refer to the representative of v in the next S-node on the path to the root by $pS(v)$.

Algorithm 1 computes the SP-tree for both input graphs G, H and implicitly extends a partial mapping ϕ between split graphs step-by-step. In each state the graphs $\overline{G^r_{uv}}$ and $\overline{H^{r'}_{u'v'}}$ are already processed and the mapping is extended to the unexplored subgraphs G^r_{uv} and $H^r_{u'v'}$, where $\phi(u) = u'$ and $\phi(v) = v'$. The main procedure starts with pairs of S-nodes, maps two edges of their skeleton graphs and roots the SP-trees accordingly.

Ongoing with this mapping the procedure Mcs-S essentially follows the cyclic skeleton graphs of the two S-nodes simultaneously and successively extends the mapping by the next unmapped vertices. Let μ be an S-node with reference edge (t, s) and a skeleton graph S_μ consisting of a cycle $(c_k, c_1, \ldots, c_k)$, where $c_k = t$ and $c_1 = s$. The function NEXT(v, S_μ) returns the edge $(v = c_i, c_j)$, where $j = i + 1 \mod k$, i.e., the next edge in the cyclic order given by the skeleton graph. The direction in which the cycle is traversed is determined by the ordering of the parameters of Mcs-S. Different cases apply based on the type of edges that are mapped. The basic case is that the edges can be matched (line 7). Then the size of the maximum common subgraph depends on (i) MATCHEDGE, i.e., the maximum common subgraph of the split graphs associated with the edges and (ii) the result of the recursive call Mcs-S with the new base vertices. Furthermore, the size of the mapping is increased by one for the vertex w, which is mapped to w'. When the next vertex corresponds to the first base vertex, i.e., the split graph consists of a single edge, the recursion ends (line 5). However, if this does not hold for both split graphs, the mapping can not be completed as indicated by the return value $-\infty$ (line 6). Assume the edge under consideration is virtual and the vertices are

---

Algorithm 1. 2-MCS

   Input   : Biconnected partial 2-trees G and H.
   Output : Size of a maximum common biconnected subgraph.
1 $\mathcal{T} \leftarrow \mathcal{T}(G); \mathcal{U} \leftarrow \mathcal{T}(H)$ ▷ Compute SP-tree decomposition
2 $mcs \leftarrow 0$
3 **forall the** $(\mu, \mu') \in V_S(\mathcal{T}) \times V_S(\mathcal{U})$ **do**   ▷ Pairs of series components
4     $r \leftarrow$ arbitrary edge (u, v) in S_μ with $(u, v) \in E(G)$
5   root $\mathcal{T}$ at r
6   **forall the** *edges* $r' = (u', v')$ *in* $S_{\mu'}$ *with* $r' \in E(H)$ **do**
7   root $\mathcal{U}$ at r'
8        $p \leftarrow \text{MCS-S}(u, v, u', v')$
9        $q \leftarrow \text{MCS-S}(u, v, v', u')$ ▷ Alternative mapping of root edges
10       $mcs \leftarrow \max\{mcs, p, q\}$
11 **return** $mcs + 2$

---

not adjacent in G, then a critical separator is reached (line 8). In this case the algorithm recursively proceeds with a representative of the second base vertex from a skeleton of a different S-node that is a child of the P-node pertinent to the edge. All possible pairs of such S-nodes are considered and the best solution is selected (line 13). Note that it is eventually required to go back to the previous S-node to finally complete a mapping. This step is realized when the reference edge is reached in line 3 and 4, respectively.

The function MATCHEDGE is called whenever the mapping is extended. Note that the parameters are edges from skeleton graphs and may be real or virtual. Assume a given edge is virtual in the skeleton than an edge between these vertices may or may not exist in G. We do not obtain a valid mapping between induced subgraphs if such an edge exists in only one of the graphs (line 1). However, we can still map a virtual edge to a real edge if the virtual edge is pertinent to a P-node with a skeleton containing a real edge (line 2). In this case the subgraph represented by the descendants of the P-node is not part of the common subgraph. Finally, assume both edges are virtual, then MWB-MATCHING is performed between the children of the pertinent P-nodes (line 6), where the edge weights w are determined by MCS-S on pairs of the associated split graphs (line 5). Note that if the result of the matching is 0 and the P-nodes do not contain real edges, the mapping is not valid since the subgraph would not be biconnected (line 7). Otherwise the result corresponds to the size of the matching (line 8). Note that the base vertices already mapped are counted in the main procedure (Algorithm 1, line 11), but not in the procedures MCS-S and MCS-P handling pairs of split graphs.

Analysis. We argue that Algorithm 1 solves 2-MCS in polynomial time. A biconnected subgraph G' can be obtained from a biconnected partial 2-tree only by deleting components of $G \setminus S$ for compulsive separators S. A separator may remain compulsive for the subgraph, either because the two vertices are adjacent in G or because $G' \setminus S$ has at least three components. These cases are handled for the two input graphs by the procedure MATCHEDGE. If the separator does not remain compulsive, there are separators of G' that do not separate G, but are potential separators of G. Algorithm 1 considers all possible potential separators of both input graphs by building all possible valid shear paths

---

Procedure MCS-S(u, v, u', v')

Input : Base vertices u, v of G, $v \in V(S_\mu)$ and u', v' of H, $v' \in V(S_{\mu'})$.

Output : Size of a 2-MCS of G_{uv}^r and $H_{u'v'}^r$ such that $u \to u', v' \to v'$.

1	$e = (v, w) \leftarrow \text{NEXT}(v, S_\mu)$	▷ next edge in S_μ
2	$e' = (v', w') \leftarrow \text{NEXT}(v', S_{\mu'})$	▷ next edge in $S_{\mu'}$
3	**if** $e = \text{ref}(\mu)$ **then return** MCS-S$(u, \text{pS}(v), u', v')$	▷ Merge S-nodes
4	**if** $e' = \text{ref}(\mu')$ **then return** MCS-S$(u, v, u', \text{pS}(v'))$	▷ Merge S-nodes
5	**if** $w = u$ *and* $w' = u'$ **then return** MATCHEDGE(e, e')	▷ Completed skeleton
6	**if** $w = u$ *xor* $w' = u'$ **then return** $-\infty$	▷ Incompletable mapping
7	$mcs \leftarrow$ MATCHEDGE(e, e') + MCS-S(u, w, u', w') + 1	
8	**if** $e \notin E(G)$ *or* $e' \notin E(H)$ **then**	▷ Consider critical separators
9	**if** $e \in E(G)$ **then** $M \leftarrow \{\mu\}$ **else** $M \leftarrow \text{cS}(e)$	
10	**if** $e' \in E(H)$ **then** $M' \leftarrow \{\mu'\}$ **else** $M' \leftarrow \text{cS}(e')$	
11	**forall the** $(\eta, \eta') \in M \times M'$ **do**	
12	$p \leftarrow$ MCS-S$(u, \eta(v), u', \eta'(v'))$	▷ Perform shear split
13	$mcs \leftarrow \max\{mcs, p\}$	
14	**return** mcs	

Procedure MATCHEDGE(e, e')

Input : Edges $e = (u, v) \in E(S_\mu)$ and $e' = (u', v') \in E(S_{\mu'})$.

Output : Size of a 2-MCS of G_{uv}^r and $H_{u'v'}^r$ such that $u \to u', v \to v'$.

1	**if** $e \in E(G)$ *xor* $e' \in E(H)$ **then return** $-\infty$	▷ Subgraph not induced
2	**if** e *is real in* S_μ *or* e' *is real in* $S_{\mu'}$ **then return** 0	▷ Valid mapping
3	$M \leftarrow \text{cS}(e); M' \leftarrow \text{cS}(e')$	
4	**forall the** $e = (\eta, \eta') \in M \times M'$ **do**	▷ Pairs of S-node children
5	$w(e) \leftarrow$ MCS-S$(\eta(u), \eta(v), \eta'(u'), \eta'(v'))$	
6	$p \leftarrow$ MWBMATCHING(M, M', w)	▷ Compute maximum matching
7	**if** $p = 0$ *and* $e \notin E(G)$, $e' \notin E(H)$ **then return** $-\infty$	▷ Not biconnected
8	**else return** p	

---

in MCS-S, cf. Lemma 3. Furthermore, it considers the associated split graphs, which is sufficient according to Corollar 1. To analyze the running time we first consider the number of possible split graphs.

Lemma 4. *Let G be a biconnected partial 2-tree and $n = |V(G)|$. The number of split graphs of G is $\mathcal{O}(n^2)$.*

Theorem 1. *2-MCS can be solved in $\mathcal{O}(n^6)$, where $n = \max\{|V(G)|, |V(H)|\}$.*

Proof (Sketch). Each call of the recursive methods MCS-S and MATCHEDGE computes 2-MCS for two split graphs defined by the parameter list. We can easily transform the methods into dynamic programming algorithms filling a table indexed by split graphs. Thus, as soon as a cell of the table has been computed every successive call for the same pair of split graphs can be answered in $\mathcal{O}(1)$. Since the number of split graphs is bounded by $\mathcal{O}(n^2)$, a table of size $\mathcal{O}(n^4)$ is sufficient. It is not hard to argue that the total running time spend in MCS-S is $\mathcal{O}(n^6)$. Moreover, we can show that the number of non-trivial matching problems to be solved is at most $\mathcal{O}(n^2)$. This leads to the total running time of $\mathcal{O}(n^5)$ for MATCHEDGE. □

Application to Outerplanar Graphs. 2-MCES was solved in the context of BBP-MCES in outerplanar graphs, but not carefully analyzed. For BBP-MCES the rough bounds of $\mathcal{O}(n^{10})$ [3] and $\mathcal{O}(n^7)$ [11] were given. A biconnected partial 2-tree G is outerplanar iff all P-nodes of $\mathcal{T}(G)$ have degree two. According to this fact and Lemma 1 there cannot be any critical separators. Therefore parts of Algorithm 1 become trivial in outerplanar graphs and we obtain the following result.

Theorem 2. 2-*MCS in outerplanar graphs can be solved in* $\mathcal{O}(n^3)$.

Proof. Since there are no critical separators, lines 9-13 of MCS-S are never reached. The sets M and M' in line 3 of MATCHEDGE always consist of single S-nodes, because all P-nodes have degree two. Therefore both functions are computed in $\mathcal{O}(1)$ except the running time required for recursive calls. Algorithm 1 causes $\mathcal{O}(n^2)$ calls of MCS-S, each again involves $\mathcal{O}(n)$ recursive calls leading to a total running time of $\mathcal{O}(n^3)$. □

References

1. Akutsu, T.: A polynomial time algorithm for finding a largest common subgraph of almost trees of bounded degree. IEICE Trans. Fundamentals E76-A(9) (1993)
2. Akutsu, T., Tamura, T.: On the complexity of the maximum common subgraph problem for partial k-trees of bounded degree. In: Chao, K.-M., Hsu, T.-S., Lee, D.-T. (eds.) ISAAC 2012. LNCS, vol. 7676, pp. 146–155. Springer, Heidelberg (2012)
3. Akutsu, T., Tamura, T.: A polynomial-time algorithm for computing the maximum common connected edge subgraph of outerplanar graphs of bounded degree. Algorithms 6(1), 119–135 (2013)
4. Bachl, S., Brandenburg, F.J., Gmach, D.: Computing and drawing isomorphic subgraphs. J. Graph Algorithms Appl. 8(2), 215–238 (2004)
5. Dessmark, A., Lingas, A., Proskurowski, A.: Faster algorithms for subgraph isomorphism of k-connected partial k-trees. Algorithmica 27, 337–347 (2000)
6. Gupta, A., Nishimura, N.: Sequential and parallel algorithms for embedding problems on classes of partial k-trees. In: Schmidt, E.M., Skyum, S. (eds.) SWAT 1994. LNCS, vol. 824, pp. 172–182. Springer, Heidelberg (1994)
7. Gupta, A., Nishimura, N.: The complexity of subgraph isomorphism for classes of partial k-trees. Theoretical Computer Science 164(1-2), 287–298 (1996)
8. Hajiaghayi, M., Nishimura, N.: Subgraph isomorphism, log-bounded fragmentation, and graphs of (locally) bounded treewidth. J. Comput. System Sci. 73(5), 755 (2007)
9. Matoušek, J., Thomas, R.: On the complexity of finding iso- and other morphisms for partial k-trees. Discrete Mathematics 108(1-3), 343–364 (1992)
10. Matula, D.W.: Subtree isomorphism in $O(n^{5/2})$. In: Algorithmic Aspects of Combinatorics. Ann. Discrete Math., vol. 2, p. 91 (1978)
11. Schietgat, L., Ramon, J., Bruynooghe, M.: A polynomial-time metric for outerplanar graphs. In: Mining and Learning with Graphs, MLG 2007, Firence, Italy, August 1-3 (2007)
12. Sysło, M.M.: The subgraph isomorphism problem for outerplanar graphs. Theoretical Computer Science 17(1), 91–97 (1982)
13. Yamaguchi, A., Aoki, K.F., Mamitsuka, H.: Finding the maximum common subgraph of a partial k-tree and a graph with a polynomially bounded number of spanning trees. Inf. Process. Lett. 92(2), 57–63 (2004)

On Coloring Resilient Graphs

Jeremy Kun and Lev Reyzin

Department of Mathematics, Statistics, and Computer Science
University of Illinois at Chicago, Chicago, IL 60607, USA
{jkun2,lreyzin}@math.uic.edu

Abstract. We introduce a new notion of resilience for constraint satisfaction problems, with the goal of more precisely determining the boundary between NP-hardness and the existence of efficient algorithms for resilient instances. In particular, we study r-resiliently k-colorable graphs, which are those k-colorable graphs that remain k-colorable even after the addition of any r new edges. We prove lower bounds on the NP-hardness of coloring resiliently colorable graphs, and provide an algorithm that colors sufficiently resilient graphs. We also analyze the corresponding notion of resilience for k-SAT. This notion of resilience suggests an array of open questions for graph coloring and other combinatorial problems.

1 Introduction and Related Work

An important goal in studying NP-complete combinatorial problems is to find precise boundaries between tractability and NP-hardness. This is often done by adding constraints to the instances being considered until a polynomial time algorithm is found. For instance, while SAT is NP-hard, the restricted 2-SAT and XOR-SAT versions are decidable in polynomial time.

In this paper we present a new angle for studying the boundary between NP-hardness and tractability. We informally define the resilience of a constraint-based combinatorial problem and we focus on the case of resilient graph colorability. Roughly speaking, a positive instance is resilient if it remains a positive instance up to the addition of a constraint. For example, an instance G of Hamiltonian circuit would be "r-resilient" if G has a Hamiltonian circuit, and G minus any r edges *still* has a Hamiltonian circuit. In the case of coloring, we say a graph G is r-resiliently k-colorable if G is k-colorable and will remain so even if any r edges are added. One would imagine that *finding* a k-coloring in a very resilient graph would be easy, as that instance is very "far" from being not colorable. And in general, one can pose the question: how resilient can instances be and have the search problem still remain hard?[1]

Most NP-hard problems have natural definitions of resiliency. For instance, resilient positive instances for optimization problems over graphs can be defined as those that remain positive instances even up to the addition or removal of

---

[1] We focus on the search versions of the problems because the decision version on resilient instances induces the trivial "yes" answer.

E. Csuhaj-Varjú et al. (Eds.): MFCS 2014, Part II, LNCS 8635, pp. 517–528, 2014.
© Springer-Verlag Berlin Heidelberg 2014

any edge. For satisfiability, we say a resilient instance is one where variables can be "fixed" and the formula remains satisfiable. In problems like set-cover, we could allow for the removal of a given number of sets. Indeed, this can be seen as a general notion of resilience for adding constraints in constraint satisfaction problems (CSPs), which have an extensive literature [24].[2]

Therefore we focus on a specific combinatorial problem, graph coloring. Resilience is defined up to the addition of edges, and we first show that this is an interesting notion: many famous, well studied graphs exhibit strong resilience properties. Then, perhaps surprisingly, we prove that 3-coloring a 1-resiliently 3-colorable graph is NP-hard – that is, it is hard to color a graph even when it is guaranteed to remain 3-colorable under the addition of any edge. Briefly, our reduction works by mapping positive instances of 3-SAT to 1-resiliently 3-colorable graphs and negative instances to graphs of chromatic number at least 4. An algorithm which can color 1-resiliently 3-colorable graphs can hence distinguish between the two. On the other hand, we observe that 3-resiliently 3-colorable graphs have polynomial-time coloring algorithms (leaving the case of 3-coloring 2-resiliently 3-colorable graphs tantalizingly open). We also show that efficient algorithms exist for k-coloring $\binom{k}{2}$-resiliently k-colorable graphs for all k, and discuss the implications of our lower bounds.

This paper is organized as follows. In the next two subsections we review the literature on other notions of resilience and on graph coloring. In Section 2 we characterize the resilience of boolean satisfiability, which is used in our main theorem on 1-resilient 3-coloring. In Section 3 we formally define the resilient graph coloring problem and present preliminary upper and lower bounds. In Section 4 we prove our main theorem, and in Section 5 we discuss open problems.

1.1   Related Work on Resilience

There are related concepts of resilience in the literature. Perhaps the closest in spirit is Bilu and Linial's notion of stability [5]. Their notion is restricted to problems over metric spaces; they argue that practical instances often exhibit some degree of stability, which can make the problem easier. Their results on clustering stable instances have seen considerable interest and have been substantially extended and improved [3,5,27]. Moreover, one can study TSP and other optimization problems over metrics under the Bilu-Linial assumption [26]. A related notion of stability by Ackerman and Ben-David [1] for clustering yields efficient algorithms when the data lies in Euclidian space.

Our notion of resilience, on the other hand, is most natural in the case when the optimization problem has natural constraints, which can be fixed or modified. Our primary goal is also different – we seek to more finely delineate the boundary between tractability and hardness in a systematic way across problems.

Property testing can also be viewed as involving resilience. Roughly speaking property testing algorithms distinguish between combinatorial structures that

---

[2] However, a resilience definition for general CSPs is not immediate because the ability to add any constraint (e.g., the negation of an existing constraint) is too strong.

satisfy a property or are very far from satisfying it. These algorithms are typically given access to a small sample depending on a parameter ε alone. For graph property testing, as with resilience, the concept of being ε-far from having a property involves the addition or removal of an arbitrary set of at most $\varepsilon\binom{n}{2}$ edges from G. Our notion of resilience is different in that we consider adding or removing a constant number of constraints. More importantly, property testing is more concerned with query complexity than with computational hardness.

1.2 Previous Work on Coloring

As our main results are on graph colorability, we review the relevant past work. A graph G is k-colorable if there is an assignment of k distinct colors to the vertices of G so that no edge is monochromatic. Determining whether G is k-colorable is a classic an NP-hard problem [19]. Many attempts to simplify the problem, such as assuming planarity or bounded degree, still result in NP-hardness [8]. A large body of work surrounds positive and negative results for explicit families of graphs. The list of families that are polynomial-time colorable includes triangle-free planar graphs, perfect graphs and almost-perfect graphs, bounded tree- and clique-width graphs, quadtrees, and various families of graphs defined by the lack of an induced subgraph [7,10,15,22,23].

With little progress on coloring general graphs, research has naturally turned to approximation. In approximating the chromatic number of a general graph, the first results were of Garey and Johnson, giving a performance guarantee of $O(n/\log n)$ colors [18] and proving that it is NP-hard to approximate chromatic number to within a constant factor less than two [11]. Further work improved this bound by logarithmic factors [4,13]. In terms of lower bounds, Zuckerman [29] derandomized the PCP-based results of Håstad [14] to prove the best known approximability lower-bound to date, $O(n^{1-\varepsilon})$.

There has been much recent interest in coloring graphs which are already known to be colorable while minimizing the number of colors used. For a 3-colorable graph, Wigderson gave an algorithm using at most $O(n^{1/2})$ colors [28], which Blum improved to $\tilde{O}(n^{3/8})$ [6]. A line of research improved this bound still further to $o(n^{1/5})$ [17]. Despite the difficulties in improving the constant in the exponent, and as suggested by Arora [2], there is no evidence that coloring a 3-colorable graph with as few as $O(\log n)$ colors is hard.

On the other hand there are asymptotic and concrete lower bounds. Khot [21] proved that for sufficiently large k it is NP-hard to color a k-colorable graph with fewer than $k^{O(\log k)}$ colors; this was improved by Huang to $2^{\sqrt[3]{k}}$ [16]. It is also known that for every constant h there exists a sufficiently large k such that coloring a k-colorable graph with hk colors is NP-hard [9]. In the non-asymptotic case, Khanna, Linial, and Safra [20] used the PCP theorem to prove it is NP-hard to 4-color a 3-colorable graph, and more generally to color a k colorable graph with at most $k + 2\lfloor k/3 \rfloor - 1$ colors. Guruswami and Khanna give an explicit reduction for $k = 3$ [12]. Assuming a variant of Khot's 2-to-1 conjecture, Dinur et al. prove that distinguishing between chromatic number K and K' is hard for constants $3 \leq K < K'$ [9]. This is the best conditional lower bound we give in Section 3.3, but it does not to our knowledge imply Theorem 2.

Without large strides in approximate graph coloring, we need a new avenue to approach the NP-hardness boundary. In this paper we consider the coloring problem for a general family of graphs which we call *resiliently colorable*, in the sense that adding edges does not violate the given colorability assumption.

2   Resilient SAT

We begin by describing a resilient version of k-satisfiability, which is used in proving our main result for resilient coloring in Section 4.

Problem 1 (resilient k-SAT). *A boolean formula φ is r-resilient if it is satisfiable and remains satisfiable if any set of r variables are fixed. We call r-resilient k-SAT the problem of finding a satisfying assignment for an r-resiliently satisfiable k-CNF formula. Likewise, r-resilient CNF-SAT is for r-resilient formulas in general CNF form.*

The following lemma allows us to take problems that involve low (even zero) resilience and blow them up to have large resilience and large clause size.

Lemma 1 (blowing up). *For all $r \geq 0$, $s \geq 1$, and $k \geq 3$, r-resilient k-SAT reduces to $[(r+1)s - 1]$-resilient (sk)-SAT in polynomial time.*

Proof. Let φ be an r-resilient k-SAT formula. For each i, let φ^i denote a copy of φ with a fresh set of variables. Construct $\psi = \bigvee_{i=1}^{s} \varphi^i$. The formula ψ is clearly equivalent to φ, and by distributing the terms we can transform ψ into (sk)-CNF form in time $O(n^s)$. We claim that ψ is $[(r+1)s - 1]$-resilient. If fewer than $(r+1)s$ variables are fixed, then by the pigeonhole principle one of the s sets of variables has at most r fixed variables. Suppose this is the set for φ^1. As φ is r-resilient, φ^1 is satisfiable and hence so is ψ.   □

As a consequence of the blowing up lemma for $r = 0, s = 2, k = 3$, 1-resilient 6-SAT is NP-hard (we reduce from this in our main coloring lower bound). Moreover, a slight modification of the proof shows that r-resilient CNF-SAT is NP-hard for all $r \geq 0$. The next lemma allows us to reduce in the other direction, shrinking down the resilience and clause sizes.

Lemma 2 (shrinking down). *Let $r \geq 1$, $k \geq 2$, and $q = \min(r, \lfloor k/2 \rfloor)$. Then r-resilient k-SAT reduces to q-resilient $(\lceil \frac{k}{2} \rceil + 1)$-SAT in polynomial time.*

Proof. For ease of notation, we prove the case where k is even. For a clause $C = \bigvee_{i=1}^{k} x_i$, denote by $C[: k/2]$ the sub-clause consisting of the first half of the literals of C, specifically $\bigvee_{i=1}^{k/2} x_i$. Similarly denote by $C[k/2 :]$ the second half of C. Now given a k-SAT formula $\varphi = \bigwedge_{j=1}^{k} C_j$, we construct a $(\frac{k}{2} + 1)$-SAT formula ψ by the following. For each j introduce a new variable z_j, and define

$$\psi = \bigwedge_{j=1}^{k} (C_j[: k/2] \vee z_j) \wedge (C_j[k/2 :] \vee \overline{z_j})$$

The formulas φ and ψ are logically equivalent, and we claim ψ is q-resilient. Indeed, if some of the original set of variables are fixed there is no problem, and each z_i which is fixed corresponds to a choice of whether the literal which will satisfy C_j comes from the first or the second half. Even stronger, we can arbitrarily *pick* another literal in the correct half and fix its variable so as to satisfy the clause. The r-resilience of φ guarantees the ability to do this for up to r of the z_i. But with the observation that there are no l-resilient l-SAT formulas, we cannot get $k/2 + 1$ resilience when $r > k/2$, giving the definition of q.   □

Combining the blowing up and shrinking down lemmas, we get a tidy characterization: r-resilient k-SAT is either NP-hard or vacuously trivial.

Theorem 1. *For all $k \geq 3$, $0 \leq r < k$, r-resilient k-SAT is NP-hard.*

Proof. We note that increasing k or decreasing r (while leaving the other parameter fixed) cannot make r-resilient k-SAT easier, so it suffices to reduce from 3-SAT to $(k-1)$-resilient k-SAT for all $k \geq 3$. For any r we can blow up from 3-SAT to r-resilient $3(r+1)$-SAT by setting $s = r+1$ in the blowing up lemma. We want to iteratively apply the shrinking down lemma until the clause size is s. If we write $s_0 = 3s$ and $s_i = \lceil s_i/2 \rceil + 1$, we would need that for some m, $s_m = s$ and that for each $1 \leq j < m$, the inequality $\lfloor s_j/2 \rfloor \geq r = s - 1$ holds.

Unfortunately this is not always true. For example, if $s = 10$ then $s_1 = 16$ and $16/2 < 9$, so we cannot continue. However, we can avoid this for sufficiently large r by artificially increasing k after blowing up. Indeed, we just need to find some $x \geq 0$ for which $a_1 = \lceil \frac{3s+x}{2} \rceil + 1 = 2(s-1)$. And we can pick $x = s - 6 = r - 5$, which works for all $r \geq 5$. For $r = 2, 3, 4$, we can check by hand that one can find an x that works.[3] For $r = 2$ we can start from 2-resilient 9-SAT; for $r = 3$ we can start from 16-SAT; and for $r = 4$ we can start from 24-SAT.   □

3   Resilient Graph Coloring and Preliminary Bounds

In contrast to satisfiability, resilient graph coloring has a more interesting hardness boundary, and it is not uncommon for graphs to have relatively high resilience. In this section we present some preliminary bounds.

3.1   Problem Definition and Remarks

Problem 2 (resilient coloring). *A graph G is called r-resiliently k-colorable if G remains k-colorable under the addition of any set of r new edges.*

---

[3] The difference is that for $r \geq 5$ we can get what we need with only two iterations, but for smaller r we require three steps.

We argue that this notion is not trivial by showing the resilience properties of some classic graphs. These were determined by exhaustive computer search. The Petersen graph is 2-resiliently 3-colorable. The Dürer graph is 1-resiliently 3-colorable (but not 2-resilient) and 4-resiliently 4-colorable (but not 5-resilient). The Grötzsch graph is 4-resiliently 4-colorable (but not 5-resilient). The Chvátal graph is 3-resiliently 4-colorable (but not 4-resilient).

There are a few interesting constructions to build intuition about resilient graphs. First, it is clear that every k-colorable graph is 1-resiliently $(k + 1)$-colorable (just add one new color for the additional edge), but for all $k > 2$ there exist k-colorable graphs which are not 2-resiliently $(k + 1)$-colorable. Simply remove two disjoint edges from the complete graph on $k + 2$ vertices. A slight generalization of this argument provides examples of graphs which are $\lfloor (k + 1)/2 \rfloor$-colorable but not $\lfloor (k + 1)/2 \rfloor$-resiliently k-colorable for $k \geq 3$. On the other hand, every $\lfloor (k + 1)/2 \rfloor$-colorable graph is $(\lfloor (k + 1)/2 \rfloor - 1)$-resiliently k-colorable, since r-resiliently k-colorable graphs are $(r + m)$-resiliently $(k + m)$-colorable for all $m \geq 0$ (add one new color for each added edge).

One expects high resilience in a k-colorable graph to reduce the number of colors required to color it. While this may be true for super-linear resilience, there are easy examples of $(k - 1)$-resiliently k-colorable graphs which are k-chromatic. For instance, add an isolated vertex to the complete graph on k vertices.

3.2 Observations

We are primarily interested in the complexity of coloring resilient graphs, and so we pose the question: for which values of k, r does the task of k-coloring an r-resiliently k-colorable graph admit an efficient algorithm? The following observations aid us in the classification of such pairs, which is displayed in Figure 1.

Observation 1. *An r-resiliently k-colorable graph is r'-resiliently k-colorable for any $r' \leq r$. Hence, if k-coloring is in P for r-resiliently k-colorable graphs, then it is for s-resiliently k-colorable graphs for all $s \geq r$. Conversely, if k-coloring is NP-hard for r-resiliently k-colorable graphs, then it is for s-resiliently k-colorable graphs for all $s \leq r$.*

Hence, in Figure 1 if a cell is in P, so are all of the cells to its right; and if a cell is NP-hard, so are all of the cells to its left.

Observation 2. *If k-coloring is in P for r-resiliently k-colorable graphs, then k'-coloring r-resiliently k'-colorable graphs is in P for all $k' \leq k$. Similarly, if k-coloring is in NP-hard for r-resiliently k-colorable graphs, then k'-coloring is NP-hard for r-resiliently k'-colorable graphs for all $k' \geq k$.*

Proof. If G is r-resiliently k-colorable, then we construct G' by adding a new vertex v with complete incidence to G. Then G' is r-resiliently $(k + 1)$-colorable, and an algorithm to color G' can be used to color G. □

Observation 2 yields the rule that if a cell is in P, so are all of the cells above it; if a cell is NP-hard, so are the cells below it. More generally, we have the following observation which allows us to apply known bounds.

Observation 3. *If it is NP-hard to $f(k)$-color a k-colorable graph, then it is NP-hard to $f(k)$-color an $(f(k) - k)$-resiliently $f(k)$-colorable graph.*

This observation is used in Propositions 2 and 3, and follows from the fact that an r-resiliently k-colorable graph is $(r + m)$-resiliently $(k + m)$-colorable for all $m \geq 0$ (here $r = 0, m = f(k) - k$).

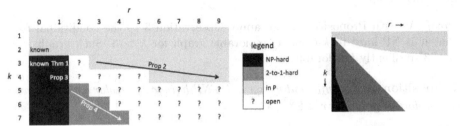

Fig. 1. The classification of the complexity of k-coloring r-resiliently k-colorable graphs. Left: the explicit classification for small k, r. Right: a zoomed-out view of the same table, with the NP-hard (black) region added by Proposition 4.

3.3 Upper and Lower Bounds

In this section we provide a simple upper bound on the complexity of coloring resilient graphs, we apply known results to show that 4-coloring a 1-resiliently 4-colorable graph is NP-hard, and we give the conditional hardness of k-coloring $(k - 3)$-resiliently k-colorable graphs for all $k \geq 3$. This last result follows from the work of Dinur et al., and depends a variant of Khot's 2-to-1 conjecture [9]; a problem is called *2-to-1-hard* if it is NP-hard assuming this conjecture holds. Finally, applying the result of Huang [16], we give an asymptotic lower bound.

All our results on coloring are displayed in Figure 1. To explain Figure 1 more explicitly, Proposition 1 gives an upper bound for $r = \binom{k}{2}$, and Proposition 2 gives hardness of the cell $(4, 1)$ and its consequences. Proposition 3 provides the conditional lower bound, and Theorem 2 gives the hardness of the cell $(3, 1)$. Proposition 4 provides an NP-hardness result.

Proposition 1. *There is an efficient algorithm for k-coloring $\binom{k}{2}$-resiliently k-colorable graphs.*

Proof. If G is $\binom{k}{2}$-resiliently k-colorable, then no vertex may have degree $\geq k$. For if v is such a vertex, one may add complete incidence to any choice of k vertices in the neighborhood of v to get K_{k+1}. Finally, graphs with bounded degree $k - 1$ are greedily k-colorable.   □

Proposition 2. *4-coloring a 1-resiliently 4-colorable graph is NP-hard.*

Proof. It is known that 4-coloring a 3-colorable graph is NP-hard, so we may apply Observation 3. Every 3-colorable graph G is 1-resiliently 4-colorable, since if we are given a proper 3-coloring of G we may use the fourth color to properly color any new edge that is added. So an algorithm A which efficiently 4-colors 1-resiliently 4-colorable graphs can be used to 4-color a 3-colorable graph.   □

Proposition 3. *For all $k \geq 3$, it is 2-to-1-hard to k-color a $(k-3)$-resiliently k-colorable graph.*

Proof. As with Proposition 2, we apply Observation 3 to the conditional fact that it is NP-hard to k-color a 3-colorable graph for $k > 3$. Such graphs are $(k-3)$-resiliently k-colorable.   □

Proposition 4. *For sufficiently large k it is NP-hard to $2^{\sqrt[3]{k}}$-color an r-resiliently $2^{\sqrt[3]{k}}$-colorable graph for $r < 2^{\sqrt[3]{k}} - k$.*

Proposition 4 comes from applying Observation 3 to the lower bound of Huang [16]. The only unexplained cell of Figure 1 is (3,1), which we prove is NP-hard as our main theorem in the next section.

4   NP-Hardness of 1-Resilient 3-Colorability

Theorem 2. *It is NP-hard to 3-color a 1-resiliently 3-colorable graph.*

Proof. We reduce 1-resilient 3-coloring from 1-resilient 6-SAT. This reduction comes in the form of a graph which is 3-colorable if and only if the 6-SAT instance is satisfiable, and 1-resiliently 3-colorable when the 6-SAT instance is 1-resiliently satisfiable. We use the colors white, black, and gray.

We first describe the gadgets involved and prove their consistency (that the 6-SAT instance is satisfiable if and only if the graph is 3-colorable), and then prove the construction is 1-resilient. Given a 6-CNF formula $\varphi = C_1 \wedge \cdots \wedge C_m$ we construct a graph G as follows. Start with a base vertex b which we may assume w.l.o.g. is always colored gray. For each literal we construct a *literal gadget* consisting of two vertices both adjacent to b, as in Figure 2. As such, the vertices in a literal gadget may only assume the colors white and black. A variable is interpreted as true iff both vertices in the literal gadget have the same color. We will abbreviate this by saying a literal is *colored true* or *colored false*.

Fig. 2. The gadget for a literal. The two single-degree vertices represent a single literal, and are interpreted as true if they have the same color. The base vertex is always colored gray. Note this gadget comes from Kun et al. [25].

We connect two literal gadgets for $x, \overline{x}$ by a *negation gadget* in such a way that the gadget for x is colored true if and only if the gadget for $\overline{x}$ is colored false. The negation gadget is given in Figure 3. In the diagram, the vertices labeled 1 and 3 correspond to x, and those labeled 10 and 12 correspond to $\bar{x}$. We start by showing that no proper coloring can exist if both literal gadgets are colored true. If all four of these vertices are colored white or all four are black, then vertices 6 and 7 must also have this color, and so the coloring is not proper. If one pair is colored both white and the other both black, then vertices 13 and 14 must be gray, and the coloring is again not proper. Next, we show that no proper coloring can exist if both literal gadgets are colored false. First, if vertices 1 and 10 are white and vertices 3 and 12 are black, then vertices 2 and 11 must be gray and the coloring is not proper. If instead vertices 1 and 12 are white and vertices 3 and 10 black, then again vertices 13 and 14 must be gray. This covers all possibilities up to symmetry. Moreover, whenever one literal is colored true and the other false, one can extend it to a proper 3-coloring of the whole gadget.

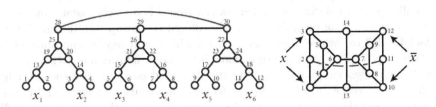

Fig. 3. Left: the gadget for a clause. Right: the negation gadget ensuring two literals assume opposite truth values.

Now suppose we have a clause involving literals, w.l.o.g., $x_1, \ldots, x_6$. We construct the *clause gadget* shown in Figure 3, and claim that this gadget is 3-colorable iff at least one literal is colored true. Indeed, if the literals are all colored false, then the vertices 13 through 18 in the diagram must be colored gray, and then the vertices 25, 26, 27 must be gray. This causes the central triangle to use only white and black, and so it cannot be a proper coloring. On the other hand, if some literal is colored true, we claim we can extend to a proper coloring of the whole gadget. Suppose w.l.o.g. that the literal in question is x_1, and that vertices 1 and 2 both are black. Then Figure 4 shows how this extends to a proper coloring of the entire gadget regardless of the truth assignments of the other literals (we can always color their branches as if the literals were false).

It remains to show that G is 1-resiliently 3-colorable when φ is 1-resiliently satisfiable. This is because a new edge can, at worst, fix the truth assignment

Fig. 4. A valid coloring of the clause gadget when one variable (in this case x_3) is true

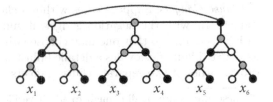

(perhaps indirectly) of at most one literal. Since the original formula φ is 1-resiliently satisfiable, G maintains 3-colorability. Additionally, the gadgets and the representation of truth were chosen so as to provide flexibility w.r.t. the chosen colors for each vertex, so many edges will have no effect on G's colorability.

First, one can verify that the gadgets themselves are 1-resiliently 3-colorable.[4] We break down the analysis into eight cases based on the endpoints of the added edge: within a single clause/negation/literal gadget, between two distinct clause/negation/literal gadgets, between clause and negation gadgets, and between negation and literal gadgets. We denote the added edge by $e = (v, w)$ and call it *good* if G is still 3-colorable after adding e.

Literal Gadgets. First, we argue that e is good if it lies within or across literal gadgets. Indeed, there is only one way to add an edge within a literal gadget, and this has the effect of setting the literal to false. If e lies across two gadgets then it has no effect: if c is a proper coloring of G without e, then after adding e either c is still a proper coloring or we can switch to a different representation of the truth value of v or w to make e properly colored (i.e. swap "white white" with "black black," or "white black" with "black white" and recolor appropriately).

Negation Gadgets. Next we argue that e is good if it involves a negation gadget. Let N be a negation gadget for the variable x. Indeed, by 1-resilience an edge within N is good; e only has a local effect within negation gadgets, and it may result in fixing the truth value of x. Now suppose e has only one vertex v in N. Figure 5 shows two ways to color N, which together with reflections along the horizontal axis of symmetry have the property that we may choose from at least two colors for any vertex we wish. That is, if we are willing to fix the truth value of x, then we may choose between one of two colors for v so that e is properly colored regardless of which color is adjacent to it.

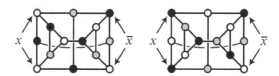

Fig. 5. Two distinct ways to color a negation gadget without changing the truth values of the literals. Only the rightmost center vertex cannot be given a different color by a suitable switch between the two representations or a reflection of the graph across the horizontal axis of symmetry. If the new edge involves this vertex, we must fix the truth value appropriately.

Clause Gadgets. Suppose e lies within a clause gadget or between two clause gadgets. As with the negation gadget, it suffices to fix the truth value of one variable suitably so that one may choose either of two colors for one end of the new edge. Figure 6 provides a detailed illustration of one case. Here, we focus on two branches of two separate clause gadgets, and add the new edge $e = (v, w)$.

---

[4] These graphs are small enough to admit verification by computer search.

The added edge has the following effect: if x is false, then neither y nor z may be used to satisfy C_2 (as w cannot be gray). This is no stronger than requiring that either x be true or y and z both be false, i.e., we add the clause $x \lor (\overline{y} \land \overline{z})$ to φ. This clause can be satisfied by fixing a single variable (x to true), and φ is 1-resilient, so we can still satisfy φ and 3-color G. The other cases are analogous.

This proves that G is 1-resilient when φ is, and finishes the proof. $\qquad\square$

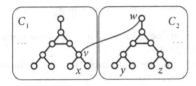

Fig. 6. An example of an edge added between two clauses C_1, C_2

5   Discussion and Open Problems

The notion of resilience introduced in this paper leaves many questions unanswered, both specific problems about graph coloring and more general exploration of resilience in other combinatorial problems and CSPs.

Regarding graph coloring, our paper established the fact that 1-resilience doesn't affect the difficulty of graph coloring. However, the question of 2-resilience is open, as is establishing linear lower bounds without dependence on the 2-to-1 conjecture. There is also room for improvement in finding efficient algorithms for highly-resilient instances, closing the gap between NP-hardness and tractability.

On the general side, our framework applies to many NP-complete problems, including Hamiltonian circuit, set cover, 3D-matching, integer LP, and many others. Each presents its own boundary between NP-hardness and tractability, and there are undoubtedly interesting relationships across problems.

Acknowledgments. We thank Shai Ben-David for helpful discussions.

References

1. Ackerman, M., Ben-David, S.: Clusterability: A theoretical study. Journal of Machine Learning Research - Proceedings Track 5, 1–8 (2009)
2. Arora, S., Ge, R.: New tools for graph coloring. In: Goldberg, L.A., Jansen, K., Ravi, R., Rolim, J.D.P. (eds.) APPROX/RANDOM 2011. LNCS, vol. 6845, pp. 1–12. Springer, Heidelberg (2011)
3. Awasthi, P., Blum, A., Sheffet, O.: Center-based clustering under perturbation stability. Inf. Process. Lett. 112(1-2), 49–54 (2012)
4. Berger, B., Rompel, J.: A better performance guarantee for approximate graph coloring. Algorithmica 5(3), 459–466 (1990)
5. Bilu, Y., Linial, N.: Are stable instances easy? Combinatorics, Probability & Computing 21(5), 643–660 (2012)
6. Blum, A.: New approximation algorithms for graph coloring. J. ACM 41(3), 470–516 (1994)

7. Cai, L.: Parameterized complexity of vertex colouring. Discrete Applied Mathematics 127(3), 415–429 (2003)
8. Dailey, D.P.: Uniqueness of colorability and colorability of planar 4-regular graphs are np-complete. Discrete Mathematics 30(3), 289–293 (1980)
9. Dinur, I., Mossel, E., Regev, O.: Conditional hardness for approximate coloring. SIAM J. Comput. 39(3), 843–873 (2009)
10. Eppstein, D., Bern, M.W., Hutchings, B.L.: Algorithms for coloring quadtrees. Algorithmica 32(1), 87–94 (2002)
11. Garey, M.R., Johnson, D.S.: The complexity of near-optimal graph coloring. J. ACM 23(1), 43–49 (1976)
12. Guruswami, V., Khanna, S.: On the hardness of 4-coloring a 3-colorable graph. SIAM J. Discrete Math. 18(1), 30–40 (2004)
13. Halldórsson, M.M.: A still better performance guarantee for approximate graph coloring. Inf. Process. Lett. 45(1), 19–23 (1993)
14. Håstad, J.: Clique is hard to approximate within $n^{1-\varepsilon}$. Acta Mathematica 182, 105–142 (1999)
15. Hoàng, C.T., Maffray, F., Mechebbek, M.: A characterization of b-perfect graphs. Journal of Graph Theory 71(1), 95–122 (2012)
16. Huang, S.: Improved hardness of approximating chromatic number. CoRR, abs/1301.5216 (2013)
17. Kawarabayashi, K.I., Thorup, M.: Coloring 3-colorable graphs with $o(n^{1/5})$ colors. In: STACS, vol. 25, pp. 458–469 (2014)
18. Johnson, D.S.: Worst case behavior of graph coloring algorithms. In: Proc. 5th Southeastern Conf. on Comb., Graph Theory and Comput., pp. 513–527 (1974)
19. Karp, R.M.: Reducibility among combinatorial problems. In: Complexity of Computer Computations, pp. 85–103 (1972)
20. Khanna, S., Linial, N., Safra, S.: On the hardness of approximating the chromatic number. Combinatorica 20(3), 393–415 (2000)
21. Khot, S.: Improved inaproximability results for maxclique, chromatic number and approximate graph coloring. In: FOCS, pp. 600–609 (2001)
22. Kobler, D., Rotics, U.: Edge dominating set and colorings on graphs with fixed clique-width. Discrete Applied Mathematics 126(2-3), 197–221 (2003)
23. Král', D., Kratochvíl, J., Tuza, Z., Woeginger, G.J.: Complexity of coloring graphs without forbidden induced subgraphs. In: Brandstädt, A., Le, V.B. (eds.) WG 2001. LNCS, vol. 2204, pp. 254–262. Springer, Heidelberg (2001)
24. Kumar, V.: Algorithms for constraint-satisfaction problems: A survey. AI Magazine 13(1), 32 (1992)
25. Kun, J., Powers, B., Reyzin, L.: Anti-coordination games and stable graph colorings. In: Vöcking, B. (ed.) SAGT 2013. LNCS, vol. 8146, pp. 122–133. Springer, Heidelberg (2013)
26. Mihalák, M., Schöngens, M., Šrámek, R., Widmayer, P.: On the complexity of the metric TSP under stability considerations. In: Černá, I., Gyimóthy, T., Hromkovič, J., Jefferey, K., Královič, R., Vukolić, M., Wolf, S. (eds.) SOFSEM 2011. LNCS, vol. 6543, pp. 382–393. Springer, Heidelberg (2011)
27. Reyzin, L.: Data stability in clustering: A closer look. In: Bshouty, N.H., Stoltz, G., Vayatis, N., Zeugmann, T. (eds.) ALT 2012. LNCS, vol. 7568, pp. 184–198. Springer, Heidelberg (2012)
28. Wigderson, A.: Improving the performance guarantee for approximate graph coloring. J. ACM 30(4), 729–735 (1983)
29. Zuckerman, D.: Linear degree extractors and the inapproximability of max clique and chromatic number. Theory of Computing 3(1), 103–128 (2007)

Document Retrieval with One Wildcard*

Moshe Lewenstein[1,**], J. Ian Munro[2], Yakov Nekrich[2],
and Sharma V. Thankachan[2]

[1] Bar-Ilan University, Israel
moshe@macs.biu.ac.il
[2] University of Waterloo, Canada
{imunro,ynekrich,thanks}@uwaterloo.ca

Abstract. In this paper we extend several well-known document listing problems to the case when documents contain a substring that approximately matches the query pattern. We study the scenario when the query string can contain a wildcard symbol that matches any alphabet symbol; all documents that match a query pattern with one wildcard must be enumerated. We describe a linear space data structure that reports all documents containing a substring P in $O(|P| + \sigma\sqrt{\log\log\log n} + \mathsf{docc})$ time, where σ is the alphabet size and docc is the number of listed documents. We also describe a succinct solution for this problem.

Furthermore our approach enables us to obtain an $O(n\sigma)$-space data structure that enumerates all documents containing both a pattern P_1 and a pattern P_2 in the special case when P_1 and P_2 differ in one symbol.

1 Introduction

The ever-growing abundance of data in databases presents unique challenges to the search community that need to be addressed with both speed and space considerations taken into account. For many years the search community was focused on offline problems. In classical *pattern matching* one is given a pattern P and text T and is required to find all occurrences of P in T. Numerous solutions for pattern matching have addressed many different variations of this problem, e.g. [7,19]. In *text indexing* one is given a text T to preprocess for subsequent pattern queries P for which one is required to find all occurrences of P in T. The practical need for text indexing was already recognized long ago. The explosion in data that stemmed from the onset of the Internet and applications in computational biology set the quest for more succinct data structures and posed new challenges in this important research area. Several broad research directions are related to extensions of the standard indexing problem.

* This research was funded in part by NSERC of Canada and the Canada Research Chairs program.
** The author is grateful for the support of the Binational Science Foundation (BSF) grant # 2010437 and for the support of the German Israel Foundation (GIF) grant # 1147/2011.

E. Csuhaj-Varjú et al. (Eds.): MFCS 2014, Part II, LNCS 8635, pp. 529–540, 2014.

First, we can keep a collection of strings (or documents) in the data structure and ask queries about documents that contain a query string. For instance, we may wish to find all documents that contain P at least once. This problem, called document listing problem, is more difficult than the standard indexing because we are interested in reporting each document only once even if it contains multiple occurrences of P. More complicated queries, that are relevant for document retrieval applications, were also studied [16,33,29]. We may be interested in documents that contain P at least k times for a query parameter k, or the k documents in which P occurs most frequently, etc. We refer to e.g. [28,13] for an overview of problems and results in this area. Second, we are frequently interested in substrings of T that approximately match the query string P. Approximate string matching is important for computational biology applications. There are several definitions of approximate matching. In wildcard pattern matching, considered in this paper, the query string P contains a symbol ϕ that matches any alphabet symbol. Indexing for patterns with wildcards and approximate indexing was considered in e.g., [2,6,9,21,35,23,18,12,20,30,17].

In this paper we consider generalizations of the document listing problem for the case when the query pattern P contains one wildcard symbol ϕ that matches any alphabet symbol, and present a linear space indexing solution.

Previous and Related Results. The suffix tree provides a linear space solution for the original indexing problem. Using suffix trees, we can report all occ occurrences of a substring P in optimal $O(|P| + \text{occ})$ time. Henceforth $|S|$ denotes the length of string S. Indexing for patterns with wildcards appears to be significantly more difficult than the standard indexing. Even in the simplest scenario when P contains only one wildcard, we either need more space or have to spend more time to answer a query.

The naive solution is to store all possible combinations of suffixes resulting from replacing an arbitrary symbol by a wildcard. The space usage of this naive solution is $O(n^2)$. Cole et al. [9] described an elegant data structure that significantly reduced the space usage of this naive approach[1]. Their data structure uses $O(n \log n)$ space and answers queries in $O(|P| + \text{occ})$ time. Another solution of one-wildcard indexing is based on reducing this problem to range reporting. This approach, introduced in [2] and used in [5], needs $O(n \log^\varepsilon n)$ space to achieve optimal query time.

Cole et al. [9] described another data structure that uses $O(n \log n)$ space and answers queries in $O(|P| + \sigma \log \log n + \text{occ})$ time. Here and further σ denotes the alphabet size. Bille et al. [6] showed how to reduce the space usage to $O(n)$ and obtained the first linear space data structure. Very recently, Lewenstein et al. [23] described a linear space data structure that supports queries in $O(|P| + \sigma \sqrt{\log \log \log n} + \text{occ})$ time. This is the fastest currently known data structure that uses linear space.

---

[1] In [9] the situation when the query string contains $k > 1$ wildcard symbols was also considered. We refer to e.g. [22] and references therein for an overview of previous results on general indexing with wildcards.

Matias et al. [26] and later Muthukrishnan [27] addressed the document listing problem. The solution of Muthukrishnan [27] uses $O(n)$ space and reports all docc documents containing P in optimal time $O(|P| + \text{docc})$. In previous works on document listing it was assumed that the pattern P contains alphabet symbols only.

Our Results. We consider the document listing problem in the case when the pattern P contains one wildcard symbol. Our solution uses linear space and reports all docc documents that contain a substring matching $P_1 \phi P_2$ in $O(|P_1| + |P_2| + \sigma \sqrt{\log \log \log n} + \text{docc})$ time, where P_1 and P_2 are arbitrary strings containing alphabet symbols only. Thus we match the complexity of the best currently known linear space data structure for reporting all occurrences of $P_1 \phi P_2$. We also describe a compact data structure that uses $|\text{CSA}| + O(n)$ bits of space, where $|\text{CSA}|$ denotes the space (in bits) needed to store a compact suffix array.

The main algorithmic challenge of document listing in the one-wildcard scenario is the requirement that each document must be reported $O(1)$ times. Our solution is based on a novel notion of unique prefixes, defined in Section 3. We believe this idea to be of independent interest and that it can find further applications.

As an additional bonus, we describe an efficient solution for a special case of the two-pattern document listing problem [8,14,10,15]. In the two-pattern document listing a query consists of two patterns P_1 and P_2; we want to report all documents that contain both P_1 and P_2. This problem is known to be very hard; the best known solution need $\tilde{O}(n^2)$ space to achieve $\tilde{O}(|P_1| + |P_2| + \text{docc})$ query time.[2] We consider the special case of this problem when P_1 and P_2 mismatch only at one position. For this special case, we describe an $O(n\sigma)$ space data structure that lists all docc relevant documents in optimal $O(|P_1| + |P_2| + \text{docc})$ time. This result also relies on the notion of unique prefixes.

2   Document Listing without Wildcards

In this section, we briefly describe the indexing solution for the document listing problem proposed by Muthukrishnan [27]. Notice that the query string P does not contain any wildcard character in this case. The data structure consists of three main components: a generalized suffix tree GST, an array $E[1..n]$ and a range minimum query data structure over E [11].

The generalized suffix tree GST of our document collection $\mathcal{D} = \{d_1, d_2, .., d_D\}$ is essentially a suffix tree [36] of the text $\mathsf{T}[1..n] = d_1 d_2 \ldots d_D$ obtained by concatenating all documents in $\mathcal{D}$. We assume that the last character of each document is \$ $\notin \Sigma$, a special symbol that does not appear anywhere else in any document in $\mathcal{D}$. Each substring $\mathsf{T}[i..n]$, with $i \in [1, n]$, is called a *suffix* of T. The GST of $\mathcal{D}$ is a lexicographic arrangement of all these n suffixes in a compact

---

[2] The notation $\tilde{O}$ ignores poly-logarithmic factors. Precisely, $\tilde{O}(f(n)) \equiv O(f(n) \log^{O(1)} n)$.

trie structure, where the ith leftmost leaf represents the ith lexicographically smallest suffix. Let $str(u, w)$ be the string obtained by concatenation of all edge labels on the path from u to w in GST and $str(w) = str(u_r, w)$, where u_r is the root node of GST. Also we use ℓ_i for $i \in [1, n]$ to denote the ith leftmost leaf in GST, and $doc(\ell)$ represents the document that contains the suffix $str(\ell)$ [3]. The locus node of a pattern P, denoted by $locus(P)$, is the highest node u such that $str(u)$ is prefixed by P. Suffix array $SA[1..n]$ is a component of GST, where $SA[i] = n + 1 - |str(\ell_i)|$ [25]. Suffix range of a pattern P is the maximal range $[sp, ep]$, such that for all $i \in [sp, ep]$, $str(\ell_i)$ is prefixed by P. Therefore, ℓ_{sp} and ℓ_{ep} represents the first and last leaves in the subtree of the locus node of P. Both the locus node and the suffix range of P can be computed in $O(|P|)$ time.

The array $E[1..n]$ is defined as $E[i] = \max\{j < i | doc(\ell_i) = doc(\ell_j)\}$; if there is no $j < i$ satisfying $doc(\ell_j) = doc(\ell_i)$, then $E[i] = -\infty$. The set $\{doc(\ell_i) | i \in [sp, ep] \text{ and } E[i] < sp\}$ represents the set of documents containing P as a substring and its cardinality is denoted by docc. The problem of enumerating elements of this set can be reduced to three-sided range reporting in two dimensions, and by maintaining a range minimum query data structure over E, optimal $O(1 + \text{docc})$ query time can be achieved. Since we initially need $O(|P|)$ time for pattern search, the total query time is $O(|P| + \text{docc})$; the data structure uses $O(n)$ words of space. See [4,32,16] for other space efficient solutions.

3   Document Listing with One Wildcard

In this section we show how all those documents in $\mathcal{D}$, containing a query string $P_1 \phi P_2$ can be reported efficiently. Our approach is based on defining the *unique prefixes*. The notion of the unique prefix will also be an important component of our other data structures. The main result is summarized in the following theorem.

Theorem 1. *Let $\mathcal{D}$ be collection of D strings (documents) of n characters in total, which are drawn from an alphabet set Σ of size σ. By maintaining $\mathcal{D}$ as an $O(n)$-word data structure, all those docc documents in $\mathcal{D}$ containing a query string $P = P_1 \phi P_2$ can be reported in $O(|P| + \sigma \sqrt{\log \log \log n} + \text{docc})$ time. Here ϕ represents a wildcard character, P_1, P_2 are strings without wildcards, and $P_1 \phi P_2$ is the concatenation of P_1, ϕ and P_2.*

The heavy-path decomposition represents an arbitrary tree as a union of *heavy paths* in such way that any leaf-to-root path intersects with $O(\log n)$ heavy paths. We refer to [34] for the definition of heavy paths. We will say that a child u_i of a node u is *heavy* if both u_i and u are on the same heavy path and *light* otherwise. A heavy (respectively, light) descendant of u is a descendant of a heavy (respectively, light) child of u. We keep the suffixes of all documents in the generalized suffix tree GST.

We now describe the notion of unique prefix. Let u be an arbitrary node in GST with $u_1, u_2, u_3, \ldots, u_x$ as its children (in the left to right order, where $x \leq \sigma$)

---

[3] To be precise, $doc(\ell_i) = d_j$, where $j = 1 +$ (the number of \$'s in $T[1..(SA[i] - 1)]$).

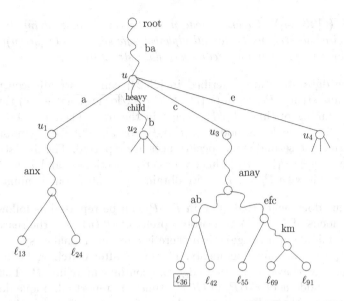

Fig. 1. Values of $\psi(l, u)$ for selected leaves in GST. All shown leaves are suffixes of the same document. We have $\psi(l_{36}, u) = 6$, because *bacan* is not a unique prefix (*baaan* appeared in the subtree u_1), but *bacana* is a unique prefix of $str(l_{36})$. In the same manner, $\psi(l_{42}, u) = 10$, $\psi(l_{55}, u) = 8$, $\psi(l_{69,u}) = 11$, $\psi(l_{91}, u) = 13$. Now, when we search for a pattern $P = ba\phi ana$, our algorithm will report $doc(l_{36})$ because $|P| \geq \psi(l_{36}, u)$. We observe that $|P| \geq \psi(\cdot, u)$ only for l_{36}, but not for the other leaves in the same document.

and u_h be its heavy child. Also let $\alpha_i \in \Sigma$ be the leading character on the edge connecting u and u_i. Let ℓ be a leaf node in the subtree of u_i for some $i \neq h$ (i.e., ℓ is a light descendant of u), then we say that a string P_i is *a unique prefix* of $str(\ell)$ with respect to u iff

1. $P_i = str(u)\alpha_i P'$ is a prefix of $str(\ell)$.
2. if w_i is the locus of P_i, then ℓ is the leftmost occurrence of $doc(\ell)$ in the subtree of w_i.
3. the document $d = doc(\ell)$ does not contain a substring $P_j = str(u)\alpha_j P'$ for any $j < i$ and $j \neq h$.

Let $\psi(\ell, u)$ be the length of the shortest unique prefix of $str(\ell)$ with respect to u. See Fig. 1 for an example illustrating the definitions of unique prefixes and $\psi(\cdot, \cdot)$. For any node u and all its *light leaf* descendants ℓ, we keep the values of $\psi(\ell, u)$ in an array A_u. That is, $A_u[k] = \psi(\ell, u)$ if ℓ is the kth leftmost light leaf descendant of u. Each leaf ℓ has at most $O(\log n)$ ancestors with ℓ as one of its light descendants. Hence all $A_{\{\cdot\}}$'s have $O(n \log n)$ entries. We also keep the linear space data structure for standard document listing (refer to Section 2), which can report all unique documents whose suffixes appear below a given node in optimal time.

We will also need the following result.

Lemma 1 ([23,24]). *Given a node u of the* GST *and a string P_2, we can find the loci of all $str(u)\alpha_i P_2$ for all alphabet symbols α_i in $O(|str(u)| + |P_2| + \sigma\sqrt{\log\log\log n})$ time using a linear space data structure.*

Proof: The data structure described in [23,24] can report all occurrences of $P_1\phi P_2$ for any strings P_1, P_2 in $O(|P_1| + |P_2| + \sqrt{\log\log\log n} + occ)$ time where occ is the number of times $P_1\phi P_2$ occurs in the text. Their method works in two stages: first the locus nodes of $P_1\alpha_i P_2$ for all alphabet symbols α_i are found. During the second stage, occurrences are reported. The first stage takes $O(\sqrt{\log\log\log n} + |P_1| + |P_2|)$ time; we refer to Sections 6 and 7 in [24]. If we apply this result with $P_2 = str(u)$, we obtain the result of our Lemma.   □

All unique documents that contain $P_1\phi P_2$ can be reported as follows. Let u denote the locus of P_1. If P_1 is a proper prefix of $str(u)$ (i.e., the search for P_1 ends in the middle of an edge), then there is only one alphabet symbol α such that $P_1\alpha$ appears in some documents. Therefore, after matching P_1 in GST, we can skip the next character on that edge and continue matching P_2. Thus we can find the locus node w of $P_1\alpha P_2$ in $O(|P|)$ time and report all unique documents corresponding to the suffixes in the subtree of w optimally using the standard document listing algorithm.

The case when $str(u) = P_1$ is more complicated. The search for P_1 ends in a node u. Hence many different strings that match $P_1\phi P_2$ may appear in documents. The difficulty arises because the same document d can contain different substrings (e.g., $P_1\alpha_t P_2$ and $P_1\alpha_s P_2$) matching $P_1\phi P_2$, but we want to report d only once. Information about unique prefixes stored in arrays A_u enables us to solve this problem. We visit all children u_i of u. Recall that α_i is the first symbol on the edge with label $str(u, u_i)$. We can find the loci w_i of $str(u)\alpha_i P_2$ for all α_i's in $O(\sigma\sqrt{\log\log\log n} + |str(u)| + |P_2|)$ time using an $O(n)$-word structure, as shown in Lemma 1. If w_i exists, let $[sp_i, ep_i]$ be the range of leaves in the subtree rooted at w_i. We visit all w_i that are light descendants of u. For every such w_i we report all indices j such that $j \in [sp_i, ep_i]$ and $A_u[map_u(j)] \leq |P|$. Here $map_u(j)$ represents the location in A_u, where $\psi(\ell_j, u)$ is stored and can be computed in constant time [4]. In other words, we report all j's corresponding to all elements within $A_u[map_u(sp_i)..map_u(ep_i)]$ which are not larger than $|P|$. All j can be enumerated in $O(1)$ time per index using a data structure for range minimum queries (RMQ) [11]. For every found j we report the document $doc(\ell_j)$ where ℓ_j is the j-th leaf in the subtree of w_i. Finally we also visit w_h that is a heavy descendant of u and report all documents whose suffixes appear in the subtree of w_h using the standard document listing algorithm, described in Section 2.

Correctness of our procedure follows from the definition of unique prefixes. Suppose that a document d contains substring $P_1\alpha_i P_2$ for one or more symbols α_i. Let α_k be the smallest such symbol. If ℓ_z is the leftmost occurrence of d in the subtree of w_k, then the shortest unique prefix of $str(\ell_z)$ is not longer

---

[4] Let $[L_u, R_u]$ represent the range of leaves in the subtree of u and $[L'_u, R'_u]$ be the range of leaves in the subtree of u_h, the heavy child of u. Then, if $j \in [L_u, L'_u - 1]$, then $map_u(j) = j - (L_u - 1)$. On the other hand, if $j \in [R'_u + 1, R_u]$, then $map_u(j) = j - R'_u$.

than $|P|$. Hence, the index of ℓ_z will be listed when a node u_k is visited and the subarray $A_u[map_u(sp_k)..map_u(ep_k)]$ is examined. When nodes u_i, $i > k$, are visited, the document d will not be reported again: Suppose that a suffix of document d is stored in a leaf ℓ' in the subtree of u_i. Then $P_1\alpha_i P_2$ is not a unique prefix of ℓ' because condition (3) is not satisfied. Hence, each document is reported at most once when light descendants u_i of u are visited. The same document can be reported at most two times: Each document d is reported at most once when light descendants u_i of u are explored and at most once when a heavy descendant is explored. We can get rid of duplicates by keeping the list of documents L found in all u_i and a binary array D. All entries of D are initially set to 0. Each document in L is reported only if the corresponding entry in D is 0; when a document d_z is reported, we set $D[z] = 1$. When the list of unique documents is generated and the query is answered, we traverse L and set all corresponding entries in D to 0.

The space usage of our data structure is $O(n \log n)$ words because all arrays A_u have $O(n \log n)$ entries and the range minima structure for an array A_u uses linear space in the number of its entries. We can reduce the space to linear by discarding arrays A_u and keeping only compact RMQ structures that need only a linear number of *bits*. Our modified procedure does not require the array A_u and works as follows: when we visit a light descendant u_i of u, we answer an RMQ query on $A_u[map_u(sp_i)..map_u(ep_i)]$ and find the index j_0 such that $A_u[map_u(j_0)] \leq A_u[map_u(j')]$ for all $j' \in [sp_i, ep_i]$. If the j_0-th leaf in the subtree u_i is a suffix of document d that was not reported before, then we recursively search in $A_u[map_u(sp_i)..map_u(j_0 - 1)]$ and $A_u[map_u(j_0 + 1)..map_u(ep_i)]$. If the document d was already reported, then we stop. Using the array D described above, we can find out whether a document d was already reported in $O(1)$ time. Correctness of our modified procedure follows from the definition of A_u. Suppose that $A_u[map_u(j)]$ is minimal on some $A_u[map_u(a)..map_u(b)]$ and the document $doc(\ell_j)$ was already reported. Then $A_u[map_u(j)] \geq |P|$ and hence $A_u[map_u(i)] \geq |P|$ for all $i \in [a, b]$. Hence, all documents corresponding to suffixes in $A_u[map_u(a)..map_u(b)]$ are already reported.

4   A Succinct Space Data Structure

In this section, we show how to obtain a space efficient version of our data structure in Theorem 3. The main result of this section is summarized in the following theorem.

Theorem 2. *There exists an* $|\mathsf{CSA}| + O(n)$ *bit structure for reporting all those* docc *documents in* $\mathcal{D}$ *containing a query string* $P = P_1 \phi P_2$ *in* $O(search(|P|) + (\sigma + \mathsf{docc})t_{SA} \log n)$ *time. Here* CSA *represents the compressed suffix array of the text obtained by concatenating all documents in* $\mathcal{D}$, $|\mathsf{CSA}|$ *denotes the space usage (in bits) of* CSA, $search(|P|)$ *is the time for finding the suffix range of* P *in* CSA *and* t_{SA} *is the time needed to compute* $SA[\cdot]$ *and* $SA^{-1}[\cdot]$.

By choosing one of the recent versions of compressed suffix array [3], where $search(|P|) = O(|P|)$ and $t_{SA} = O(\log n)$, the following can be obtained.

Corollary 1. *There exists an $nH_k + o(nH_k) + O(n)$ bit structure for reporting all those docc documents in $\mathcal{D}$ containing a query string $P = P_1\phi P_2$ in $O(|P| + (\sigma + \mathsf{docc})\log^2 n)$ time. Here $H_k \leq \log\sigma$ represents the kth order empirical entropy of $\mathcal{D}$ for $k \leq \epsilon\log_\sigma n$ for any $0 \leq \epsilon < 1$.*

As the first step, we replace the GST by its compressed version that uses $|\mathsf{CSA}| + O(n)$ bits. The total space over all RMQ's in our earlier structure is $O(n\log n)$ bits. In order to compress this, we replace each RMQ structure by sampled RMQ structures as follows: We partition each A_u into contiguous blocks of size $(\log n)$. Then we obtain another array A'_u, where the i-th entry in A'_u is the minimum element in the i-th block of A_u. We now maintain RMQ structures over A'_u's and the total space can be bounded by $O(n)$ bits. We may also maintain a bit vector $B[1..n]$ with constant time rank/select support [31], where $B[i] = 1$ iff the i-th character in T is $\$$. Using this, we can compute $j = doc(\ell_i)$ in $O(t_{SA})$ time, where j is one plus number of 1's in $B[1..(SA[i]-1)]$.

Next we describe how to handle a query with $P_1\phi P_2$ as the input string. The case where $P_1 \neq str(u)$ for any node u in GST can be handled using a standard document listing algorithm. We shall use an $o(n)$-bit (in addition to CSA) solution with query time $O(search(|P|) + \mathsf{docc} \times t_{SA}\log^\epsilon n)$ for handling this case [16]. We now describe the complicated case, i.e., $P_1 = str(u)$ using the same notations defined in Section 3. Our first task is to find the suffix ranges $[sp_i, ep_i]$ corresponding to the patterns $P_1\alpha_i P_2$ for all possible $\alpha_i \in \Sigma$. As we cannot afford to keep the $O(n)$-word structure serving this purpose efficiently, we use the following alternative solution: first find the suffix range of P_1 in $search(|P_1|)$ time and then find the suffix range of P_2 in $search(|P_2|)$ time. We may also find the suffix ranges corresponding to all characters in another $O(\sigma \times search(1))$ time: in fact, we can precompute and store these σ ranges explicitly. We now make use of the following result from [17].

Lemma 2 ([17]). *Let $[sp', ep']$ and $[sp'', ep'']$ be the suffix ranges of strings S' and S'' in CSA. If $[sp', ep']$ and $[sp'', ep'']$ are known, then the suffix range $[sp, ep]$ of the concatenated string $S'S''$ can be computed in $O(t_{SA}\log n)$ time.*

Proof: Clearly, $sp' \leq sp \leq ep \leq ep'$. Moreover for any suffix $str(\ell_i)$ with $i \in [sp, ep]$, the suffix obtained by deleting its first $|S'|$ characters must be within the range $[sp'', ep'']$. Also, the lexicographical ordering of two suffixes within the suffix range $[sp, ep]$ will not alter even if we are comparing them after deleting their first $|S'|$ characters. This essentially means, sp (resp., ep) is the minimum (resp., maximum) k satisfying the following conditions: $k \in [sp', ep']$ and $SA^{-1}[SA[k] + |S'|] \in [sp'', ep'']$. Thus we can obtain $[sp, ep]$ in $O(t_{SA}\log n)$ time via a binary search for k in $[sp', ep']$. $\qquad\square$

Using the above lemma, we can compute the suffix range $[sp_i, ep_i]$ of $P_1\alpha_i P_2$ in $O(t_{SA}\log n)$ time, provided the suffix ranges of $|P_1|$ and $|P_2|$ are already calculated. Thus we need $O(|P| + \sigma t_{SA}\log n)$ time for the initial phase of computing the locus nodes. The next step is to report the elements within $A_u[map_u(sp_i)..map_u(ep_i)]$ based on RMQ's. Notice that we do not have an RMQ structure over

A_u, instead an RMQ structure over the minimum element of each block of A_u. Therefore, we slightly modify the remaining part of the search algorithm as follows: after each RMQ, we extract all documents within the corresponding block and we stop recursing iff all documents within an extracted block are already reported. Notice that when each new document in the output set is reported, we may report another $O(\log n)$ documents, which we might have already reported. Also, computing $str(\ell)$ require $O(t_{SA})$ time. Therefore, by putting all pieces together, the query time can be bounded by $O(search(|P|) + (\sigma + \mathsf{docc})t_{SA}\log n)$.

5 Multiple Patterns in a Document

We consider the problem of reporting all documents that contain both a string P and a string P', such that P and P' differ in only position. The main result is summarized below.

Theorem 3. *There exists an $O(n\sigma)$-word data structure that reports all docc documents containing both $P = P_1\alpha P_2$ and $P_1\alpha' P_2$ in $O(|P_1| + |P_2| + 1 + \mathsf{docc})$ time, where α and α' are characters in Σ.*

Our method is similar to the approach of Section 3. Let ℓ be a light leaf descendant of a node u such that $str(\ell)$ is prefixed by $str(u)\alpha_i$. For an alphabet symbol $\alpha_j \neq \alpha_i$, a string $P_i = str(u)\alpha_i P_i'$ is an α_j-unique prefix of $str(\ell)$ with respect to u iff the following conditions are satisfied:

1. P_i is a prefix of $str(\ell)$
2. if w_i is the locus of P_i, then ℓ is the leftmost occurrence of $doc(\ell)$ in the subtree of w_i
3. the document $d = doc(\ell)$ does not contain a substring $P_j = str(u)\alpha_j P_i'$.

We denote by $\psi_\alpha(\ell, u)$ the length of the shortest α-unique prefix of $str(\ell)$ with respect to u. For every node u in the generalized suffix tree GST and for every alphabet symbol α, we keep an array $A_{u,\alpha}$ that contains values $\psi_\alpha(\ell, u)$ for all light leaf descendants ℓ of u. The total number of entries in all arrays is $O(n\sigma\log n)$ where σ is the size of the alphabet.

Suppose that $P = P_1\alpha P_2$ and $P' = P_1\alpha' P_2$. If there is at least one document that contains both P and P', then GST contains a node u such that $str(u) = P_1$. We descend to the node u in $O(|P|)$ time. If w and w' are the loci of P and P', then at least one of them is a light descendant of u. Suppose that w is a light descendant of u and let $[l_w, r_w]$ be the range of leaves in the subtree of w. We report all indices i such that $l_w \leq i \leq r_w$ and $A_{u,\alpha'}[map_{u,\alpha'}(i)] \geq |P|$. The definition of $map_{u,\alpha}(\cdot)$ with respect to $A_{u,\alpha}$ is the same as the definition of $map_u(\cdot)$ with respect to A_u in Section 3. For every found index i, we report the corresponding document. If ℓ_i is a leaf below the node w and $\psi_{\alpha'}(\ell_i, u) \geq |P|$, then the document $doc(\ell_i)$ also contains the string $P_1\alpha P_2$. Each document is reported exactly once because only the leftmost occurrence of a document in the subtree of w is reported.

We can reduce the space usage to $O(n\sigma)$ by storing only compact data structures for answering range maxima queries on $A_{u,\alpha}$ and discarding the arrays $A_{u,\alpha}$. We also need reporting data structures R_d for all documents d. R_d contains indices of all leaves in GST in which suffixes of the document d are stored. Using the result of [1], we can keep R_d in linear space and answer existential range reporting queries in $O(1)$ time. Thus we can determine whether a suffix from a document d appears in the subtree of a node u in $O(1)$ time for any u and any d.

The modified procedure for listing documents is almost the same as in Section 3. We find the index j where $A_{u,\alpha'}[l_w..r_w]$ reaches its maximum. If the document d_j corresponding to $A_{u,\alpha'}[map_{u,\alpha'}(j)]$ occurs in the subtree of w', we split the range $[l_w, r_w]$ and recursively explore $A_{u,\alpha'}[l_w..j-1]$ and $A_{u,\alpha'}[j+1..r_w]$. If d_j does not appear in the subtree of w', we stop the recursion.

6   Conclusions

In this paper we showed that a collection of documents can be stored in a linear space data structure so that all documents containing a query pattern $P_1\phi P_2$ can be enumerated. This is the first non-trivial result for the case when the query pattern contains a wildcard symbol ϕ. The query time of our data structure matches the query time of the fastest currently known linear space data structure for listing all occurrences of $P_1\phi P_2$. We also described efficient compact solutions for this problem.

The research area of document retrieval contains many important questions related to approximate pattern matching. The first question is existence of a linear space data structure that reports all documents containing a pattern P, where P contains $g > 1$ wildcards. There are linear space data structures that report all occurrences of such P [6,23]. But in the document listing problem each document has to be reported only once; this requirement makes the latter problem significantly more difficult. Unfortunately our definition of uniqueness is not efficient in the case of $g > 1$ wildcards. A straightforward application of our technique would significantly increase the space usage because a superlinear number of patterns has to be taken into account: in addition to document suffixes we would also have to consider the "approximate suffixes" obtained by changing $g - 1$ arbitrary symbols in every suffix. This would blow up the space usage of our data structure. Extending our method to patterns with many wildcards in a space-efficient way presents an algorithmic challenge.

Document listing for other variants of approximate matching appears to be even more challenging. It would be interesting to design a data structure that lists all documents containing a pattern $\tilde{P}$, such that the Hamming distance between $\tilde{P}$ and the query pattern P does not exceed a threshold value g. Even for $g = 1$ no efficient solution for this problem is known. Other proximity measures, e.g., the edit distance between P and $\tilde{P}$ can also be considered. Designing efficient data structures for these queries or proving that no such data structure exists is an interesting and important open problem. Yet another open problem concerns counting documents; it would be interesting to design a $O(n)$-space index that counts the number of documents containing $P_1 \phi P_2$ in $O(|P| + \sigma \texttt{polylog}(n))$ time.

References

1. Alstrup, S., Brodal, G.S., Rauhe, T.: Optimal static range reporting in one dimension. In: Proc. 33rd Annual ACM Symposium on Theory of Computing (STOC), pp. 476–482 (2001)
2. Amir, A., Keselman, D., Landau, G.M., Lewenstein, M., Lewenstein, N., Rodeh, M.: Text indexing and dictionary matching with one error. J. Algorithms 37(2), 309–325 (2000)
3. Belazzougui, D., Navarro, G.: Alphabet-independent compressed text indexing. In: Demetrescu, C., Halldórsson, M.M. (eds.) ESA 2011. LNCS, vol. 6942, pp. 748–759. Springer, Heidelberg (2011)
4. Belazzougui, D., Navarro, G., Valenzuela, D.: Improved compressed indexes for full-text document retrieval. J. Discrete Algorithms 18, 3–13 (2013)
5. Bille, P., Gørtz, I.L.: Substring range reporting. In: Giancarlo, R., Manzini, G. (eds.) CPM 2011. LNCS, vol. 6661, pp. 299–308. Springer, Heidelberg (2011)
6. Bille, P., Gørtz, I.L., Vildhøj, H.W., Vind, S.: String indexing for patterns with wildcards. In: Fomin, F.V., Kaski, P. (eds.) SWAT 2012. LNCS, vol. 7357, pp. 283–294. Springer, Heidelberg (2012)
7. Boyer, R.S., Moore, J.S.: A Fast String Searching Algorithm. Communications of the ACM 20(10), 762–772 (1977)
8. Cohen, H., Porat, E.: Fast set intersection and two-patterns matching. Theoretical Computer Science 411(40-42), 3795–3800 (2010)
9. Cole, R., Gottlieb, L.-A., Lewenstein, M.: Dictionary matching and indexing with errors and don't cares. In: Proc. 36th Annual ACM Symposium on Theory of Computing (STOC 2004), pp. 91–100 (2004)
10. Fischer, J., Gagie, T., Kopelowitz, T., Lewenstein, M., Mäkinen, V., Salmela, L., Välimäki, N.: Forbidden patterns. In: Fernández-Baca, D. (ed.) LATIN 2012. LNCS, vol. 7256, pp. 327–337. Springer, Heidelberg (2012)
11. Fischer, J., Heun, V.: Space-efficient preprocessing schemes for range minimum queries on static arrays. SIAM J. Comput. 40(2), 465–492 (2011)
12. Hon, W.-K., Ku, T.-H., Shah, R., Thankachan, S.V., Vitter, J.S.: Compressed dictionary matching with one error. In: Proc. 2011 Data Compression Conference (DCC 2011), pp. 113–122 (2011)
13. Hon, W.-K., Patil, M., Shah, R., Thankachan, S.V., Vitter, J.S.: Indexes for document retrieval with relevance. In: Brodnik, A., López-Ortiz, A., Raman, V., Viola, A. (eds.) Ianfest-66. LNCS, vol. 8066, pp. 351–362. Springer, Heidelberg (2013)
14. Hon, W.-K., Shah, R., Thankachan, S.V., Vitter, J.S.: String retrieval for multi-pattern queries. In: Chavez, E., Lonardi, S. (eds.) SPIRE 2010. LNCS, vol. 6393, pp. 55–66. Springer, Heidelberg (2010)
15. Hon, W.-K., Shah, R., Thankachan, S.V., Vitter, J.S.: Document listing for queries with excluded pattern. In: Kärkkäinen, J., Stoye, J. (eds.) CPM 2012. LNCS, vol. 7354, pp. 185–195. Springer, Heidelberg (2012)
16. Hon, W.-K., Shah, R., Vitter, J.S.: Space-efficient framework for top-k string retrieval problems. In: FOCS, pp. 713–722. IEEE Computer Society (2009)
17. Huynh, T.N.D., Hon, W.-K., Lam, T.-W., Sung, W.-K.: Approximate string matching using compressed suffix arrays. Theoretical Comp. Science 352(1), 240–249 (2006)

18. Iliopoulos, C.S., Rahman, M.S.: Indexing factors with gaps. Algorithmica 55(1), 60–70 (2009)
19. Knuth, D.E., Morris, J.H., Pratt, V.B.: Fast Pattern Matching in Strings. SIAM Journal on Computing 6(2), 323–350 (1977)
20. Lam, T.-W., Sung, W.-K., Tam, S.-L., Yiu, S.-M.: Space efficient indexes for string matching with don't cares. In: Tokuyama, T. (ed.) ISAAC 2007. LNCS, vol. 4835, pp. 846–857. Springer, Heidelberg (2007)
21. Lewenstein, M.: Indexing with gaps. In: Grossi, R., Sebastiani, F., Silvestri, F. (eds.) SPIRE 2011. LNCS, vol. 7024, pp. 135–143. Springer, Heidelberg (2011)
22. Lewenstein, M.: Orthogonal range searching for text indexing. In: Brodnik, A., López-Ortiz, A., Raman, V., Viola, A. (eds.) Ianfest-66. LNCS, vol. 8066, pp. 267–302. Springer, Heidelberg (2013)
23. Lewenstein, M., Nekrich, Y., Vitter, J.S.: Space-efficient string indexing for wildcard pattern matching. In: Proc. 31st International Symposium on Theoretical Aspects of Computer Science (STACS 2014), pp. 506–517 (2014)
24. Lewenstein, M., Nekrich, Y., Vitter, J.S.: Space-efficient string indexing for wildcard pattern matching. CoRR, abs/1401.0625 (2014)
25. Manber, U., Myers, E.W.: Suffix arrays: A new method for on-line string searches. SIAM J. Comput. 22(5), 935–948 (1993)
26. Matias, Y., Muthukrishnan, S., Şahinalp, S.C., Ziv, J.: Augmenting suffix trees, with applications. In: Bilardi, G., Pietracaprina, A., Italiano, G.F., Pucci, G. (eds.) ESA 1998. LNCS, vol. 1461, pp. 67–78. Springer, Heidelberg (1998)
27. Muthukrishnan, S.: Efficient algorithms for document retrieval problems. In: Proc. 13th Annual ACM-SIAM Symposium on Discrete Algorithms (SODA 2002), pp. 657–666 (2002)
28. Navarro, G.: Spaces, trees and colors: The algorithmic landscape of document retrieval on sequences. CoRR, abs/1304.6023 (2013)
29. Navarro, G., Nekrich, Y.: Top-k document retrieval in optimal time and linear space. In: Rabani, Y. (ed.) SODA, pp. 1066–1077. SIAM (2012)
30. Rahman, M.S., Iliopoulos, C.S.: Pattern matching algorithms with don't cares. In: Proc. 33rd Conference on Current Trends in Theory and Practice of Computer Science (SOFSEM 2007), pp. 116–126 (2007)
31. Raman, R., Raman, V., Satti, S.R.: Succinct indexable dictionaries with applications to encoding k-ary trees, prefix sums and multisets. ACM Transactions on Algorithms 3(4) (2007)
32. Sadakane, K.: Succinct data structures for flexible text retrieval systems. J. Discrete Algorithms 5(1), 12–22 (2007)
33. Shah, R., Sheng, C., Thankachan, S.V., Vitter, J.S.: Top-k document retrieval in external memory. In: Bodlaender, H.L., Italiano, G.F. (eds.) ESA 2013. LNCS, vol. 8125, pp. 803–814. Springer, Heidelberg (2013)
34. Sleator, D.D., Tarjan, R.E.: A data structure for dynamic trees. J. Comput. Syst. Sci. 26(3), 362–391 (1983)
35. Tam, A., Wu, E., Lam, T.-W., Yiu, S.-M.: Succinct text indexing with wildcards. In: Karlgren, J., Tarhio, J., Hyyrö, H. (eds.) SPIRE 2009. LNCS, vol. 5721, pp. 39–50. Springer, Heidelberg (2009)
36. Weiner, P.: Linear pattern matching algorithms. In: SWAT (FOCS), pp. 1–11. IEEE Computer Society (1973)

An $H_{n/2}$ Upper Bound on the Price of Stability of Undirected Network Design Games

Akaki Mamageishvili, Matúš Mihalák, and Simone Montemezzani

Department of Computer Science, ETH Zurich, Switzerland

Abstract. In the network design game with n players, every player chooses a path in an edge-weighted graph to connect her pair of terminals, sharing costs of the edges on her path with all other players fairly. We study the price of stability of the game, i.e., the ratio of the social costs of a best Nash equilibrium (with respect to the social cost) and of an optimal play. It has been shown that the price of stability of any network design game is at most H_n, the n-th harmonic number. This bound is tight for directed graphs. For undirected graphs, the situation is dramatically different, and tight bounds are not known. It has only recently been shown that the price of stability is at most $H_n \left(1 - \frac{1}{\Theta(n^4)} \right)$, while the worst-case known example has price of stability around 2.25. In this paper we improve the upper bound considerably by showing that the price of stability is at most $H_{n/2} + \epsilon$ for any ϵ starting from some suitable $n \geq n(\epsilon)$.

1 Introduction

Network design game was introduced by Anshelevich et al. [1] together with the notion of price of stability (PoS), as a formal model to study and quantify the strategic behavior of non-cooperative agents in designing communication networks. Network design game with n players is given by an edge-weighted graph G (where n does not stand for the number of vertices), and by a collection of n terminal (source-target) pairs $\{s_i, t_i\}$, $i = 1, \ldots, n$. In this game, every player i connects its terminals s_i and t_i by an s_i-t_i path P_i, and pays for each edge e on the path a fair share of its cost (i.e., all players using the edge pay the same amount totalling to the cost of the edge). A Nash equilibrium of the game is an outcome $(P_1, \ldots, P_n)$ in which no player i can pay less by changing P_i to a different path P_i'.

Nash equilibria of the network design game can be quite different from an optimal outcome that could be created by a central authority. To quantify the difference in quality of equilibria and optima, one compares the total cost of a Nash equilibrium to the cost of an optimum (with respect to the total cost). Taking the worst-case approach, one arrives at the *price of anarchy*, which is the ratio of the maximum cost of any Nash equilibrium to the cost of an optimum. Price of anarchy of network design games can be as high as n (but not higher) [1]. Taking the slightly less pessimistic approach leads to the notion of the *price of*

E. Csuhaj-Varjú et al. (Eds.): MFCS 2014, Part II, LNCS 8635, pp. 541–552, 2014.

stability, which is the ratio of the smallest cost of any Nash equilibrium to the cost of an optimum. The motivation behind this is that often a central authority exists, but cannot force the players into actions they do not like. Instead, a central authority can suggest to the players actions that correspond to a best Nash equilibria. Then, no player wants to deviate from the action suggested to her, and the overall cost of the outcome can be lowered (when compared to the worst case Nash equilibria).

Network design games belong to the broader class of congestion games for which a function (called a *potential function*) $\Phi(P_1, \ldots, P_n)$ exists, with the property that $\Phi(P_1 \ldots, P_i, \ldots, P_n) - \Phi(P_1, \ldots, P_i', \ldots, P_n)$ exactly reflects the changes of the cost of any player i switching from P_i to P_i'. This property implies that a collection of paths $(P_1, \ldots, P_n)$ minimizing Φ necessarily needs to be a Nash equilibrium. Up to an additive constant, every congestion game has a unique potential function of a concrete form, which can be used to show that the price of stability of any network design game is at most $H_n := \sum_{i=1}^{n} \frac{1}{i}$, the n-th harmonic number, and this is tight for directed graphs (i.e., there is a network design game for which the price of stability is arbitrarily close to H_n) [1].

Obtaining tight bounds on the price of stability for undirected graphs turned out to be much more difficult. The worst case known example is an involved construction of a game by Bilò et al. [4] achieving in the limit the price of stability of around 2.25. While the general upper bound of H_n applies also for undirected graphs, it has not been known for a long time whether it can be any lower, until the recent work of Disser et al. [7] who showed that the price of stability of any network design game with n players is at most $H_n \cdot \left(1 - \frac{1}{\Theta(n^4)}\right)$. Improved upper bounds have been obtained for special cases. For the case where all terminals t_i are the same, Li showed [10] that the price of stability is at most $O\left(\frac{\log n}{\log \log n}\right)$ (note that H_n is approximately $\ln n$). If, additionally, every vertex of the graph is a source of a player, a series of papers by Fiat et al. [9], Lee and Ligett [12], and Bilò et al. [5] showed that the price of stability is in this case at most $O(\log \log n)$, $O(\log \log \log n)$, and $O(1)$, respectively. Fanelli et al. [8] restrict the graphs to be rings, and prove that the price of stability is at most $3/2$. Further special cases concern the number of players. Interestingly, tight bounds on price of stability are known only for $n = 2$ (we do not consider the case $n = 1$ as a game) [1,6], while for already 3 players there are no tight bounds; for the most recent results for the case $n = 3$, see [7] and [3].

All obtained upper bounds on the price of stability use the potential function in one way or another. Our paper is not an exception in that aspect. Bounding the price of stability translates effectively into bounding the cost of a best Nash equilibrium. A common approach is to bound this cost by the cost of the potential function minimizer $(P_1^\Phi, \ldots, P_n^\Phi) := \arg\min_{(P_1, \ldots, P_n)} \Phi(P_1, \ldots, P_n)$, which is (as we argued above) also a Nash equilibrium. Using just the inequality $\Phi(P_1^\Phi, \ldots, P_n^\Phi) \le \Phi(P_1^O, \ldots, P_n^O)$, where $(P_1^O, \ldots, P_n^O)$ is an optimal outcome (minimizing the total cost of having all pairs of terminals connected), one obtains the original upper bound H_n on the price of stability [1]. In [7,6] authors consider other inequalities obtained from the property that potential optimizer is also a

Nash equilibrium to obtain improved upper bounds. In this paper, we consider n different specifically chosen strategy profiles $(P_1^i, \ldots, P_n^i)$, $i = 1, \ldots, n$, in which players use only edges of the optimum $(P_1^O, \ldots, P_n^O)$ and of the Nash equilibrium $(P_1^\Phi, \ldots, P_n^\Phi)$. This idea is a generalization of the approach used by Bilò and Bove [3] to prove an upper bound of $286/175 \approx 1.634$ for Shapley network design games with 3 players. Clearly, the potential of each of the considered strategy profile is at least the potential of $(P_1^\Phi, \ldots, P_n^\Phi)$. Summing all these n inequalities and combining it with the original inequality $\Phi(P_1^\Phi, \ldots, P_n^\Phi) \leq \Phi(P_1^O, \ldots, P_n^O)$ gives an asymptotic upper bound of $H_{n/2} + \epsilon$ on the price of stability. Our result thus shows that the price of stability is strictly lower than H_n by an additive constant (namely, by $\log 2$).

Albeit the idea is simple, the analysis is not. It involves carefully chosen strategy profiles for various possible topologies of the optimum solution. These considerations can be of independent interest in further attempts to improve the bounds on the price of stability of network design games.

2 Preliminaries

Shapley network design game is a strategic game of n players played on an edge-weighted graph $G = (V, E)$ with non-negative edge costs c_e, $e \in E$. Each player i, $i = 1, \ldots, n$, has a *source* node s_i and a *target* node t_i. All s_i-t_i paths form the set $\mathcal{P}_i$ of the *strategies* of player i. A vector $P = (P_1, \ldots, P_n) \in \mathcal{P}_1 \times \cdots \times \mathcal{P}_n$ is called a *strategy profile*. Let $E(P) := \bigcup_{i=1}^n P_i$ be the set of all edges used in P. The *cost of player* i in a strategy profile P is $\text{cost}_i(P) = \sum_{e \in P_i} c_e/k_e(P)$, where $k_e(P) = |\{j | e \in P_j\}|$ is the number of players using edge e in P. A strategy profile $N = (N_1, \ldots, N_n)$ is a *Nash equilibrium* if no player i can unilaterally switch from her strategy N_i to a different strategy $N_i' \in \mathcal{P}_i$ and decrease her cost, i.e., $\text{cost}_i(N) \leq \text{cost}_i(N_1, \ldots, N_i', \ldots, N_n)$ for every $N_i' \in \mathcal{P}_i$.

Shapley network design games are exact potential games. That is, there is a so called *potential function* $\Phi : \mathcal{P}_1 \times \cdots \times \mathcal{P}_n \to \mathbb{R}$ such that, for every strategy profile P, every player i, and every alternative strategy P_i', $\text{cost}_i(P) - \text{cost}_i(P_1, \ldots, P_i', \ldots, P_n) = \Phi(P) - \Phi(P_1, \ldots, P_i', \ldots, P_n)$. Up to an additive constant, the potential function is unique [13], and is defined as

$$\Phi(P) = \sum_{e \in E(P)} \sum_{i=1}^{k_e(P)} c_e/i = \sum_{e \in E(P)} H_{k_e(P)} \, c_e \ .$$

To simplify the notation (e.g., to avoid writing $H_{\lceil n/2 \rceil}$), we extend H_k also for non-integer values of k by setting $H(k) := \int_0^1 \frac{1-x^k}{1-x} dx$, which is an increasing function, and which agrees with the (original) k-th harmonic number whenever k is an integer.

The *social cost* of a strategy profile P is defined as the sum of the player costs: $\text{cost}(P) = \sum_{i=1}^n \text{cost}_i(P) = \sum_{i=1}^n \sum_{e \in P_i} c_e/k_e(P) = \sum_{e \in E(P)} k_e(P) c_e/k_e(P) = \sum_{e \in E(P)} c_e$. A strategy profile $O(G)$ that minimizes the social cost of a game

G is called a *social optimum*. Observe that a social optimum $O(G)$ so that $E(O(G))$ induces a forest always exists (if there is a cycle, we could remove one of its edges without increasing the social cost). Let $\mathcal{N}(G)$ be the set of Nash equilibria of a game G. The *price of stability of a game G* is the ratio $\text{PoS}(G) = \min_{N \in \mathcal{N}(G)} \text{cost}(N)/\text{cost}(O(G))$.

Let $\mathcal{M}(G)$ be the set of Nash equilibria that are also global minimizers of the potential function Φ of the game. The *potential-optimal price of anarchy* of a game G, introduced by Kawase and Makino [11], is defined as $\text{POPoA}(G) = \max_{N \in \mathcal{M}(G)} \text{cost}(N)/\text{cost}(O(G))$. Properties of potential optimizers were earlier observed and exploited by Asadpour and Saberi in [2] for other games.

Since $\mathcal{M}(G) \subset \mathcal{N}(G)$, it follows that $\text{PoS}(G) \leq \text{POPoA}(G)$. Let $\mathcal{G}(n)$ be the set of all Shapley network design games with n players. The *price of stability of Shapley network design games* is defined as $\text{PoS}(n) = \sup_{G \in \mathcal{G}(n)} \text{PoS}(G)$. The quantity $\text{POPoA}(n)$ is defined analogously, and we get that $\text{PoS}(n) \leq \text{POPoA}(n)$.

3   The $\approx H_{n/2}$ Upper Bound

The main result of the paper is the new upper bound on the price of stability, as stated in the following theorem.

Theorem 1. $PoS(n) \leq H_{n/2} + \epsilon$, *for any* $\epsilon > 0$ *given that* $n \geq n(\epsilon)$ *for some suitable* $n(\epsilon)$.

We consider a Nash equilibrium N that minimizes the potential function Φ. For each player i we construct a strategy profile S^i as follows. Every player $j \neq i$, whenever possible (the terminals of players i and j lie in the same connected component of the optimum O), uses edges of $E(O(G))$ to reach s_i, from there it uses the Nash equilibrium strategy (a path) of player i to reach t_i, and from there it again uses edges of $E(O(G))$ to reach the player j's other terminal node. From the definition of N, we then obtain the inequality $\Phi(N) \leq \Phi(S^i)$. We then combine these n inequalities in a particular way with the inequality $\Phi(N) \leq \Phi(O(G))$, and obtain the claimed upper bound on the cost of N.

The proof of Theorem 1 is structured in the following way. We first prove the theorem for the special case where an optimum $O(G)$ contains an edge that is used by every player. We then extend the proof of this special case, first to the case where $E(O(G))$ is a tree, but with no edge used by every player, and, second, to the case where $E(O(G))$ is a general forest (i.e., not one connected component).

We will use the following notation. For a strategy profile $P = (P_1, \ldots, P_n)$ and a set $U \subset \{1, \ldots, n\}$, we denote by P_U the set of edges $e \in E$ for which $\{j | e \in P_j\} = U$ and by P^l the set of edges $e \in E$ for which $|\{j | e \in P_j\}| = l$. That is, P_U is the set of edges used in P by exactly the players U, and $P^l = \bigcup_{\substack{U \subset \{1, \ldots, n\} \\ |U| = l}} P_U$ is the set of edges used by exactly l many players. Then the edges used by player i in P are $\bigcup_{\substack{U \subset \{1, \ldots, n\} \\ i \in U}} P_U$. We stress that for every player $i \in U$, the edges of P_U

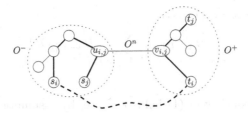

Fig. 1. The non dashed lines are the edges of $E(O)$, the dashed line is the Nash strategy N_i. The path S_j^i from s_j to t_j is given by the thicker dashed and non dashed lines.

are part of the strategy P_i; this implies that, whenever $E(P)$ induces a forest, the source s_i and the target t_i are in two different connected components of $E(P) \setminus P_U$. For any set of edges $F \subset E$, let $|F| := \sum_{e \in F} c_e$. We then have, for instance, that the cost of player i in P is given by $\mathrm{cost}_i(P) = \sum_{\substack{U \subset \{1,\dots,n\} \\ i \in U}} \frac{|P_U|}{|U|}$.

From now on, G is an arbitrary Shapley network design game with n players, $N = (N_1, \dots, N_n)$ is a Nash equilibrium minimizing the potential function and $O = (O_1, \dots, O_n)$ is an arbitrary social optimum so that $E(O)$ has no cycles.

3.1   Case O^n Is Not Empty

In this section we assume that O^n is not empty. In this case, $E(O)$ is actually a tree. Then, $E(O) \setminus O^n$ is formed by two disconnected trees, which we call O^- and O^+, such that each player has the source node in one tree and the target node in the other tree (see also Fig. 1). Without loss of generality, assume that all source nodes s_i are in O^-. Given two players i and j, let $u_{i,j}$ be the first[1] edge of $O_i \cap O_j$ and $v_{i,j}$ be the last edge of $O_i \cap O_j$. Notice that every edge between s_i and $u_{i,j}$ is used in O by player i but not by player j. That is, each edge e between s_i and $u_{i,j}$ satisfies $e \in O_i$ and $e \notin O_j$, or equivalently, $e \in \bigcup_{\substack{U \subset \{1,\dots,n\} \\ i \in U, j \notin U}} O_U$. An analogous statement holds for each edge e between t_i and $v_{i,j}$.

For every player i, we define a strategy profile S^i, where player $j = 1, \dots, n$ uses the following s_j-t_j path S_j^i (see Fig. 1 for an example.):

1. From s_j to $u_{i,j}$, it uses edges of O^-.
2. From $u_{i,j}$ to s_i, it uses edges of O^-.
3. From s_i to t_i, it uses edges of N_i.
4. From t_i to $v_{i,j}$, it uses edges of O^+.
5. From $v_{i,j}$ to t_j, it uses edges of O^+.

If S_j^i contains cycles, we skip them to obtain a simple path from s_j to t_j. This can be the case if N_i is not disjoint from $E(O)$, so that an edge appears both in step 3 and in one of the steps $1, 2, 4$ or 5. Observe that the path S_j^i uses exactly the edges of O_U for $i \in U, j \notin U$ (in steps 2 and 4), the edges of O_U for $i \notin U, j \in U$ (in steps 1 and 5) and the edges of N_U for $i \in U$ (in step 3). We now can prove the following lemma.

---

[1] The edges are ordered naturally along the path from s_i to t_i.

Lemma 1. *For every* $i \in \{1, \ldots, n\}$,

$$\Phi(N) \leq \Phi(S^i) \leq \sum_{\substack{U \subset \{1,\ldots,n\} \\ i \in U}} H_n |N_U| + \sum_{\substack{U \subset \{1,\ldots,n\} \\ i \in U}} H_{n-|U|} |O_U| + \sum_{\substack{U \subset \{1,\ldots,n\} \\ i \notin U}} H_{|U|} |O_U|.$$

$$(1)$$

Proof. The first inequality of (1) holds because, by assumption, N is a global minimum of the potential function Φ.

To prove the second inequality, recall that for any strategy profile P we can write $\Phi(P) = \sum_{e \in P} H_{k_e(P)} c_e = \sum_{U \subset \{1,\ldots,n\}} H_{|U|} |P_U|$. In our case, every edge $e \in S^i$ belongs either to N_U, $U \subset \{1, \ldots, n\}$, $i \in U$, or to O_U, and we therefore sum only over these terms. We now show that, in our sum, the cost c_e of every edge e in S^i is accounted for with at least coefficient $H_{k_e(S^i)}$.

For the first sum in the right hand side of (1), obviously at most n players can use an edge of $N_U, i \in U$, i.e., $k_e(S^i) \leq n$. To explain the second and third sums, notice that if an edge $e \in O_U$ that is present in S^i also belongs to N_i, its cost is already accounted for in the first sum. So, we just have to look at edges that are only present in steps $1, 2, 4$ and 5 of the definition of S^i_j.

To explain the second sum, let $i \in U$. Then, as we already noted, in the definition of S^i_j, player j uses edges of O_U with $i \in U$ only if $j \notin U$ (in steps 2 and 4). Since there are exactly $n - |U|$ players that satisfy $j \notin U$, this explains the second sum.

Finally, to explain the third sum, let $i \notin U$. Similarly to the previous argument, in the definition of S^i_j, player j uses edges of O_U with $i \notin U$ only if $j \in U$ (in steps 1 and 5). Since there are exactly $|U|$ players that satisfy $j \in U$, this explains the third sum. $\qquad \square$

We now show how to combine Lemma 1 with the inequality $\Phi(N) \leq \Phi(O)$ to prove Theorem 1, whenever $O^n \neq \emptyset$.

Lemma 2. *Suppose that Inequality (1) holds for every i. Then, for $x = \frac{n - H_n}{H_n - 1}$,*

$$PoS(G) \leq \frac{n + x}{n + x - H_n} H_{\frac{n+x}{2}} \leq H_{n/2} + \epsilon$$

holds for any $\epsilon > 0$, given that $n \geq n(\epsilon)$ for some suitable $n(\epsilon)$.

Proof. We sum (1) for $i = 1, \ldots, n$ to obtain

$$n\Phi(N) \leq \sum_{i=1}^{n} \left(\sum_{\substack{U \subset \{1,\ldots,n\} \\ i \in U}} H_n |N_U| + \sum_{\substack{U \subset \{1,\ldots,n\} \\ i \in U}} H_{n-|U|} |O_U| + \sum_{\substack{U \subset \{1,\ldots,n\} \\ i \notin U}} H_{|U|} |O_U| \right) =$$

$$= \sum_{U \subset \{1,\ldots,n\}} |U| H_n |N_U| + \sum_{U \subset \{1,\ldots,n\}} |U| H_{n-|U|} |O_U| + \sum_{U \subset \{1,\ldots,n\}} (n - |U|) H_{|U|} |O_U| =$$

$$= \sum_{l=1}^{n} l H_n |N^l| + \sum_{l=1}^{n} (l H_{n-l} + (n - l) H_l) |O^l| .$$

Since $\Phi(N) = \sum_{l=1}^{n} H_l |N^l|$, by putting all terms relating to N on the left hand side we obtain

$$\sum_{l=1}^{n} (nH_l - lH_n)|N^l| \leq \sum_{l=1}^{n} (lH_{n-l} + (n-l)H_l)|O^l| . \qquad (2)$$

On the other hand, we have $\Phi(N) \leq \Phi(O)$, which we can write as

$$\sum_{l=1}^{n} H_l |N^l| \leq \sum_{l=1}^{n} H_l |O^l| . \qquad (3)$$

If we multiply (3) by $x = \frac{n - H_n}{H_n - 1}$ and sum it with (2) we get

$$\sum_{l=1}^{n} ((n+x)H_l - lH_n)|N^l| \leq \sum_{l=1}^{n} (lH_{n-l} + ((n+x) - l)H_l)|O^l| . \qquad (4)$$

Let $\alpha(l) = (n+x)H_l - lH_n$ and $\beta(l) = lH_{n-l} + ((n+x) - l)H_l$. We will show that $\min_{l \in \{1,...,n\}} \alpha(l) = n + x - H_n$ and that $\max_{l \in \{1,...,n\}} \beta(l) \leq (n+x)H_{\frac{n+x}{2}}$. This will allow us to bound the left and right hand side of (4), giving us the desired bound on the price of stability.

To prove $\min_{l \in \{1,...,n\}} \alpha(l) = n + x - H_n$, we observe that $\alpha(l)$ first increases and then decreases and that $\alpha(1) = \alpha(n)$. By the choice of x the values at the two extremes coincide, the minimum is $\alpha(1) = n + x - H_n$ by inserting 1 in the formula of $\alpha(l)$.

To prove $\max_{l \in \{1,...,n\}} \beta(l) \leq (n+x)H_{\frac{n+x}{2}}$, we first show that $\theta(l) = lH_{n-l} + (n-l)H_l$ has maximum $nH_{n/2}$. Since θ is symmetric around $n/2$, we just have to show that the difference $\theta(l+1) - \theta(l)$ is always positive for $l+1 \leq n/2$. This proves that θ reaches at $l = n/2$ the maximum value of $\frac{n}{2}H_{n/2} + \frac{n}{2}H_{n/2} = nH_{n/2}$. We have that

$$\theta(l+1) - \theta(l) = (l+1)H_{n-(l+1)} + (n-(l+1))H_{l+1} - (lH_{n-l} + (n-l)H_l) =$$

$$= lH_{n-l} + H_{n-l} - \frac{l+1}{n-(l+1)} + (n-l)H_l - H_l + \frac{n-(l+1)}{l+1} - lH_{n-l} - (n-l)H_l =$$

$$= \frac{n-(l+1)}{l+1} - \frac{l+1}{n-(l+1)} + H_{n-l} - H_l.$$

The term $\frac{n-(l+1)}{l+1} - \frac{l+1}{n-(l+1)}$ is positive if $n - (l+1) \geq l+1$, that is if $l+1 \leq n/2$. Since H is an increasing function, $H_{n-l} - H_l$ is positive if $l \leq n/2$, in particular if $l+1 \leq n/2$. This proves our claim that $\theta(l) = lH_{n-l} + (n-l)H_l$ has maximum $nH_{n/2}$.

Since H is an increasing function, we then have the bound

$$\beta(l) = lH_{n-l} + ((n+x) - l)H_l \leq lH_{(n+x)-l} + ((n+x) - l)H_l \leq (n+x)H_{\frac{n+x}{2}} .$$

We can now finally prove Lemma 2. We know that

$$(n+x - H_n) \operatorname{cost}(N) = (n+x - H_n) \sum_{l=1}^{n} |N^l| \leq \sum_{l=1}^{n} ((n+x)H_l - lH_n)|N^l| , \qquad (5)$$

$$\sum_{l=1}^{n}(lH_{n-l}+((n+x)-l)H_l)|O^l| \le (n+x)H_{\frac{n+x}{2}}\sum_{l=1}^{n}|O^l| = (n+x)H_{\frac{n+x}{2}}\,\text{cost}(O)\ ,$$

(6)

which together with (4) proves that $\text{PoS}(G) \le \frac{\text{cost}(N)}{\text{cost}(O)} \le \frac{n+x}{n+x-H_n}H_{\frac{n+x}{2}}$.

Now observe that for any ϵ there is an $n(\epsilon)$ so that $\frac{n+x}{n+x-H_n}H_{\frac{n+x}{2}} \le H_{n/2}+\epsilon$ whenever $n \ge n(\epsilon)$, because $x = \frac{n-H_n}{H_n-1} \in o(n)$ and $(H_n)^2 \in o(n)$. □

3.2   Case O^n Is Empty

In the previous section we proved Theorem 1 if $O^n \ne \emptyset$ by constructing for every pair of players i and j a particular path S_j^i that uses edges of $E(O)$ to go from s_j to s_i and from t_j to t_i.

If $E(O)$ is not connected, then there is a pair of players i, j for which s_i and s_j are in different connected components of $E(O)$, and we cannot define the path S_j^i. Even if $E(O)$ is connected, but $O^n = \emptyset$, there might be a pair of players i and j for which the path S_j^i exists, but this path is not optimal. See Fig. 6 for an example: the path S_j^i (before cycles are removed to make S_j^i a simple path) traverses some edges of $E(O)$ twice, including the edge denoted by e in the figure. The same holds even if we exchange the labeling of s_i and t_i. Thus, we may need to define a new path T_j^i for some players i and j.

To define the new path T_j^i, let us introduce some notation. Given two players i, j and two nodes $x_i \in \{s_i, t_i\}, x_j \in \{s_j, t_j\}$ in the same connected component of $E(O)$, let $O(x_i, x_j)$ be the unique path in $E(O)$ between x_i and x_j. If s_i and s_j are in the same connected component of $E(O)$, let $(T_j^i)'$ (respectively $(T_j^i)''$) be the following s_j-t_j path:

1′. From s_j to s_i (respectively t_i), it uses edges of $O(s_i, s_j)$ (respectively $O(t_i, s_j)$).
2′. From s_i (respectively t_i) to t_i (respectively s_i), it uses edges of N_i.
3′. From t_i (respectively s_i) to t_j, it uses edges of $O(t_i, t_j)$ (respectively $O(s_i, t_j)$).

If $(T_j^i)'$ or $(T_j^i)''$ contain cycles, we skip them to obtain a simple path from s_j to t_j. See Fig. 2 for an example of $(T_j^i)'$ and Fig. 4 for an example of $(T_j^i)''$.

Notice that in the previous section, we had $S_j^i = (T_j^i)'$ (where steps 1 and 2 are now step 1′; steps 4 and 5 are now step 3′) and $O(s_i, s_j) \cap O(t_i, t_j) = \emptyset$, since $O(s_i, s_j) \subset O^-$ and $O(t_i, t_j) \subset O^+$. This ensured that there was no edge that is traversed both in step 1′ and 3′, which would make Lemma 1 not hold. In general, $O(s_i, s_j) \cap O(t_i, t_j) = \emptyset$ does not have to hold; for example in Fig. 6 we have $e \in O(s_i, s_j) \cap O(t_i, t_j)$. We call the path $(T_j^i)'$ (respectively $(T_j^i)''$) *O-cycle free* if $O(s_i, s_j) \cap O(t_i, t_j) = \emptyset$ (respectively if $O(s_i, t_j) \cap O(t_i, s_j) = \emptyset$). For instance, in Fig. 6 both $(T_j^i)'$ and $(T_j^i)''$ are not *O-cycle free*.

We are now ready to define the path T_j^i for two players i and j. If s_i and s_j are in the same connected component of $E(O)$, we set $T_j^i = (T_j^i)'$ (respectively $T_j^i = (T_j^i)''$) if $(T_j^i)'$ (respectively $(T_j^i)''$) is *O-cycle free*. Otherwise, we set $T_j^i = O_j$. Similar to the previous section, let $T^i = (T_1^i, \ldots, T_n^i)$. That is, in T^i a player

j uses the optimal path O_j if the paths $(T_j^i)'$ and $(T_j^i)''$ are not defined (meaning that s_i and s_j are in different connected components of $E(O)$), or if they are not O-cycle free (meaning that they use some edges of $E(O)$ twice). Otherwise, player j uses the O-cycle free path.

The following lemma shows that the paths T^i satisfy the requirements of Lemma 2 if $E(O)$ is connected but $O^n = \emptyset$. A subsequent lemma will then show that the requirements of Lemma 2 are satisfied even if $E(O)$ is not connected.

Lemma 3. *If $E(O)$ is connected, then for every $i \in \{1, \ldots, n\}$*

$$\Phi(N) \le \Phi(T^i) \le \sum_{\substack{U \subset \{1,\ldots,n\} \\ i \in U}} H_n |N_U| + \sum_{\substack{U \subset \{1,\ldots,n\} \\ i \in U}} H_{o_i(U)} |O_U| + \sum_{\substack{U \subset \{1,\ldots,n\} \\ i \notin U}} H_{|U|} |O_U| \ ,$$

(7)

with $o_i(U) \le n - |U|$.

Proof. Since the initial part of the proof is exactly the same as the proof of Lemma 1, we only prove that the cost c_e of every edge e in T^i is accounted for with at least coefficient $H_{k_e(T^i)}$ in the right hand side of (7). In particular, we just look at edges that are only present in steps 1' and 3' of the definition of T_j^i, since an edge $e \in O_U$ that also belongs to N_i has its cost already accounted for in the first sum.

To explain the second and third sum, let $U \subset \{1, \ldots, n\}$ and $e \in O_U$. We will look at all the possibilities of where the nodes s_i, s_j, t_i and t_j can be in the tree $E(O)$ and see whether e can be traversed in the path T_j^i. Denote by e^- and e^+ the two distinct connected components of $E(O) \setminus \{e\}$. Then, by the definition of O_U, each player $k \in U$ has $s_k \in e^-$ and $t_k \in e^+$, or viceversa. Always by the definition of O_U, each player $k \notin U$ has either $s_k, t_k \in e^-$ or $s_k, t_k \in e^+$.

To explain the third sum of (7), let $i \notin U$. For illustration purposes, assume without loss of generality that $s_i, t_i \in e^-$. Then, the only possibilities are that

- $j \in U$. Then e can be traversed, since T_j^i has to go from e^- to e^+ to connect s_j and t_j. See Fig. 2 for an illustration in the case $T_j^i \ne O_j$ and Fig. 3 for the case $T_j^i = O_j$.
- $j \notin U, s_j, t_j \in e^-$. Then e cannot be traversed, since all terminal nodes are in e^- and there is no need to traverse e. See Fig. 4 for an illustration in the case $T_j^i \ne O_j$ and Fig. 5 for the case $T_j^i = O_j$.
- $j \notin U, s_j, t_j \in e^+$. Then e cannot be traversed, since both $(T_j^i)'$ and $(T_j^i)''$ traverse e twice, so we must have $T_j^i = O_j$. See Fig. 6 for an illustration.

As we can see, e can be traversed only if $j \in U$, that is, at most $|U|$ times. This explains the third sum of (7).

Finally, to explain the second sum of (7), let $i \in U$. The only possibilities are that

- $j \in U$. Then e cannot be traversed, since at least one of $(T_j^i)'$ or $(T_j^i)''$ is a O-cycle free path that does not traverse e. See Fig. 7 for an illustration.

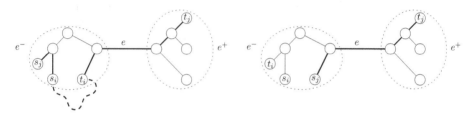

Fig. 2. $i \notin U, j \in U$ and $T_j^i \neq O_j$. Then e can be traversed in the path T_j^i.

Fig. 3. $i \notin U, j \in U$ and $T_j^i = O_j$. Then e can be traversed in the path T_j^i.

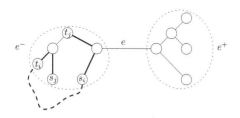

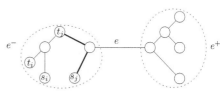

Fig. 4. $i \notin U, j \notin U$, $s_j, t_j \in e^-$ and $T_j^i \neq O_j$. Then e cannot be traversed in the path T_j^i.

Fig. 5. $i \notin U, j \notin U$, $s_j, t_j \in e^-$ and $T_j^i = O_j$. Then e cannot be traversed in the path T_j^i.

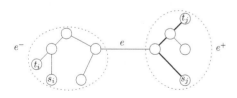

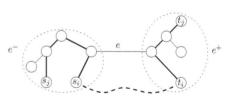

Fig. 6. $i \notin U, j \notin U$, $s_j, t_j \in e^+$ and $T_j^i = O_j$. Then e cannot be traversed in the path T_j^i.

Fig. 7. $i \in U, j \in U$ and $T_j^i \neq O_j$. Then e cannot be traversed in the path T_j^i.

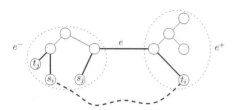

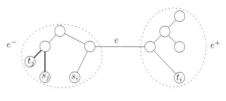

Fig. 8. $i \in U, j \notin U$, and $T_j^i \neq O_j$. Then e can be traversed in the path T_j^i.

Fig. 9. $i \in U, j \notin U$, and $T_j^i = O_j$. Then e cannot be traversed in the path T_j^i.

  – $j \notin U$ and $T^i_j \neq O_j$. Then e can be traversed, since s_j and t_j are in the same connected component of $E(O) \setminus \{e\}$, but s_i and t_i are in different ones. See Fig. 8 for an illustration.
  – $j \notin U$ and $T^i_j = O_j$. Then e cannot be traversed, since s_j and t_j are in the same connected component of $E(O) \setminus \{e\}$ and we just take the direct path between them, which does not traverse e. See Fig. 9 for an illustration.

Let $o_i(U)$ be the number of $j \notin U$ with $T^i_j \neq O_j$. Then, as we can see, e is traversed at most $o_i(U) \leq n - |U|$ times. This explain the second sum of (7) and finishes the proof of Lemma 3. □

Theorem 1 follows directly if $E(O)$ is connected but O^n is empty by Lemma 3 and Lemma 2. The following lemma handles the last case we have left to analyze, which is when $E(O)$ is not a connected tree. This, together with Lemma 2, finishes the proof of Theorem 1.

Lemma 4. Let $E(O) = C_1 \sqcup \cdots \sqcup C_q$, with each C_m being a connected component of $E(O)$. Let R_m be the set of players j with $s_j, t_j \in C_m$. Then for a player $i \in R_k$

$$\Phi(N) \leq \Phi(T^i) \leq \sum_{\substack{U \subset \{1,\ldots,n\} \\ i \in U}} H_n |N_U| + \sum_{\substack{U \subset R_k \\ i \in U}} H_{o_i(U)} |O_U| + \sum_{\substack{U \subset R_m \text{ for some } m \\ i \notin U}} H_{|U|} |O_U|,$$

(8)

with $o_i(U) \leq |T_k| - |U| \leq n - |U|$.

Proof. Since the initial part of the proof is exactly the same as the proof of Lemma 1 and Lemma 3, we only prove that the cost c_e of every edge e in T^i is accounted for with at least coefficient $H_{k_e(T^i)}$ in the right hand side of (8). In particular, we just look at edges that are only present in steps $1'$ and $3'$ of the definition of T^i_j, since an edge $e \in O_U$ that also belongs to N_i has its cost already accounted for in the first sum.

To explain the second and third sum, let $U \subset \{1, \ldots, n\}$ and $e \in O_U$. Notice that if $U \not\subset R_m$ for every m, then O_U is the empty set and e does not contribute anything to $\Phi(T^i)$. We begin by looking at the second sum.

Notice that since $i \in R_k$, the only possibility to have $i \in U$ is that $U \subset R_k$. By the definition of T^i the players $j \in R_m$, $m \neq k$ use the path O_j, which does not traverse e. With the exact same reasoning of Lemma 3, by looking at all the possibilities of where s_i, t_i, s_j and t_j can be in C_k, we can see that e can be traversed by player $j \in R_k$ only if $j \notin U$ and $T^i \neq O_j$. If we then define the number of players $j \in T_k$ with this property to be $o_i(U) \leq |T_k| - |U| \leq n - |U|$, the second sum in the right hand side of (8) is explained.

Finally, for the third sum, we fix $i \notin U$ and look at the cases $U \subset R_k$ and $U \subset R_m$, $m \neq k$ separately.

Suppose first that $U \subset R_k$. By the definition of T^i the players $j \in R_m$, $m \neq k$ use the path O_j, which does not traverse e. With the exact same reasoning of Lemma 3, by looking at all the possibilities of where s_i, t_i, s_j and t_j can be in C_k, we can see that e can be traversed by player $j \in R_k$ only if $j \in U$. That is, by at most $|U|$ players. This explains the third sum for the case $U \subset R_k$.

We now look at the case $U \subset R_m$, $m \neq k$. By the definition of T^i, players $j \in R_l$, $l \neq m$ do not traverse e, since they only use edges of C_l (if $l \neq k$) or edges of C_k and of N_i (if $l = k$). Players $j \in R_m$ use the path O_j, and by the definition of O_U exactly $|U|$ players traverse e. This explains the third sum for the case $U \subset R_m$, $m \neq k$, which finishes the proof. $\square$

Acknowledgements. We are grateful to Rati Gelashvili for valuable discussions and remarks. This work has been partially supported by the Swiss National Science Foundation (SNF) under the grant number 200021_143323/1.

References

1. Anshelevich, E., Dasgupta, A., Kleinberg, J.M., Tardos, É., Wexler, T., Roughgarden, T.: The price of stability for network design with fair cost allocation. In: FOCS, pp. 295–304 (2004)
2. Asadpour, A., Saberi, A.: On the inefficiency ratio of stable equilibria in congestion games. In: Leonardi, S. (ed.) WINE 2009. LNCS, vol. 5929, pp. 545–552. Springer, Heidelberg (2009)
3. Bilò, V., Bove, R.: Bounds on the price of stability of undirected network design games with three players. Journal of Interconnection Networks 12(1-2), 1–17 (2011)
4. Bilò, V., Caragiannis, I., Fanelli, A., Monaco, G.: Improved lower bounds on the price of stability of undirected network design games. Theory Comput. Syst. 52(4), 668–686 (2013)
5. Bilò, V., Flammini, M., Moscardelli, L.: The price of stability for undirected broadcast network design with fair cost allocation is constant. In: FOCS, pp. 638–647 (2013)
6. Christodoulou, G., Chung, C., Ligett, K., Pyrga, E., van Stee, R.: On the price of stability for undirected network design. In: Bampis, E., Jansen, K. (eds.) WAOA 2009. LNCS, vol. 5893, pp. 86–97. Springer, Heidelberg (2010)
7. Disser, Y., Feldmann, A.E., Klimm, M., Mihalák, M.: Improving the H_k-bound on the price of stability in undirected shapley network design games. In: Spirakis, P.G., Serna, M. (eds.) CIAC 2013. LNCS, vol. 7878, pp. 158–169. Springer, Heidelberg (2013)
8. Fanelli, A., Leniowski, D., Monaco, G., Sankowski, P.: The ring design game with fair cost allocation. In: Goldberg, P.W. (ed.) WINE 2012. LNCS, vol. 7695, pp. 546–552. Springer, Heidelberg (2012)
9. Fiat, A., Kaplan, H., Levy, M., Olonetsky, S., Shabo, R.: On the price of stability for designing undirected networks with fair cost allocations. In: Bugliesi, M., Preneel, B., Sassone, V., Wegener, I. (eds.) ICALP 2006. LNCS, vol. 4051, pp. 608–618. Springer, Heidelberg (2006)
10. Li, J.: An upper bound on the price of stability for undirected shapley network design games. Information Processing Letters 109, 876–878 (2009)
11. Kawase, Y., Makino, K.: Nash equilibria with minimum potential in undirected broadcast games. Theor. Comput. Sci. 482, 33–47 (2013)
12. Lee, E., Ligett, K.: Improved bounds on the price of stability in network cost sharing games. In: EC, pp. 607–620 (2013)
13. Monderer, D., Shapley, L.S.: Potential games. Games and Economic Behavior 14(1), 124–143 (1996)

Traveling Salesman Problems in Temporal Graphs[*]

Othon Michail[1] and Paul G. Spirakis[1,2]

[1] Computer Technology Institute & Press "Diophantus" (CTI), Patras, Greece
[2] Department of Computer Science, University of Liverpool, UK
michailo@cti.gr, P.Spirakis@liverpool.ac.uk

Abstract. In this work, we introduce the notion of time to some well-known combinatorial optimization problems. In particular, we study problems defined on *temporal graphs*. A temporal graph $D = (V, A)$ may be viewed as a time-sequence $G_1, G_2 \ldots, G_l$ of static graphs over the same (static) set of nodes V. Each $G_t = D(t) = (V, A(t))$ is called the *instance of D at time t* and l is called the *lifetime of D*. Our main focus is on analogues of *traveling salesman problems* in temporal graphs. A sequence of time-labeled edges (e.g. a tour) is called *temporal* if its labels are strictly increasing. We begin by considering the problem of exploring the nodes of a temporal graph as soon as possible. In contrast to the positive results known for the static case, we prove that, it cannot be approximated within cn, for some constant $c > 0$, in general temporal graphs and within $(2 - \varepsilon)$, for every constant $\varepsilon > 0$, in the special case in which $D(t)$ is connected for all $1 \leq t \leq l$, both unless $\mathbf{P} = \mathbf{NP}$. We then study the temporal analogue of TSP(1,2), abbreviated TTSP(1,2), where, for all $1 \leq t \leq l$, $D(t)$ is a complete weighted graph with edge-costs from $\{1, 2\}$ and the cost of an edge may vary from instance to instance. The goal is to find a minimum cost temporal TSP tour. We give several *polynomial-time approximation algorithms* for TTSP(1,2). Our best approximation is $(1.7 + \varepsilon)$ for the generic TTSP(1,2) and $(13/8 + \varepsilon)$ for its interesting special case in which the lifetime of the temporal graph is restricted to n. In the way, we also introduce temporal versions of other fundamental combinatorial optimization problems, for which we obtain polynomial-time approximation algorithms and hardness results.

1 Introduction

A *temporal graph* is, informally speaking, a graph that changes with time. A great variety of both modern and traditional networks such as information and communication networks, social networks, transportation networks, and several physical systems can be naturally modeled as temporal graphs.

[*] Supported in part by the (i) project **FOCUS**, "ARISTEIA" Action, OP EdLL, EU and Greek National Resources, (ii) FET EU IP project **MULTIPLEX** under contract no 317532, and (iii) School of EEE/CS of the Univ. of Liverpool. Full version: http://ru1.cti.gr/aigaion/?page=publication&kind=single&ID=1051

E. Csuhaj-Varjú et al. (Eds.): MFCS 2014, Part II, LNCS 8635, pp. 553–564, 2014.
© Springer-Verlag Berlin Heidelberg 2014

In this work, we restrict attention to *discrete time*. This is totally plausible when the dynamicity of the system is inherently discrete, which is for example the case in synchronous mobile distributed systems that operate in discrete rounds, but can also satisfactory approximate a wide range of continuous-time systems. Moreover, this choice gives to the resulting models and problems a purely combinatorial flavor. We also restrict attention to systems in which only the relationships between the participating entities may change and not the entities themselves. Therefore, in this paper, a temporal graph $D = (V, A)$ may always be viewed as a sequence $G_1, G_2 \ldots, G_l$ of static graphs over the same (static) set of nodes V. Each $G_t = D(t) = (V, A(t))$ is called the *instance of D at time t* and l is called the *lifetime of D*.

Though static graphs have been extensively studied, for their temporal generalization we are still far from having a concrete set of structural and algorithmic principles. Additionally, it is not yet clear how is the complexity of combinatorial optimization problems affected by introducing to them a notion of time. In an early but serious attempt to answer this question, Orlin [Orl81] observed that many dynamic languages derived from **NP**-complete languages can be shown to be **PSPACE**-complete. Among the other few things that we do know, is that the max-flow min-cut theorem holds with unit capacities for time-respecting paths [Ber96]. Additionally, Kempe *et al.* [KKK00] proved that, in temporal graphs, the classical formulation of Menger's theorem is violated and the computation of the number of node-disjoint *s-t* paths becomes **NP**-complete. In a very recent work [MMCS13], among other things, the authors achieved a reformulation of Menger's theorem which is valid for all temporal graphs and introduced several interesting cost minimization parameters for optimal temporal network design. Dutta *et al.* [DPR+13], working on a distributed online dynamic network model, presented offline centralized algorithms for the k-gossip problem.

We make here one more step towards the direction of revealing the algorithmic principles of temporal graphs. In particular, we introduce the study of traveling salesman problems in temporal graphs, which, to the best of our knowledge, have not been considered before in the literature. Our main focus is on the TEMPORAL TRAVELING SALESMAN PROBLEM WITH COSTS ONE AND TWO abbreviated TTSP(1,2) throughout the paper. In this problem, we are given a temporal graph $D = (V, A)$ every instance of which is a complete graph, i.e. $D(t) = (V, A(t))$ is complete for all $1 \le t \le l$. Moreover, the edges of every $D(t)$ are weighted according to a cost function $c : A \to \{1, 2\}$. Observe that A is a set of *time-edges* (e, t), where e is an edge and t is the time at which e appears. So, the cost function c is allowed to assign different cost values to different instances of the same edge, therefore, in this model, *costs are dynamic* in nature. We are asked to find a *Temporal TSP tour* (abbreviated *TTSP tour*) of minimum total cost. A TSP tour $(u_1, t_1, u_2, t_2, \ldots, t_{n-1}, u_n, t_n, u_1)$ is temporal if $t_i < t_{i+1}$ for all $1 \le i \le n - 1$. The cost of such a TSP tour is $\sum_{1 \le i \le n} c((u_i, u_{i+1}), t_i)$, where $u_{n+1} = u_1$. We should remark that, in general, the lifetime of D is not restricted and therefore it can be much greater than n. Whenever we restrict attention to limited lifetime, this will be explicitly stated. We should also emphasize that, throughout this work, *the entire temporal graph is provided to the centralized algorithms in advance.* It is useful to observe that TTSP(1,2) seems to be naturally closer to the ASYMMETRIC TSP(1,2) and this seems to hold independent of whether the temporal graph has directed or undirected instances. In most places we assume directed edge-sets, however keep in mind that the undirected case is not expected to be any simpler. Finally, note that ATSP(1,2) is a special case of TTSP(1,2) which implies that TTSP(1,2) is also **APX**-hard [PY93] and cannot be approximated within any factor less than 207/206 [KS13].

1.1   Our Approach-Contribution

We now summarize our approach to approximate TTSP(1,2). Note that all the approximation algorithms that we present in this work are *polynomial-time algorithms* on a binary encoding of the temporal graph, i.e. on $|\langle D \rangle|$. In the static case, one easily obtains a (3/2)-factor approximation for ATSP(1,2) by computing a perfect matching maximizing the number of ones and then patching the edges together arbitrarily. This works well, because such a minimum cost perfect matching can be computed in polynomial time in the static case. This was one of the first algorithms known for ATSP(1,2). Other approaches have improved the factor to the best currently known 5/4 [Blä04]. Unfortunately, as we shall see, even the apparently simple task of computing a matching maximizing the number of ones is not that easy in temporal graphs. In particular, we prove that computing a matching maximizing the number of ones and additionally satisfying the temporal condition that all its edges appear at distinct times is **NP**-hard. The reason that we insist on distinct times is that we can form a temporal TSP tour by patching the edges of a matching only if the edges of the matching can be strictly ordered in time. In fact, an additional requirement is that the edges of the matching should have time differences of at least two, so that we can always fit a patching-edge between two time-consecutive edges of the matching. We call the corresponding problem Max-TEM(≥ 2).

We naturally then search for good approximations for Max-TEM(≥ 2). We follow two main approaches. One is to reduce the problem to MAXIMUM INDEPENDENT SET (MIS) in $(k+1)$-claw free graphs and the other is to reduce it to k'-SET PACKING for some k and k' to be determined. The first approach gives a $(7/4 + \varepsilon)$-approximation ($= 1.75 + \varepsilon$) for the generic TTSP(1,2) and a $(12/7 + \varepsilon)$-approximation ($\approx 1.71 + \varepsilon$) for the special case of TTSP(1,2) in which the lifetime is restricted to n. The latter is obtained by approximating a temporal path packing instead of a matching. The second approach improves these to $1.7 + \varepsilon$ for the general case and to $13/8 + \varepsilon = 1.625 + \varepsilon$ when the lifetime is n. In all cases, $\varepsilon > 0$ is a small constant (not necessarily the same in all cases) adopted from the factors of the approximation algorithms for independent set and set packing. We leave as an interesting open problem whether a (3/2)-factor for TTSP(1,2) or its special case with lifetime restricted to n is within reach.

Apart from TTSP(1,2) we also consider the TEMPORAL (NODE) EXPLORATION (abbreviated TEXP) problem, in which we are given a temporal graph (unweighted and non-complete) and the goal is to visit all nodes of the temporal graph, by possibly revisiting nodes, minimizing the arrival time (in the static case, appears as GRAPHIC TSP in the literature). Though, in the static case, the decision version of the problem, asking whether a given graph is explorable, can be solved in linear time, we show that in the temporal case it becomes **NP**-complete. Additionally, in the static case, there is a $(3/2 - \varepsilon)$-approximation for undirected graphs [GSS11] and a $O(\log n / \log \log n)$ for directed [AGM^{+}10]. In contrast to these, we prove that there exists some constant $c > 0$ such that TEXP cannot be approximated within cn unless **P** = **NP**. Additionally, we prove that even the special case in which every instance of the temporal graph is connected, cannot be approximated within $(2 - \varepsilon)$, for every constant $\varepsilon > 0$, unless **P** = **NP**. On the positive side, we show that TEXP can be approximated within the *dynamic diameter* (definition in Section 2) of the temporal graph.

Finally, in the way to approaching the above two main problems, we also obtain several interesting side-results, such as a $[3/(5+\varepsilon)]$-approximation for Max-TEM(≥ 2), a $[1/(7/2+\varepsilon)]$-approximation for TEMPORAL PATH PACKING (TPP) when the lifetime is restricted to n, and in the full paper a $(1/5)$-approximation for Max-TTSP and an inapproximability result stating that for any polynomial time computable function $\alpha(n)$, TEMPORAL CYCLE COVER cannot be approximated within $\alpha(n)$, unless **P** = **NP**. To the best of our knowledge, all the aforementioned temporal problems are first studied in this work.

In Section 2, we formally define the model of temporal graphs under consideration and provide all further necessary definitions. Section 2.1 presents formal definitions of all temporal problems that we consider in this work. In Section 3, we consider the TEMPORAL EXPLORATION problem. Then, in Section 4 we introduce and study the TTSP(1,2) problem in weighted temporal graphs.

2   Preliminaries

Definition 1. *A* temporal graph *(or* dynamic graph*)* D *is an ordered pair of disjoint sets* (V, A) *such that* $A \subseteq \binom{V}{2} \times \mathbb{N}$ *(*$V^2 \setminus \{(u, u) : u \in V\}$ *in case of a digraph). The set* V *is the set of* nodes *and the set* A *is the set of* time-edges.

A temporal (di)graph $D = (V, A)$ can be also viewed as a static (underlying) graph $G_D = (V, E)$, where $E = \{e : (e, t) \in A \text{ for some } t \in \mathbb{N}\}$ contains all edges that appear at least once, together with a labeling $\lambda_D : E \to 2^{\mathbb{N}}$ defined as $\lambda_D(e) = \{t : (e, t) \in A\}$ (we omit the subscript D when no confusion can arise). We denote by $\lambda(E)$ the multiset of all labels assigned by λ to G_D and by $\lambda_{\min} = \min\{l \in \lambda(E)\}$ ($\lambda_{\max} = \max\{l \in \lambda(E)\}$) the minimum (maximum) label of D. We define the *lifetime* (or *age*) of a temporal graph D as $\alpha(D) = \lambda_{\max} - \lambda_{\min} + 1$. Note that in case $\lambda_{\min} = 1$ we have $\alpha(D) = \lambda_{\max}$.

For every time $t \in \mathbb{N}$, we define the *t-th instance of a temporal graph* $D = (V, A)$ as the static graph $D(t) = (V, A(t))$, where $A(t) = \{e : (e, t) \in A\}$ is the (possibly empty) set of all edges that appear in D at time t. A temporal graph $D = (V, A)$ may be also viewed as a *sequence of static graphs* $(G_1, G_2, \ldots, G_{\alpha(D)})$, where $G_i = D(\lambda_{\min} + i - 1)$ for all $1 \leq i \leq \alpha(D)$. Another, often convenient, representation of a temporal graph is the following.

The *static expansion* of a temporal graph $D = (V, A)$ is a DAG $H = (S, E)$ defined as follows. If $V = \{u_1, u_2, \ldots, u_n\}$ then $S = \{u_{ij} : \lambda_{\min} - 1 \leq i \leq \lambda_{\max}, 1 \leq j \leq n\}$ and $E = \{(u_{(i-1)j}, u_{ij'}) : \text{if } (u_j, u'_j) \in A(i) \text{ for some } \lambda_{\min} \leq i \leq \lambda_{\max}\}$.

A *temporal* (or *time-respecting*) *walk* W of a temporal graph $D = (V, A)$ is an alternating sequence of nodes and times $(u_1, t_1, u_2, t_2, \ldots, u_{k-1}, t_{k-1}, u_k)$ where $(u_i u_{i+1}, t_i) \in A$, for all $1 \leq i \leq k - 1$, and $t_i < t_{i+1}$, for all $1 \leq i \leq k - 2$. We call $t_{k-1} - t_1 + 1$ the *duration* (or *temporal length*) of the walk W, t_1 its *departure time* and t_{k-1} its *arrival time*. A *journey* (or *temporal/time-respecting path*) J is a temporal walk with pairwise distinct nodes. A (u, v)-journey J is called *foremost from time* $t \in \mathbb{N}$ if it departs after time t and its arrival time is minimized. The *temporal distance* from a node u at time t (also called *time-node* (u, t)) to a node v is defined as the duration of a foremost (u, v)-journey from time t. We say that a temporal graph $D = (V, A)$ has *dynamic diameter* d, if d is the minimum integer for which it holds that the temporal distance from every time-node $(u, t) \in V \times \{0, 1, \ldots, \alpha(D) - d\}$ to every node $v \in V$ is at most d. A *temporal matching* of a temporal graph $D = (V, A)$ is a set of time-edges $M = \{(e_1, t_1), (e_2, t_2), \ldots, (e_k, t_k)\}$, such that $(e_i, t_i) \in A$, for all $1 \leq i \leq k$, $t_i \neq t_j$, for all $1 \leq i < j \leq k$, and $\{e_1, e_2, \ldots, e_k\}$ is a matching of G_D.

Similarly to weighted graphs we may define *weighted temporal graphs* by introducing a (temporal) cost function $c : A \to C$, where C denotes the range of the costs, e.g. $C = \mathbb{N}$. A temporal graph $D = (V, A)$ is called *complete* (*continuously connected*) if $D(t)$ is complete (connected, resp.) for all $1 \leq t \leq \alpha(D)$. In these cases, we may also say that D *has complete/connected instances*.

Throughout the text, unless otherwise stated, we denote by n the number of nodes of (temporal) (di)graphs. When no confusion may arise, we use the term *edge* for both undirected edges and arcs. Finally, a δ-factor (polynomial-time) approximation algorithm for a problem Π satisfies $\delta \geq 1$ if Π is a minimization problem and $\delta \leq 1$ if Π is a maximization problem.

2.1 Problem Definitions

TEMPORAL EXPLORATION - TEXP. Given a temporal graph $D = (V, A)$ and a source node $s \in V$, find a temporal walk that begins from s and visits all nodes minimizing the arrival time.

TTSP(1,2). Given a complete temporal graph $D = (V, A)$ and a cost function $c : A \to \{1, 2\}$ find a temporal TSP tour of minimum total cost.

Max-TEM($\geq k$). Given a temporal graph $D = (V, A)$ find a maximum cardinality temporal matching $M = \{(e_1, t_1), (e_2, t_2), \ldots, (e_h, t_h)\}$ satisfying that there is a permutation $t_{i_1}, t_{i_2}, \ldots, t_{i_h}$ of the t_js s.t. $t_{i_{(l+1)}} \geq t_{i_l} + k$ for all $1 \leq l \leq h - 1$.

TEMPORAL PATH PACKING - TPP. We are given a temporal graph and we want to find time and node disjoint time-respecting paths maximizing the number of edges used. By time disjoint we require that they correspond to distinct intervals that differ by ≥ 2 in time.

3 Exploration of Temporal Graphs

In this section, we study the TEMPORAL EXPLORATION (TEXP) problem in (unweighted) temporal graphs. In contrast to several positive results known for the static case, we show that in temporal graphs the problem is quite hard. In particular, we show that the decision version of the problem is **NP**-complete and we give two hardness of approximation results for the optimization version, one for the generic case and another for the special case in which the temporal graph is continuously connected. On the positive side, we approximate the optimum of the generic instances within the dynamic diameter of the temporal graph.

3.1 Deciding Explorability is Hard in Temporal Graphs

Note that a walk in the (TEMPORAL) EXPLORATION is allowed to revisit nodes several times. Let us first focus on static graphs. Consider the decision version DEXP of EXPLORATION in which the goal is to decide whether a given graph is explorable. DEXP and finding an arbitrary solution can be solved in linear time for both undirected and directed static graphs. On the other hand, we prove that its temporal version, abbreviated DTEXP, is **NP**-complete.

3.2 Hardness of Approximate Temporal Exploration

Theorem 1. *There exists some constant $c > 0$ such that TEXP cannot be approximated within cn unless* **P** = **NP**.

The reason that we managed to obtain such a strong inapproximability result was that we were free to totally break at some point the connectivity of the temporal graph. This freedom is lost in continuously connected temporal graphs.

Theorem 2. *For every constant $\varepsilon > 0$, there is no $(2 - \varepsilon)$-approximation for TEXP in continuously (strongly) connected temporal graphs unless* **P** = **NP**.

Proof. The reduction is from HAMILTONIAN PATH (abbreviated HAMPATH). We prove that a $(2 - \varepsilon)$-factor approximation for TEXP in continuously connected temporal graphs could be used to decide HAMPATH. Let (G, s) be an instance of HAMPATH. We construct an instance of TEXP consisting of a continuously strongly connected temporal graph $D = (V, A)$ and a source node s'. D consists of three static graphs T_1, T_2, and T_3 as illustrated in Figure 1. The first graph T_1 (Figure 1(a)) consists of $G_1 = G$ and a set V_2 of additional nodes, i.e. $V = V_1 \cup V_2$. Denote by n, n_1, and n_2 the cardinalities of V, V_1, and V_2, respectively. For the time being it suffices to assume that $n_2 > n_1$. We set $s' = s$. We connect every node of V_1 to the leftmost node of V_2, then continue with a directed path spanning V_2 (i.e. a hamiltonian path on V_2), and finally we connect the rightmost node of V_2 to each node of V_1. T_1 persists until time $n_1 - 1$, that is $D(t) = T_1$ for all $1 \leq t \leq n_1 - 1$. Then, at time n_1, D changes to the second graph T_2 (Figure 1(b)) which is the same as T_1 without the internal edges of set V_1

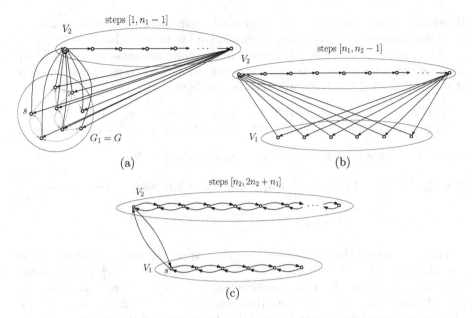

steps $[1, n_1 - 1]$

V_2

s

$G_1 = G$

(a)

steps $[n_1, n_2 - 1]$

V_2

V_1

(b)

steps $[n_2, 2n_2 + n_1]$

V_2

V_1 s

(c)

Fig. 1. The temporal graph constructed by the reduction. (a) T_1 (b) T_2 (c) T_3

(those are the edges of G that were present in T_1). T_2 persists until time $n_2 - 1$, that is $D(t) = T_2$ for all $n_1 \leq t \leq n_2 - 1$. Finally, at time n_2, D changes to the third graph T_3 (Figure 1(c)) in which each of V_1 and V_2 has its nodes connected by a line of 2-cycles and the left endpoints of the two sets are also connected by a 2-cycle. T_3 is preserved up to the lifetime of D, that is $D(t) = T_3$ for all $n_2 \leq t \leq \alpha(D)$. To ensure explorability of D, it suffices to set $\alpha(D) = 2n_2 + n_1$. Note that D is a continuously strongly connected temporal graph because T_1, T_2, and T_3 are strongly connected graphs.

($\Rightarrow$) If G is hamiltonian, then the hamiltonian path of G_1, followed by an edge leading from V_1 to V_2, and finally followed by the hamiltonian path on V_2 gives a hamiltonian journey of D and thus V can be explored optimally in $n_1 + n_2 - 1$ steps.

($\Leftarrow$) If G is not hamiltonian, then we prove that in this case the optimum exploration needs at least $2n_2 + 1$ steps. Observe that by time $n_1 - 1$ the exploration cannot have visited all nodes of V_1 because G_1 is not hamiltonian from s (Figure 1(a)). This remains true until time $n_2 - 1$, because in the interval $[n_1, n_2 - 1]$ the only edges that lead to nodes in V_1 cannot have been reached before time n_2 (Figure 1(b)). So, by time $n_2 - 1$ there is an unvisited node in V_1. Moreover, by the same time the rightmost node of V_2 is also unvisited because the temporal distance from $(s, 0)$ to it is n_2. Then, even if at time n_2 the exploration hits one of them, the other is at distance $\geq n_2 + 1$ because the leftmost node of V_1 in Figure 1(c) is s. So, in total, at least $2n_2 + 1$ steps are needed to explore V.

It remains to prove that the above reduction can be adjusted to introduce the claimed gap. As ε is a constant, we can restrict attention to instances of

HAMPATH of order at least $2/\varepsilon$ and provide a gap introducing reduction from those instances (which obviously still remain hard to decide), that is $n_1 \geq 2/\varepsilon \Rightarrow \varepsilon \geq 2/n_1 \Rightarrow 2 - \varepsilon \leq 2 - (2/n_1)$. Moreover, in the above reduction set $n_2 = n_1^2 + n_1$ (observe that we can set n_2 equal to any polynomial-time computable function of n_1). So, by what has been proved so far, we have that:

- If G is hamiltonian, then $OPT = n_1 + n_2 - 1 = n_1^2 + 2n_1 - 1$.
- If G is not hamiltonian, then $OPT \geq 2n_2 + 1 = 2(n_1^2 + n_1) + 1 > 2(n_1^2 + n_1)$.

Consider the hamiltonian case. As $2 - \varepsilon \leq 2 - (2/n_1)$ we have

$$(2 - \varepsilon)(n_1^2 + 2n_1 - 1) \leq (2 - \frac{2}{n_1})(n_1^2 + 2n_1 - 1) = 2n_1^2 + 4n_1 - 2 - 2n_1 - 4 + \frac{2}{n_1}$$

$$= 2(n_1^2 + n_1) + (\frac{2}{n_1} - 6) \leq 2(n_1^2 + n_1).$$

Thus, whenever G is hamiltonian, the $(2-\varepsilon)$-approximation algorithm returns a solution of cost $\leq (2 - \varepsilon)OPT = (2 - \varepsilon)(n_1^2 + 2n_1 - 1) \leq 2(n_1^2 + n_1)$. On the other hand, whenever G is not hamiltonian, $OPT > 2(n_1^2 + n_1)$ and thus also the solution returned by the algorithm must have cost $> 2(n_1^2 + n_1)$. Thus a $(2-\varepsilon)$-approximation algorithm would decide instances of HAMPATH of order at least $2/\varepsilon$ in polynomial time, by comparing the solution to the polynomial-time computable $2(n_1^2 + n_1)$ threshold. This cannot be the case unless **P = NP**.   □

On the positive side:

Theorem 3. *We provide a d-approximation algorithm for TEXP restricted to temporal graphs with dynamic diameter $\leq d$ and lifetime $\geq (n - 1)d$.*

4   Temporal Traveling Salesman with Costs One and Two

In this section, we deal with TTSP(1,2) which is a generalization of the well known ATSP(1,2) to weighted temporal graphs. Recall that in TTSP(1,2) we are given a complete temporal graph $D = (V, A)$, with its time-edges weighted according to a cost function $c : A \to \{1, 2\}$, and we are asked to find a temporal TSP tour of minimum total cost. Our approach is to compute a temporal matching using many 1s and then extend it to a TTSP tour. Unfortunately:

Theorem 4. *Max-TEM($\geq k$) is **NP**-hard for every independent of the lifetime polynomial-time computable $k \geq 1$.*

4.1   Approximating TTSP(1,2) via Maximum Independent Sets

Clearly, by taking an arbitrary temporal TSP tour, one obtains a trivial 2-factor approximation for TTSP(1,2). In the worst case, its cost is $2n$ (paying always 2s) while the cost of the optimum TSP tour is at least n (paying always 1s). Can we do better? Recall that, in ATSP(1,2) it is known that we can do much better as there is a (5/4)-factor approximation [Blä04]. In this section, we provide our first approximation algorithms for both the generic TTSP(1,2) and its special case with lifetime restricted to n. To do this, we first show that, though by Theorem 4 Max-TEM(≥ 2) is **NP**-hard, it can still be approximated within some constant via a reduction to MAXIMUM INDEPENDENT SET (MIS) in 5-claw free graphs. Recall that a graph is k-claw free if there is no k-independent set in the neighborhood of any node. We then translate this to an approximation for TTSP(1,2). For the restricted lifetime case we follow in Section 4.1.1 a similar approach by approximating a temporal path packing this time.

We begin by showing that a constant factor approximation algorithm for Max-TEM(≥ 2) translates to a constant approximation algorithm for TTSP(1,2) with factor strictly smaller than 2. This then naturally motivates us to search for constant approximations for temporal matchings.

Lemma 1. *An $(1/c)$-factor approximation for Max-TEM(≥ 2) implies a $(2 - \frac{1}{2c})$-factor approximation for TTSP(1,2).*

We now present a constant factor approximation for Max-TEM(≥ 1).

Theorem 5. *There is a $(3/5)$-approximation algorithm for Max-TEM(≥ 1).*

Proof. We are given a temporal graph $D = (V, A)$ and our goal is to return a temporal matching M of maximum cardinality. To simplify the description let us consider the static expansion $H = (S, E)$ of D. Now given an edge $e = (u_{(i-1)j}, u_{ij'})$ of the static expansion we may think of it as having the following positions for conflicts with other edges, i.e. edges that cannot be taken together with e in a temporal matching: (1) Edges of the same row as e, i.e. all edges of the form $(u_{(i-1)l}, u_{il'})$, (2) edges of the same column as $u_{(i-1)j}$, i.e. all edges that have one endpoint of the form u_{kj}, and (3) edges of the same column as $u_{ij'}$, i.e. all edges that have one endpoint of the form $u_{kj'}$. Consider now the graph $G = (E, K)$ where $(e_1, e_2) \in K$ iff e_1 and e_2 satisfy some of the above three constraints. Observe that the set of nodes E of G is the set of edges of the static expansion H. It is straightforward to observe that temporal matchings of D are now equivalent to independent sets of G. Observe now that G is 4-claw free which means that there is no 4-independent set in the neighborhood of any node. To see this take any $e \in E$ and any set $\{e_1, e_2, e_3, e_4\}$ of four neighbors of e in G. As there are only three constraints at least two of the neighbors, say e_i and e_j, must be connected to e by the same constraint. Finally, observe that if e_i and e_j both satisfy the same constraint with e (e.g. belong to the same row as e) then they must satisfy the same constraint with each other (e.g. if e_i and e_j belong to the same row as e then e_i belongs to the same row as e_j) implying that e_i and e_j are also connected by an edge in G. From [Hal95] we have a factor of 3/5 for MIS in 4-claw free graphs. $\square$

The following lemma makes a slight modification to the proof of Theorem 5 to obtain a constant approximation for Max-TEM(≥ 2).

Lemma 2. *There is a $\frac{1}{2+\varepsilon}$-approximation algorithm for Max-TEM(≥ 2).*

Theorem 6. *There is a $(7/4 + \varepsilon)$-approximation algorithm for TTSP(1, 2).*

4.1.1   Lifetime Restricted to n

We now restrict our attention to temporal graphs with lifetime $\alpha(D)$ restricted to n. In this case, we show that an extension of the above ideas provides us with an improved $12/7 \approx 1.71$-factor approximation for TTSP(1, 2). A difference now is that instead of approximating a temporal matching we approximate a temporal path packing.

Lemma 3. *An $(1/c)$-factor approximation for TPP implies a $(2 - \frac{1}{c})$-factor approximation for TTSP(1, 2).*

Lemma 4. *There is a $\frac{1}{(7/2)+\varepsilon}$-factor approximation for TPP when $\alpha(D) = n$.*

Proof. We directly express a TPP as an independent set of time-edges in the static expansion $H = (S, E)$. Given an edge $e = (u_{ij}, u_{(i+1)j'}) \in E$ we add the following constraints. (1) All edges with tail u_{ik} (i.e. for all $1 \leq k \leq n$), (2) all edges $(u_{(i-1)k}, u_{il})$ such that $l \neq j$ or $k = j'$, (3) all edges $(u_{(i+1)k}, u_{(i+2)l})$ such that $k \neq j'$ or $l = j$, (4) all edges (with tails) in $[1, i-2] \cup [i+2, n]$ that have an endpoint in the same column as the tail of e, and (5) all edges (with tails) in $[1, i-2] \cup [i+2, n]$ that have an endpoint in the same column as the head of e. Note now that the resulting graph of constraints is $(7 + 1)$-claw free. From [Hal95], in $(h + 1)$-claw free graphs, for all $h \geq 4$, MIS can be approximated within $1/(h/2 + \varepsilon)$. As in our case $h = 7$ we have a $[1/(7/2 + \varepsilon)]$-factor approximation for MIS and thus for TPP. □

Theorem 7. *There is a $(12/7 + \varepsilon)$-factor approximation algorithm for TTSP(1, 2) when $\alpha(D) = n$.*

4.2   Improved Approximations for TTSP(1,2) via Set Packing

We now present a different reduction idea, from Max-TEM(≥ 2) to k-SET PACK-ING this time, that gives improved approximations for TTSP(1,2).

Lemma 5. *There is a $\frac{3}{5+\varepsilon}$-approximation algorithm for Max-TEM(≥ 2).*

Proof. We express the temporal matching problem as a 4-SET PACKING. Then, from [Cyg13], we have that k-SET PACKING can be approximated within $3/(k + 1 + \varepsilon)$ yielding $3/(5 + \varepsilon)$ for $k = 4$. In k-SET PACKING we are given a family $F \subseteq 2^U$ of sets of size at most k, where U is some universe of elements, and we are asked to find a maximum size subfamily of F of pairwise disjoint sets. Given $D = (V, A)$, we set $U = V \cup \{1, 2, \ldots, \alpha(D)\}$. Let $H = (S, E)$ be the static expansion of D. Construct now F as follows. For every $(u_{ij}, u_{(i+1)j'}) \in E$ set $F \leftarrow F \cup \{\{u_j, u_{j'}, i - 1, i\}\}$. Clearly, $\{u_j, u_{j'}, (i - 1), i\} \in 2^U$ because $u_j, u_{j'}$, $i - 1$, and i are pairwise distinct elements, thus indeed $F \subseteq 2^U$. Note that every set contains 4 elements, thus we have created an instance of 4-SET PACKING. The claim follows by observing that there is a temporal matching of size h in D iff there is a packing of F of size h. □

Theorem 8. *There is a $(1.7 + \varepsilon)$-approximation algorithm for TTSP(1, 2).*

4.2.1   Lifetime Restricted to n

Now assume again that the lifetime $\alpha(D)$ of the temporal graph is restricted to n. In this case, we devise via a reduction to 3-SET PACKING an improved $13/8 = 1.625$-factor approximation for TTSP$(1, 2)$.

Theorem 9. *There is a $(13/8 + \varepsilon)$-factor approximation algorithm for TTSP$(1,2)$ when $\alpha(D) = n$.*

Proof. Every TTSP tour, including the optimum tour, must necessarily use precisely the time-labels $1, 2, \ldots, n$ because otherwise it cannot cover all nodes in n steps. So, the optimum TTSP tour can be partitioned into two temporal matchings, M_O and M_E, both with time differences ≥ 2 between consecutive labels. M_O is the *odd matching* using labels $1, 3, 5, \ldots$ and M_E is the *even matching* using labels $2, 4, 6, \ldots$. So, if we denote by OPT$_{TTSP}$ the cost of the optimum TTSP tour and by $o(D')$ the number of edges of cost one of a single-label subgraph D' of the temporal graph D, we have $o(M_O) + o(M_E) = 2n - \text{OPT}_{TTSP}$.

We now approximate the maximum odd and maximum even matchings of the temporal graph D (counting the number of edges of cost one). Assume, for example, that we want to approximate the maximum matching that uses only odd labels (the even labels case is symmetric). We express it as a 3-SET PACKING as follows. Recall that in 3-SET PACKING we are given a family $F \subseteq 2^U$ of sets of size at most 3, where U is some universe of elements, and we are asked to find a maximum size subfamily of F of pairwise disjoint sets. We set $U = V \cup L_O$, where $L_O = \{1, 3, 5, \ldots\} \subset \{1, 2, \ldots, n\}$ is the set of all odd labels. Now consider the subgraph $H = (S, E)$ of the static expansion of D consisting only of the edges of cost one that appear at odd times and construct F as follows. For every $(u_{ij}, u_{(i+1)j'}) \in E$ set $F \leftarrow F \cup \{\{u_j, u_{j'}, i\}\}$. Clearly, $\{u_j, u_{j'}, i\} \in 2^U$ because u_j, $u_{j'}$, and i are pairwise distinct elements, thus indeed the constructed $F \subseteq 2^U$. Note that every set contains 3 elements, thus we have created an instance of 3-SET PACKING. It is not hard to show that there is an odd temporal matching of size h iff there is a packing of size h. The reason is that two sets $\{u, v, t\}$ and $\{u', v', t'\}$ do not conflict and can be picked together in the packing iff the corresponding edges can be picked at the same time in an odd temporal matching. Now, from [Cyg13], we have that k-SET PACKING can be approximated within $3/(k + 1 + \varepsilon)$ yielding $3/(4 + \varepsilon')$ for $k = 3$. We omit ε' in the sequel and add it in the end. So, if we denote by OPT$_O$ and ALG$_O$ (OPT$_E$ and ALG$_E$) the size of the optimum odd (even) matching and of the odd (even) matching produced by the above algorithm, respectively, we have ALG$_O \geq \frac{3}{4}$OPT$_O$ and ALG$_E \geq \frac{3}{4}$OPT$_E$. Now from the two computed matchings we keep the one with maximum cardinality. Denote its cardinality by ALG$_M$. Clearly, $2\text{ALG}_M \geq \text{ALG}_O + \text{ALG}_E$, so we have

$$\text{ALG}_M \geq \frac{1}{2}(\text{ALG}_O + \text{ALG}_E) \geq \frac{1}{2} \cdot \frac{3}{4}(\text{OPT}_O + \text{OPT}_E) = \frac{3}{8}(\text{OPT}_O + \text{OPT}_E)$$

$$\geq \frac{3}{8}[o(M_O) + o(M_E)] = \frac{3}{8}(2n - \text{OPT}_{TTSP}) = \frac{6}{8}n - \frac{3}{8}\text{OPT}_{TTSP}$$

Now, we complete the produced matching arbitrarily with the missing edges to obtain a TTSP tour. This is feasible because the matching has time differences ≥ 2 between its edges. Denote by ALG_{TTSP} the cost of the produced TTSP tour. As in the worst case, every added edge has cost 2, we have

$$\mathrm{ALG}_{TTSP} \leq 2n - \mathrm{ALG}_M \leq 2n - \frac{6}{8}n + \frac{3}{8}\mathrm{OPT}_{TTSP} = \frac{10}{8}n + \frac{3}{8}\mathrm{OPT}_{TTSP}$$

$$\leq \frac{10}{8}\mathrm{OPT}_{TTSP} + \frac{3}{8}\mathrm{OPT}_{TTSP} = \frac{13}{8}\mathrm{OPT}_{TTSP}. \qquad \square$$

References

[AGM+10] Asadpour, A., Goemans, M.X., Madry, A., Gharan, S.O., Saberi, A.: An $O(\log n/\log\log n)$-approximation algorithm for the asymmetric traveling salesman problem. In: Proceedings of the Twenty-First Annual ACM-SIAM Symposium on Discrete Algorithms (SODA), pp. 379–389 (2010)

[Ber96] Berman, K.A.: Vulnerability of scheduled networks and a generalization of Menger's theorem. Networks 28(3), 125–134 (1996)

[Blä04] Bläser, M.: A 3/4-approximation algorithm for maximum ATSP with weights zero and one. In: Jansen, K., Khanna, S., Rolim, J.D.P., Ron, D. (eds.) APPROX and RANDOM 2004. LNCS, vol. 3122, pp. 61–71. Springer, Heidelberg (2004)

[Cyg13] Cygan, M.: Improved approximation for 3-dimensional matching via bounded pathwidth local search. In: Proceedings of the IEEE 54th Annual Symposium on Foundations of Computer Science, FOCS (2013)

[DPR+13] Dutta, C., Pandurangan, G., Rajaraman, R., Sun, Z., Viola, E.: On the complexity of information spreading in dynamic networks. In: Proc. of the 24th Annual ACM-SIAM Symp. on Discrete Algorithms (SODA), pp. 717–736 (2013)

[GSS11] Gharan, S.O., Saberi, A., Singh, M.: A randomized rounding approach to the traveling salesman problem. In: Proceedings of the IEEE 52nd Annual Symposium on Foundations of Computer Science (FOCS), pp. 550–559. IEEE Computer Society Press, Washington, DC (2011)

[Hal95] Halldórsson, M.M.: Approximating discrete collections via local improvements. In: Proceedings of the Sixth Annual ACM-SIAM Symposium on Discrete Algorithms (SODA), pp. 160–169 (1995)

[KKK00] Kempe, D., Kleinberg, J., Kumar, A.: Connectivity and inference problems for temporal networks. In: Proceedings of the 32nd Annual ACM Symposium on Theory of Computing (STOC), pp. 504–513 (2000)

[KS13] Karpinski, M., Schmied, R.: On improved inapproximability results for the shortest superstring and related problems. In: Proc. 19th CATS, pp. 27–36 (2013)

[MMCS13] Mertzios, G.B., Michail, O., Chatzigiannakis, I., Spirakis, P.G.: Temporal network optimization subject to connectivity constraints. In: Fomin, F.V., Freivalds, R., Kwiatkowska, M., Peleg, D. (eds.) ICALP 2013, Part II. LNCS, vol. 7966, pp. 657–668. Springer, Heidelberg (2013)

[Orl81] Orlin, J.B.: The complexity of dynamic languages and dynamic optimization problems. In: Proceedings of the 13th Annual ACM Symposium on Theory of Computing (STOC), pp. 218–227. ACM (1981)

[PY93] Papadimitriou, C.H., Yannakakis, M.: The traveling salesman problem with distances one and two. Mathematics of Operations Research 18(1), 1–11 (1993)

Inferring Strings from Lyndon Factorization

Yuto Nakashima[1], Takashi Okabe[1], Tomohiro I[2], Shunsuke Inenaga[1],
Hideo Bannai[1], and Masayuki Takeda[1]

[1] Department of Informatics, Kyushu University, Japan
{yuto.nakashima,takashi.okabe,inenaga,bannai,takeda}@inf.kyushu-u.ac.jp
[2] Department of Computer Science, TU Dortmund, Germany
tomohiro.i@cs.tu-dortmund.de

Abstract. The Lyndon factorization of a string w is a unique factorization $\ell_1^{p_1}, \ldots, \ell_m^{p_m}$ of w s.t. $\ell_1, \ldots, \ell_m$ is a sequence of Lyndon words that is monotonically decreasing in lexicographic order. In this paper, we consider the *reverse-engineering problem on Lyndon factorization*: Given a sequence $S = ((s_1, p_1), \ldots, (s_m, p_m))$ of ordered pairs of positive integers, find a string w whose Lyndon factorization corresponds to the input sequence S, i.e., the Lyndon factorization of w is in a form of $\ell_1^{p_1}, \ldots, \ell_m^{p_m}$ with $|\ell_i| = s_i$ for all $1 \le i \le m$. Firstly, we show that there exists a simple $O(n)$-time algorithm if the size of the alphabet is unbounded, where n is the length of the output string. Secondly, we present an $O(n)$-time algorithm to compute a string over an alphabet of the smallest size. Thirdly, we show how to compute only the size of the smallest alphabet in $O(m)$ time. Fourthly, we give an $O(m)$-time algorithm to compute an $O(m)$-size representation of a string over an alphabet of the smallest size. Finally, we propose an efficient algorithm to enumerate all strings whose Lyndon factorizations correspond to S.

1 Introduction

A string ℓ is said to be a *Lyndon word*, if ℓ is lexicographically smallest among its circular permutations of characters of ℓ. Lyndon words have various and important applications in, e.g., musicology [4], bioinformatics [8], string matching [6], and free Lie algebras [23]. The *Lyndon factorization* of a string w, denoted LF_w, is a factorization $\ell_1^{p_1}, \ldots, \ell_m^{p_m}$ of w such that ℓ_i is a Lyndon word and p_i is a positive integer for each $1 \le i \le m$, ℓ_i is lexicographically larger than ℓ_{i+1} for each $1 \le i < m$, and $w = \ell_1^{p_1} \cdots \ell_m^{p_m}$. It is known that for any string w, this factorization is unique [5]. Lyndon factorizations are used in a bijective variant of Burrows-Wheeler transform [22,16] and a digital geometry algorithm [3]. Several efficient algorithms to compute Lyndon factorization exist: Duval [9] proposed an elegant on-line algorithm to compute LF_w of a given string w of length n in $O(n)$ time. Efficient parallel algorithms to compute the Lyndon factorization are also known [1,7]. Recently, algorithms to compute Lyndon factorization from compressed strings were proposed [20,21].

In this paper, we consider the *reverse-engineering problem on Lyndon factorization*: Given a sequence $S = ((s_1, p_1), \ldots, (s_m, p_m))$ of ordered pairs of positive

E. Csuhaj-Varjú et al. (Eds.): MFCS 2014, Part II, LNCS 8635, pp. 565–576, 2014.

integers, find a string w whose Lyndon factorization corresponds to the input sequence S (formal definitions of the problems will be given in Section 3). Firstly, we show that there exists a simple $O(n)$-time algorithm if the size of the alphabet is unbounded, where n is the length of the output string. Secondly, we present an $O(n)$-time algorithm to compute a string over an alphabet of the smallest size. Thirdly, we show how to compute only the size of the smallest alphabet in $O(m)$ time, where $m \leq n$ is the size of the input sequence S. Fourthly, we give an $O(m)$-time algorithm to compute an $O(m)$-size representation of a string over an alphabet of the smallest size. Finally, we propose an efficient algorithm to enumerate all strings whose Lyndon factorizations correspond to S.

The problems we consider in this paper belong to a well-studied class of reverse-engineering problems. There exist efficient reverse-engineering algorithms for, e.g., border arrays [14,11] suffix arrays [13,2], KMP failure tables [12,15], palindromic structures [18], suffix trees on binary alphabets [19], and directed acyclic word graphs [2], while hardness results are known for previous factor tables [17] and runs on a finite alphabet of size at least 4 [24]. Counting and enumerating versions of some of the above-mentioned problems have also been studied in the literature [25,26].

2   Preliminaries

Let Σ be an ordered finite alphabet, and let $\sigma = |\Sigma|$. An element of Σ^* is called a *string*. The length of a string w is denoted by $|w|$. The empty string ε is a string of length 0. Let Σ^+ be the set of non-empty strings, i.e., $\Sigma^+ = \Sigma^* - \{\varepsilon\}$. For a string $w = xyz$, x, y and z are called a *prefix*, *substring*, and *suffix* of w, respectively. A prefix x of w is called a *proper prefix* of w if $x \neq w$. The i-th character of a string w is denoted by $w[i]$ for $1 \leq i \leq |w|$. For a string w and two integers $1 \leq i \leq j \leq |w|$, let $w[i..j]$ denote the substring of w that begins at position i and ends at position j. For convenience, let $w[i..j] = \varepsilon$ when $i > j$. For any string w let $w^0 = \varepsilon$, and for any integer $k \geq 1$ let $w^k = ww^{k-1}$, i.e., w^k is a k-time repetition of w.

An integer $p \geq 1$ is said to be a *period* of a string w if $w[i] = w[i + p]$ for all $1 \leq i \leq |w| - p$. If p is a period of a string w with $p < |w|$, then $|w| - p$ is said to be a *border* of w. If w has no borders, then w is said to be *border-free*.

If character c is lexicographically smaller than another character c', then we write $c \prec c'$. For any $1 \leq i \leq \sigma$, let c_i denote the i-th "largest" element of Σ. Namely, $c_{i+1} \prec c_i$ for any $1 \leq i < \sigma$. For any non-empty strings $x, y \in \Sigma^+$, let $lcp(x, y)$ be the length of the longest common prefix of x and y, namely, $lcp(x, y) = \max(\{j \mid x[i] = y[i] \text{ for all } 1 \leq i \leq j\} \cup \{0\})$. For any non-empty strings $x, y \in \Sigma^+$, we write $x \prec y$ iff either $lcp(x, y) + 1 \leq \min\{|x|, |y|\}$ and $x[lcp(x, y) + 1] \prec y[lcp(x, y) + 1]$, or x is a proper prefix of y.

For any non-empty string x and integer $2 \leq i \leq |x|$, let $\mathbf{cs}_i(x)$ denote the i-th cyclic shift of x, namely, $\mathbf{cs}_i(x) = x[i..|x|]x[1..i - 1]$, and let $\mathbf{cs}_1(x) = x$. A string x is said to be a *Lyndon word*, if $x \prec cs_i(x)$ for all $2 \leq i \leq |x|$. Notice that any Lyndon word is border-free. The following lemma is also useful.

Lemma 1 (Proposition 1.3 [9]). *For any Lyndon words u and v, uv is a Lyndon word iff $u \prec v$.*

Definition 1. *The* Lyndon factorization *of a non-empty string $w \in \Sigma^+$, denoted LF_w, is the factorization $\ell_1^{p_1}, \ldots, \ell_m^{p_m}$ of w, such that each $\ell_i \in \Sigma^+$ is a Lyndon word, $p_i \geq 1$, and $\ell_i \succ \ell_{i+1}$ for all $1 \leq i < m$. Each $\ell_i^{p_i}$ is called a* Lyndon factor, *and can be represented by two positive integers $(|\ell_i|, p_i)$. ℓ_i is called a decomposed Lyndon factor.*

The Lyndon factorization is unique for each string w. Also, the Lyndon factorization of any string w of length n can be computed in $O(n)$ time [9].

3   Inferring a String with Given Lyndon Factorization

3.1   Computing String on Alphabet of Arbitrary Size

The simplest variant of our reverse-engineering problem is the following:

Problem 1. Given a sequence $S = ((s_1, p_1), \ldots, (s_m, p_m))$ of ordered pairs of positive integers, compute a string $w \in \Sigma^+$ such that $LF_w = \ell_1^{p_1}, \ldots, \ell_m^{p_m}$ and $|\ell_i| = s_i$.

The length n of w is clearly $n = \sum_{i=1}^{m} s_i p_i$. In Problem 1, there is no restriction on the size of the alphabet from which the output string w is drawn. A solution always exists, and the problem can be solved in $O(n)$-time by, basically just assigning decreasingly smaller characters to the first character of each factor.

Proposition 1. *Problem 1 can be solved in $O(n)$ time, where n is the length of an output string.*

3.2   Computing String on Alphabet of Smallest Size

Now, we consider a more interesting variant of our reverse-engineering problem, where a string on an alphabet of the smallest size is to be computed. For any $1 \leq j \leq \sigma$, let $\Sigma_j = \{c_1, \ldots, c_j\}$ denote the set of the j largest characters of Σ.

Problem 2. Given a sequence $S = ((s_1, p_1), \ldots, (s_m, p_m))$ of ordered pairs of positive integers, compute a string $w \in \Sigma_k^+$ such that $LF_w = \ell_1^{p_1}, \ldots, \ell_m^{p_m}$, $|\ell_i| = s_i$, and k is the smallest possible.

In what follows, we present an $O(n)$-time algorithm to solve Problem 2. This algorithm computes factors from left to right, and is based on the lemma below.

Lemma 2. *Let $S = ((s_1, p_1), \ldots, (s_m, p_m))$ be a sequence of ordered pairs of positive integers. If for some string $w \in \Sigma^+$, $LF_w = \ell_1^{p_1}, \ldots, \ell_m^{p_m}$, where ℓ_1 is the lexicographically largest Lyndon word of length s_1 and for all $2 \leq i \leq m$, ℓ_i is the lexicographically largest Lyndon word of length s_i which is lexicographically smaller than ℓ_{i-1}, then, $w \in \Sigma_k^+$ where $\Sigma_k = \{c_1, \ldots, c_k\} \subseteq \Sigma$ and k is the smallest possible.*

Proof. Let $L_1 \succ \cdots \succ L_\alpha$ be the decreasing sequence of Lyndon words on Σ of length at most $\max\{s_i \mid 1 \leq i \leq m\}$. Then the sequence $\ell_1, \ldots, \ell_m$ is a subsequence of $L_1, \ldots, L_\alpha$, i.e., there exist $1 \leq i_1 < \cdots < i_m \leq \alpha$ s.t. $\ell_1 = L_{i_1}, \ldots, \ell_m = L_{i_m}$. If $L_i \in \Sigma_k^+ - \Sigma_{k-1}^+$ for some i, then it must be that $L_i[1] = c_k$ or else L_i cannot be a Lyndon word. Thus, for any Lyndon words $L_i \in \Sigma_{k-1}^+$ and $L_j \in \Sigma_k^+ - \Sigma_{k-1}^+$, $L_i \succ L_j$, and thus $i < j$ holds. As the condition on w indicates that $i_1 = \min\{i \mid |L_i| = s_1\}$ and $i_j = \min\{i \mid L_{i_{j-1}} \succ L_i, |L_i| = s_j\}$ for any $1 < j \leq m$, i_m is the smallest possible, and thus k is the smallest possible. □

The string w of Lemma 2 is the lexicographically largest string whose Lyndon factorization corresponds to input S. We compute w as defined in Lemma 2.

Duval [10] proposed a linear time algorithm which, given a Lyndon word, computes the next Lyndon word (i.e., the lexicographical successor) of the same length. Although our algorithm to be shown in this section is somewhat similar to his algorithm, ours is more general in that it can compute the previous Lyndon word (i.e., the lexicographical predecessor) of a *given length*, in linear time.

If $s_1 = 1$, then $\ell_1 = c_1$. If $s_1 \geq 2$, then $\ell_1 = c_2 c_1^{s_1-1}$. Assume that we have already computed $\ell_1, \ldots, \ell_{i-1}$ for $1 < i \leq m$, and we are computing ℓ_i. In so doing, we will need the two following lemmas.

Lemma 3. *Let x be any Lyndon word such that $|x| \geq 2$ and $x[i..|x|] = c_q c_1^{|x|-i}$ for some $1 < i \leq |x|$ and some $1 < q \leq \sigma$. Then, for any $1 \leq p < q$, $y = x[1..i-1]c_p c_1^{|x|-i}$ is a Lyndon word.*

Proof. Firstly, we show that $y[1..i]$ is a Lyndon word. Assume on the contrary that $y[1..i]$ is not a Lyndon word. Then, there exists $2 \leq j \leq i$ satisfying $y[1..i] \succ y[j..i]$. Since $x[1..i-1] = y[1..i-1]$ and $x[i] = c_q \prec c_p = y[i]$, $y[1..i] \succ x[1..i]$ and $y[j..i] \succ x[j..i]$. Since $2 \leq j$, $|y[j..i]| = i - j + 1 \leq i - 1$. Since $y[1..i-1] = x[1..i-1]$, we get $x[1..i] \succ y[j..i]$, which implies that $x[1..i] \succ x[j..i]$. However, this contradicts that x is a Lyndon word. Hence $y[1..i]$ is a Lyndon word.

Now we show y is a Lyndon word by induction on k, where $i \leq k \leq |y|$. The case where $k = i$ has already been shown. Assume $y[1..k]$ is a Lyndon word for $i \leq k < |y|$. As $2 \leq i \leq k$, $y[1..k] \prec y[k+1] = c_1$. Since $y[k+1] = c_1$ is a Lyndon word, by Lemma 1, $y[1..k+1]$ is a Lyndon word. This completes the proof. □

Lemma 4. *For any Lyndon word x with $|x| \geq 2$ and any $1 \leq i \leq |x|$, $y = x[1..i]c_1^{|x|-i}$ is a Lyndon word.*

Proof. Let $k = |x| - i$. We prove the lemma by induction on k. If $k = 0$, i.e. $i = |x|$, then $y = x$ and hence the lemma trivially holds. Assume the lemma holds for some $0 \leq k < |x| - 1$, i.e., $x[1..|x| - k]c_1^k = x[1..i]c_1^{|x|-i}$ is a Lyndon word. Then, by Lemma 3, $x[1..|x| - (k+1)]c_1^{k+1} = x[1..i-1]c_1^{|x|-i+1}$ is also a Lyndon word. Hence the lemma holds. □

Computing ℓ_i from ℓ_{i-1} When $s_i = s_{i-1}$. Here, we describe how to compute ℓ_i from ℓ_{i-1} when $s_i = s_{i-1}$, namely, $|\ell_i| = |\ell_{i-1}|$. The following is a key lemma:

Algorithm 1. Compute next smaller Lyndon word of same length.

Input: String x.

Output: The lexicographically largest Lyndon word y of length $|x|$ which is lexicographically smaller than x.

1   compute C_x;

2   $k \leftarrow 1$, $h \leftarrow 2$, $i \leftarrow 0$;

3   **if** $x[1] \neq c_\sigma$ **then** $i \leftarrow 1$;

4   **while** $h \leq |x|$ **do**

5   **if** $x[k] \neq c_\sigma$ **then**

6   **if** $x[k] \prec c_{r+1}$ **then** $i \leftarrow h$;   // $c_r = x[h]$

7   **else if** $x[k] = c_{r+1}$ **then**

8            $len \leftarrow \min\{C_x[k+1], h - k - 1\}$;

9   **if** $len < |x| - h$ **then** $i \leftarrow h$;

     // Operation of Duval's algorithm.

10   **if** $x[k] \prec x[h]$ **then** $k \leftarrow 1$, $h \leftarrow h + 1$;

11   **else if** $x[k] = x[h]$ **then** $k \leftarrow k + 1$, $h \leftarrow h + 1$;

12   **else** break;

13 **output** $y = x[1..i-1]c_{j+1}c_1^{|x|-i}$;   // $c_j = x[i]$

Lemma 5. *For any non-empty string x, let y be the lexicographically largest Lyndon word of length $|x|$ that is lexicographically smaller than x. Then $y = x[1..i-1]c_{j+1}c_1^{|x|-i}$ where $x[i] = c_j$ and i is the largest position s.t. $x[1..i-1]c_{j+1}c_1^{|x|-i}$ is a Lyndon word.*

Proof. Assume on the contrary that there is a Lyndon word z of length $|x|$ s.t. $y \prec z \prec x$. As $x[1..i-1] = y[1..i-1]$ and $c_j = x[i] \succ y[i] = c_{j+1}$, there is a position $i' > i$ s.t. $z[1..i'-1] = x[1..i'-1]$ and $z[i'] \prec x[i']$. By Lemma 4, $z[1..i']c_1^{|x|-i'} = x[1..i'-1]z[i']c_1^{|x|-i'}$ is a Lyndon word. By Lemma 3 $x[1..i'-1]c_{j'+1}c_1^{|x|-i'}$ is a Lyndon word, where $x[i'] = c_{j'} \succ c_{j'+1} \succeq z[i']$. This contradicts that i is the largest position in x s.t. $x[1..i-1]c_{j+1}c_1^{|x|-i}$ is a Lyndon word.   □

Algorithm 1 shows a pseudo-code of our linear-time algorithm to find the lexicographically largest Lyndon word y of length $|x|$ that is lexicographically smaller than x. To efficiently compute i of Lemma 5, we use, as a sub-routine, Duval's algorithm [9] which computes the Lyndon factorization of a string.

Lemma 6. *For any non-empty string x, Algorithm 1 computes the lexicographically largest Lyndon word y of length $|x|$ which is lexicographically smaller than x in $O(|x|)$ time.*

Proof. Let C_x be an array of length $|x|$ such that, for any $1 \leq i \leq |x|$, $C_x[i] = \max\{q \mid x[i..i+q-1] = c_1^q\}$. Namely, $C_x[i]$ represents the number of consecutive c_1's starting at position i in x. Algorithm 1 firstly computes C_x.

For $1 \leq h \leq |x|$, let $x_h = x[1..h-1]c_{r+1}c_1^{|x|-h}$, where $c_r = x[h]$. Namely, x_h is the concatenation of the prefix of x of length $h - 1$, the lexicographically next character c_{r+1} to the character $c_r = x[h]$, and the repetition of c_1 of length $|x| - h$. The algorithm checks whether x_h is a Lyndon word for all $1 \leq h \leq |x|$ in increasing order of h, based on Duval's algorithm [9]. For each h, our algorithm maintains a variable k to be the largest integer satisfying $x_h[1..k-1] = x_h[h-k+1..h-1]$. To check if x_h is a Lyndon word, we compare $x_h[k] = x[k]$ and $x_h[h] = c_{r+1}$ (lines 6 and 7). There are the three following cases:

- If $x[k] \prec c_{r+1}$, then we know that $x[1..h-1]c_{r+1}$ is a Lyndon word, due to Duval's algorithm [9]. It follows from Lemma 1 that $x[1..h-1]c_{r+1}c_1^{|x|-h}$ is a Lyndon word. The value of i is replaced by h (line 6).
- If $x[k] \succ c_{r+1}$, then we know that $x[1..h-1]c_{r+1}c_1^{|x|-h}$ is not a Lyndon word, due to Duval's algorithm [9].
- If $x[k] = c_{r+1}$, then $x[1..k] = x[j-k+1..j-1]c_{r+1}$. In this case, we compare the substrings immediately following $x[1..k]$ and $x[h-k+1..h-1]c_{r+1}$, respectively. Since $x_h[h+1..|x|] = c_1^{|x|-h}$, if we know the number of consecutive c_1's from position $k+1$ in x, then we can efficiently check whether or not x_h is a Lyndon word. Let len be the number of consecutive c_1's from position $k+1$ in x_h which can be calculated by $\min\{C_x[k+1], h-k-1\}$. We compare len with the number of consecutive c_1's from position $h+1$ in x_h, which is clearly $|x| - h$. If $len < |x| - h$, then x_h is a Lyndon word. Otherwise, x_h is not a Lyndon word since it has a border.

In lines 10-12 we update the values of k and h using Duval's algorithm [9].

After the **while** loop, the variable i stores the largest integer s.t. $x[1..i-1]c_{j+1}c_1^{|x|-i}$ is a Lyndon word, which is the output of the algorithm (line 13).

It is easy to see that C_x can be computed in $O(|x|)$ time. The **while** loop repeats at most $|x|$ times, and each operation in the **while** loop takes constant time. Therefore the overall time complexity is $O(|x|)$.   □

Computing ℓ_i from ℓ_{i-1} When $s_i \neq s_{i-1}$. Here we show how to compute ℓ_i from ℓ_{i-1} when their lengths s_i and s_{i-1} are different. Firstly, we consider the case where $s_i < s_{i-1}$, namely $|\ell_i| < |\ell_{i-1}|$:

Lemma 7. *For any non-empty string x and positive integer $k < |x|$, let y be the lexicographically largest Lyndon word of length k that is lexicographically smaller than x. If $x[1..k]$ is a Lyndon word, then $y = x[1..k]$. Otherwise, $y = x[1..i-1]c_{j+1}c_1^{k-i}$, where $x[i] = c_j$ and i is the largest position s.t. $x[1..i-1]c_{j+1}c_1^{k-i}$ is a Lyndon word.*

Proof. Let $x' = x[1..k]$. No string of length k which is lexicographically smaller than x and larger than x' exists. Thus, if x' is a Lyndon word, $y = x'$. Otherwise, y is the lexicographically largest string of length k that is lexicographically smaller than x'. Since $|x'| = |y| = k$, the statement follows from Lemma 5.   □

Secondly, we consider the case where $s_i > s_{i-1}$, namely $|\ell_i| > |\ell_{i-1}|$. The following lemma can be shown in a similar way to Lemma 5.

Lemma 8. *For any non-empty string x and positive integer $k > |x|$, let y be the lexicographically largest Lyndon word of length k that is lexicographically smaller than x. Then $y = x[1..i-1]c_{j+1}c_1^{k-i}$, where $x[i] = c_j$ and i is the largest position s.t. $x[1..i-1]c_{j+1}c_1^{k-i}$ is a Lyndon word and $1 \leq i \leq |x|$.*

Due to the two above lemmas, ℓ_i can be computed from ℓ_{i-1} in a similar way to the case where $|\ell_i| = |\ell_{i-1}|$, using a slightly modified version of Algorithm 1, as described in the following theorem.

Theorem 1. *Problem 2 can be solved in $O(n)$ time, where n is the length of an output string.*

Proof. Assume we have already computed $\ell_1, \ldots, \ell_{i-1}$ and are computing ℓ_i.

- If $|\ell_{i-1}| > |\ell_i|$, then let x be the prefix of ℓ_{i-1} of length s_i, namely, $x = \ell_{i-1}[1..s_i]$. By Lemma 7, if x is a Lyndon word, then $\ell_i = x$. We can check whether x is a Lyndon word or not in $O(|x|)$ time, by using Duval's algorithm [9]. Otherwise, ℓ_i can be computed from x by Algorithm 1. This takes $O(|\ell_i|) = O(s_i)$ time by Lemma 6.
- If $|\ell_{i-1}| = |\ell_i|$, then ℓ_i can be computed from ℓ_{i-1} in $O(s_i)$ time, by Lemma 6.
- If $|\ell_{i-1}| < |\ell_i|$, then let $x = \ell_{i-1}c_1^{|\ell_i|-|\ell_{i-1}|}$. We take x as input to Algorithm 1, with a slight modification to the algorithm. Since $\ell_{i-1} \prec x \prec \ell_i$ must hold, we are only interested in positions from 1 to $|\ell_{i-1}|$ in x. Hence, as soon as the value of h in Algorithm 1 exceeds ℓ_{i-1}, we exit from the **while** loop, and the resulting string is ℓ_{i-1}. The above modification clearly does not affect the time complexity of the algorithm, and hence it takes $O(s_i)$ time.

Thus we can compute the output string in $O(\sum_{i=1}^m s_i p_i) = O(n)$ time. $\square$

We can remark that for computing $\ell_1, \ldots, \ell_m$ we do not use $p_1, \ldots, p_m$, and hence, the following corollaries are immediate from Theorem 1.

Corollary 1. *We can compute the Lyndon factorization $\ell_1^{p_1}, \ldots, \ell_m^{p_m}$ of a string which is a solution to Problem 2 in $O(\sum_{i=1}^m s_i)$ time.*

Corollary 2. *Given a sequence $S = ((s_1, p_1), \ldots, (s_m, p_m))$ of ordered pairs of integers and an integer $k' \geq 1$, we can determine in $O(\sum_{i=1}^m s_i)$ time if there exists a string w over an alphabet of size at most k' s.t. $LF_w = \ell_1^{p_1}, \ldots, \ell_m^{p_m}$ and $|\ell_i| = s_i$.*

3.3   Computing the Smallest Alphabet Size

In this subsection we consider the following problem.

Problem 3. Given a sequence $S = ((s_1, p_1), \ldots, (s_m, p_m))$ of ordered pairs of positive integers, compute the smallest integer k for which there exists a string $w \in \Sigma_k^+$ such that $LF_w = \ell_1^{p_1}, \ldots, \ell_m^{p_m}$ and $|\ell_i| = s_i$.

Clearly, this problem can be solved in $O(n)$ time by Theorem 1. However, since only the smallest alphabet size is of interest, a string does not have to be computed in this problem. To this end, we present an optimal $O(m)$-time algorithm to solve Problem 3, where $m \leq n$ is the size of the input sequence S. The basic strategy is the same as the previous algorithm, i.e., we simulate the algorithm of computing ℓ_i from ℓ_{i-1}, for all $1 < i \leq m$. The difficulty is that, in order to achieve an $O(m)$-time algorithm, we cannot afford to store ℓ_i's explicitly. Hence, we simulate the previous algorithm on a compact representation of ℓ_i's.

We introduce *the largest character block encoding (LCBE)* X of a non-empty string x. Consider factorizing x into blocks according to the following rules; the b-th block X_b is the longest prefix of $x[pos(X_b)..|x|]$ s.t. $X_b = c_j c_1^q$ for some $j \geq 1$ and $q \geq 0$, where $pos(X_b) = 1$ if $b = 1$, and $pos(X_b) = pos(X_{b-1}) + |X_{b-1}|$ otherwise. Let $\|X\|$ denote the number of blocks of X, i.e., $x = X = X_1 X_2 \ldots X_{\|X\|}$. For any $1 < b \leq \|X\|$, $X_b[1] \neq c_1$. Notice that X can be encoded in $O(\|X\|)$ space by storing $X_b[1]$ and $pos(X_b)$ for each block.

Let X and X' be the *LCBE* of ℓ_{i-1} and ℓ_i, respectively. It holds that $\|X'\| \leq \|X\| + 1$ and $\|X\| \leq m$. To compute X' from X efficiently, each block X_b stores the information about the position k s.t. $X[k]$ and $X[pos(X_b)]$ are supposed to be compared in Algorithm 1. Since X is a Lyndon word, $X[k] \preceq X[pos(X_b)] \prec c_1$ and there exists a block starting at k. Also, $X[1..k-1]$ is the longest prefix which is a suffix of $X[1..pos(X_b)-1]$. Thus we let X_b have the value $pbi(X_b) = \max\{b' \mid 1 \leq b' < b, X[1..pos(X_{b'}) - 1] = X[pos(X_b) - pos(X_{b'}) + 1..pos(X_b) - 1]\}$, that is, $b' = pbi(X_b)$ is the block index s.t. $k = pos(X_{b'})$. We can let $pbi(X_1)$ remain undefined since we will never use it. An example of *LCBE* follows.

Example 1. Let $\Sigma = \{c, b, a\}$ and $\ell_{i-1} = \mathtt{acaccacbacaccbcc}$. Then the *LCBE* of ℓ_{i-1} is $X = \mathtt{ac}, \mathtt{acc}, \mathtt{ac}, \mathtt{b}, \mathtt{ac}, \mathtt{acc}, \mathtt{bcc}$ and $pbi(X_2) = 1, pbi(X_3) = 1, pbi(X_4) = 2, pbi(X_5) = 1, pbi(X_6) = 2, pbi(X_7) = 3$.

Lemma 9 shows how to efficiently compute X' from X using pbi. Since X and X' share at least the first $\|X'\| - 2$ blocks, we do not build X' from scratch.

Lemma 9. *Given LCBE X of ℓ_{i-1} with pbi, we can compute LCBE X' of ℓ_i in $O(\max\{1, \|X\| - \|X'\|\})$ time.*

Proof. Since it is trivial when $X = c_1$, we consider the case where $X \neq c_1$ and $X[1] \prec c_1$. We simulate the task described in Theorem 1. First, we adjust the length of X to s_i, i.e., add $c_1^{s_i - s_{i-1}}$ if $s_{i-1} < s_i$, or truncate X to represent $X[1..s_i]$ if $s_{i-1} > s_i$. A major difference from the algorithm of Theorem 1 is that we process the blocks from right to left, checking whether each block contains the position h s.t. $X[1..h-1]c_{j+1}c_1^{|X|-h}$ is a Lyndon word, where $c_j = X[h]$. We show each block X_b can be investigated in $O(1)$ time by using *LCBE* and *pbi*.

For any $1 < h \leq |X|$, let $p(h)$ be the position k s.t. $X[1..k-1]$ is the longest prefix of X which is a suffix of $X[1..h-1]$. In Algorithm 1, $X[h]$ is compared with $X[p(h)]$. As described in Lemma 6, for any $1 < h \leq |X|$ with $X[p(h)] \prec X[h] = c_j$, $X[1..h-1]c_{j+1}c_1^{|X|-h}$ is a Lyndon word iff $X[p(h)] \prec c_{j+1}$ or $|X| - h > d$, where d is the maximum repeat of c_1's as a prefix of $X[p(h) + 1..h - 1]$.

Consider the case where $b \neq 1$. Let $b' = pbi(X_b)$. Since we know $p(pos(X_b)) = pos(X_{b'})$, position $g = pos(X_b) + lcp(X_b, X_{b'})$ is the leftmost position inside X_b s.t. $X[p(g)] \prec X[g]$ if $|X_b| > lcp(X_b, X_{b'})$. By the definition of $LCBE$ and that X is a prefix of a Lyndon word, $lcp(X_b, X_{b'}) = 0$ if $X_b[1] \neq X_{b'}[1]$, and $lcp(X_b, X_{b'}) = |X_{b'}|$ otherwise. For any h with $g < h \leq pos(X_b) + |X_b| - 1$, $X[p(h)] \prec X[h] = c_1$ holds since $X[1] \prec c_1$ and $p(h) = 1$. Since we can compute d from the information of $LCBE$ in $O(1)$ time, we can check if $X[1..g-1]c_j{+}c_1^{|X|-g}$ is a Lyndon word or not in $O(1)$ time. Since $p(g'') = 1$ and $X[g''] = c_1$ for any g'' with $g < g'' \leq pos(X_b) + |X_b| - 1$, if $X[1] \prec c_2$ or $|X| - g'' > d'$, then $X[1..g'' - 1]c_{j+1}c_1^{|X|-g''}$ is a Lyndon word, where d' is the maximum integer s.t. $c_1^{d'}$ is a prefix of $X[2..g'' - 1]$. Hence by a simple arithmetic operation we can compute in $O(1)$ time the largest position g' with $g < g' \leq pos(X_b) + |X_b| - 1$ s.t. $X[1..g' - 1]c_2c_1^{|X|-g'}$ is a Lyndon word. A minor remark is that we add a constraint for g' not to exceed s_{i-1} when $s_{i-1} < s_i$ and $b = \|X\|$.

The case where $b = 1$ can be managed in a similar way to the case where $g < g'' \leq pos(X_b) + |X_b| - 1$ described above, and hence it takes $O(1)$ time.

We check each block from right to left until we find the largest position h s.t. $X[1..h-1]c_{j+1}c_1^{|X|-h}$ is a Lyndon word, where $c_j = X[h]$. Since each block can be checked in $O(1)$ time, the whole computational time is $O(\max\{1, \|X\| - \|X'\|\})$.

Since pbi for the blocks in X' other than the last block remain unchanged from pbi for X, it suffices to calculate pbi for the last block of X'. Let $b = \|X'\|$. Assume $b > 1$ since no updates are needed when $b = 1$. Then $pbi(X_b') = pbi(X_{b-1}') + 1$ if $b - 1 \geq 2$ and $X'_{pbi(X'_{b-1})} = X'_{b-1}$, and $pbi(X_b') = 1$ otherwise. □

Theorem 2. *Problem 3 can be solved in $O(m)$ time and $O(m)$ space.*

Proof. We begin with $LCBE$ of ℓ_1 and transform it to $LCBE$ of $\ell_2, \ell_3, \ldots, \ell_m$ in increasing order, using Lemma 9. Finally we get $LCBE$ of ℓ_m and we can obtain the alphabet size by looking into the first character of ℓ_m.

Let $B_1, B_2, \ldots, B_m$ denote the number of blocks in $LCBEs$ of $\ell_1, \ell_2, \ldots, \ell_m$, respectively. Clearly $B_1 = 1$. By Lemma 9, the total time complexity to get $LCBE$ of ℓ_m is $O(\sum_{i=2}^{m} \max\{1, B_{i-1} - B_i\}) = O(m + B_1 - B_m) = O(m)$. □

We also remark that an $O(m)$-size compact representation of the lexicographically largest solution for Problem 2 can be computed in $O(m)$ time through the algorithm described above. To do so, we store all $LCBE$'s of $\ell_1, \ell_2, \ldots, \ell_m$ as a tree where the common prefix blocks are shared. Using this representation we can obtain the desired string in $O(n)$ time.

4 Enumerate Strings with Given Lyndon Factorization

In this section we consider the problem of enumerating all strings whose Lyndon factorizations correspond to a given sequence of integer pairs:

Problem 4. Given a sequence $S = ((s_1, p_1), \ldots, (s_m, p_m))$ of ordered pairs of positive integers and $\Sigma_k = \{c_1, \ldots, c_k\}$, compute all strings $w \in \Sigma_k^+$ such that $LF_w = \ell_1^{p_1}, \ldots, \ell_m^{p_m}$ and $|\ell_i| = s_i$.

Let K be the set of output strings for Problem 4. We consider a tree T defined as follows. Let $\texttt{root}$ be the root of T.

- $\texttt{root}$ is $w_{\texttt{root}}$, which is the lexicographically largest string in K, computed by the algorithm of Section 3.2;
- Each child v of any node u is a pair (w_v, j) of a string and integer j, such that w_v is the string obtained by replacing the j-th factor $(\ell_j)^{p_j}$ of the Lyndon factorization of w_u with $(\ell'_j)^{p_j}$, where ℓ'_j is a Lyndon word of length s_j satisfying $\ell_j \succ \ell'_j \succ \ell_{j+1}$ (ℓ_{m+1} denotes ε for convenience);
- For any non-root node $u = (w_u, i)$ and any child $v = (w_v, j)$ of u, $i > j$.

The next lemma shows that T represents all and only the strings in K.

Lemma 10. *Let $S = ((s_1, p_1), \ldots, (s_m, p_m))$ be a sequence of ordered pairs of positive integers. Then T contains all and only strings $w \in \Sigma_k^+$ s.t. $LF_w = \ell_1^{p_1}, \ldots, \ell_m^{p_m}$ with $|\ell_i| = s_i$ for all $1 \le i \le m$.*

Proof. If w is the lexicographically largest string in K, then it is represented by $\texttt{root}$, i.e., $w = w_{\texttt{root}}$. Otherwise, let $LF_w = \ell_1^{p_1}, \ldots, \ell_m^{p_m}$ and $LF_{w_{\texttt{root}}} = r_1^{p_1}, \ldots, r_m^{p_m}$. Let $J = \{j \mid r_j \succ \ell_j\}$ and $g = |J|$. For any $1 \le i \le g$, let j_i denote the i-th smallest element of J. The node u that corresponds to w can be located from $\texttt{root}$, as follows. By the definition of T, (ℓ_{j_g}, j_g) is a child of $\texttt{root}$. Assume we have arrived at a non-root node $v_{j_i} = (\ell_{j_i}, j_i)$ with $1 < i \le g$. Let $LF_{w_v} = x_1^{p_1}, \ldots, x_m^{p_m}$. Then, for any $k < j_i$, $x_k = r_k$. Thus we have that $r_{j_{i-1}} = x_{j_{i-1}} \succ \ell_{j_{i-1}} \succ \ell_{j_{i-1}+1} = x_{j_{i-1}+1}$. This implies that $v_{j_{i-1}} = (\ell_{j_{i-1}}, j_{i-1})$ is a child of v_{j_i}. Note that for any $k' > j_{i-1}$, $x_{k'} = \ell_{k'}$. Hence, $v_{j_1} = u$, the desired node which corresponds to w.

By the definition of T, any string corresponding to a node of T is in K. $\qquad\square$

A naïve representation of T requires $O(|K|n)$ space. To reduce the output size of Problem 4, we introduce the following compact representation of T:

- $\texttt{root}$ is $LF_{w_{\texttt{root}}} = r_1^{p_1}, \ldots, r_m^{p_m}$, where $w_{\texttt{root}}$ is the lexicographically largest string in K, computed by the algorithm of Section 3.2;
- Each child v of any node u is a pair (ℓ'_j, j) of a Lyndon word ℓ'_j of length s_j and integer j, such that LF_{w_v} is obtained by replacing the j-th factor $(\ell_j)^{p_j}$ of the Lyndon factorization LF_{w_u} of w_u with $(\ell'_j)^{p_j}$, where ℓ'_j satisfies $\ell_j \succ \ell'_j \succ \ell_{j+1}$ (ℓ_{m+1} denotes ε for convenience);
- For any non-root node $u = (\ell'_i, i)$ and any child $v = (\ell''_j, j)$ of u, $i > j$.

Let $s_{\max} = \max\{s_i \mid 1 \le i \le m\}$ and $n' = \sum_{i=1}^m s_i$. Then, this compact representation of T requires only $O(|K|s_{\max} + n')$ space. In the sequel, we mean by T the compact representation of T.

We show how to construct T in linear time. Let $LF_{w_{\texttt{root}}} = r_1^{p_1}, \ldots, r_m^{p_m}$. Let H be the set of integers h ($1 \le h \le m$) s.t. there exists a Lyndon word r'_h satisfying $r_h \succ r'_h \succ r_{h+1}$, where r_{m+1} is the empty string ε for convenience.

Lemma 11. *Given a sequence $S = ((s_1, p_1), \ldots, (s_m, p_m))$ of ordered pairs of integers, H can be computed in $O(n')$ time, where $n' = \sum_{i=1}^m s_i$.*

Proof. We compute $LF_{w_{root}} = r_1^{p_1}, \ldots, r_m^{p_m}$ in $O(n')$ time by Corollary 1. For each $1 \le h \le m$, we apply Algorithm 1 to r_h and compute the lexicographically largest Lyndon word r_h' of length $|r_h| = s_h$ that is lexicographically smaller than r_h. If $h < m$, then we lexicographically compare r_h' with r_{h+1}, and $h \in H$ only if $r_h' \succ r_{h+1}$. This takes $O(s_h)$ time for each h. If $h = m$, then $m \in H$ only if r_m' contains at most $k = |\Sigma_k|$ distinct characters. This can be easily checked in $O(s_m)$ time. Hence, it takes a total of $O(n')$ time for all $1 \le h \le m$.   □

Theorem 3. *An $O(|K|s_{max} + n')$-size representation of the solution to Problem 4 can be computed in $O(|K|s_{max} + n')$ time with $O(n')$ extra working space, where $|K|$ is the number of strings which corresponds to a given S, $s_{max} = \max\{s_i \mid 1 \le i \le m\}$ and $n' = \sum_{i=1}^{m} s_i$.*

Proof. We first check if there is a string over a given alphabet Σ_k whose Lyndon factorization corresponds to the input sequence S, in $O(n')$ time by Corollary 2. If there exist such strings, then $K = \emptyset$ and thus the theorem holds.

Assume $K \ne \emptyset$. The root node **root** of T is the Lyndon factorization $LF_{w_{root}} = r_1^{p_1}, \ldots, r_m^{p_m}$, which can be computed in $O(n')$ by Corollary 1. Then, we compute the children of **root** as follows. For all $h \in H$, we compute all the Lyndon words r_h' of length $|r_h| = s_h$ that satisfy $r_h \succ r_h' \succ r_{h+1}$, over alphabet Σ_k. Each of these Lyndon words can be computed in $O(s_h) = O(s_{max})$ time by Lemma 6.

Given a non-root node $u = (\ell_j, j)$, we compute the children of u as follows. If $j = 1$, then u is a leaf and has no children. Otherwise, we first compute all the Lyndon words ℓ_{j-1}' of length $|r_{j-1}| = s_{j-1}$ that satisfy $r_{j-1} \succ \ell_{j-1}' \succ \ell_j$, over alphabet Σ_k. Then, for all $h \in H \cap \{1, .., j-2\}$, we compute all the Lyndon words ℓ_h' of length $|r_h| = s_h$ that satisfy $r_h \succ \ell_h' \succ r_{h+1}$, over alphabet Σ_k. Each of these Lyndon words can be computed in $O(s_{max})$ time as well.

We can compute H in $O(n')$ time by Lemma 11. Since each node can be computed in $O(s_{max})$ time as above, the total running time for constructing T is $O(|K|s_{max} + n')$. We need extra $O(n')$ working space to store H.

We show the correctness of the algorithm. Clearly, all the children of **root** are computed by the above algorithm. Consider any non-root node $u = (\ell_j, j)$ of T. Let w_u be the string that corresponds to node u. Since $r_j \succ \ell_j$, if $j \ge 2$, then there may exist some Lyndon words ℓ_{j-1}' of length $|r_{j-1}|$ with $r_{j-1} \succ \ell_{j-1}' \succ \ell_j$. All such Lyndon words over Σ_k are computed by the above algorithm. Consider the other children of u. Since the first $j - 1$ factors of LF_{w_u} are $r_1^{p_1}, \ldots, r_{j-1}^{p_{j-1}}$, all the Lyndon words ℓ_h' of length $|r_h|$ satisfying $r_h \succ \ell_h' \succ r_{h+1}$ with $h \in H \cap \{1, ..., j-2\}$ correspond to the children of u. All these Lyndon words over Σ_k are also computed by the above algorithm. This completes the proof.   □

References

1. Apostolico, A., Crochemore, M.: Fast parallel Lyndon factorization with applications. Mathematical Systems Theory 28(2), 89–108 (1995)
2. Bannai, H., Inenaga, S., Shinohara, A., Takeda, M.: Inferring strings from graphs and arrays. In: Rovan, B., Vojtáš, P. (eds.) MFCS 2003. LNCS, vol. 2747, pp. 208–217. Springer, Heidelberg (2003)

3. Brlek, S., Lachaud, J.O., Provençal, X., Reutenauer, C.: Lyndon + Christoffel = digitally convex. Pattern Recognition 42(10), 2239–2246 (2009)
4. Chemillier, M.: Periodic musical sequences and Lyndon words. Soft Comput. 8(9), 611–616 (2004)
5. Chen, K.T., Fox, R.H., Lyndon, R.C.: Free differential calculus. IV. the quotient groups of the lower central series. Annals of Mathematics 68(1), 81–95 (1958)
6. Crochemore, M., Perrin, D.: Two-way string matching. J. ACM 38(3), 651–675 (1991)
7. Daykin, J.W., Iliopoulos, C.S., Smyth, W.F.: Parallel RAM algorithms for factorizing words. Theor. Comput. Sci. 127(1), 53–67 (1994)
8. Delgrange, O., Rivals, E.: STAR: an algorithm to search for tandem approximate repeats. Bioinformatics 20(16), 2812–2820 (2004)
9. Duval, J.P.: Factorizing words over an ordered alphabet. J. Algorithms 4(4), 363–381 (1983)
10. Duval, J.P.: Génération d'une section des classes de conjugaison et arbre des mots de Lyndon de longueur bornée. Theor. Comput. Sci. 60, 255–283 (1988)
11. Duval, J.P., Lecroq, T., Lefebvre, A.: Border array on bounded alphabet. Journal of Automata, Languages and Combinatorics 10(1), 51–60 (2005)
12. Duval, J.P., Lecroq, T., Lefebvre, A.: Efficient validation and construction of border arrays and validation of string matching automata. RAIRO - Theoretical Informatics and Applications 43(2), 281–297 (2009)
13. Duval, J.P., Lefebvre, A.: Words over an ordered alphabet and suffix permutations. Theoretical Informatics and Applications 36, 249–259 (2002)
14. Franek, F., Gao, S., Lu, W., Ryan, P.J., Smyth, W.F., Sun, Y., Yang, L.: Verifying a border array in linear time. J. Comb. Math. and Comb. Comp. 42, 223–236 (2002)
15. Gawrychowski, P., Jeż, A., Jeż, Ł.: Validating the Knuth-Morris-Pratt failure function, fast and online. Theory Comput. Syst. 54(2), 337–372 (2014)
16. Gil, J.Y., Scott, D.A.: A bijective string sorting transform. CoRR abs/1201.3077 (2012)
17. He, J., Liang, H., Yang, G.: Reversing longest previous factor tables is hard. In: Dehne, F., Iacono, J., Sack, J.-R. (eds.) WADS 2011. LNCS, vol. 6844, pp. 488–499. Springer, Heidelberg (2011)
18. I, T., Inenaga, S., Bannai, H., Takeda, M.: Counting and verifying maximal palindromes. In: Chavez, E., Lonardi, S. (eds.) SPIRE 2010. LNCS, vol. 6393, pp. 135–146. Springer, Heidelberg (2010)
19. I, T., Inenaga, S., Bannai, H., Takeda, M.: Inferring strings from suffix trees and links on a binary alphabet. In: Proc. PSC 2011, pp. 121–130 (2011)
20. I, T., Nakashima, Y., Inenaga, S., Bannai, H., Takeda, M.: Efficient Lyndon factorization of grammar compressed text. In: Fischer, J., Sanders, P. (eds.) CPM 2013. LNCS, vol. 7922, pp. 153–164. Springer, Heidelberg (2013)
21. I, T., Nakashima, Y., Inenaga, S., Bannai, H., Takeda, M.: Faster lyndon factorization algorithms for SLP and LZ78 compressed text. In: Kurland, O., Lewenstein, M., Porat, E. (eds.) SPIRE 2013. LNCS, vol. 8214, pp. 174–185. Springer, Heidelberg (2013)
22. Kufleitner, M.: On bijective variants of the Burrows-Wheeler transform. In: Proc. PSC 2009, pp. 65–79 (2009)
23. Lyndon, R.C.: On Burnside's problem. Transactions of the American Mathematical Society 77, 202–215 (1954)
24. Matsubara, W., Ishino, A., Shinohara, A.: Inferring strings from runs. In: Proc. PSC 2010, pp. 150–160 (2010)
25. Moore, D., Smyth, W.F., Miller, D.: Counting distinct strings. Algorithmica 23(1), 1–13 (1999)
26. Schürmann, K.B., Stoye, J.: Counting suffix arrays and strings. Theoretical Computer Science 395(2-3), 220–234 (2008)

Betweenness Centrality – Incremental and Faster*

Meghana Nasre[1], Matteo Pontecorvi[2], and Vijaya Ramachandran[2]

[1] Indian Institute of Technology Madras, India
meghana@cse.iitm.ac.in
[2] University of Texas at Austin, USA
{cavia,vlr}@cs.utexas.edu

Abstract. We present an incremental algorithm that updates the betweenness centrality (BC) score of all vertices in a graph G when a new
edge is added to G, or the weight of an existing edge is reduced. Our
incremental algorithm runs in $O(\nu^* \cdot n)$ time, where ν^* is bounded by
m^*, the number of edges that lie on a shortest path in G. We achieve
the same bound for the more general incremental vertex update problem.
Even for a single edge update, our incremental algorithm is the first algorithm that is provably faster on sparse graphs than recomputing with the
well-known static Brandes algorithm. It is also likely to be much faster
than Brandes on dense graphs since m^* is often close to linear in n.

Our incremental algorithm is very simple, and we give an efficient
cache-oblivious implementation that incurs $O(n \cdot sort(\nu^*))$ cache misses,
where *sort* is a well-known measure for caching efficiency.

1 Introduction

Betweenness centrality (BC) of vertices is a widely-used measure in the analysis
of large complex networks. As a classical measure, BC is widely used in sociology
[6,22], biology [9], physics [17] and network analysis [32,34]. BC is also useful
for critical applications such as identifying lethality in biological networks [31],
identifying key actors in terrorist networks [20] and finding attack vulnerability
of complex networks [14]. In recent years, BC also had a wide impact in the
analysis of social networks [11,33], wireless [25] and mobile networks [4], P2P
networks [18] and more.

In this paper we present incremental BC algorithms that are provably faster
on sparse graphs than current algorithms for the problem. By an *incremental
update on edge* (u, v) we mean the addition of a new edge (u, v) with finite weight
if (u, v) is not present in the graph, or a decrease in the weight of an existing
edge (u, v); in an *incremental vertex update*, incremental updates can occur on
any subset of edges incident to v, and this includes adding new edges.

We now define BC, and describe the widely-used Brandes algorithm [3] for
this problem. We then describe our contributions and related work.

---

* This work was supported in part by NSF grants CCF-0830737 and CCF-1320675.

E. Csuhaj-Varjú et al. (Eds.): MFCS 2014, Part II, LNCS 8635, pp. 577–588, 2014.
© Springer-Verlag Berlin Heidelberg 2014

Betweenness Centrality (BC) and the Brandes Algorithm. Let $G = (V, E)$ be a (directed or undirected) graph with positive edge weights $\mathbf{w}(e)$, $e \in E$. The distance $d(s, t)$ from s to t is the weight of a shortest path from s to t. Let σ_{st} be the number of shortest paths from s to t in G (with $\sigma_{ss} = 1$) and let $\sigma_{st}(v)$ be the number of shortest paths from s to t that pass through v. Thus, $\sigma_{st}(v) = \sigma_{sv} \cdot \sigma_{vt}$ if $d(s, t) = d(s, v) + d(v, t)$, and $\sigma_{st}(v) = 0$ otherwise.

The *pair dependency* of s, t on an intermediate vertex v is $\delta_{st}(v) = \frac{\sigma_{st}(v)}{\sigma_{st}}$ [3]. For $v \in V$, the *betweenness centrality* $\mathrm{BC}(v)$ is defined by Freeman [6] as:

$$\mathrm{BC}(v) = \sum_{s \neq v, t \neq v} \frac{\sigma_{st}(v)}{\sigma_{st}} = \sum_{s \neq v, t \neq v} \delta_{st}(v) \qquad (1)$$

Let $P_s(v)$ denote the predecessors of v on shortest paths from s. Brandes [3] defined the dependency of a vertex s on a vertex v as $\delta_{s\bullet}(v) = \sum_{t \in V \setminus \{v, s\}} \delta_{st}(v)$, and observed that

$$\delta_{s\bullet}(v) = \sum_{w : v \in P_s(w)} \frac{\sigma_{sv}}{\sigma_{sw}} \cdot (1 + \delta_{s\bullet}(w)) \quad \text{and} \quad \mathrm{BC}(v) = \sum_{s \neq v} \delta_{s\bullet}(v) \qquad (2)$$

Alg. 1 gives Brandes' algorithm to compute $BC(v)$ for all $v \in V$. This algorithm runs in $O(mn + n^2 \log n)$ time, where $|V| = n$ and $|E| = m$.

Algorithm 1. Betweenness-centrality$(G = (V, E))$ (from Brandes [3])

1: **for** every $v \in V$ **do** $\mathrm{BC}(v) \leftarrow 0$.
2: **for** every $s \in V$ **do**
3:      Run Dijkstra's SSSP from s and compute σ_{st} and $P_s(t), \forall\, t \in V \setminus \{s\}$.
4:      Store the explored nodes in a stack S in non-increasing distance from s.
5:      Accumulate dependency of s on all $t \in S$ using Eqn. 2.

1.1   Our Contributions

Let E^* be the set of edges in G that lie on shortest paths, let $m^* = |E^*|$, and let ν^* be the maximum number of edges that lie on shortest paths through any single vertex. Here is our main result:

Theorem 1. *After an incremental update on an edge or a vertex in a directed or undirected graph with positive edge weights, the betweenness centrality of all vertices can be recomputed in:*
   1. $O(\nu^* \cdot n)$ time using $O(\nu^* \cdot n)$ space;
   2. $O(m^* \cdot n)$ time using $O(n^2)$ space.

Since $\nu^* \leq m^*$ and $m^* \leq m$, the worst case time for both results is bounded by $O(mn + n^2)$, which is a $\log n$ factor improvement over Brandes' algorithm on sparse graphs. Our results also have benefits for dense graphs (when $m = \omega(n \log n)$) similar to the Hidden Paths algorithm of Karger et al. [15] for the all pairs shortest paths (APSP) problem (see also McGeoch [26]), although our techniques are different. This is through the use of ν^* or m^* in place of m, and we comment more on this below. Our algorithms are simple, and only use stack, queue and linked list data structures. We also give an efficient cache-oblivious implementation which avoids the high caching cost of Dijkstra's algorithm that is present in Alg. 1 (its bound is given in Section 5).

Both ν^* and m^* are typically much smaller than m in dense graphs. For instance, it is known [7,13,15,23] that $m^* = O(n \log n)$ with high probability in a complete graph where edge weights are chosen from a large class of probability distributions, including the uniform distribution on integers in $[1, n^2]$ or reals in $[0, 1]$. For such graphs, both results in Theorem 1 imply an $O(n^2 \log^2 n)$ algorithm for an incremental update. For the random real weights, the first result would in fact give $O(n^2)$ time and space since shortest paths are unique with probability 1 in this setting, hence $\nu^* = O(n)$.

We observe that Alg. 1 (Brandes) can be made to run faster: In a directed graph, by using the Pettie [29] or the Hidden Paths algorithm in place of Dijkstra in Step 3 of Alg. 1, we can compute BC scores in $O(mn + n^2 \log \log n)$ or $O(m^*n + n^2 \log n)$ time, respectively. In an undirected graph, we can obtain $O(mn \cdot \log \alpha(m, n))$ time, where α is an inverse-Ackermann function, using [30]. Our incremental bounds are better than any of these bounds for sparse graphs.

There are several results on dynamic BC algorithms and heuristics [10,12,19,21], but our time bounds are better than any of these on sparse graphs. In fact, ours is the first incremental BC algorithm that gives a provable improvement over Brandes' algorithm for sparse graphs, which are the type of graphs that typically occur in practice. While the space used by our algorithms is higher than Brandes', which uses only linear space, our second result matches the best space bound obtained by any of these other dynamic BC algorithms and heuristics.

We consider only incremental updates in this paper. Computing decremental and fully dynamic updates efficiently appears to be more challenging (as is the case for APSP [5]). In recent work [28], we have developed decremental and fully dynamic BC algorithms that build on techniques in [5], and run in amortized time $O(\nu^{*2} \cdot polylog(n))$.

Organization. In Section 2 we discuss related work on dynamic BC. Since the algorithm for a single edge update is simpler than that for a vertex update, we first present our edge update result in Section 3. We describe the $O(n \cdot \nu^*)$ algorithm, and then the simple changes needed to obtain the second $O(n^2)$ space result. We present the vertex update result in Section 4. In Section 5 we sketch our efficient cache-oblivious BC algorithm, and mention some preliminary experimental results.

Step 5 of Alg. 1: For completeness, Alg. 2 below gives the algorithm for Step 5 in Brandes' algorithm (Alg. 1). We will use Alg. 2 unchanged for our first result of Theorem 1, and modified (to eliminate the use of predecessor lists $P_s(t)$) for the second result of Theorem 1.

2   Related Work

Approximation and parallel algorithms for BC have been considered in [2,8], [24] respectively. More recently, the problem of dynamic betweenness centrality has

---

Algorithm 2. Accumulate-dependency(s, S) (from [3])

Input: For every $t \in V$: $\sigma_{st}, P_s(t)$.
        A stack S containing $v \in V$ in a suitable order (non-increasing $d(s,v)$ in [3]).
1: **for** every $v \in V$ **do** $\delta_{s\bullet}(v) \leftarrow 0$.
2: **while** $S \neq \emptyset$ **do**
3:    $w \leftarrow \text{pop}(S)$.
4:    **for** $v \in P_s(w)$ **do** $\delta_{s\bullet}(v) \leftarrow \delta_{s\bullet}(v) + \frac{\sigma_{sv}}{\sigma_{sw}} \cdot (1 + \delta_{s\bullet}(w))$.
5:    **if** $w \neq s$ **then** $\text{BC}(w) \leftarrow \text{BC}(w) + \delta_{s\bullet}(w)$.

---

received attention, and these results for incremental and in some cases, decremental, BC are listed in the table below. All of these results except [16] deal with unweighted graphs as opposed to our results, which are for the weighted case. Further, while all give encouraging experimental results or match the Brandes worst-case time complexity, none prove any worst-case improvement. As mentioned in the Introduction, BC is also widely used in weighted networks (see [4,18,31,32]); however, only the heuristic in Kas et al. [16], which has no worst-case bounds, addresses this version.

Paper	Year	Space	Time	Weights	Update Type
Brandes static [3]	2001	$O(m+n)$	$O(mn)$	NO	Static Alg.
Lee et al. [21]	2012	$O(n^2+m)$	Heuristic	NO	Single Edge
Green et al. [12]	2012	$O(n^2+mn)$	$O(mn)$	NO	Single Edge
Kourtellis+ [19]	2014	$O(n^2)$	$O(mn)$	NO	Single Edge
Singh et al. [10]	2013	–	Heuristic	NO	Vertex update
Brandes static [3]	2001	$O(m+n)$	$O(mn + n^2 \log n)$	YES	Static Alg.
Kas et al. [16]	2013	$O(n^2+mn)$	Heuristic	YES	Single Edge
This paper	2014	$O(\nu^* \cdot n)$	$O(\nu^* \cdot n)$	YES	Vertex Update
This paper	2014	$O(n^2)$	$O(m^* \cdot n)$	YES	Vertex Update

Our first algorithm, which takes time $O(\nu^* \cdot n)$ in a weighted graph even for a vertex update, improves on all previous results when $\nu^* = o(m)$. By slightly relaxing the time complexity to $O(m^* \cdot n)$, we are also able to match the best space complexity in any of the previous results, while matching their time complexities and improving on all of them when $m^* = o(m)$.

3   Incremental Edge Update

In this section we present our algorithm to recompute BC scores of all vertices in a directed graph $G = (V, E)$ after an incremental edge update (i.e., adding an edge or decreasing the weight of an existing edge). Let $G' = (V, E')$ denote the graph obtained after an edge update to $G = (V, E)$. A path π_{st} from s to t in G has *weight* $\mathbf{w}(\pi_{st}) = \sum_{e \in \pi_{st}} \mathbf{w}(e)$. Let $d(s,t), \sigma_{st}, \delta_{s\bullet}(t)$ and $\text{DAG}(s)$ denote the distance from s to t in G, the number of shortest paths from s to t in G, the dependency of s on t and the SSSP DAG rooted at s in G respectively; let $d'(s,t), \sigma'_{st}, \delta'_{s\bullet}(t)$ and $\text{DAG}'(s)$ denote these parameters in G'.

Lemma 1. *If weight of edge (u, v) in G is decreased to obtain G', then for any $x \in V$, the set of shortest paths from x to u and from v to x is the same in G and G', and $d'(x, u) = d(x, u)$, $d'(v, x) = d(v, x)$; $\sigma'_{xu} = \sigma_{xu}$, $\sigma'_{vx} = \sigma_{vx}$.*

Proof. Since edge weights are positive, the edge (u, v) cannot lie on a shortest path to u or from v. The lemma follows. □

By Lemma 1, $DAG(v) = DAG'(v)$ after the decrease of weight on edge (u, v). The next lemma shows that after the weight of (u, v) is decreased we can efficiently obtain the updated values $d'(s, t)$ and σ'_{st} for any $s, t \in V$.

Lemma 2. *Let the weight of edge (u, v) be decreased to $\mathbf{w}'(u, v)$, and for any given pair of vertices s, t, let $D(s, t) = d(s, u) + \mathbf{w}'(u, v) + d(v, t)$. Then,*

1. *If $d(s, t) < D(s, t)$, then $d'(s, t) = d(s, t)$ and $\sigma'_{st} = \sigma_{st}$.*
   The shortest paths from s to t in G' are the same as in G.
2. *If $d(s, t) = D(s, t)$, then $d'(s, t) = d(s, t)$ and $\sigma'_{st} = \sigma_{st} + (\sigma_{su} \cdot \sigma_{vt})$.*
   The shortest paths from s to t in G' are a superset of the shortest paths G.
3. *If $d(s, t) > D(s, t)$, then $d'(s, t) = D(s, t)$ and $\sigma'_{st} = \sigma_{su} \cdot \sigma_{vt}$.*
   The shortest paths from s to t in G' are new (shorter distance).

Proof. Case 1 holds because the shortest path distance from s to t remains unchanged and no new shortest path is created in this case. In case 2, the shortest path distance from s to t remains unchanged, but there are $\sigma_{su} \cdot \sigma_{vt}$ new shortest paths from s to t created via edge (u, v). In case 3, the shortest path distance from s to t decreases and all new shortest paths pass through (u, v). □

By Lemma 2, the updated values $d'(s, t)$ and σ'_{st} can be computed in constant time for each pair s, t. Once we have the updated $d'(\cdot)$ and $\sigma'_{(\cdot)}$ values, we need the updated predecessors $P'_s(t)$ for every s, t pair for Alg. 2. The SSSP $DAG(s)$ rooted at a source s is the union of all the $P_s(t), \forall \, t \in V$. Thus, obtaining $DAG'(s)$ after the edge update is equivalent to computing the $P'_s(t), \forall \, t \in V$. The next section gives a simple algorithm to maintain the SSSP DAGs rooted at every source $s \in V$, after an incremental edge update.

3.1   Updating an SSSP DAG

For each pair s, t we define $flag(s, t)$ to indicate the specific case of Lemma 2 that is applicable.

$$flag(s, t) = \begin{cases} \text{UN-changed} & \text{if } d'(s, t) = d(s, t) \text{ and } \sigma'_{st} = \sigma_{st} \text{ (Lemma 2-1)} \\ \text{NUM-changed} & \text{if } d'(s, t) = d(s, t) \text{ and } \sigma'_{st} > \sigma_{st} \text{ (Lemma 2-2)} \\ \text{WT-changed} & \text{if } d'(s, t) < d(s, t) \text{ (Lemma 2-3)} \end{cases}$$

By Lemma 2, $flag(s, t)$ can be computed in constant time for each pair s, t. Given an input s and the updated edge (u, v), Alg. 3 (Update-DAG) constructs a set of edges H using these $flag$ values, together with $DAG(s)$ and $DAG(v)$. We will show that H contains exactly the edges in $DAG'(s)$. The algorithm considers edges in DAG(s) (Steps 3–5) and edges in DAG(v) (Steps 6–8), and for each edge (a, b) in either DAG, it decides whether to include it in H based on the value of $flag(s, b)$. For the updated edge (u, v) there is a separate check (Steps 9–10). The algorithm clearly takes time linear in the size of DAG(s) and DAG(v), i.e., $O(\nu^*)$ time.

Algorithm 3. Update-DAG($s, \mathbf{w}'(u, v)$)

Input: DAG(s), DAG(v), and $flag(s, t), \forall t \in V$.
Output: An edge set H after decrease of weight on edge (u, v), and $P_s'(t), \forall t \in V - \{s\}$.
1: $H \leftarrow \emptyset$.
2: **for** each $v \in V$ **do** $P_s'(v) = \emptyset$.
3: **for** each edge $(a, b) \in$ DAG(s) and $(a, b) \neq (u, v)$ **do**
4:      **if** $flag(s, b) = $ UN-changed or $flag(s, b) = $ NUM-changed **then**
5:          $H \leftarrow H \cup \{(a, b)\}$ and $P_s'(b) \leftarrow P_s'(b) \cup \{a\}$.
6: **for** each edge $(a, b) \in$ DAG(v) **do**
7:      **if** $flag(s, b) = $ NUM-changed or $flag(s, b) = $ WT-changed **then**
8:          $H \leftarrow H \cup \{(a, b)\}$ and $P_s'(b) \leftarrow P_s'(b) \cup \{a\}$.
9: **if** $flag(s, v) = $ NUM-changed or $flag(s, v) = $ WT-changed **then**
10:      $H \leftarrow H \cup \{(u, v)\}$ and $P_s'(v) \leftarrow P_s'(v) \cup \{u\}$.

Lemma 3. *Let H be the set of edges output by Alg. 3. An edge $(a, b) \in H$ if and only if $(a, b) \in DAG'(s)$.*

Proof. Since the update is an incremental update on edge (u, v), we note that for any b, a shortest path π'_{sb} from s to b in G' can be of two types:
(i) π'_{sb} is a shortest path in G. Therefore every edge on such a path is present in DAG(s) and each such edge is added to H in Steps 3–5 of Alg. 3.
(ii) π'_{sb} is not a shortest path in G. However, since π'_{sb} is a shortest path in G', therefore π'_{sb} is of the form $s \rightsquigarrow u \rightarrow v \rightsquigarrow b$. Since shortest paths from s to u in G and G' are unchanged (by Lemma 1), the edges in the sub-path $s \rightsquigarrow u$ are present in DAG(s) and are added to H in Steps 3–5 of Alg. 3. Finally, shortest paths from v to any b in G and G' remain unchanged. Thus, the edges in the sub-path $v \rightsquigarrow b$ are present in DAG(v) and are added to H in Steps 6–8 of Alg. 3.

For the other direction, if the edge (a, b) is added to H by Step 5, this implies that the edge $(a, b) \in$ DAG(s). Thus, there exists a shortest path $\pi_{sb} = s \rightsquigarrow a \rightarrow b$ in G. We execute Step 5 when $flag(s, b) = $ UN-changed or $flag(s, b) = $ NUM-changed. Thus every shortest path from s to b in G is also shortest path in G'. Therefore, $(a, b) \in$ DAG$'(s)$. If the edge (a, b) is added to H by Step 8, then the edge $(a, b) \in$ DAG(v). Thus, there exists a shortest path $\pi_{vb} = v \rightsquigarrow a \rightarrow b$ in G. Since decreasing the weight of the edge (u, v) does not change shortest paths from v to any other vertex, π_{vb} is in G'. We execute Step 8 when $flag(s, b) = $ NUM-changed or $flag(s, b) = $ WT-changed. Therefore, there exists at least one shortest path from s to b in G' that uses the updated edge (u, v). Hence the path $\pi'_{sb} = \pi'_{su} \cdot (u, v) \cdot \pi_{vb}$ is shortest in G', and this establishes that $(a, b) \in$ DAG$'(s)$. Finally, edge (u, v) is added to H by Step 10 only if $flag(s, v)$ is NUM-changed or WT-changed, and in either case, there is at least a new shortest path from s to v through (u, v). Hence $(u, v) \in$ DAG$'(s)$. □

3.2 Updating Betweenness Centrality Scores

The algorithm for updating the BC scores after an edge update (Alg. 4) is similar to Alg. 1, but with the following changes: an extended Step 1 also computes, for every s, t pair, the updated $d'(s, t)$ and σ'_{st}, as well as $flag(s, t)$. Using Lemma 2, we spend constant time for each s, t pair, hence $O(n^2)$ time for all pairs. In Step 3, instead of Dijkstra's algorithm, we run Alg. 3 to obtain the updated predecessor lists $P'_s(t)$, for all s, t. This step requires time $O(\nu^*)$ for a source s, and $O(\nu^* \cdot n)$ over all sources. The last difference is in Step 4: we place in the stack S the vertices in reverse topological order in $DAG'(s)$, instead of non-increasing distance from s. This requires time linear in the size of the updated DAG. Thus the time complexity of Edge-Update is $O(\nu^* \cdot n)$.

Algorithm 4. Edge-Update$(G = (V, E), \mathbf{w}'(u, v))$

Input: updated edge with $\mathbf{w}'(u, v)$, $d(s, t)$ and σ_{st}, $\forall\, s, t \in V$; $DAG(s), \forall\, s \in V$.
Output: $BC'(v), \forall\, v \in V$; $d'(s, t)$ and $\sigma'_{st}\ \forall\, s, t \in V$; $DAG'(s), \forall\, s \in V$.
1: **for** every $v \in V$ **do** $BC'(v) \leftarrow 0$.
    for every $s, t \in V$ **do** compute $d'(s, t), \sigma'_{st}, flag(s, t)$.   // use Lemma 2
2: **for** every $s \in V$ **do**
3:      Update-DAG$(s, (u, v))$.   // use Alg. 3
4:      Stack $S \leftarrow$ vertices in V in a reverse topological order in $DAG'(s)$.
5:      Accumulate-dependency(s, S).   // use Alg. 2

Undirected Graphs. For an undirected G, we construct the corresponding directed graph G_D in which every undirected edge is replaced with 2 directed edges. An incremental update on an undirected edge (u, v) is equivalent to two edge updates on (u, v) and (v, u) in G_D. Thus, Theorem 1 holds for undirected graphs.

Space Efficient Implementation. In order to obtain $O(n^2)$ space complexity, we do not store the SSSP DAGs rooted at every source. Instead, we only store the edge set E^*. After an incremental update on edge (u, v) we first construct the updated set E'^* in $O(m^* \cdot n)$ time as follows. For each edge $(a, b) \in E^*$, if $d'(s, b) = d(s, a) + \mathbf{w}(a, b)$ for some source $s \in V$, then $(a, b) \in E'^*$. Using the updated E'^* we can construct $DAG'(s)$ in $O(m^*)$ time, by using the fact that an edge $(a, b) \in E'^*$ belongs to $DAG'(s)$ iff $d(s, b) = d(s, a) + \mathbf{w}(a, b)$. Since the construction of each updated DAG takes $O(m^*)$ time and there are n DAGs to be constructed, the $O(m^* \cdot n)$ time complexity follows. The space used is $O(m^* + n^2)$ to store E^* and $d(s, t)$, σ_{st}, for all $s, t \in V$.

4 Incremental Vertex Update

We now consider an incremental update to a vertex v in $G = (V, E)$, which allows an incremental edge update on any subset of edges incoming to and outgoing from v. In this algorithm, we use the graph G and the graph $G_R = (V, E_R)$, which is obtained by reversing every edge in G, i.e., $(a, b) \in E_R$ iff $(b, a) \in E$. Thus, for every $s \in V$, we also maintain $DAG_R(s)$, the SSSP DAG rooted at s in G_R. We will obtain the same time bound as in Section 3.

4.1   Overview

Let $E_i(v)$ and $E_o(v)$ denote the set of updated edges incoming to v and outgoing from v respectively. Our algorithm is a natural extension, with some new features, of the algorithm for a single edge update, and works as follows. We process $E_i(v)$ in G in Step 1 to form G', G'_R, $\mathrm{DAG}'(s)$ and $\mathrm{DAG}'_R(s)$; we then process $E_o(v)$ in G'_R in a complementary Step 2 to obtain the updated G'', $\mathrm{DAG}''(s)$ and $\mathrm{DAG}''_R(s)$. Step 1, which processes $E_i(v)$, consists of two phases.

Step 1, Phase 1: Constructing the $\mathrm{DAG}'(s)$ for updates in $E_i(v)$.
Since $E_i(v)$ contains updated edges incoming to v, $\mathrm{DAG}(v) = \mathrm{DAG}'(v)$ (as in the single edge update case). In order to handle updates to several edges incoming to v, we strengthen Lemma 2 by introducing $\hat{\sigma}$, which keeps track of new shortest paths from s to v that go through any of the updated edges in $E_i(v)$. This allows us to efficiently recompute the number of shortest paths from a source to any node in G', and thus update all the $\mathrm{DAG}'(s)$ using an algorithm similar to Alg. 3. Parts (A), (B), (C) in Section 4.2 describe Phase 1 in detail.

Step 1, Phase 2: Constructing the $\mathrm{DAG}'_R(s)$ for updates in $E_i(v)$.
We present an efficient algorithm to construct the $\mathrm{DAG}'_R(s)$ for all s in G'. We construct these reverse graphs because the edges in $E_o(v)$ are in fact incoming edges to v in G'_R. Hence our method to maintain DAGs when incoming edges are updated can be applied to G'_R with E_o to obtain $\mathrm{DAG}''_R(s)$, for every s, in Phase 1 of Step 2 (and then we can obtain the $\mathrm{DAG}''(s)$ in Phase 2 of Step 2).

Let (t, a) be the first edge on a shortest path from t to v in G'. Then (t, a) is an outgoing edge from t in $\mathrm{DAG}'(t)$, and its reverse (a, t) is on a shortest path from v to t in G'_R. Further an edge (a, t) is on a new shortest path from v to t in G'_R if and only if its reverse is on a new shortest path from t to v in G'. These edges on new shortest paths are the ones we need to keep track of in order to update the reverse DAGs, and to facilitate this we define a collection of sets R_t, $t \in V$. The set R_t is the set of (reversed) outgoing edges from t in $\mathrm{DAG}'(t)$ that lie on a shortest path from t to v in G' (see also Eqn. 5 in the next section). Thus, if a new shortest path π_{sb} is present in $\mathrm{DAG}'_R(s)$ (π_{sb} must pass through v), its last edge (a, b) is present in R_b. Using the sets $R_t, \forall\, t \in V$, it is possible to quickly build the $\mathrm{DAG}'_R(t)$ after Phase 1 as shown in part (D) in section 4.2.

Step 2: After applying Phase 1 and 2 on the initial DAGs using E_i to obtain the $\mathrm{DAG}'_R(s)$ and G'_R, Step 2 re-applies Phase 1 and Phase 2 on these updated graphs using E_o in order to complete all of the updates to vertex v. We can then apply Alg. 2 to the $\mathrm{DAG}''(s)$ to obtain the BC scores for the updated graph G''.

4.2   Vertex Update Algorithm

We now give details of each phase of our algorithm starting with the graph G.

Step 1, Phase 1
(A) **Compute $d'(s, v)$ and σ'_{sv} for any s.** We show how to compute in G' the distance and number of shortest paths to v from any s. Let $(u_j, v) \in E_i(v)$

and let $D_j(s,v) = d(s,u_j) + \mathbf{w}'(u_j,v)$. Since the updates on edges in $E_i(v)$ are incremental, it follows that:

$$d'(s,v) = \min\{d(s,v), \min_{j:(u_j,v)\in E_i(v)}\{D_j(s,v)\}\} \tag{3}$$

Further, if $d'(s,v) = d(s,v)$, we define:

$$\widehat{\sigma}'_{sv} = |\{\pi'_{sv} : \pi'_{sv} \text{ is a shortest path in } G' \text{ and } \pi'_{sv} \text{ uses } e \in E_i(v)\}| \tag{4}$$

We also need to compute σ'_{sv}, the number of shortest paths from s to v in G'. It is straightforward to compute $d'(s,v)$, σ'_{sv}, and $\widehat{\sigma}'_{sv}$ in $O(|E_i(v)|)$ time. Alg. 5 gives the details of this step.

Algorithm 5. Dist-to-v $(s, E_i(v))$	**Algorithm 6.** Upd-Rev-DAG$(s, E_i(v))$
Input: $E_i(v)$ with updated weights $\mathbf{w}'$. $d(s,t)$ and σ_{st}, $\forall s,t \in V$.	**Input:** $\text{DAG}_R(s)$; $R_t, flag(s,t), \forall t \in V$.
Output: $d'(s,v), \sigma'_{sv}, \widehat{\sigma}'_{sv}$.	**Output:** An edge set X after update on edges in $E_i(v)$.
1: $\widehat{\sigma}'_{sv} \leftarrow 0$, $\sigma'_{sv} \leftarrow \sigma_{sv}$, $D' \leftarrow d(s,v)$.	1: $X \leftarrow \emptyset$.
2: **for** each edge $(u_i,v) \in E_i(v)$ **do**	2: **for** each edge $(a,b) \in \text{DAG}_R(s)$ **do**
3: **if** $D' = d(s,u_i) + \mathbf{w}'(u_i,v)$ **then**	3: **if** $flag(b,s) = $ UN-changed or $flag(b,s) = $ NUM-changed **then**
4: $\sigma'_{sv} \leftarrow \sigma'_{sv} + \sigma_{su_i}$.	4: $X \leftarrow X \cup (a,b)$.
5: $\widehat{\sigma}'_{sv} \leftarrow \widehat{\sigma}'_{sv} + \sigma_{su_i}$.	5: **for** each $b \in V \setminus \{s\}$ **do**
6: **else if** $D' > d(s,u_i) + \mathbf{w}'(u_i,v)$ **then**	6: **if** $flag(b,s) = $ NUM-changed or $flag(b,s) = $ WT-changed **then**
7: $D' \leftarrow d(s,u_i) + \mathbf{w}'(u_i,v)$.	7: $X \leftarrow X \cup R_b$.
8: $\sigma'_{sv} \leftarrow \sigma_{su_i}$.	
9: $d'(s,v) \leftarrow D'$.	

(B) **Compute $d'(s,t)$ and $\sigma'(s,t)$ for all s,t.** After computing $d'(s,v), \sigma'_{sv}$ and $\widehat{\sigma}'_{sv}$, we show that the values $d'(s,t)$ and $\sigma'(s,t)$ can be computed efficiently. We state Lemma 4 which captures this computation. The proof of this lemma is similar to Lemma 2 in the edge update case.

Lemma 4. Let $E_i(v)$ be the set of updated edges incoming to v. Let G' be the graph obtained by applying the updates in $E_i(v)$ to G. For any $s \in V$ and $t \in V \setminus \{v\}$, let $D(s,t) = d'(s,v) + d(v,t)$, $\Sigma_{st} = \sigma_{st} + \widehat{\sigma}'_{sv} \cdot \sigma_{vt}$, $\Sigma'_{st} = \sigma_{st} + \sigma'_{sv} \cdot \sigma_{vt}$.

1. If $d(s,t) < D(s,t)$, then $d'(s,t) = d(s,t)$ and $\sigma'_{st} = \sigma_{st}$.
2. If $d(s,t) = D(s,t)$ and $d(s,v) = d'(s,v)$, then $d'(s,t) = d(s,t)$ and $\sigma'_{st} = \Sigma_{st}$.
3. If $d(s,t) = D(s,t)$ and $d(s,v) > d'(s,v)$, then $d'(s,t) = d(s,t)$ and $\sigma'_{st} = \Sigma'_{st}$.
4. If $d(s,t) > D(s,t)$, then $d'(s,t) = D(s,t)$ and $\sigma'_{st} = \sigma'_{sv} \cdot \sigma_{vt}$.

The value $flag(s,t)$ for every s,t can be computed using the updated distances and number of shortest paths ($flag(s,t)$ is UN-changed for 1, NUM-changed for both 2 and 3, and WT-changed for 4, in Lemma 4).

(C) **Compute DAG$'(s)$ for every s.** Given $d'(s,t)$ and $\sigma'(s,t)$ updated for all $s,t \in V$, the algorithm to compute DAG$'(s)$ for any $s \in V$ is similar to Alg. 3 in the edge update case. The only modification we need is in Steps 9–10 where instead of a single edge (u,v), we consider every edge $(u_1,v) \in E_i(v)$.

Step 1, Phase 2

(D) **Compute DAG$'_R(s)$ for every** s. We update DAG$_R(s)$, for every s, for which we use Alg. 6. Recall the sets $R_t, \forall\, t \in V$ defined as:

$$R_t = \{(a,t) \mid (t,a) \in \text{DAG}'(t) \text{ and } \mathbf{w}'(t,a) + d'(a,v) = d'(t,v)\} \qquad (5)$$

The set R_t is the set of (reversed) outgoing edges from t in DAG$'(t)$ that lie on a shortest path from t to v in G'. Consider an edge $e = (a,b)$ in the updated DAG$'_R(s)$. If e is in DAG$_R(s)$, it is added to DAG$'_R(s)$ by Steps 2–4. If e lies on a new shortest path present only in G'_R, its reverse must also lie on a shortest path that goes through v in G', and it will be added to DAG$'_R(s)$ by the R_b during Steps 5–7 (R_b could also contain edges on old shortest paths through v already processed in Steps 2–4, but even in that case each edge is added to DAG$'_R(s)$ at most twice by Alg. 6). Note that we do not need to process edges (u_j, v) in E_i separately (as with edge (u,v) in Alg. 2), because these edges will be present in the relevant R_{u_j}. The correctness of Alg. 6 follows from Lemma 5, whose proof is similar to Lemma 3, and is omitted.

Lemma 5. *In Alg. 6, an edge (a,b) is placed in X if and only if $(a,b) \in DAG'_R(s)$ after the incremental update of the set $E_i(v)$.*

Step 2: To process the updates in $E_o(v)$, we re-apply Phase 1 and 2 over G'_R. Since we are processing incoming edges in G'_R, our earlier steps apply unchanged, and we obtain modified values for $d''(\cdot), \sigma''_{(\cdot)}$, and DAG$''_R(s)$ for every s. Then, using Alg. 6 we obtain the DAG$''(s)$ for every s. Finally, to compute the updated BC values, we apply Alg. 2.

Performance: Computing $d'(s,v), \sigma'_{sv}$ and $\hat{\sigma}'_{sv}$ requires time $O(|E_i(v)|) = O(n)$ for each s, and hence $O(n^2)$ time for all sources. Applying Lemma 4 to all pairs of vertices takes time $O(n^2)$. The complexity of modified Alg. 3 applied to all DAGs is again $O(\nu^* \cdot n)$. Creating set R_t requires at most $O(E^* \cap \{\text{outgoing edges of } t\})$, so the overall complexity for all the sets is $O(m^*)$. Finally, we bound the complexity of Algorithm 6: the algorithm adds (a,b) in a reverse DAG edge set X at most twice. Since $\sum_{s \in V} |E(\text{DAG}'(s))| = \sum_{s \in V} |E(\text{DAG}'_R(s))|$, at most $O(\nu^* \cdot n)$ edges can be inserted into all the sets X when Algorithm 6 is executed over all sources. Finally, applying the updates in $E_o(v)$ requires a symmetric procedure starting from the reverse DAGs, the final complexity bound of $O(\nu^* \cdot n)$ follows.

5 Efficient Cache Oblivious Algorithm

We give a cache-oblivious implementation with $O(n \cdot sort(\nu^*))$ cache misses. Here, for a size M cache that can hold B blocks, $sort(r) = \frac{r}{B} \cdot \log_M r$ when $M \geq B^2$; $sort$ is a measure of good caching performance (even though $sort(r)$ performs $r \log r$ operations, the base of M in the log makes $sort(r)$ preferable to, say, r cache misses). In contrast, the Brandes algorithm calls Dijkstra's algorithm, which is affected by unstructured accesses to adjacency lists that lead to large caching costs (see, e.g., [27]).

We consider the basic edge update algorithm. The main change is in the cache-oblivious (CO) implementation of Alg. 2, which is the last step of Alg. 4.

Instead of the stack S, we use an optimal CO max-priority queue Z [1], that is initially empty. Each element in Z has an ordered pair $(d'(s,v),v)$ as its key value, and also has auxiliary data as described below. Consider the execution of Step 4 in Alg. 2 for vertices $v \in P_s(w)$. Instead of computing the contribution of w to $\delta_{s\bullet}(v)$ for each $v \in P_s(w)$ when w is processed, we insert an element into Z with key value $(d'(s,v),v)$ and auxiliary data $(w, \sigma_{sw}, \delta_{s\bullet}(w))$. With this scheme, entries will be extracted from Z in nonincreasing values of $d'(s,v)$, and all entries for a given v will be extracted consecutively. We compute $\delta_{s\bullet}(v)$ as these extractions for v occur from Z, and also update $BC(v)$. Initially, for each sink t in DAG(s), we insert an element with key value $(d'(s,t),t)$ and NIL auxiliary data. Using [1], Alg. 2 (which is Step 6 in Alg. 4) takes $sort(\nu^*)$ cache misses for source s, and hence $O(n \cdot sort(\nu^*))$ over all sources. The earlier steps in Alg. 4 can be performed in $O(n \cdot sort(\nu^*))$ cache misses by suitably storing and rearranging data for cache-efficiency.

Preliminary experimental results for our basic edge update algorithm (in Section 3) on random graphs generated using the Erdős-Rényi model give 2 to 15 times speed-up over Brandes' algorithm for graphs with 256 to 2048 nodes, with the larger speed-ups on dense graphs.

Acknowledgment. We thank Varun Gangal and Aritra Ghosh at IIT Madras for implementing the algorithms.

References

1. Arge, L., Bender, M.A., Demaine, E.D., Holland-Minkley, B., Munro, J.I.: An optimal cache-oblivious priority queue and its application to graph algorithms. SIAM J. Comput. 36(6), 1672–1695 (2007)
2. Bader, D.A., Kintali, S., Madduri, K., Mihail, M.: Approximating betweenness centrality. In: Bonato, A., Chung, F.R.K. (eds.) WAW 2007. LNCS, vol. 4863, pp. 124–137. Springer, Heidelberg (2007)
3. Brandes, U.: A faster algorithm for betweenness centrality. J. of Mathematical Sociology 25(2), 163–177 (2001)
4. Catanese, S., Ferrara, E., Fiumara, G.: Forensic analysis of phone call networks. Social Network Analysis and Mining 3(1), 15–33 (2013)
5. Demetrescu, C., Italiano, G.F.: A new approach to dynamic all pairs shortest paths. J. ACM 51(6), 968–992 (2004)
6. Freeman, L.C.: A set of measures of centrality based on betweenness. Sociometry 40(1), 35–41 (1977)
7. Frieze, A., Grimmett, G.: The shortest-path problem for graphs with random arc-lengths. Discrete Applied Mathematics 10(1), 57–77 (1985)
8. Geisberger, R., Sanders, P., Schultes, D.: Better approximation of betweenness centrality. In: Proc. ALENEX, pp. 90–100 (2008)
9. Girvan, M., Newman, M.E.J.: Community structure in social and biological networks. Proc. the National Academy of Sciences 99(12), 7821–7826 (2002)
10. Goel, K., Singh, R.R., Iyengar, S., Sukrit: A faster algorithm to update betweenness centrality after node alteration. In: Bonato, A., Mitzenmacher, M., Prałat, P. (eds.) WAW 2013. LNCS, vol. 8305, pp. 170–184. Springer, Heidelberg (2013)
11. Goh, K.-I., Oh, E., Kahng, B., Kim, D.: Betweenness centrality correlation in social networks. Phys. Rev. E 67, 017101 (2003)

12. Green, O., McColl, R., Bader, D.A.: A fast algorithm for streaming betweenness centrality. In: Proc. PASSAT, pp. 11–20 (2012)
13. Hassin, R., Zemel, E.: On shortest paths in graphs with random weights. Mathematics of Operations Research 10(4), 557–564 (1985)
14. Holme, P., Kim, B.J., Yoon, C.N., Han, S.K.: Attack vulnerability of complex networks. Phys. Rev. E 65, 056109 (2002)
15. Karger, D.R., Koller, D., Phillips, S.J.: Finding the hidden path: Time bounds for all-pairs shortest paths. SIAM J. Comput. 22(6), 1199–1217 (1993)
16. Kas, M., Wachs, M., Carley, K.M., Carley, L.R.: Incremental algorithm for updating betweenness centrality in dynamically growing networks. In: Proc. ASONAM, pp. 33–40. ACM (2013)
17. Kitsak, M., Havlin, S., Paul, G., Riccaboni, M., Pammolli, F., Stanley, H.E.: Betweenness centrality of fractal and nonfractal scale-free model networks and tests on real networks. Phys. Rev. E 75, 056115 (2007)
18. Kourtellis, N., Iamnitchi, A.: Leveraging peer centrality in the design of socially-informed peer-to-peer systems. CoRR, abs/1210.6052 (2012)
19. Kourtellis, N., Morales, G.D.F., Bonchi, F.: Scalable online betweenness centrality in evolving graphs. CoRR, abs/1401.6981 (2014)
20. Krebs, V.: Mapping networks of terrorist cells. Connections 24(3), 43–52 (2002)
21. Lee, M.-J., Lee, J., Park, J.Y., Choi, R.H., Chung, C.-W.: Qube: a quick algorithm for updating betweenness centrality. In: Proc. WWW, pp. 351–360 (2012)
22. Leydesdorff, L.: Betweenness centrality as an indicator of the interdisciplinarity of scientific journals. J. Am. Soc. Inf. Sci. Technol. 58(9), 1303–1319 (2007)
23. Luby, M., Ragde, P.: A bidirectional shortest-path algorithm with good average-case behavior. Algorithmica 4(1-4), 551–567 (1989)
24. Madduri, K., Ediger, D., Jiang, K., Bader, D.A., Chavarría-Miranda, D.G.: A faster parallel algorithm and efficient multithreaded implementations for evaluating betweenness centrality on massive datasets. In: Proc. IPDPS, pp. 1–8 (2009)
25. Maglaras, L., Katsaros, D.: New measures for characterizing the significance of nodes in wireless ad hoc networks via localized path-based neighborhood analysis. Social Network Analysis and Mining 2(2), 97–106 (2012)
26. McGeoch, C.C.: All-pairs shortest paths and the essential subgraph. Algorithmica 13(5), 426–441 (1995)
27. Mehlhorn, K., Meyer, U.: External-memory breadth-first search with sublinear I/O. In: Möhring, R.H., Raman, R. (eds.) ESA 2002. LNCS, vol. 2461, pp. 723–735. Springer, Heidelberg (2002)
28. Nasre, M., Pontecorvi, M., Ramachandran, V.: Decremental and fully dynamic all pairs all shortest paths and betweenness centrality. Manuscript (2014)
29. Pettie, S.: A new approach to all-pairs shortest paths on real-weighted graphs. Theoretical Computer Science 312(1), 47–74 (2004)
30. Pettie, S., Ramachandran, V.: A shortest path algorithm for real-weighted undirected graphs. SIAM J. Comput. 34(6), 1398–1431 (2005)
31. Pinney, J.W., McConkey, G.A., Westhead, D.R.: Decomposition of biological networks using betweenness centrality. In: Proc. RECOMB. Poster session (2005)
32. Puzis, R., Altshuler, Y., Elovici, Y., Bekhor, S., Shiftan, Y., Pentland, A.S.: Augmented betweenness centrality for environmentally aware traffic monitoring in transportation networks. J. of Intell. Transpor. Syst. 17(1), 91–105 (2013)
33. Ramírez: The social networks of academic performance in a student context of poverty in Mexico. Social Networks 26(2), 175–188 (2004)
34. Singh, B.K., Gupte, N.: Congestion and decongestion in a communication network. Phys. Rev. E 71, 055103 (2005)

Deterministic Parameterized Algorithms for the Graph Motif Problem

Ron Y. Pinter, Hadas Shachnai, and Meirav Zehavi

Department of Computer Science, Technion, Haifa 32000, Israel
{pinter,hadas,meizeh}@cs.technion.ac.il

Abstract. We study the classic GRAPH MOTIF problem: given a graph $G = (V, E)$ with a set of colors for each node, and a multiset M of colors, we seek a subtree $T \subseteq G$, and a coloring of the nodes in T, such that T carries exactly (also with respect to multiplicity) the colors in M. GRAPH MOTIF plays a central role in the study of pattern matching problems, primarily motivated from the analysis of complex biological networks.

Previous algorithms for Graph Motif and its variants either rely on techniques for developing randomized algorithms that, if derandomized, render them inefficient, or the algebraic narrow sieves technique for which there is no known derandomization. In this paper, we present fast *deterministic* parameterized algorithms for GRAPH MOTIF and its variants. Specifically, we give such an algorithm for the more general GRAPH MOTIF WITH DELETIONS problem, followed by faster algorithms for GRAPH MOTIF and other well-studied special cases. Our algorithms make non-trivial use of *representative families*, and a novel tool that we call *guiding trees*, together enabling the efficient construction of the output tree.

1 Introduction

With the advent of network biology and complex network analysis in general, the study of pattern matching problems in graphs has become of major importance [12,16]. Indeed, the term "graph motif" plays a central role in this context, with different node colors used to model different functionalities of the network (see, e.g., [17,7]). Due to the generic nature of the GRAPH MOTIF (GM) problem (also known as the TOPOLOGY-FREE NETWORK QUERY problem), the so called *motif analysis* approach has become useful also in the study of social networks (see, e.g., [23] and the references therein).

The GM problem is a natural variant of classic pattern matching problems, where the topology of the pattern M is unknown or of lesser importance. Given a graph $G = (V, E)$ with a set of colors for each node, and a multiset M of colors, we seek a subtree $T \subseteq G$, and a coloring of the nodes in T, such that T carries exactly (also with respect to multiplicity) the colors in M. We call T an *occurrence* of M in G. To allow more flexibility in the definition of an occurrence, and since biological network data often contains noise, a generalized version of GM allows *deleting* colors from M.

Parameterized algorithms solve NP-hard problems by confining the combinatorial explosion to a parameter k. More precisely, a problem is *fixed-parameter*

E. Csuhaj-Varjú et al. (Eds.): MFCS 2014, Part II, LNCS 8635, pp. 589–600, 2014.
© Springer-Verlag Berlin Heidelberg 2014

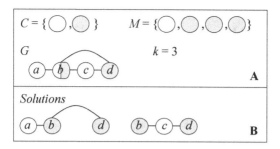

Fig. 1. An input for GM_D (A), and two possible solutions (B)

tractable (FPT) with respect to a parameter k if it can be solved in time $O^*(f(k))$ for some function f, where O^* hides factors polynomial in the input size. Since GM is NP-complete [17], there is a growing body of literature studying its parameterized complexity (see the excellent survey in [26]). In this paper, we present fast *deterministic* parameterized algorithms for GM and its variants.

1.1 Problem Statement

The most general variant considered in this paper is GRAPH MOTIF WITH DELETIONS (GM_D): the input is a set of colors C, a multiset M of colors from C, and an undirected graph $G = (V, E)$. The nodes in V are associated with colors via a (set-)coloring $Col : V \to 2^C$. We are also given a parameter $k \leq |M|$.

We need to decide if there exists a subtree $T = (V_T, E_T)$ of G on k nodes, and a coloring $col : V_T \to C$ that assigns a color from $Col(v)$ to each node $v \in V_T$, such that

$$\forall c \in C : |\{v \in V_T : col(v) = c\}| \leq occ(c), \tag{1}$$

where $occ(c)$ is the number of occurrences of a color c in M (see Fig. 1).[1]

Special Cases: RESTRICTED GM_D (RGM_D) is the special case of GM_D where for any node $v \in V$, $|Col(v)| = 1$. Also, GM and RGM are the special cases of GM_D and RGM_D, respectively, where deletions are not allowed (i.e., the inequality in (1) is replaced by equality, and $k = |M|$).

1.2 Known Results and Our Contribution

GM_D has received considerable attention since it was introduced by Lacroix et al. [17]. The paper [17] also shows that RGM is NP-hard when M is a set and G is a tree. Even seemingly simpler cases of RGM are known to be NP-hard (see [11,2,8]). Moreover, a natural optimization version of RGM_D, minimizing the number of deletions from M, is hard to approximate within factor $|V|^{\frac{1}{3}-\epsilon}$ [24].

---

[1] Some papers define GM_D as a problem where one seeks a connected subgraph S of G, which is equivalent to our definition (simply consider some spanning tree T of S).

On the positive side, using techniques for developing randomized parameterized algorithms, many such algorithms have been obtained for GM_D and its variants [3,4,6,7,9,10,14,15,21,22]. Some of these algorithms can be derandomized, resulting, however, in inefficient algorithms. In particular, Fellows et al. [10] gave a deterministic algorithm for RGM that runs in time $O^*(87^k)$, based on a derandomization of the color coding technique [1]. Currently, the best randomized algorithm for GM_D runs in time $O^*(2^k)$, due to Björklund et al. [6]. This algorithm is based on the narrow sieves technique [5], for which there is no known derandomization. Thus, previous studies left open the existence of a *fast* deterministic parameterized algorithm for GM_D.

In this paper, we present fast deterministic parameterized algorithms for GM_D and its variants. In particular, we develop an $O^*(6.86^k)$ time algorithm for GM_D, an $O^*(5.22^k)$ time algorithm for GM, and an $O^*(5.18^k)$ time algorithm for RGM_D.

Due to space constraints, some of the proofs are omitted. The detailed results appear in [20].

1.3 Techniques

Our algorithms make non-trivial use of *representative families*, and a novel tool that we call *guiding trees*, together enabling the efficient construction of the output tree. Informally, a guiding tree is a constant-size rooted tree which provides some structural information about the solution tree. To efficiently compute a family S of partial solutions, we first construct a polynomial number of *suitable* guiding trees. We then use these trees to generate S, by combining previously computed families of partial solutions. Thus, we avoid iterating over all $O^*(2^k)$ possible topologies for the solution tree.

The efficiency of our algorithms is further improved via replacement of each family of partial solutions, S, by a subfamily $\widehat{S} \subseteq S$, which represents S. Each representative family $\widehat{S}$ contains enough sets from S, thus, we preserve the correctness of the algorithm while improving its running time.

Building on the powerful technique of Fomin et al. [13], for efficient construction of representative families, we tailor the definitions of these sets to the problem at hand. This also leads to replacing *uniform* matroid (often used for fast computation of representative families) by *partition* matroid, which captures more closely the restricted variants of GM.

2 Preliminaries

Given a graph H, let V_H and E_H denote its node-set and edge-set, respectively. **Matroids:** In deriving our results, we use two types of matroids.[2] Given a constant k, the first is defined by a pair $M = (E, \mathcal{I})$, where E is an n-element set, and $\mathcal{I} = \{S \subseteq E : |S| \leq k\}$. Such a pair is called a *uniform matroid*, denoted by $U_{n,k}$.

---

[2] For a broader overview of matroids, see, e.g., [19].

Given some constants ℓ and $k_1, k_2, \ldots, k_\ell$, the second is defined by a pair $(E, \mathcal{I})$, where E is an n-element set partitioned into disjoint sets $E_1, E_2, \ldots, E_\ell$, and $\mathcal{I} = \{S \subseteq E : |S \cap E_1| \leq k_1, |S \cap E_2| \leq k_2, \ldots, |S \cap E_\ell| \leq k_\ell\}$. Such a pair is called a *partition matroid*. Note that, when $\ell = 1$, the definitions for the two types of matroids coincide.

Representative Families: Given a family $\mathcal{S}$ of sets that are partial solutions, we would like to replace $\mathcal{S}$ by a smaller subfamily $\widehat{\mathcal{S}} \subseteq \mathcal{S}$. If there is a partial solution in $\mathcal{S}$ that can be extended to a solution, it is clearly necessary that there would also be a partial solution in $\widehat{\mathcal{S}}$ that can be extended to a solution. The following definition captures such a family $\widehat{\mathcal{S}}$.

Definition 1. *Given a matroid $M = (E, \mathcal{I})$, and a family $\mathcal{S}$ of subsets of size p of E, we say that a subfamily $\widehat{\mathcal{S}} \subseteq \mathcal{S}$ q-represents $\mathcal{S}$ if for every pair of sets $X \in \mathcal{S}$, and $Y \subseteq E \setminus X$ such that $|Y| \leq q$ and $X \cup Y \in \mathcal{I}$, there is a set $\widehat{X} \in \widehat{\mathcal{S}}$ disjoint from Y such that $\widehat{X} \cup Y \in \mathcal{I}$.*

The next two results enable the efficient construction of small representative families.

Theorem 1 ([13,25]). *Given a parameter $c \geq 1$, a uniform matroid $U_{n,k} = (E, \mathcal{I})$, and a family $\mathcal{S}$ of subsets of size p of E, a family $\widehat{\mathcal{S}} \subseteq \mathcal{S}$ of size at most $\dfrac{(ck)^k}{p^p(ck-p)^{k-p}} 2^{o(k)} \log n$ that $(k-p)$-represents $\mathcal{S}$ can be found in time $O(|\mathcal{S}|(ck/(ck-p))^{k-p} 2^{o(k)} \log n)$.*

Theorem 2 ([13,18]). *Given constants $\ell, k_1, k_2, \ldots, k_\ell$ and $k \leq \sum_{i=1}^{\ell} k_i$, a corresponding partition matroid $M = (E, \mathcal{I})$, and a family $\mathcal{S}$ of subsets of size p of E, a family $\widehat{\mathcal{S}} \subseteq \mathcal{S}$ of size at most $\binom{k}{p} n^{O(1)}$ that $(k-p)$-represents $\mathcal{S}$ can be found in time $O(|\mathcal{S}|\binom{k}{p}^{\widetilde{w}-1} n^{O(1)})$, where $\widetilde{w} < 2.3727$ is the matrix multiplication exponent [27].*

Let $\mathsf{UniRep}(c, U_{n,k}, \mathcal{S})$ and $\mathsf{ParRep}(k, M, \mathcal{S})$ be the algorithms implied by Theorems 1 and 2, respectively.

Guiding Trees: Recall that $G = (V, E)$ is the input graph, and let $2 \leq d \leq k/2$ be a constant (to be determined).[3] Given a rooted tree T and a node $v \in V_T$ that is not the root of T, let $f_T(v)$ be the father of v in T. Given nodes $v, u \in V$, we say that a tree T rooted at v is a (v, u)-*tree* if $u \in V_T$. Furthermore, a (v, u)-tree R is a (v, u)-*guide* if $3 \leq |V_R| \leq 2d$ and $V_R \subseteq V$ (E_R may not be contained in E). Let $\mathcal{G}_{v,u}$ be the set of (v, u)-guides. Finally, let $\mathcal{T}_{v,u,\ell}$ be the set of (v, u)-trees on ℓ nodes, that, when unrooted, are subtrees of G.

We now define which subtrees of G listen to the instructions of a given guide (see Fig. 2).

---

[3] The choice of d concerns the analysis of the running times of our algorithms.

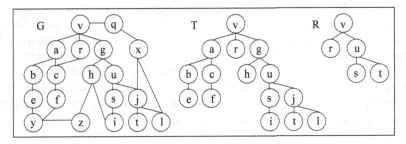

Fig. 2. A (v, u)-tree T, and a (v, u)-guide R, where $d = 3$, $k = 12$, and T listens to R

Definition 2. *Given $v, u \in V$ and $\ell \leq k$, we say that $T \in \mathcal{T}_{v,u,\ell}$ listens to $R \in \mathcal{G}_{v,u}$ if the following two conditions are fulfilled.*

1. *$\forall v', u' \in V_R$: v' is an ancestor of u' in R iff v' is an ancestor of u' in T.*
2. *For each tree X in the forest obtained by removing V_R from T, let $N_X = \{v' \in V_R : \{v', u'\} \in E_T$ for some $u' \in V_X\}$.*
   Then, $|N_X| \leq 2$, and $[N_X \neq \{v\}] \to (|V_X \cup N_X| \leq k/d)$.

The next lemma, which asserts that none of the subtrees of G relevant to solving GM_D is completely undisciplined, is implicit in [13].

Lemma 3. *For any rooted tree $T \in \mathcal{T}_{v,u,\ell}$, where $v, u \in V$ and $3 \leq \ell \leq k$, there exists $R \in \mathcal{G}_{v,u}$ to whom T listens.*

Feasible Colorings: Given $U \subseteq V$, we say that a coloring $col : U \to C$ is *feasible* if $[\forall v \in U : col(v) \in Col(v)]$ and $[\forall c \in C : |\{v \in U : col(v) = c\}| \leq occ(c)]$. Denote by $ima(col)$ the image of col.

3 An Algorithm for GM_D

In this section we solve GM_D in time $O^*(6.86^k)$. Since in GM_D each node is assigned a set of colors whose size can be greater than 1, we may assume w.l.o.g that M is a set equal to C (a formal proof is given, e.g., in [22]).

The main idea of the algorithm is to iterate over all pairs of nodes $v, u \in V$, and all values $1 \leq \ell \leq k$. When we reach such v, u and ℓ, we have already computed, for all $v', u' \in V$ and $1 \leq \ell' < \ell$, representative families for families of corresponding "partial solutions". Each such partial solution is a union of a set A containing exactly ℓ' nodes, and a set B containing exactly ℓ' colors. The sets A and B correspond to a pair of a rooted tree $T \in \mathcal{T}_{v',u',\ell'}$ satisfying $A = V_T$, and a feasible coloring $col : A \to B$.

To compute a family of partial solutions corresponding to v, u and ℓ, we iterate over all (v, u)-guides in $\mathcal{G}_{v,u}$. We follow the instructions of the current guide R by using another, internal dynamic programming-based computation. At each stage of this computation, we have a family of partial solutions listening to a certain subtree of R. We unite these partial solutions with other *small* partial solutions, according to the instructions of R, thus *efficiently* constructing a family of partial solutions listening to a greater subtree of R. For this family, we compute a smaller representative family, so that the following stage can be executed efficiently. After iterating over all relevant guides, we find a family representing the union of the families returned by the internal dynamic programming-based computations. This family includes enough, but not too many, partial solutions corresponding to v, u and ℓ, which ensures the correctness of the algorithm.

3.1 The Algorithm

We now describe GM_D-Alg, our algorithm for GM_D (see the pseudocode below). GM_D-Alg first generates a matrix M, where each entry $[v, u, c_v, c_u, \ell]$ holds a family that represents $Sol_{v,u,c_v,c_u,\ell}$, the family of every set $(X \cup Y)$ satisfying $|X| = |Y| = \ell$, for which there exist $T \in \mathcal{T}_{v,u,\ell}$ such that $X = V_T$, and a feasible $col : X \to Y$ satisfying $col(v) = c_v$ and $col(u) = c_u$.

Algorithm 1. GM_D-Alg$(C, G = (V, E), Col, k)$

---

1. let M be a matrix that has an entry $[v, u, c_v, c_u, \ell]$ for all $v, u \in V$, $c_v \in Col(v)$, $c_u \in Col(u)$, and $1 \leq \ell \leq k$, initialized to $\emptyset$.
2. $M[v,v,c,c,1] \Leftarrow \{\{v,c\}\}$ for all $v \in V$ and $c \in Col(v)$.
3. $M[v,v,c,c,2] \Leftarrow \{\{v,u,c,c'\} : \{v,u\} \in E, c' \in Col(u) \setminus \{c\}\}$ for all $v \in V$ and $c \in Col(v)$.
4. $M[v,u,c,c',2] \Leftarrow \{\{v,u,c,c'\}\}$ for all $\{v, u\} \in E$, $c \in Col(v)$ and $c' \in Col(u) \setminus \{c\}$.
5. **for all** $v, u \in V$, $c_v \in Col(v)$, $c_u \in Col(u)$, and $\ell = 3, \ldots, k$ **do**
6.   let N be a matrix that has an entry $[R, col_R]$ for all $R \in \mathcal{G}_{v,u}$, and feasible $col_R : V_R \to C$ satisfying $col_R(v) = c_v$ and $col_R(u) = c_u$, initialized to $\emptyset$.
7.   **for all** $[R, col_R] \in N$ **do**
8.   let $w_1, \ldots, w_{|V_R|}$ be a preorder on V_R, where $w_1 = v$.
9.   let L be a matrix that has an entry $[i, \ell']$ for all $1 \leq i \leq |V_R|$ and $1 \leq \ell' \leq \ell$, initialized to $\emptyset$.
10. $L[1, \ell'] \Leftarrow M[v, v, c_v, c_v, \ell']$ for all $1 \leq \ell' < \ell$.
11.   **for** $i = 2, \ldots, |V_R|$, and $\ell' = 2, \ldots, \ell$ **do**
12.   let $\mathcal{A}$ include all sets $(U \cup W)$ for which there exists $2 \leq \ell'' \leq \min\{\ell', \ell - 1, k/d\}$ satisfying (1) or (2):
            (1) $U \cap W = \{f_R(w_i), col_R(f_R(w_i))\}$,
               $U \in M[f_R(w_i), w_i, col_R(f_R(w_i)), col_R(w_i), \ell'']$ and $W \in L[i-1, \ell' - \ell'' + 1]$.
            (2) $U \cap W = \{w_i, col_R(w_i)\}$,
               $U \in M[w_i, w_i, col_R(w_i), col_R(w_i), \ell'']$ and $W \in L[i, \ell' - \ell'' + 1]$.
13. $L[i, \ell'] \Leftarrow \text{UniRep}(1.447, U_{(|V|+|C|), 2k}, \mathcal{A})$.
14.   **end for**
15. $N[R, col_R] \Leftarrow L[|V_R|, \ell]$.
16.   **end for**
17. $M[v, u, c_v, c_u, \ell] \Leftarrow \text{UniRep}(1.447, U_{(|V|+|C|), 2k}, \bigcup_{[R, col_R] \in N} N[R, col_R])$.
18. **end for**
19. accept iff $(\bigcup_{v \in V, c_v \in Col(v)} M[v, v, c_v, c_v, k]) \neq \emptyset$.

Then, in Steps 2–4, GM_D-Alg computes all "basic" entries of M, i.e., entries of the form $[v, u, c_v, c_u, \ell]$, where $\ell \leq 2$. Next, in Step 5, GM_D-Alg iterates over all values v, u, c_v, c_u and ℓ that define an entry of M that is not basic, in an order that guarantees that when we reach an entry [\$] of M, we have already computed entries of M that are relevant to [\$]. Now, consider a specific iteration of Step 5, and note that the goal of this iteration is to compute $M[v, u, c_v, c_u, \ell]$.

GM_D-Alg, in Step 6, generates a matrix N. Each entry $[R, col_R]$ holds a family that represents a subfamily of $Sol_{v,u,c_v,c_u,\ell}$. A set $(X \cup Y) \in Sol_{v,u,c_v,c_u,\ell}$ belongs to this subfamily if its corresponding (v, u)-tree $T \in \mathcal{T}_{v,u,\ell}$ and feasible coloring col also satisfy the requirements that T listens to R, and col colors the nodes in V_R exactly as col_R colors them. Now, consider a specific iteration of Step 7, and note that the goal of this iteration is to compute $N[R, col_R]$. To this end, GM_D-Alg executes an internal dynamic programming-based computation, which takes place in Steps 9–14.

First, in Step 9, GM_D-Alg generates a matrix L. Almost every entry $[i, \ell']$ holds a family that represents $Sol_{i,\ell'}$,[4] the family including every set $(X \cup Y)$ satisfying $|X| = |Y| = \ell'$, for which there exist a (v, w_i)-tree $T \in \mathcal{T}_{v,w_i,\ell'}$ and a feasible coloring $col : X \to Y$, satisfying the following conditions. The subtree T listens to the subtree of R induced by $\{w_1, \ldots, w_i\}$, $X = V_T$, and col colors the nodes in $\{w_1, \ldots, w_i\}$ exactly as col_R colors them. Note that the subgraph of R induced by $\{w_1, \ldots, w_i\}$ is a tree because of the preorder defined in Step 8. Then, in Step 10, GM_D-Alg computes all "basic" entries of L, i.e., entries of the form $[1, \ell']$. Next, in Step 11, GM_D-Alg iterates over all values i and ℓ' that define an entry of L that is not basic, in an order that guarantees that when we reach an entry [\$] of L, we have already computed entries of L that are relevant to [\$]. Now, consider a specific iteration of Step 11, and note that the goal of this iteration is to compute $L[i, \ell']$.

GM_D-Alg, in Step 12, computes a family $\mathcal{A}$ that represents $Sol_{i,\ell'}$. The computation involves uniting sets U, found in previous stages of the external dynamic programming-based computation (i.e., U belongs to an entry of M), with sets W, found in previous stages of the internal dynamic programming-based computation (i.e., W belongs to an entry of L). It is easy to verify that the restrictions posed on the choices of U and W gaurantee that their union indeed belongs to $Sol_{i,\ell'}$, noting the following observations. The restriction $\ell'' \leq k/d$ concerns Condition 2 in Definition 2, whose relevance follows from the requirement of existence of a (v, w_i)-tree T as defined above. The first line in each of the options (1) and (2) ensures that we do not use any node or color more than once. The other line of option (1) ensure that $U \in Sol_{f_R(w_i),w_i,col_R(f_R(w_i)),col_R(w_i),\ell''}$ and $W \in Sol_{i-1,\ell'-\ell''+1}$, and the other line of option (2) ensures that $U \in Sol_{w_i,w_i,col_R(w_i),col_R(w_i),\ell''}$ and $W \in Sol_{i,\ell'-\ell''+1}$.

After computing $\mathcal{A}$, GM_D-Alg computes $L[i, \ell']$ (in Step 13) by finding a smaller family that represents $\mathcal{A}$. Upon completing the computation of L, since $V_R = \{w_1, \ldots, w_{|V_R|}\}$, GM_D-Alg can compute $N[R, col_R]$ (in Step 15) by a simple assignment. Then, the union of the families stored in N is a family that represents

---

[4] More precisely, here we refer to all entries $[i, \ell']$ such that $(\ell' = \ell \to i = |V_R|)$.

$Sol_{v,u,c_v,c_u,\ell}$, a claim supported by Lemma 3. Therefore, in Step 19, $\mathsf{GM_D}$-Alg can compute $M[v, u, c_v, c_u, \ell]$ by simply finding a family that represents this union.

Finally, $\mathsf{GM_D}$-Alg accepts iff $\bigcup_{v \in V, c_v \in Col(v)} M[v, v, c_v, c_v, k] \neq \emptyset$. Indeed, note that the input is a yes-instance iff $\bigcup_{v \in V, c_v \in Col(v)} Sol_{v,v,c_v,c_v,k} \neq \emptyset$.

3.2   Correctness

Recall that $Sol_{v,u,c_v,c_u,\ell}$ is the family of every set $(X \cup Y)$ satisfying $|X| = |Y| = \ell$, for which there exist $T \in \mathcal{T}_{v,u,\ell}$ such that $X = V_T$, and a feasible $col : X \to Y$ satisfying $col(v) = c_v$ and $col(u) = c_u$.

The correctness of the algorithm follows directly from the next lemma.

Lemma 4. *Every entry* $M[v, u, c_v, c_u, \ell]$ $(2k - 2\ell)$*-represents* $Sol_{v,u,c_v,c_u,\ell}$.

Proof (Lemma 4). By Steps 1–4, the lemma holds for any entry $[v, u, c_v, c_u, \ell]$ in M such that $\ell \leq 2$. Now, consider some $v, u \in V$, $c_v \in Col(v)$, $c_u \in Col(u)$ and $3 \leq \ell \leq k$, and assume that the lemma holds for all $v', u' \in V$, $c'_v \in Col(v')$, $c'_u \in Col(u')$ and $1 \leq \ell' < \ell$.

For an entry $N[R, col_R]$, let $Sol(R, col_R)_{v,u,c_v,c_u,\ell}$ include every set $(X \cup Y) \in Sol_{v,u,c_v,c_u,\ell}$ whose corresponding (v, u)-tree $T \in \mathcal{T}_{v,u,\ell}$ and feasible coloring col also satisfy the requirements that T listens to R, and col colors the nodes in V_R exactly as col_R colors them.

Towards proving the main inductive claim, we need the following claim.

Claim 1. *Every entry* $N[R, col_R]$ $(2k - 2\ell)$*-represents* $Sol(R, col_R)_{v,u,c_v,c_u,\ell}$.

We first show that Claim 1 implies the correctness of the main inductive claim. Since representation is a transitive relation, it is enough to prove that $\mathcal{B} = \bigcup_{[R,col_R] \in N} N[R, col_R]$ $(2k - 2\ell)$-represents $Sol_{v,u,c_v,c_u,\ell}$. By Claim 1, $\mathcal{B} \subseteq \bigcup_{[R,col_R] \in N} Sol(R, col_R)_{v,u,c_v,c_u,\ell} \subseteq Sol_{v,u,c_v,c_u,\ell}$.

Consider some sets $A \in Sol_{v,u,c_v,c_u,\ell}$, and $B \subseteq (V \cup C) \setminus A$ such that $|B| \leq 2k - 2\ell$. Since $A \in Sol_{v,u,c_v,c_u,\ell}$, we have that A is of the form $(X_A \cup Y_A)$, where $|X_A| = |Y_A| = \ell$, for which there exist $T \in \mathcal{T}_{v,u,\ell}$ such that $X_A = V_T$, and a feasible $col : X_A \to Y_A$ satisfying $col(v) = c_v$ and $col(u) = c_u$. By Lemma 3, there exists $R \in \mathcal{G}_{v,u}$ such that T listens to R. Let col_R be defined as col when restricted to the domain V_R. We get that $A \in Sol(R, col_R)_{v,u,c_v,c_u,\ell}$. By Claim 1, there is $\widehat{A} \in N[R, col_R] \subseteq \mathcal{B}$ such that $\widehat{A} \cap B = \emptyset$. Thus, $\mathcal{B}$ $(2k - 2\ell)$-represents $Sol_{v,u,c_v,c_u,\ell}$. $\square$

We now turn to prove Claim 1.

Proof (Claim 1). Consider an iteration of Step 7, corresponding to an entry $N[R, col_R]$. For an entry $L[i, \ell']$, let $R(i)$ be the subtree of R induced by $\{w_1, \ldots, w_i\}$. Moreover, let $Sol_{i,\ell'}$ be the family including every set $(X \cup Y)$ satisfying $|X| = |Y| = \ell'$, for which there exist a (v, w_i)-tree $T \in \mathcal{T}_{v,w_i,\ell'}$ and a feasible coloring $col : X \to Y$, satisfying the following conditions. The subtree T listens to $R(i)$, $X = V_T$, and col colors the nodes in $\{w_1, \ldots, w_i\}$ exactly as col_R colors them.

Towards proving Claim 1, we need the following claim.

Claim 2. *Every entry $L[i, \ell']$, where $(\ell' = \ell \to i = |V_R|)$, $(2k\text{–}2\ell')$-represents $Sol_{i,\ell'}$.*

Since $N[R, col_R] = L[|V_R|, \ell]$ and $Sol(R, col_R)_{v,u,c_v,c_u,\ell} = Sol_{|V_R|,\ell}$, Claim 2 implies the correctness of Claim 1. □

Finally, we turn to prove Claim 2, concluding the correctness of the algorithm.

Proof (Claim 2). By Steps 9 and 10, and the induction hypothesis concerning the matrix M, the claim holds for $(i = 1$ and all $1 \leq \ell' < \ell)$ and (all $1 \leq i \leq |V_R|$ and $\ell' = 1$). Now, consider some $2 \leq i \leq |V_R|$ and $2 \leq \ell' \leq \ell$, and assume that the claim holds for all $1 \leq i' \leq i$ and $1 \leq \ell'' < \ell'$. Since representation is a transitive relation, it is enough to prove that $\mathcal{A}$ $(2k - 2\ell')$-represents $Sol_{i,\ell'}$.

By definition, a set A belongs to $Sol_{i,\ell'}$ iff there are sets U and W whose union is A, for which there exists $2 \leq \ell'' \leq \min\{\ell, \ell-1, k/d\}$ satisfying (1) or (2):

1. $U \cap W = \{f_R(w_i), col_R(f_R(w_i))\}$,
   $U \in Sol_{f_R(w_i),w_i,col_R(f_R(w_i)),col_R(w_i),\ell''}$ and $W \in Sol_{i-1,\ell'-\ell''+1}$.
2. $U \cap W = \{w_i, col_R(w_i)\}$,
   $U \in Sol_{w_i,w_i,col_R(w_i),col_R(w_i),\ell''}$ and $W \in Sol_{i,\ell'-\ell''+1}$.

Thus, by Step 12 and the inductive hypotheses for the matrices M and L, $\mathcal{A} \subseteq Sol_{i,\ell'}$. Now, consider some $A \in Sol_{i,\ell'}$, and $B \subseteq (V \cup C) \setminus A$ such that $|B| \leq 2k - 2\ell'$. Since $A \in Sol_{i,\ell'}$, there are U, W, and ℓ'' as mentioned above.

First, suppose that U, W, and ℓ'' correspond to the first option. Note that $|(W \setminus \{f_R(w_i), col_R(f_R(w_i))\}) \cup B| = |W| - 2 + |B| \leq 2(\ell' - \ell'' + 1) - 2 + (2k - 2\ell') = 2k - 2\ell''$. Therefore, by the inductive hypothesis concerning M, there is a set $\widehat{U} \in M[f_R(w_i), w_i, col_R(f_R(w_i)), col_R(w_i), \ell'']$ such that $\widehat{U} \cap ((W \setminus \{f_R(w_i), col_R(f_R(w_i))\}) \cup B) = \emptyset$. Moreover, $|(\widehat{U} \setminus \{f_R(w_i), col_R(f_R(w_i))\}) \cup B| = |\widehat{U}| - 2 + |B| \leq (2\ell'') - 2 + (2k - 2\ell') = 2k - 2(\ell' - \ell'' + 1)$. Therefore, by the inductive hypothesis concerning L, there is a set $\widehat{W} \in L[i - 1, \ell' - \ell'' + 1]$ such that $\widehat{W} \cap ((\widehat{U} \setminus \{f_R(w_i), col_R(f_R(w_i))\}) \cup B) = \emptyset$.

Now, suppose that U, W, and ℓ'' correspond to the second option. Note that $|(W \setminus \{w_i, col_R(w_i)\}) \cup B| = |W| - 2 + |B| \leq 2(\ell' - \ell'' + 1) - 2 + (2k - 2\ell') = 2k - 2\ell''$. Therefore, by the inductive hypothesis concerning M, there is a set $\widehat{U} \in M[w_i, w_i, col_R(w_i), col_R(w_i), \ell'']$ such that $\widehat{U} \cap ((W \setminus \{w_i, col_R(w_i)\}) \cup B) = \emptyset$. Moreover, $|(\widehat{U} \setminus \{w_i, col_R(w_i)\}) \cup B| = |\widehat{U}| - 2 + |B| \leq (2\ell'') - 2 + (2k - 2\ell') = 2k - 2(\ell' - \ell'' + 1)$. Therefore, by the inductive hypothesis concerning L, there is a set $\widehat{W} \in L[i, \ell' - \ell'' + 1]$ such that $\widehat{W} \cap ((\widehat{U} \setminus \{w_i, col_R(w_i)\}) \cup B) = \emptyset$. □

3.3 Running Time

Let $0 < \epsilon < 1$ be some constant, $c = 1.447$, and $q = 2k$. Choose a constant $d \geq 2$ satisfying, for any integer n, $\binom{cn}{n/d} = O(2^{\epsilon n})$ and $1/d \leq \epsilon$.

For any $0 \leq r^* \leq q$ and call $\mathsf{UniRep}(c, U_{|V|+|C|,q}, \mathcal{S})$ executed by $\mathsf{GM_D\text{-}Alg}$, where $\mathcal{S}$ is a family of subsets of size r^* of $V \cup C$, there exists $0 \leq r' \leq \min\{r^*, q/d\}$ such that

$$|S| \leq 2^{o(q)}|V|^{O(d)}\Big(\frac{(cq)^q}{(r^*-r')^{r^*-r'}(cq-(r^*-r'))^{q-(r^*-r')}}\Big)\Big(\frac{(cq)^q}{r'^{r'}(cq-r')^{q-r'}}\Big).$$

We get that $\mathsf{GM_D}$-Alg runs in time

$$O\Big(2^{o(q)}|V|^{O(d)}\max_{r=0}^{q}\max_{r'=0}^{\min\{q-r,q/d\}}\Big\{\Big(\frac{(cq)^q}{r^r(cq-r)^{q-r}}\Big)\Big(\frac{(cq)^q}{r'^{r'}(cq-r')^{q-r'}}\Big)\Big(\frac{cq}{cq-(r+r')}\Big)^{q-(r+r')}\Big\}\Big)$$

$$=O\Big(2^{o(q)}|V|^{O(1)}\max_{r=0}^{q}\max_{r'=0}^{\min\{q-r,q/d\}}\Big\{\Big(\frac{(cq)^q}{r^r(cq-r)^{q-r}}\Big)\binom{cq}{r'}\Big(\frac{cq}{cq-(r+q/d)}\Big)^{q-r}\Big\}\Big)$$

$$=O\Big(2^{o(q)}|V|^{O(1)}\max_{r=0}^{q}\Big\{\Big(\frac{(cq)^q}{r^r(cq-r)^{q-r}}\Big)\binom{cq}{q/d}\Big(\frac{cq}{cq-r-(1/d)q}\Big)^{q-r}\Big\}\Big)$$

$$=O\Big(2^{\epsilon q+o(q)}|V|^{O(1)}\max_{r=0}^{q}\Big\{\Big(\frac{(cq)^q}{r^r(cq-r)^{q-r}}\Big)\Big(\frac{cq}{cq-r-\epsilon q}\Big)^{q-r}\Big\}\Big).$$

By choosing a small enough $\epsilon > 0$, the maximum is obtained at $r = \alpha q$, where $\alpha \cong 0.55277$. Thus, $\mathsf{GM_D}$-Alg runs in time $O(6.85414^k|V|^{O(1)})$.

4   An Algorithm for GM

In this section we develop algorithm GM-Alg, proving the following result.

Theorem 5. GM-Alg *solves* GM *in time* $O^*(5.21914^k)$.

Algorithm GM-Alg computes families of "partial solutions" that contain only nodes, and handles colors by adding a parameter to the matrices holding these families. More precisely, given a pair of nodes $v, u \in V$, and a subset of colors $D \subseteq C$, we compute families of partial solutions of the following form. A partial solution is a subset $U \subseteq V$ of $|D|$ nodes, for which there exist a (v,u)-tree $T \in \mathcal{T}_{v,u,|D|}$ satisfying $U = V_T$, and a feasible coloring $col : U \rightarrow D$. Having a family of such partial solutions, we compute a family that represents it, calling algorithm UniRep. Such computations of representative families are embedded in a dynamic programming-based framework, whose progress is governed by guiding trees. Note that, since we iterate over every subset $D \subseteq C$, the running time of GM-Alg crucially relies on the fact that deletions are not allowed in GM.

5   An Algorithm for $\mathrm{RGM_D}$

In this section we develop algorithm $\mathsf{RGM_D}$-Alg, proving the following result.

Theorem 6. $\mathsf{RGM_D}$-Alg *solves* $\mathrm{RGM_D}$ *in time* $O^*(5.1791^k)$.

To efficiently compute representative families, we define a partition matroid $P = P(C, M, G, Col) = (E, \mathcal{I})$ as follows. Denote $C = \{c_1, \ldots, c_{|C|}\}$. Now, let $E = V$ be partitioned into sets $E_1, \ldots, E_{|C|}$, where $E_i = \{v \in V : c_i \in Col(v)\}$, for all $1 \leq i \leq |C|$. The sets $E_1, \ldots, E_{|C|}$ are disjoint because $|Col(v)| = 1$, for all $v \in V$. Now, let $k_i = occ(c_i)$ for all $1 \leq i \leq |C|$ (recall that $occ(c)$ is the number of occurences of a color c in M). Accordingly, define $\mathcal{I} = \mathcal{I}(C, M, G, Col) = \{S \subseteq E : |S \cap E_1| \leq k_1, \ldots, |S \cap E_{|C|}| \leq k_{|C|}\}$.

Intuitively, this definition ensures that $U \in \mathcal{I}$ iff U can be colored without using any color "too many" times, i.e., there exists a feasible coloring $col \colon U \to C$.

Algorithm RGM$_D$-Alg computes families of "partial solutions" that contain only nodes, and handles colors by computing representative families with respect to the partition matroid P. More precisely, when we now consider a pair of nodes $v, u \in V$, and a value $1 \leq \ell \leq k$, we compute families of partial solutions of the following form. A partial solution is a set of nodes $U \in \mathcal{I}$, for which there exists a (v, u)-tree $T \in \mathcal{T}_{v,u,\ell}$ satisfying $U = V_T$. Having a family of such partial solutions, we compute a family that represents it with respect to the matroid P, calling algorithm ParRep. Such computations of representative families are embedded in a dynamic programming-based framework, whose progress is governed by guiding trees.

References

1. Alon, N., Yuster, R., Zwick, U.: Color coding. J. Assoc. Comput. Mach. 42(4), 844–856 (1995)
2. Ambalath, A.M., Balasundaram, R., Rao H., C., Koppula, V., Misra, N., Philip, G., Ramanujan, M.S.: On the kernelization complexity of colorful motifs. In: Raman, V., Saurabh, S. (eds.) IPEC 2010. LNCS, vol. 6478, pp. 14–25. Springer, Heidelberg (2010)
3. Betzler, N., Bevern, R., Fellows, M.R., Komusiewicz, C., Niedermeier, R.: Parameterized algorithmics for finding connected motifs in biological networks. IEEE/ACM Trans. Comput. Biol. Bioinf. 8(5), 1296–1308 (2011)
4. Betzler, N., Fellows, M.R., Komusiewicz, C., Niedermeier, R.: Parameterized algorithms and hardness results for some graph motif problems. In: Ferragina, P., Landau, G.M. (eds.) CPM 2008. LNCS, vol. 5029, pp. 31–43. Springer, Heidelberg (2008)
5. Björklund, A., Husfeldt, T., Kaski, P., Koivisto, M.: Narrow sieves for parameterized paths and packings. CoRR abs/1007.1161 (2010)
6. Björklund, A., Kaski, P., Kowalik, L.: Probably optimal graph motifs. In: STACS, pp. 20–31 (2013)
7. Bruckner, S., Hüffner, F., Karp, R.M., Shamir, R., Sharan, R.: Topology-free querying of protein interaction networks. J. Comput. Biol. 17(3), 237–252 (2010)
8. Dondi, R., Fertin, G., Vialette, S.: Finding approximate and constrained motifs in graphs. In: Giancarlo, R., Manzini, G. (eds.) CPM 2011. LNCS, vol. 6661, pp. 388–401. Springer, Heidelberg (2011)
9. Dondi, R., Fertin, G., Vialette, S.: Maximum motif problem in vertex-colored graphs. In: Kucherov, G., Ukkonen, E. (eds.) CPM 2009. LNCS, vol. 5577, pp. 221–235. Springer, Heidelberg (2009)
10. Fellows, M.R., Fertin, G., Hermelin, D., Vialette, S.: Sharp tractability borderlines for finding connected motifs in vertex-colored graphs. In: Arge, L., Cachin, C., Jurdziński, T., Tarlecki, A. (eds.) ICALP 2007. LNCS, vol. 4596, pp. 340–351. Springer, Heidelberg (2007)

11. Fellows, M.R., Fertin, G., Hermelin, D., Vialette, S.: Upper and lower bounds for finding connected motifs in vertex-colored graphs. J. Comput. Syst. Sci. 77(4), 799–811 (2011)
12. Fionda, V., Palopoli, L.: Biological network querying techniques: Analysis and comparison. J. Comput. Biol. 18(4), 595–625 (2011)
13. Fomin, F.V., Lokshtanov, D., Saurabh, S.: Efficient computation of representative sets with applications in parameterized and exact agorithms. In: SODA (see also: CoRR abs/1304.4626), pp. 142–151 (2014)
14. Guillemot, S., Sikora, F.: Finding and counting vertex-colored subtrees. In: Hliněný, P., Kučera, A. (eds.) MFCS 2010. LNCS, vol. 6281, pp. 405–416. Springer, Heidelberg (2010)
15. Koutis, I.: Constrained multilinear detection for faster functional motif discovery. Inf. Process. Lett. 112(22), 889–892 (2012)
16. Koyutürk, M.: Algorithmic and analytical methods in network biology. Wiley Interdiscip. Rev. Syst. Biol. Med. 2(3), 277–292 (2010)
17. Lacroix, V., Fernandes, C.G., Sagot, M.F.: Motif search in graphs: Application to metabolic networks. IEEE/ACM Trans. Comput. Biol. Bioinf. 3(4), 360–368 (2006)
18. Lokshtanov, D., Misra, P., Panolan, F., Saurabh, S.: Deterministic truncation of linear matroids. CoRR abs/1404.4506 (2014)
19. Oxley, J.G.: Matroid theory. Oxford University Press (2006)
20. Pinter, R.Y., Shachnai, S., Zehavi, M.: Deterministic parameterized algorithms for the graph motif problem, http://www.cs.technion.ac.il/~hadas/PUB/Graph_Motif_full.pdf
21. Pinter, R.Y., Zehavi, M.: Partial information network queries. In: Lecroq, T., Mouchard, L. (eds.) IWOCA 2013. LNCS, vol. 8288, pp. 362–375. Springer, Heidelberg (2013)
22. Pinter, R.Y., Zehavi, M.: Algorithms for topology-free and alignment network queries. J. Discrete Algorithms (to appear, 2014)
23. Pinter-Wollman, N., Hobson, E.A., Smith, J.E., Edelman, A.J., Shizuka, D., de Silva, S., Waters, J.S., Prager, S.D., Sasaki, T., Wittemyer, G., Fewell, J., McDonald, D.B.: The dynamics of animal social networks: analytical, conceptual, and theoretical advances. Behavioral Ecology 25(2), 242–255 (2014)
24. Rizzi, R., Sikora, F.: Some results on more flexible versions of graph motif. In: Hirsch, E.A., Karhumäki, J., Lepistö, A., Prilutskii, M. (eds.) CSR 2012. LNCS, vol. 7353, pp. 278–289. Springer, Heidelberg (2012)
25. Shachnai, H., Zehavi, M.: Faster computation of representative families for uniform matroids with applications. CoRR abs/1402.3547 (2014)
26. Sikora, F.: An (almost complete) state of the art around the graph motif problem. Université Paris-Est Technical reports (2012)
27. Williams, V.V.: Multiplying matrices faster than Coppersmith-Winograd. In: STOC, pp. 887–898 (2012)

The Two Queries Assumption
and Arthur-Merlin Classes

Vyas Ram Selvam

International Institute of Information Technology, Hyderabad, India
vyasram.s@research.iiit.ac.in

Abstract. We explore the implications of the two queries assumption, $P^{SAT[1]} = P_{||}^{SAT[2]}$, with respect to the polynomial hierarchy (PH) and Arthur-Merlin classes. We prove the following results under the assumption $P^{SAT[1]} = P_{||}^{SAT[2]}$:

1. $AM = MA$
2. There exists no relativizable proof for $PH \subseteq AM$
3. Every problem in PH can be solved by a non-uniform variant of an Arthur-Merlin(AM) protocol where Arthur(the verifier) has access to one bit of advice.
4. $PH = P_{||}^{SAT[1], MA[1]}$

Under the two queries assumption, Chakaravarthy and Roy showed that PH collapses to NO_2^P [5]. Since $NP \subseteq MA \subseteq NO_2^P$ unconditionally, our result on relativizability improves upon the result by Buhrman and Fortnow that we cannot show that $PH \subseteq NP$ using relativizable proof techniques [3]. However, we show a containment of PH in a non-uniform variant of AM where Arthur has one bit of advice. This also improves upon the result by Kadin that $PH \subset NP_{/poly}$ [11]. Our fourth result shows that simulating MA in a $P^{SAT[1]}$ machine is as hard as collapsing PH to $P^{SAT[1]}$.

Keywords: Computational complexity, two queries assumption, SAT oracle, Arthur-Merlin classes.

1 Introduction

The assumption that a single query to a SAT oracle is as powerful as two queries leads to some interesting implications such as the collapse of the polynomial hierarchy (PH). The question has a long history in the field of structural complexity theory with several efforts [11,13,6,1] for a deeper collapse of PH. For the functional class analogous to P, FP, Krentel showed that, $FP^{SAT[1]} = FP^{SAT||[2]} \implies P = NP$ [12]. Such a nice collapse of PH, even to the class NP, is not known under the assumption $P^{SAT[1]} = P_{||}^{SAT[2]}$. Throughout this paper, we shall refer to the assumption, $P^{SAT[1]} = P_{||}^{SAT[2]}$, as *two queries assumption*. Buhrman and Fortnow [3] showed a relativized world in which $P^{SAT[1]} = P_{||}^{SAT[2]}$ and $NP \neq coNP$. However, Chang and Purini

E. Csuhaj-Varjú et al. (Eds.): MFCS 2014, Part II, LNCS 8635, pp. 601–612, 2014.
© Springer-Verlag Berlin Heidelberg 2014

[7] proved that $NP = coNP$ under NP machine hypothesis and two queries assumption. For higher levels of PH, Hemaspaandra et al. [10] showed a downward collapse of PH to Σ_k^P if $P^{\Sigma_k^P[1]} = P_{||}^{\Sigma_k^P[2]}$, for $k \geq 2$.

Using the hard/easy argument, Kadin showed that $coNP \subset$ NP/poly under the two queries assumption [11]. This implies a collapse of PH to Σ_3^P using Yap's theorem [14]. Wagner improved the collapse to $P^{\Sigma_2^P}$ [13]. Later, Chang and Kadin [6] brought down the collapse to $P^{\Sigma_2^P[1],NP[1]}$. Buhrman and Fortnow extended the hard/easy argument of Kadin to arrive at the following results under the two queries assumption [3]:

1. $P^{SAT[1]} = P^{SAT}$.
2. Locally (at every input length n), either $NP = coNP$ or NP has polynomial sized circuits.

Fortnow, Pavan and Sengupta [9] showed that $PH \subseteq S_2^P$ under the two queries assumption by making use of the second part of the above result. Chakaravarthy and Roy [5] improved this collapse to their newly defined class $NO_2^P \cap YO_2^P$. Later, Chang and Purini [8] showed a collapse of PH to $ZPP_{1/2-1/exp}^{SAT[1]}$. The last two results are incomparable since the relationship between the two classes is not known independent of the two queries assumption.

1.1 Our Results

We first show that $AM = MA$ under the two quereis assumption. Since $NP \subseteq MA \subseteq NO_2^P$, the next natural target for a PH collapse under two queries assumption is the class MA. In this paper, we show that no relativizable proof technique can achieve such a collapse. Applying our first result, we can extend this non-relativizability result to the class AM. However, we prove that PH is contained in a non-uniform variant of AM where Arthur has access to one bit of advice. We also show that $PH \subseteq P_{||}^{SAT[1],MA[1]}$. It can be observed that any assumption which derandomizes MA gives a flat collapse to $P^{SAT[1]}$. The intractability assumptions for derandomizing MA are closer to those for BPP rather than AM or ZPP^{NP}. The following is a summary of results we obtain under the two queries assumption.

2 Preliminaries

Definition 1: $P^{SAT[f(n)]}$ is the class of all languages decidable by a polynomial time Turing machine asking at most $f(n)$ queries to the SAT oracle where n is the input length. $P_{||}^{SAT[f(n)]}$ is similar to $P^{SAT[f(n)]}$ except that all the queries are asked in parallel. Here, P is the base class and SAT is the oracle. Similar definitions hold for other base classes and other oracles as well.

Definition 2: $ZPP_{p(n)}^{SAT[f(n)]}$ is the class of all languages decidable by a zero-error probabilistic polynomial-time machine which can ask at most $f(n)$ SAT

queries and succeed with probability $p(n)$ where n is the input length. The terms *poly* and *exp* are used in place of $p(n)$ to represent polynomial and exponential functions on the input length respectively.

Definition 3: We define $AM_{/Arthur-Advice=1}$ as a non-uniform variant of the class MA where Arthur is provided with an advice of length 1. Using operator algebra, we can define this class as $AM_{/Arthur-Advice=1} = BP.(\exists.(P_{/1}))$. This is not the same as the class $AM_{/1} = (BP.(\exists.(P)))_{/1}$.

Definition 4: For any complexity class B, $pr(B)$ is defined as the promise version of the class and $pr_A(B)$ is defined as the promise version of the decision class B where the promise is restricted to be in the decision class A.

3 Main Results

3.1 AM = MA

Theorem 5: If $P^{SAT[1]} = P_{||}^{SAT[2]}$, then $AM = MA$.

Proof. Let the language $L \in AM$. Then, L can be characterized as

$$x \in L \implies Pr_r[\phi_{x,r} \in SAT] \geq 2/3$$
$$x \notin L \implies Pr_r[\phi_{x,r} \in \overline{SAT}] \geq 2/3$$

Using the fact that $AM \subseteq \Pi_2^P$ and $\Sigma_2^P = \Pi_2^P$ [9] under the two queries assumption, we can also characterize L as $x \in L \iff \exists u \ \psi_{x,u} \in \overline{SAT}$.

In the above characterizations, ϕ and ψ are some polynomial time computable functions and the lengths of r and u are bounded by some polynomials on the input length. Without loss of generality, we assume that the formulae $\phi_{x,r}$ and $\psi_{x,u}$ have the same length $p_1(|x|)$ for some polynomial p_1. Under the two queries assumption, Buhrman and Fortnow [3, Definition 5.4] showed that $\overline{SAT}^{=n}$ can be partitioned into what are called $Easy - IV$ and $Hard - IV$ formulae. The exact definitions for these are complicated but the crux of our proof relies on the fact that $Easy - IV \in NP$ and the fact that the presence of a $Hard - IV$ formula at a particular length implies the presence of a polynomial sized circuit for SAT at this length [3, Theorem 5.3(1)]. Let R be a polynomial time verifier such that $F \in Easy\text{-}IV \iff \exists v \ R(F, v) = 1$ where $|v| = p_2(|F|)$ for some polynomial p_2.

Case 1: There exists a $Hard - IV$ formula at length $p_1(|x|)$. This means that there is a polynomial sized circuit, C', for SAT at this length. Using this and the self-reducibility property of SAT, we can construct another polynomial sized circuit C which outputs a satisfying assignment for satisfiable inputs. Let V be a polynomial time verifier such that $V(F, u)$ is equal to 1 if and

only if the variable assignment u satisfies the boolean formula F. In this case,
$$x \in L \implies \exists C \; Pr_r[V(\phi_{x,r}, C(\phi_{x,r})) = 1] \geq 2/3$$

Case 2: All the $\overline{SAT}$ formulae of length $p_1(|x|)$ for an input string x are $Easy - IV$. In this case, $x \in L \implies \exists u \; \psi_{x,u} \in \overline{SAT} \implies \exists u, v \; R(\psi_{x,u}, v) = 1$
Regardless of which of the two cases holds,
$$x \notin L \implies \forall C \; Pr_r[V(\phi_{x,r}, C(\phi_{x,r})) = 1] < 1/3$$
$$x \notin L \implies \forall u \; \psi_{x,u} \in SAT \implies \forall u, v \; R(\psi_{x,u}, v) = 0$$
Finally, by combining these cases as follows, we can prove that $L \in MA$:

$$x \in L \implies \exists C, u, v \; Pr_r[V(\phi_{x,r}, C(\phi_{x,r})) = 1 \vee R(\psi_{x,u}, v) = 1] \geq 2/3$$
$$x \notin L \implies \forall C, u, v \; Pr_r[V(\phi_{x,r}, C(\phi_{x,r})) = 1 \vee R(\psi_{x,u}, v) = 1] < 1/3$$

$\square$

3.2   Relativizability of PH Collapse to AM

Definition 6: The towers of two set ω consists of natural numbers $w_0 = 1$ and $w_i = 2^{w_{i-1}}$, $\forall i > 0$.

Definition 7: A oracle A is said to be *gappy* if both these conditions are satisfied i) $\forall n \geq 1$, $|A^{=n}| \leq 1$ and ii) $x \in A \implies |x|$ is a tower of 2. [1]

Theorem 8: There exists an oracle B such that $NP^B \not\subseteq coMA^B$ and $P^{SAT[1]^B} = PSPACE^B$.

Proof. The oracle B is defined as $B = A \oplus TQBF$, where A is a *gappy* oracle, the construction of which constitutes the main part of the proof and $TQBF$ is the language of true quantified boolean formulae. Buhrman and Fortnow [3] showed that if A is gappy and $B = A \oplus TQBF$, then $P^{SAT[1]^B} = PSPACE^B$. This trivially implies that $P^{SAT[1]^B} = P_{||}^{SAT[2]^B}$. We now construct a gappy oracle A such that $NP^B \not\subseteq coMA^B$. Towards that direction, let us define the unary language L_A as follows:

$$L_A = \{ \; 1^n \mid \exists x, \; |x| = n \text{ and } x \in A \; \}$$

It is easy to see that an NP^A machine can decide the language L_A. Initially, oracle A is empty. We construct the oracle A in such a way that $L_A \notin coMA^B$. In our construction here, we set $2/3$ as the required probability threshold for the $coMA$ protocols.

Let $M_1^B, M_2^B, \cdots$ be an enumeration of $coMA$ protocols with oracle access to B. This includes $coMA^B$ protocols which are invalid (protocols which accept some input x with probability between $1/3$ and $2/3$). For the protocol M_i^B, let $q_i(n)$ be the running time of the verifier, $r_i(n)$ be the number of random bits

---

[1] The conventional definition of *gappy* only has the condition that strings are absent at non-tower lengths. In our definition, we include the additional condition that it contains at most one string at tower lengths.

used and $p_i(n)$ be the length of the proof supplied by the prover, where n is the input length. We first define an integer series $t_i, \forall i \geq 0$ as follows. Let $t_0 = 1$ and $\forall i > 0$, let t_i be the smallest integer satisfying the following constraints:

1. $t_i \in \omega$ is a tower of 2.
2. $\forall j \in \mathbb{N}, \ j < i \implies t_i > q_j(t_j)$
3. $2^{t_i} > 6.q_i(t_i)$

The construction of the oracle A happens in stages. In the ith stage of the construction we diagonalize against M_i^B on input 1^{t_i}. The first condition ensures that the oracle is *gappy*. The second condition ensures that the protocols already diagonalized stay diagonalized. The third condition ensures that the next protocol (i^{th} protocol) to be diagonalized can be diagonalized at that length.

By definition, M_i^B accepts an input of length n iff for all proofs of length $p_i(n)$, the verifier accepts with $>= 2/3$ probability. And, M_i rejects an input of length n iff there exists a proof of length $p_i(n)$ such that the verifier rejects with $>= 2/3$ probability.

We first check whether the protocol M is valid. If not, we have already diagonalized against it. If M_i^B does not reject 1^{t_i}, we include no string of length t_i into A. This means $1^{t_i} \notin L_A$ and yet M_i^B does not reject. If M_i^B rejects 1^{t_i}, we would like to include a string x into A at length t_i, such that M_i^B does not accept 1^{t_i} even after the inclusion of x to A. This again means that $1^{t_i} \in L_A$ and yet M_i^B does not accept. by including 1^{t_i} in L_A. We now prove that there exists such a string x at length t_i satisfying the required properties.

Since 1^{t_i} is rejected by M_i^B, there exists a proof of length $p_i(t_i)$ such that $2/3^{rd}$ of the random paths of the verifier reject. Under this proof path, there are $2^{r_i(t_i)}$ random paths and in each random path, the number of queries asked by the verifier to the oracle A is bounded by $q_i(t_i)$ which is the running time of the verifier. Hence, a total of $q_i(t_i)2^{r_i(t_i)}$ queries to A could have been asked under this proof path. There are 2^{t_i} strings at length t_i which could be queried to the oracle A. By a simple counting argument, we get that there exists a string of length t_i that is queried to oracle A by at most $m = q_i(t_i)2^{r_i(t_i)}/2^{t_i}$ different random paths under this proof path. Since $2^{t_i} > 6q_i(t_i)$ by definition, we get that $m < 2^{r_i(t_i)}/6$. In other words, there exists a string x of length 1^{t_i}, the inclusion of which into A reduces the reject probability along the proof path to at most $1/2$. This is due to the fact that at most $1/6$ fraction of random paths could turn into accepting paths from previously rejecting paths due to the inclusion of the string x into A. So, M_i^B does not accept 1^{t_i}.

Finally, since we are adding at most one string only at lengths which are towers of 2, the language A remains *gappy* and by construction, no $coMA^B$ protocol decides L_A. □

Since we have shown a relativizable proof that $AM = MA$ (Theorem 5) under the two queries assumption, a relativizable collapse to AM isn't possible either.

3.3 $PH \subset AM_{/Arthur-Advice=1}$

Definition 9: The languages $\overline{SAT} \wedge SAT \in P_{||}^{SAT[2]}$ and $SAT \oplus \overline{SAT} \in P^{SAT[1]}$ are defined as follows:

$$\overline{SAT} \wedge SAT = \{ \langle x_1, x_2 \rangle \mid x_1 \in \overline{SAT} \text{ and } x_2 \in SAT \}$$
$$SAT \oplus \overline{SAT} = \{ \langle +, x \rangle \mid x \in SAT \} \cup \{ \langle -, x \rangle \mid x \in \overline{SAT} \}$$

Since $P_{||}^{SAT[2]} = P^{SAT[1]}$, there exists a polynomial time reduction h such that

$$\langle x_1, x_2 \rangle \in \overline{SAT} \wedge SAT \iff h(\langle x_1, x_2 \rangle) \in SAT \oplus \overline{SAT}.$$

Definition 10: $x \in \overline{SAT}$ is said to be *Easy* if there exists a string u, $|u| = |x|$, such that $h(\langle x, u \rangle) = \langle +, y \rangle$ and $y \in SAT$.

Definition 11: x is *Hard* if and only if $x \in \overline{SAT}$ and x is not *Easy*.

$\overline{SAT}$ formulae which are *Easy* have a short proof of unsatisfiability whereas it need not be the case for *Hard* formulae.

Lemma 12: If y is *Hard*, then for all x of length $|y|$, $x \in SAT \iff h(\langle y, x \rangle) = \langle -, z \rangle$ with $z \in \overline{SAT}$ [3, Lemma 5.6]

We now mention a technique to classify a set of formuale into satisfiable and unsatisfiable formulae when it is promised that at least one *hard* formula exists in the given set by using a non-deterministic polynomial time machine. This technique is used by Buhrman and Fortnow in [3, Lemma 5.10]. A similar technique is used by Buhrman, Chang and Fortnow in [2, Theorem 2].

Lemma 13: Given a set W of boolean formulae which are of length n where $|W|$ is of polynomial size in n and a promise that there exists a *Hard* formula in W, then the set W can be partitioned into three sets W_{SAT}, W_{easy} and W_{hard} such that

1. $f \in W_{SAT} \implies f \in SAT$
2. $f \in W_{easy} \implies f$ is *easy*
3. f is *hard* $\implies f \in W_{hard}$
4. $f \in W_{easy} \cup W_{hard} \iff f \in \overline{SAT}$
5. An NP machine can guess a partition and verify if it satisfies the above four properties.

Proof. An NP machine initially guesses the partition. Along each nondeterministic path it verifies whether the guessed partition satisfies the stated conditions. The NP machine can easily verify the properties of the formulae in W_{SAT} and W_{easy} as they have polynomial time verifiable proofs.

Now, consider the set W_{hard}. It consists of at least one *Hard* formula, say y. If there exists a formula x such that $x \in W_{hard} \cap SAT$, we would have that

$h(\langle y, x \rangle) = \langle -, z \rangle$ and $z \in \overline{SAT}$ (by Lemma 12). Since we do not know which of the formulae in W_{hard} is the promised $Hard$ formula; we check for all pairs (x, y) such that $x, y \in W_{hard}$ whether $h(\langle y, x \rangle) = \langle +, z \rangle$ or $(h(\langle y, x \rangle) = \langle -, z \rangle$ and $z \in SAT)$. If it is true for all pairs, it is verified that no satisfiable formula is in W_{hard} and we have already seen that no satisfiable formula can be present in W_{easy}. $\qquad\square$

We note that the above non-deterministic polynomial time algorithm classifies satisfiable and unsatisfiable formulae. We will make use of this later.

Definition 14: $HardBits = \{ \langle 1^n, 0 \rangle \mid$ there exists a Hard formula of length $n \} \cup \{ \langle 1^n, i \rangle \mid i^{th}$ bit of the lexically smallest Hard string of length n is 1 $\}$. Here, the strings are represented in such a way that, for a given n, $|\langle 1^n, 0 \rangle| = |\langle 1^n, i \rangle|$ for all $i <= n$.

Theorem 15: If $P^{SAT[1]} = P_{||}^{SAT[2]}$, then $coNP \subset AM_{/Arthur-Advice=1}$.

Proof. Let M be a $ZPP_{1/2-1/exp}^{SAT[1]}$ machine deciding HardBits. This is possible since $HardBits$ is in PH and $PH \subseteq ZPP_{1/2-1/exp}^{SAT[1]}$ under the two queries assumption [8]. We say that a random path in the computation of M on an input is *simple*, if that path leads to a definite answer (accept or reject) and the query at the end of the path is either in SAT or is a $\overline{SAT}$ formula which is $Easy$. Consider the execution of machine M on inputs $\langle 1^n, i \rangle$, for $0 \leq i \leq n$. We would like to find if there exists a $Hard$ string of length n and if so construct the lexically smallest string by observing the computations of machine M on those input. The following are the two possible cases and the one bit of advice directs us to the appropriate case:

Case 1: There exists no $Hard$ formula at length n or there exists a *simple* random path in each of the computations $M(\langle 1^n, i \rangle)$, for $0 \leq i \leq n$, in which case we can construct a polynomial time verifiable proof for the lexically smallest $Hard$ formula.

Case 2: There exists a *hard* string at length n and there exists a computation $M(\langle 1^n, i \rangle)$ which does not contain any *simple* random path. Since the success probability of machine M is $1/2 - 1/exp$, a uniform random sampling of that computation yields a $Hard$ formula with almost $1/2$ probability. A polynomial number of random samples contain a $Hard$ formula with $1 - 1/exp$ probability. In this case, we construct a query sample set W by collecting polynomial number of boolean queries from each computation $M(\langle 1^n, i \rangle)$, for $0 \leq i \leq n$, by uniform random sampling. The set W contains a $Hard$ formula with exponentially high probability. Without loss of generality, we assume that all the strings in W are of the same length. Then by using the Lemma 13, we can partition the set W into polynomial time verifiable sets W_{SAT}, W_{easy} and W_{hard}. Using these sets again as answers to the queries, we can constitute a proof for the lexically smallest $Hard$ formula.

We note that it is necessary to sample as many paths for each computation of $M(\langle 1^n, i \rangle)$, (where $0 \le i \le n$) so that we have at least one rejecting path or one accepting path for each of these computations with high probability. Since the success probability of these computations is greater than $1/3$, we observe that the number of paths that must be sampled to achieve this is bounded by some polynomial. In fact, we can show that sampling $(n+1)^2$ paths uniformly at random for each computation is enough. If we sample $(n+1)^2$ random paths for each computation, the probability that there does not exist a $Hard$ string in the queries sampled is less than $(2/3)^{(n+1)^2}$. The probability that one of the computations does not have an accepting or rejecting path from the sampled random paths is less than $1 - (1 - (2/3)^{(n+1)^2})^{n+1}$. So, the total probability of success is $(1 - (2/3)^{(n+1)^2})^{n+1} - (2/3)^{(n+1)^2}$, which tends to 1.

Now, given an input string x, we have to construct a $AM_{/Arthur-Advice=1}$ protocol to decide if $x \in \overline{SAT}$. The one bit of advice directs the AM protocol to one of the two cases as discussed above. The AM protocol handles the two cases as follows:

Case 1: If all the $\overline{SAT}$ formulae at length n are $Easy$, then Merlin directly provides a proof of unsatisfiability of x. If there exists a $Hard$ formula at length n, then Merlin supplies the lexically smallest $Hard$ formula, let it be y, and a verifiable proof for it. Then, Merlin shows that either $h(\langle y, x \rangle) = \langle +, z \rangle$ or $(h(\langle y, x \rangle) = \langle -, z \rangle$ and $z \in SAT)$ to prove that $x \in \overline{SAT}$ (by Lemma 12).

Case 2: In this case, Arthur constructs the query sample set W as discussed before. With exponentially high probability it contains a $Hard$ formula. The query sample set is given to Merlin. Then Merlin constructs the lexically smallest $Hard$ formula, let it be y, and provides a proof for it through the verifiable partition W_{SAT}, W_{easy} and W_{hard} of the queries sent by Arthur. This y is a $Hard$ formula with high probability. Then, Merlin shows that either $h(\langle y, x \rangle) = \langle +, z \rangle$ or $(h(\langle y, x \rangle) = \langle -, z \rangle$ and $z \in SAT)$ to prove that $x \in \overline{SAT}$ (by Lemma 12).

If the input formula x is in $\overline{SAT}$, then it will be accepted with probability 1 or $1 - 1/exp$, depending on whether we fall in Case 1 or 2 respectively. Whereas if $x \in SAT$, then it will be accepted with probability 0 or at most $1/exp$, depending on whether we fall in Case 1 or 2 respectively. $\qquad\square$

3.4 Towards a Flat Collapse to $\mathbf{P^{SAT[1]}}$

While a relativizable collapse to MA is not possible, we do not know of any such result for the longstanding question of whether PH collapses to $P^{SAT[1]}$ under the two queries assumption. The following result due to Chang and Purini [8] gives a possible approach towards achieving such a flat PH collapse: $P^{SAT[1]} = P_{||}^{SAT[2]} \implies P^{SAT[1]} = P^{SAT} \subseteq ZPP_{1-1/exp}^{SAT[1]} = ZPP_{1/2+1/poly}^{SAT[1]} \subseteq ZPP_{1/2-1/exp}^{SAT[1]} = PH$. We can attempt to show that $ZPP_{1/2-1/exp}^{SAT[1]} \subseteq P^{SAT}$ by eliminating the randomness using SAT queries. In this section, we show that

$ZPP^{SAT[1]}_{1/2-1/exp}$ is contained in $P^{SAT[1],MA[1]}_{||}$ under the two queries assumption. This gives us a PH collapse to $P^{SAT[1]}$ under the derandomization assumptions for MA. We first start with an unconditional result.

Theorem 16: $ZPP^{SAT[1]}_{1/poly} \subseteq P^{SAT[1],AM[1]}_{||}$

Proof. Since $ZPP^{SAT[1]}_{1/poly} = ZPP^{SAT[1]}_{1/4-1/exp}$ [8, Theorem 6], we just have to prove our theorem for $ZPP^{SAT[1]}_{1/5}$. Let M be a $ZPP^{SAT[1]}_{1/5}$ machine deciding a language L. For a string $x \notin L$, We say that a rejecting random path in the computation M(x) is *simple* if the SAT query asked in that path is satisfiable. Let us define the language $L_{simpleR}$ as follows:

$$L_{simpleR} = \{\ x \in \overline{L} \mid x \text{ has a } simple \text{ reject path in M}(x)\ \}$$

We can easily observe that $L_{simpleR} \in NP$. Let us define a new language L_{ext} by extending the language L as $L_{ext} = L \cup L_{simpleR}$. Since $L_{simpleR} \subseteq \overline{L}$, we have that $L = \overline{L_{simpleR}} \cap L_{ext}$. We now show that $L_{ext} \in AM$ by providing a protocol which proves the theorem statement.

On an input x, Arthur uniformly randomly samples 100 computation paths of the machine M(x) through simulation and collects the SAT query in each of the paths in a set R. For a query $f \in R$, let r_f denotes the random path from which it is collected. Define the formula set R' as follows:

$$R' = \{\ f \in R \mid f \in \overline{SAT} \implies \text{M}(x) \text{ outputs rejects along } r_f\ \}$$

Arthur asks Merlin to provide proof for either $x \in L_{simpleR}$ or $\forall f \in R'$, $f \in SAT$. Arthur accepts if Merlin provides such a proof; otherwise Arthur rejects the input string. If $x \in L_{simpleR}$, then Merlin can provide an *simple* reject path as a proof along with a satisfying assignment for the query along the path. If $x \in L$, then every formula in R' is indeed present in SAT. So a satisfying assignment for all the formulae can be presented. If $x \notin L_{ext}$, then no *simple* reject path proof exists. Further this forces the machine M to reject x through the random paths corresponding to the formulae in R'. So with high probability ($> 1 - \frac{4}{5}^{100}$) there exists an $f \in R'$ which is in $\overline{SAT}$. Merlin will not be able to provide a satisfying assignment for such a boolean formula.

We can ask the two queries in parallel and decide accordingly and so $L \in P^{SAT[1],AM[1]}_{||}$. $\qquad\square$

Theorem 17: $P^{SAT[1]} = P^{SAT[2]}_{||} \implies PH = P^{SAT[1],MA[1]}_{||}$

Proof. We know that under two queries assumptions PH collapses to $ZPP^{SAT[1]}_{1/2-1/exp}$ [8]. Then from Theorem 16, we get that $PH \subseteq P^{SAT[1],AM[1]}_{||}$. Since $AM = MA$ under two queries assumption from Theorem 5, finally we have that $PH \subseteq P^{SAT[1],MA[1]}_{||}$. The other direction is trivial. $\qquad\square$

Results Based on Promise Oracles. A promise oracle is an oracle based on the promise version of a decision problem. The oracle is expected to provide the correct answers to queries that fulfill the promise and can give arbitrary answers to queries that do not fulfill the promise. We now show a couple of results based on promise oracles starting with an unconditional result.

Theorem 18: $ZPP^{SAT[1]}_{1/poly} \subseteq P^{SAT[2],pr_{coNP}(AM \cap coAM)[1]}_{||}$

Proof. The proof of this is similar to the proof of Theorem 16 and uses the technique used by Cai and Chakaravarthy [4] to show that $ZPP^{SAT[1]}_{1/2+1/poly} \subseteq S^P_2$. Since $ZPP^{SAT[1]}_{1/poly} = ZPP^{SAT[1]}_{1/4-1/exp}$ [8, Theorem 6], we just have to prove our theorem for $ZPP^{SAT[1]}_{1/5}$. Let M be a $ZPP^{SAT[1]}_{1/5}$ machine deciding a language L. For a given input x, We say that a rejecting or accepting random path in the computation $M(x)$ is *simple* if the SAT query asked in that path is satisfiable. Let us define the languages $L_{simpleR}$ and $L_{simpleA}$ as follows:

$$L_{simpleR} = \{ x \in \overline{L} \mid x \text{ has a } simple \text{ reject path in } M(x) \}$$

$$L_{simpleA} = \{ x \in L \mid x \text{ has a } simple \text{ accept path in } M(x) \}$$

We can easily observe that $L_{simpleR} \in NP$ and $L_{simpleA} \in NP$. We want to decide L using a $P^{SAT[2],pr_{coNP}(AM \cap coAM)}_{||}$ machine. We make use of the first two queries to check whether $x \in L_{simpleA}$ or $x \in L_{simpleR}$ and accept or reject accordingly. Under the promise that both these queries are false, we can reduce the problem to one in $AM \cap coAM$. So, $\overline{L_{simpleR} \cup L_{simpleA}}$ which is in $coNP$ is the promise. We now define the AM protocol that decides L under the promise. Arthur uniformly samples 100 computation paths of the machine $M(x)$ and collects the SAT query in each of the paths in a set R. For a query $f \in R$, let r_f denotes the random path from which it is collected. Define the formula set R' as follows:

$$R' = \{ f \in R \mid f \in \overline{SAT} \implies M(x) \text{ outputs rejects along } r_f \}$$

Arthur asks Merlin to prove that $\forall f \in R'$, $f \in SAT$. Arthur accepts if Merlin provides such a proof; otherwise Arthur rejects the input string. If $x \in L$, then every formula in R' is indeed present in SAT. So a satisfying assignment for all the formulae can be presented. If $x \notin L$, then with high probability $(> 1 - \frac{4}{5}^{100})$ there exists an $f \in R'$ which is in $\overline{SAT}$. This is because all the rejecting paths are not *simple* and M succeeds with $1/5$ probability. Merlin will not be able to provide a satisfying assignment for such a boolean formula.

We can have a similar AM protocol for $\overline{L}$ under the same promise. So, the language L is reduced to a problem in $pr_{coNP}(AM \cap coAM)$. We can ask these three queries in parallel and decide accordingly and so $L \in P^{SAT[2],pr_{coNP}(AM \cap coAM)}_{||}$.

$\square$

Theorem 19: $P^{SAT[1]} = P_{||}^{SAT[2]} \implies PH = P_{||}^{SAT[2],pr_{coNP}(MA\cap coMA)[1]}$

Proof. We know that under two queries assumptions PH collapses to $ZPP_{1/2-1/exp}^{SAT[1]}$ [8]. Then from Theorem 18 and Theorem 5, we get that $PH \subseteq P_{||}^{SAT[2],pr_{coNP}(MA\cap coMA)[1]}$.

Also, $P_{||}^{SAT[2],pr_{coNP}(MA\cap coMA)[1]} \subseteq PH$ unconditionally because the promise oracle can be replaced with a $coMA$ decision oracle (by combining the promise with the query) and $P_{||}^{SAT[2],coMA[1]} \subseteq PH$. □

Taking into account the results in this section related to the two queries assumption, we observe that it is possible to collapse PH to $P^{SAT[1]}$ if and only if MA is contained in $P^{SAT[1]}$ under the two queries assumption. One can view this observation as a reduction to the hard-core of the problem.

4 Conclusions and Open Problems

Whether PH collapses to $P^{SAT[1]}$ under the two queries assumption is a long standing open problem. Our results point out that such a flat collapse can be achieved by showing that either $MA \subseteq P^{SAT[1]}$ or by simulating $pr_{coNP}(MA \cap coMA)$ in $P^{SAT[1]}$ under the two queries assumption. Even showing that either BPP or $MA \cap coMA$ is contained in $P^{SAT[1]}$ under the two queries assumption might be a step in that direction.

Buhrman, Chang and Fortnow [2] showed that if $coNP \subset NP_{/k}$ then $PH \subseteq P^{SAT}$, where k is any constant. Our result which claims that $coNP \subset AM_{/Arthur-Advice=1}$ under two queries assumption make some progress in showing that $coNP \subset NP_{/k}$. This leaves an interesting open question if $AM \subseteq NP_{/k}$ under the two queries assumption. Another open question is whether $PH \subset AM_{/1}$ under the two queries assumption.

A related open question asked by Buhrman and Fortnow [3] is whether $\overline{SAT}$ can be described as the union/intersection of an NP set and a BPP/1 set under the two queries assumption. The question remains open but we have made some progress by showing that $\overline{SAT} \in AM_{/Arthur-Advice=1}$.

Acknowledgments. I thank Suresh Purini for providing valuable suggestions and proof-reading the paper. I thank the anonymous reviewers for providing valuable feedback and suggestions.

References

1. Beigel, R., Chang, R., Ogiwara, M.: A relationship between difference hierarchies and relativized polynomial hierarchies. Mathematical Systems Theory 26(3), 293–310 (1993)

2. Buhrman, H., Chang, R., Fortnow, L.: One Bit of Advice. In: Alt, H., Habib, M. (eds.) STACS 2003. LNCS, vol. 2607, pp. 547–558. Springer, Heidelberg (2003)
3. Buhrman, H., Fortnow, L.: Two Queries. J. Comput. Syst. Sci. 59(2), 182–194 (1999)
4. Cai, J., Chakaravarthy, V.: A note on zero error algorithms having oracle access to one NP query. In: Proceedings of the 11th Annual International Conference on Computing and Combinatorics, pp. 339–348 (2005)
5. Chakaravarthy, V.T., Roy, S.: Oblivious Symmetric Alternation. In: Durand, B., Thomas, W. (eds.) STACS 2006. LNCS, vol. 3884, pp. 230–241. Springer, Heidelberg (2006)
6. Chang, R., Kadin, J.: The Boolean Hierarchy and the Polynomial Hierarchy: A closer connection. SIAM J. Comput. 25(2), 340–354 (1996)
7. Chang, R., Purini, S.: Bounded Queries and the NP Machine Hypothesis. In: Proceedings of the Twenty-Second Annual IEEE Conference on Computational Complexity, CCC 2007, pp. 52–59. IEEE Computer Society, Washington, DC (2007)
8. Chang, R., Purini, S.: Amplifying ZPP^SAT[1] and the Two Queries Problem. In: Proceedings of the 2008 IEEE 23rd Annual Conference on Computational Complexity, CCC 2008, pp. 41–52. IEEE Computer Society Press, Washington, DC (2008)
9. Fortnow, L., Pavan, A., Sengupta, S.: Proving SAT does not have small circuits with an application to the two queries problem. J. Comput. Syst. Sci. 74(3), 358–363 (2008)
10. Hemaspaandra, E., Hemaspaandra, L.A., Hempel, H.: A Downward Collapse within the Polynomial Hierarchy. SIAM Journal on Computing 28(2), 383–393 (1999)
11. Kadin, J.: The Polynomial Time Hierarchy Collapses if the Boolean Hierarchy Collapses. SIAM J. Comput. 17(6), 1263–1282 (1988)
12. Krentel, M.W.: The Complexity of Optimization Problems. J. Comput. Syst. Sci. 36(3), 490–509 (1988)
13. Wagner, K.W.: Bounded query computations. In: Proceedings of the Third Annual Structure in Complexity Theory Conference, pp. 260–277 (June 1988)
14. Yap, C.-K.: Some Consequences of Non-Uniform Conditions on Uniform Classes. Theor. Comput. Sci. 26, 287–300 (1983)

Flexible Bandwidth Assignment
with Application to Optical Networks
(Extended Abstract)

Hadas Shachnai, Ariella Voloshin, and Shmuel Zaks

Department of Computer Science, Technion, Haifa 32000, Israel
{hadas,variella,zaks}@cs.technion.ac.il

Abstract. We introduce two scheduling problems, the *flexible bandwidth allocation problem* (FBAP) and the *flexible storage allocation problem* (FSAP). In both problems, we have an available resource, and a set of requests, each consists of a minimum and a maximum resource requirement, for the duration of its execution, as well as a profit accrued per allocated unit of the resource. In FBAP the goal is to assign the available resource to a feasible subset of requests, such that the total profit is maximized, while in FSAP we also require that each satisfied request is given a contiguous portion of the resource. Our problems generalize the classic *bandwidth allocation problem* (BAP) and *storage allocation problem* (SAP) and are therefore NP-Hard.

Our main results are a 3-approximation algorithm for FBAP and a $(3 + \epsilon)$-approximation algorithm for FSAP, for any fixed $\epsilon > 0$. These algorithms make non-standard use of the *local ratio* technique. Furthermore, we present a $(2 + \epsilon)$-approximation algorithm for SAP, for any fixed $\epsilon > 0$, thus improving the best known ratio of $\frac{2e-1}{e-1} + \epsilon$. Our study is motivated also by critical resource allocation problems arising in all-optical networks.

Keywords: Approximation algorithms, local ratio, resource allocation, all-optical networks.

1 Introduction

1.1 Background

Scheduling activities with resource demands arise in a wide range of applications. In these problems we have a set of activities competing for a reusable resource. Each activity utilizes a certain amount of the resource for the duration of its execution and frees it upon completion. The problem is to find a feasible schedule for a subset of the activities which satisfies certain constraints, including the requirement that the total amount of resource allocated simultaneously for executing activities never exceeds the amount of available resource. Two classic problems that fit in this scenario are the *Bandwidth Allocation Problem* (BAP) and the *Storage Allocation Problem* (SAP). In BAP the goal is to assign the available resource to a feasible subset of activities, such that the total profit is

E. Csuhaj-Varjú et al. (Eds.): MFCS 2014, Part II, LNCS 8635, pp. 613–624, 2014.
© Springer-Verlag Berlin Heidelberg 2014

maximized, while in SAP it is also required that any satisfied activity is given the same contiguous portion of the resource for its entire duration (for references and further discussion see Section 1.3). We introduce two variants of these problems where each activity has a minimum and a maximum possible request size, as well as a profit per unit of the resource allocated to it. We refer to these variants as the *Flexible Bandwidth Allocation Problem* (FBAP) and the *Flexible Storage Allocation Problem* (FSAP).

1.2   Problem Statement

In this work we study FBAP and FSAP on path network. In graph-theoretical terms, the input for FBAP and FSAP consists of a path $P = (V, E)$ and a set $\mathcal{I}$ of n intervals on P. Each interval $I \in \mathcal{I}$ requires the utilization of a given, limited, resource. The amount of resource available, denoted by $W > 0$, is fixed over P. Each interval $I \in \mathcal{I}$ is defined by the following parameters. (i) A left endpoint, $l(I) \geq 0$, and a right endpoint, $r(I) > l(I)$. Thus, I is associated with the half-open interval $[l(I), r(I))$ on P. (ii) The amount of resource range required by each interval I, where $a(I)$, $b(I)$ are integers satisfying $0 \leq a(I) \leq b(I) \leq W$. Thus I can take any integer value in the *possible range* for I, given by $[a(I), b(I)]$ or 0. (iii) The profit $w(I)$ gained for each unit of the resource allocated to I.[1]

A solution has to satisfy the following conditions. (i) Each assigned interval $I \in \mathcal{I}$ is allocated an amount of the resource in its possible range or is not allocated at all. (ii) The specific resources allocated to an interval are fixed along the interval. (iii) The total amount of the resource allocated at any time does not exceed the available amount W. In FBAP we seek a feasible allocation which maximizes the total profit accrued by the intervals. In FSAP we add the requirement that the allocation to each interval is a contiguous block of the resource for the entire duration. We give an example for FSAP and FSAP in the full version of this paper [17]. A primary application of FBAP and FSAP is spectrum assignment for connection in all-optical networks.

Approximation Algorithms. We develop approximation algorithms and analyze their worst case performance. For $\rho \geq 1$, a ρ-*approximation* algorithm for optimization problem Π yields in polynomial time a solution whose value is within a factor ρ of the optimum, for any input for Π.

1.3   Related Works

Bandwidth Allocation Problem (BAP). We are given a network having some available bandwidth, and a set of connection requests. Each request consists of a path in the network, a bandwidth requirement, and a weight. The goal is to feasibly assign bandwidth to a maximum weight subset of requests. BAP is strongly NP-Hard even for uniform profit on a path network [9]. [3] presents a 3-approximation algorithm for the problem. The best known result is a deterministic $(2 + \epsilon)$-approximation algorithm of [7].

---

[1] We note that our results hold when $a(I)$, $b(I)$ and $w(I)$ are non-negative rational numbers.

Storage Allocation Problem (SAP). The SAP is a special case of BAP in which we require that each activity is allocated a single contiguous block of resource for all of its edges. This problem is NP-hard. SAP was first studied in [3,15]. In [3] an approximation algorithm is presented, that yields a ratio of 7. The authors of [8] study the special case in which all rectangle heights are multiple of $1/K$, for some integer $K \geq 1$. They present an $O(n(nK)^K)$ time dynamic programming algorithm to solve this special case of SAP, and also give an approximation algorithm with ratio $\frac{e}{e-1} + \epsilon$, for any $\epsilon > 0$, assuming that the maximum height of any rectangle is $O(1/K)$. In [5] a randomized $(2 + \epsilon)$-approximation algorithm for SAP is presented, together with a deterministic $(\frac{2e-1}{e-1} + \epsilon)$-approximation algorithm for the problem, for any fixed $\epsilon > 0$.

Flex Non-Contiguous (FNC) and Flex Contiguous (FC). The problems FNC and FC are restricted variants of the FBAP and FSAP, respectively, in which all intervals have to be assigned an amount of resource in their required range, i.e., for each interval $I \in \mathcal{I}$ the amount of the assigned resource is at least $a(I)$. We note that the special case of FNC and FC in which $a(I) = 0$, for all $I \in \mathcal{I}$, is also a special case of the FBAP and FSAP, respectively. Papers [14,18,19] consider these problems. In [19] the authors show that FNC is polynomially solvable. For a contrast, in [18] it is observed that FC cannot be approximated within any bounded ratio, unless $P = NP$. The authors of [18] show that FC is NP-Complete for the subclass of instances where $a(I) = 0$ for all $I \in \mathcal{I}$, and present a $(2 + \epsilon)$-approximation algorithm for such instances, for any fixed $\epsilon > 0$. For this special subclasses of inputs, the authors of [14] present a $(5/4+\epsilon)$-approximation algorithm for instances where the input graph is a proper interval graph, for any fixed $\epsilon > 0$. Finally, the authors of [14] show that when for all $I \in \mathcal{I}$, $a(I) = 0, b(I) = \text{Max}$ for some $1 \leq \text{Max} \leq W$, and $w(I) = 1$, the problem is NP-Hard. For this subclass, they obtain a $(\frac{2k}{2k-1})$-approximation algorithm, where $k = \lceil \frac{W}{Max} \rceil$, and show that this subclass admits a PTAS.

1.4   Our Results

We study the scheduling problems FBAP and FSAP. We note that both problems are NP-Hard, and show that in fact, FSAP is NP-Hard in the strong sense for any instance $\mathcal{I}$ where $a(I) \neq b(I)$ for all $I \in \mathcal{I}$. In Section 3, we give a 3-approximation algorithm for FBAP (based on the local ratio technique). We then show (in Section 4) that this algorithm can be extended to yield a $(3 + \epsilon)$-approximation for FSAP, for any fixed $\epsilon > 0$. Finally, in Section 5 we consider SAP, the special case of FSAP where $a(I) = b(I)$ for all $I \in \mathcal{I}$. We present a $(2 + \epsilon)$-approximation algorithm, for any fixed $\epsilon > 0$, thus improving the best known ratio of $\frac{2e-1}{e-1} + \epsilon$, due to [5]. In Section 6 we present applications for the problems, particularly, the primary application in spectrum assignment in all-optical networks. We conclude with summary and future work in Section 7.

Techniques. In developing our approximation algorithm for FBAP, we make non-standard use of the *local ratio* technique. In particular, we formulate a lemma which shows that the classic Local Ratio Theorem holds also for instances associated with profit *per unit* vectors, where the solution vector specifies the amount of resource allocated to the input elements. To the best of our knowledge this interpretation of the local ratio technique is given here for the first time. Some of the proofs are omitted in this Extended Abstract. They can be found in the full version of this paper [17].

2 Preliminaries

Throughout the paper we use graph-theoretical coloring terminology. Specifically, the requests $\mathcal{I}$ are represented as a set of n intervals on a path $P = (V, E)$. For an interval I we denote $l(I)$ and $r(I)$ as its left-endpoint and right-endpoint, respectively.

The amount of available resource, W, can be viewed as the amount of available distinct colors. Each interval $I \in \mathcal{I}$ is also defined by a minimum $a(I)$ and a maximum $b(I)$ color requirements, $0 \leq a(I) \leq b(I) \leq W$, and a positive profit per allocated color (or, profit per unit) $w(I)$, where $a(I)$, $b(I)$ and $w(I)$ are non-negative integers.

The set of available colors is $\Lambda = \{1, \ldots, W\}$. A contiguous range of colors is any set $\Lambda_i^j = \{t : 1 \leq i \leq t \leq j \leq W\}$, and is termed an *interval* of colors. A *(multi)coloring* is a function $\sigma : \mathcal{I} \mapsto 2^\Lambda$ that assigns to each interval $I \in \mathcal{I}$ a subset of the set Λ of colors. A coloring σ is *feasible* if for every $I \in \mathcal{I}$ $a(I) \leq |\sigma(I)| \leq b(I)$ or $|\sigma(I)| = 0$, and for any two intervals $I, I' \in \mathcal{I}$ such that $I \cap I' \neq \emptyset$ we have $\sigma(I) \cap \sigma(I') = \emptyset$. A *contiguous color assignment* is a coloring σ that assigns a contiguous range of colors. For any disjoint subsets $\mathcal{I}', \mathcal{I}'' \subseteq \mathcal{I}$, a coloring function σ for $\mathcal{I}'$ can be expanded to a coloring function $\overline{\sigma}$ for $\mathcal{I}' \cup \mathcal{I}''$ such that $\overline{\sigma}(I) = \sigma(I)$ for each $I \in \mathcal{I}'$ and $\overline{\sigma}(I) = \emptyset$ for each $I \notin \mathcal{I}'$. The total profit of a feasible coloring σ of $\mathcal{I}' \subseteq \mathcal{I}$ is $profit^\sigma(\mathcal{I}') \stackrel{def}{=} \sum_{I \in \mathcal{I}'} |\sigma(I)| \cdot w(I)$. When there is no ambiguity regarding the set of intervals we simply write $profit^\sigma$.

The Problems. We first introduce the following coloring problem:

FBAP$(\mathcal{I}, W)$
Input: A tuple $(\mathcal{I}, W)$, where W is a positive integer, and $\mathcal{I}$ is a set of intervals together with the integers $a(I)$, $b(I)$, and $w(I)$ for each $I \in \mathcal{I}$, such that $0 \leq a(I) \leq b(I) \leq W$ and $0 < w(I)$.
Output: A feasible color assignment σ for $\mathcal{I}$.
Objective: Maximize $profit^\sigma(\mathcal{I})$.

A solution S for the FBAP consists of the intervals that were assigned at least one color, i.e., a set of pairs, where the first entry of each pair is an interval $I \in \mathcal{I}$ and the second entry is the interval coloring size $|\sigma(I)|$.

The second problem is the contiguous color assignment variant, FSAP, in which the goal is to achieve a feasible contiguous coloring function σ for $\mathcal{I}$ that maximizes $profit^\sigma(\mathcal{I})$. A solution S for FSAP consists of the intervals that were assigned at least one color; i.e., a set of triples, where the first entry of each triple is an interval $I \in \mathcal{I}$, the second entry is the interval coloring size $|\sigma(I)|$, and the third entry is its lower color index. Given a solution S to the FSAP (or FBAP), we denote by $\mathcal{I}_S$ the intervals of S and by σ_S their coloring function. Note that for the FBAP it is impossible to return a solution of the exact coloring function σ in polynomial time. For example, suppose we are given a path and a set of intervals $\mathcal{I} = \{I\}$ where $a(I) = 0$ and $b(I) = W$. Presenting a coloring solution σ such that $|\sigma(I)| > \Omega(\log W)$ is not polynomial in the input size. Therefore, following the description of the algorithm for FBAP, we explain how to achieve the exact coloring function.

We note that BAP and SAP are special cases of FBAP and FSAP, respectively (where $a(I) = b(I)$ for every $I \in \mathcal{I}$). BAP and SAP are NP-Hard since they include the Knapsack problem as a special case, where all intervals share the same edge; thus we have that FSAP and FBAP are NP-Hard.

In addition, following the hardness result of [14], FSAP is NP-Hard for the subclass in which for all $I \in \mathcal{I}$, $a(I) = 0, b(I) = $ Max for some $1 \leq$ Max $\leq W$, and $w(I) = 1$. For another subclass, where $a(I) \neq b(I)$ for all $I \in \mathcal{I}$, we show (see the full version of this paper [17]) that the problem remains NP-Hard, even where all intervals have the same unit profit, as follows

Lemma 1. FSAP *is* NP-Hard *in the strong sense, even for uniform profit instances, where* $a(I) \neq b(I)$ *for all* $I \in \mathcal{I}$.

Narrow and Wide Intervals. In deriving our approximation results, for a given set of interval requests, we form two new interval sets. We solve separately the problem for each set, and then the solution of larger profit is expanded into a solution for the original instance. Formally, given a set of intervals $\mathcal{I}$ and a parameter $\delta \in (0, 1)$, we form sets of intervals $\mathcal{I}_{wide}$ and $\mathcal{I}_{narrow}$ as follows. For any $I \in \mathcal{I}$ for which $b(I) > \delta W$, we define an interval I_{wide} with the same left and right endpoint as I, $a(I_{wide}) = \max\{a(I), \delta W + 1\}$, $b(I_{wide}) = b(I)$, and $w(I_{wide}) = w(I)$. This set of intervals is termed $\mathcal{I}_{wide}$. For any $I \in \mathcal{I}$ for which $a(I) \leq \delta W$, we define an interval I_{narrow} with the same left and right endpoint as I, $a(I_{narrow}) = a(I)$, $b(I_{narrow}) = \min\{b(I), \delta W\}$, and $w(I_{narrow}) = w(I)$. We call this set of intervals $\mathcal{I}_{narrow}$.

Given a feasible coloring function σ_{narrow} (or σ_{wide}) of the instance $(\mathcal{I}_{narrow}, W)$ (or $(\mathcal{I}_{wide}, W)$) we expand it to a feasible coloring function σ_{narrow}^{expand} (or σ_{wide}^{expand}) for $(\mathcal{I}, W)$ such that for each $I \in \mathcal{I}$ if $a(I) \leq \delta W$ (or $b(I) \geq \delta W$) then $\sigma_{narrow}^{expand}(I) = \sigma_{narrow}(I)$ (or $\sigma_{wide}^{expand}(I) = \sigma_{wide}(I)$), otherwise, $\sigma_{narrow}^{expand}(I) = \emptyset$.

By the above definition of $\mathcal{I}_{narrow}$ and $\mathcal{I}_{wide}$, we do not necessarily have that $\mathcal{I}_{narrow} \subseteq \mathcal{I}$ (or $\mathcal{I}_{wide} \subseteq \mathcal{I}$), as for some intervals the lower or upper bound for the range of possible colors were changed. Therefore, we need the following lemma in order to claim an approximation ratio for the returned result (for proof see the full version of this paper [17]).

Lemma 2. *Let* $(\mathcal{I}, W)$ *be an instance of* FBAP *(or* FSAP*). For any* $\delta \in (0, 1)$, *let* σ_{narrow} *and* σ_{wide} *be a* (ρ_{narrow})-*approximate solution for the instance* $(\mathcal{I}_{narrow}, W)$ *and a* (ρ_{wide})-*approximate solution for the instance* $(\mathcal{I}_{wide}, W)$, respectively, for $\rho_{narrow}, \rho_{wide} \geq 1$. Then, the solution of larger profit can be expanded to a $(\rho_{narrow} + \rho_{wide})$-approximate solution for the instance $(\mathcal{I}, W)$.

The Local Ratio Technique. The local ratio technique (see [2–4,6]) is based on the Local Ratio Theorem. Let $\mathbf{w} \in \mathbb{R}^n$ be a profit per unit vector, and let F be a set of feasibility constraints on vectors $\mathbf{x} \in \mathbb{R}^n$. A vector $\mathbf{x} \in \mathbb{R}^n$, which specifies the amount of resource units allocated to the input elements, is a *feasible solution* to a given problem instance $(F, \mathbf{w})$ if it satisfies all of the constraints in F. Its value is the inner product $\mathbf{w} \cdot \mathbf{x}$. We now state the Local Ratio Theorem (of [3]) using profit per unit vectors. The proof of the lemma is immediate.

Lemma 3. *Let F be a set of constraints and let $\mathbf{w}$, $\mathbf{w_1}$, and $\mathbf{w_2}$ be profit per unit vectors such that $\mathbf{w} = \mathbf{w_1} + \mathbf{w_2}$. Then, if $\mathbf{x}$ is an r-approximate solution with respect to $(F, \mathbf{w_1})$ and with respect to $(F, \mathbf{w_2})$, then it is an r-approximate solution with respect to $(F, \mathbf{w})$.*

3   A 3-Approximation Algorithm for FBAP

In this section we present a polynomial-time 3-approximation algorithm for the problem FBAP. Given an instance $(\mathcal{I}, W)$ of FBAP, the algorithm starts by forming two sets of intervals: $\mathcal{I}_{wide}$ and $\mathcal{I}_{narrow}$, using $\delta = 1/2$ as defined in Section 2. For the wide intervals, $\mathcal{I}_{wide}$, the algorithm reduces the problem to the *Maximum Weight Independent Set* (MWIS) in interval graphs, which has a optimal polynomial-time algorithm [12]. Since each interval requires at least more than half of the resource, no pair of intersecting intervals can be colored together, therefore by reducing it to an interval with a width of its maximal resource requirement we get an optimal solution. For the narrow intervals, $\mathcal{I}_{narrow}$, we present a 2-approximation algorithm. The algorithm returns expansion to $\mathcal{I}$ of the color assignment of larger profit. By Lemma 2, we obtain a 3-approximation for FBAP$(\mathcal{I}, W)$.

FBAP on Narrow Paths. In the following we show algorithm NARROWFBAP and prove that it yields a 2-approximation for the narrow intervals. The input for the algorithm is a tuple $(\mathcal{I}, \mathbf{w})$, where $\mathcal{I}$ is a set of intervals and $\mathbf{w} \in \mathbb{R}^n$ is the profit per unit vector of $\mathcal{I}$. Algorithm NARROWFBAP uses the local ratio technique, it is recursive and works as follows. The algorithm starts by removing all intervals of non-positive profit per unit value as they don't change the optimum value. If no intervals remain, then it returns $\emptyset$. Otherwise, it chooses an interval $\tilde{I}$ with the minimum right endpoint. It constructs a new profit per unit vector $\mathbf{w_1}$, which assign profit only to intervals which intersect $\tilde{I}$ and solves the problem recursively on $\mathbf{w_2} = \mathbf{w} - \mathbf{w_1}$. Then, if the solution that was computed recursively has at least $a(\tilde{I})$ colors available for $\tilde{I}$, it adds $\tilde{I}$ to the solution with the maximal amount of colors such that the feasibility is maintained. For a profit per unit vector $\mathbf{w}$, the total profit of a feasible coloring σ of a subset $\mathcal{I}' \subseteq \mathcal{I}$ is denoted by $profit^{\sigma}(\mathcal{I}', \mathbf{w})$. Given a solution S, the load of an edge e in S is defined as $load(S, e) \overset{def}{=} \sum_{I \in \mathcal{I}_e \cap \mathcal{I}_S} |\sigma_S(I)|$.

We note that Algorithm NARROWFBAP returns the number of colors assigned to each interval. The algorithm can be easily modified to return the coloring of the intervals, by changing Line 7 to add to the solution the list of assigned allocated colors, instead of the coloring size.

Algorithm 1. NARROWFBAP$(\mathcal{I}, \mathbf{w})$

1: $\mathcal{I} \leftarrow \mathcal{I} \setminus \{I \in \mathcal{I} : w(I) \leq 0\}$
2: **If** $\mathcal{I} = \emptyset$ return $\emptyset$
3: Select an interval $\tilde{I} \in \mathcal{I}$ with a minimum right endpoint
4: Define for each $I \in \mathcal{I}$

$$w_1(I) = w(\tilde{I}) \cdot \begin{cases} 1 & I = \tilde{I}, \\ \frac{b(\tilde{I})}{W-a(\tilde{I})} & I \neq \tilde{I}, I \cap \tilde{I} \neq \emptyset, \\ 0 & \text{otherwise.} \end{cases}$$

   and $\mathbf{w_2} = \mathbf{w} - \mathbf{w_1}$
5: $S' \leftarrow$ NARROWFBAP$(\mathcal{I}, \mathbf{w_2})$
6: Let $\tilde{e}$ be the rightmost edge of $\tilde{I}$
7: **If** $a(\tilde{I}) \leq W - load(S', \tilde{e})$, **then** $S \leftarrow S' \cup \{(\tilde{I}, \min\{b(\tilde{I}), W - load(S', \tilde{e})\}\}$
8: **Else** $S \leftarrow S'$
9: **Return** S

Theorem 1. *Algorithm* NARROWFBAP$(\mathcal{I}, \mathbf{w})$ *computes in polynomial time a 2-approximate solution for any* FBAP *instance in which* $b(I) \leq W/2$ *for all* $I \in \mathcal{I}$.

Proof. Clearly, the first step, in which intervals of non-positive profit are deleted, does not change the optimum value. Thus, it is sufficient to show that S is a 2-approximation with respect to the remaining intervals. The proof is by induction on the number of recursive calls. At the basis of the recursion, the solution returned is optimal and is a 2-approximation, since $\mathcal{I} = \emptyset$. For the inductive step, we show that the returned solution S is a 2-approximation with respect to $\mathbf{w_1}$ and $\mathbf{w_2}$, and thus, by the Lemma 2, it is 2-approximation with respect to $\mathbf{w}$. Assuming that S' is a 2-approximation with respect to $\mathbf{w_2}$, we have to show that S is a 2-approximation with respect to $\mathbf{w_2}$. Since $w_2(\tilde{I}) = 0$ and $S' \subseteq S$, it follows that S is a 2-approximation with respect to $\mathbf{w_2}$.

We now show that S is a 2-approximation with respect to $\mathbf{w_1}$. In order to prove this, we need the following claims.

Claim 1. *For the solution* S, $profit^{\sigma_S}(\mathcal{I}_S, \mathbf{w_1}) \geq w_1(\tilde{I}) \cdot b(\tilde{I})$.

Proof. The claim holds since, either $\tilde{I} \in \mathcal{I}_S$ and $a(\tilde{I}) \leq \left|\sigma_S(\tilde{I})\right| \leq b(\tilde{I})$, or $S' \cup \{(\tilde{I}, a(\tilde{I}))\}$ is infeasible. For the case that $\tilde{I} \in \mathcal{I}_S$, if $\left|\sigma_S(\tilde{I})\right| = b(\tilde{I})$, then $profit^{\sigma_S}(\mathcal{I}_S, \mathbf{w_1}) \geq w_1(\tilde{I}) \cdot b(\tilde{I})$; else, $\tilde{I} \in \mathcal{I}_S$ and thus the profit accrued by the intervals intersecting $\tilde{I}$ is $w_1(\tilde{I}) \cdot \frac{b(\tilde{I})}{W-a(\tilde{I})} \cdot (W - \left|\sigma_I(\tilde{I})\right|)$. In addition, $\left|\sigma_S(\tilde{I})\right| <$ $b(\tilde{I})$, and since $b(\tilde{I}) \leq W/2$, we have that $a(\tilde{I}) + b(\tilde{I}) \leq W$, and thus

$$profit^{\sigma_S}(\mathcal{I}_S, \mathbf{w_1}) = w_1(\tilde{I}) \cdot \left|\sigma_S(\tilde{I})\right| + w_1(\tilde{I}) \cdot \frac{b(\tilde{I})}{W-a(\tilde{I})} \cdot (W - \left|\sigma_S(\tilde{I})\right|)$$
$$= w_1(\tilde{I}) \cdot \left|\sigma_S(\tilde{I})\right| \cdot (1 - \frac{b(\tilde{I})}{W-a(\tilde{I})}) + w_1(\tilde{I}) \cdot b(\tilde{I}) \cdot \frac{W}{W-a(\tilde{I})}$$
$$\geq w_1(\tilde{I}) \cdot b(\tilde{I}).$$

Consider now the case where $\tilde{I} \notin \mathcal{I}_S$. Since $S' \cup \{(\tilde{I}, a(\tilde{I}))\}$ is infeasible, it follows that $profit^{\sigma_S}(\mathcal{I}_S, \mathbf{w_1}) \geq w_1(\tilde{I}) \cdot \frac{b(\tilde{I})}{W-a(\tilde{I})} \cdot (W - a(\tilde{I}) + 1) \geq w_1(\tilde{I}) \cdot b(\tilde{I})$. Therefore we conclude that $profit^{\sigma_S}(\mathcal{I}_S, \mathbf{w_1}) \geq w_1(\tilde{I}) \cdot b(\tilde{I})$. $\square$

Claim 2. *For any optimal solution* S^*, $profit^{\sigma_{S^*}}(\mathcal{I}_{S^*}, \mathbf{w_1}) \leq 2 \cdot w_1(\tilde{I}) \cdot b(\tilde{I})$.

Proof. The claim holds since if $\left|\sigma_{S^*}(\tilde{I})\right| \geq a(\tilde{I})$, then

$$profit^{\sigma_{S^*}}(\mathcal{I}, \mathbf{w_1}) \leq w_1(\tilde{I}) \cdot \left|\sigma_{S^*}(\tilde{I})\right| + w_1(\tilde{I}) \cdot \frac{b(\tilde{I})}{W-a(\tilde{I})} \cdot (W - \left|\sigma_{S^*}(\tilde{I})\right|)$$
$$\leq w_1(\tilde{I}) \cdot b(\tilde{I}) + w_1(\tilde{I}) \cdot b(\tilde{I}) \cdot (\frac{W - \left|\sigma_{S^*}(\tilde{I})\right|}{W-a(\tilde{I})})$$
$$\leq 2 \cdot w_1(\tilde{I}) \cdot b(\tilde{I}).$$

Else, $\left|\sigma_{S^*}(\tilde{I})\right| = 0$, and thus we have that $profit^{\sigma_{S^*}}(\mathcal{I}, \mathbf{w_1}) \leq w_1(\tilde{I}) \cdot \frac{b(\tilde{I})}{W-a(\tilde{I})} \cdot W$, and since $a(\tilde{I}) \leq \frac{W}{2}$ we get that $profit^{\sigma_{S^*}}(\mathcal{I}_{S^*}, \mathbf{w_1}) \leq 2 \cdot w_1(\tilde{I}) \cdot b(\tilde{I})$. $\square$

Combining Claim 1 and Claim 2, we have that S is 2-approximate solution with respect to $\mathbf{w_1}$. By the Lemma 3, S is also a 2-approximate solution with respect to $\mathbf{w}$. The running time is polynomial, since the number of recursive call is at most the number of input intervals, and each call requires linear time. $\square$

Combining the exact algorithm for MWID in interval graphs of [12], Theorem 1, and Lemma 2, we conclude

Theorem 2. *There exists a polynomial-time 3-approximation algorithm for the problem* FBAP.

4   A $(3 + \epsilon)$-Approximation Algorithm for FSAP

We now show that our result for FBAP can be extended to yield (almost) the same bound for FSAP. Given an instance $\mathcal{I}$ of FSAP, we form two sets of intervals: $\mathcal{I}_{narrow}$ and $\mathcal{I}_{wide}$, as defined in Section 2, using a parameter $\delta > 0$ (to be determined). For the wide intervals, $\mathcal{I}_{wide}$, we present a $(1+\epsilon)$-approximation algorithm for any fixed $\epsilon > 0$. For the narrow intervals, $\mathcal{I}_{narrow}$, we give a $(2 + \epsilon)$-approximation algorithm for any fixed $\epsilon > 0$. The algorithm selects color assignment of larger profit among the assignments found for $\mathcal{I}_{narrow}$ and $\mathcal{I}_{wide}$. This assignment is then expanded into a solution for the original instance, $\mathcal{I}$. By Lemma 2, we obtain a $(3 + \epsilon)$-approximate solution for FSAP, for any fixed $\epsilon > 0$. We note that any future improvements in the approximation ratio of FBAP would improve also the approximation ratio of our algorithm for FSAP.

FSAP on Wide Intervals. In the full version of this paper [17] we show a $(1+\epsilon)$-approximation algorithm for the wide instance of FSAP. Specifically, we prove

Lemma 4. *Given an instance $\mathcal{I}$ of FSAP, for any fixed $\epsilon > 0$, there exists $\delta > 0$, such that there is a polynomial-time $(1 + \epsilon)$-approximation algorithm for the wide intervals in $\mathcal{I}$.*

FSAP on Narrow Intervals. We now show how to obtain a $(2 + \epsilon)$-approximate solution for the $\mathcal{I}_{narrow}$ intervals. Recall, than in BAP, we are given a path having one unit of available resource, and a set $\mathcal{I}$ of intervals on the path. Each interval $I \in \mathcal{I}$ consists of an arrival time and departure time, a resource requirement $s(I) \in [0, 1]$, and a profit $p(I) \in \mathbb{Z}$. The goal is to assign the resource to a maximum weight subset of requests. A solution S for BAP consists of the assigned intervals. The profit S is given by $profit(S) \overset{def}{=} \sum_{I \in S} p(I)$. SAP is a special case of BAP in which we require that each interval is allocated a single contiguous block of resource for its entire duration.

We use as subroutines Algorithm NARROWFBAP (of Section 3) and a subroutine of an algorithm for SAP due to [5], which transforms in polynomial-time a BAP solution into a SAP solution, as formulated in the following lemma.

Lemma 5. ([5]) *There exists a constant $\delta_0 \in (0, 1]$, such that if S is a solution for BAP on intervals $\mathcal{I}$ for which $s(I) \leq \delta$ for all $I \in \mathcal{I}$, where $\delta \in (0, \delta_0)$, then S can be transformed in polynomial time into a solution for SAP with profit at least $(1 - 4\delta)profit(S)$.*

Combining Theorem 1 and Lemma 5, we have (for proof see the full version of this paper [17])

Lemma 6. *Given an instance $\mathcal{I}$ of FSAP, for any fixed $\epsilon > 0$, there exists $\delta > 0$, such that there is a polynomial-time $(2 + \epsilon)$-approximation algorithm for the narrow intervals in $\mathcal{I}$.*

Combining Lemmas 4, 6, and 2, we obtain

Theorem 3. *There exists a polynomial-time $(3 + \epsilon)$-approximation algorithm for FSAP, for any fixed $\epsilon > 0$.*

5   A $(2 + \epsilon)$-Approximation Algorithm for SAP

In this section we consider SAP, the special case of FSAP where $a(I) = b(I)$ for all $I \in \mathcal{I}$. We present a polynomial-time $(2 + \epsilon)$-approximation algorithm for any fixed $\epsilon > 0$.

In deriving the algorithm we use a technique similar to the one used for solving the general instance of FSAP. Given an instance $\mathcal{I}$ of SAP, we partition the intervals into two sets: *narrow* and *wide*, using a parameter $\delta > 0$ (to be determined). Specifically, *narrow* intervals are those for which $|s(I)| \leq \delta W$ and *wide* intervals are those for which $|s(I)| > \delta W$. For the *wide* intervals, the algorithm runs an optimal polynomial-time dynamic programming algorithm of [5] for SAP on *wide* intervals. For the *narrow* intervals we show how to obtain a $(1 + \epsilon)$-approximate solution. The algorithm returns the color assignment of greater profit. By Lemma 2, this yield a $(2+\epsilon)$-approximation algorithm for SAP.

SAP on Narrow Intervals. In the following we show how to obtain a $(1 + \epsilon)$-approximate solution for the *narrow* intervals. We rely on two subroutines of known algorithms for BAP and SAP. For BAP, the paper [7] presents a $(2 + \epsilon)$-approximation algorithm, for any $\epsilon > 0$. The authors obtain this result by dividing the input intervals into *wide* and *narrow* intervals, for some $\delta \in (0, 1)$. They use dynamic programming to compute an optimal solution for the *wide* intervals, and LP-based algorithm to obtain a $(1+0(1)\sqrt{\delta})$-approximate solution for the *narrow* intervals, as states in the next result.

Lemma 7. *([7]) There exist constants $\delta_1 \in (0,1)$ and $C > 0$, such that for any $\delta \in (0, \delta_1)$ there exists a $(1 + C\sqrt{\delta})$-approximation algorithm for the* narrow *intervals of* BAP.

The second subroutine that we use is an algorithm for SAP of [5], which transforms in polynomial-time a BAP solution into a SAP solution (Lemma 5). Combining Lemma 7 and Lemma 5, we have (for proof see the full version of this paper [17])

Lemma 8. *For any fixed $\epsilon > 0$, there exists $\delta > 0$, such that there is a polynomial-time $(1 + \epsilon)$-approximation algorithm for the narrow intervals of* SAP.

Summarizing the above discussion, we have (for proof see the full version of this paper [17])

Theorem 4. *For any fixed $\epsilon > 0$, there is a polynomial-time $(2+\epsilon)$-approximation algorithm for* SAP.

6   Applications of FSAP and FBAP

The problems FSAP and FBAP have important applications in real-time scheduling. Consider, for example, a reusable resource of fixed size and activities that have a minimum and a maximum range for contiguous or non-contiguous resource requirement. The resource may be memory, computing units, servers in a Cloud, or network bandwidth. The allocated amount of resource for the activities actually determine it performance, quality-of-service, or processing time. In the following we present the application of FBAP and FSAP in optimizing spectrum assignment in all-optical networks.

Spectrum Assignment in All-Optical Networks. In modern optical networks several high-speed signals are sent through a single optical fiber. A signal transmitted optically from some source node to some destination node over a wavelength is termed *lightpath* (for comprehensive surveys on optical networks, see, e.g., [1,16]). Traditionally, the spectrum of light that can be transmitted through the fiber has been divided into frequency intervals of *fixed* width, with a gap of unused frequencies between them. In this context, the term wavelength refers to each of these predefined frequency intervals. An emerging architecture, which moves away from the rigid model towards a flexible model, was suggested in [10,13]. In this model, the usable frequency intervals are of *variable* width (even within the same link). Every lightpath has to be assigned a frequency interval (sub-spectrum), which remains fixed through all of the links it traverses. As in the traditional model, two different lightpaths using the same link must be assigned disjoint sub-spectra. This technology is termed *flex-grid* (or, *flex-spectrum*), as opposed to the *fixed-grid* (or, *fixed-spectrum*) traditional technology. The network implications of this new architecture are described in detail in [11]. The following spectrum assignment problems arising in the fixed-grid and flex-grid technology correspond to FBAP and FSAP, respectively. We are given a set of flexible connection requests, each with a lower and upper bound on the width of its spectrum request, as well as an associated positive profit per allocated spectrum unit. In the fixed-grid (or flex-grid) technology the goal is to find a non-contiguous (or contiguous) spectrum assignment for a subset of requests that maximizes the total profit.

7   Summary and Future Work

In this paper we studied the FBAP and FSAP problems. We observed that both problems are NP-Hard even for highly restricted inputs. We also presented a 3-approximation and a $(3 + \epsilon)$-approximation algorithms for general input of FBAP and FSAP, respectively.

We point to a few of the many problems that remain open. (a) We showed that FSAP is NP-Hard for the subclass of instances where $a(I) \neq b(I)$ for all $I \in \mathcal{I}$. For these cases it would be interesting to obtain a better approximation ratio than the one derived for the general instance. (b) Our results for intervals on a line call for the study FBAP and FSAP in other graph, especially those that are relevant in optical networks. (c) The flex-grid technology enables to combine non-contiguous and contiguous spectrum assignment; thus, a request can be assigned either a contiguous or non-contiguous set of colors. In this setting, a non-contiguous color assignment for any request requires accumulatively more spectrum than contiguous color assignment of the same request (due to the gap of unused frequencies between wavelengths). This setting opens up an unexplored terrain for future study. (d) Finally, as stated above, FSAP and FBAP are the flexible variants of the classic SAP and BAP, respectively. There is much importance in exploring the implications of these new problems and our results in the context of resource allocation, emerging computing, and network technologies.

Acknowledgment. We thank Dror Rawitz for valuable discussions. We also thank the anonymous referees for valuable comments.

References

1. Ali Norouzi, A.Z., Ustundag, B.B.: An integrated survey in optical networks: Concepts, components and problems. IJCSNS International Journal of Computer Science and Network Security 11(1), 10–26 (2011)

2. Bafna, V., Berman, P., Fujito, T.: A 2-approximation algorithm for the undirected feedback vertex set problem. SIAM Journal on Discrete Mathematics 12(3), 289–297 (1999)
3. Bar-Noy, A., Bar-Yehuda, R., Freund, A., Naor, J.S., Schieber, B.: A unified approach to approximating resource allocation and scheduling. Journal of the ACM, 735–744 (2000)
4. Bar-Yehuda, R.: One for the price of two: A unified approach for approximating covering problems. Algorithmica 27(2), 131–144 (2000)
5. Bar-Yehuda, R., Beder, M., Cohen, Y., Rawitz, D.: Resource allocation in bounded degree trees. Algorithmica 54(1), 89–106 (2009)
6. Bar-Yehuda, R., Even, S.: A local-ratio theorem for approximating the weighted vertex cover problem. Annals of Discrete Mathematics 25, 27–46 (1985)
7. Chekuri, C., Mydlarz, M., Shepherd, F.B.: Multicommodity demand flow in a tree and packing integer programs. ACM Transaction on Algorithms 3(3) (2007)
8. Chen, B., Hassin, R., Tzur, M.: Allocation of bandwidth and storage. IIE Transactions 34(5), 501–507 (2002)
9. Chrobak, M., Woeginger, G.J., Makino, K., Xu, H.: Caching is hard - even in the fault model. Algorithmica 63(4), 781–794 (2012)
10. Gerstel, O.: Flexible use of spectrum and photonic grooming. In: Photonics in Switching, OSA (Optical Society of America)Technical Digest (2010)
11. Gerstel, O.: Realistic approaches to scaling the ip network using optics. In: Optical Fiber Communication Conference and Exposition (OFC/NFOEC), 2011 and the National Fiber Optic Engineers Conference, pp. 1–3 (2011)
12. Golumbic, M.C.: Algorithmic graph theory and perfect graphs. Academic Press, New York (1980)
13. Jinno, M., Takara, H., Kozicki, B., Tsukishima, Y., Sone, Y., Matsuoka, S.: Spectrum-efficient and scalable elastic optical path network: architecture, benefits, and enabling technologies. IEEE Communications Magazine 47(11), 66–73 (2009)
14. Katz, D., Schieber, B., Shachnai, H.: The flexible storage allocation problem (submitted, 2014)
15. Leonardi, S., Marchetti-Spaccamela, A., Vitaletti, A.: Approximation algorithms for bandwidth and storage allocation problems under real time constraints. In: Kapoor, S., Prasad, S. (eds.) FST TCS 2000. LNCS, vol. 1974, pp. 409–420. Springer, Heidelberg (2000)
16. Ramaswami, R., Sivarajan, K., Sasaki, G.: Optical Networks: A Practical Perspective, 3rd edn. Morgan Kaufmann Publishers Inc., San Francisco (2009)
17. Shachnai, H., Voloshin, A., Zaks, S.: Flexible bandwidth assignment with application to optical networks, http://www.cs.technion.ac.il/~hadas/PUB/flex.pdf
18. Shachnai, H., Voloshin, A., Zaks, S.: Optimizing bandwidth allocation in flex-grid optical networks with application to scheduling. In: 28th IEEE International Parallel and Distributed Processing Symposium (IPDPS), Phoenix, USA (May 2014)
19. Shalom, M., Wong, P.W.H., Zaks, S.: Profit maximization in flex-grid all-optical networks. In: Moscibroda, T., Rescigno, A.A. (eds.) SIROCCO 2013. LNCS, vol. 8179, pp. 249–260. Springer, Heidelberg (2013)

Approximation Algorithms for Bounded Color Matchings via Convex Decompositions[*]

Georgios Stamoulis[1,2]

[1] LAMSADE, CNRS UMR 7243, Universitè Paris-Dauphine, France
[2] Universitá della Svizzera Italiana (USI),
Lugano, Switzerland
stamoulis.georgios@dauphine.fr

Abstract. We study the following generalization of the maximum matching problem in general graphs: Given a simple non-directed graph $G = (V, E)$ and a partition of the edges into k classes (i.e. $E = E_1 \cup \cdots \cup E_k$), we would like to compute a matching M on G of maximum cardinality or profit, such that $|M \cap E_j| \leq w_j$ for every class E_j. Such problems were first studied in the context of network design in [17]. We study the problem from a linear programming point of view: We provide a polynomial time $\frac{1}{2}$-approximation algorithm for the weighted case, matching the integrality gap of the natural LP formulation of the problem. For this, we use and adapt the technique of *approximate convex decompositions* [19] together with a different analysis and a polyhedral characterization of the natural linear program to derive our result. This improves over the existing $\frac{1}{2}$, but with additive violation of the color bounds, approximation algorithm [14].

1 Introduction

In modern optical fiber network systems, we encode the information as an electromagnetic signal and we transfer it through the optical fiber as a beam of light in a specified frequency. Typically, at most one beam of light is allowed to travel through the fiber at any given time. In WDM[1] optical networks we allow *multiplexing* of a number of different light beams to travel simultaneously through the optical fiber as follows: We partition the electromagnetic spectrum into a number of k non-overlapping intervals. For each interval f_i we have an upper bound on how many different beams of light that have frequencies within this interval can travel at the same time through the optical carrier. This constraint is imposed since, otherwise, we would have quantum phenomena as *interference* of the light beams of a given interval. Naturally, if we allow a large number of beams of light with frequencies within a small interval to travel through the optical fiber, we can expect with very high probability two or more beams to be interfered. Our goal in an optical network is to establish communication between an as large as

---

[*] Part of this work was done while the author was a PhD student at IDSIA.
[1] WDM stands for Wavelength-Division Multiplexing

E. Csuhaj-Varjú et al. (Eds.): MFCS 2014, Part II, LNCS 8635, pp. 625–636, 2014.

possible number of pairs that want to communicate in their own frequency, such that in a given interval of frequencies f_i we allow no more than w_i connections to be established. This naturally reduces to the following problem:

Bounded Color Matching: We are given a (simple, non-directed) graph $G = (V, E)$. The edge set is partitioned into k sets $E_1 \cup E_2 \cup \cdots \cup E_k$ i.e. every edge e has color c_j if $e \in E_j$ and a profit $p_e \in \mathbb{Q}^+$. We are asked to find a maximum (weighted) matching M such that in M there are no more that w_j edges of color c_j, where $w_j \in \mathbb{Z}^+$ i.e. a matching M such that $|M \cap E_j| \leq w_j, \forall j \in [k]$.

In the following, we denote as $\mathcal{C}$ the collection of all the color classes. In other words, $\mathcal{C} = \{c_j\}_{j \in [k]}$. Moreover, for a given edge $e \in E(G)$, we denote by $c^{-1}(e)$ its color i.e. $c^{-1}(e) = c_j \Leftrightarrow e \in c_j$.

Bounded Color Matching is a budgeted version of the classical matching problem: For a given instance G, let $\mathcal{F}$ be the set of all feasible solutions. Associated with every feasible solution $M \in \mathcal{F}$ we are given a set of ℓ linear *cost* functions $\{\alpha_i\}_{i \in [\ell]}$ and a linear profit function π such that $\pi, \alpha_i : \mathcal{F} \to \mathbb{Q}^+$ and for every cost function α_i a *budget* $\beta_i \in \mathbb{Q}^+$. The goal is to find $M \in \mathcal{F} : \alpha_i(F) \leq \beta_i, \forall i \in [\ell]$ that also maximizes $\pi(M)$. Budgeted versions of the maximum matching problem have been recently studied intensively. When G is bipartite there is a PTAS for the case where $\ell = 1$ [2] and the case where $\ell = \mathcal{O}(1)$ [11]. For general graphs there is a PTAS for the 2-budgeted maximum matching problem [12] and a bicriteria PTAS for $\ell = \mathcal{O}(1)$ [7] (where the returned solution might violate the budgets by a factor of $(1 + \epsilon)$). This approach works also for *unbounded* number of budgets albeit a *logarithmic* overflow of the budgets.

Bounded Color Matching (BCM) not only is **NP**-hard even in bipartite graphs where $w_j = 1$, $\forall c_j \in \mathcal{C}$ [10] but also **APX**-hardness can be deduced even in 2-regular bipartite graphs from [15]. In [13] the BCM was considered from a bi-criteria point of view: given a parameter $\lambda \in [0, 1]$ there is an $(\frac{2}{3+\lambda})$ approximation algorithm for BCM which might violate the budgets w_j by a factor of at most $(\frac{2}{1+\lambda})$.

To the best of our knowledge, the first case where matching problems with cardinality (disjoint) budgets were considered, was in [17] where the authors defined and studied the *blue-red* Matching problem: compute a maximum (cardinality) matching that has at most w blue and at most w red edges, in a blue-red colored (multi)-graph. A $\frac{3}{4}$ polynomial time combinatorial approximation algorithm and an **RNC²** algorithm (that computes the maximum matching that respects both budget bounds with high probability) were presented. This was motivated by network design problems, in particular they showed how *blue-red* Matching can be used in approximately solving the Directed Maximum Routing and Wavelength Assignment problem (DirMRWA) [16] in *rings* which is a fundamental network topology, see [17] (also [4] for alternative and slightly better approximation algorithms and [1] for combinatorial algorithms). Here, approximately solving means that an (asymptotic) α-approximation algorithm for blue-red results in an (asymptotic) $\frac{\alpha+1}{\alpha+2}$-approximation algorithm for DirMRWA in rings.

We note that the exact complexity of the blue-red matching problem is not known: it is only known that blue-red matching is at least as hard as the Exact

Matching problem [18] whose complexity is open for more than 30 years. A polynomial time algorithm for the blue-red matching problem will imply that Exact Matching is polynomial time solvable. On the other hand, blue-red matching is probably not **NP**-hard since it admits an **RNC²** algorithm. We note that this algorithm can be extended to a constant number of color classes with arbitrary bounds w_j. Using the results of [20] one can deduce an "almost" optimal algorithm for blue-red matching, i.e. an algorithm that returns a matching of maximum cardinality that violates the two color bounds by at most one edge. This is the best possible, unless of course blue-red matching (and, consequently, exact matching) is in **P**.

If we formulate BCM as a linear program, the polyhedron $\mathcal{M}_c$ containing all feasible matchings M for the BCM is

$$\mathcal{M}_c = \left\{ \boldsymbol{y} \in \{0,1\}^{|E|} : \ \boldsymbol{y} \in \mathcal{M} \bigwedge \sum_{e \in E_j} y_e \leq w_j, \ \forall j \in [k] \right\} \tag{1}$$

where $\mathcal{M}$ is the usual matching polyhedron: $\mathcal{M} = \{x \in \{0,1\}^{|E|} : \ \sum_{e \in \delta(v)} x_e \leq 1, \ \forall v \in V\}$. We would like to find $\boldsymbol{y} \in \{0,1\}^{|E|}$ such that $\boldsymbol{y} = \max_{x \in \mathcal{M}_c}\{\boldsymbol{p}^T \boldsymbol{x} = \sum_{e \in E} p_e x_e\}$, $\boldsymbol{p} \in \mathbb{Q}_{\geq 0}^{|E|}$. As usual, we relax the integrality constraints $\boldsymbol{y} \in \{0,1\}^{|E|}$ to $\boldsymbol{y} \in [0,1]^{|E|}$ and we solve the corresponding linear relaxation efficiently to obtain a *fractional* $|E|$-dimensinal vector $\boldsymbol{y}$. It is not hard to show (see later section) that the *integrality gap* of $\mathcal{M}_c$ is essentially 2 and this is true even if we add the blossom inequalities i.e. if instead of $\mathcal{M}$ as defined here, we use the well known Edmond's LP [8].

Our Contribution: In this work we study the BCM problem with *unbounded* number of budgets: we provide a deterministic $\frac{1}{2}$ approximation algorithm based on the concept of *approximate convex decompositions* from [19] together with a different analysis and an extra step based on polyhedral properties of extreme point solutions of $\mathcal{M}_c$. This might be helpful also in the context of k-uniform b-matching problem. This result improves over the $\frac{1}{2}$ but with an additive violation of the color bounds w_j from [13] and matches the integrality gap of 2 of the natural linear formulation of the problem (captured by 1) which implies that a $\frac{1}{2}$ approximation algorithm is the best we can hope using this natural linear relaxation.

We note that the BCM problem can be easily seen as a special case of the 3-hypergraph β-matching problem [19] or 3-set packing. Using the existing LP-based results for these problems which guarantee a $\frac{1}{k-1+\frac{1}{k}}$ for the k-hypergraph β-matching, we can only guarantee a $\frac{3}{7}$-approximation algorithm [19]. We show that by taking advantage of the special structure of the problem we can do better than this. For 3-set packing there there exists a $\frac{1}{2} - \epsilon$ approximation algorithm for the *weighted* case [3] and a recent $\frac{3}{4}$ for the non-weighted case [9]. But these tell us nothing about the strengths (and limitations) of linear programming techniques for the problem, which is our main motivation. See also [6] for a similar study on the effect of linear programming techniques on k-dimensional matching problems.

2   A $\frac{1}{2}$ Approximation Based on Approximate Convex Decompositions

In this section we will provide a polynomial time $\frac{1}{2}$-approximation algorithm for BCM based on the technique of approximate convex decompositions from [19]. Given x^*, an optimal solution of the LP for the BCM, the main idea of the algorithm is to construct a collection of *feasible* (for the BCM problem) matchings $\mu_1, \mu_2, \ldots, \mu_\rho$ for some ρ, such that an approximate version of $p^T x$ can be written as a *convex combination* of these matchings μ_i. Recall the famous Carathéodory's theorem:

Theorem 1 (Carathéodory[5]). *Let $P = \{x \mid Ax \leq b\}$ where $x \in \mathbb{R}^n, A \in \mathbb{R}^{m \times n}$ and $b \in \mathbb{R}^m$ be a polyhedron in $\mathbb{R}^n$. Assume a point $z \in P$ and assume that this point satisfies r of the m inequalities with equality. Then z can be written as a convex combination of at most $n - r + 1$ vertices of the polyhedron P.*

As an immediate corollary, we have that any point z belonging in (the convex hull of) a bounded convex polyhedron $P \subseteq \mathbb{R}^n$ can be written as a convex combination of at most $n + 1$ vertices of P.

Let x^* be an optimal (fractional) solution of the relaxation of $\mathcal{M}_c$. Ideally we would like to use Carathéodory's theorem to write x^* as a *convex* combination of feasible integral extreme point solutions (vertices) of $\mathcal{M}_c$. But unfortunately this is not always the case, meaning that x^* might not belong in the convex hull of all feasible integral vertices of $\mathcal{M}_c$. Instead of that, we will settle for an *approximate* convex combination of x^* by vertices of $\mathcal{M}_c$. An approximate convex combination of x^* is a convex combination of ρ extreme points μ_i of $\mathcal{M}_c$ satisfying the following:

$$\alpha p^T x^* = \sum_{i \in [\rho]} \lambda_i \mu_i, \quad \alpha \in (0,1], \quad \sum_{i \in [\rho]} \lambda_i = 1, \quad \mu_i \in \mathcal{M}_c \qquad (2)$$

The fact that we insist for convex combination directly implies that $\lambda_i \geq 0, \forall i$. An important feature of the above convex combination is that it gives us immediately an α approximation algorithm. Indeed, since all points μ_i, for $i \in [\rho]$, are feasible and they constitute a *convex* combination of $p^T x^*$ then, by a standard averaging argument, at least one of the μ_is will have profit at least $\alpha \cdot p^T x^*$, and so we have the following:

Lemma 1. *Given an optimal fractional solution x^* of the relaxation of $\mathcal{M}_c$ and assume that x^* can be written as in (2) then we can retrieve a feasible integral solution for $\mathcal{M}_c$ with total profit at least α times $p^T x^*$.*

We will inductively construct an α approximate convex combination of $\bar{x}^*$ where $\bar{x}^*$ is x^* without a specific edge (i.e. after removing an edge e in $\mathsf{supp}(x^*)$ where we define for any vector $y \in \mathbb{R}^n$ $\mathsf{supp}(y) = \{j \in [n] : y_j \neq 0\}$). Then, if we can add e in a α fraction of the points $\bar{\mu}_i$ constituting the α approximate convex combination of $\bar{x}^*$, we will have a convex combination with the desired properties. In other words, we require that $\sum_{j \in \kappa} \lambda_j \geq \alpha x_e^*$, where $\kappa = \{\mu_i : \mu_i \cup e \in \mathcal{M}_c\}$ is the set of all the solutions (that constitute the approximate version of $\bar{x}^*$) that can facilitate e preserving feasibility.

The inductive process of obtaining (constructing) μ_is is roughly the following: at the "bottom" of the induction process we start with the trivial empty solution ($\mu_1 = \emptyset$). At each next step we try to pack any of the edges e_j into exactly an α fraction of the current set of the matchings. This is the step where we may create new solutions in order to maintain the invariant "pack any edge into exactly α fraction". Usually this process in its basic form greedily packs e_j and, moreover, is oblivious to any ordering of the edges. In our case however, this cannot lead to any meaningful approximation guarantee. We will show how we can carefully select an edge at the current step i such that edge e_i to be packed can fit into an $\frac{1}{2}$ fraction of the current set of solutions $\{\mu_j\}$. In other words, we will define an ordering of the edge set $(e_1, \ldots, e_m)$ ($m = |E(G)|$) such that at step $i \in [m]$, assuming that we have an α-approximate convex decomposition for $(e_1, \ldots e_{i-1})$ (characterized by a set of feasible matchings $\{\mu_j\}$), then the current edge e_i can be successfully inserted into at least an α fraction of the μ_js. We will show that in our case we can in fact select $\alpha = \frac{1}{2}$, thus giving us the desired approximation guarantee.

Now we will define the ordering on the edges $\{e_i\}_{i \in [m]}$. Assume that we are in some inductive step where the remaining edges are $R_i = \{e_1, e_2, \ldots, e_i\}$ for some i, i.e. we have removed $m - i$ edges to iteratively obtain an α-approximate decomposition for the remaining solution. How do we choose which edge to select from R_i? Intuitively, the larger the fractional value of x_e is, the larger the fraction of matchings μ_i that can be added is. This is because high fractional value of x_e implies low fractional values of the other "blocking" components. And low value of these components means "few" matchings μ_j that are actually blocking e. So, a good starting strategy is to select at each step edges with high fractional value. Unfortunately, we cannot always guarantee that such edges are present in x^*. But there is hope: let x' be x^* restricted to the set of edges R_i. Now, it is not hard to see (we omit the easy proof) that x' is an extreme point solution for the *reduced* instance where we set $w_j := w_j - \sum_{e \in E_j \setminus R_i} x_e^*$, $\forall C_j$ and $\beta_v := \sum_{e \in \delta_v \setminus R_i} x_e^*$, $\forall v \in V$. Such extreme point solutions have very nice properties. To see that, we use the following slight generalization of a result due to [13] where we consider a version where for each vertex $v \in V$ we have a bound $\beta_v \leq 1$ and where w_j are no longer integers such that, when we say "tight" vertex (with respect to x^*) we mean a vertex v such that $\sum_{e \in \delta(v)} x_e = \beta_v$. The same for tight color classes. Initially, all w_j are integers and $\beta_v = 1$, $\forall v$. Define the residual graph with respect to a solution vector x to be the graph where we include an edge e only if $x_e > 0$.

Lemma 2. *Take any basic feasible solution x of $\mathcal{M}_c$ (where we no longer require the degree bounds on vertices and the bounds on color classes be integer anymore) such that $0 < x_e < 1$, $\forall e$ (i.e. remove all integer variables reducing the bounds appropriately if necessary). Then one the following must be true:*

1. *either there is a tight color class $c_j \in Q$ such that $|E_j| \leq \lceil w_j \rceil + 1$ in the residual graph,*
2. *or there is a tight vertex $v \in F$ such that the degree of v in the residual graph $\leq \lceil \beta_v \rceil + 1$.*

The proof follows similar lines as in [13] and we omit it from the current version. In our lemma, F, Q are the linearly independent set of rows of the LP ($\mathcal{M}_c$) that characterize the basic solution x. This lemma will give us what we need in order to successfully select the right edge from R_i and prove that we can successfully pack it into $\frac{1}{2}$ of the matchings for the inductively obtained approximate convex decomposition for the rest of the edges i.e. for R_{i-1}:

---

1. Given x^* select e according to Lemma 2:
   (a) if $\exists v \in F: |\mathsf{supp}(x^* \cap \delta(v))| \leq 2$ then select as e the edge $\in \delta(v): x_e \geq \frac{\beta_v}{2}$.
   (b) else $\exists c_j \in Q: |\mathsf{supp}(x^*) \cap E_j| \leq \lceil w_j \rceil + 1$. Select in this iteration an edge $e \in c_j$ with $x_e^* \geq \frac{w_j}{\lceil w_j \rceil + 1}$.
2. Zero out the coordinate of x^* corresponding to e and let x' the resulting vector.
3. Iteratively obtain an approximate convex decomposition for x'.
4. Add e to the convex decomposition of x' obtained in the previous step.

Using Lemma 2 we will show now how, in each iteration of our algorithm, we can successfully pack an edge $e = \{u, v\} \in \mathsf{supp}(x^*)$ into a large number (i.e. half) of the solutions that constitute the approximate convex decomposition of the residual solution vector x'. In order to insert e into a large number of the solutions μ_i that constitute the approximate convex decomposition of the residual solution vector x', we need to see in what fraction of the μ_i's the edge e *cannot* be added. These are all μ_i's such that

(1): $\exists e' \in \mu_i: u \vee v \in e'$, or
(2): $|\mu_i \cap E_j| = w_j$, where $j \in [k]$ is the color of edge e.

The first condition says that e *cannot* be added to those matchings μ_i (which constitute that approximate convex decomposition of x') that have edges incident to either of the endpoints of e. The second condition says that, additionally, e cannot be added to all μ_is that are "full" of color c_j. All these matchings are blocking the insertion of e, meaning that for such a μ_i, $\mu_i \cup e$ is not feasible anymore for either of the previous two reasons. In order to guarantee that e can be added to exactly α fraction of the matchings we may need to double a current solution μ_i and break its multiplier λ_i appropriately (see appendix).

We will distinguish between two cases (one for each case of the algorithm, i.e. step 1.(a) or 1.(b)) and we will prove the result inductively. The base case of the induction inside the algorithm is the trivial case of the empty graph. Focus, w.l.o.g., at the first execution of the algorithm and assume that we have an α-approximate convex decomposition of x' (x^* without edge e) i.e. $\alpha p^T x' = \sum_i \lambda_i \mu_i$ where μ_i's $\in \mathcal{M}_c$ and $\mathbf{1}^T \mu = 1$. Moreover, by Carathéodory's Theorem, this collection of matchings is *sparse* (at most $|\mathsf{supp}(x')| + 1$ matchings μ_i). Let u, v the endpoints of e selected in the first step of the algorithm and assume that $c^{-1}(e) = j \in [k]$. Firstly, we will handle the case where the edge e has been selected according to the rule 1.(b) of the algorithm, which is slightly easier.

According to the algorithm (using Lemma 2), we know that there must exist a tight color class $c_j \in C$ with the property that

$$|\mathsf{supp}(x^*) \cap E_j| \leq \lceil w_j \rceil + 1 \Rightarrow \exists e \in \mathsf{supp}(x^*) \cap c_j : x_e^* \geq \frac{w_j}{\lceil w_j \rceil + 1}$$

Lemma 3. *If we select to pack an edge e of color c_j according to rule 1.(b) of the algorithm, then the fraction of the solutions μ_i that e can be added is at least $\frac{1}{2}$ i.e. $\sum_{j \in \kappa_e} \lambda_j \geq \frac{1}{2}$.*

Proof. We will show that at least $\frac{1}{2}$ fraction of the solutions μ_i can facilitate such an e. Assume that we can identify a color class $c_j \in Q$ in the *reduced instance* induced by the current set of edges such that for these reduced color bounds we have that $|\mathsf{supp}(x^*) \cap E_j| \leq \lceil w_j \rceil + 1$. Let $e = \{u, v\}$ be an edge from $\mathsf{supp}(x^*) \cap E_j$. Let $\xi \in \mathsf{supp}(x^*)$ be the index of e. Observe that in the reduced solution x' (defined as x^* without e), there are exactly $\lceil w_j \rceil$ edges of color c_j (summing up to $w_j - x_e^*$) and, moreover, this number $\lceil w_j \rceil$ is *at most* the initial (integer) color bound for E_j.

All the solutions μ_i that constitute the α-approximate decomposition for the residual solution x' that block the insertion of e (i.e. the solutions μ_i such that $\mu_i \cup \{e\}$ is not feasible anymore) are these μ_is that have edges adjacent to either of u, v and those that have $\lceil w_j \rceil$ edges of color j and so we have that in an α-approximate decomposition, the fraction of μ_i's that block e is

$$\alpha(1 - x_e) + \alpha(1 - x_e) + \alpha\left(\frac{w_j - x_e}{\lceil w_j \rceil}\right) = A$$

and so if we would like to pack edge e to an α fraction of the μ_i's that constitute the α approximate decomposition of x', then we must require that $1 - A \geq \alpha x_e^*$. From this we conclude that the fraction of the solutions that e can be added is

$$1 - \left(\alpha(1 - x_e) + \alpha(1 - x_e) + \alpha\left(\frac{w_j - x_e}{\lceil w_j \rceil}\right)\right)$$

from which we get that $\alpha \leq \frac{1}{2 + \frac{w_j}{\lceil w_j \rceil} - x_e(1 + \frac{1}{\lceil w_j \rceil})} = \frac{1}{\sigma}$. We will deliver an upper bound on σ. Using the bound on the variable x_e, we have that

$$\sigma \leq 2 + \left(\frac{w_j}{\lceil w_j \rceil} - \frac{w_j}{\lceil w_j \rceil + 1}\left(1 + \frac{1}{\lceil w_j \rceil}\right)\right)$$

$$= 2 + \left(\frac{w_j}{\lceil w_j \rceil} - \frac{w_j}{\lceil w_j \rceil + 1} - \frac{w_j}{\lceil w_j \rceil} \cdot \frac{1}{\lceil w_j \rceil + 1}\right)$$

$$= 2 + \left(\frac{(\lceil w_j \rceil + 1)w_j - w_j\lceil w_j \rceil - w_j}{\lceil w_j \rceil(\lceil w_j \rceil + 1)}\right) = 2$$

and so $1/\sigma \geq 1/2$ so we can select $\alpha = \frac{1}{2}$ to satisfy the bound on α delivered above i.e. $\alpha \leq \frac{1}{\sigma}$.

Now we move on to prove the second case: if we select an edge e (with endpoints $u, v \in V$ and color $j \in [k]$) according to rule 1.(a) of the algorithm, then this edge can be packed again into an α fraction of the solutions (matchings) that constitute an α-approximate decomposition of the subgraph induced by the current residual solution x' without the edge e. For this, we need to define the set of the solutions μ_i that block the insertion of e slightly more carefully.

Let B_e be the set of all solutions μ_i that constitute an α-approximate decomposition of x' (current solution without edge e) that block the insertion of e. In B_e, as before, we add all μ_i that have edges adjacent to either of the endpoints u, v of e. We need to describe which solutions μ_i are blocking solutions for e with respect to its color class c_j. The natural way is to consider the solutions that are "full" of color c_j i.e. have w_j edges of color c_j (condition (2)). Unfortunately, if we follow this rule, the result would be a slightly worse approximation guarantee than our goal i.e. we can guarantee that e can be packed into a $\frac{2}{5}$ fraction of the μ_i's resulting in a $\frac{2}{5}$ approximation guarantee.

Instead, we will define blocking solutions of edge e, with respect to color c_j, all the solutions μ_i such that $|\mu_i \cap E_j| = \lceil \sum_{e' \in E_j} x_{e'}^* \rceil$ (in the subgraph induced by the current solution vector x^*).

Lemma 4. *Let $x \in [0,1]^E$ be any fractional feasible solution of the natural LP for the BCM problem and let $\theta_j = \lceil \sum_{e \in E_j} x_e \rceil \leq w_j$, for every color class $c_j \in C$. Assume that we have an α-approximate decomposition I for x, i.e. a collection of feasible matchings $\{\mu_i\}_{i \in I}$ together with their multipliers such that $\alpha \cdot x = \sum_i \lambda_i \mu_i$. For a color class c_j define $W(j) = \{\mu_i : |\mu_i \cap E_j| = \theta_j\}$. Then we have that*

$$\Lambda(W_j) = \sum_{\mu_i \in W(j)} \lambda_i \leq \alpha\left(1 + \mathcal{F}\left(\sum_{e' \in E_j} x_{e'}\right)\right)$$

where $\mathcal{F}(y) = y - \lceil y \rceil$ for $y > 0$ is the fractional part of y. Moreover, let $e = \{u, v\}$ be an edge with $c(e) = c_j, j \in [k]$ such that $x_e > 0$. Let $G[x'] := G[x] \setminus \{e\}$ and assume that we have an α approximate convex decomposition for $G[x']$ for some $\alpha \in (0, 1)$. Define the set of blocking solutions (with respect to the approximate decomposition of x') for e due to color c_j as:

$$B_e(j) = \left\{\mu_i : |\mu_i \cap E_j| = \left\lceil \sum_{e' \in E_j} x_{e'} \right\rceil\right\}$$

Then we have that

$$\sum_{i : \mu_i \in B_e(j)} \lambda_i \leq \alpha(1 - x_e)$$

Proof. We will prove the first claim with induction on x. The second claim will follow easily from the first one. For the base case assume that $x = 0^E$. Then the condition is automatically satisfied (take any empty matching with multiplier equal to 1). In this case $W_j = \emptyset$, for all color classes c_j.

Assume that the claim is true for all but the *first* edge of color c_j to be removed (first with respect to the inductive process of obtaining approximate decompositions), i.e., assume that it is true for the subgraph $G[x'] := G[x] \setminus \{e\}$. Let e_f be that edge. Define $X_{\bar{e}_f} = \sum_{e \in E_j \setminus \{e_f\}} x_e$ in the current subgraph. We will distinguish between two cases:

First case: $\lceil X_{\bar{e}_f} \rceil = \lceil X_{\bar{e}_f} + x_{e_f} \rceil$: In this case we have that

$$\sum_{j \in W(j)} \lambda_j \leq \sum_{\mu_\gamma \in \Gamma} \lambda_{\mu_\gamma} + \alpha x_{\bar{e}_f} \leq \alpha \left(1 + \mathcal{F}\left(X_{\bar{e}_f} \right) \right) + \alpha x_{\bar{e}_f} =$$

$$= \alpha \left(1 + \sum_{e \in E_j \setminus \{e_f\}} x_e - \left\lceil \sum_{e' \in E_j \setminus \{e_f\}} x_{e'} \right\rceil + x_{e_f} \right)$$

$$= \alpha \left(1 + \sum_{e \in E_j} x_e - \lceil X_{\bar{e}_f} \rceil \right)$$

$$= \alpha \left(1 + \sum_{e \in E_j} x_e - \lceil X_{\bar{e}_f} + x_{e_f} \rceil \right) = \alpha \left(1 + \sum_{e \in E_j} x_e - \left\lceil \sum_{e \in E_j} x_e \right\rceil \right)$$

$$= \alpha \left(1 + \mathcal{F}\left(\sum_{e' \in E_j} x_{e'} \right) \right)$$

where $\Gamma = \{\mu_\gamma : |\mu_i \cap E_j| = \lceil X_{\bar{e}_f} \rceil\}$ for the feasible matchings μ_γ that constitute the approximate decomposition of the current subgraph $G[x']$ (without the edge e_f). The first inequality is true because we have inserted an α fraction of e_f into the current approximate decomposition. The second inequality follows by the inductive hypothesis.

Second case: $\lceil X_{\bar{e}_f} \rceil \neq \lceil X_{\bar{e}_f} + x_{e_f} \rceil$: Observe that in this case we have that $\lceil X_{\bar{e}_f} \rceil = \lceil X_{\bar{e}_f} + x_{e_f} \rceil - 1$. Assume that we have an α-approximate decomposition for the subgraph induced by all the edges except e_f. Now, the set of matchings that block the insertion of edge e_f to the current approximate decomposition contains all the matchings μ_i such that $|\mu_i \cap E_j| = \lceil \sum_{e \in E_j} x_e \rceil$ (for the current set of edges of color c_j). But observe that since $\lceil X_{\bar{e}_f} \rceil \neq \lceil X_{\bar{e}_f} + x_{e_f} \rceil = \lceil \sum_{e \in E_j} x_e \rceil$, no matching μ_i from the current approximate decomposition has this property (that $|\mu_i \cap E_j| = \lceil \sum_{e \in E_j} x_e \rceil$). If either of αx_{e_f} or $\Lambda(W_j)$ (for the subgraph induced by all edges but e_f) is less than $\alpha(1 + \mathcal{F}(\sum_{e' \in E_j} x_{e'}))$, then we are done. Otherwise we might need to duplicate some solutions μ_i that constitute the approximate convex decomposition of x' to make sure that $\Lambda(W_j) = \alpha(1 + \mathcal{F}(\sum_{e' \in E_j} x_{e'}))$.

We need to prove the second claim i.e., $\sum_{i : \mu_i \in B_e(j)} \lambda_i \leq \alpha(1 - x_e)$. In the first case ($\lceil X_{\bar{e}_f} \rceil = \lceil X_{\bar{e}_f} + x_{e_f} \rceil$), observe that $B_e(j)$ does not change after the removal of e_f, i.e., in the subgraph induced by the remaining edges, the matchings μ_i with the property $|\mu_i \cap E_j| = \lceil \sum_{e' \in E_j} x_{e'} \rceil$ are the same in both cases. So we have that

$$\sum_{i:\mu_i\in B_e(j)} \lambda_i = \Lambda(W_j) = \sum_{\mu_i\in W(j)} \lambda_i \leq \alpha\left(1+\mathcal{F}\left(\sum_{e'\in E_j\setminus\{x_{e_f}\}} x_{e'}\right)\right)$$

$$= \alpha\left(1+\sum_{e\in E_j\setminus\{e_f\}} x_{e'} - \left\lceil \sum_{e\in E_j\setminus\{e_f\}} x_{e'}\right\rceil\right)$$

$$= \alpha\left(1+\sum_{e\in E_j} x_{e'} - x_{e_f} - \lceil X_{e_f} + x_{e_f}\rceil\right)$$

$$= \alpha\left(1 - x_{e_f} + \underbrace{\sum_{e\in E_j} x_{e'} - \lceil X_{e_f} + x_{e_f}\rceil}_{\leq 0}\right)$$

$$\leq \alpha(1 - x_{e_f})$$

where the quantities $\Lambda(W_j)$ and $\sum_{\mu_i\in W(j)}\lambda_i$ are defined in the subgraph without e_f and with respect to the α-approximate decomposition defined by x'.

As for the second case, we already claimed that in an α-approximate decomposition I for $G[x'] = G[x]\setminus\{e_f\}$, no solution $\mu_i\in I$ can block the insertion of e_f because in $G[x']$ we have that $\lceil \bar{X}_{e_f}\rceil = \lceil X_{\bar{e}_f} + x_{e_f}\rceil - 1$ and by construction in I we do not have any μ_i such that $|\mu_i\cap E_j| = \lceil \bar{X}_{e_f}\rceil + 1$ to block the insertion of e_f. Now, as we already argued, if $\Lambda(W_j)$ (in $G[x']$ which to avoid confusion we denote as $\Lambda_{x'}(W_j)$)is less than $\alpha(1+\mathcal{F}(\sum_{e'\in E_j} x_{e'}))$ then $B_{e_f}(j) = \emptyset$ in this case and we are done. But in the case that $\Lambda(W_j)$ (in $G[x']$)is strictly more than $\alpha(1+\mathcal{F}(\sum_{e'\in E_j} x_{e'}))$, then we should set $B_{e_f}(j)\subseteq \Lambda_{x'}(W_j)$ such that

$$\sum_{i\in B_{e_f}(j)} = \Lambda_{x'}(W_j) - \alpha(1+\mathcal{F}(\sum_{e'\in E_j} x_{e'}))$$

$$\leq \alpha\left(1+\bar{X}_{e_f} - \lceil\bar{X}_{e_f}\rceil\right) - \alpha\left(1+\bar{X}_{e_f} + x_{e_f} - \lceil\bar{X}_{e_f} + x_{e_f}\rceil\right)$$

$$= \alpha\left(-x_{e_f}\underbrace{-\lceil\bar{X}_{e_f}\rceil + \lceil\bar{X}_{e_f} + x_{e_f}\rceil}_{=1}\right)$$

$$= \alpha\left(1 - x_{e_f}\right)$$

Now, with the help of the previous claim, we will show that when we apply rule 1.(a) of our algorithm, we can add the selected edge e into an $\alpha = \frac{1}{2}$ of the matchings μ_i that constitute an α approximate convex decomposition I of the residual solution. Remember that the rule 1.(a) says that

$$\exists v\in F : |\mathrm{supp}(x^*\cap\delta(v))|\leq 2 \Rightarrow \exists e\in\delta(v) : x_e \geq \frac{\beta_v}{2}$$

As before, we want to calculate the fraction of the matchings μ_i that $e = \{u,v\}$ of color $c_j\in\mathcal{C}$ can be inserted preserving feasibility (i.e., μ_i is still a matching) and the above rule (that in the resulting matching μ_i after the addition of e we have that $|\mu_i\cap E_j|\leq\theta_j$). For this, we will calculate the fraction of μ_i that block the insertion of e: these are all the matchings μ_i that have edges adjacent to either u or v, and all the matchings μ_j that have θ_j edges of color c_j. In the residual solution vector x' (x without e) we have that (1) $\sum_{e' \in \delta(u)} x'_{e'} \leq 1 - x_e$, (2) the single edge e_2 adjacent to v has $x_{e_2} = \beta_v - x_e \leq \beta_v/2$ (remember that we have selected v such that the degree of v is equal to 2), and since we have an α-approximate convex decomposition of x', this means that the fraction of solutions that block the insertion of e (using also the previous claim) is at most

$$\alpha(1 - x_e) + \alpha(\beta_v - x_e) + \alpha(1 - x_e) = B$$

In clear analogy with the previous case (Lemma 3), since we want to insert e into an α fraction of the matchings in I, we want that $1 - B \geq \alpha x_e$ from which we conclude that the fraction of the matchings μ_i of I that e can be inserted is at least

$$1 - \alpha\left(1 - x_e + \beta_v - x_e + 1 - x_e\right) \geq \alpha x_e \Rightarrow \alpha \leq \frac{1}{1 + \beta_v - x_e + 1 - x_e}$$

and using the fact that $x_e \geq \frac{\beta_v}{2} \Rightarrow \beta_v - 2x_e \leq 0$ we conclude that $\alpha \leq \frac{1}{2}$ such that we can select $\alpha = \frac{1}{2}$ in this case as well. And so, in clear analogy with Lemma 3 we have proved the following:

Lemma 5. *If we select to insert an edge e of color c_j according to rule 1.(a) of the algorithm, then the fraction of the solutions μ_i of an α-approximate convex decomposition of the residual solution x' that e can be added is at least $\frac{1}{2}$.*

Theorem 2. *We can, in polynomial time, construct an $\frac{1}{2}$-approximate convex decomposition of x^*, resulting in a polynomial time $\frac{1}{2}$-approximation algorithm for BCM in general graphs.*

Acknowledgements. The author would like to thank Monaldo Mastrolilli for his support during the development of this work, Christos Nomikos and Panagiotis Cheilaris for discussions various issues in preliminary versions of this work.

References

1. Bampas, E., Pagourtzis, A., Potika, K.: An experimental study of maximum profit wavelength assignment in wdm rings. Networks 57(3), 285–293 (2011)
2. Berger, A., Bonifaci, V., Grandoni, F., Schäfer, G.: Budgeted matching and budgeted matroid intersection via the gasoline puzzle. In: Lodi, A., Panconesi, A., Rinaldi, G. (eds.) IPCO 2008. LNCS, vol. 5035, pp. 273–287. Springer, Heidelberg (2008)
3. Berman, P.: A $d/2$ approximation for maximum weight independent set in d-claw free graphs. In: Halldórsson, M.M. (ed.) SWAT 2000. LNCS, vol. 1851, pp. 214–219. Springer, Heidelberg (2000)
4. Caragiannis, I.: Wavelength management in wdm rings to maximize the number of connections. SIAM J. Discrete Math. 23(2), 959–978 (2009)

5. Carathéodory, C.: Über den variabilitätsbereich der fourierschen konstanten von positiven harmonischen funktionen. Rendiconti del Circolo Matematico di Palermo 32, 193–217 (1911)
6. Chan, Y.H., Lau, L.C.: On linear and semidefinite programming relaxations for hypergraph matching. Math. Program. 135(1-2), 123–148 (2012)
7. Chekuri, C., Vondrák, J., Zenklusen, R.: Multi-budgeted matchings and matroid intersection via dependent rounding. In: SODA, pp. 1080–1097 (2011)
8. Edmonds, J.: Maximum matching and a polyhedron with 0, 1 vertices. J. of Res. the Nat. Bureau of Standards 69B, 125–130 (1965)
9. Fürer, M., Yu, H.: Approximate the k-set packing problem by local improvements. In: ISCO-3rd International Symbosium on Combinatorial Optimization, Lisboa, Portugal, March 5-7, Lisboa, Portugal, March 5-7 (2014)
10. Garey, M.R., Johnson, D.S.: Computers and Intractability: A Guide to the Theory of NP-Completeness. W. H. Freeman (1979)
11. Grandoni, F., Ravi, R., Singh, M.: Iterative rounding for multi-objective optimization problems. In: Fiat, A., Sanders, P. (eds.) ESA 2009. LNCS, vol. 5757, pp. 95–106. Springer, Heidelberg (2009)
12. Grandoni, F., Zenklusen, R.: Approximation schemes for multi-budgeted independence systems. In: de Berg, M., Meyer, U. (eds.) ESA 2010, Part I. LNCS, vol. 6346, pp. 536–548. Springer, Heidelberg (2010)
13. Mastrolilli, M., Stamoulis, G.: Constrained matching problems in bipartite graphs. In: Mahjoub, A.R., Markakis, V., Milis, I., Paschos, V.T. (eds.) ISCO 2012. LNCS, vol. 7422, pp. 344–355. Springer, Heidelberg (2012)
14. Mastrolilli, M., Stamoulis, G.: Bi-criteria approximation algorithms for restricted matchings. Theoretical Computer Science 540-541, 115–132 (2014)
15. Monnot, J.: The labeled perfect matching in bipartite graphs. Inf. Process. Lett. 96(3), 81–88 (2005)
16. Nomikos, C., Pagourtzis, A., Zachos, S.: Minimizing request blocking in all-optical rings. In: IEEE INFOCOM (2003)
17. Nomikos, C., Pagourtzis, A., Zachos, S.: Randomized and approximation algorithms for blue-red matching. In: Kučera, L., Kučera, A. (eds.) MFCS 2007. LNCS, vol. 4708, pp. 715–725. Springer, Heidelberg (2007)
18. Papadimitriou, C.H., Yannakakis, M.: The complexity of restricted spanning tree problems. J. ACM 29(2), 285–309 (1982)
19. Parekh, O.: Iterative packing for demand and hypergraph matching. In: Günlük, O., Woeginger, G.J. (eds.) IPCO 2011. LNCS, vol. 6655, pp. 349–361. Springer, Heidelberg (2011)
20. Yuster, R.: Almost exact matchings. Algorithmica 63(1-2), 39–50 (2012)

Author Index